HOW THE
BODY WORKS

HOW THE
BODY WORKS

GENERAL EDITOR:
DR. PETER ABRAHAMS

amber
BOOKS

First published in 2007

Published by
Amber Books Ltd
74–77 White Lion Street
London N1 9PF
www.amberbooks.co.uk
www.twitter.com/amberbooks
www.facebook.com/amberbooks

Copyright © 2007 Bright Star Publishing plc

Reprinted 2008, 2012, 2013, 2014, 2015 (twice), 2016, 2017

ISBN 978-1-78274-435-1

Contributors: Claire Cross
Project Editor: Michael Spilling
Design: Colin Hawes
Jacket design: Jenny Dawe

Printed in China

Contents

Introduction

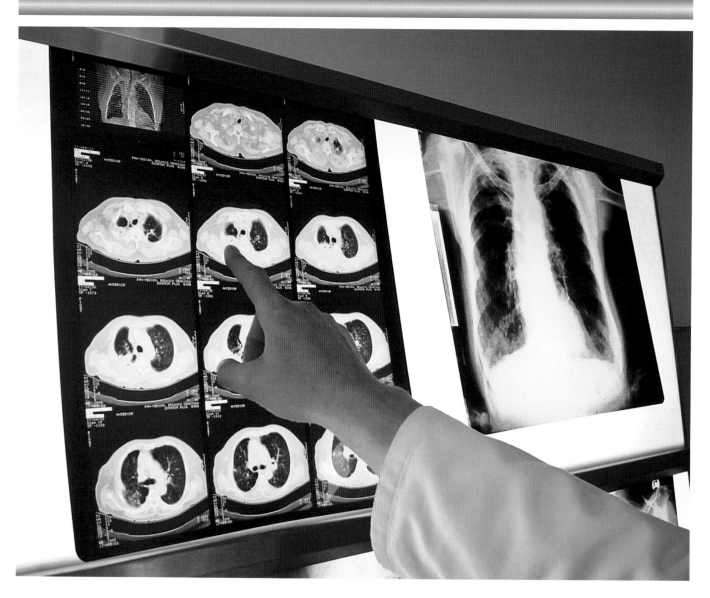

The human body, with its complex network of interdependent systems and structures, has provided mankind with an enduring challenge: to discover and unravel the intricate workings of each of its parts and to use this knowledge to enhance and develop ever-new ways to conquer illness and disease. Akin to the most masterly machine, the human body is comprised of billions of microscopic units, each with their own unique function yet all working together to create one smoothly operating entity. The study of exactly how each tissue, organ, cell and system in the body operates, both on its own and as part of this interlinking structure, is known as physiology. It is this science that has occupied some of the greatest minds throughout history.

THE BEGINNINGS OF KNOWLEDGE

There is evidence that points to a keen interest in the mechanics of the human body even in the ancient civilizations. Numerous preserved papyruses record the ancient Egyptians' knowledge of human anatomy and the conclusions they drew about its workings – knowledge drawn, it is thought, from their practice of embalming dead bodies. The Egyptians also held supernatural beliefs about illness, maintaining that sickness resulted from the interference of malevolent spirits, which could be driven away with the use of potions and spells.

A doctor examines a series of x-rays for abnormalities in a patient's chest.

The Greek physician Hippocrates, born in Cos in 460 BC, was renowned for his methods and approach to healing. His fundamental principles of observation and reasoning inspired fellow physicians, both during his lifetime and beyond, earning him the title of 'the father of medicine'. He looked for natural causes for illnesses and placed great importance

8

on the recording of symptoms and constant observation of a patient – fundamental principles that are still adhered to today and which are honoured in the taking of the 'Hippocratic oath'.

From this point, medical knowledge steadily evolved. Born in AD 131, Claudius Galen was a Greek physician and ardent follower of Hippocrates. He supported the popular, though inaccurate, theory that the body contains four systems, or humours – blood, phlegm, yellow bile and red bile – and that illness was caused by an imbalance of these humours. Despite such primitive beliefs, he made important advances in physiological understanding by dissecting animals and drawing parallels with the

This computer graphic shows strands of DNA in the classic helix pattern.

inner workings of the human body. His work was hugely influential, even hundreds of years after his death.

STEADY ADVANCES

With the Middle Ages came a period of relative stagnation. Medicine continued to be studied, at universities under a master physician, but students relied heavily on ancient texts and the observations of physicians such as Galen. The influence of the Church was such that illness and disease were seen as a punishment from God. An uneasy mix of secular and spiritual beliefs defined the reasons given for illness, with astrology, destiny and sin all held to play a significant part. Bloodletting and crude mixes of herbs and potions were popular remedies. The dawning of the Renaissance saw a revival of the Classical

ideals of rationality and reasoning. The archetypal Renaissance man, Leonardo da Vinci, recognized the importance of anatomical investigation in the furtherance of medical knowledge. He believed that to obtain this knowledge the dissection of human bodies was necessary. Anatomists Andreas Vesalius and William Harvey made hugely significant contributions to the science of physiology. In 1543, Vesalius challenged previously held beliefs in his text De Humani Corporis Fabrica. Based upon dissections of human bodies, it demonstrated how human anatomy differed from that of animals.

Harvey's discoveries on circulation were equally astounding, based on dissections of both animals and humans. In 1628, his revolutionary text An Anatomical Study of the

Motion of the Heart and of the Blood in Animals described how the heart worked as a muscle, causing arteries to pulsate and constantly move blood around the body. It also demonstrated the importance of valves in the blood's circulation.

A new era in medical technology began, laying the foundation for major breakthroughs in the treatment of disease well into the twentieth century. In the late seventeenth century, the microscope was invented; in 1896, Wilhelm Rontgen discovered X-rays. Both offered previously undreamt of opportunities for examining the inner workings of the human body. From 1800, surgical operations were transformed by the creation of anaesthetics, and in 1901 the discovery by Karl Landsteiner of the four blood groups paved the way

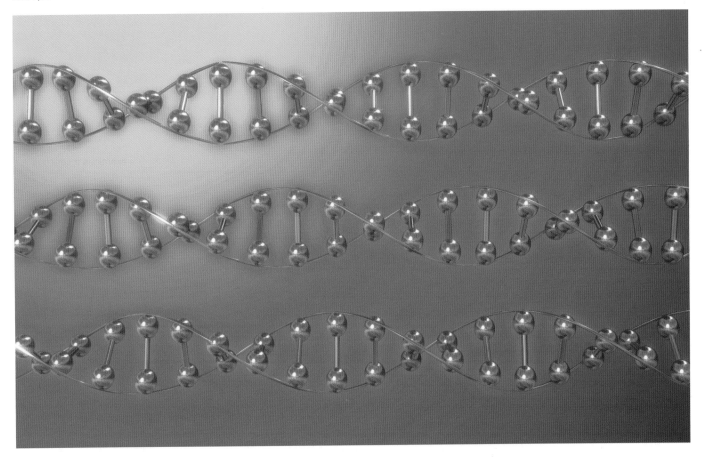

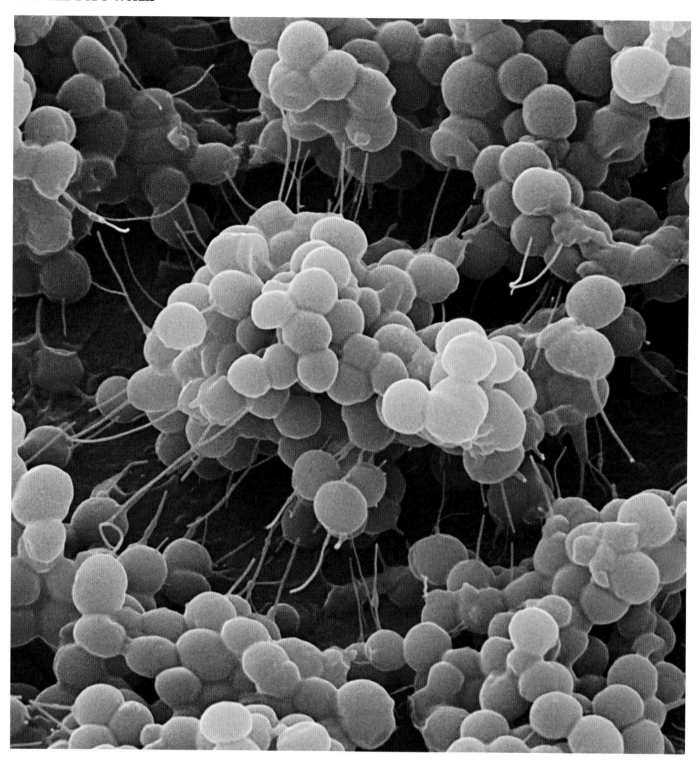

for some of the most important advances in the twentieth century, ultimately leading to transplant surgery. However, some of the greatest innovations were due to basic improvements in hygiene. In the nineteenth century, Louis Pasteur established the link between micro-organisms and disease. He influenced Joseph Lister, the professor of surgery at Glasgow University, who transformed practices in hospitals in 1865, when he made the link between disease-causing microorganisms that could be passed from patient to patient via medical staff. Through the simple expedient of using carbolic acid as a ward antiseptic, he saw a dramatic drop in fatality figures among patients. These first crude attempts at improving hospital cleanliness earned Lister the title of the 'Father of Antiseptic Surgery'.

MEDICINE TRANSFORMED

The dawn of the twentieth century heralded an unprecedented period in terms of medical breakthroughs. Simultaneous

Staphylococcus epidermidis bacteria. This normally harmless bacteria is part of the normal skin flora, but can cause infection on broken skin or wounds.

advances in medical research and technological developments meant that huge strides were made in medicine, surgery and pharmacology. The two World Wars provided additional incentives to discover new methods and techniques. Improvements in Rontgen's X-ray machine made it an increasingly powerful diagnostic tool, and there were significant developments in the use of sutures. The groundbreaking discovery of penicillin in 1928 by Alexander Fleming made a huge contribution during World War II, preventing the death of thousands of soldiers.

As well as improvements in diagnostic tools, this period saw the dawn of a new era in preventative medicines. In the 1950s and 1960s, vaccines were developed to combat debilitating and sometimes fatal childhood diseases, such as polio, rubella and measles. Childhood health in the developed world was now better than at any time in history.

The survival rate of infants was given a further boost by significant improvements in healthcare during pregnancy and childbirth. The advent of targeted anaesthetics known as epidurals helped to ease difficult labours, and the ability to induce labour increased the options for problematic pregnancies.

In the 1950s, Ian Donald, the Professor of Midwifery at the University of Glasgow, Scotland, discovered the ultrasound scanner, using sound waves rather than potentially harmful radiation to visualize internal organs. This discovery revolutionized the world of diagnostic medicine and become a fundamental tool during pregnancy.

Pharmacology witnessed staggering transformations in the twentieth century as

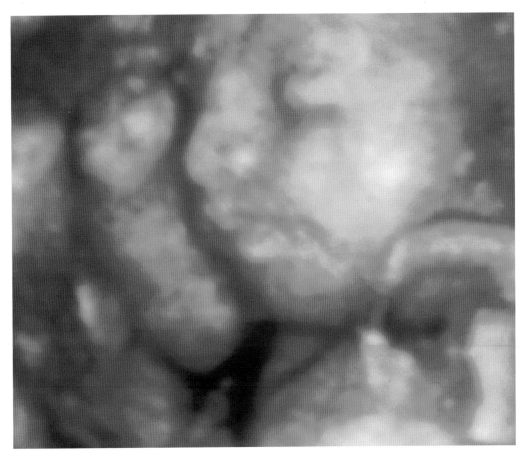

ABOVE: This tree-dimensional sonogram shows a six months fetus by ultrasound.

RIGHT: An asthmatic boy uses his ventolin inhaler to open up his bronchial airways.

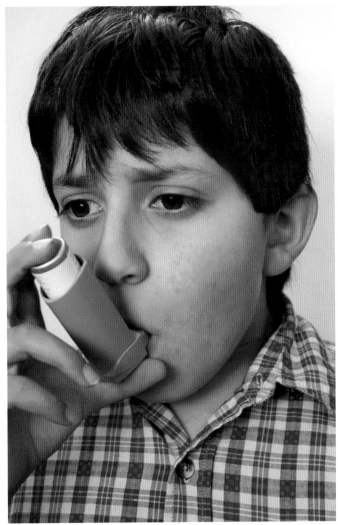

technologies and our understanding of the effects of medicine on the human body improved. The phenomenal success of penicillin sparked a search to discover new microorganisms to fight bacterial infections. In the 1940s, the scientists Ernst Chain and Howard Florey were inspired by Fleming to undertake further research into natural antibiotics. After World War II, the United States stepped up this research, leading to the discovery of a mould called streptomycin, which successfully treated the highly prevalent disease tuberculosis. As throughout the history of medicine, one discovery opened the door to others. This was the case with the development after 1945 of the group of drugs known as steroids. The steroid cortisone was used primarily to relieve pain and inflammation, proving hugely beneficial in the management of diseases affecting the joints. An unforeseen side effect was that cortisone lowered the body's immune system. However, this discovery prompted the use of cortisone in some transplant surgery, reducing the body's chances of rejecting its new organ.

UP CLOSE AND PERSONAL

One of the most monumental and important breakthroughs in the history of medicine came in 1953, when James Watson and Francis Crick discovered the structure of DNA. Both men realized that they had found 'the secret of life' – a claim that is by no means farfetched since the double helix structure of a DNA cell contains the genes that form the basis of all living tissue. The effects of this discovery continue to resonate into the twenty-first century. As has often been the case throughout the centuries, it was not until decades later that the significance of this discovery truly came into its own, with the advent of the human genome project.

One of the twentieth century's most significant contributions to medicine was the constant refinement of imaging technologies. These have allowed scientists to observe the inner workings of the body and have revolutionised the diagnosis and treatment of many serious conditions. In addition to the X-ray and ultrasound, scientists developed the magnetic resonance imaging (MRI) scan, which uses magnetic fields and radio waves to create an image; and the computed tomography (CT) scan, which is a highly accurate, three-dimensional form of X-ray. These imaging techniques are used widely today to provide amazingly accurate and detailed pictures, from which physicians can form often life-saving diagnoses. Further innovations in the field of imaging include the use of the endoscope, a tubular viewing instrument that allows a telescopic internal examination without the need for surgery.

SURGICAL INTERVENTION

Improvements in surgery continued apace throughout the twentieth century. Operations began to be performed on parts of the body previously deemed untouchable. The most famous surgical triumph was achieved in 1967, when Christian Barnard successfully carried out the first heart transplant.

Although the patient died shortly afterwards from pneumonia, this pioneering operation inspired others to follow suit, culminating in the first heart-lung

The human head has a detailed and complex muscle structure that allows us to communicate using a wide range of facial expressions.

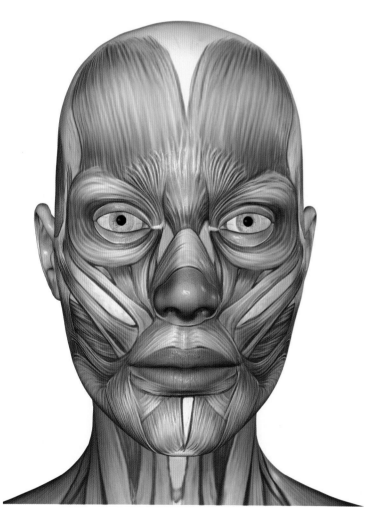

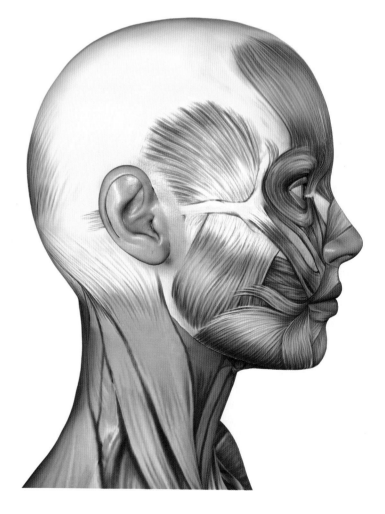

A diabetic takes blood from her finger to test her glucose level using a finger-level glucose tester.

transplant in 1982 and the first brain tissue transplant in 1987.

With increasingly sophisticated drugs, the survival time following transplant surgery was greatly extended. Other surgical innovations included the invention of a heart-lung machine that temporarily took over the work of the heart and lungs, allowing ever-more complex surgery to take place and resulting in heart bypass surgery becoming a commonplace procedure. Developments in replacement surgery, particularly hip and knee replacements, offered a new lease of life to many.

Ongoing technical innovations have allowed increasingly intricate surgical operations to be performed with the minimum of inconvenience to the patient. The advent of keyhole surgery, combining the use of a telescopic endoscope with small incisions, has vastly improved post-operative recovery rates. The development of microsurgery, using powerful magnification and fine instruments, has allowed operations to be carried out on minute or delicate tissues. The introduction of lasers opened up a whole new branch of surgery, bringing high-precision techniques to such delicate procedures as the removal of birthmarks.

CURING DISEASE

With increasingly complex innovations, medical practitioners are being offered a whole new range of tools to combat some of the most serious diseases facing the world. The result of such a rapid change in the world of medicine has meant that many diseases considered untreatable in the first half of the twentieth century are today treatable and often curable, or may be in the near future.

A major challenge facing the world of medicine has been the fight against the deadly and prevalent disease of cancer. Over the past few decades, scientists have battled to conquer this most feared of diseases, resulting in huge strides forward in the understanding and treatment of it. Today, many types of cancer are successfully treated thanks to increasingly effective drugs. Meanwhile, surgery and research has established, often conclusively, the link between certain risk factors, such as tobacco smoking, and the onset of cancer.

Further radical improvements in the

The hair, skin and nails make up the integumentary system. As well as keeping the body waterproof, they protect the body against microbial attack.

treatment of cancer are forecast. Advancements in radiotherapy, as well as the emergence of a new group of 'smart' drugs, are already offering promising results. Radiotherapy, one of the most effective means for fighting cancer, traditionally caused unwanted and sometimes serious side effects as it attacked not just cancerous cells but the surrounding healthy tissues. Technological advances have led to the development of a new, highly precise form of radiation, with rays concentrated more exactly on the target, allowing physicians to deliver higher doses with fewer side effects.

'Smart' drugs, resulting from DNA technology, work on a similar principle. While chemotherapy affects healthy and diseased cells alike (causing nausea, hair loss and other side effects), smart drugs work as targeted missiles, binding to the surface of cancer cells to stop them reproducing, and at the same time causing limited damage to surrounding tissue. The outcome is a reduction in trauma to the patient and improved results. Another prediction currently exciting scientists is that viruses will be found to play a far more significant role in the development of cancer than originally thought. Already the medical world has produced a vaccine to combat cervical cancer, and it is predicted that there will be many more cancer-preventing vaccines in the next few decades.

Transformations in the treatment of heart disease are similarly predicted. Heart surgery using robotics allows surgeons to make small incisions in the patient's chest and then insert robotic arms that have cameras and tiny instruments to carry out bypass surgery. This procedure eliminates the need to open up the patient's chest and saw through bones, thus dramatically reducing the pain experienced and improving the recovery rate. The treatment of blood clots is also set to improve dramatically with state-of-the-art diagnostic scanning devices that enable doctors to monitor the status of a blood clot and deliver clot-busting drugs into the affected artery as required.

UNLOCKING THE CODE OF LIFE

Of all medical advances, the ones that seem to offer the greatest possibilities are those of gene therapy and stem cell research. In 1990, the international Human Genome Project was spearheaded by scientists in the United States. The project's aim was to identify the 25,000 genes in human DNA and to unlock the sequence of the 3 billion chemical bases that make up DNA. And in 2003, it was completed. The potential this offers is incredible.

With every gene in the human body mapped out, scientists are able to compare the genes of healthy and non-healthy individuals and thus pinpoint problem genes. The idea behind gene therapy is that a defective disease-causing gene, often inherited, would be replaced with a normal therapeutic gene, thus eliminating the disease. More controversially, gene therapy has been considered for replacing defective genes in egg and sperm cells to prevent conditions being passed on to future generations.

However, gene therapy has been beset with practical, ethical and social issues, and is still very much in its infancy, with trials experimental, costs prohibitive and success stories limited. There are many obstacles to be overcome, including how to deal with the body's immune response to the introduction of new genes and the short-lived nature of any treatment. Nonetheless, some scientists predict that individual gene profiling may one day become available, and that this will enable individuals to take precautions to optimize their health and even avoid certain diseases entirely.

Stem cell research offers treatment possibilities that seem almost miraculous, signalling hope in conditions and illnesses previously deemed untreatable. Stem cells are primal 'blank' cells that can renew themselves. Taken from an early-stage embryo, they can be isolated and then cultured into specialized cell types, such as muscles or nerves. This offers remarkable potential for the treatment of diseases such as Parkinson's, Alzheimer's and cancer. They may also be used to counter spinal cord damage and muscle injury, replacing diseased or injured tissue with healthy, renewable tissue. Ultimately, it is thought that stem cells could be used to grow entirely new organs or body parts, and thus play a significant role in transplant surgery.

Throughout history, the thirst for medical knowledge and the ability to control disease has led to both innovation and controversy, often in equal measure. Today, this still holds true, perhaps more so than ever. Gene therapy and stem cell research are both emerging technologies that raise

enormous ethical issues and have thus been at the heart of much debate. The destruction of human embryos involved in stem cell research is seen by many to threaten the sanctity of human life, putting stem cell research at the heart of the abortion debate.

This research also raises the possibility of human cloning – a possibility that the majority of scientists regard as morally reprehensible. Gene therapy in turn raises the question of the rights and wrongs of genetic engineering, and

some people worry about the possibility of 'designer' babies and are concerned by the potential for discrimination based on an individual's genetic profiling. It also presents ethical considerations about the nature of disability, creating comparisons between what is 'normal' and what is not. And do all disabilities even require a cure? This medical step into the future has sparked a debate that looks set to continue.

How the Body Works takes the reader on an incredible journey through the human

body, offering us a glimpse into the extraordinary world of medical science. It charts the body's anatomy, with detailed artworks illustrating the structure of each body part, and it also explains the physiology of each body system, providing a fascinating insight into exactly how our bodies operate. Working from head to toe, the book offers a fascinating introduction to the human body.

This x-ray shows the human pelvis. The socket joint where the femur attaches to the hip bone can clearly be seen.

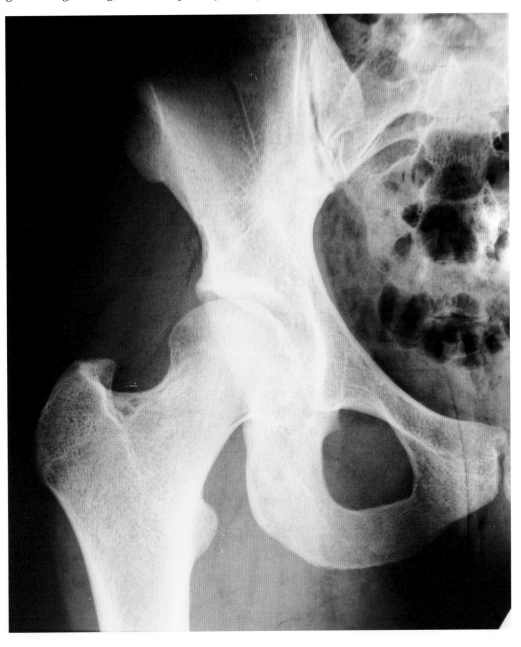

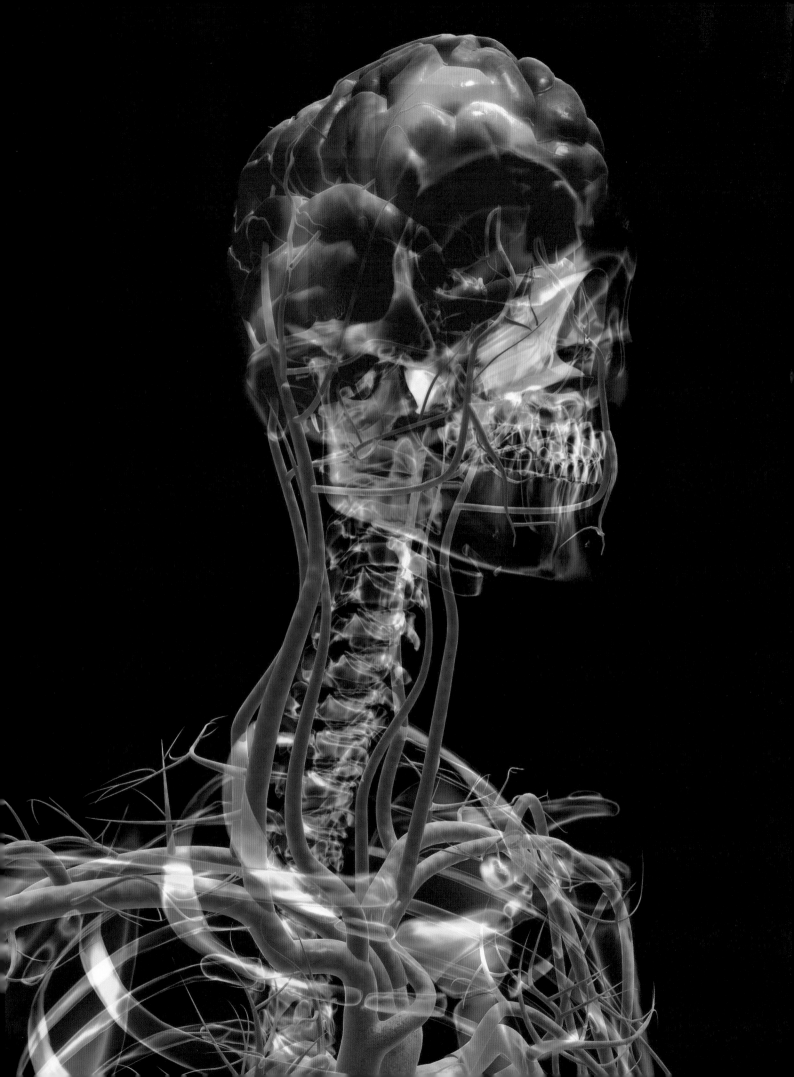

Chapter One

The Head

The head is undoubtedly one of the most complex, multifaceted parts of the human body. Within the protective casing of the skull bones lies the brain, the part of the central nervous system that acts as the control centre for most of the body. Our memory and ability to communicate and to experience thoughts, emotions and creativity are all made possible by the activities of the brain.

Hundreds of bodily processes, including simple unconscious actions such as breathing, are controlled here too. In this chapter, you can also explore other elements of the head, including the facial muscles, which are pivotal in establishing our identity and allowing us to communicate, and the teeth, which play a crucial part in the body's digestive system.

LEFT: This computer graphic shows the cardiovascular system within the head, neck and chest.

Front of the skull

The skull is the head's natural crash helmet, protecting the brain and sense organs from damage. It is made up of 28 separate bones and is the most complex element of the human skeleton.

The skull is the skeleton of the face and head. Its basic role is protecting the brain, the organs of special sense such as the eyes, and the cranial parts of the breathing and digestive system. It also provides attachment for many of the muscles of the neck and head.

Although often thought of as a single bone, the skull is made up of 28 separate bones. For convenience, it is often divided into two main sections: the cranium and the mandible. The basis for this is that, whereas most of the bones of the skull articulate by relatively fixed joints, the mandible (jawbone) is easily detached. The cranium is then subdivided into a number of smaller regions, including:

- cranial vault (upper dome part of the skull)
- cranial base
- facial skeleton
- upper jaw
- acoustic cavities (ears)
- cranial cavities (interior of skull housing the brain).

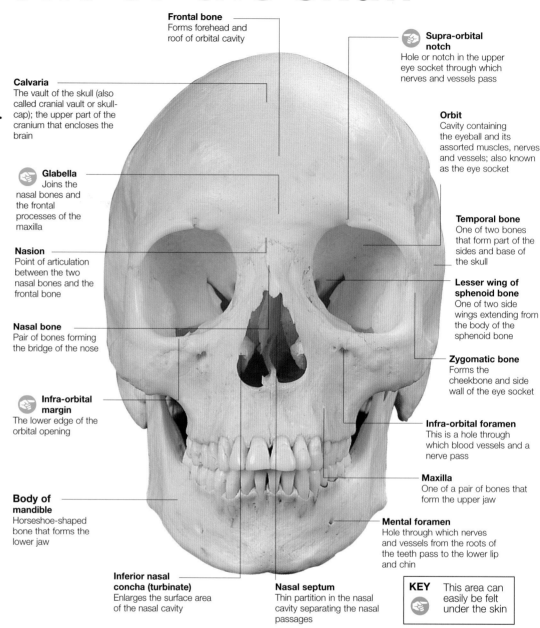

Frontal bone
Forms forehead and roof of orbital cavity

Supra-orbital notch
Hole or notch in the upper eye socket through which nerves and vessels pass

Orbit
Cavity containing the eyeball and its assorted muscles, nerves and vessels; also known as the eye socket

Calvaria
The vault of the skull (also called cranial vault or skull-cap); the upper part of the cranium that encloses the brain

Glabella
Joins the nasal bones and the frontal processes of the maxilla

Nasion
Point of articulation between the two nasal bones and the frontal bone

Nasal bone
Pair of bones forming the bridge of the nose

Infra-orbital margin
The lower edge of the orbital opening

Body of mandible
Horseshoe-shaped bone that forms the lower jaw

Temporal bone
One of two bones that form part of the sides and base of the skull

Lesser wing of sphenoid bone
One of two side wings extending from the body of the sphenoid bone

Zygomatic bone
Forms the cheekbone and side wall of the eye socket

Infra-orbital foramen
This is a hole through which blood vessels and a nerve pass

Maxilla
One of a pair of bones that form the upper jaw

Mental foramen
Hole through which nerves and vessels from the roots of the teeth pass to the lower lip and chin

Inferior nasal concha (turbinate)
Enlarges the surface area of the nasal cavity

Nasal septum
Thin partition in the nasal cavity separating the nasal passages

KEY This area can easily be felt under the skin

Sinuses of the skull

In the general sense, sinuses are cavities or hollow spaces in the body. In the skull, there are four sinuses, more accurately known as 'paranasal sinuses'. They are named after the bones in which they lie:

- Frontal
- Ethmoidal
- Maxillary
- Sphenoidal

This exploded skull shows three of the paranasal sinuses: frontal (1), ethmoidal (2) and maxillary (3). The fourth, sphenoidal, is not clear in this view as it is inside the skull, behind the eyes. All paranasal sinuses are connected to the nasal cavity.

The paranasal sinuses are air-containing sacs connected to the nasal cavity through narrow – and therefore easily blocked – channels. Their usefulness is limited to adding resonance to the voice, and possibly lightening the skull. The same tissue that lines the nasal cavities lines the sinuses, so they easily become infected (resulting in sinusitis).

The most commonly infected sinus is the maxillary. When this happens, the mucous membrane lining the sinuses becomes inflamed, resulting in a blocked-up nose, a loss of sense of smell and the discharge of pus and mucus from the nose. The main treatment is drainage with or without antibiotics.

Illuminated skull

Most of the bones of the skull are connected by sutures – immovable fibrous joints. These, and the bones inside the skull, can be seen most clearly using a brightly illuminated skull.

The areas where skull bones meet are called 'sutures'. The coronal suture, for example, occurs between the frontal and parietal bones, and the sagittal suture connects the two parietal bones. It is important to learn the position of these joints, because they can be confused with fractures on X-rays.

In babies, there are relatively large gaps between skull bones, allowing the head to squeeze through the birth canal without fracturing. The gaps are covered in fibrous membranes called 'fontanelles'. In most 'head-first' births, the fontanelles can be palpated (examined using the fingertips) during vaginal examinations to determine the position of the head.

CHANGE OF FACE

Because children have only rudimentary teeth and sinuses, their faces are smaller proportionally to adults'. (The skull of a newborn, however, is one-quarter of its body size.) As we get older, the relative size of the face diminishes as our gums shrink and we lose our teeth and the bony sockets.

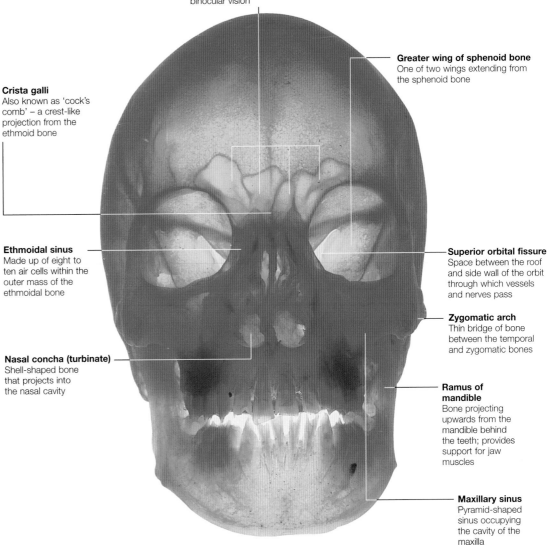

Frontal sinuses
Pockets of air connected to the nasal passage; not fully understood, but believed to help shape the orbitals and provide binocular vision

Greater wing of sphenoid bone
One of two wings extending from the sphenoid bone

Crista galli
Also known as 'cock's comb' – a crest-like projection from the ethmoid bone

Superior orbital fissure
Space between the roof and side wall of the orbit through which vessels and nerves pass

Ethmoidal sinus
Made up of eight to ten air cells within the outer mass of the ethmoidal bone

Zygomatic arch
Thin bridge of bone between the temporal and zygomatic bones

Nasal concha (turbinate)
Shell-shaped bone that projects into the nasal cavity

Ramus of mandible
Bone projecting upwards from the mandible behind the teeth; provides support for jaw muscles

Maxillary sinus
Pyramid-shaped sinus occupying the cavity of the maxilla

Painted skull

The front view of the skull reveals about nine of the major bones of the head. The painted skull (right) shows these areas clearly:
1 Frontal bone
2 Parietal bone
3 Temporal bone
4 Nasal bone
5 Sphenoid bone
6 Lacrimal bone
7 Zygomatic bone
8 Maxilla
9 Mandible
The other principal features of the skull are the orbits (eye sockets), nasal cavity and teeth.

Some bones in the skull, such as those surrounding the orbital, are relatively thin and prone to fracturing. However, the large number of overlapping bones makes it difficult for doctors to see fractures in X-rays.

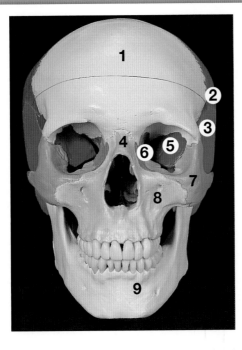

The colours of this painted skull identify the major bones of the head as seen from the front. In this view – known as an anterior view – some bones, such as the occipital (back of the head) and palatine (plate of the upper mouth) cannot be seen.

Skull X-rays clearly show the sutures between the bones. However, the appearance of these sutures makes it difficult for doctors to assess fractures to the skull. In order to identify broken bones, doctors look for five black lines in the white bone. If, however an area of white is seen inside a sinus, this may suggest fluid such as pus or blood inside the cavity.

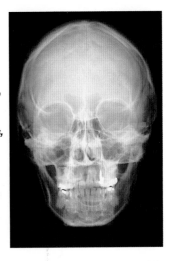

Side of the skull

A lateral or side view of the skull clearly reveals the complexity of the structure, with many separate bones and the joints between them.

Several of the bones of the skull are paired, with one on either side of the midline of the head. The nasal, zygomatic, parietal and temporal bones all conform to this symmetry. Others, such as the ethmoid and sphenoid bones, occur singly along the midline. Some bones develop in two separate halves and then fuse at the midline, namely the frontal bone and the mandible (lower jaw).

The bones of the skull constantly undergo a process of remodelling: new bone develops on the outer surface of the skull, while the excess on the inside is reabsorbed into the bloodstream. This dynamic process is facilitated by the presence of numerous cells and a good blood supply.

Occasionally, a deficiency in the cells responsible for reabsorption upsets the bone metabolism, which can result in severe thickening of the skull – osteopetrosis, or Paget's disease – and deafness or blindness may follow.

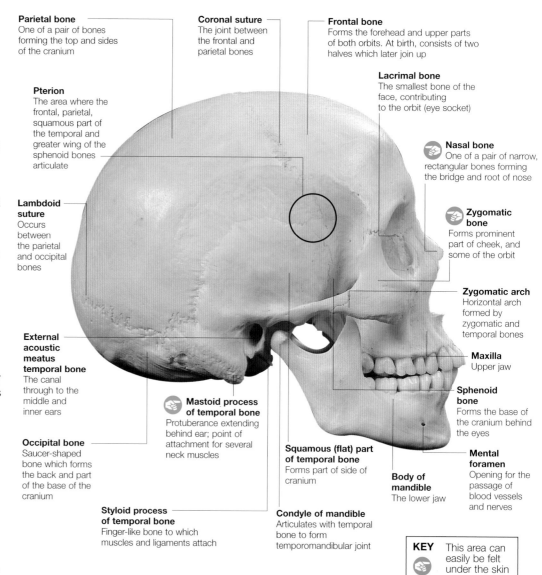

Parietal bone
One of a pair of bones forming the top and sides of the cranium

Pterion
The area where the frontal, parietal, squamous part of the temporal and greater wing of the sphenoid bones articulate

Lambdoid suture
Occurs between the parietal and occipital bones

External acoustic meatus temporal bone
The canal through to the middle and inner ears

Occipital bone
Saucer-shaped bone which forms the back and part of the base of the cranium

Styloid process of temporal bone
Finger-like bone to which muscles and ligaments attach

Mastoid process of temporal bone
Protuberance extending behind ear; point of attachment for several neck muscles

Coronal suture
The joint between the frontal and parietal bones

Frontal bone
Forms the forehead and upper parts of both orbits. At birth, consists of two halves which later join up

Lacrimal bone
The smallest bone of the face, contributing to the orbit (eye socket)

Nasal bone
One of a pair of narrow, rectangular bones forming the bridge and root of nose

Zygomatic bone
Forms prominent part of cheek, and some of the orbit

Zygomatic arch
Horizontal arch formed by zygomatic and temporal bones

Maxilla
Upper jaw

Sphenoid bone
Forms the base of the cranium behind the eyes

Mental foramen
Opening for the passage of blood vessels and nerves

Body of mandible
The lower jaw

Condyle of mandible
Articulates with temporal bone to form temporomandibular joint

Squamous (flat) part of temporal bone
Forms part of side of cranium

KEY This area can easily be felt under the skin

Joints of the skull: sutures

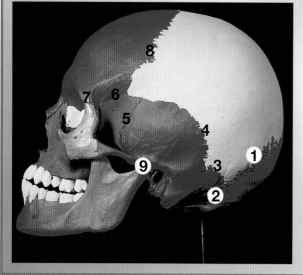

1 Lambdoid suture
2 Occipitomastoid suture
3 Parietomastoid suture
4 Squamosal suture
5 Sphenosquamosal suture
6 Sphenofrontal suture
7 Frontozygomatic suture
8 Coronal suture
9 Temporomandibular joint

The only moveable skull joint is the temporomandibular joint (where the jaw hinges against the cranium) allowing all the actions of chewing and speech.

All the other bones are fixed to each other by joints known as sutures,

This painted skull shows the location of the 11 major bones and the sutures that join them.

which are only found in the skull. In the adult, these comprise thin zones of unmineralized fibrous tissue bonding the irregular, interlocking margins of adjacent bones.

The purpose of sutures in the skull of the developing infant is to allow for growth at right angles to their alignment. For example, the coronal suture allows growth in length and the squamosal allows for increase in height of the skull.

During the rapid period of cranial growth, from baby to child, the enlarging brain forces the bones apart at their sutures, and new bone is then deposited at the edge of the sutures, stabilizing the skull at its new size. By the age of seven, the sutural growth has slowed, and the skull enlarges at a slower rate by bone remodelling.

Inside the skull

The inside of the left half of the skull shows the large cranial vault (calvaria) and facial skeleton in section.

Comparing this photograph with the one of the skull's exterior, many of the same bones can be seen, as well as additional structures. The bony part of the nasal septum (the dividing wall of the nasal cavity) consists of the vomer and the perpendicular plate of the ethmoid bone.

In this skull, the sphenoidal air sinuses are large. The pituitary fossa, containing the pea-sized, hormone-producing pituitary gland, projects down into the sinus. The circle marks the pterion, corresponding to the position marked on the external photograph.

The skull covers the brain, and skull fractures can lead to potentially life-threatening situations. If the side of the skull is fractured in the region of the temporal bone, the blood vessel of the middle meningeal artery may be damaged (extra dural haemorrhage). This vessel supplies the skull bones and the meninges (outer coverings of the brain), and if ruptured, the escaping blood may cause pressure on vital centres in the brain. If not relieved, this can rapidly cause death. The artery is accessible to the surgeon if entry is made near the pterion.

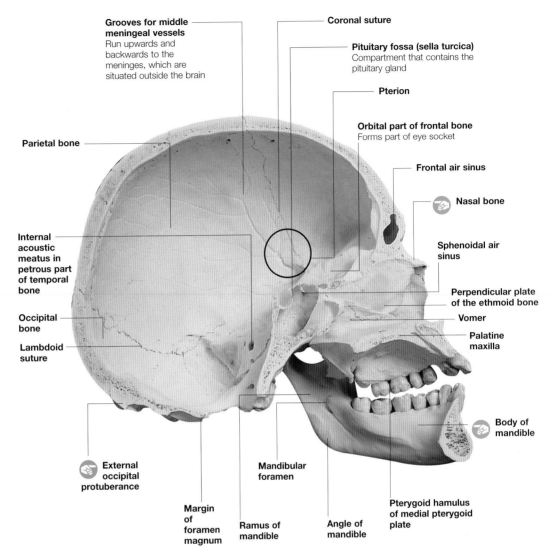

Grooves for middle meningeal vessels
Run upwards and backwards to the meninges, which are situated outside the brain

Coronal suture

Pituitary fossa (sella turcica)
Compartment that contains the pituitary gland

Pterion

Orbital part of frontal bone
Forms part of eye socket

Frontal air sinus

Nasal bone

Sphenoidal air sinus

Perpendicular plate of the ethmoid bone

Vomer

Palatine maxilla

Body of mandible

Parietal bone

Internal acoustic meatus in petrous part of temporal bone

Occipital bone

Lambdoid suture

External occipital protuberance

Margin of foramen magnum

Ramus of mandible

Mandibular foramen

Angle of mandible

Pterygoid hamulus of medial pterygoid plate

Types of bone in the skull

Bone is a hard, dense, mineralized connective tissue comprising three components:
- an organic matrix (about 25 per cent by weight) mostly of the fibrous protein collagen
- mineralized crystals of calcium phosphate and calcium carbonate (65 per cent by weight), known as hydroxyapatite
- approximately 10 per cent water.

The combination of mineral and organic material ensures that it has strength and rigidity, as well as the flexibility to absorb loads without being brittle.

The bones of the cranium – that is the frontal, parietals,

occipital and temporals – are 'flat bones', consisting of two thin tables (layers) of compact bone enclosing a looser type called diplöe, or cancellous bone. This is spongy, lattice-like bone

containing the marrow.

Blood cells are produced within the marrow, while the bone itself – as elsewhere in the body – is a source of the calcium ions essential for the normal working

of nerves and muscles.

The diplöe is unique to the skull, and allows for large, yet light and strong, areas of bone to protect and nourish the brain and vital sense organs.

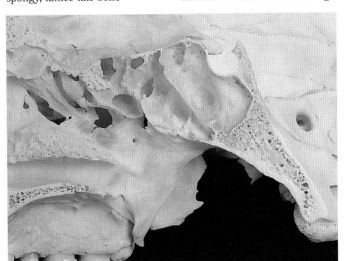

This cross-section through the upper jaw reveals the honeycombed nature of the paranasal air sinuses – this makes them lighter in weight, but no less strong.

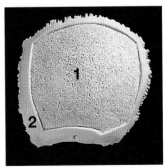

The right parietal bone, dissected to reveal the diplöe (1), beneath the outer table of compact bone (2). Its honeycomb of cancellous (spongy) bone is visible; under it will be an inner table of compact bone.

Top and base of the skull

The calvaria, or vault of the skull, is the upper section of the cranium, surrounding and protecting the brain.

The four bones that make up the calvaria are the frontal bone, the two parietals and a portion of the occipital bone.

These bones are formed by a process in which the original soft connective tissue membrane ossifies (hardens) into bone substance, without going through the intermediate cartilage stage, as happens with some other bones of the skull.

Points of interest in the calvaria include:
■ The sagittal suture running longitudinally from the lambdoid suture at the back of the head to the coronal suture.
■ The vertex (highest point) of the skull; the central uppermost part, along the sagittal suture.
■ The distance between the two parietal tuberosities is the widest part of the cranium.
■ The complex, interlocking nature of the sutures which enable substantial skull growth in the formative years, and provide strength and stability in the adult skull.

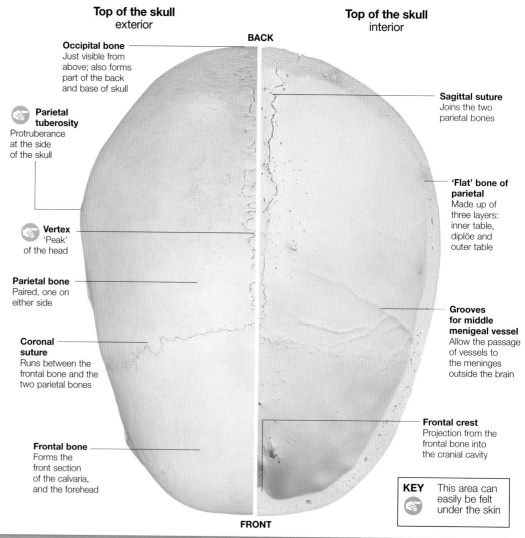

Top of the skull
exterior

Top of the skull
interior

BACK

Occipital bone — Just visible from above; also forms part of the back and base of skull

Parietal tuberosity — Protruberance at the side of the skull

Vertex — 'Peak' of the head

Parietal bone — Paired, one on either side

Coronal suture — Runs between the frontal bone and the two parietal bones

Frontal bone — Forms the front section of the calvaria, and the forehead

Sagittal suture — Joins the two parietal bones

'Flat' bone of parietal — Made up of three layers: inner table, diplöe and outer table

Grooves for middle menigeal vessel — Allow the passage of vessels to the meninges outside the brain

Frontal crest — Projection from the frontal bone into the cranial cavity

KEY This area can easily be felt under the skin

FRONT

Skull defects

Because of the unique way in which the skull develops – the growing brain forcing the bones apart at the sutures – any defect either in the component bones themselves or in the sutures may result in drastic changes in the shape and appearance of the baby's or child's head.

Isolated premature fusion of the sutures (where individual joints become fixed and closed before the brain has reached its full size) is called craniostenosis. This greatly reduces the capacity for expansion in the direction normally allowed by the suture.

The brain continues to expand, however, in whatever direction is available, distorting the usual shape, depending upon which suture is affected:
■ **scaphocephaly** is an elongated, boat-shaped skull resulting from stenosis (premature closure) of the sagittal suture

■ **brachycephaly** gives a markedly pointed, short skull, caused by bilateral stenosis of the coronal suture
■ **plagiocephaly,** in which the head takes on the 'twisted' appearance of asymmetrical deformity, is caused by stenosis of one half of the coronal suture
■ **oxycephaly** produces an abnormally high, sloping skull, usually because of early fusion of the sagittal and coronal sutures.
Disorders in bone production may also result in skull deformation. With achondroplasia (dwarfism of the whole body), the cartilaginous bones are affected. This means that the bones of the base of the skull are foreshortened, while the vault of the skull is normal (the intramembranous bones being unaffected). Hydrocephalus ('water on the brain') is a serious

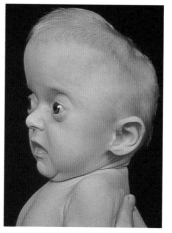

A baby with oxycephaly displays a peaked crown, and an under-developed vault. The condition may also result in poor eyesight.

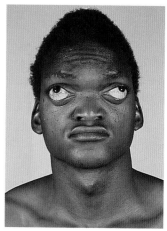

This man is displaying the classic signs of brachycephaly – a high, pointed head and bulging eyes – due to a fused coronal suture.

condition caused by a build-up of the cerebrospinal fluid surrounding the brain. This

enlarges the skull from the inside, putting immense pressure on the brain.

Base of the skull

This unusual view of the skull is from below. The upper jaw and the hole through which the spinal cord goes can be seen.

The bones found in the midline region of the base of the skull (the ethmoid, sphenoid and part of the occipital bone) develop in a different way from those of the vault of the skull. They are derived from an earlier cartilaginous structure in a process called endochondral ossification.

The maxillae are the two tooth-bearing bones of the upper jaw, one on each side. The palatine processes of the maxillae and the horizontal plates of the palatine bones form the hard palate.

PALATE DEFECTS
A cleft palate occurs when the structures of the palate do not fuse as normal before birth, creating a gap in the roof of the mouth. This links the oral and nasal cavities. If the gap extends through to the upper jaw, a harelip will become apparent on the upper lip. However, surgery can often improve the defect.

Children with narrow palates and crowded teeth can have an orthodontic appliance fitted which gradually increases tension across the longitudinally running midline palatine.

Over a period of months, the edges of the suture are forced apart, allowing for the growth of new bone, and extra space for the teeth.

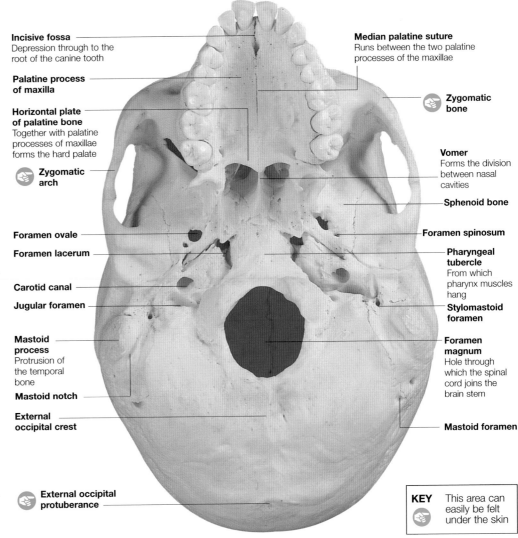

Incisive fossa
Depression through to the root of the canine tooth

Palatine process of maxilla

Horizontal plate of palatine bone
Together with palatine processes of maxillae forms the hard palate

Zygomatic arch

Foramen ovale

Foramen lacerum

Carotid canal

Jugular foramen

Mastoid process
Protrusion of the temporal bone

Mastoid notch

External occipital crest

External occipital protuberance

Median palatine suture
Runs between the two palatine processes of the maxillae

Zygomatic bone

Vomer
Forms the division between nasal cavities

Sphenoid bone

Foramen spinosum

Pharyngeal tubercle
From which pharynx muscles hang

Stylomastoid foramen

Foramen magnum
Hole through which the spinal cord joins the brain stem

Mastoid foramen

KEY This area can easily be felt under the skin

Foramina – channels through the skull

'Foramina' is the plural form of the Latin word *foramen*, meaning simply a hole or opening. These openings are the numerous canals through the bones of the skull that allow blood vessels and the 24 cranial nerves (12 on each side) to pass into and out of the cranial cavity.

Other small, less regular, channels link the external veins of the skull with those on the inside. They are termed emissary veins, and the openings are emissary foramina. Such pathways can allow the spread of an infection from outside the skull to a more serious infection inside. The most important foramina are:

- **foramen magnum**, where the spinal cord joins the brain stem
- **foramen lacerum**, between the petrous part of the temporal bone and the sphenoid
- **foramen ovale** (one on either side), for the mandibular branch of the tregiminal nerve
- **foramen spinosum**, allows the middle menigeal artery to pass into the interior of the cranial cavity
- **stylomastoid foramen**, transmits the seventh cranial nerve
- **jugular foramen**, aperture for the sigmoid and inferior petrosal sinus and three of the cranial nerves
- **carotid canal**, for the passage of the carotid artery (chief artery of the neck) and associated nerve fibres.

This close-up picture shows the left half of the middle cranial fossa (one of three depressions of the floor of the cranial cavity) from the inside, and four of the foramina: (1) foramen spinosum, (2) foramen ovale, (3) foramen lacerum, (4) foramen magnum.

Baby's skull

The newborn's skull has the same bones as an adult's skull, although it is much smaller. There are, however, significant differences in the proportions of the skull, the size and shape of the bones, and how they are joined.

If all the bones of the child's skull were to grow equally until reaching their adult size, then we would end up looking very different. What actually happens is that there is a marked change in the proportions of the skull, with the bones growing at different rates. The face in particular changes radically.

PROPORTIONS

In the newborn skull, the neurocranium (that houses and protects the brain) is about eight times as large as the viscerocranium, or face (which includes the jaws), whereas, in the adult it is only three times as large. This is because the brain develops rapidly and reaches adult proportions early in life, while the jaws, teeth and associated musculature develop over a longer period of time.

At birth, the circumference of the skull is, on average, about 33 cm and the capacity of the brain case is 400 ml. By two years of age, the circumference is about 47 cm and the brain capacity already nearly 1,000 ml, compared with the adult skull circumference of about 55 cm and a capacity of about 1,400 ml.

The orbit (eye socket) in the newborn is comparatively large and its floor is almost level with that of the nasal cavity.

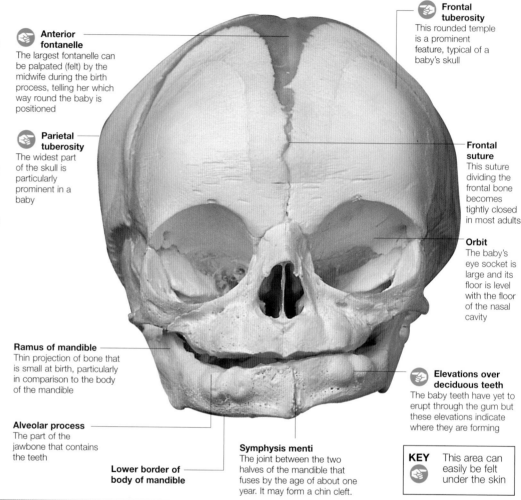

Anterior fontanelle
The largest fontanelle can be palpated (felt) by the midwife during the birth process, telling her which way round the baby is positioned

Parietal tuberosity
The widest part of the skull is particularly prominent in a baby

Frontal tuberosity
This rounded temple is a prominent feature, typical of a baby's skull

Frontal suture
This suture dividing the frontal bone becomes tightly closed in most adults

Orbit
The baby's eye socket is large and its floor is level with the floor of the nasal cavity

Ramus of mandible
Thin projection of bone that is small at birth, particularly in comparison to the body of the mandible

Alveolar process
The part of the jawbone that contains the teeth

Lower border of body of mandible

Symphysis menti
The joint between the two halves of the mandible that fuses by the age of about one year. It may form a chin cleft.

Elevations over deciduous teeth
The baby teeth have yet to erupt through the gum but these elevations indicate where they are forming

KEY This area can easily be felt under the skin

Development of the fetal skull

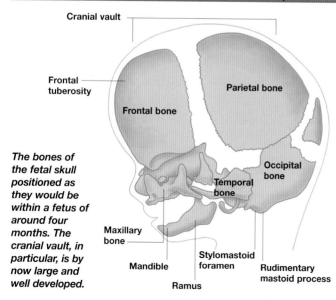

Cranial vault

Frontal tuberosity

Frontal bone

Parietal bone

Occipital bone

Temporal bone

Maxillary bone

Mandible

Stylomastoid foramen

Rudimentary mastoid process

Ramus

The bones of the fetal skull positioned as they would be within a fetus of around four months. The cranial vault, in particular, is by now large and well developed.

The bones of the skull of the newborn are smaller than their adult counterparts, the exceptions being the ossicles of the middle ear (malleus, incus and stapes), which are adult size at birth. Many of the bones are slightly different shapes from those in the adult due to their relative proportions.

In the baby, the bones of the cranial vault are more curved and the frontal and parietal tuberosities (at the temple, and above and behind the ear) are especially prominent.

The mandibular fossa, a depression in the temporal bone where the lower jaw hinges (temporomandibular joint), is flat. The mastoid process behind the ear canal is poorly developed. Consequently, the stylomastoid foramen of the temporal bone, is superficial. One of the nerves that supplies the facial muscles passes through this foramen. An occasional problem is a forceps delivery in which the baby's head is held behind the ears. This may compress the nerve and result in temporary facial paralysis.

The newborn's mandible has no defined chin, but is composed mostly of the alveolar process containing the developing teeth. Each maxillary (upper jaw) bone also consists mainly of its alveolar process.

The first teeth (deciduous) do not begin to erupt until about six months after birth. These teeth have all erupted by three years. The second (permanent) set may not be complete until the age of 20.

Joints of the baby's skull

Fontanelles are a prominent feature of the skull of the newborn.

Fontanelles are the fibrous membranes that fill in the gaps between the growing bones of the vault of the skull. These, and the wide sutures, permit sliding and overriding of the bones of the cranium during the passage of the head through the narrow birth canal. This often leads to temporary distortion of the skull at birth.

There are six fontanelles, each one located at the corners of the parietal bones.

FONTANELLE POSITIONS
Along the midline at the top of the cranium are the anterior and posterior fontanelles. The anterior is the largest of all the fontanelles and is diamond-shaped. It lies between the frontal bone and the parietal bones. At the back of the head is the small, triangular posterior fontanelle.

On each side of the skull are the paired sphenoidal (anterolateral) and mastoid (posterolateral) fontanelles. Both are small and irregular in shape. The posterior and sphenoidal fontanelles close up within three months of birth, the mastoid fontanelle closes at about one year, and the anterior fontanelle at about one-and-a-half years.

PALPATING FONTANELLES
During labour, when the head is engaged in the birth canal, the anterior and posterior fontanelles can be palpated (felt) and identified by the doctor or midwife. In the ideal labour position, the anterior fontanelle should be in front, and a reversal of this position indicates rotation of the baby within the uterus. This may result in a difficult labour.

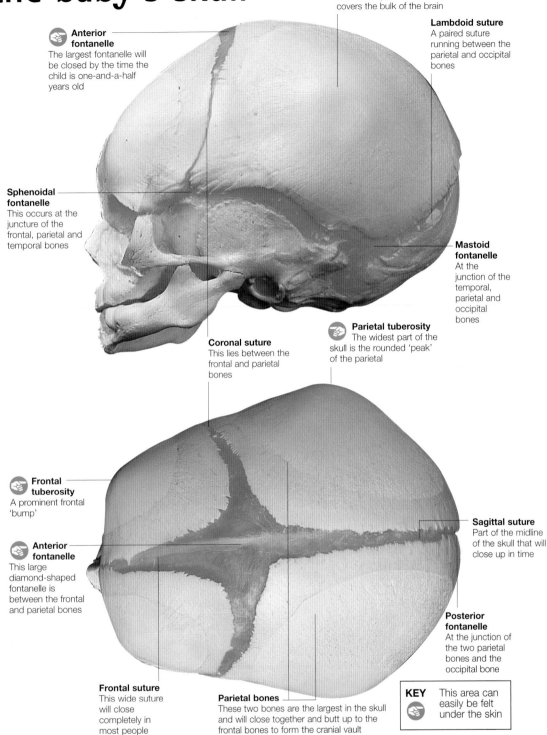

Parietal bone
One of the pair of bones that covers the bulk of the brain

Anterior fontanelle
The largest fontanelle will be closed by the time the child is one-and-a-half years old

Lambdoid suture
A paired suture running between the parietal and occipital bones

Sphenoidal fontanelle
This occurs at the juncture of the frontal, parietal and temporal bones

Mastoid fontanelle
At the junction of the temporal, parietal and occipital bones

Coronal suture
This lies between the frontal and parietal bones

Parietal tuberosity
The widest part of the skull is the rounded 'peak' of the parietal

Frontal tuberosity
A prominent frontal 'bump'

Anterior fontanelle
This large diamond-shaped fontanelle is between the frontal and parietal bones

Sagittal suture
Part of the midline of the skull that will close up in time

Posterior fontanelle
At the junction of the two parietal bones and the occipital bone

Frontal suture
This wide suture will close completely in most people

Parietal bones
These two bones are the largest in the skull and will close together and butt up to the frontal bones to form the cranial vault

KEY This area can easily be felt under the skin

The ways in which a child's skull grows

The bones of the growing skull develop in one of two ways. Some bones, such as those of the vault of the skull and face, may develop directly from a soft connective tissue membrane by the process of intramembranous ossification. Other bones, such as those found in the midline of the base of the skull (the ethmoid and parts of the sphenoid and occipital bones) are derived from pre-existing cartilage, and this is known as endochondral ossification.

The bones of the skull are joined by fibrous joints called sutures that allow for growth. The enlarging brain and eyeballs both generate a force sufficient to separate the bones at their sutures. Bone is then deposited at the edges of the sutures, stabilizing the skull at its new size.

When the brain stops its main phase of growth after about seven years, sutural growth also slows down and the skull enlarges at a slower rate by bone remodelling. Bone is deposited on the outer surface of the skull and reabsorbed on the inner surface. This gradually changes the shape of bones during continued growth.

In the child's skull, certain bones take some years to fuse and form a single bone, because the ossification occurs at more than one site.

For example, the frontal bone is initially separated into two by a midline suture which disappears at about the age of four. The mandible is also in two halves, separated by a midline mandibular symphysis, which fuses between the ages of one and two. At birth, the occipital bone is divided into four parts, and complete fusion does not occur until around six years.

Teeth

Teeth are designed for biting and chewing up food, and each has a particular function.

The teeth are specialized hardened regions of gum tissue, partly embedded in the jaw bones. They break up solid foods by the actions of biting and chewing.

The visible part of a tooth is the crown. This is composed of a shell of hard, calcified material called dentine (similar to compact bone but without blood vessels), which is covered by a thin layer of even harder calcified material called enamel.

The hidden part (the root) is embedded in a socket of the jaw bone (the alveolus). It is also made of dentine, covered by a layer of cementum which, with dense, collagen-rich periodontal ligaments, anchors the root to the bone of the alveolar socket.

INSIDE THE TEETH

Inside is an internal pulp cavity, containing soft connective tissue, blood vessels and nerves. The pulp is linked to the jaw via the root.

The layout of adult teeth is the same in the upper and lower jaws. Each side (quadrant) has eight teeth: two incisors, one canine, two premolars and three molars, making 32 in total. Children have 20 milk teeth, with only one molar in each quadrant.

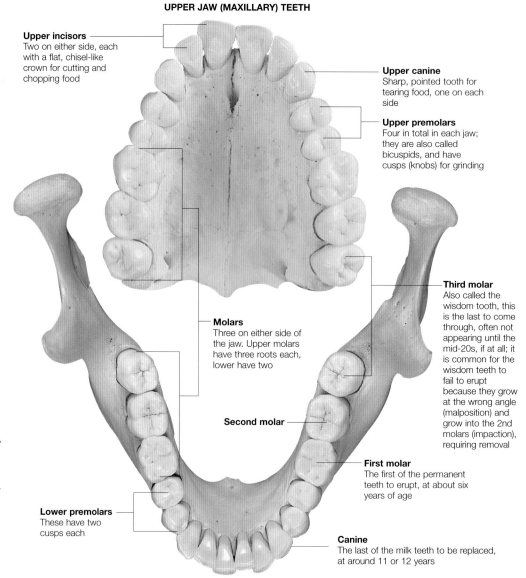

UPPER JAW (MAXILLARY) TEETH

Upper incisors
Two on either side, each with a flat, chisel-like crown for cutting and chopping food

Upper canine
Sharp, pointed tooth for tearing food, one on each side

Upper premolars
Four in total in each jaw; they are also called bicuspids, and have cusps (knobs) for grinding

Molars
Three on either side of the jaw. Upper molars have three roots each, lower have two

Second molar

Third molar
Also called the wisdom tooth, this is the last to come through, often not appearing until the mid-20s, if at all; it is common for the wisdom teeth to fail to erupt because they grow at the wrong angle (malposition) and grow into the 2nd molars (impaction), requiring removal

First molar
The first of the permanent teeth to erupt, at about six years of age

Lower premolars
These have two cusps each

Canine
The last of the milk teeth to be replaced, at around 11 or 12 years

LOWER JAW (MANDIBULAR) TEETH

Tooth shape and function

The teeth are a variety of shapes, and each one is specialized for a particular function. The incisors at the front of the mouth have a flat, chisel-like crown for cutting. Behind them, the canines are sharp and pointed, and are used to tear tough food.

The surface of the crowns of the premolars and molars is broader, with cusps to aid grinding. The premolars (also known as the bicuspids) have two cusps each, and the molars have four or five cusps.

The arrangement of the teeth in the upper jaw (maxillary teeth) and the lower jaw (mandibular teeth) is essentially identical, although there are some differences in size and shape. The maxillary incisors, for instance, are typically wider than the mandibular incisors.

The roots of the teeth vary; incisors have one root, while lower molars have two roots and upper molars have three roots.

Human dentition, like that of other primates, is thought to have been originally suited to a diet of fruits, nuts and roots. It has, however, proved to be flexible, by adapting to a wider, omnivorous diet.

- Cusp
- Enamel
- Pulp
- Dentine
- Cementum

This cross-section of a lower molar shows typical tooth structure. The two large roots need secure support from the gums to remain snug in their jaw socket.

This X-ray is an orthopantomograph, showing all the teeth of both jaws. Such images are taken using a special machine that moves horizontally around the face.

Development of teeth

There are two major phases of tooth development during childhood. This is to allow the head to grow and adult teeth to develop.

The teeth begin to develop in the human embryo around the sixth week of pregnancy. Six to eight months after birth, root growth pushes the tooth crown through the gum in the process of eruption called teething.

This first set are the primary or deciduous teeth (milk teeth). These erupt in a specific order, usually the lower central incisors first, then the upper central incisors. The deciduous teeth do not include premolars.

ADULT TOOTH GROWTH

The tooth buds for the second wave of tooth production develop at the same time. These permanent teeth remain dormant until the ages of five to seven, when they begin to grow, causing the roots of the deciduous teeth to break down.

This breakdown, together with the pressure of the underlying permanent teeth, results in the shedding of the deciduous teeth. The new teeth then start to appear and continue to do so until the ages of 10 to 12.

Eruption of the permanent set follows a similar pattern to the earlier growth (although premolars erupt between the canines and the molars). The permanent set has additional, third molars (wisdom teeth) that tend to appear after between 15 and 25 years.

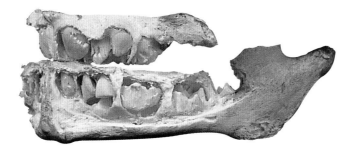

Jaws of a newborn
Unerupted deciduous (primary or milk) teeth can be be seen in the dental follicles (tooth-bearing capsules) of both jaws. They will start to emerge at about six months of age.

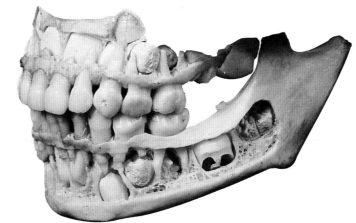

Jaws of a six-year-old
The deciduous teeth have all erupted. Beneath them in the alveoli (tooth sockets) are the permanent teeth ready to come through. This process continues until the early to mid teens.

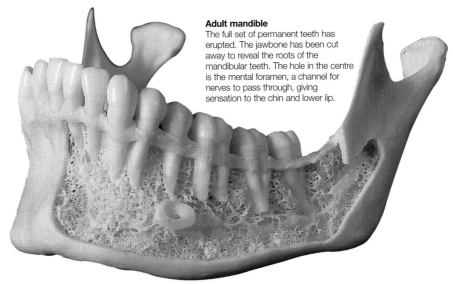

Adult mandible
The full set of permanent teeth has erupted. The jawbone has been cut away to reveal the roots of the mandibular teeth. The hole in the centre is the mental foramen, a channel for nerves to pass through, giving sensation to the chin and lower lip.

Dental decay and other problems

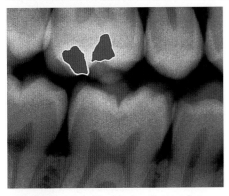

This coloured X-ray of the teeth reveals two metal fillings in one of the upper premolars. The pink areas indicate the central pulp within each tooth.

Dental caries (tooth decay) is caused by the formation of plaque, a combination of saliva, food residue and acid-producing bacteria that can eat into tooth enamel and dentine. Once the decay goes deep enough, infection and inflammation of the dental pulp inside may result. The pulp is living tissue, so this causes much pain. If untreated, the tooth will die. Infection can also result in a dentoalveolar abscess (gumboil), when the gum erupts.

Dentists can treat cavities by root canal work, removing the pulp (including the vessels and nerves), cleaning out the canal, and packing with a suitable material.

Gum disease is a major problem. Infections of the gum (gingivitis) can result in loosening or loss of teeth; there may even be a loss of bone from the jaws. This is probably due to a lack of mechanical force (chewing) on the bones.

Tartar (also known as calculus) is a chalky residue from saliva. If not regularly removed, it can harm the gums and shelter bacteria, increasing the likelihood of tooth decay.

When the internal pulp is infected, the tooth can be saved by drilling out the root canal with a fine drill. The cavity is then filled, to avoid recurrence of the problem.

How teeth develop

Teeth are the body's toughest and most durable organs. They play a vital role in the digestion of food by helping to break it down into smaller fragments by biting and mastication (chewing).

The teeth are used to chew and grind food into smaller pieces. The action of chewing increases the surface area of food exposed to digestive enzymes, thus speeding up the process of digestion.

Teeth also play an important role in speech – the teeth, lips and tongue allow the formation of words by controlling airflow through the mouth. In addition, teeth provide structural support for muscles in the face as well as helping to form the smile.

ANATOMY OF THE TEETH

Each tooth is composed of the crown and the root. The crown is the visible part of the tooth which emerges from the gingiva or gum (which helps to hold the tooth firmly in place). The crown of each premolar and molar includes projections or cusps which facilitate the chewing and grinding of food.

Teeth provide structural support for the muscles of the face. In addition, they play an important role in speech, and help to form the smile.

The root is the portion of the tooth embedded in the jawbone.

STRUCTURE

Teeth are composed of four distinct types of tissue:
- Enamel – the clear outer layer and the hardest substance in the body. Composed of a densely packed structure, heavily mineralized with calcium salts, this layer helps to protect the inner layers of the teeth from harmful bacteria, and changes in temperature caused by hot or cold food and drink
- Dentine – encloses and protects the inner core of the tooth, and is similar in composition to bone. It is composed of odontoblast cells, which secrete and maintain dentine throughout adult life
- Pulp – contains blood vessels which supply the tooth with oxygen and nutrients. It also contains nerves responsible for the transmission of pain and temperature sensations to the brain
- Cementum – covers the outer surface of the root. It is a calcium-containing connective tissue that attaches the tooth to the periodontal ligament, which anchors the tooth firmly in the tooth socket (alveolus) located in the jawbone.

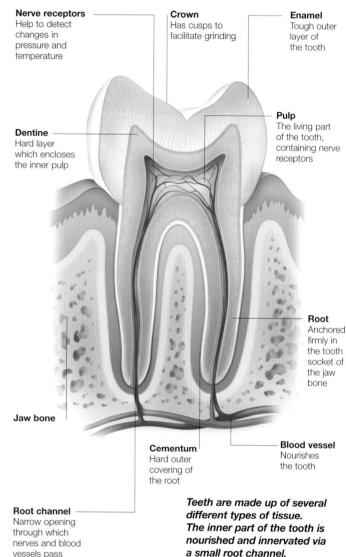

Nerve receptors
Help to detect changes in pressure and temperature

Dentine
Hard layer which encloses the inner pulp

Jaw bone

Root channel
Narrow opening through which nerves and blood vessels pass

Crown
Has cusps to facilitate grinding

Enamel
Tough outer layer of the tooth

Pulp
The living part of the tooth, containing nerve receptors

Root
Anchored firmly in the tooth socket of the jaw bone

Cementum
Hard outer covering of the root

Blood vessel
Nourishes the tooth

Teeth are made up of several different types of tissue. The inner part of the tooth is nourished and innervated via a small root channel.

Development of milk teeth

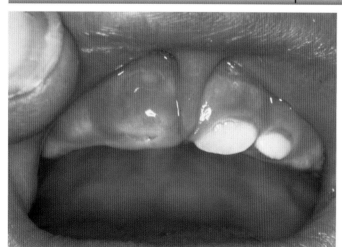

Humans develop two sets of teeth during their lives. The first set, known as the deciduous or milk teeth, starts developing in the fetus at around two months after conception and will consist of 20 teeth in total.

STAGES OF DEVELOPMENT

The dentine of these teeth forms while the fetus is still in the uterus. After birth, the tooth enamel develops in stages.

Tooth enamel begins to develop soon after birth. The milk teeth emerge in stages, with the front teeth erupting first and the second molars last.

Front tooth enamel develops first and is usually complete at around one month after birth, while the enamel on the second (back) molars is not completely developed until a year and a half later.

Once the enamel is fully developed the tooth begins to emerge (erupt). Front teeth usually erupt at between six and 12 months of age, while second molars emerge between 13 and 19 months and canines at 19 months or older. The final stage of root development is root completion, a slow process that continues until the child is over three years old.

Development of permanent teeth

Over several years, the milk teeth are replaced by a set of permanent adult teeth. There are 32 adult teeth, including the third molars, or wisdom teeth.

Around the age of six, the roots of the deciduous teeth are slowly eroded by the pressure of erupting permanent teeth and by the action of specialized bone cells in the jaws. This process, called resorption, allows the permanent teeth to emerge from beneath. If a permanent tooth is missing – a relatively common condition, the corresponding milk tooth is retained.

COMPLETE SET
As the milk teeth are replaced, the mouth and jaw lose their childhood shape and take on a more pronounced and adult appearance. Adult teeth are usually darker in colour and differ in size and proportion from the milk teeth. A full complement of permanent teeth is generally present by the end of adolescence, with the exception of the third molars (wisdom teeth) which tend to emerge at around 18–25 years.

By the age of six most children have begun to lose their milk teeth. These become loose and are eventually replaced by the permanent teeth.

Types of teeth

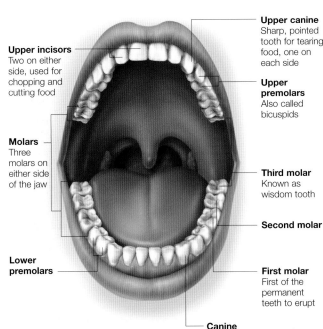

Upper incisors
Two on either side, used for chopping and cutting food

Molars
Three molars on either side of the jaw

Lower premolars

Upper canine
Sharp, pointed tooth for tearing food, one on each side

Upper premolars
Also called bicuspids

Third molar
Known as wisdom tooth

Second molar

First molar
First of the permanent teeth to erupt

Canine

Adults generally have 32 teeth – 16 on the upper jaw and 16 on the lower – which fit together to bite and chew food. Humans are referred to as heterodonts as they have a variety of different tooth types, with a specific size, shape and function:

■ Incisors – adults have eight incisors located at the front of the mouth – four in the upper jaw, and four in the lower jaw. Incisors have a sharp edge used to cut up food

■ Canines – on either side of the incisors are the canines, so-called because of their resemblance to the sharp fangs of dogs. There are two canines on each jaw, and their primary role is to pierce and tear food

■ Bicuspids – also known as premolars, these are flat teeth with pronounced cusps that grind and mash food. There are four bicuspids in each jaw

■ Molars – behind the bicuspids are the molars, where the most vigorous chewing occurs. There are twelve molars, referred to as the first, second and third molars. Third molars are commonly called wisdom teeth.

Humans are heterodonts – that is, their teeth are not of a uniform size or shape. The various types of teeth are formed differently to fulfil specific roles.

WISDOM TEETH
Wisdom teeth are remnants from thousands of years ago when the human diet consisted of raw foods that required the extra chewing and grinding power of a third set of molars. Today, wisdom teeth are not required for chewing and, as they can crowd other teeth and cause them to become impacted, they are often removed by dentists.

Chewing

The muscles of the jaws allow the teeth to close together (occlude) and open in a vertical plane, and also to slide over one another in a horizontal plane. The former action is necessary in biting through food, the latter in grinding.

Pressure sensors
The periodontal ligament of each tooth contains sensory nerve receptors that protect the teeth and supporting tissues from excessive chewing and biting forces, which can cause damage. These nerve receptors transmit impulses to the central nervous system in response to sensation and stimulation and give information about jaw movement and position, and pressure being exerted on the teeth.

Brain control
The brain responds by sending nerve impulses which control the position of the jaw and attached muscles (and thus the position of the tooth) allowing the process of chewing (mastication) to occur in an appropriate manner.

Nerve receptors in each tooth convey information to the brain about the position of the jaw. The brain then transmits nerve impulses, controlling the jaw.

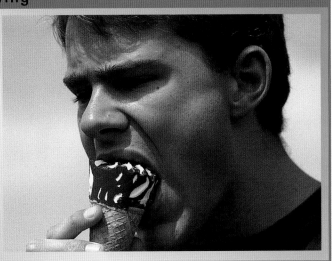

Scalp

The scalp is composed of five layers of tissue that cover the bones of the skull. The skin is firmly attached to the muscles of the scalp by connective tissue which also carries numerous blood vessels.

The scalp is the covering of the top of the head which stretches from the hairline at the back of the skull to the eyebrows at the front. It is a thick, mobile, protective covering for the skull, and it has five distinct layers, the first three of which are bound tightly together.

PROTECTION

The skin of the scalp is the thickest in the body and the hairiest. As well as its functions of hair-bearing and protection of the skull, the skin of the front of the scalp in particular has an important role in facial expression. This is because many of the fibres of the scalp muscles are attached to the skin, allowing it to move backwards and forwards.

DENSE CONNECTIVE TISSUE

Under the skin, and attached firmly to it, is a layer of dense tissue which carries numerous arteries and veins. The arteries are branches of the external and internal carotid arteries, which interconnect to give a rich blood supply to all areas of the scalp.

This layer of connective tissue is also attached firmly to the underlying layer of muscle. The connective tissue binds the skin to the muscle in such a way that even if the scalp is torn from the head in an accident, these three layers will remain together.

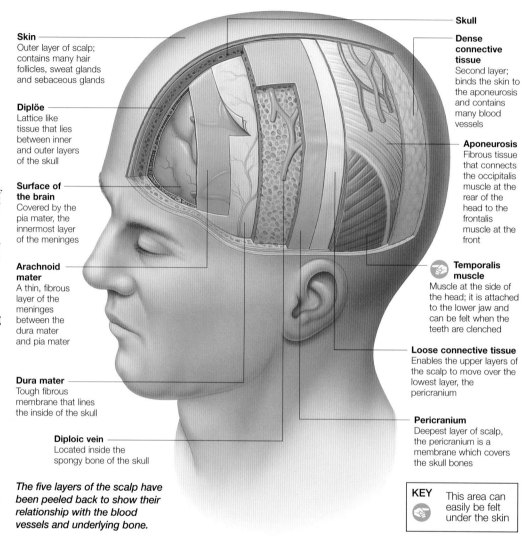

Skin
Outer layer of scalp; contains many hair follicles, sweat glands and sebaceous glands

Diplöe
Lattice like tissue that lies between inner and outer layers of the skull

Surface of the brain
Covered by the pia mater, the innermost layer of the meninges

Arachnoid mater
A thin, fibrous layer of the meninges between the dura mater and pia mater

Dura mater
Tough fibrous membrane that lines the inside of the skull

Diploic vein
Located inside the spongy bone of the skull

The five layers of the scalp have been peeled back to show their relationship with the blood vessels and underlying bone.

Skull

Dense connective tissue
Second layer; binds the skin to the aponeurosis and contains many blood vessels

Aponeurosis
Fibrous tissue that connects the occipitalis muscle at the rear of the head to the frontalis muscle at the front

Temporalis muscle
Muscle at the side of the head; it is attached to the lower jaw and can be felt when the teeth are clenched

Loose connective tissue
Enables the upper layers of the scalp to move over the lowest layer, the pericranium

Pericranium
Deepest layer of scalp, the pericranium is a membrane which covers the skull bones

KEY This area can easily be felt under the skin

Hair follicles of the scalp

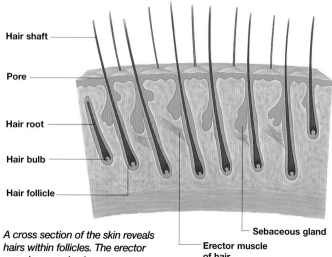

Hair shaft

Pore

Hair root

Hair bulb

Hair follicle

Sebaceous gland

Erector muscle of hair

A cross section of the skin reveals hairs within follicles. The erector muscles can also be seen.

The hair-bearing part of the scalp is the hairiest region of the body. Scalp hair provides the head with insulation against cold and protection from sunlight and, to some extent, acts to cushion the head from minor injury.

Each hair consists of a root embedded in the skin and the shaft which protrudes from the scalp. Within the scalp, the root is enclosed in a hair follicle. The hair projects from the follicle at an angle of less than 90 degrees and thus covers the skin effectively.

The hair follicles of the scalp go through a cycle of growth and rest phases. After an active growth stage, the follicle and hair bulb rest for a short time.

Hair is shed during the rest phase, but because the growth and rest phases of the individual scalp follicles are staggered, hair loss is not normally noticeable.

Sebaceous glands, which produce an oily secretion called sebum, are also found in the skin connected to the follicles. The sebum lubricates the hair shaft and plays a role in protecting the skin against bacteria and fungi.

Attached to the follicle is a muscle which pulls the hair into an erect position when it contracts, producing 'goose pimples'. The contraction of this muscle also squeezes the sebaceous gland, which in turn releases sebum.

Muscles of the scalp

The muscles of the scalp lie below the skin and a layer of connective tissue. They act to move the skin of the forehead and the jaw while chewing.

The occipitofrontalis is a large muscle formed by two sections at the front and the back of the scalp, which are connected by a thin, tough, fibrous sheet (aponeurosis). The frontalis is the section of muscle over the forehead, arising from the skin overlying the eyebrow and passing back to become continuous with the aponeurosis. This muscle acts to raise the eyebrows, thus wrinkling the forehead or pulling the scalp forward, as when frowning.

The occipitalis is the section of muscle that arises from the top of the back of the neck and passes forward to the aponeurosis. It acts to pull the scalp backwards.

The temporalis muscle lies at the side of the scalp, above the ears, and runs from the skull down to the lower jaw. It is involved in the action of chewing.

LOOSE CONNECTIVE TISSUE

The fourth layer, underlying the muscle and aponeurosis, is a layer of loose connective tissue which allows the layers above to move relatively freely over the layer below. It is at this level that the scalp may be torn away during accidents, such as the head going forward through the windscreen of a car.

The pericranium, the fifth layer of the scalp, is the tough membrane covering the bone of the skull itself.

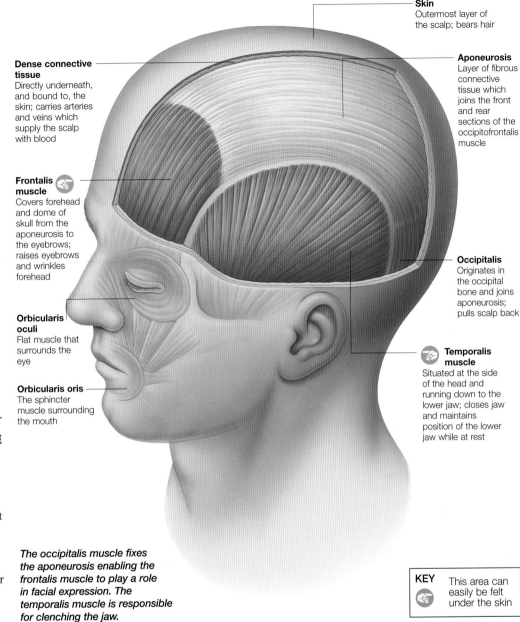

Skin
Outermost layer of the scalp; bears hair

Aponeurosis
Layer of fibrous connective tissue which joins the front and rear sections of the occipitofrontalis muscle

Dense connective tissue
Directly underneath, and bound to, the skin; carries arteries and veins which supply the scalp with blood

Frontalis muscle
Covers forehead and dome of skull from the aponeurosis to the eyebrows; raises eyebrows and wrinkles forehead

Orbicularis oculi
Flat muscle that surrounds the eye

Orbicularis oris
The sphincter muscle surrounding the mouth

Occipitalis
Originates in the occipital bone and joins aponeurosis; pulls scalp back

Temporalis muscle
Situated at the side of the head and running down to the lower jaw; closes jaw and maintains position of the lower jaw while at rest

The occipitalis muscle fixes the aponeurosis enabling the frontalis muscle to play a role in facial expression. The temporalis muscle is responsible for clenching the jaw.

KEY This area can easily be felt under the skin

Trauma to the scalp

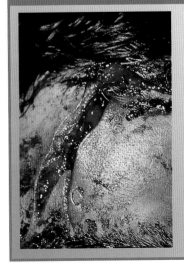

Major wounds to the scalp bleed excessively as the arteries do not constrict after injury. This hampers the clotting response.

Damage to the scalp causes profuse bleeding, even from a relatively minor cut. Two factors together explain why the scalp bleeds so profusely and for so long when it is cut.

To nourish its many hair follicles, the scalp has a much more profuse blood supply compared to the skin of the rest of the body. Blood is supplied to the scalp via several arteries which anastomose (interconnect) freely within the dense connective tissue layer underlying the skin. These interconnecting vessels provide a good blood supply for the whole of the scalp.

The fibrous tissue within this dense layer acts to hold these blood vessels open, so that they are unable to contract as arteries in other sites would in the event of an injury. If the artery cannot react to damage by narrowing its aperture, then blood clotting is hampered. To stop the bleeding, pressure applied to the area of the wound is needed.

Most of the blood that serves the scalp travels through vessels that lie on the surface of the muscles, just below the skull.

Facial muscles

One of the features that distinguishes humans from animals is
our ability to communicate using a wide range of facial expressions.
The power behind this ability is a complex system of facial muscles.

Just under the skin of the scalp and face lies a group of very thin muscles, which are collectively known as the muscles of facial expression. These muscles play a vital role in a number of ways, in addition to their physiological function. They alter facial expression – providing a means of non-verbal communication by transmitting a range of emotional information – and are also one of the means of articulating speech.

Apart from this, the facial muscles also form sphincters that open and close the orifices of the face – the eyes and mouth.

SKIN AND BONE

The majority of facial muscles are attached to the skull bone at one end and to the deep layer of skin (dermis) at the other. From these attachments, it can be seen see how the numerous muscles alter facial expression, and also how they eventually cause creases and wrinkles in the overlying skin.

A number of small muscles called 'dilators' open the mouth. They radiate out from the corners of the mouth and lips, where they have an attachment to bone. The mouth and lips can be pulled up, pushed down and moved from side to side.

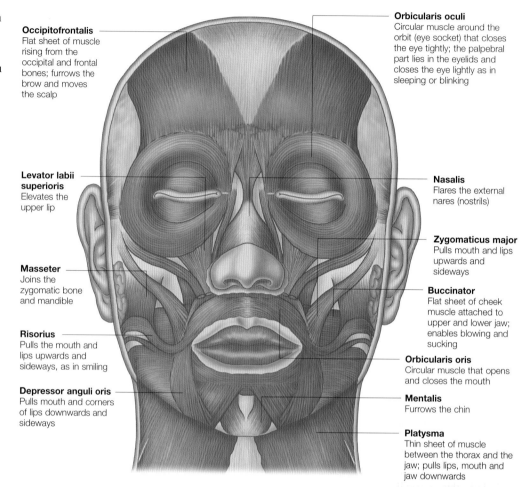

Occipitofrontalis
Flat sheet of muscle rising from the occipital and frontal bones; furrows the brow and moves the scalp

Levator labii superioris
Elevates the upper lip

Masseter
Joins the zygomatic bone and mandible

Risorius
Pulls the mouth and lips upwards and sideways, as in smiling

Depressor anguli oris
Pulls mouth and corners of lips downwards and sideways

Orbicularis oculi
Circular muscle around the orbit (eye socket) that closes the eye tightly; the palpebral part lies in the eyelids and closes the eye lightly as in sleeping or blinking

Nasalis
Flares the external nares (nostrils)

Zygomaticus major
Pulls mouth and lips upwards and sideways

Buccinator
Flat sheet of cheek muscle attached to upper and lower jaw; enables blowing and sucking

Orbicularis oris
Circular muscle that opens and closes the mouth

Mentalis
Furrows the chin

Platysma
Thin sheet of muscle between the thorax and the jaw; pulls lips, mouth and jaw downwards

Platysma

The large, flat platysma muscle is a superficial (close to the surface) muscle extending from below the collar bone to the skin and muscle of the mouth and lips. Its main role is lowering the lower lip and jaw, as in a snarl.

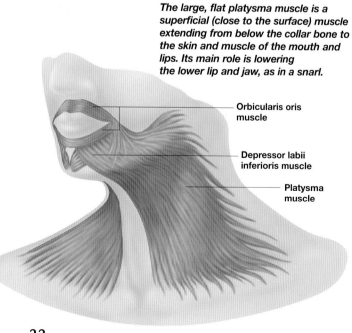

Orbicularis oris muscle

Depressor labii inferioris muscle

Platysma muscle

Although not strictly a muscle of the head, the platysma muscle plays an important role in facial expression. This thin sheet of muscle extends from below the collar bone up to the mandible. It covers the front of the neck, where it tightens the skin, and connects to the muscle and skin at the corners of the mouth.

The platysma's role in altering facial expression is to pull the neck skin out and depress the mandible (lower the jaw). This pulls the mouth down, as in an expression of disgust. It also assists in the movement of the lower lip.

In addition, the platysma is the muscle that is cut and pinned behind the ear during plastic surgery to reduce 'double chins'. This is, arguably, another way in which the platysma can alter the facial expression.

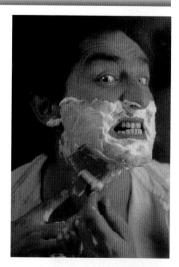

The platysma muscle is the large sheet of muscle that is tensed while shaving under the chin. This action stretches and tightens the skin of the neck.

Opening and closing the eye

Whether fluttered alluringly or squeezed tightly shut for protection, the eyelids communicate a range of non-verbal signals. The eyelids are also vital for cleaning and lubricating the eyes.

The orbicularis oculi is the muscle responsible for the closing of the eye. This flat sphincter muscle lines the rim of the orbit (eye socket), and various sections of it can be manipulated individually.

Part of the orbicularis oculi lies in the eyelid (the palpebral part). This section of the muscle closes the eye lightly, as in sleeping or in routine blinking. This action also aids the flow of lacrimal secretion (tears) across the conjunctiva (the membrane covering the eyes) to keep it clean, free of foreign bodies and lubricated.

OPENING THE LIDS

A larger part of the orbicularis oculi consists of concentrically arranged fibres that cover the front of the eye socket. The role of this part of the muscle is to close the eye tightly (screw up the eyes) to protect against a blow or bright light.

The second orbital muscle is the levator palpebrae superioris. As its name suggests, this small muscle pulls on the upper lid to open the eye. Unlike the larger orbicularis, this muscle lies within the eye socket.

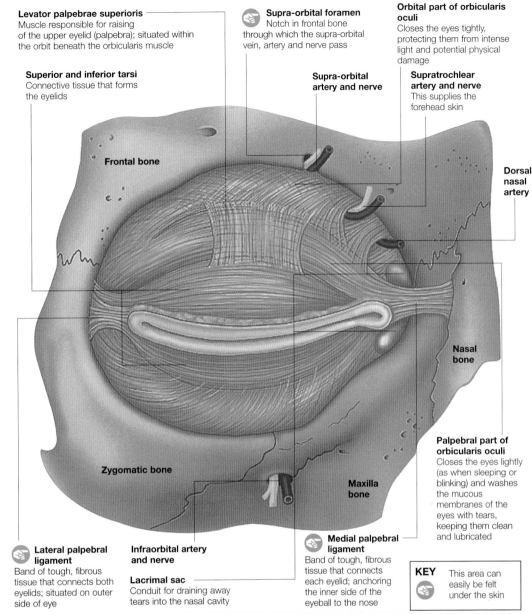

Levator palpebrae superioris
Muscle responsible for raising of the upper eyelid (palpebra); situated within the orbit beneath the orbicularis muscle

Superior and inferior tarsi
Connective tissue that forms the eyelids

Frontal bone

Supra-orbital foramen
Notch in frontal bone through which the supra-orbital vein, artery and nerve pass

Supra-orbital artery and nerve

Orbital part of orbicularis oculi
Closes the eyes tightly, protecting them from intense light and potential physical damage

Supratrochlear artery and nerve
This supplies the forehead skin

Dorsal nasal artery

Nasal bone

Palpebral part of orbicularis oculi
Closes the eyes lightly (as when sleeping or blinking) and washes the mucous membranes of the eyes with tears, keeping them clean and lubricated

Zygomatic bone

Maxilla bone

Lateral palpebral ligament
Band of tough, fibrous tissue that connects both eyelids; situated on outer side of eye

Infraorbital artery and nerve

Lacrimal sac
Conduit for draining away tears into the nasal cavity

Medial palpebral ligament
Band of tough, fibrous tissue that connects each eyelid; anchoring the inner side of the eyeball to the nose

KEY This area can easily be felt under the skin

Muscles of the cheek and lips

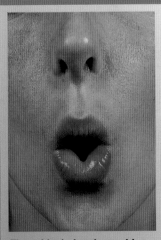

The orbicularis oris provides sphincteric control of the mouth. When the muscles contract, the orifice closes up.

The buccinator muscle forms the cheek on the side of the face. Its upper margin is attached to the maxilla bone along a line just above the upper tooth sockets, and its lower to the mandible below the lower tooth sockets. Its main action occurs during eating.

The buccinator pushes food that has collected in the cheek during chewing back on to the tongue and between the surfaces of the teeth ready for the next bite. The buccinator is also used in activities such as blowing up a balloon or playing a musical instrument, such as a trumpet (the word 'buccinator' is derived from the Latin for trumpet).

The muscle within the upper and lower lips surrounding the opening of the mouth is the orbicularis oris. It consists of concentrically arranged muscle fibres which curve around the angles of the mouth and are attached to bone beneath the nose at one end and to a region above the point of the chin at the other.

The front margin of buccinator contributes muscle fibres that crisscross at the corners of the mouth, and blend with the orbicularis oris.

The orbicularis oris is the muscle that closes the mouth. It is in continuous use to hold the lips firmly together and prevent saliva, which is being produced all the time, escaping from the mouth and also to retain food in the mouth during eating. It is also used to purse the lips, as in whistling or blowing a kiss.

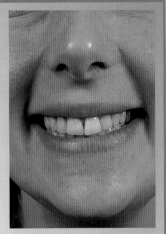

The versatility of the ring of muscle fibres around the lips helps humans to articulate a wide range of sounds.

Facial nerves

The facial muscles, and the involuntary functions such as tear formation, are served by the facial nerve, which transmits signals to and from the brain.

The muscles of facial expression are supplied by left and right facial nerves, each supplying muscle on their respective side of the face. Each nerve emerges through a hole in the skull (the stylomastoid foramen) next to the lower part of the ear, and reaches the facial muscles by branching through a salivary gland (the parotid) located on the side of the face.

Nerves are bundles of fibres that transmit electrical impulses from the brain or spinal column to the muscles, or from the sense organs to the brain or spinal column. Most nerves – including the facial nerves – are composed of a mixture of the two types, sending and receiving data to and from the brain.

NERVE DAMAGE

There are 12 pairs of cranial nerves serving various functions from moving the eyeballs to maintaining balance. The facial nerves are the seventh pair, and their principle task is providing motor impulses to the muscles of facial expression. (The muscles of mastication, used for chewing food, are served by the fifth cranial nerve, the trigeminal.)

As well as innervating the muscles (transmitting impulses to them), the facial nerves serve the autonomic functions, such as the production of tears and saliva. They also convey sensory impulses from the taste buds.

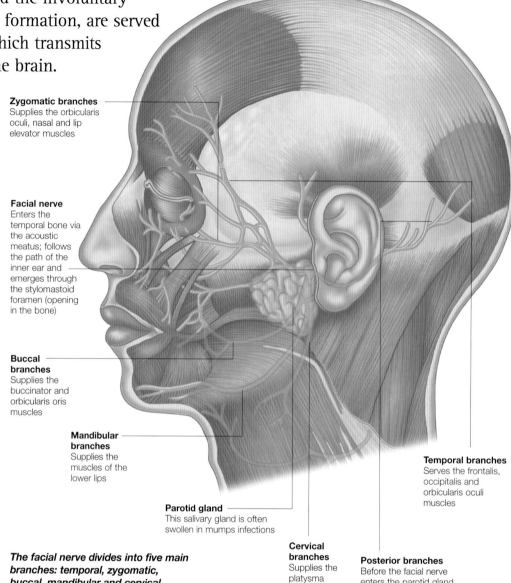

Zygomatic branches
Supplies the orbicularis oculi, nasal and lip elevator muscles

Facial nerve
Enters the temporal bone via the acoustic meatus; follows the path of the inner ear and emerges through the stylomastoid foramen (opening in the bone)

Buccal branches
Supplies the buccinator and orbicularis oris muscles

Mandibular branches
Supplies the muscles of the lower lips

Parotid gland
This salivary gland is often swollen in mumps infections

Temporal branches
Serves the frontalis, occipitalis and orbicularis oculi muscles

Cervical branches
Supplies the platysma muscles

Posterior branches
Before the facial nerve enters the parotid gland and divides, it serves the occipitalis (scalp) and auricular (ear) muscles

The facial nerve divides into five main branches: temporal, zygomatic, buccal, mandibular and cervical. These five branches fan out over the face and further divide, serving the muscles of facial expression.

Facial nerve disorders

With a relaxed gaze, a Bell's palsy sufferer displays the characteristic facial droop (in this case the left side).

When asked to grin and screw up the eyes, the patient cannot move the left side of the mouth or eye because of paralysis.

The facial nerve can be damaged by direct trauma to the side of the face or by inflammation, which causes the nerve to swell where it lies in the skull bone (facial canal). This may result in paresis (weakness) or paralysis of the facial muscles, giving that side of the face a droopy expression.

In a person with facial nerve damage, the eye is permanently open, leaving the cornea and conjunctiva in danger of drying. Speech is slurred as the lips cannot articulate clearly and the mouth cannot form an effective seal.

Consequently, saliva and food often spill out of the mouth.

Bell's palsy is a paralysis most commonly associated with the swelling of the facial nerve. The symptoms are numerous, affecting hearing, taste, vision and muscle strength.

The facial nerve can occasionally be damaged if an infant is delivered by forceps. The bony mastoid process (protuberance) at the back of the ear is undeveloped in the new born, leaving the nerve unprotected. Such damage would leave the facial muscles paralysed, preventing suckling.

Muscles of mastication

The muscles that help us chew our food also play a part in speech, breathing and yawning.

The muscles of mastication are the muscles that move the mandible (jaw bone) up and down, and forwards and backwards, resulting in the opening and closing of the mouth.

This action is used in activities such as speaking, breathing through the mouth and in yawning. The closing action is also used very powerfully in the movements necessary for biting off and chewing up food (mastication), when side-to-side slewing of the jaw is also employed.

MOVING THE JAW

All jaw movements take place at the pair of temporo-mandibular joints, which lie in front of the ears.

The bones forming the joint are the head of the mandible (the rounded section at the top of the jaw bone) and the mandibular fossa of the temporal bone (the hollow in the skull in which the head of the mandible sits).

The hinge-like action allows up and down movements of the jaw. Additionally, the head of the mandible is covered with a closely fitting disc of cartilage, which allows forward and backward rocking movements. This latter movement enables the lower jaw to be slewed across the upper jaw on opening, and so provides the sideways forces necessary to grind up hard food on closing the mouth and chewing.

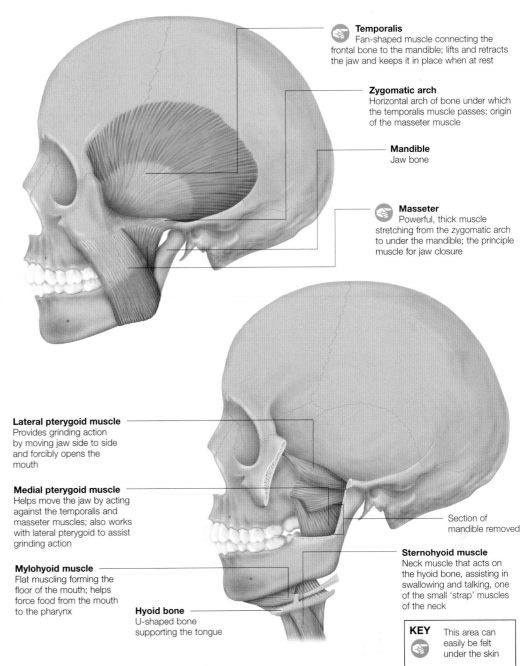

Temporalis
Fan-shaped muscle connecting the frontal bone to the mandible; lifts and retracts the jaw and keeps it in place when at rest

Zygomatic arch
Horizontal arch of bone under which the temporalis muscle passes; origin of the masseter muscle

Mandible
Jaw bone

Masseter
Powerful, thick muscle stretching from the zygomatic arch to under the mandible; the principle muscle for jaw closure

Lateral pterygoid muscle
Provides grinding action by moving jaw side to side and forcibly opens the mouth

Medial pterygoid muscle
Helps move the jaw by acting against the temporalis and masseter muscles; also works with lateral pterygoid to assist grinding action

Mylohyoid muscle
Flat muscling forming the floor of the mouth; helps force food from the mouth to the pharynx

Hyoid bone
U-shaped bone supporting the tongue

Section of mandible removed

Sternohyoid muscle
Neck muscle that acts on the hyoid bone, assisting in swallowing and talking, one of the small 'strap' muscles of the neck

KEY This area can easily be felt under the skin

Dislocated jaw

As the jaw is opened, the head of the mandible and its disc of cartilage moves forward out of the joint socket and on to a tubercle (small protrusion) in front. This forward movement can be easily seen and felt just in front of the ear hole.

If this movement goes too far, as in a wide yawn or enthusiastic laughter, the mandible may slip in front of the tubercle and become lodged under the zygomatic bone. This jams the mouth open, necessitating medical treatment. The same action can result from a blow to the side of the jaw, and is the reason boxers are taught to keep

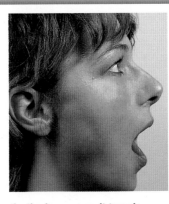

As the jaw opens, it travels forward on to a tubercle, and can be felt as a swelling in front of the ear.

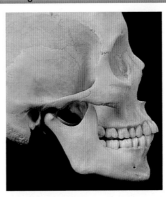

If the jaw is opened too wide, or is knocked out of place, it can cause a spasm of the temporalis muscle, locking the jaw open.

the mouth shut and teeth tightly clamped on a gum shield.

Unlocking the jaw and closing the mouth involves pushing the jaw downwards, against the pull of the temporalis, masseter and medial pterygoid muscles. This forces the head of the mandible back over the tubercle, and the joint snaps back into place.

When performing this manoeuvre, it is important not to press down on the teeth using the thumbs. Instead, downwards force should be applied to the mandible, below the tooth line. This prevents the molars biting the doctor's fingers when the jaw snaps back into place.

Arteries of the face and neck

The pulse you feel in your neck is blood being pumped to the head via the carotid artery.

The head and neck are supplied with blood from the two common carotid arteries that ascend either side of the neck. They are encased, along with the internal jugular vein and the vagus nerve, in a protective covering of connective tissue called the carotid sheath. They have slightly different origins at the base of the neck with the left common carotid arising directly from the arch of the aorta while the right arises from the brachiocephalic trunk.

BRANCHING ARTERIES

The common carotid arteries divide at the level of the upper border of the thyroid cartilage (Adam's apple) to form the internal and external carotid arteries. The former enters the skull and supplies the brain and the latter provides branches that supply the face and scalp.

Many of the branches of the external carotid artery have a wavy or looped course. This flexibility ensures that when the mouth, larynx or pharynx are moved, during swallowing for example, the vessels are not stretched and damaged.

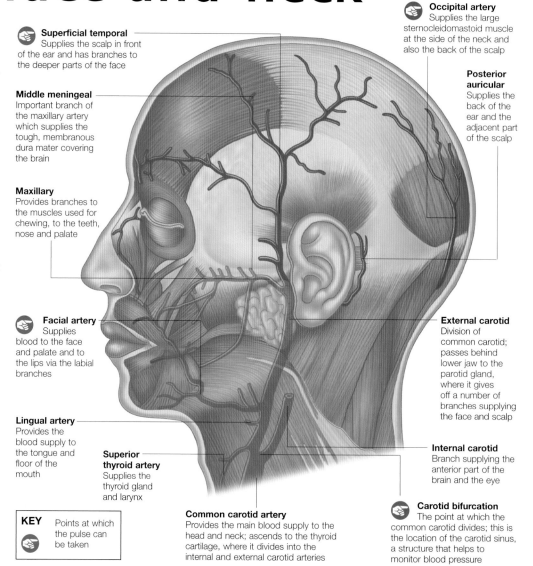

Superficial temporal
Supplies the scalp in front of the ear and has branches to the deeper parts of the face

Middle meningeal
Important branch of the maxillary artery which supplies the tough, membranous dura mater covering the brain

Maxillary
Provides branches to the muscles used for chewing, to the teeth, nose and palate

Facial artery
Supplies blood to the face and palate and to the lips via the labial branches

Lingual artery
Provides the blood supply to the tongue and floor of the mouth

Superior thyroid artery
Supplies the thyroid gland and larynx

Occipital artery
Supplies the large sternocleidomastoid muscle at the side of the neck and also the back of the scalp

Posterior auricular
Supplies the back of the ear and the adjacent part of the scalp

External carotid
Division of common carotid; passes behind lower jaw to the parotid gland, where it gives off a number of branches supplying the face and scalp

Internal carotid
Branch supplying the anterior part of the brain and the eye

Carotid bifurcation
The point at which the common carotid divides; this is the location of the carotid sinus, a structure that helps to monitor blood pressure

Common carotid artery
Provides the main blood supply to the head and neck; ascends to the thyroid cartilage, where it divides into the internal and external carotid arteries

KEY Points at which the pulse can be taken

Carotid angiography

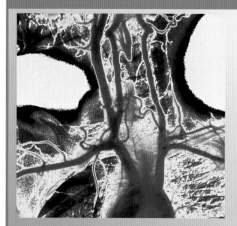

A false-colour angiogram of the arteries running from the aorta (the large vessel at the bottom of the image) to the head.

The injection of a contrast medium, and a rapid series of X-ray recordings, can illustrate the branches of the common carotid artery. This process is called angiography.

Angiography is used to study blood vessels and to look for any abnormalities, such as blockages in the carotid arteries. A build-up of fatty deposits in the artery wall, associated with atherosclerosis, can sometimes be found where the common carotid artery divides. It is possible for surgeons to operate and carefully remove the fatty deposit without damaging the vessel walls. This procedure, known as carotid endarterectomy, effectively improves the blood supply to the head and neck, and reduces the risk of a subsequent stroke.

Another abnormality which can be detected using angiography is aneurysm of the artery, where a balloon-like swelling is found in the wall of the vessel.

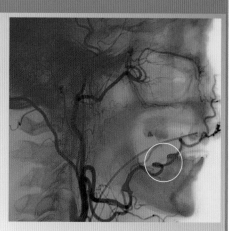

This digitally subtracted arteriogram reveals branching of the carotid artery. The looped facial artery is circled.

Veins of the face and neck

The veins have a similar distribution around the face and neck as the arteries. Many of the veins also share the same names.

Blood drains from the head and neck back to the heart via the internal jugular veins that lie on either side of the neck. As with the common carotid arteries, the veins are protected by the carotid sheath.

Unlike in the rest of the body, the veins of this region generally lack valves, and the return of blood to the heart is by gravity and negative pressure in the thorax (chest).

The superficial (close to the surface) veins are often visible during exertion, and may be seen standing out on the necks of singers, for example.

JUGULAR VEIN

There is very little variation in the position of the internal jugular vein. Because of this, the vein is used for monitoring central venous pressure (blood pressure within the right atrium of the heart). A cannula (hollow tube) is inserted into the vein and passed to the heart. The other end of the cannula is attached to a transducer, an instrument that records pressure. The blood volume may then be assessed.

As well as the veins draining the face, there is a series of emissary veins communicating between the venous sinuses (which drain blood from the brain) and the veins of the scalp. Along with the diploic veins (found in the bones of the skull), these provide a potential route for infection from the scalp into the brain.

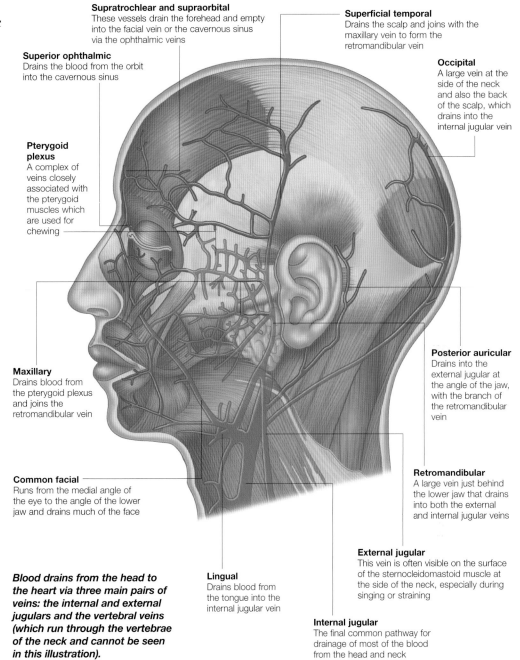

Supratrochlear and supraorbital
These vessels drain the forehead and empty into the facial vein or the cavernous sinus via the ophthalmic veins

Superior ophthalmic
Drains the blood from the orbit into the cavernous sinus

Pterygoid plexus
A complex of veins closely associated with the pterygoid muscles which are used for chewing

Maxillary
Drains blood from the pterygoid plexus and joins the retromandibular vein

Common facial
Runs from the medial angle of the eye to the angle of the lower jaw and drains much of the face

Superficial temporal
Drains the scalp and joins with the maxillary vein to form the retromandibular vein

Occipital
A large vein at the side of the neck and also the back of the scalp, which drains into the internal jugular vein

Posterior auricular
Drains into the external jugular at the angle of the jaw, with the branch of the retromandibular vein

Retromandibular
A large vein just behind the lower jaw that drains into both the external and internal jugular veins

External jugular
This vein is often visible on the surface of the sternocleidomastoid muscle at the side of the neck, especially during singing or straining

Lingual
Drains blood from the tongue into the internal jugular vein

Internal jugular
The final common pathway for drainage of most of the blood from the head and neck

Blood drains from the head to the heart via three main pairs of veins: the internal and external jugulars and the vertebral veins (which run through the vertebrae of the neck and cannot be seen in this illustration).

Interconnections

There are numerous interconnections – called anastomoses – between the arteries on the left and right side of the face and between the branches of the external and internal carotid. This has implications for treating a cut lip, for example, when pressure may need to be applied to both left and right facial arteries in order to stop the bleeding.

The extensive array of blood vessels in the scalp also means that injuries in this area may bleed profusely. This is not only because of the plentiful blood supply but

A resin cast of the veins and arteries serving the face and neck clearly shows the extensive branching networks. Note the particular density of vessels at the front of the neck (the thyroid gland) and the tongue.

also because the vessels are prevented from contracting rapidly due to the fibrous connective tissue under the skin.

The highly interconnected nature of the veins in the head and neck also means that they are an important potential route for the spread of infection. Boils or spots around the side of the nose may cause thrombosis (clot formation) in the facial vein. This in turn may transmit thrombotic material via the ophthalmic vein to the cavernous sinus (the paired sinus in the sphenoid bone of the skull, into which blood drains from the brain, eyes and nose). The resulting cavernous sinus thrombosis is fatal if not treated with antibiotics. Treatment of this condition was one of the first recorded uses of penicillin in the 1940s.

Infratemporal fossa

The infratemporal fossa (a fossa is a depression or hollow) is a region at the side of the head which contains a number of important nerves, blood vessels and muscles involved in mastication (chewing).

The infratemporal fossa is located below the base of the skull, between the pharynx and the ramus (side) of the mandible (lower jawbone). The region is of particular importance to dental surgeons, not only because many of its components are essential to the process of mastication, but also as many of the nerves and blood vessels supplying the mouth are transmitted through it.

ANATOMY OF THE FOSSA
The region is largely defined by the skeletal boundaries of the infratemporal fossa. The anterior boundary is the posterior surface of the maxillary bone, and the posterior boundary is the styloid process of the temporal bone and the carotid sheath. The midline boundary is formed by the lateral pterygoid plate of the sphenoid bone; the lateral boundary is the ramus of the mandible and the roof is the base of the greater wing of the sphenoid bone. The infratemporal fossa has no floor, and is continuous with the neck.

CONTENTS OF THE FOSSA
The fossa contains the pterygoid muscles, branches of the mandibular nerve, the chorda tympani branch of the facial nerve, the otic ganglion (part of the autonomic nervous system), the maxillary artery and the pterygoid venous plexus (vessels surrounding pterygoid muscles).

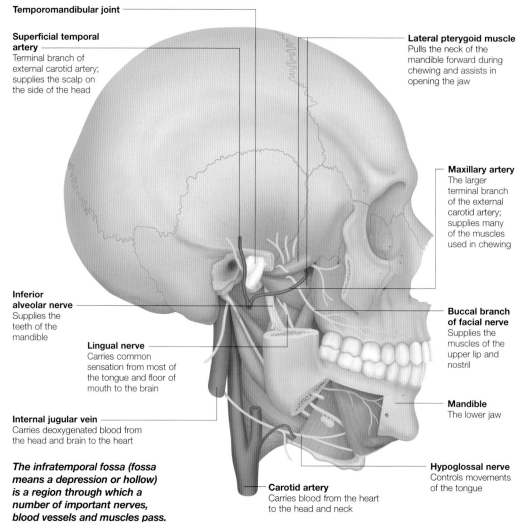

Temporomandibular joint

Superficial temporal artery
Terminal branch of external carotid artery; supplies the scalp on the side of the head

Inferior alveolar nerve
Supplies the teeth of the mandible

Lingual nerve
Carries common sensation from most of the tongue and floor of mouth to the brain

Internal jugular vein
Carries deoxygenated blood from the head and brain to the heart

Lateral pterygoid muscle
Pulls the neck of the mandible forward during chewing and assists in opening the jaw

Maxillary artery
The larger terminal branch of the external carotid artery; supplies many of the muscles used in chewing

Buccal branch of facial nerve
Supplies the muscles of the upper lip and nostril

Mandible
The lower jaw

Hypoglossal nerve
Controls movements of the tongue

Carotid artery
Carries blood from the heart to the head and neck

The infratemporal fossa (fossa means a depression or hollow) is a region through which a number of important nerves, blood vessels and muscles pass.

Chorda tympani

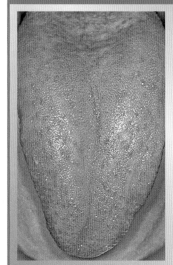

Most of the taste buds are located on the tongue. However, taste is a special sensation which is not in the repertoire of the trigeminal nerve.

A branch of the facial nerve, the chorda tympani, joins the lingual nerve in the infratemporal fossa to provide the sensation of taste. The chorda tympani also carries nervous information telling the submandibular and sublingual salivary glands when to secrete saliva.

The chorda tympani is a branch of the facial nerve which carries the sensation of taste from the front of the tongue.

Pterygoid muscles

The pterygoid muscles, which are contained within the infratemporal fossa, are two of the four muscles known collectively as the muscles of mastication. All these muscles have the same developmental origin and therefore the same nerve supply, which is the mandibular nerve – a branch of the trigeminal nerve.

CHEWING
Chewing involves movements of the mandible (lower jaw) relative to the cranium. Thus, the muscles of mastication have their origin in the cranial part of the skull and their insertions in the mandible or the joint between the two.

As the muscle fibres are directed backwards, contraction of the muscle (shortening of the fibres) on both sides of the head results in the the mandible protruding forwards.

Acting alternately, the left and right muscles will produce side to side motions of the jaws. Muscle fibres of the medial pterygoid insert into the medial surface of the angle of the mandible. Since the muscle fibres are orientated downwards and backwards, contraction of these fibres will result in elevation of the mandible and closing of the jaws. Acting alternately, the left and right muscles will produce a grinding action of the teeth.

Mandibular nerve

The mandibular nerve leaves the skull (through the foramen ovale) to enter directly into the infratemporal fossa, where it divides into its many branches.

The mandibular nerve has motor branches which supply all of the muscles of mastication, allowing contractions of these muscles to take place. The nerve also supplies sensation to the skin of the temple and around the ear via the auriculotemporal nerve and to the skin on the outside of the cheek and the tissue lining the inside via the buccal nerve.

BRANCHES OF THE MANDIBULAR NERVE

The branch called the inferior alveolar nerve (or inferior dental nerve) travels downwards and then forwards to enter the body of the mandible. It supplies sensation to all of the lower teeth, but it also has a branch called the mental nerve which leaves the mandible through a foramen in the lower premolar region. This supplies sensation to the lower lip.

The lingual nerve branch supplies common sensation (for example touch, temperature and pain) to most of the tongue and the floor of the mouth.

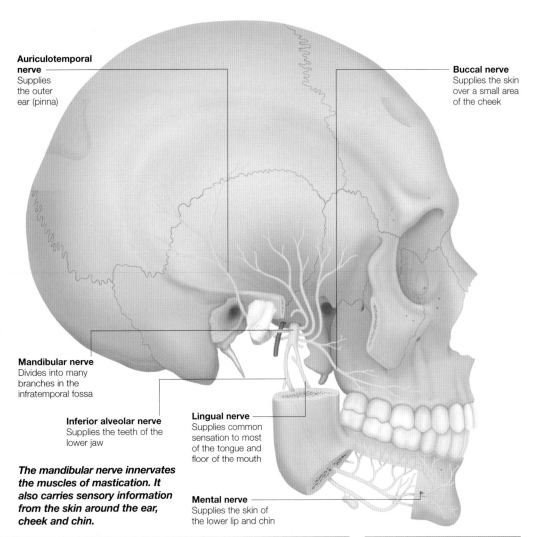

Auriculotemporal nerve
Supplies the outer ear (pinna)

Buccal nerve
Supplies the skin over a small area of the cheek

Mandibular nerve
Divides into many branches in the infratemporal fossa

Inferior alveolar nerve
Supplies the teeth of the lower jaw

Lingual nerve
Supplies common sensation to most of the tongue and floor of the mouth

Mental nerve
Supplies the skin of the lower lip and chin

The mandibular nerve innervates the muscles of mastication. It also carries sensory information from the skin around the ear, cheek and chin.

Dental anaesthetics

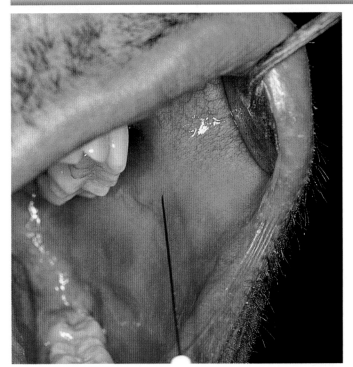

When a dentist needs to work on the teeth and gums of the lower jaw, it is common practice to use an inferior alveolar nerve block. The procedure involves injecting a local anaesthetic (LA) at the point where the nerve enters the body of the mandible on the side of interest. The anaesthetic desensitizes the nerve including its mental nerve branch and hence the lower lip becomes numb – often the first sign that the anaesthetic is working properly.

To desensitize the upper teeth, LA is injected either side of the individual tooth, and sometimes into the palate; the bone of the upper jaw is thin enough for the LA to filter through.

Local anaesthetic (LA) is used to numb the area of the mouth that needs to be worked on. In order to desensitize the lower teeth and gums, the inferior alveolar nerve is numbed with an LA injection.

Tumours

Tumours affecting the infratemporal fossa are usually benign. The commonest are pleomorphic adenoma of the parotid salivary gland and schwannoma (tumour around a nerve). These often present as a fullness in the neck or a swelling at the back of the throat.

MR and CT scanning delineates the soft tissue extent of the mass and its relationship to the internal carotid artery. The scan images will also show any erosion of the skull base.

Treatment is by surgical removal. Small tumours may be approached from the neck and removed from behind the lower jaw. Larger or more vascular tumours may require some excision of bone to give better access to the internal carotid artery, thus preventing life-threatening haemorrhage.

Inside the base of the skull

The floor of the skull is divided into three cranial fossae, which accommodate the brain.

The three fossae (cup-like depressions) have a marked step-like appearance, with the floor of the anterior cranial fossa the highest level and the floor of the posterior fossa the lowest

The nose and eyes lie under the anterior cranial fossae. The floor is formed by the frontal bone (orbital plates), the ethmoid bone and part of the sphenoid bone. Extending upwards from the cribriform plate is a process (projection) called the crista galli.

The floor of the 'butterfly-shaped' middle cranial fossa is formed by the body of the sphenoid bone centrally, and the greater wings of the sphenoid and the temporal bones laterally. The depressions in the floor contain the temporal lobes of the brain.

HOUSING THE BRAIN
The posterior fossa is the largest of the cranial fossae. It contains the cerebellum pons and medulla oblongata of the brain. The floor and posterior wall of the posterior cranial fossa are formed mainly by the occipital bone.

The anterior wall of the posterior crania fossa leading up to the middle cranial fossa is formed by the basilar part of the occipital bone, the temporal bones (petrous and mastoid parts) and the sphenoid bone.

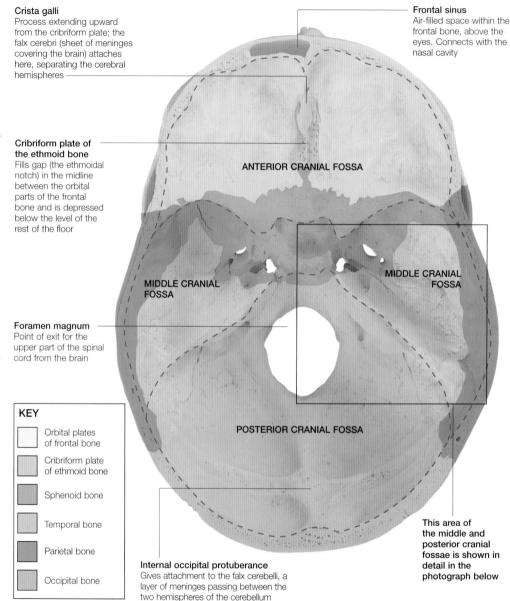

Crista galli
Process extending upward from the cribriform plate; the falx cerebri (sheet of meninges covering the brain) attaches here, separating the cerebral hemispheres

Cribriform plate of the ethmoid bone
Fills gap (the ethmoidal notch) in the midline between the orbital parts of the frontal bone and is depressed below the level of the rest of the floor

Foramen magnum
Point of exit for the upper part of the spinal cord from the brain

Frontal sinus
Air-filled space within the frontal bone, above the eyes. Connects with the nasal cavity

ANTERIOR CRANIAL FOSSA

MIDDLE CRANIAL FOSSA

MIDDLE CRANIAL FOSSA

POSTERIOR CRANIAL FOSSA

KEY

☐	Orbital plates of frontal bone
☐	Cribriform plate of ethmoid bone
☐	Sphenoid bone
☐	Temporal bone
☐	Parietal bone
☐	Occipital bone

This area of the middle and posterior cranial fossae is shown in detail in the photograph below

Internal occipital protuberance
Gives attachment to the falx cerebelli, a layer of meninges passing between the two hemispheres of the cerebellum

Foramina of the cranial fossae

The foramina (see also Sheet 3) are channels in the skull for the passage of blood vessels and nerves. Several important foramina pass through the middle cranial fossa.

The superior orbital fissure lies between the greater and lesser wings of the sphenoid bone. The fissure transmits many structures from the orbit (eye socket). The nerves passing through it include the oculomotor, trochlear and abducent cranial nerves, branches of the trigeminal nerve, and filaments from the internal carotid plexus. The fissure also transmits the ophthalmic veins.

The foramen rotundum lies within the greater wing of the sphenoid. Passing

through it is the maxillary division of the trigeminal nerve (fifth cranial nerve). The foramen ovale is present in the greater wing of the sphenoid, linking the middle cranial fossa to the infratemporal fossa. The major structure passing through it is the mandibular division of the trigeminal nerve.

The foramen spinosum lies just behind the foramen ovale. It transmits the middle meningeal vessels (see right).

Four of the major foramina of the right lateral part of the middle cranial fossa are shown here: (1) foramen magnum, (2) foramen lacerum, (3) foramen ovale, (4) foramen spinosum.

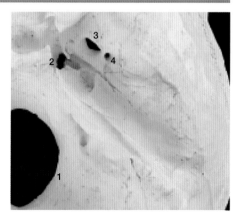

Inside the skull in detail

There are several foramina (openings for blood vessels and nerves) within each of the three cranial fossa.

There are three openings in the floor of the anterior cranial fossa: the two cribriform plates and the foramen caecum.

The anterior ethmoidal nerve (a branch of the fifth cranial nerve) enters the cranial cavity between the cribriform plate and the orbital part of the frontal bone. The anterior ethmoidal vessels from the ophthalmic artery accompany the nerve.

The openings of the middle cranial fossa are described overleaf. With the exception of the optic canals, the lateral regions of the middle cranial fossa contain all the foramina.

FORAMEN MAGNUM

The most prominent structure in the floor of the posterior cranial fossa is the foramen magnum. Passing through the foramen are the medulla oblongata (the upper end of the spinal cord, at the point where it becomes the lower brain stem) the vertebral and spinal arteries and the spinal parts of the accessory nerves. The hypoglossal canals which transmit the hypoglossal nerve (12th cranial nerve) lie close to the foramen magnum.

The jugular foramen lies between the clivus and the petrous process and transmits the glossopharyngeal, vagus and accessory nerves (9th, 10th and 11th).

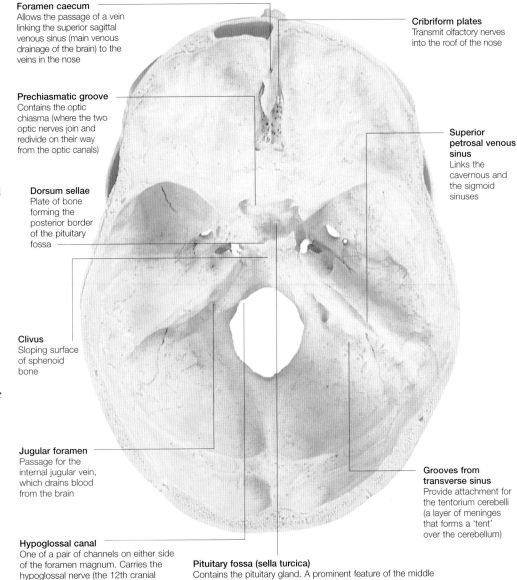

Foramen caecum
Allows the passage of a vein linking the superior sagittal venous sinus (main venous drainage of the brain) to the veins in the nose

Prechiasmatic groove
Contains the optic chiasma (where the two optic nerves join and redivide on their way from the optic canals)

Dorsum sellae
Plate of bone forming the posterior border of the pituitary fossa

Clivus
Sloping surface of sphenoid bone

Jugular foramen
Passage for the internal jugular vein, which drains blood from the brain

Hypoglossal canal
One of a pair of channels on either side of the foramen magnum. Carries the hypoglossal nerve (the 12th cranial nerve), which supplies the muscles of the tongue

Cribriform plates
Transmit olfactory nerves into the roof of the nose

Superior petrosal venous sinus
Links the cavernous and the sigmoid sinuses

Grooves from transverse sinus
Provide attachment for the tentorium cerebelli (a layer of meninges that forms a 'tent' over the cerebellum)

Pituitary fossa (sella turcica)
Contains the pituitary gland. A prominent feature of the middle fossa, lying above the sphenoidal sinuses; a sheet of meninges (diaphragma sellae) forms a roof over the fossa

Disorders of the base of the skull

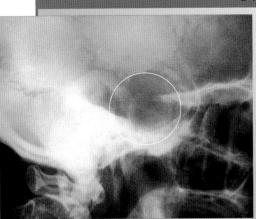

A pituitary tumour is visible as a dark mass on this X-ray taken from the side of the head. As it grows, the tumour erodes away the bone of the pituitary fossa, leading to a characteristic appearance.

A tumour of the pituitary gland may result in enlargement of the pituitary fossa, the bony compartment in which it is located. On an X-ray image, the tumour may be seen to depress the floor of the pituitary fossa, giving the appearance of a double floor and a narrowing of the sphenoidal air sinus. A pituitary tumour may also encroach on the optic chiasma (the structure formed where the two optic nerves meet as they pass back from the eyeballs to the midline of the brain), with the result that early symptoms of this condition may involve vision.

Tumours of nerves as they pass through bone may compress the nerves, with obvious symptoms. For example, a benign tumour of the vestibulocochlear nerve (acoustic neuroma or neurofibroma) in the internal acoustic meatus (the passage from the inner ear to the brain) may cause some deafness. Such a tumour may also compress the facial nerve which runs in the same canal. X-ray examination can reveal enlargement of the canal.

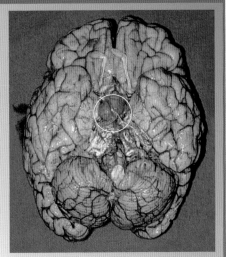

The red structure (circled) in the midline of this inverted brain is a pituitary tumour.

The brain

The brain is the part of the central nervous system that lies inside the skull.
It controls many body functions including our heart rate, the ability to walk
and run, and the creation of our thoughts and emotions.

The brain comprises three major
parts: forebrain, midbrain and
hindbrain. The forebrain is
divided into two halves, forming
the left and right cerebral
hemispheres.

HEMISPHERES

The cerebral hemispheres form
the largest part of the forebrain.
Their outer surface is folded
into a series of gyri (ridges) and
sulci (furrows) that greatly
increases its surface area. Most
of the surface of each
hemisphere is hidden in the
depths of the sulci.

Each hemisphere is divided
into frontal, parietal, occipital
and temporal lobes, named after
the closely related bones of the
skull. Connecting the two
hemispheres is the corpus
callosum, a large bundle of
fibres deep in the longitudinal
fissure.

GREY AND WHITE MATTER

The hemispheres consist of an
outer cortex of grey matter and
an inner mass of white matter.
- Grey matter contains nerve
cell bodies, and is found in the
cortex of the cerebral and
cerebellar hemispheres and in
groups of sub-cortical nuclei.
- White matter comprises nerve
fibres found below the cortex.
They form the communication
network of the brain, and can
project to other areas of the
cortex and spinal cord.

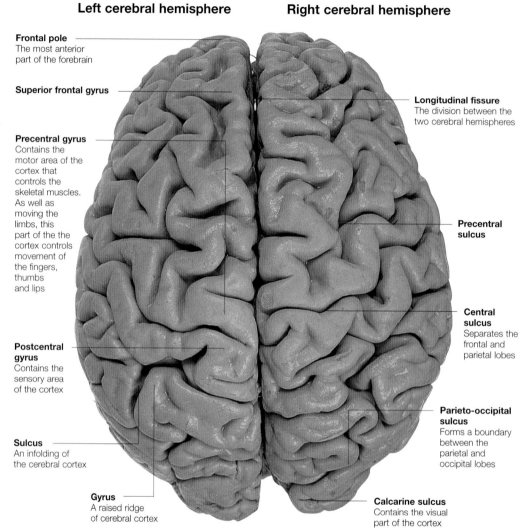

Left cerebral hemisphere　　**Right cerebral hemisphere**

Frontal pole
The most anterior
part of the forebrain

Superior frontal gyrus

Precentral gyrus
Contains the
motor area of the
cortex that
controls the
skeletal muscles.
As well as
moving the
limbs, this
part of the the
cortex controls
movement of
the fingers,
thumbs
and lips

**Postcentral
gyrus**
Contains the
sensory area
of the cortex

Sulcus
An infolding of
the cerebral cortex

Gyrus
A raised ridge
of cerebral cortex

Longitudinal fissure
The division between the
two cerebral hemispheres

**Precentral
sulcus**

**Central
sulcus**
Separates the
frontal and
parietal lobes

**Parieto-occipital
sulcus**
Forms a boundary
between the
parietal and
occipital lobes

Calcarine sulcus
Contains the visual
part of the cortex

Ridges and furrows

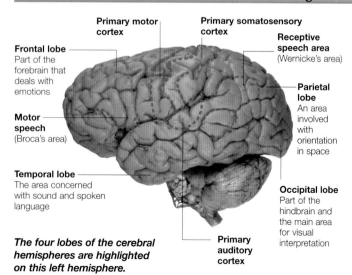

**Primary motor
cortex**

**Primary somatosensory
cortex**

**Receptive
speech area**
(Wernicke's area)

Frontal lobe
Part of the
forebrain that
deals with
emotions

**Motor
speech**
(Broca's area)

Temporal lobe
The area concerned
with sound and spoken
language

**Parietal
lobe**
An area
involved
with
orientation
in space

Occipital lobe
Part of the
hindbrain and
the main area
for visual
interpretation

**Primary
auditory
cortex**

*The four lobes of the cerebral
hemispheres are highlighted
on this left hemisphere.*

The central sulcus runs from the
longitudinal fissure to the lateral
fissure, and marks the boundary
between the frontal and parietal
lobes. The precentral gyrus runs
parallel to and in front of the
central sulcus and contains the
primary motor cortex, where
voluntary movement is initiated.
The postcentral gyrus contains
the primary somatosensory
cortex that perceives bodily
sensations. The parieto-occipital
sulcus (on the medial surface of
both hemispheres) marks the
border between the parietal and
occipital lobes.

The calcarine sulcus marks
the position of the primary
visual cortex, where visual
images are perceived. The

primary auditory cortex is
located towards the posterior
(back) end of the lateral fissure.

On the medial surface of the
temporal lobe, at the rostral
(front) end of the most superior
gyrus, lies the primary
olfactory cortex, which is
involved with smell. Internal to
the parahippocampal gyrus lies
the hippocampus, which is part
of the limbic system and is
involved in memory formation.
The areas responsible for
speech are located in the
dominant hemisphere (usually
the left) in each individual. The
motor speech area (Broca's
area) lies in the inferior frontal
gyrus and is essential for the
production of speech.

Inside the brain

A midline section between the two cerebral hemispheres reveals the main structures that control a vast number of activities in the body. While particular areas monitor sensory and motor information, others control speech and sleep.

SPEECH, THOUGHT AND MOVEMENT

The receptive speech area (Wernicke's area) lies behind the primary auditory cortex and is essential for understanding speech. The prefrontal cortex has high-order cognitive functions, including abstract thinking, social behaviour and decision-making ability.

Within the white matter of the cerebral hemispheres are several masses of grey matter, known as the basal ganglia. This group of structures is involved in aspects of motor function, including movement programming, planning and motor programme selection and motor memory retrieval.

DIENCEPHALON

The medial part of the forebrain comprises the structures surrounding the third ventricle. These form the diencephalon which includes the thalamus, hypothalamus, epithalamus and subthalamus of either side. The thalamus is the last relay station for information from the brainstem and spinal cord before it reaches the cortex.

The hypothalamus lies below the thalamus in the floor of the diencephalon. It is involved in

Corpus callosum
A thick band of nerve fibres, found in the depths of the longitudinal fissure that connects the cerebral hemispheres

Right cerebral hemisphere
One of two hemispheres that form the largest part of the forebrain

Ventricle
Fluid-filled cavity

Thalamus
Directs sensory information from the sense organs to the correct part of the cerebral cortex

Optic nerve
Carries visual information from the eye to the brain

Pituitary stalk
The pituitary gland is not included when the brain is removed from the skull

Hypothalamus
Concerned with emotions and drives, such as hunger and thirst; it also helps to control body temperature and the water-salt balance in the blood

Precentral gyrus

Central sulcus

Postcentral gyrus

Pineal gland
Part of the epithalamus that synthesizes melatonin

Parieto-occipital sulcus
Divides the occipital and parietal lobes

Calcarine sulcus
Where most of the primary visual cortex lies

Cerebellum
Controls body movement and maintains balance; consists of grey matter on the outside and white matter on the inside

Medulla oblongata
Contains vital centres that control breathing, heart-beat and blood supply

Midbrain
Important in vision; links the forebrain to the hindbrain

Pons
Part of the brainstem that contains numerous nerve tracts

Spinal cord

a variety of homeostatic mechanisms, and controls the pituitary gland which descends from its base. The anterior (front) lobe of the pituitary secretes substances that influence the thyroid and adrenal glands, and the gonads and produces growth factors. The posterior lobe produces

hormones that increase blood pressure, decrease urine production and cause uterine contraction. The hypothalamus also influences the sympathetic and parasympathetic nervous systems and controls body temperature, appetite and wakefulness. The epithalamus is a relatively small part of the

dorso-caudal diencephalon that includes the pineal gland, which synthesizes melatonin and is involved in the control of the sleep/wake cycle.

The subthalamus lies beneath the thalamus and next to the hypothalamus. It contains the subthalamic nucleus which controls movement.

Brainstem and cerebellum

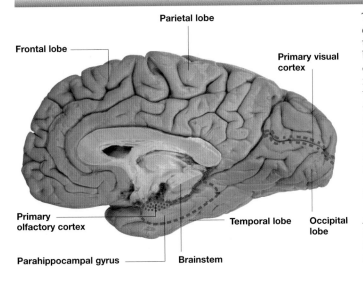

Parietal lobe

Frontal lobe

Primary visual cortex

Primary olfactory cortex

Temporal lobe

Occipital lobe

Parahippocampal gyrus

Brainstem

The posterior part of the diencephalon is connected to the midbrain, which is followed by the pons and medulla oblongata of the hindbrain. The midbrain and hindbrain contain the nerve fibres connecting the cerebral hemispheres to the cranial nerve nuclei, to lower centres within the brainstem and to the spinal cord. They also contain the cranial nerve nuclei.

Most of the reticular formation, a network of nerve

A view of the medial surface of the right hemisphere, with the brainstem removed, allowing the lower hemisphere to be seen.

pathways, lies in the midbrain and hindbrain. This system contains the important respiratory, cardiac and vasomotor centres.

The cerebellum lies posterior to the hindbrain and is attached to it by three pairs of narrow stalk-like structures called peduncles. Connections with the rest of the brain and spinal cord are established via these peduncles. The cerebellum functions at an unconscious level to co-ordinate movements initiated in other parts of the brain. It also controls the maintenance of balance and influences posture and muscle tone.

Meninges

The meninges are three membranes that protect the brain and spinal cord.

The meninges cover the brain and spinal cord, and serve to protect these important structures. They comprise three layers.

THE DURA MATER

The dura mater is a thick fibrous tissue that lines the inner layer of the skull. It deviates from the contours of the skull by forming a double fold (falx cerebri) that dips down between the cerebral hemispheres, and into the gaps between the cerebral hemispheres and the cerebellum (tentorium cerebelli).

THE ARACHNOID MATER

The arachnoid mater is an impermeable membrane that follows the contours of the dura mater. It is separated from the dura mater by a small gap called the subdural space. The arachnoid mater is connected to the pia mater by bands of web-like tissue (called trabeculae).

THE PIA MATER

The pia mater covers the surface of the brain and spinal cord. The space between the arachnoid and pia (subarachnoid space) is filled with cerebrospinal fluid (CSF). The brain and spinal cord are suspended in this fluid, which provides the most important means of physical protection for the brain.

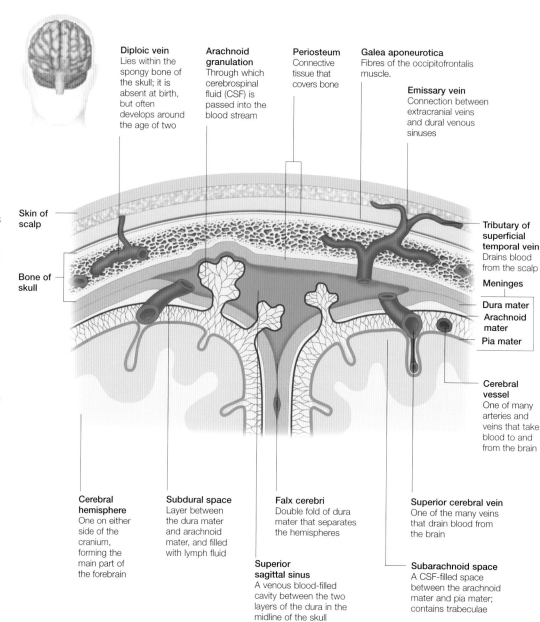

Diploic vein
Lies within the spongy bone of the skull; it is absent at birth, but often develops around the age of two

Arachnoid granulation
Through which cerebrospinal fluid (CSF) is passed into the blood stream

Periosteum
Connective tissue that covers bone

Galea aponeurotica
Fibres of the occipitofrontalis muscle.

Emissary vein
Connection between extracranial veins and dural venous sinuses

Skin of scalp

Bone of skull

Tributary of superficial temporal vein
Drains blood from the scalp

Meninges
Dura mater
Arachnoid mater
Pia mater

Cerebral vessel
One of many arteries and veins that take blood to and from the brain

Cerebral hemisphere
One on either side of the cranium, forming the main part of the forebrain

Subdural space
Layer between the dura mater and arachnoid mater, and filled with lymph fluid

Falx cerebri
Double fold of dura mater that separates the hemispheres

Superior sagittal sinus
A venous blood-filled cavity between the two layers of the dura in the midline of the skull

Superior cerebral vein
One of the many veins that drain blood from the brain

Subarachnoid space
A CSF-filled space between the arachnoid mater and pia mater; contains trabeculae

Potential sites of bleeding

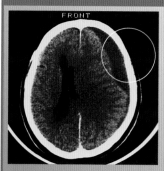

A subdural haemorrhage (circled on this CT scan) can be extensive, compressing a large area of the brain.

Although arteries are found in the arachnoid membrane, the most important of the meningeal arteries is the middle meningeal artery.

This artery is found outside the dura mater (the extradural space), and lies close to the inner layer of the periosteum. It is set in the bone, therefore reducing pressure on the meninges, but if the artery becomes too enlarged, for example through increased blood pressure, the person will suffer from a headache.

The middle meningeal artery runs across the junction of three skull bones. Here, it is susceptible

to injury from fracture. If the artery ruptures, blood will leak into the extradural space, and may result in a quick death if left untreated.

Following a head injury, such as violent shaking of the head, bleeding may occur from the intra-cranial venous sinuses – blood-filled cavities between layers of dura (see over). The blood will leak into the subdural space, and may compress brain tissue. If this occurs slowly, gradual loss of brain function or a change in the individual's personality will result. Sometimes, it can occur rapidly, and may cause death.

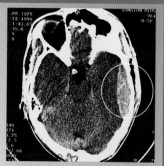

This is the characteristic appearance of an extradural haemorrhage. The leakage of blood into the blood is circled.

Dural venous sinuses

The sinuses play a role in the circulation and drainage of the blood and fluids that protect and bathe the brain.

There are 15 dural venous sinuses – blood-filled cavities between double folds of dura mater. Venous sinuses are lined by endothelium, but unlike other veins, they have no muscular layer. They are therefore very delicate, relying on surrounding tissues for support. The network of sinuses can be seen on the illustrations.

VENOUS CIRCULATION

There are two sets of dural venous sinuses, those in the upper part of the skull and those on the floor of the skull. They receive blood draining from the brain via the cerebral and cerebellar veins, the red bone marrow of the skull via diploic veins, and the scalp via the emissary veins. They are crucial to the reabsorption of CSF.

ROUTES OF BRAIN INFECTION

The sinuses are valveless and provide no resistance to the spread of infection. The connections between the veins of the face, and the dural venous sinuses allow the possibility of infection on the face spreading to the brain, which can be life-threatening.

Valveless connections between the veins around the spinal cord and dural venous sinuses allow the spread of infection or cancer cells between the body and the brain.

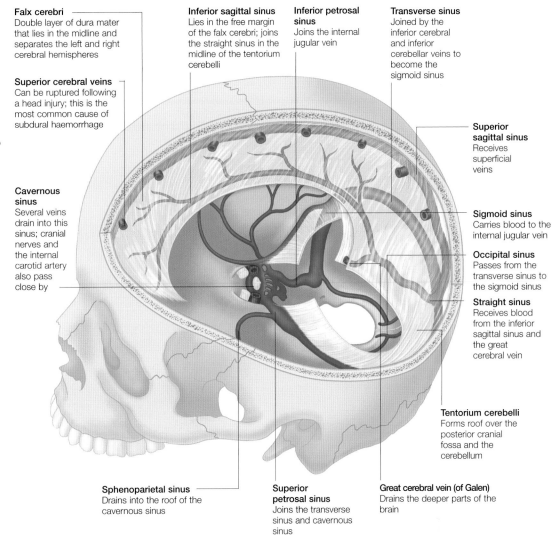

Falx cerebri
Double layer of dura mater that lies in the midline and separates the left and right cerebral hemispheres

Superior cerebral veins
Can be ruptured following a head injury; this is the most common cause of subdural haemorrhage

Cavernous sinus
Several veins drain into this sinus; cranial nerves and the internal carotid artery also pass close by

Inferior sagittal sinus
Lies in the free margin of the falx cerebri; joins the straight sinus in the midline of the tentorium cerebelli

Inferior petrosal sinus
Joins the internal jugular vein

Transverse sinus
Joined by the inferior cerebral and inferior cerebellar veins to become the sigmoid sinus

Superior sagittal sinus
Receives superficial veins

Sigmoid sinus
Carries blood to the internal jugular vein

Occipital sinus
Passes from the transverse sinus to the sigmoid sinus

Straight sinus
Receives blood from the inferior sagittal sinus and the great cerebral vein

Tentorium cerebelli
Forms roof over the posterior cranial fossa and the cerebellum

Sphenoparietal sinus
Drains into the roof of the cavernous sinus

Superior petrosal sinus
Joins the transverse sinus and cavernous sinus

Great cerebral vein (of Galen)
Drains the deeper parts of the brain

Sinuses in the base of the skull

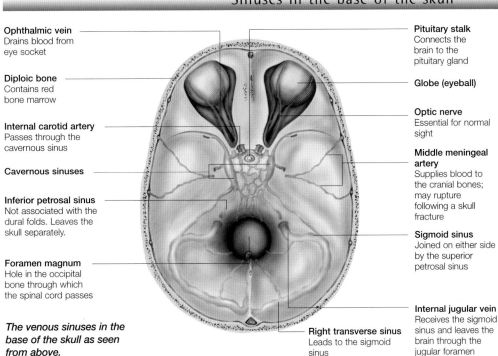

Ophthalmic vein
Drains blood from eye socket

Diploic bone
Contains red bone marrow

Internal carotid artery
Passes through the cavernous sinus

Cavernous sinuses

Inferior petrosal sinus
Not associated with the dural folds. Leaves the skull separately.

Foramen magnum
Hole in the occipital bone through which the spinal cord passes

The venous sinuses in the base of the skull as seen from above.

Pituitary stalk
Connects the brain to the pituitary gland

Globe (eyeball)

Optic nerve
Essential for normal sight

Middle meningeal artery
Supplies blood to the cranial bones; may rupture following a skull fracture

Sigmoid sinus
Joined on either side by the superior petrosal sinus

Internal jugular vein
Receives the sigmoid sinus and leaves the brain through the jugular foramen

Right transverse sinus
Leads to the sigmoid sinus

There are seven pairs of sinuses on the floor of the skull. These include the transverse, inferior petrosal sinus, superior petrosal sinus, cavernous, sigmoid, sphenoparietal and occipital sinuses.

CAVERNOUS SINUSES

The cavernous sinuses lie on either side of the pituitary gland. The roof of each sinus is continuous with the dural sheet (diaphragma sellae) that covers the pituitary gland, surrounding the pituitary stalk.

Several important structures lie close to the cavernous sinuses. These include the internal carotid artery, three nerves supplying eye movement, and branches of the trigeminal nerve, which supplies sensation to the skin of the face and enables movement of the muscles of mastication.

Blood vessels of the brain

The arteries provide the brain with a rich supply of oxygenated blood.

The brain weighs about 1.4 kg and accounts for two per cent of our total body weight. However, it requires 15–20 per cent of the cardiac output to be able to function properly. If the blood supply to the the brain is cut for as little as 10 seconds we lose consciousness and, unless blood flow is quickly restored, it takes only a matter of minutes before the damage is irreversible.

THE ARTERIAL NETWORK

Blood reaches the brain via two pairs of arteries. The internal carotid arteries originate from the common carotid arteries in the neck, enter the skull via the carotid canal and then branch to supply the cerebral cortex. The two main branches of the internal carotid are the middle and anterior cerebral arteries.

The vertebral arteries arise from the subclavian arteries, enter the skull via the foramen magnum and supply the brainstem and cerebellum. They join, forming the basilar artery which then divides to produce the two posterior cerebral arteries that supply, among other things, the occipital or visual cortex at the back of the brain.

These two sources of blood to the brain are linked by other arteries to form a circuit at the base of the brain called the 'circle of Willis'.

Inferior (from below) view of the brain

Right hemisphere **Left hemisphere**

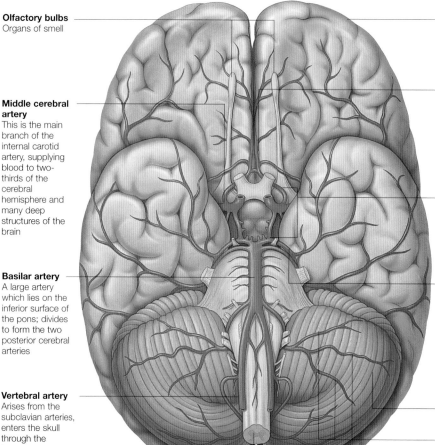

Olfactory bulbs
Organs of smell

Middle cerebral artery
This is the main branch of the internal carotid artery, supplying blood to two-thirds of the cerebral hemisphere and many deep structures of the brain

Basilar artery
A large artery which lies on the inferior surface of the pons; divides to form the two posterior cerebral arteries

Vertebral artery
Arises from the subclavian arteries, enters the skull through the foramen magnum to supply the brainstem, then fuses with its opposite number to form the basilar artery

Spinal cord

Cerebrum

Anterior cerebral artery
Supplies blood to the frontal lobe and to the medial surface of the cerebral hemisphere

Circle of Willis
Circle of communicating arteries at the base of the brain

Posterior cerebral artery
Supplies blood to the inferior part of the temporal lobe and to the occipital lobe at the back of the brain

Cerebellum

Cerebellar arteries
Branches from the vertebral and basilar arteries that provide the blood supply to the cerebellum

What happens if blood supply stops

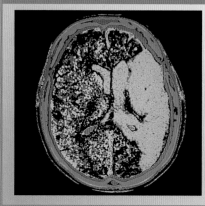

The importance of the blood supply to the brain becomes very clear when that supply is lost, as is seen in a stroke.

Strokes may result from blockage of (ischaemic stroke) or bleeding from (haemorrhagic stroke) an artery. This results in the death

This false-colour CT scan shows an area of dead tissue (blue) caused by a blockage in a cerebral artery. This may be due to a blood clot.

of the brain tissue supplied by that particular vessel.

The precise effects on the patient depend on which vessel is affected. In the case of a 'classic stroke' it tends to be the middle cerebral artery that is affected (see scan), resulting in paralysis down the opposite side of the body. This is because the motor cortex, which controls the voluntary movement of muscles on the opposite side of the body, is damaged.

Other symptoms that may be associated with damage to this artery are:
■ Loss of sensation down the opposite side of the body
■ Visual disturbances
■ Language problems (if the damage caused by the stroke is in the dominant left hemisphere)

The extent of damage, and degree of recovery, depends on the extent of the infarct (area of dead tissue). In some cases, the paralysis persists.

Veins of the brain

Deep and superficial veins drain blood from the brain into a complex system of sinuses. These sinuses rely on gravity to return blood to the heart as, unlike other veins, they do not possess valves.

The veins of the brain can be divided into deep and superficial groups. These veins, none of which have valves, drain into the venous sinuses of the skull.

The sinuses are formed between layers of dura mater, the tough outer membrane covering the brain, and are unlike the veins in the rest of the body in that they have no muscular tissue in their walls.

The superficial veins have a variable arrangement on the surface of the brain and many of them are highly interconnected. Most superficial veins drain into the superior sagittal sinus. By contrast, most of the deep veins, associated with structures within the body of the brain, drain into the straight sinus via the great cerebral vein (vein of Galen).

FUNCTIONS OF THE SINUSES

The straight sinus and the superior sagittal sinus converge. Blood flows through the transverse and sigmoid sinuses and exits the skull through the internal jugular vein before flowing back towards the heart.

Beneath the brain, on either side of the sphenoid bone, are the cavernous sinuses. These drain blood from the orbit (eye socket) and deep parts of the face. This provides a potential route of infection into the skull.

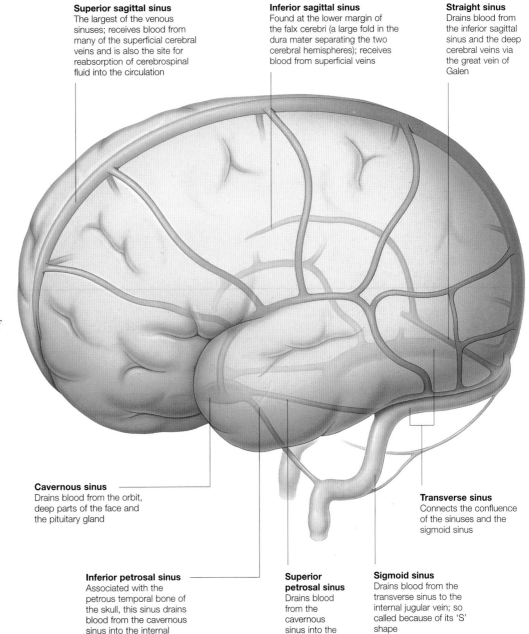

Superior sagittal sinus
The largest of the venous sinuses; receives blood from many of the superficial cerebral veins and is also the site for reabsorption of cerebrospinal fluid into the circulation

Inferior sagittal sinus
Found at the lower margin of the falx cerebri (a large fold in the dura mater separating the two cerebral hemispheres); receives blood from superficial veins

Straight sinus
Drains blood from the inferior sagittal sinus and the deep cerebral veins via the great vein of Galen

Cavernous sinus
Drains blood from the orbit, deep parts of the face and the pituitary gland

Transverse sinus
Connects the confluence of the sinuses and the sigmoid sinus

Inferior petrosal sinus
Associated with the petrous temporal bone of the skull, this sinus drains blood from the cavernous sinus into the internal jugular vein

Superior petrosal sinus
Drains blood from the cavernous sinus into the transverse sinus

Sigmoid sinus
Drains blood from the transverse sinus to the internal jugular vein; so called because of its 'S' shape

---

Visualizing the veins of the brain

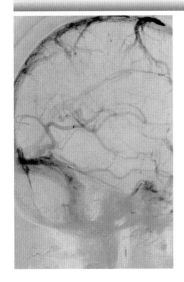

The cerebral veins and venous sinuses can be clearly seen using the technique of angiography.

The procedure involves injecting a radio-opaque contrast medium into the internal carotid artery. After about seven seconds the medium has had an opportunity to reach the venous circulation. A rapid series of X-rays is then taken, and the details of abnormalities or problems with

the venous drainage of the brain can be readily visualized.

This technique can be used to detect venous thrombosis (blood clots) and congenital abnormalities in the connections between arteries and veins (arteriovenous malformations, or AVMs).

However, problems with the cerebral venous system are far less common than those associated with the cerebral arteries.

The veins of the cerebrum are visible on this carotid arteriogram (venous phase). The radio-opaque medium shows up as black.

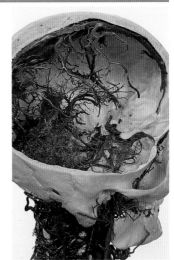

This cast shows the venous sinuses of the brain, which drain deoxygenated (no longer containing oxygen) blood back to the heart.

Ventricles of the brain

The brain 'floats' in a protective layer of cerebrospinal fluid – the watery liquid produced in a system of cavities within the brain and brainstem.

The brain contains a system of communicating (connected) cavities known as the ventricles. There are four ventricles within the brain and brainstem, each secreting cerebrospinal fluid (CSF), the fluid that surrounds and permeates the brain and spinal cord, protecting them from injury and infection.

Three of the ventricles – namely the two (paired) lateral ventricles and the third ventricle – lie within the forebrain. The lateral ventricles are the largest, and lie within each cerebral hemisphere. Each consists of a 'body' and three 'horns' – anterior (situated in the frontal lobe), posterior (occipital lobe) and inferior (temporal lobe). The third ventricle is a narrow cavity between the thalamus and hypothalamus.

HINDBRAIN VENTRICLE

The fourth ventricle is situated in the hindbrain, beneath the cerebellum. When viewed from above, it is diamond-shaped, but in sagittal section (see right) it is triangular. It is continuous with the third ventricle via a narrow channel called the cerebral aqueduct of the midbrain. The roof of the fourth ventricle is incomplete, allowing it to communicate with the subarachnoid space (see over).

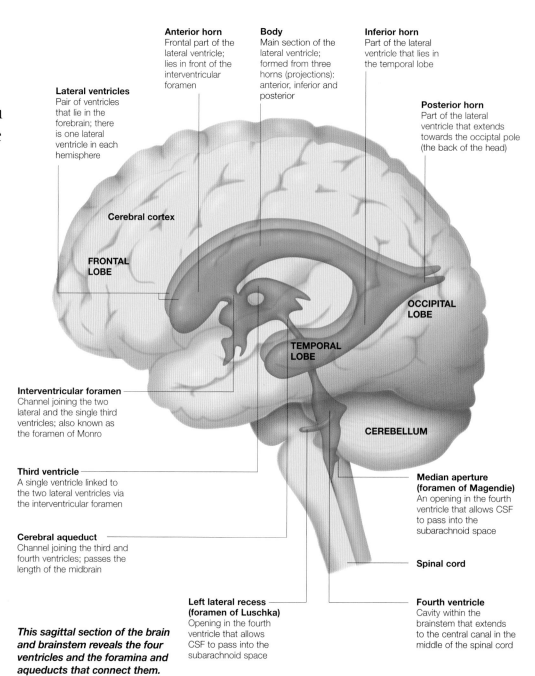

Anterior horn
Frontal part of the lateral ventricle; lies in front of the interventricular foramen

Body
Main section of the lateral ventricle; formed from three horns (projections): anterior, inferior and posterior

Inferior horn
Part of the lateral ventricle that lies in the temporal lobe

Lateral ventricles
Pair of ventricles that lie in the forebrain; there is one lateral ventricle in each hemisphere

Posterior horn
Part of the lateral ventricle that extends towards the occipital pole (the back of the head)

Cerebral cortex

FRONTAL LOBE

OCCIPITAL LOBE

TEMPORAL LOBE

CEREBELLUM

Interventricular foramen
Channel joining the two lateral and the single third ventricles; also known as the foramen of Monro

Third ventricle
A single ventricle linked to the two lateral ventricles via the interventricular foramen

Cerebral aqueduct
Channel joining the third and fourth ventricles; passes the length of the midbrain

Left lateral recess (foramen of Luschka)
Opening in the fourth ventricle that allows CSF to pass into the subarachnoid space

Median aperture (foramen of Magendie)
An opening in the fourth ventricle that allows CSF to pass into the subarachnoid space

Spinal cord

Fourth ventricle
Cavity within the brainstem that extends to the central canal in the middle of the spinal cord

This sagittal section of the brain and brainstem reveals the four ventricles and the foramina and aqueducts that connect them.

Cerebrospinal fluid in the ventricles

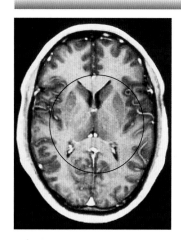

Within each ventricle is a network of blood vessels known as the choroid plexus. This is where the cerebrospinal fluid is produced. CSF fills the ventricles and also the subarachnoid space surrounding the brain and spinal cord, where it acts a protective buffer. CSF is also believed to remove waste products into the venous system. The appearance of CSF can often provide clues to infection.

The symmetrical arrangement of the ventricles can be seen on this MR scan (circled).

CSF can be taken from various locations, usually along the spinal cord, in a procedure known as a 'tap'. A small hole is made in the dural sac (the outermost layer of meninges surrounding the brain and spinal cord) and into the lumbar subarachnoid space, where fluid can be aspirated (sucked out).

As CSF is normally a clear, colourless fluid, any change from this can indicate disease. A red appearance, for example, might suggest that the CSF contains blood from a recent haemorrhage.

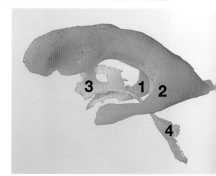

The ventrical system consists of four communicating cavities, as seen on this resin cast.

Circulation of cerebrospinal fluid

Cerebrospinal fluid (CSF) is produced by the choroid plexus within the lateral, third and fourth ventricles.

The choroid plexuses are a rich system of blood vessels originating from the pia mater, the innermost tissue surrounding the brain. The plexuses contain numerous folds (villous processes) projecting into the ventricles, from which cerebrospinal fluid is produced.

From the choroid plexuses in the two lateral ventricles, CSF passes to the third ventricle via the interventricular foramen. Together with additional fluid produced by the choroid plexus in the third ventricle, CSF then passes through the cerebral aqueduct of the midbrain and into the fourth ventricle. Additional fluid is produced by the choroid plexus in the fourth ventricle.

SUBARACHNOID SPACE
From the fourth ventricle, CSF finally passes out into the subarachnoid space surrounding the brain. It does this through openings in the fourth ventricle – a median opening (foramen of Magendie) and two lateral ones (foramina of Luschka). Once in the subarachnoid space, the CSF circulates to surround the central nervous system.

As CSF is produced constantly, it needs to be drained continuously to prevent any build-up of pressure. This is achieved by passage of the CSF into the venous sinuses of the brain through protrusions known as arachnoid granulations. These are particularly evident in the region of the superior sagittal sinus.

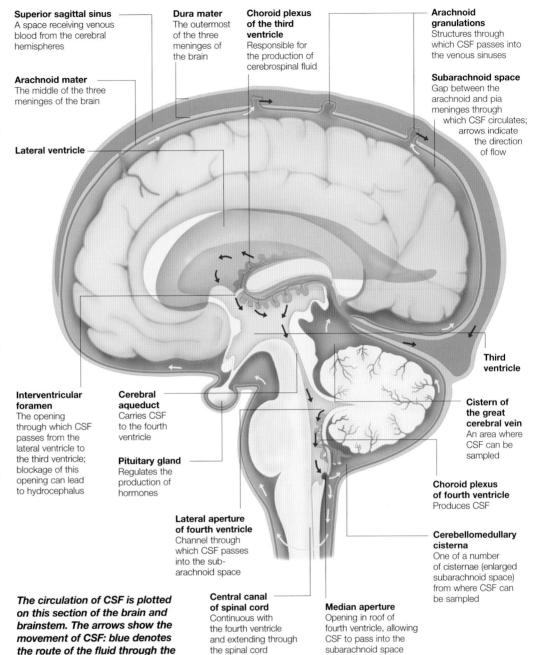

Superior sagittal sinus
A space receiving venous blood from the cerebral hemispheres

Arachnoid mater
The middle of the three meninges of the brain

Lateral ventricle

Dura mater
The outermost of the three meninges of the brain

Choroid plexus of the third ventricle
Responsible for the production of cerebrospinal fluid

Arachnoid granulations
Structures through which CSF passes into the venous sinuses

Subarachnoid space
Gap between the arachnoid and pia meninges through which CSF circulates; arrows indicate the direction of flow

Interventricular foramen
The opening through which CSF passes from the lateral ventricle to the third ventricle; blockage of this opening can lead to hydrocephalus

Cerebral aqueduct
Carries CSF to the fourth ventricle

Pituitary gland
Regulates the production of hormones

Lateral aperture of fourth ventricle
Channel through which CSF passes into the subarachnoid space

Central canal of spinal cord
Continuous with the fourth ventricle and extending through the spinal cord

Median aperture
Opening in roof of fourth ventricle, allowing CSF to pass into the subarachnoid space

Third ventricle

Cistern of the great cerebral vein
An area where CSF can be sampled

Choroid plexus of fourth ventricle
Produces CSF

Cerebellomedullary cisterna
One of a number of cisternae (enlarged subarachnoid space) from where CSF can be sampled

The circulation of CSF is plotted on this section of the brain and brainstem. The arrows show the movement of CSF: blue denotes the route of the fluid through the ventricular system; yellow the route through the subarachnoid space.

CSF analysis

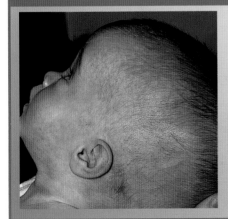

As CSF is continually being secreted, raised intracranial pressure will result if there is interference in the circulation. Such a situation will result from blockage of the interventricular foramen, the cerebral aqueduct or the apertures in the roof of the fourth ventricle, and will

Hydrocephalus is a condition that results from an obstruction of the flow of CSF within the ventricular system, or in the flow of CSF through the subarachnoid space. A blockage in the ventricular system may occur because of a tumour, and an obstruction in the subarachnoid space may develop after a head injury or be due to infection from meningitis.

produce the condition known as hydrocephalus (water on the brain). A patient with this condition presents with headaches, unsteadiness and mental impairment.

In the newborn, hydrocephalus may produce a tensed and raised anterior fontanelle and an enlarged skull, and will require immediate treatment to relieve the pressure. If a sample of CSF is required for analysis in an adult, a lumbar puncture may be performed. In this procedure, a needle may be inserted into the subarachnoid space between the bones of the fourth and fifth lumbar vertebrae. This does not damage nervous tissue, as the spinal cord normally terminates at a higher level (between the first and second lumbar vertebrae).

PET brain imaging

PET is a non-invasive imaging technique that reveals
local changes in brain activity. It is very useful at showing changes
associated with mental processes and nervous system diseases.

PET (positron emission tomography) is a technique for imaging brain function which makes use of small amounts of radiation from positron (sub-atomic particle) emitting isotopes or tracers.

The isotope is injected into the bloodstream and taken up into the brain. Depending on the particular isotope used, the tracer becomes most concentrated in certain areas, such as those with the highest rates of metabolism.

CROSS-SECTIONS OF THE BRAIN

The scanner uses a set of sensors to detect radiation emitted from the head after the injection and uses this data to construct detailed cross-sections of the brain. The level of radiation coming from each area can be colour-coded to produce a map of activity in the different brain regions.

While PET does not show brain structure clearly (as an MRI would), it gives unique information on functional activity. PET scans can now be superimposed onto MRI scans of the same person to combine the strengths of both techniques.

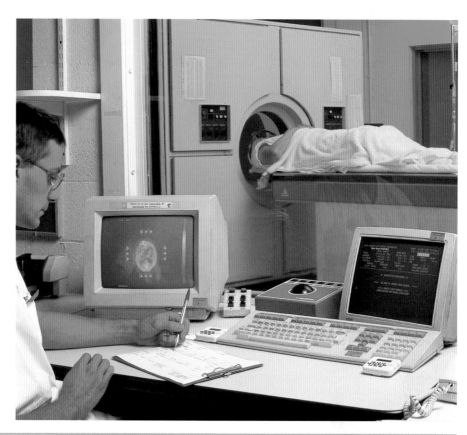

PET scans yield immediate results, as the technician receives images of cross-sections of the brain during the scan (screen on centre left). Metabolic activity shows up as 'hot spots'.

Applications

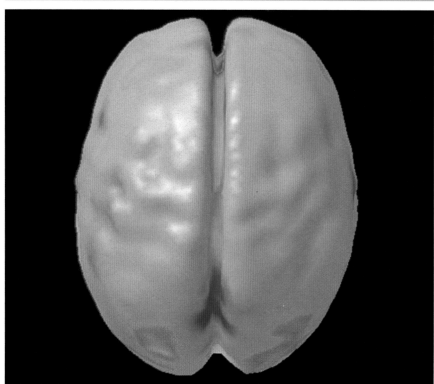

In healthy subjects, the short half-life of the isotopes used in PET scanning means that several scans can be carried out in the same session, monitoring changes in the brain during different activities such as movement, speaking or memorizing. PET scans have shown increases in blood flow in specific regions of the brain during different tasks, providing new information about the organization of brain function.

Other uses of PET scanning include:
■ Helping to discover which area of the brain is active at the beginning of an epileptic seizure. Although highly active during a seizure, the abnormal brain region will also often show reduced glucose metabolism between seizures
■ Estimating the rate of metabolic activity in brain tumours: more malignant tumours show higher levels of metabolism
■ In patients previously treated for brain tumours, helping to decide whether changes seen on MRI following surgery are due to the effects of the treatment or recurrence of the tumour.

PET imaging shows activity within different areas of the brain. This scan of the brain at rest demonstrates that there are still small areas of activity (shown in green) at the front and back of the brain.

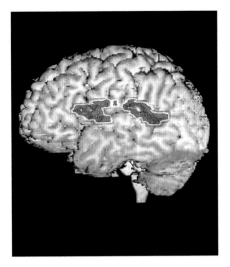

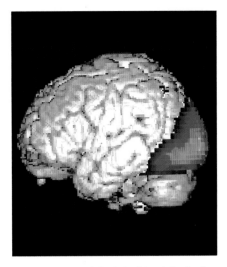

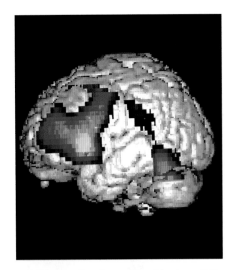

A PET scan showing areas of brain activity associated with turning thoughts into speech. The smaller area is involved in checking for cognition (understanding).

Generating speech

PET scanning can demonstrate the brain areas involved in speech generation. A region known as Broca's area in the frontal cortex is particularly involved. In most people, this is located in the left frontal lobe and so the left hemisphere is said to be dominant for speech.

Other areas are also activated during speaking, including the supplementary motor area in the frontal lobe, part of the motor cortex and the cerebellum.

In verbal short-term memory there are two areas of high activity (shown in green/red): Broca's area (left) and the inferior parietal/superior temporal cortex (right).

Remembering words

Verbal short-term memory involves several different components. When a person is trying to keep a string of letters in mind for a few seconds, at least two processes are active. The items are first circulated in a 'rehearsal system' (with associated activity in Broca's area), and then repeatedly put into a buffer store for verbal material (associated activity in the parietal lobe), which has a limited capacity and rapidly decays over a few seconds.

This PET scan of the left side of the brain shows the visual area of the occipital cortex being activated by visual stimuli (shown in red/orange).

Viewing images

The principal visual areas are in the occipital lobe at the back of the brain. PET scans have shown that the pattern of activation depends on the degree of complexity of the visual image. Simple white light activates only a limited area of the occipital lobes, while a more complicated visual stimulus activates a more extensive region. Colour perception involves a specific sub-region of the visual cortex, known as V4.

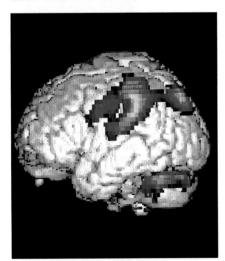

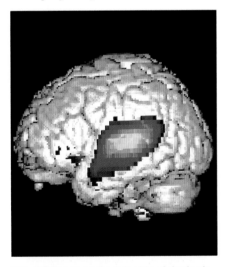

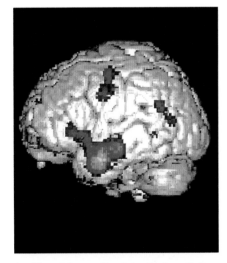

Tactile areas (upper) of the parietal cortex are activated as a blind person's fingers feel Braille dots. A cognitive area also becomes active (lower right).

Reading Braille

People who have been blind from birth show activation in the tactile areas of the parietal cortex as their fingers feel and make sense of the Braille dots. People with acquired blindness, however, show activity in the visual areas of the occipital lobe, as a sighted person would when reading a book. This suggests that they may be using visual imagery to interpret the touch sensations from their fingers as they use their newly acquired skill.

This PET scan of the left side of the brain shows the auditory area in the superior temporal lobe (part of the cerebrum) being activated by hearing.

Hearing sound

Listening to sound produces activation in the primary auditory cortex of the temporal lobes. Speech and non-speech sounds produce different patterns of activation. Hearing speech is associated with Wernicke's area in the left temporal lobe, while non-verbal sounds, such as a door slamming or the melody of music, produce activation in corresponding regions of the other side of the brain, in the right temporal lobe.

Looking at faces stimulates brain activity in an area of the occipital lobe. Face recognition, however, takes place mainly in the temporal lobe (orange/red).

Viewing faces

The perception of faces involves several different brain regions working together. A pathway running from the visual areas in the occipital lobe to the pre-frontal cortex is activated during facial perception, while an area known as the fusiform face area is activated during facial recognition.

Information also passes to the amygdala, a part of the limbic system concerned with emotion. This identifies the face as known or unknown, and as friend or foe.

Cerebral hemispheres

The cerebral hemispheres are the largest part of the brain.
In humans, they have developed out of proportion to the other
regions, distinguishing our brains from those of other animals.

The left and right cerebral hemispheres are separated from each other by the longitudinal fissure which runs between them. Looking at the surface of the hemispheres from the top and side, there is a prominent groove running downwards, beginning about 1 cm behind the midpoint between the front and back of the brain.

This is the central sulcus or rolandic fissure. Further down on the side of the brain there is a second large groove, the lateral sulcus or sylvian fissure.

LOBES OF THE BRAIN

The cerebral hemispheres are divided into lobes, named after the bones of the skull which lie over them:

■ The frontal lobe lies in front of the rolandic fissure and above the sylvian fissure

■ The parietal lobe lies behind the rolandic fissure and above the back part of the sylvian fissure; it extends back as far as the parieto-occipital sulcus, a groove separating it from the occipital lobe, which is at the back of the brain

■ The temporal lobe is the area below the sylvian fissure and extends backward to meet the occipital lobe.

Lobes of the cerebral hemispheres

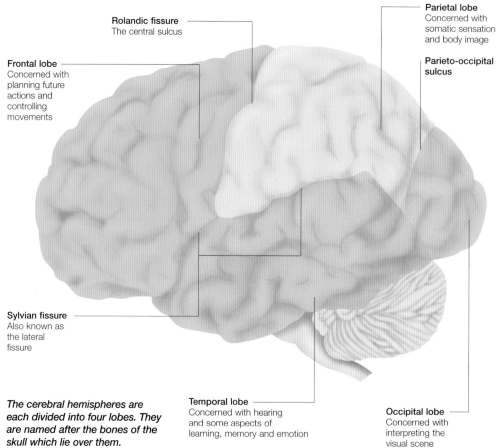

Rolandic fissure
The central sulcus

Frontal lobe
Concerned with planning future actions and controlling movements

Parietal lobe
Concerned with somatic sensation and body image

Parieto-occipital sulcus

Sylvian fissure
Also known as the lateral fissure

Temporal lobe
Concerned with hearing and some aspects of learning, memory and emotion

Occipital lobe
Concerned with interpreting the visual scene

The cerebral hemispheres are each divided into four lobes. They are named after the bones of the skull which lie over them.

Gyri and sulci

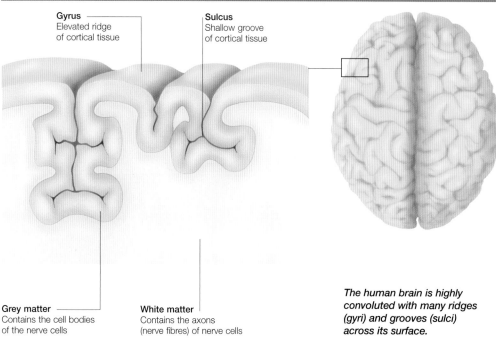

Gyrus
Elevated ridge of cortical tissue

Sulcus
Shallow groove of cortical tissue

Grey matter
Contains the cell bodies of the nerve cells

White matter
Contains the axons (nerve fibres) of nerve cells

The human brain is highly convoluted with many ridges (gyri) and grooves (sulci) across its surface.

As the brain grows rapidly before birth, the cerebral cortex folds in on itself, producing the characteristic appearance that resembles a walnut. The folds are known as gyri and the shallow grooves between them are the sulci.

Certain sulci are found in the same position in all human brains and as a result are used as landmarks to divide the cortex into the four lobes.

DEVELOPMENT OF GYRI AND SULCI

Gyri and sulci begin to appear about the third or fourth month after conception. Before this time, the surface of the brain is smooth, like the brains of birds or reptiles. This complicated folding of its surface allows a larger area of cerebral cortex to be contained within the confined space of the skull.

52

Functions of the cerebral hemispheres

Different regions of the cortex have distinct and highly specialized functions.

The cerebral cortex is divided into:
■ Motor areas, which initiate and control movement. The primary motor cortex controls voluntary movement of the opposite side of the body. Just in front of the primary motor cortex is the pre-motor cortex and a third area, the supplementary motor area, lies on the inner surface of the frontal lobe. All of these areas work with the basal ganglia and cerebellum to allow us to perform complex sequences of finely controlled movements.
■ Sensory areas, which receive and integrate information from sensory receptors around the body. The primary somatosensory area receives information from sensory receptors on the opposite side of the body about touch, pain, temperature and the position of joints and muscles (proprioception).
■ Association areas, which are involved with the integration of more complex brain functions – the higher mental processes of learning, memory, language, judgment and reasoning, emotion and personality.

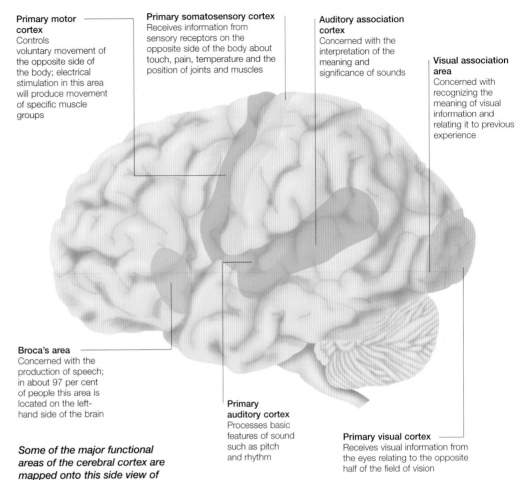

Primary motor cortex
Controls voluntary movement of the opposite side of the body; electrical stimulation in this area will produce movement of specific muscle groups

Primary somatosensory cortex
Receives information from sensory receptors on the opposite side of the body about touch, pain, temperature and the position of joints and muscles

Auditory association cortex
Concerned with the interpretation of the meaning and significance of sounds

Visual association area
Concerned with recognizing the meaning of visual information and relating it to previous experience

Broca's area
Concerned with the production of speech; in about 97 per cent of people this area is located on the left-hand side of the brain

Primary auditory cortex
Processes basic features of sound such as pitch and rhythm

Primary visual cortex
Receives visual information from the eyes relating to the opposite half of the field of vision

Some of the major functional areas of the cerebral cortex are mapped onto this side view of the human brain.

Motor and sensory body map

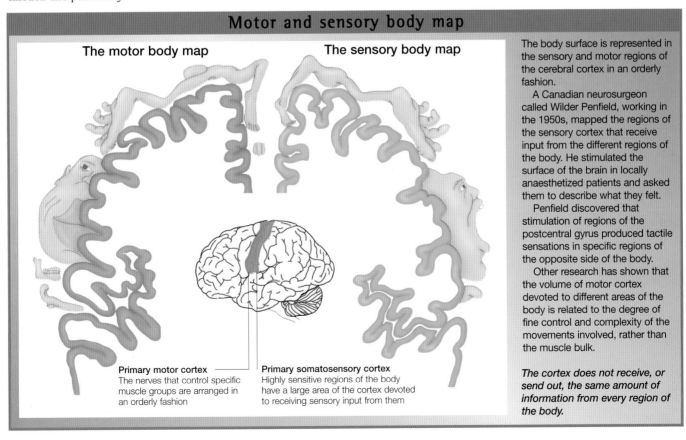

The motor body map

The sensory body map

Primary motor cortex
The nerves that control specific muscle groups are arranged in an orderly fashion

Primary somatosensory cortex
Highly sensitive regions of the body have a large area of the cortex devoted to receiving sensory input from them

The body surface is represented in the sensory and motor regions of the cerebral cortex in an orderly fashion.

A Canadian neurosurgeon called Wilder Penfield, working in the 1950s, mapped the regions of the sensory cortex that receive input from the different regions of the body. He stimulated the surface of the brain in locally anaesthetized patients and asked them to describe what they felt.

Penfield discovered that stimulation of regions of the postcentral gyrus produced tactile sensations in specific regions of the opposite side of the body.

Other research has shown that the volume of motor cortex devoted to different areas of the body is related to the degree of fine control and complexity of the movements involved, rather than the muscle bulk.

The cortex does not receive, or send out, the same amount of information from every region of the body.

Structure of the cerebral hemispheres

The cerebral cortex is made of two distinct layers: grey matter,
a thin layer of nerve and glial cells about 2-4 mm in thickness;
and white matter, consisting of nerve fibres (axons) and glial cells.

On the surface of the hemispheres is the grey matter, which ranges in thickness from about 2 to 4 mm. The grey matter is made up of nerve cells (neurones) together with supporting glial cells. In most parts of the cortex, six separate layers of cells can be distinguished under a microscope.

CORTICAL NEURONES
The cell bodies (which contain the cell's nucleus) of cortical neurones differ markedly in shape, though two main types of cell can be distinguished:
■ Pyramidal cells are so called because their cell body is shaped like a pyramid; their axons (nerve fibres) project out of the cortex, carrying information to other regions of the brain
■ Non-pyramidal cells, in contrast, have a smaller and rounder cell body and are involved in receiving and analysing input from other sources.

The grey matter of the cerebral hemispheres can be subdivided into six layers of cells, based on the type of brain cell present.

The six layers of cells of the cortex

I: Molecular
Contains mostly axons that run laterally and glial cells

II: External granular
Contains mostly small pyramidal neurones

III: External pyramidal
Contains larger pyramidal cells which provide output to other cortical regions

IV: Internal granular
Rich in non-pyramidal cells which receive afferent input from the thalamus

V: Internal pyramidal
Contains the largest pyramidal cells whose long axons leave the cortex and descend to the brainstem and spinal cord

VI: Multiform
Contains pyramidal cells, some of which project back to the thalamus

Brodmann's area

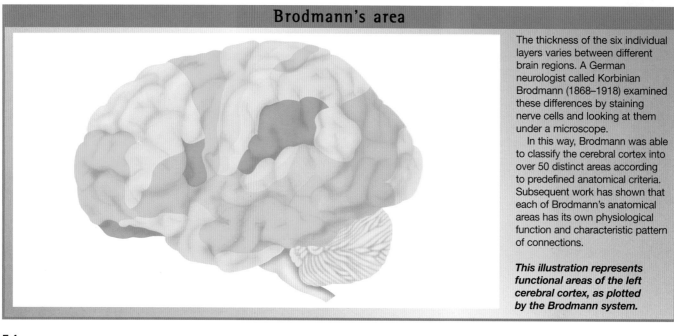

The thickness of the six individual layers varies between different brain regions. A German neurologist called Korbinian Brodmann (1868–1918) examined these differences by staining nerve cells and looking at them under a microscope.

In this way, Brodmann was able to classify the cerebral cortex into over 50 distinct areas according to predefined anatomical criteria. Subsequent work has shown that each of Brodmann's anatomical areas has its own physiological function and characteristic pattern of connections.

This illustration represents functional areas of the left cerebral cortex, as plotted by the Brodmann system.

White matter

The white matter is composed of nerve fibres (axons) which connect different regions of brain.

Underneath the cerebral cortex (grey matter) is the white matter, which makes up the bulk of the inside of the cerebral hemispheres. It is arranged into bundles or tracts of three types:

■ **Commissural fibres**
These cross between the hemispheres, connecting corresponding regions on the two sides. The corpus callosum is the largest of these tracts.

■ **Association fibres**
These connect different areas within the same hemisphere. Short association fibres connect adjacent gyri and long association fibres interconnect more widely separated regions of the cortex.

■ **Projection fibres**
These connect the cerebral cortex with deeper underlying regions of the brain, the brainstem and the spinal cord. These fibres enable the cortex to receive incoming information from the rest of the body and to send out instructions controlling movement and other bodily functions.

Nerve bundles can be classified into three groups – commissural fibres (green), association fibres (blue) and projection fibres (red).

Distribution of the major nerve fibre tracts

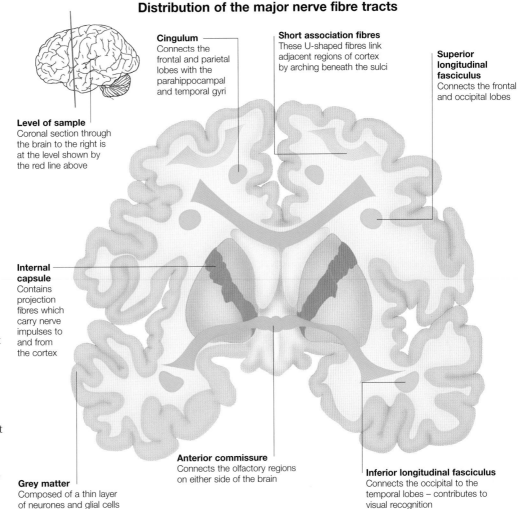

Level of sample
Coronal section through the brain to the right is at the level shown by the red line above

Cingulum
Connects the frontal and parietal lobes with the parahippocampal and temporal gyri

Short association fibres
These U-shaped fibres link adjacent regions of cortex by arching beneath the sulci

Superior longitudinal fasciculus
Connects the frontal and occipital lobes

Internal capsule
Contains projection fibres which carry nerve impulses to and from the cortex

Anterior commissure
Connects the olfactory regions on either side of the brain

Grey matter
Composed of a thin layer of neurones and glial cells

Inferior longitudinal fasciculus
Connects the occipital to the temporal lobes – contributes to visual recognition

Brain damage and the case of Phineas Gage

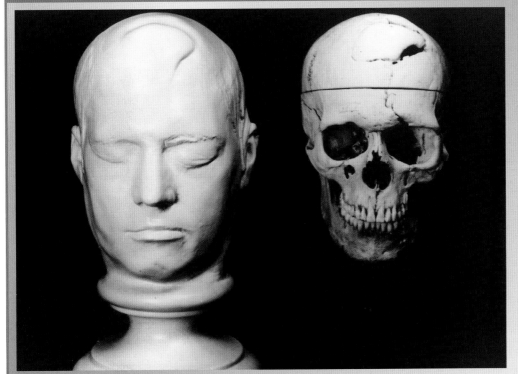

Neurologists have learnt a great deal about the brain by examining the behaviour of patients with brain damage.

One of the most notorious case histories was that of an American railway construction worker called Phineas Gage. Gage survived a blasting accident in 1848 in which a metal rod was driven through his left cheek and out of the top of his head.

After the accident, Gage's personality changed considerably – he became inconsiderate, moody, foul-mouthed and unable to plan ahead. This provided evidence that the frontal cortex, which was severely damaged during the accident, is involved in forward planning and self-image.

Phineas Gage's death mask and skull demonstrate the damage caused by the metre-long, 4 cm-wide metal rod.

Thalamus

The thalamus is a major sensory relay and integrating centre in the brain, lying deep within its central core. It consists of two halves, and receives sensory inputs of all types, except smell.

The thalamus is made up of paired egg-shaped masses of grey matter (cell bodies of nerve cells) 3–4 cm long and 1.5 cm wide, located in the deep central core of the brain known as the diencephalon, or 'between brain'.

The thalamus makes up about 80 per cent of the diencephalon and lies on either side of the fluid-filled third ventricle. The right and left parts of the thalamus are connected to each other by a bridge of grey matter – the massa intermedia, or interthalamic adhesion.

NEUROANATOMY

The front end of the thalamus is rounded and is narrower than the back, which is expanded into the pulvinar. The upper surface of the thalamus is covered with a thin layer of white matter – the stratum zonale. A second layer of white matter – the external medullary lamina – covers the lateral surface.

Its structure is very complex and it contains more than 25 distinct nuclei (collections of nerve cells with a common function). These thalamic nuclear groups are separated by a vertical Y-shaped sheet of white matter – the internal medullary lamina. The anterior nucleus lies in the fork of the Y, and the tail divides the medial and lateral nuclei and splits to enclose the intralaminar nuclei.

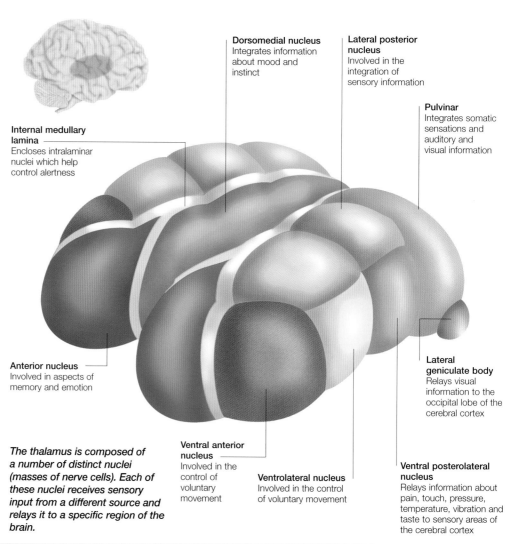

Dorsomedial nucleus
Integrates information about mood and instinct

Lateral posterior nucleus
Involved in the integration of sensory information

Pulvinar
Integrates somatic sensations and auditory and visual information

Internal medullary lamina
Encloses intralaminar nuclei which help control alertness

Anterior nucleus
Involved in aspects of memory and emotion

Lateral geniculate body
Relays visual information to the occipital lobe of the cerebral cortex

Ventral anterior nucleus
Involved in the control of voluntary movement

Ventrolateral nucleus
Involved in the control of voluntary movement

Ventral posterolateral nucleus
Relays information about pain, touch, pressure, temperature, vibration and taste to sensory areas of the cerebral cortex

The thalamus is composed of a number of distinct nuclei (masses of nerve cells). Each of these nuclei receives sensory input from a different source and relays it to a specific region of the brain.

Higher brain control

Side view of brain

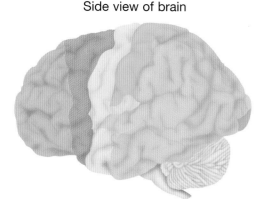

Internal view of brain

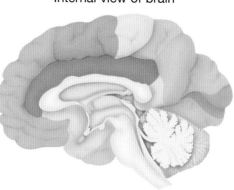

Each thalamic nucleus is connected to a specific region of the cerebral cortex (outer tissue of the brain). The two illustrations above are colour-coded to the artwork of the thalamus at the top of the page.

Each thalamic nucleus is linked to a distinct region of the cerebral cortex. These connections are made via a nerve fibre bundle called the internal capsule.

Some thalamic nuclei relay information received from different sensory modalities, including somatic (physical) sensation, vision and hearing, to the somatosensory cortex.

Others are involved in transmitting information about movement from the cerebellum and basal ganglia to the motor regions of the frontal cortex.

The thalamus is also involved in autonomic (unconscious) functions, including the maintenance of consciousness.

Hypothalamus

The hypothalamus is a complex structure located in the deep core of the brain. It regulates fundamental aspects of body function, and is critical for homeostasis – the maintenance of equilibrium in the body's internal environment.

The hypothalamus is a small region of the diencephalon; it is the size of a thumbnail and weighs only about four grams. It lies below the thalamus and is separated from it by a shallow groove, the hypothalamic sulcus. The hypothalamus is just behind the optic chiasm, the point where the two optic nerves cross over as they travel from the eyes towards the visual area at the back of the brain.

Several distinct structures stand out on its undersurface:
■ The mammillary bodies – two small, pea-like projections which are involved in the sense of smell
■ The infundibulum or pituitary stalk – a hollow structure connecting the hypothalamus with the posterior part of the pituitary gland (neurohypophysis) which lies below it
■ The tuber cinereum or median eminence – a greyish-blue, raised region surrounding the base of the infundibulum.

Hypothalamic nuclei

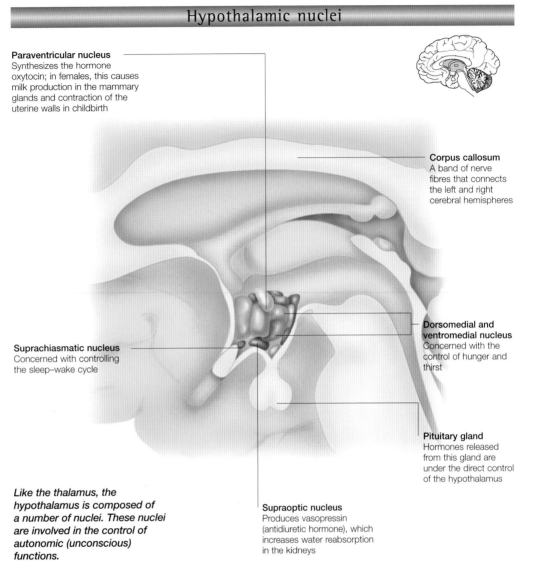

Paraventricular nucleus
Synthesizes the hormone oxytocin; in females, this causes milk production in the mammary glands and contraction of the uterine walls in childbirth

Corpus callosum
A band of nerve fibres that connects the left and right cerebral hemispheres

Dorsomedial and ventromedial nucleus
Concerned with the control of hunger and thirst

Suprachiasmatic nucleus
Concerned with controlling the sleep–wake cycle

Pituitary gland
Hormones released from this gland are under the direct control of the hypothalamus

Supraoptic nucleus
Produces vasopressin (antidiuretic hormone), which increases water reabsorption in the kidneys

Like the thalamus, the hypothalamus is composed of a number of nuclei. These nuclei are involved in the control of autonomic (unconscious) functions.

Hypothalamic control of other functions

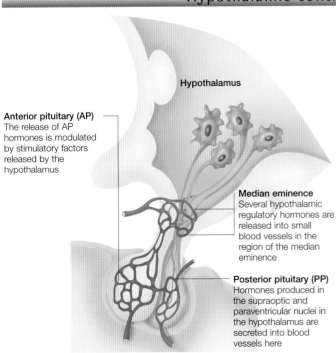

Hypothalamus

Anterior pituitary (AP)
The release of AP hormones is modulated by stimulatory factors released by the hypothalamus

Median eminence
Several hypothalamic regulatory hormones are released into small blood vessels in the region of the median eminence

Posterior pituitary (PP)
Hormones produced in the supraoptic and paraventricular nuclei in the hypothalamus are secreted into blood vessels here

The hypothalamus regulates a wide range of basic processes:
■ **The pituitary gland**
The hypothalamus is the main link between the central nervous system and the endocrine system, controlling pituitary gland function
■ **The autonomic nervous system**
Nerve fibres travel from the hypothalamus to the autonomic control centres in the brainstem. By this pathway, the hypothalamus can influence heart rate and blood pressure; contraction of the gut and bladder; sweating; and salivation
■ **Eating and drinking behaviour**
Stimulation of the lateral

The hypothalamus controls the pituitary gland via nerve fibres innervating the posterior pituitary, and blood capillaries supplying the anterior pituitary.

hypothalamus increases hunger and thirst. In contrast, activation of the ventromedial hypothalamus reduces hunger and food intake
■ **Body temperature**
Certain areas of the hypothalamus monitor the temperature of the blood and act as a thermostat
■ **Control of emotional behaviour**
The hypothalamus is involved, along with other brain regions, in the expression of fear and aggression, as well as in the control of sexual behaviour
■ **Control of sleep cycles**
The suprachiasmatic nucleus contributes to the daily patterns of sleeping and waking
■ **Memory**
Damage to the mammillary bodies is associated with impairment of the ability to learn and retain new information.

Limbic system

The limbic system is a ring of interconnected structures that lies deep within the brain. It makes connections with other parts of the brain, and is associated with mood and memory.

The limbic system is a collection of structures deep within the brain that is associated with the perception of emotions and the body's response to them.

The limbic system is not one, discrete part of the brain. Rather it is a ring of interconnected structures surrounding the top of the brainstem. The connections between these structures are complex, often forming loops or circuits and, as with much of the brain, their exact role is not fully understood.

STRUCTURE

The limbic system is made up from all or parts of the following brain structures:

■ Amygdala – this almond-shaped nucleus appears to be linked to feelings of fear and aggression

■ Hippocampus – this structure seems to play a part in learning and memory

■ Anterior thalamic nuclei – these collections of nerve cells form part of the thalamus. One of their roles seems to lie in the control of instinctive drives

■ Cingulate gyrus – this connects the limbic system to the cerebral cortex, the part of the brain that carries conscious thoughts

■ Hypothalamus – this regulates the body's internal environment, including blood pressure, heart rate and hormone levels. The limbic system generates its effects on the body by sending messages to the hypothalamus.

Medial view of the limbic system within the brain

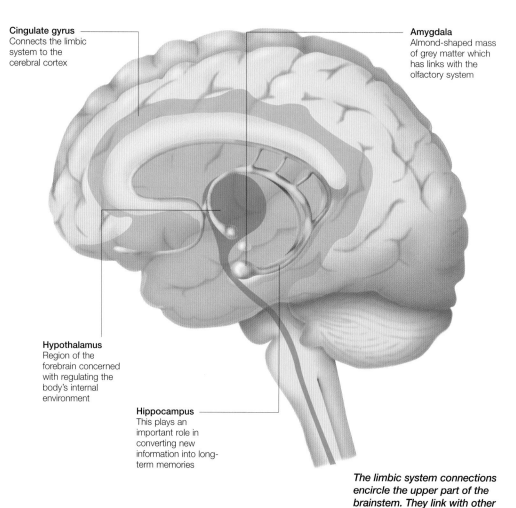

Cingulate gyrus
Connects the limbic system to the cerebral cortex

Amygdala
Almond-shaped mass of grey matter which has links with the olfactory system

Hypothalamus
Region of the forebrain concerned with regulating the body's internal environment

Hippocampus
This plays an important role in converting new information into long-term memories

The limbic system connections encircle the upper part of the brainstem. They link with other parts of the brain and are associated with emotion.

The limbic system and the sense of smell

Our sense of smell is strongly linked to memories of the past or emotions. For example, the smell of a new baby can trigger maternal affection.

The olfactory system (responsible for the sense of smell) is often included with the limbic system. There is certainly a close connection between the two.

EMOTIONS

Nerve fibres carrying information to the brain from the sensory receptors in the nose connect with structures in the limbic system, especially the amygdala. These connections mean that different smells are often associated with a variety of emotions and feelings. Examples might be the disgust that accompanies the smell of excreta, or the maternal affection that is associated with the smell of a tiny child.

MEMORIES

Smell is also linked to memory; it is not uncommon for a passing scent to suddenly evoke memories which are believed to be long forgotten. This may be explained by the role of the limbic system, and especially the hippocampus, in learning and memory.

Connections of the limbic system

The limbic system has connections with the higher centres of the brain in the cortex, and with the more primitive brainstem. It not only allows the emotions to influence the body, but also enables the emotional response to be regulated.

The human brain can be considered to be made up of three parts. These parts have evolved one after another over the millennia.

BRAINSTEM

The 'oldest' part of the brain, in evolutionary terms, is the brainstem, which is concerned largely with unconscious control of the internal state of the body. The brainstem can be seen as a sort of 'life support system'.

LIMBIC SYSTEM

With the evolution of mammals came another 'layer' of brain, the limbic system. The limbic system allowed the development of feelings and emotions in response to sensory information. It is also associated with the development of newer – in evolutionary terms – behaviours, such as closeness to offspring (maternal bonding).

CEREBRAL CORTEX

The final layer of the human brain is shared to some extent with higher mammals. It is the cerebral cortex, the part of the brain that allows humans to think and reason. With this part of the brain, individuals perceive the outside world and make

conscious decisions about their behaviour and actions.

ROLE OF THE LIMBIC SYSTEM

The limbic system lies between the cortex and the brainstem and makes connections with both. Through its connections

with the brainstem, the limbic system provides a way in which an individual's emotional state can influence the internal state of the body. This may prepare the body perhaps for an act of self-preservation such as running away in fear, or for a sexual encounter.

The extensive connections between the limbic system and the cerebral cortex allow human beings to use their knowledge of the outside world to regulate their response to emotions. The cerebral cortex can thus 'override' the more primitive limbic system when necessary.

The developing brain

The cerebral cortex
This outer layer of the brain evolved last, and is related to higher intellect

The brainstem
This part of the brain evolved first, and is responsible for self-preservation and aggression

The limbic system
This system developed secondly and enabled the emotions necessary for mammalian existence, which include caring for offspring

The three layers of the brain evolved one by one over thousands of years. Each is responsible for different bodily and intellectual functions.

Disorders of the limbic system

Temporal lobe epilepsy may involve the limbic system. Electroencephalography shows any abnormal electrical activity in particular areas of the brain.

As the limbic system is associated with emotions, mood and memory, damage to the structures of this system may have effects in these areas.

WERNICKE'S ENCEPHALOPATHY

Wernicke's encephalopathy is a disorder of the brain in which bleeding from tiny capillaries occurs in the upper brainstem and limbic system.

The condition is caused by

long-term alcohol abuse associated with a poor diet; affected individuals experience confusion and may eventually fall into a coma. If a person recovers, there is usually a degree of amnesia and an inability to learn new facts.

TEMPORAL LOBE EPILEPSY

In temporal lobe epilepsy, seizures arise in the temporal lobe of the brain, close to the limbic system. If the amygdala or hippocampus are involved, the patient may report complex experiences of smell, mood and memory during the seizure. These may even be severe enough to mimic schizophrenia.

Basal ganglia

The basal ganglia lie deep within the white matter of the cerebral hemispheres. They are collections of nerve cell bodies that are involved in the control of movement.

The common term basal ganglia is, in fact, a misnomer, as the term ganglion refers to a mass of nerve cells in the peripheral nervous system rather than the central nervous system, as here. The term basal nuclei is anatomically more appropriate.

COMPONENTS

There are a number of component parts to the basal nuclei which are all anatomically and functionally closely related to each other. The parts of the basal nuclei include:

■ Putamen. Together with the caudate nucleus, the putamen receives input from the cortex
■ Caudate nucleus. Named for its shape, as it has a long tail, this nucleus is continuous with the putamen at the anterior (front) end
■ Globus pallidus. This nucleus relays information from the putamen to the pigmented area of the midbrain known as the substantia nigra, with which it bears many similarities.

GROUPING

Various names are associated with different groups of the basal nuclei. The term corpus striatum (striped body) refers to the whole group of basal nuclei, whereas the striatum includes only the putamen and caudate nuclei. Another term, the lentiform nucleus, refers to the putamen and the globus pallidus which, together, form a lens-shaped mass.

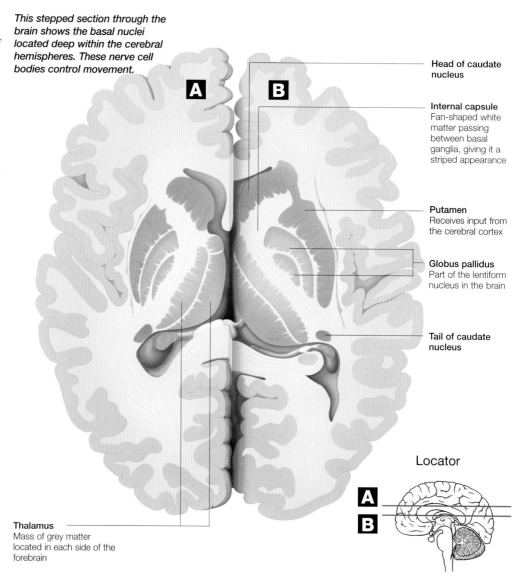

This stepped section through the brain shows the basal nuclei located deep within the cerebral hemispheres. These nerve cell bodies control movement.

Head of caudate nucleus

Internal capsule
Fan-shaped white matter passing between basal ganglia, giving it a striped appearance

Putamen
Receives input from the cerebral cortex

Globus pallidus
Part of the lentiform nucleus in the brain

Tail of caudate nucleus

Thalamus
Mass of grey matter located in each side of the forebrain

Locator

Coronal section through brain

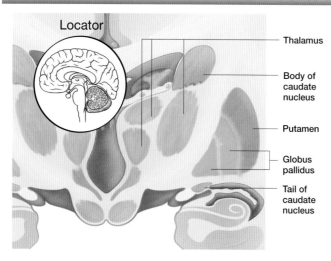

Locator

Thalamus

Body of caudate nucleus

Putamen

Globus pallidus

Tail of caudate nucleus

A coronal section through the brain reveals two anatomical factors: the shape of the basal ganglia and the location of the ganglia in relation to other structures in the area.

SHAPE OF GANGLIA

The coronal section shows that the lentiform nucleus is the shape of a brazil nut or an orange segment.

The putamen lies on the lateral (outer) side of the paler

A coronal section through the brain is shown. It reveals the relationship of the basal ganglia to other structures in the brain.

globus pallidus, which tapers to a blunt point. Just lateral to the putamen lies a streak of grey matter known as the claustrum (not shown), which is sometimes included under the heading of basal nuclei.

CAPSULE

The basal nuclei lie close to the thalamus, an important area of the brain, with which they make many connections. They are separated from the thalamus by the internal capsule, which is an area of white matter consisting of nerve fibres that pass from the cortex down to the spinal cord.

Structure and role of the basal ganglia

The overall shape of the basal ganglia (nuclei) is complex, and is hard to imagine by looking at two-dimensional cross-sections.

When seen in a three-dimensional view, the size and shape of the basal nuclei, together with their position within the brain as a whole, can be appreciated more easily.

In particular, the shape of the caudate nucleus can now be understood – it connects at its head with the putamen, then bends back to arch over the thalamus before turning forwards again. The tip of the tail of the caudate nucleus ends as it merges with the amygdala, part of the limbic system (concerned with unconscious, autonomic functions).

ROLE OF THE BASAL NUCLEI

The functions of the basal nuclei have been difficult to study because they lie deep within the brain and are therefore relatively inaccessible. Much of what is known of their function derives from the study of those patients who have disorders of the basal nuclei that lead to particular disruptions of movement and posture, such as Parkinson's disease.

A summary of what is currently known about the function of the basal nuclei is that: they help to produce movements which are appropriate; and they inhibit unwanted or inappropriate movements.

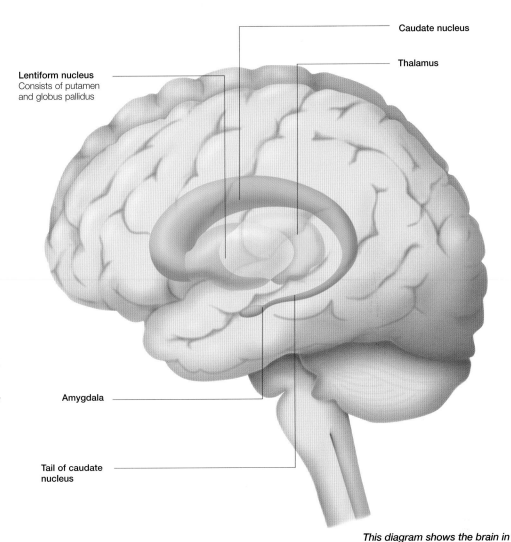

Lentiform nucleus
Consists of putamen and globus pallidus

Caudate nucleus

Thalamus

Amygdala

Tail of caudate nucleus

This diagram shows the brain in three dimensions. The size and shape of the basal ganglia can be seen in relation to other structures.

Disorders of basal ganglia

Parkinson's disease may result from damage to the basal nuclei. Symptoms of the disorder include a stooping posture, tremor and a shuffling walk.

A range of movement disorders result from damage to the basal nuclei. These disorders include Parkinson's disease, Huntington's chorea and Wilson's disease.

MOVEMENT DISORDERS

Parkinson's disease is a disease of unknown cause that mainly affects the elderly. It results in various combinations of slowness of movement, increased muscle tone, tremor and a bent posture. Affected individuals have difficulty in starting and finishing a movement and may also have a mask-like facial expression.

Studies of the basal nuclei of those individuals with Parkinson's disease have shown a lack of the chemical dopamine, a substance that allows neurones to communicate with each other.

Huntington's chorea is an inherited disease, which does not become apparent until an individual reaches mid life. It is associated with progressive degeneration of parts of the basal nuclei and cortex, leading to abnormal movements and dementia

Wilson's disease is an inherited disorder associated with damage to the basal nuclei and progressive dementia in the young.

Cerebellum

The cerebellum, which means 'little brain', lies under the occipital lobes of the cerebral cortex at the back of the brain. It is important to the subconscious control of movement.

The part of the brain known as the cerebellum lies under the occipital lobes of the cerebral cortex at the back of the head. The vital roles of the cerebellum include the co-ordination of movement and the maintenance of balance and posture. The cerebellum works subconsciously and so an individual is not aware of its functioning.

STRUCTURE
The cerebellum is composed of two hemispheres which are bridged in the midline by the vermis. The hemispheres extend laterally (sideways) and posteriorly (backwards) from the midline to form the bulk of the cerebellum.

The surface of the cerebellum has a very distinctive appearance. In contrast to the large folds of the cerebral hemispheres, the surface of the cerebellum is made up of numerous fine folds (folia).

LOBES
Between the folia of the cerebellar surface lie deep fissures which divide it into three lobes:
- Anterior lobe
- Posterior lobe
- Flocculonodular lobe.

The cerebellum has two hemispheres, one on either side of the worm-like vermis. The surface of the cerebellum is made up of thin folds (folia).

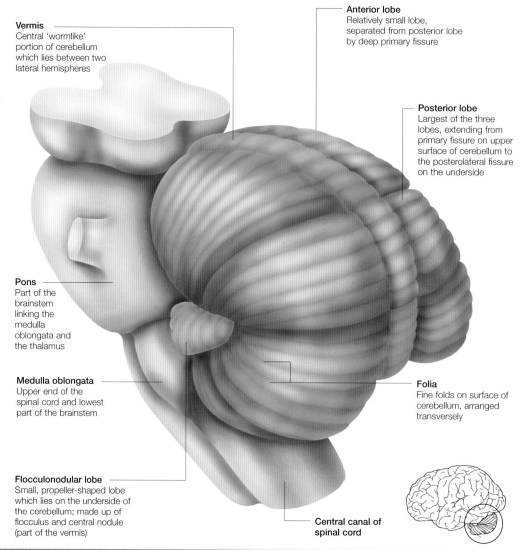

Vermis
Central 'wormlike' portion of cerebellum which lies between two lateral hemispheres

Anterior lobe
Relatively small lobe, separated from posterior lobe by deep primary fissure

Posterior lobe
Largest of the three lobes, extending from primary fissure on upper surface of cerebellum to the posterolateral fissure on the underside

Pons
Part of the brainstem linking the medulla oblongata and the thalamus

Medulla oblongata
Upper end of the spinal cord and lowest part of the brainstem

Folia
Fine folds on surface of cerebellum, arranged transversely

Flocculonodular lobe
Small, propeller-shaped lobe which lies on the underside of the cerebellum; made up of flocculus and central nodule (part of the vermis)

Central canal of spinal cord

Cerebellar peduncles

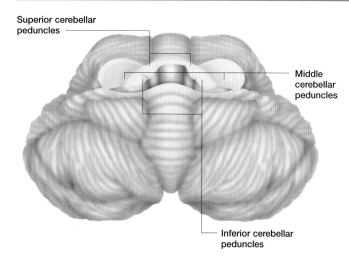

Superior cerebellar peduncles

Middle cerebellar peduncles

Inferior cerebellar peduncles

The cerebellum is connected to the brainstem, and thus to the rest of the brain, by three pairs of nerve fibre tracts which make up the cerebellar peduncles, or stalks. These can be seen on the inferior surface of the cerebellum, where they emerge together. The three tracts are:
- The superior cerebellar peduncles, connecting the cerebellum to the midbrain
- The middle cerebellar peduncles, connecting the

The three pairs of cerebellar peduncles serve to anchor the cerebellum to the brainstem. The peduncles consist of bundles of nerve fibres.

cerebellum to the pons
- The inferior cerebellar peduncles, connecting the cerebellum to the medulla.

There are no direct connections between the cerebellum and the cerebral cortex. All information to and from the cerebellum goes through the peduncles.

Unlike the cerebral cortex, where each side controls the opposite (contralateral) side of the body, each half of the cerebellum controls the same (ipsilateral) side of the body. This means that any damage to one side of the cerebellum will cause symptoms in the same side of the body.

Internal structure of the cerebellum

The cerebellum has an outer grey cortex and a core of nerve fibres, or white matter. Deep within the white matter lie four pairs of cerebellar nuclei: the fastigial, globose, emboliform and dentate nuclei.

The cerebellum is composed of a surface layer of nerve cell bodies, or grey matter, which overlies a core of nerve fibres, or white matter. Deep within the white matter lie the cerebellar nuclei.

CEREBELLAR CORTEX
Due to the presence of the numerous fine folia (folds) in the surface of the cerebellum, the cortex is very extensive. It is made up of the cell bodies and dendrites (cell processes) of the vast majority of cerebellar neurones.

The cells of the cortex receive information from outside the cerebellum via the cerebellar peduncles and make frequent connections between themselves within the cortex.

SIGNALS
In most cases, signals from the cerebellar cortex are conveyed in the fibres of the white matter down to the cerebellar nuclei. It is from here that information leaves the cerebellum to be carried to the rest of the central nervous system.

CEREBELLAR NUCLEI
There are four pairs of cerebellar nuclei which, from the midline outwards, are known as the:
- Fastigial nuclei
- Globose nuclei
- Emboliform nuclei
- Dentate nuclei.

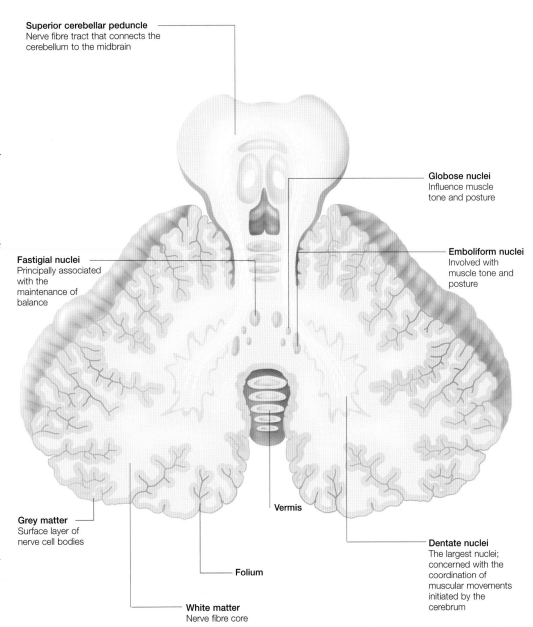

Cross-section through cerebellum

Superior cerebellar peduncle
Nerve fibre tract that connects the cerebellum to the midbrain

Globose nuclei
Influence muscle tone and posture

Emboliform nuclei
Involved with muscle tone and posture

Fastigial nuclei
Principally associated with the maintenance of balance

Vermis

Grey matter
Surface layer of nerve cell bodies

Folium

White matter
Nerve fibre core

Dentate nuclei
The largest nuclei; concerned with the coordination of muscular movements initiated by the cerebrum

Layers of the cerebellar cortex

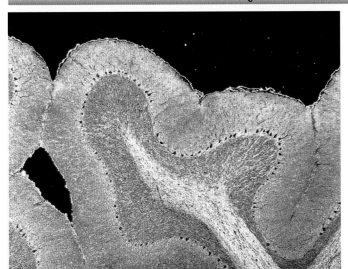

If the cerebellar cortex is stained and studied under the microscope, its structure can be seen to form a quite distinctive pattern of layers:
- Molecular layer (the outermost layer) – contains nerve cell bodies and is rich in nerve fibres from the deeper layers of cells
- Purkinje cell layer (the next layer) – although only one-cell

A micrograph of a stained section of the cerebellar cortex reveals its complex layered structure. Each layer contains distinctive cell types.

thick, this is relatively easy to visualize due to the size of the Purkinje cells. These specialized neurones are very important to cerebellar function. They receive signals via their dendrites which lie mainly in the layer above, and send information down to the cerebellar nuclei
- Granular layer (the innermost layer) – this contains the cell bodies of numerous granule cells. These cells receive information via the peduncles and send signals themselves through their axons up into the molecular layer.

Cranial nerves (1)

There are twelve pairs of cranial nerves which leave the brain to supply structures mainly of the head and neck. The cranial nerves carry information to and from the brain.

Nerves are the routes by which information passes between the central nervous system (CNS) and the rest of the body. From the neck down, these nerves emerge from the spinal cord, passing out through openings in the bony spinal column. However, the cranial nerves emerge directly from the brain.

There are twelve pairs of cranial nerves, which are named and numbered with Roman numerals. The first two pairs attach to the forebrain while the rest come from the brainstem. The cranial nerves serve structures of the head and neck. To reach these they must pass through special openings, or foramina, in the bony skull.

CRANIAL NERVE FIBRES

Cranial nerves are made up of sensory and motor nerve fibres and so carry information to and from the CNS:
■ Sensory nerve fibres bring information such as pain, touch and temperature sensations from the face as well as the senses of taste, vision and hearing
■ Motor fibres send instructions to head, neck and face muscles, allowing various facial expressions and eye movements
■ Autonomic nerve fibres allow the subconscious control of internal structures such as the salivary glands, the iris and some of the major organs of the chest and abdomen.

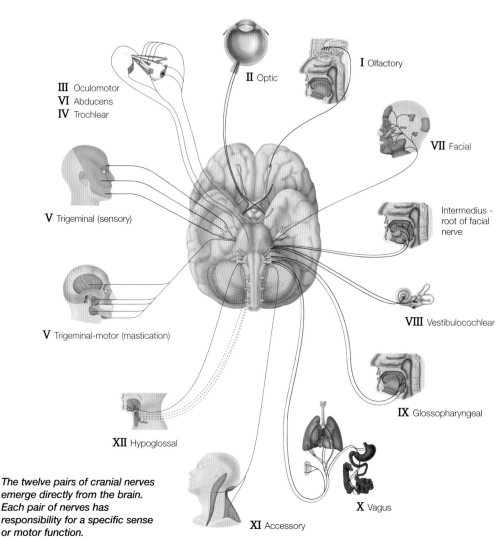

The twelve pairs of cranial nerves emerge directly from the brain. Each pair of nerves has responsibility for a specific sense or motor function.

Functions of the 12 cranial nerves		
Number	Name	Comments
I	Olfactory nerve	The sensory nerve of smell
II	Optic nerve	The sensory nerve of vision
III	Oculomotor nerve	Supplies four of the six muscles that move the eyeball
IV	Trochlear nerve	Supplies a muscle that moves the eyeball
V	Trigeminal nerve	Carries sensation from the face and moves the muscles in chewing
VI	Abducens nerve	Supplies a muscle that moves the eyeball
VII	Facial nerve	Moves the muscles of facial expression
VIII	Vestibulocochlear nerve	The sensory nerve for hearing and balance
IX	Glossopharyngeal nerve	Helps to innervate the tongue and pharynx (gullet)
X	Vagus nerve	Supplies many structures including organs in the thorax and abdomen
XI	Accessory nerve	Supplies structures in the throat and some neck muscles
XII	Hypoglossal nerve	Supplies the tongue muscles.

The olfactory nerves

The olfactory nerves are the tiny sensory nerves of smell. They run from the nasal mucosa to the olfactory bulbs.

The olfactory nerves carry the special sense of smell from the receptor cells in the nasal cavity to the brain above.

OLFACTORY EPITHELIUM
The olfactory epithelium is the part of the lining of the nasal cavity that carries special receptor cells for the sense of smell. It is found in the upper part of the nasal cavity and the septum, the partition between the two sides of the cavity.

The olfactory receptors are specialized neurones, or nerve cells, and are able to detect odorous substances which are present in the form of minute droplets in the air.

OLFACTORY NERVES
Information from the olfactory receptor neurones is passed up to the brain through their long processes, or axons, which group together to form about 20 bundles. These bundles are the true olfactory nerves, which pass up through the thin perforated layer of bone, the cribriform plate of the ethmoid bone, to reach the olfactory bulbs in the cranial cavity.

The fibres of the olfactory nerves make connections (synapse) with the neurones within the olfactory bulb.

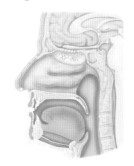

Locator

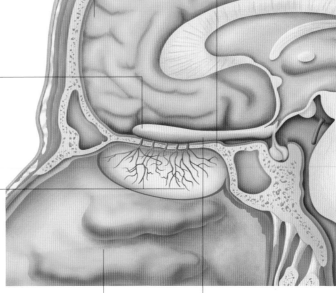

Frontal lobe of cerebral hemisphere

Cribriform plate of the ethmoid bone

Olfactory bulb
Two olfactory bulbs receive and process information from the olfactory nerve fibres

Olfactory nerve fibres
These fibres arise from the receptor cells in the nasal mucosa

Nasal lining
The olfactory epithelium in the lining of the nose contains receptor cells

Olfactory tract
These tracts extend out from the brain to culminate in the olfactory bulbs

The olfactory nerve fibres pass from the nasal lining to the brain. Information passes along these fibres and is interpreted in the olfactory centre.

OLFACTORY BULBS
The paired olfactory bulbs are actually part of the brain, extended out on stalks, the olfactory tracts, which contain fibres linking them to the cerebral hemispheres.

Large specialized neurones, known as mitral cells, connect with the olfactory nerves within the olfactory bulb. This connection permits information about smell to be passed on from the olfactory nerves.

The axons of these mitral cells then carry this information to the olfactory centre of the brain via the olfactory tracts.

Loss of the sense of smell

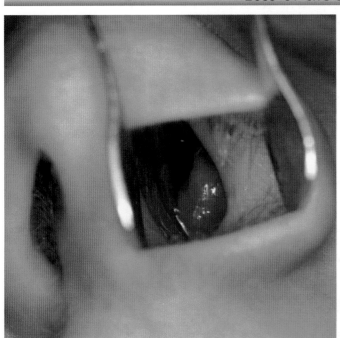

Loss of the sense of smell, known as anosmia, may affect one or both sides of the nose and may be permanent or temporary.

The chief complaint of people with anosmia is of a lack of the sense of taste rather than smell. This is because much of what we think of as the sensation of taste is actually the detection of odours within our food. Without the sense of smell the tongue can only discern four flavours: sweet, bitter, sour and salt.

EFFECTS OF AGEING
As the body ages there is a progressive loss of olfactory fibres. Elderly people, therefore,

Rhinitis causes the turbinate bone in the nasal cavity to become swollen. The resulting inflammation affects the smell mechanism.

often complain that their food has 'lost its flavour' due to a loss of their sense of smell.

HEAD INJURIES
Injuries to the head can lead to tearing of the olfactory bulbs or breaking of the delicate cribriform plate. This can lead to anosmia, usually affecting one nostril only.

RHINITIS
Rhinitis, or inflammation of the mucous lining of the nose, may occur as a result of a viral infection such as the common cold or with allergies such as hay fever. This usually causes a blocked or runny nose and sneezing.

In this situation it is common for people to experience a temporary loss of smell due to involvement of the olfactory epithelium.

Cranial nerves (2)

Cranial nerves II, III, IV and VI are responsible
for visual processes as well as movements of the eyes.
The optic nerve (cranial nerve II) is the nerve of sight.

The optic nerve carries information from the retina at the back of the eye to the brain. Unlike some of the other cranial nerves, the optic nerve is sensory only, which means that it only takes information to the brain, not from it.

OPTIC NERVE STRUCTURE

The optic nerve is formed from the axons, or long processes, of the retinal cells at the back of the eye. These join together to form the nerve, which leaves the back of the eyeball at a point known as the optic disc.

As it is an outgrowth of the brain, the optic nerve is covered by surrounding layers of meninges, the membranes which protect the brain. Running within the optic nerve are blood vessels which serve the retina.

OPTIC NERVE FIBRES

The nerve fibres that originate in the retina enter the optic nerve, which then passes back through the eye socket to the optic canal, an opening in the skull. Entering the cranial cavity through this opening, the optic nerve fibres converge to form the optic chiasma where some of them cross to the other side. This exchange of nerve fibres allows for binocular vision.

The optic nerve fibres continue until they reach the lateral geniculate body of the brain. From here optic nerve fibres radiate back to the visual cortex where their information is processed.

Optic nerve (from below)

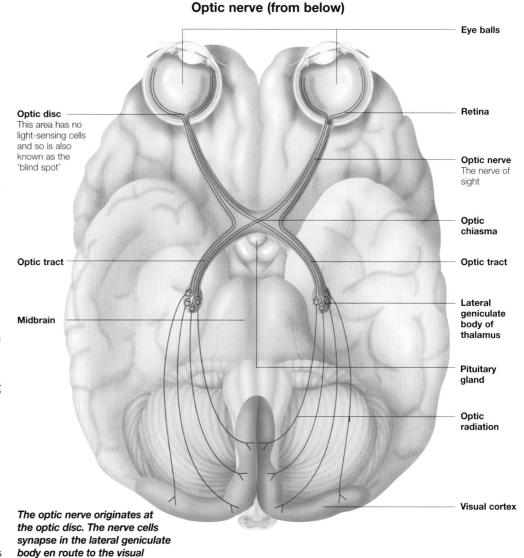

Eye balls

Optic disc
This area has no light-sensing cells and so is also known as the 'blind spot'

Retina

Optic nerve
The nerve of sight

Optic chiasma

Optic tract

Optic tract

Lateral geniculate body of thalamus

Midbrain

Pituitary gland

Optic radiation

Visual cortex

The optic nerve originates at the optic disc. The nerve cells synapse in the lateral geniculate body en route to the visual cortex.

Nerve cells of the retina

Structure of the retina

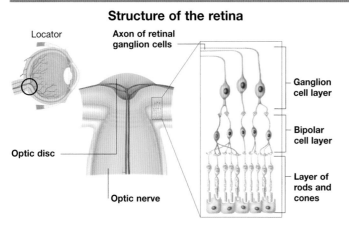

Locator

Axon of retinal ganglion cells

Optic disc

Optic nerve

Ganglion cell layer

Bipolar cell layer

Layer of rods and cones

The optic nerve carries information from the retina, the delicate innermost layer at the back of the eyeball that responds to the presence of light.

Beneath a layer of pigmented cells lies the neural layer. This structure contains cells that can convert light energy into information in a form that the brain can process and understand.

Before reaching the rods and cones, light has to pass through two layers of cells. These are the ganglion and bipolar cell layers.

LAYERS OF NERVE CELLS

The retina has three layers of nerve cells. Photoreceptor cells (rods and cones) lie at the deepest level and are stimulated by light. Above these cells is a layer of bipolar cells that make connections between the rods and cones and the uppermost layer of ganglion cells.

Ganglion cells receive information from the rods and cones via the bipolar cells. The axons of these ganglion cells run along the surface of the neural layer and converge to form the optic nerve.

Cranial nerves III, IV and VI

The oculomotor, trochlear and abducent nerves are usually
considered together because between them they supply the six muscles that move
the eye. These nerves do not carry any information relating to sight.

There are three cranial nerves which supply the muscles responsible for eye movement.

OCULOMOTOR NERVE

The oculomotor nerve mainly carries motor fibres to the eye muscles, but it also has fibres that carry sensory information back to the brain concerning the position of those muscles. In addition, it contains some fibres from the autonomic nervous system that constrict the pupil and alter the lens shape.

Fibres of the oculomotor nerve originate in the midbrain, part of the brainstem, and leave the cranial cavity to enter the eye socket by passing through the superior orbital fissure. The nerve then splits into superior and inferior divisions.

TROCHLEAR NERVE

The small trochlear nerve supplies only one of the muscles of eye movement, carrying both motor information to the muscle and sensory information from it. The fibres originate in the midbrain and then take a long course around the brainstem to enter the eye socket through the superior orbital fissure.

ABDUCENT NERVE

The abducent nerve supplies the lateral rectus muscle of eye movement and has both motor fibres and sensory fibres. Its fibres originate in the pons, a part of the brainstem. The nerve then arrives at the eye socket by passing through the superior orbital fissure.

Oculomotor, trochlear and abducent nerves

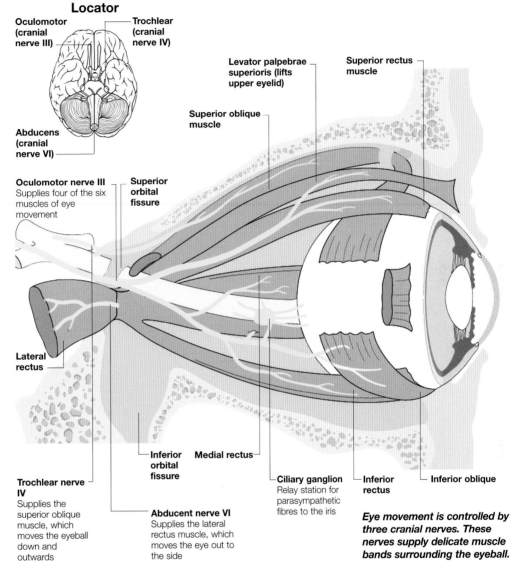

Locator

Oculomotor (cranial nerve III)

Trochlear (cranial nerve IV)

Abducens (cranial nerve VI)

Levator palpebrae superioris (lifts upper eyelid)

Superior rectus muscle

Superior oblique muscle

Oculomotor nerve III
Supplies four of the six muscles of eye movement

Superior orbital fissure

Lateral rectus

Trochlear nerve IV
Supplies the superior oblique muscle, which moves the eyeball down and outwards

Inferior orbital fissure

Medial rectus

Abducent nerve VI
Supplies the lateral rectus muscle, which moves the eye out to the side

Ciliary ganglion
Relay station for parasympathetic fibres to the iris

Inferior rectus

Inferior oblique

Eye movement is controlled by three cranial nerves. These nerves supply delicate muscle bands surrounding the eyeball.

Injuries to cranial nerves III, IV and VI

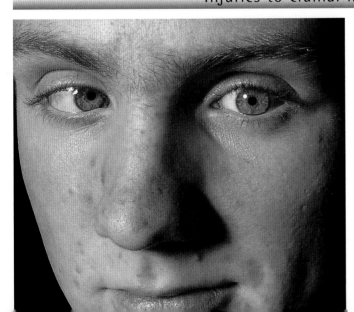

A damaged abducent nerve can cause paralysis of the lateral rectus muscle. This man's left eye cannot move sideways to look at the red pen.

Damage to one of the nerves that supplies the muscles of eye movement will affect that eye's ability to perform a full range of movements. Such damage may be obvious on looking at the eye.

■ **Oculomotor nerve damage**
If the oculomotor nerve is damaged, all but two of the muscles of eye movement will be paralysed on that side. There will be drooping of the eyelid, or ptosis, and dilatation of the pupil, which will no longer be able to constrict in response to bright light. The eye will be held in a position where it appears to look down and outwards.

■ **Trochlear nerve damage**
It is rare for the trochlear nerve to be damaged alone. If this does occur the affected person will complain of double vision when looking down.

■ **Abducent nerve damage**
Damage to this nerve leads to paralysis of the lateral rectus muscle. The affected eye will appear to look inwards and there will be double vision.

Cranial nerves (3)

The trigeminal nerve is the fifth and largest of the cranial nerves, and is the main sensory nerve of the face. It has three branches: the ophthalmic, maxillary and mandibular nerves.

As its name implies, the trigeminal nerve has three branches, each of which supplies a part of the face:

■ Ophthalmic branch – this carries purely sensory fibres and supplies structures in the upper part of the face and scalp on each side. The nerve emerges from the cranium through a gap (superior orbital fissure) in the back of the eye socket

■ Maxillary branch – the maxillary nerve exits the cranium through a small hole (foramen rotundum) and runs forward to supply the central part of the face on each side

■ Mandibular branch – the mandibular nerve has both sensory and motor fibres. It passes out of the cranium through the foramen ovale to reach the lower third of the face and jaw.

TRIGEMINAL GANGLION

Sensory information carried in the three branches of the trigeminal nerve passes back to the pons (part of the brainstem) to be processed.

On the way, information passes through the trigeminal ganglion, an expansion of the trigeminal nerve. This ganglion houses the cell bodies of the nerve cells whose peripheral processes make up the ophthalmic, maxillary and mandibular nerves. The central processes of these nerve cells then leave the ganglion to pass back to the brainstem as the solitary trigeminal nerve.

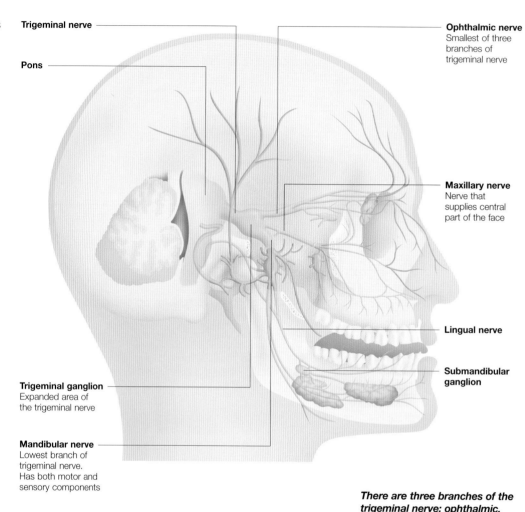

Trigeminal nerve

Pons

Ophthalmic nerve
Smallest of three branches of trigeminal nerve

Maxillary nerve
Nerve that supplies central part of the face

Lingual nerve

Submandibular ganglion

Trigeminal ganglion
Expanded area of the trigeminal nerve

Mandibular nerve
Lowest branch of trigeminal nerve. Has both motor and sensory components

There are three branches of the trigeminal nerve: ophthalmic, maxillary and mandibular. The three branches merge in the trigeminal ganglion.

Sensory supply to the skin

Distribution of sensory fibres of face and scalp

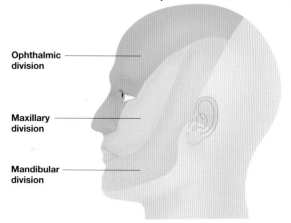

Ophthalmic division

Maxillary division

Mandibular division

Information which passes back through the trigeminal nerve includes the sensations of touch, pressure, pain and temperature from the face and scalp, cornea and the nasal and oral cavities.

THREE AREAS

The sensory supply to the face and scalp can be divided neatly into three areas, corresponding to the three main branches of

The sensory supply to the face and scalp can be divided into three separate areas. These are the ophthalmic, maxillary and mandibular divisions.

the trigeminal nerve:

■ The ophthalmic nerve supplies the scalp over the top of the head and forehead, and the skin of the upper eyelid and centre of the nose

■ The maxillary nerve supplies a central area including the lower eyelid, side of the nose, upper lip and cheek

■ The mandibular nerve supplies the skin over the chin, lower lip and side of the face in front of and above the ear.

The skin of the rest of the face and head is supplied by branches of the first few spinal nerves of the cervical region.

Motor branches of the trigeminal nerve

As well as receiving sensory information from the face, the trigeminal nerve provides the motor supply for some important muscles located around the jaw. Some of these muscles include those involved in chewing.

The large trigeminal nerve is predominantly a sensory nerve, carrying information from the face and scalp back to the brain.

MUSCLE CONTROL
The trigeminal nerve also has a role to play in the control of some important muscles, including those involved in mastication (chewing). The majority of the muscles of the face, many of which are involved in facial expression, are supplied by another nerve, the facial nerve (cranial nerve VII).

NERVE SUPPLY
The trigeminal nerve has three branches; only the lowest branch, the mandibular nerve, carries motor fibres.

The muscles which receive their nerve supply from the motor fibres of the mandibular nerve include:
■ Masseter – this strong muscle runs just under the skin, from the angle of the jaw up to the cheek. It lifts and extends the lower jaw during chewing and biting, and closes the jaw
■ Temporalis – this large fan-shaped muscle can be felt in front of and above the ear as it contracts during chewing
■ Medial and lateral pterygoids

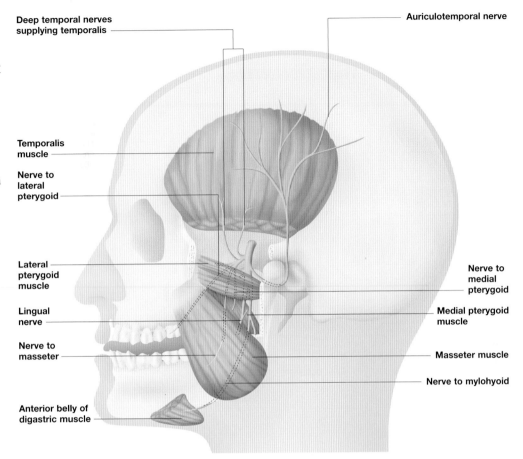

Deep temporal nerves supplying temporalis

Auriculotemporal nerve

Temporalis muscle

Nerve to lateral pterygoid

Lateral pterygoid muscle

Lingual nerve

Nerve to masseter

Anterior belly of digastric muscle

Nerve to medial pterygoid

Medial pterygoid muscle

Masseter muscle

Nerve to mylohyoid

– these strong muscles run between the lower jaw and the skull and are involved in producing chewing movements, including side-to-side grinding
■ Tensor veli palatini – this is one of the muscles of the soft palate
■ Mylohyoid – arising from the

back of the mandible, the mylohyoid muscles of each side unite in the midline to form the floor of the mouth
■ Anterior belly of digastric – this muscle acts with others to pull the hyoid bone and the larynx forwards and upwards during swallowing
■ Tensor tympani – this tiny

The trigeminal nerve plays an important role in controlling a number of muscles around the jaw. The muscles receive a nerve supply from the motor fibres of the mandibular nerve.

muscle tenses the eardrum in order to protect it from loud sounds.

Herpes zoster and trigeminal neuralgia

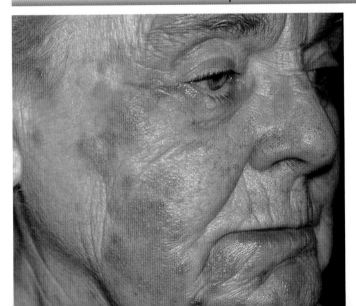

This red rash has been caused by herpes zoster infection. In this patient, the virus has affected the mandibular branch of the trigeminal nerve.

The trigeminal nerve provides a sensory supply for the skin of the face, each of the three branches supplying a well-defined area. This may be demonstrated in cases of herpes zoster infection (shingles) of the head.

After an attack of chickenpox the virus can hide away in sensory nerve cells. When conditions are right for the virus to emerge, it produces skin vesicles (blisters) and pain in the area of the skin supplied by that

sensory nerve. In the case of the trigeminal nerve, the area of the face affected by herpes zoster depends upon whether the virus is infecting the ophthalmic, maxillary or mandibular nerve.

PAINFUL EPISODES
Trigeminal neuralgia is an unpleasant condition in which a person suffers repeated attacks of excruciating pain in the area of the head supplied by the maxillary or mandibular branches of the trigeminal nerve. These episodes, which may only last a few seconds, can be triggered by mild stimuli such as brushing the teeth or even just a gentle touch in the affected area.

Cranial nerves (4)

The facial nerve (seventh cranial nerve) supplies the muscles of facial expression. It also carries autonomic parasympathetic information to the lacrimal and salivary glands and conveys sensory signals.

The facial nerve emerges from the brainstem at the junction of the pons and medulla. There are two roots: the larger root carries motor fibres which transmit instructions to muscles, and the smaller root (nervus intermedius) carries sensory information and parasympathetic fibres.

The facial nerve then travels through the temporal bone, enclosed within the facial canal. The nerve comes into very close contact with the inner ear, before it emerges from the skull through a gap known as the stylomastoid foramen. The nerve then passes through the parotid gland (the salivary gland of the cheek), where it divides into its terminal branches.

MUSCLES

The facial nerve contains three types of fibres: motor, sensory and autonomic. The motor fibres control a number of important muscles, which include the:

■ Muscles of facial expression, including those which allow smiling and frowning
■ Scalp muscles, including the occipitalis and auricular muscle, which give some degree of mobility to the scalp
■ Posterior belly of digastric (helps to raise the hyoid bone during swallowing and speaking)
■ Stylohyoid, a small muscle which also lifts the hyoid bone
■ Stapedius muscle (which lies within the middle ear).

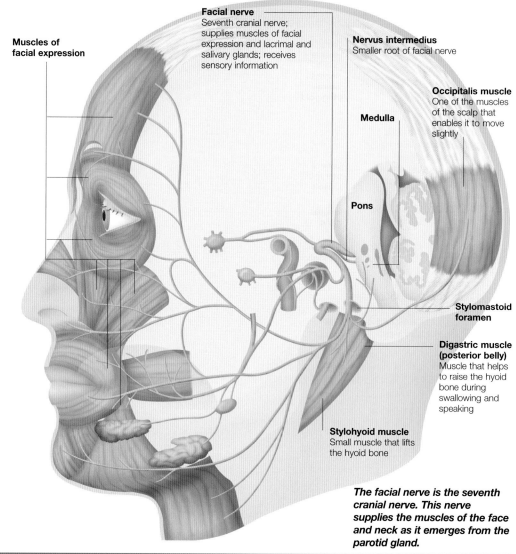

Muscles of facial expression

Facial nerve
Seventh cranial nerve; supplies muscles of facial expression and lacrimal and salivary glands; receives sensory information

Nervus intermedius
Smaller root of facial nerve

Occipitalis muscle
One of the muscles of the scalp that enables it to move slightly

Medulla

Pons

Stylomastoid foramen

Digastric muscle (posterior belly)
Muscle that helps to raise the hyoid bone during swallowing and speaking

Stylohyoid muscle
Small muscle that lifts the hyoid bone

The facial nerve is the seventh cranial nerve. This nerve supplies the muscles of the face and neck as it emerges from the parotid gland.

Branches of the facial nerve

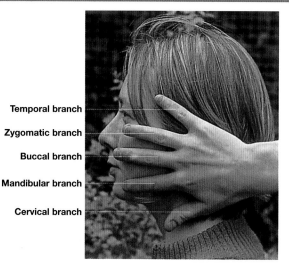

Temporal branch
Zygomatic branch
Buccal branch
Mandibular branch
Cervical branch

The facial nerve emerges from the cranial cavity via the stylomastoid foramen. At that point, the nerve lies within the parotid gland and here it divides into six branches, each named after the region it supplies.

BRANCHES

The six branches of the facial nerve comprise:
■ Posterior auricular – runs up behind the ear
■ Temporal – passes up over the

The hand placed against the face illustrates the radiation of five of the branches of the facial nerve. Each branch supplies a different area of the face.

temples towards the forehead and upper eyelid
■ Zygomatic – this takes its name from the zygoma (cheekbone)
■ Buccal – runs forward to supply the region around the mouth, or buccal cavity
■ Mandibular – runs along the mandible, or lower jaw
■ Cervical – runs down to the neck area, and is the lowest of the branches of the facial nerve.

POSITIONS

The positions of these nerves can be recalled by imagining a hand spread out across the side of the face (see picture), with the thumb pointing down.

The sensory and autonomic branches

As well as providing many muscles of the face with a nerve supply, the facial nerve carries sensory information, and transports fibres of the autonomic nervous system to the lacrimal and salivary glands.

The facial nerve carries sensory information back to the brain from:
■ The tongue – the sense of taste from the front two thirds of the tongue and soft palate is carried by fibres which become part of the facial nerve
■ A small area of skin around the external auditory meatus, the entrance to the ear canal.

The cell bodies of these sensory neurones lie in the geniculate ganglion, a swelling of the facial nerve. Infection of these nerve fibres by the virus *Herpes zoster*, which causes shingles, may occur. This condition, in which the shingles rash is found around the opening of the ear canal and on the soft palate, is known as Ramsay Hunt syndrome.

PARASYMPATHETIC COMPONENTS OF THE FACIAL NERVE

The parasympathetic system is the part of the autonomic nervous system that is concerned with the unconscious control and regulation of the body's internal environment.

The facial nerve contains some important fibres, which provide a parasympathetic

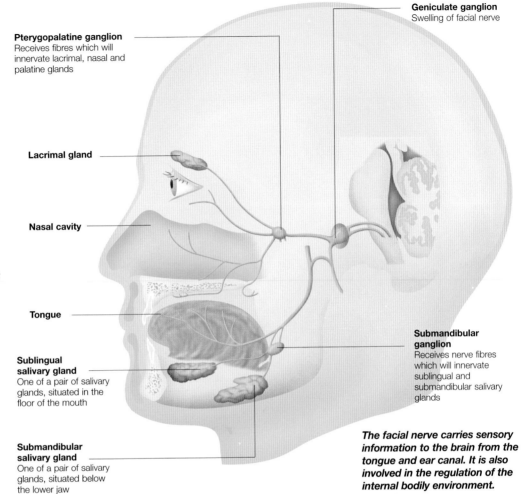

Pterygopalatine ganglion
Receives fibres which will innervate lacrimal, nasal and palatine glands

Geniculate ganglion
Swelling of facial nerve

Lacrimal gland

Nasal cavity

Tongue

Sublingual salivary gland
One of a pair of salivary glands, situated in the floor of the mouth

Submandibular salivary gland
One of a pair of salivary glands, situated below the lower jaw

Submandibular ganglion
Receives nerve fibres which will innervate sublingual and submandibular salivary glands

The facial nerve carries sensory information to the brain from the tongue and ear canal. It is also involved in the regulation of the internal bodily environment.

supply for the lacrimal (tear) glands and the sublingual (under the tongue) and submandibular (under the jaw) salivary glands.

GANGLIA
Parasympathetic fibres which leave the brain in the facial nerve run forward to form connections in two swellings, or ganglia. These ganglia comprise:
■ Pterygopalatine ganglion. Lying at the level of the cheekbone, this ganglion receives fibres which will go on to innervate the lacrimal, nasal and palatine glands
■ Submandibular ganglion. This swelling, lying nearer to the angle of the jaw, receives fibres which will innervate the sublingual and submandibular salivary glands.

Disorders of the facial nerve

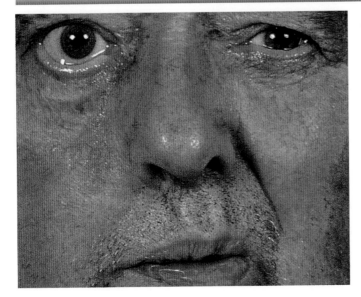

Of all the motor cranial nerves, the facial nerve is the most likely to suffer paralysis. If this happens on one side, the resulting loss of facial expression may be very obvious. Degrees of facial paralysis can occur in a number of disorders.

BELL'S PALSY
This relatively common disorder can develop suddenly with no apparent cause. The facial muscles, which receive their nerve supply from the facial nerve, become paralysed, taste

Paralysis of the facial nerve prevents normal facial movement. This man has Bell's palsy of the right side of his face and is attempting to smile.

sensation from part of the tongue is lost and the ear may be troubled by loud sounds. The disorder sometimes resolves quickly or may last for months.

UPPER MOTOR NEURONES
The motor nerve fibres of the facial nerve receive instructions through upper motor neurones (UMNs). The muscles of the lower face are controlled by UMNs from the opposite side of the head, while those of the upper face are controlled by both sides. If there is damage to the UMNs on one side, only the lower face is paralysed. If both upper and lower face muscles are paralysed, the damage must be in the lower motor neurones of the facial nerve itself.

71

Cranial nerves (5)

The vestibulocochlear nerve is the eighth cranial nerve. It is responsible for relaying information about balance and hearing from the inner ear to the brain.

The vestibulocochlear nerve is the eighth of the 12 cranial nerves. It is a sensory nerve which carries information about balance and hearing from the inner ear to the brain. The nerve is made up of two parts, the vestibular and cochlear nerves, which correspond to these two functional areas of the inner ear.

VESTIBULAR NERVE
The vestibular nerve carries information about the position and movement of the head from the semicircular canals and vestibule of the inner ear.

These delicate structures contain hair cells that are sensitive to head movement. Information from the hair cells is relayed via the vestibular nerve to the vestibular nuclei – four areas of grey matter within the brainstem.

COCHLEAR NERVE
The cochlear nerve carries information from hearing receptors in the cochlea of the inner ear. Fibres of the cochlear nerve receive information from the hair cells of the organ of Corti within the cochlea. Information about sound is then carried away from the inner ear to the brainstem where it is processed further.

As they leave the inner ear, the vestibular and cochlear nerves join together to form the vestibulocochlear nerve. This nerve then passes through the internal auditory meatus to reach the brainstem.

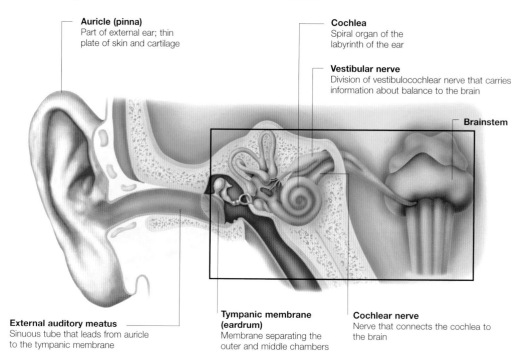

Auricle (pinna)
Part of external ear; thin plate of skin and cartilage

Cochlea
Spiral organ of the labyrinth of the ear

Vestibular nerve
Division of vestibulocochlear nerve that carries information about balance to the brain

Brainstem

External auditory meatus
Sinuous tube that leads from auricle to the tympanic membrane

Tympanic membrane (eardrum)
Membrane separating the outer and middle chambers

Cochlear nerve
Nerve that connects the cochlea to the brain

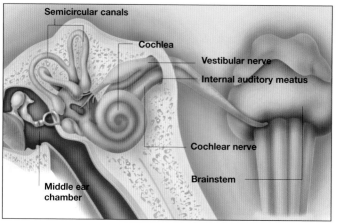

Semicircular canals

Cochlea

Vestibular nerve

Internal auditory meatus

Cochlear nerve

Brainstem

Middle ear chamber

The vestibulocochlear nerve is vital to hearing, balance and posture. The nerve comprises two parts: the vestibular and cochlear nerves.

The information from the inner ear is transmitted through the internal auditory meatus to the brainstem. Sound is then analysed in the auditory cortex.

The vestibular nerve and posture

The information from the vestibular apparatus is essential for balance, posture and co-ordination. Such information reaches the vestibular nuclei, where numerous connections are made with other parts of the central nervous system.

The information from the inner ear is combined with other sensory information and used for:
■ Maintaining correct balance

Dancers can spin without becoming dizzy. They do this by fixing the eyes on a single spot, thus reducing the movement of fluid in the semicircular canals.

and posture
■ Co-ordinating head and eye movements so that the eyes can be fixed on an object despite movement of the head and neck
■ Allowing consciousness of orientation and movement.

MOTION SICKNESS
Some people are sensitive to stimulation of the vestibular apparatus, especially during motion in a vehicle. This may be due to a mismatch of information received from the inner ear and the eye, and may be helped by medicines which depress the input from the vestibular nerve.

Auditory pathway

The perception of sound involves the passage of information along a fairly complex path. This auditory pathway runs from the inner ear to the highest level of the brain, the cortex.

The auditory pathway runs from the inner ear to the brain. In this diagram, arrows indicate the direction of flow of information to the brain.

The pathway of hearing begins at the cochlea of the inner ear, which is stimulated by sound. The information is then carried to the brain along the vestibulo-cochlear nerve. Within the brain, the information passes through a number of connections before reaching the auditory cortex, where sound is analysed.

THE AUDITORY PATHWAY
Points along the pathway include:
■ Cochlea – the hair cells of the organ of Corti within the cochlea perceive sounds and convert them into electrical information
■ Vestibulocochlear nerve – nerve fibres receive data from the cochlea and pass it back to the brainstem in the cochlear part of the vestibulocochlear nerve. The nerve cell bodies lie in the spiral ganglion
■ Cochlear nuclei – the fibres of the cochlear nerve make connections in the dorsal and ventral cochlear nuclei
■ Superior olivary nucleus – from the cochlear nuclei, auditory information passes up to the superior olivary nucleus, from where some fibres pass back to the cochlea to influence its perception of sound
■ Lateral lemniscus – nerve fibres from the superior olivary nucleus ascend the tract of the lateral lemniscus to reach the inferior colliculus. Some of the fibres in this area are thought to make connections which lead to

the reflex contraction of small muscles in the ear in response to loud sounds
■ Medial geniculate nucleus of thalamus – fibres from the inferior colliculus ascend to the thalamus to end in the medial

Cortex

Acoustic area of temporal lobe cortex (auditory cortex)

Medial geniculate nucleus or body

Inferior colliculus

Midbrain

Tract of lateral lemniscus

Nucleus of lateral lemniscus

Dorsal cochlear nucleus

Ventral cochlear nucleus

Medulla oblongata

Spiral ganglion

Superior olivary complex

Hair cells

geniculate nucleus, the last staging point along the pathway of sound to the cortex
■ Auditory cortex – the highest level within the brain is the cortex; the part of the cortex which receives information

about sound is the primary auditory cortex in the temporal lobe. The area around the auditory cortex is known as Wernicke's area and it is here that information about sound is analysed and interpreted.

Acoustic neuromas

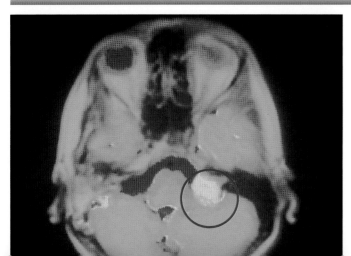

An acoustic neuroma is a slow-growing benign tumour of the vestibulocochlear nerve. It arises from a Schwann cell, which provides the insulating myelin sheath around the nerve. Tumour growth causes compression of both the vestibular and cochlear parts of the nerve, as well as of other cranial nerves.

This scan shows an acoustic neuroma – a benign tumour on the vestibulocochlear nerve. The tumour appears as a yellow area on the right (circled).

SYMPTOMS
Symptoms and signs of an acoustic neuroma relate to the nerve tissue which is being damaged. There are usually attacks of dizziness, deafness, and tinnitus, followed by paralysis of other cranial nerves supplying the face and eye muscles.

Acoustic neuromas may be associated with an inherited disease known as neuro-fibromatosis, in which similar tumours form throughout the body. They are accompanied by disfiguring skin pigmentation.

Cranial nerves (6)

The vagus nerve is the largest of the cranial nerves, stretching down from the head to the abdomen. Its role in monitoring and controlling breathing and digestion makes it vital to life.

The vagus nerve (cranial nerve X) is the most extensive of the 12 paired cranial nerves.

FUNCTIONS OF THE VAGUS

The vagus is a mixed nerve, containing both sensory and motor fibres. These provide:
- Sensation from the lower pharynx (throat), larynx and the organs of the chest and abdomen
- Taste from the root of the tongue and back of the throat
- Motor innervation of the muscles of the soft palate, pharynx and internal larynx
- The parasympathetic nerve supply to the internal organs of the chest and abdomen, which helps to monitor and control these organs subconsciously.

The vagus nerve has an important role in swallowing and speaking, as well as a vital part to play in the control of the heart, breathing and digestion. The role of the vagus is so important that a person would die if both nerves were destroyed.

BRANCHES OF THE VAGUS

Numerous branches include:
- A small branch from the base of the skull to the back of the brain, and a branch to a small area of skin around the ear
- Branches from the neck to the pharynx, larynx and the heart
- Branches from the thorax which form the plexuses serving the heart, lungs and oesophagus
- The anterior and posterior vagal trunks, which supply the stomach and digestive tract.

Schematic representation of branches of the vagus nerve

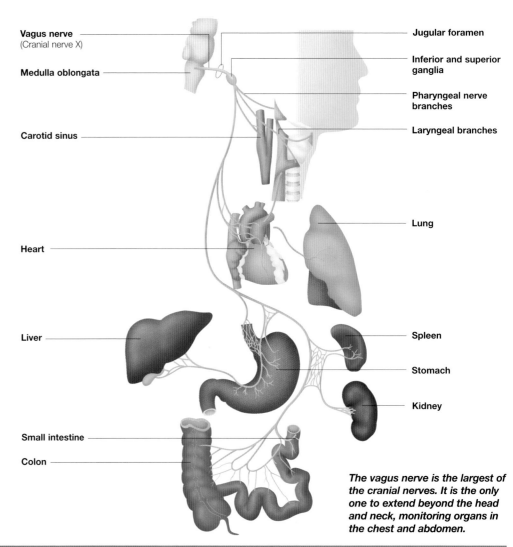

Vagus nerve (Cranial nerve X)
Medulla oblongata
Carotid sinus
Heart
Liver
Small intestine
Colon

Jugular foramen
Inferior and superior ganglia
Pharyngeal nerve branches
Laryngeal branches
Lung
Spleen
Stomach
Kidney

The vagus nerve is the largest of the cranial nerves. It is the only one to extend beyond the head and neck, monitoring organs in the chest and abdomen.

Course of the vagus in the neck

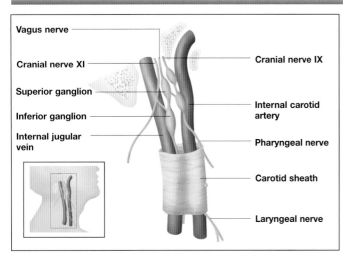

Vagus nerve
Cranial nerve XI
Superior ganglion
Inferior ganglion
Internal jugular vein

Cranial nerve IX
Internal carotid artery
Pharyngeal nerve
Carotid sheath
Laryngeal nerve

The vagus nerve travels through the neck between the internal jugular vein and the carotid artery. They are all protected by the carotid sheath.

Fibres of the paired vagus nerves leave the medulla oblongata and exit the cranial cavity by passing through the jugular foramen together with cranial nerves IX and XI.

At this level the vagus has two ganglia (swellings containing nerve cell bodies): a small superior and a larger inferior ganglion. There are also connections here with cranial nerves IX and XI, and with fibres of the sympathetic nervous system.

THROUGH THE NECK

As the vagus emerges from the jugular foramen, it lies between the large carotid blood vessels within the carotid sheath (a protective sleeve of connective tissue). It continues down with them, through the neck to enter the thoracic cavity below.

Within the neck the vagus provides part of the nerve supply to the carotid sinus and the carotid bodies – specialized areas that detect changes in the pressure and chemical composition of blood.

The vagus in the thorax and abdomen

The vagus is the only cranial nerve to extend beyond the head and neck. Reaching into the thoracic and abdominal cavities, it has a vital role to play in the control of the internal organs of the chest and abdomen, including the heart.

The right and left vagal nerves follow a similar course in the head and neck. However, as they reach the root of the neck and enter the thoracic cavity their courses begin to differ:

■ On the right, the vagus passes in front of the right subclavian artery, where it gives off the right recurrent laryngeal nerve. This runs back up to the larynx. The vagus then continues downwards, passing behind the large superior vena cava and the root of the right lung

■ On the left side, the vagus nerve enters the thoracic cavity, lying between the left common carotid and the left subclavian arteries. As it reaches and curves around the large aortic arch, it gives off the left recurrent laryngeal nerve, which loops under the aortic arch before returning to the larynx. The left vagus then descends behind the root of the left lung.

VAGAL TRUNKS
As they descend, the vagus nerves become part of a series of nerve networks, or plexuses, which supply the lungs, heart and oesophagus.

Fibres from the oesophageal plexus converge to form two nerves, the anterior and posterior vagal trunks:

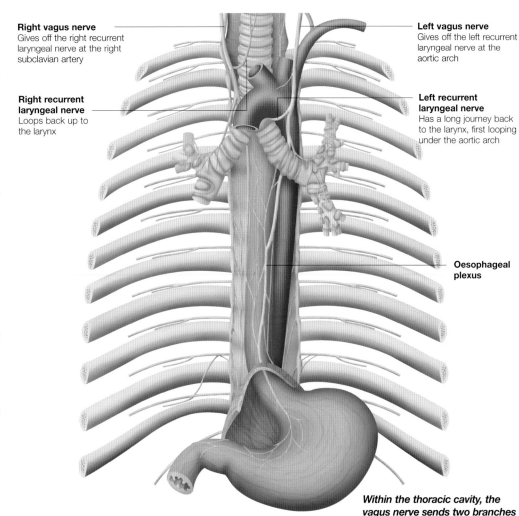

Right vagus nerve
Gives off the right recurrent laryngeal nerve at the right subclavian artery

Right recurrent laryngeal nerve
Loops back up to the larynx

Left vagus nerve
Gives off the left recurrent laryngeal nerve at the aortic arch

Left recurrent laryngeal nerve
Has a long journey back to the larynx, first looping under the aortic arch

Oesophageal plexus

Within the thoracic cavity, the vagus nerve sends two branches back up to the larynx. It also continues down to form part of the plexuses of the abdomen.

■ The anterior vagal trunk is derived mainly from the left vagus nerve. It lies on the anterior surface of the oesophagus and extends towards the lesser curvature of the stomach, giving off branches to the stomach, liver and duodenum

■ The larger posterior vagal trunk is derived mainly from the right vagus. It gives off numerous branches to both sides of the stomach as well as contributing to the coeliac plexus.

The vagus helps to control and encourage the digestive processes through these nerves.

Damage to branches of the vagus

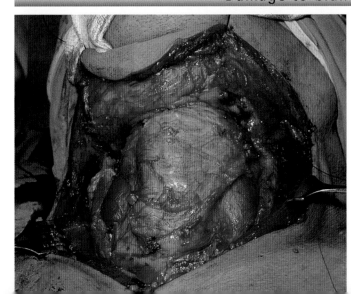

As it is vital for life, it is fortunate that the vagus nerve itself is rarely damaged. However, damage to branches may occur:

■ Injuries to the branches serving the pharynx (throat) may cause dysphagia (difficulty in swallowing)

■ Lesions of the superior laryngeal nerve (within the neck) weaken and tire the voice

■ Damage to the recurrent

The recurrent laryngeal nerve may be damaged during surgery to treat cancer of the larynx. Damage to the nerve results in an impaired ability to speak.

laryngeal nerve is the most well-known lesion. This paralyses the vocal cord on the affected side, leading to hoarseness and difficulty in speaking. When both right and left recurrent nerves are damaged, aphonia (a complete inability to speak) develops.

CAUSES OF DAMAGE
The recurrent laryngeal nerve may be damaged by cancer of the larynx or thyroid, or during surgical operations on the neck (especially the thyroid gland) or thorax.

Its longer course makes the left recurrent laryngeal nerve more vulnerable than the right.

Cranial nerves (7)

The glossopharyngeal nerve (IX) carries sensory information from the throat and tongue to the brain. The accessory (XI) and hypoglossal (XII) nerves supply muscles within the throat and mouth.

The term 'glossal' relates to the tongue, and the glossopharyngeal nerve is named according to the main areas it serves: the tongue and the pharynx (throat).

FUNCTIONS

The glossopharyngeal nerve is mixed, which means that it carries both motor and sensory nerve fibres. It also carries fibres of the parasympathetic branch of the autonomic nervous system.

Sensory information carried back to the brain by the glossopharyngeal nerve includes:
■ Taste from the back third of the tongue
■ Sensation from the lining of the pharynx, back third of the tongue and auditory (Eustachian) tube
■ Blood oxygen and carbon dioxide levels from the carotid body (tissue within the carotid artery), and blood pressure monitored by the carotid sinus.

Motor fibres of the nerve carry impulses to the stylopharyngeus muscle – one of the longitudinal muscles in the pharynx used in swallowing and speaking.

COURSE

The glossopharyngeal nerve emerges from the medulla and runs forward to leave the skull through the jugular foramen together with the tenth and eleventh cranial nerves. There are two swellings of the nerve at this point (superior and inferior ganglia). The nerve then travels downwards alongside the stylopharyngeus muscle towards the pharynx and the back of the tongue.

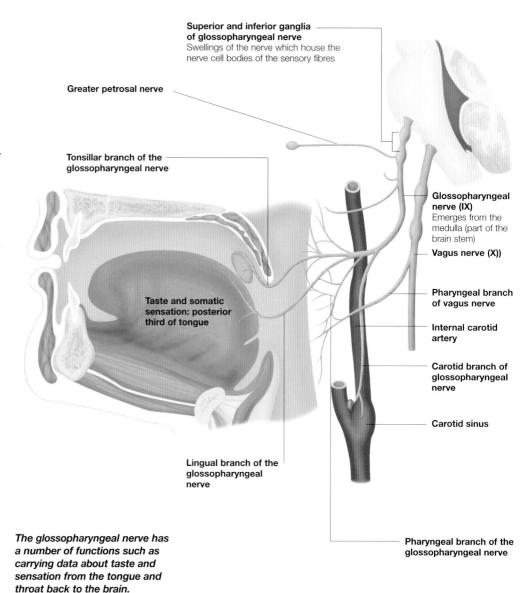

Superior and inferior ganglia of glossopharyngeal nerve
Swellings of the nerve which house the nerve cell bodies of the sensory fibres

Greater petrosal nerve

Tonsillar branch of the glossopharyngeal nerve

Taste and somatic sensation: posterior third of tongue

Lingual branch of the glossopharyngeal nerve

Glossopharyngeal nerve (IX)
Emerges from the medulla (part of the brain stem)

Vagus nerve (X))

Pharyngeal branch of vagus nerve

Internal carotid artery

Carotid branch of glossopharyngeal nerve

Carotid sinus

Pharyngeal branch of the glossopharyngeal nerve

The glossopharyngeal nerve has a number of functions such as carrying data about taste and sensation from the tongue and throat back to the brain.

The accessory nerve

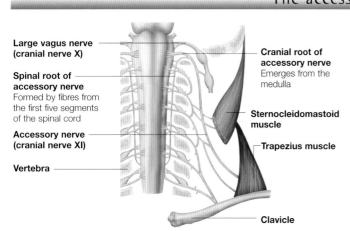

Large vagus nerve (cranial nerve X)

Spinal root of accessory nerve
Formed by fibres from the first five segments of the spinal cord

Accessory nerve (cranial nerve XI)

Vertebra

Cranial root of accessory nerve
Emerges from the medulla

Sternocleidomastoid muscle

Trapezius muscle

Clavicle

The accessory nerve is unique among the cranial nerves in that it has a spinal root as well as a cranial root. Together, these form the accessory nerve, which exits the skull through the jugular foramen.

SEPARATE FUNCTIONS

Once through the jugular foramen, the two roots of the

The accessory nerve has both spinal and cranial roots. These join to exit the skull and then separate to perform separate functions in the body.

accessory nerve separate again to fulfil their different functions.

Fibres from the cranial root then join with the large vagus nerve and go on to supply the muscles of the soft palate, pharynx, larynx and oesophagus.

Fibres from the spinal root run down as the accessory nerve, lying alongside the internal carotid artery to reach the sternocleidomastoid muscle, which they supply. The spinal accessory nerve then continues on its journey to supply the large trapezius muscle at the back of the neck.

The hypoglossal nerve (XII)

The twelfth cranial nerve, whose name 'hypoglossal' means 'under the tongue', supplies the muscles of the tongue. It has an important role in the actions of chewing, swallowing and speaking.

The hypoglossal nerve provides a motor nerve supply for many of the muscles of the tongue including the three extrinsic muscles:
■ The styloglossus muscle
■ The hyoglossus muscle
■ The genioglossus muscle.
   The hypoglossal nerve is also joined by fibres from the first cervical nerve which go on to supply other structures. These include muscles attached to the hyoid bone in the neck, which provides a base for tongue movements. They also carry sensory information from the dura (membranes) lining the rear part of the brain.

COURSE OF THE HYPOGLOSSAL NERVE
The hypoglossal nerve arises from each side of the medulla of the brainstem, usually as four separate roots which unite as they pass through an opening in the skull called the hypoglossal canal.
   The paired nerves then pass down and outwards, between the internal carotid artery and the internal jugular vein, to the angle of the jaw. Here, the nerves curve forward to run under the tongue, as the name of the nerve implies. The hypoglossal nerves end in a series of branches within the substance of the tongue.

The hypoglossal nerve joins with the first cervical nerve to supply the muscles of the tongue. It arises in the medulla of the brain stem and ends within the tongue.

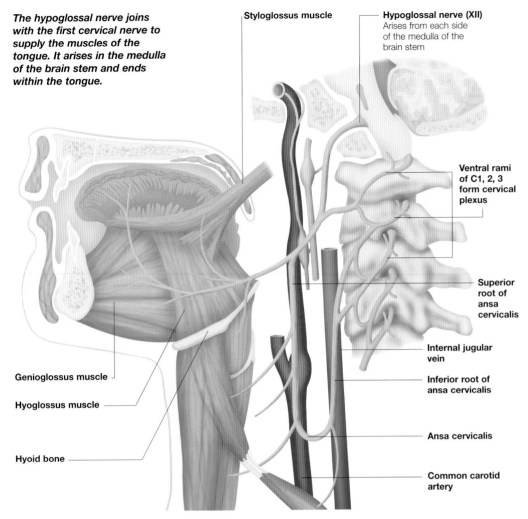

Styloglossus muscle

Hypoglossal nerve (XII)
Arises from each side of the medulla of the brain stem

Ventral rami of C1, 2, 3 form cervical plexus

Superior root of ansa cervicalis

Internal jugular vein

Inferior root of ansa cervicalis

Ansa cervicalis

Common carotid artery

Genioglossus muscle

Hyoglossus muscle

Hyoid bone

BRANCHES OF THE HYPOGLOSSAL NERVE
The branches of the hypoglossal nerve, together with the fibres from the cervical nerve which join it, include:

■ A meningeal branch that returns through the hypoglossal canal to innervate (supply) the dura of the brain (not shown)
■ A descending branch that joins the ansa cervicalis (a loop of nerves that supply the muscles below the hyoid)
■ Terminal branches that provide a nerve supply to all the intrinsic and most of the extrinsic muscles of the tongue.

Damage to cranial nerves IX, XI and XII

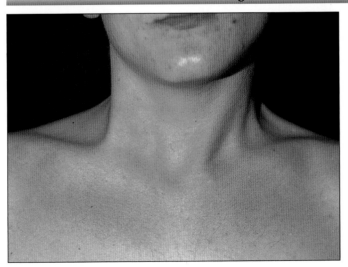

The glossopharyngeal nerve is very rarely damaged as it lies so deep within the body. However, if damage does occur, taste sensation is lost from the back third of the tongue, and the 'gag reflex' is absent on that side.
   The accessory nerve has a longer, more superficial, course and is more likely to become damaged. This occasionally occurs as a result of surgery on the neck or inflammation of the neck's lymph nodes in children.

This 20 year-old-man is suffering from torticollis (twisted neck). This condition can be caused by damage to cranial nerve XI – the accessory nerve.

When the accessory nerve is damaged there may be paralysis of:
■ The sternocleidomastoid muscle on that side, which leads to a torticollis (twisted neck)
■ The trapezius muscle, which impairs shrugging of the shoulder.

TONGUE PARALYSIS
Damage to the hypoglossal nerve is uncommon but, if it does occur, that side of the tongue will become paralysed and eventually waste and become shrunken. If the tongue is then stuck out it will be seen to deviate, or lean, towards the affected side.

How memory works

Memory is the brain's ability to store and access information.
Short-term memory stores only small amounts of information, while
greater amounts of data are kept in long-term memory.

Memory is the ability to store and retrieve information. Remembering is a vital function, since learning, thought and reasoning could not occur without it. We learn not to touch hot objects, for example, from a very early age as we remember that they burn us, causing pain. In addition, our memories, the sum of our experiences, play a huge part in the development of our personalities.

BRAIN

Memory is regarded as a function of the brain, often likened to the way in which a computer stores and processes information.

Whereas a computer can only store one billion bits of information, however, the brain can store up to 100 trillion. Moreover, the word 'store' is misleading, as, unlike a computer, there is no single centre in which memories are filed away. Remembering appears to be a function of many parts of the brain, rather than any one structure.

MEMORY INPUT

The storage of memory is very complex, and our sensory experiences suggest that there may be many different kinds of memory: visual (sight), auditory

(hearing), olfactory (smell), gustatory (taste) and tactile (touch). Information is never presented in one simple form, but tends to be embedded in a complex context – we know from daily experience how important context and associations are for effective memorizing. For example, a

single item of information conveyed to us by speech will be set in the context of other data such as the speaker's face, voice and displays of emotion.

TWO FORMS

There are two forms of memory: short-term and long-term. Short-term memory stores small

Individuals carry with them a huge amount of information which has been amassed over the years. This data is stored in the long-term memory.

amounts of information and the contents are quickly lost. Long-term memory stores larger quantities of information.

Short-term memory

Research has shown that the short-term memory is able to hold around five to seven items at a time for a maximum of one minute.

For example, you are able to remember a telephone number while you dial the number. However, if the number is engaged and you redial, you will have to look up the number again – the result of having no memory trace for it in the brain.

The reason for this inability to remember in the short term is that complex data cannot be stored the moment it is perceived. It appears that some

If an individual is dialling a telephone number for the first time, they remember it only briefly. Short-term memory can only store a few items at a time.

form of analysis and selection process is necessary for the brain to determine which information is assimilated, and which is discarded. It seems that this process cannot occur without first storing the data temporarily.

CONSOLIDATION

To last, a memory has to be recorded via the short-term memory first, before being consolidated. This process requires repetition or study, and usually classification (organization into a category of related items).

Consolidation moves a fact from the short-term memory to long-term storage. Consolidation is believed to result in an alteration in the structure of the brain, as a memory trace is formed.

Long-term memory

Large amounts of information are carried in the long-term memory. This data is stored and accessed through large collections of nerve cells located within the brain – the amygdala and hippocampus.

Every individual carries an immeasurable amount of data, often preserved for life in the long-term memory. It is now known exactly where in the brain sensory data must pass in order to be stored in the memory.

CEREBRAL HEMISPHERES

The interface circuits, through which long-term memory is recorded and recalled, are located in large structures on the inner surfaces of the temporal lobes of each cerebral hemisphere of the brain.

These huge collections of nerve cells are known as the amygdala and the hippocampus, and together they make up the limbic system. Both structures are connected to all the sensory areas of the cortex (outer layer of the brain), and damage to these structures, for example

through a stroke or brain trauma, will lead to profound memory loss.

MEMORY SITES

The exact physical basis for the long-term memory is unknown. There is evidence, however, that the sites of memory are the same as the areas of the brain in which the corresponding sensory impressions are processed in the cortex.

It appears that, during recall, the amygdala and the hippocampus play back the neurological activity that occurs during sensory activity to the appropriate part of the cerebral cortex.

The amygdala and hippocampus are structures in the brain associated with memory. They convert new information into long-term memories.

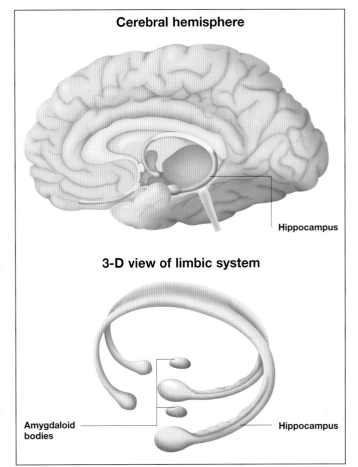

Cerebral hemisphere

Hippocampus

3-D view of limbic system

Amygdaloid bodies

Hippocampus

Amnesia

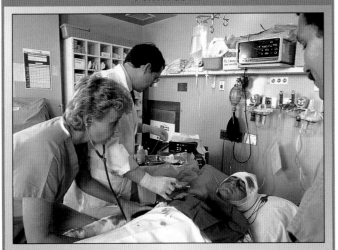

Amnesia is a failure to remember recent or past events. Most cases are caused by physical damage to the brain, although in rare instances it may be induced by emotional trauma, whereby an experience is too painful to remember.

Two types
Amnesia may take one of two forms:
■ Retrograde amnesia – this is most commonly caused by a blow to the head. The patient fails to remember what happened several hours prior to the accident, as the brain did

Patients with head injuries may experience amnesia. This can affect memories before (retrograde) or after (anterograde) the incident.

not have the chance to process that information.
■ Anterograde amnesia – this is caused by damage to the hippocampus, whereby memory of events occurring after the injury is impaired. Memory of the past remains intact, but everyday life becomes very difficult since the patient has no recollection of events from one moment to the next.

Memory loss

Most of us have no recollection of the first few years of our lives, after which time memories are fragmented and vague until we reach the age of around 10. This is probably because, in the first few years of life, the brain is not yet developed enough to process and store information.

DEGENERATION

Likewise, in the latter years of life, the brain undergoes a natural degeneration, and so memory may become impaired.

Interestingly, in older people, it is the short-term memory that is usually affected. For example, an older person may be able to

remember the exact details of a journey made 50 years before, but may be unable to recall what they did yesterday. This is because the ability to process new information often declines with age due to physical and chemical changes in the brain. Moreover, the regular recollection of long-term memories sharpens them, leaving a permanent memory trace in the brain.

Older people often have a good memory for past events, but have trouble recalling more recent memories. This is due to age-related changes in the brain.

How we feel emotion

External stimuli received through the senses arrive in the brain
as nervous impulses. Their emotional significance is determined by
the limbic system before producing a physiological response.

Experiencing an emotion involves a combination of physical and mental processes, which produce both physiological and psychological sensations.

RESPONDING TO STIMULI
To a large extent emotion is produced in response to external stimuli. The emotion experienced depends on the nature of stimuli and the individual's interpretation of those stimuli.

The physical aspects of emotional experience can be divided into two main elements:
■ The neurological processes produced by environmental or psychological stimuli
■ The physiological arousal that results from the stimuli.

ROLE OF THE AMYGDALA
Nervous impulses from the senses arrive in the brain at the thalamus, a mid-brain structure, where they are processed and passed on in a number of ways. Their emotional significance is believed to be determined by the limbic system within the brain, and in particular by the amygdala, an almond-shaped structure near the brainstem.

The amygdala assigns emotional content and value to incoming stimulus to provide a rapid initial assessment of its significance. This helps to determine quickly whether something is dangerous. The stimulus of encountering someone unexpectedly in a dark room, for instance, is labelled by the amygdala as a potential threat, and so produces an initial emotional response of fear.

ROLE OF THE CORTEX
Higher brain centres in the cortex can override the amygdala, integrating data from other sources, such as memory and context, to make a more accurate and considered determination of emotional significance.

In the example above, the cortex uses memory to identify the encountered person as a friend, and overrides the initial amygdala-produced emotion.

Emotional reactions may be complicated by culture or context. Emotions stimulated by fictional events, as in the theatre, produce real physical responses.

Physiology of emotion

A lie detector test operates on the presumption that telling lies produces a stress response in the person talking. This can be identified by physical changes.

Physiological processes are responsible for what are known as the visceral sensations involved in emotion. These include responses such as dry mouth, dilated pupils or unsettled stomach.

One purpose of emotion is to elicit an active response, and these visceral sensations are part of the process by which emotion readies the body for a physical response. For example, a stimulus that causes fear will also prepare the body to act on that fear.

THE ENDOCRINE SYSTEM
The level and type of physiological arousal is determined by the autonomic nervous system (ANS). This in turn is regulated by the endocrine system as it releases hormones in response to emotional triggers. These hormones produce many of the visceral sensations associated with emotion. Two of the most important hormones in this context are adrenaline and noradrenaline, which set up the sympathetic division of the ANS for a 'fight or flight' response.

PATTERNS OF AROUSAL
Hormones produce widespread physiological sensations that are common to different types of emotion. Specific emotions, however, produce more specific patterns of physiological arousal according to their effect on another type of chemical messenger: neurotransmitters.

These work in conjunction with hormones to produce a range of distinctive response patterns, each with its own heart rate, finger temperature, galvanic skin response (electrical conductivity of skin) and so on.

Psychological and cultural factors

More complex emotions, such as shame, involve input from brain centres that control learning and memory. Cultural factors also affect the final response.

Emotion involves more than just visceral sensations. As a subjective experience, emotion is as much psychological as it is physiological. Experiments show that drugs or hormones can produce the physiological correlates of emotional arousal, without producing the conscious sensation of emotion.

In particular, more complex, less visceral emotions, such as guilt, involve more input from higher brain centres and processes such as learning, memory and self-image.

CONTEXT AND CULTURE

Factors that influence the psychology of emotion include context and culture. For instance, in the context of a theme-park ride, 'scary' stimuli can produce a mixture of terror and pleasure.

Cultural influence on emotions includes culture-specific emotions, such as the Chinese 'sad love'. Whereas in the West love is characterized as a positive emotion, in China love is not always positive and can be a negative or mixed emotion.

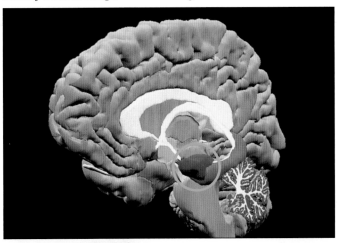

Nervous impulses send external stimuli from the senses to the brain. Part of the limbic system, the amygdala (circled), then determines the response.

A roller-coaster ride produces a variety of emotional responses. Our higher brain centres intervene to determine whether the experience is fun or terrifying.

Expression of emotion

It has long been believed that some emotions are physically linked to the way they are expressed, especially in terms of the muscular activity of facial expression. To put it simply, some emotions and facial expressions are 'hard-wired' together.

This has led in turn to the belief that the process can be reversed, and that adopting an expression can induce a particular emotion.

For example, it has been found that smiling primes individuals to interpret stimuli as more positive. Suggested mechanisms to explain this finding include:
■ Smiling induces the release of endorphins (naturally occurring mood-enhancing opiates)
■ Some sort of feedback effect occurs with smiling that primes the brain centres involved with happiness and positive emotions.

Brain hemispheres

The two hemispheres of the brain play different roles in the recognition of faces, facial expressions and even the

recognition and experience of positive and negative emotions.
For instance:
■ Some areas of the left hemisphere are specialized for the recognition and processing of positive emotions, such as happiness
■ Some areas of the right hemisphere are specialized for negative emotions, such as sadness and fear.

The face on the left is a mirror image of that on the right but is perceived as happier. This is because of the way that the hemispheres process emotion.

Brain damage can impair the experience of some emotions and exaggerate others. Left-hemisphere damage can produce excessive fear or depression, while on the right, damage can produce uncontrollable laughter or mania. Clinically depressed people may show reduced function in the left frontal lobe.

This hemispheric lateralization can be demonstrated by looking

at the picture below. Which face looks happier?

Most people choose the left-hand picture. This is because in the left-hand picture the smiley face appears in the left visual field of the observer, and is therefore processed primarily by the right hemisphere, where recognition of expression primarily occurs. The faces are actually mirror images of each other.

How laughter occurs

Laughter is the body's response to happiness and comprises both gestures and sound. Although laughter is not essential to survival, it is thought to act as a type of relief mechanism.

Laughter is a physiological response to happiness and humour and appears to be a distinctly human response. It consists of two components: a set of gestures and the production of sound. When a person finds something humorous, the brain triggers both of these responses to occur simultaneously, at the same time causing changes to occur throughout the body.

FACIAL MUSCLES

Laughter involves the contraction of 15 facial muscles as well as stimulation of the zygomatic major muscle, which causes the upper lip to lift. Meanwhile, the epiglottis partially blocks the larynx, resulting in irregular air intake and causing a person to gasp. In extreme cases the tear ducts may be activated, causing tears of laughter to stream down the face. A person may even turn red in the face as they continue to gasp for air.

RANGE OF NOISES

A range of characteristic noises, which range from a gentle giggle to a loud guffaw, accompanies this response. In fact, research into the sonic structure of laughter (the pattern of sound waves produced when a person laughs) shows that all human laughter consists of varying patterns of a basic form, which is short vowel-like notes repeated every 210 milliseconds. It has also been revealed that laughter triggers other neural circuits in the brain, which in turn generate more laughter. This explains why laughter can be contagious.

The average person laughs around 17 times a day. Laughter clubs (one in Bombay shown) encourage people to get together and laugh.

The role of laughter

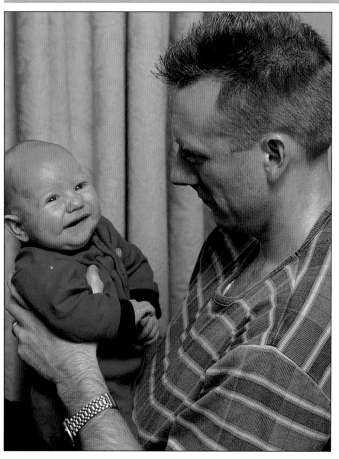

Smiling and laughter may be a sign of trust in a friend or relative. Babies often communicate with their parents by smiling and laughing.

As with much of human behaviour, it is difficult to determine the exact purpose of laughter, although there appear to be a variety of physical and psychological benefits. Many theories suggest that laughter may have evolved as a relief mechanism, whereby our ancestors would use laughter as a gesture of shared relief at the passing of danger. What is more, because laughter inhibits the 'fight or flight' response – designed to protect the body in danger – laughter may also indicate trust in a companion.

COMMUNICATION

In this way, many scientists believe that the purpose of laughter is one of communication, a social signal that strengthens human connections. Research by cultural anthropologists demonstrates that people tend to laugh together when they feel comfortable with each other, and that the more people laugh together, the greater the bonding between them. Interestingly, research shows that people are 30 times more likely to laugh in a group than when alone and that even the use of nitrous oxide (laughing gas) elicits less laughter when taken in solitude.

RESULTS OF STUDIES

Studies have also shown that laughter conforms to a social hierarchy whereby dominant individuals tend to use humour more than their subordinates; for example, when a boss laughs it is not uncommon for all his or her employees to laugh too. By controlling the emotional climate of the group, the boss exercises power.

In fact, it appears that laughter, like most human behaviour, has evolved as a means of influencing the behaviour of others. Studies have shown that in an embarrassing or threatening situation, laughter may serve as a conciliatory gesture and a means to deflect anger. If a threatening person can be made to laugh, then the risk of confrontation is reduced.

The role of the brain in laughter

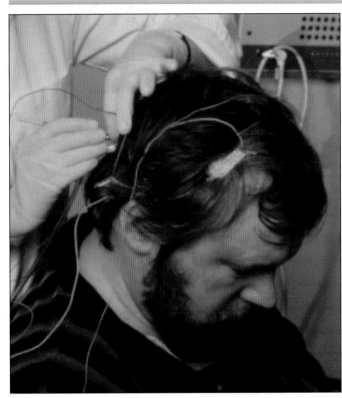

Gelotology is the study of the physiological responses that occur during laughter, and much research has been conducted to work out how laughter is triggered. While other emotional responses seem to be confined to one part of the brain (frontal lobe), laughter appears to involve a circuit running through several brain areas.

EEG RESEARCH

This has been demonstrated by research in which human subjects were connected to an electroencephalograph (EEG) (an instrument that allows the electrical activity of the brain to be measured). The subjects were then told a joke and the consequent electrical activity of their brains was observed.

An EEG can help to study response to laughter. When subjects are connected to an EEG and told a joke, the brain's electrical activity is measured.

Within less than a second, a wave of electrical activity was seen to pass through the cerebral cortex (the largest part of the brain). It was found that if this wave had a negative charge then laughter resulted, but if it was positive, laughter did not occur.

ELECTRICAL ACTIVITY

The electrical activity of the brain during laughter appears to take the following path:

1 The left side of the cortex (the layer of cells that covers the surface of the forebrain) is stimulated as the structure of the joke is analysed

2 The frontal lobe, usually involved with emotional responses, is activated

3 The right hemisphere of the cortex is stimulated – intellectual analysis of the joke occurs here, determining whether or not the joke is funny

4 The sensory processing area of the occipital lobe (the area located at the back of the head that is associated with the processing of visual signals) is activated as nerve impulses from the right hemisphere are interpreted and converted into a sensory response

5 Various motor sections (responsible for movement) of the brain are stimulated causing a physical response to the joke.

THE LIMBIC SYSTEM

As with any emotional response, the limbic system in the brain appears to be central to laughter. The limbic system is a network of complex structures that lies beneath the cerebral cortex, and controls behaviour that is essential to survival.

While this area of the brain in other animals is heavily involved in defending territory and hunting, in humans it has evolved to become more involved in emotional behaviour and memory.

EMOTIONAL RESPONSE

Indeed, research has shown that the amygdala. which controls anxiety and fear, and the hippocampus, which plays a role in learning and memory, seem to be the main areas of the brain involved with emotional responses.

The amygdala interacts with the hippocampus and thalamus (the part of the brain that relays information from the senses to the cortex), playing a key role in the expression of emotions.

In addition, the hypo-thalamus has been identified by researchers as a major contributor to the production of loud, uncontrollable laughter.

The benefits of laughter

While people have always known that laughter makes them feel good, there is now scientific evidence that it promotes health in a number of ways.

Health benefits

Laughter has many benefits for the general health of an individual. These include:

■ Immune system – laughter inhibits the 'fight or flight' response, by reducing levels of certain stress hormones responsible. This is beneficial to health since these hormones suppress the immune system and raise blood pressure. Laughter actually boosts the immune system by causing an increase in white blood cells

■ Blood pressure – laughter lowers the blood pressure, while

Laughter is more than an expression of happiness; it can actually promote health. Here, a stroke patient and speech therapist laugh together.

increasing vascular blood flow and oxygenation of the blood. This in turn aids healing

■ Saliva – laughter leads to increased production of salivary immunoglobulin A, which helps to prevent pathogens (disease-causing organisms) invading the body via the respiratory tract

■ Exercise – it has been estimated that laughing 100 times is the equivalent of 15 minutes' workout on an exercise bike. Laughter exercises the diaphragm and abdominal, respiratory, facial, leg and back muscles, which explains why people often feel exhausted after laughing a lot

■ Mental health – laughter provides a way for negative emotions, such as anger or frustration, to be released. Ever since the pioneering work of Patch Adams (a physician who recognized the benefits of humour when treating patients) doctors have become increasingly aware of the therapeutic benefits of laughter.

How we sleep

The body enters an altered state of consciousness during sleep.
While it used to be believed that the sole function of sleep was rest,
studies show that the brain is far from inactive during this time.

Sleep is defined as a state of relative unconsciousness and reduced body movement. Unlike coma, subjects can be aroused from sleep by external stimulation. Relatively little is known as to the exact function of the phenomenon of sleep, despite the fact that the average person spends around a third of their lifetime asleep.

RESTORATIVE FUNCTION

In the past it was believed that sleep served a restorative function only. More recently however, sleep studies with electroencephalography (using electrodes attached to the head which measure the electrical activity of the brain) suggest otherwise. While motor activity is inhibited by sleep, it seems the brain is far from inactive during this time. Although the functioning of the conscious part of the brain is depressed, brain stem functions such as control of respiration, heart rate, and blood pressure are maintained.

PHYSIOLOGICAL CHANGES

While sleeping, humans close their eyes, and adopt a sleeping posture – typically lying down. Hormonal changes cause heart, respiration, and breathing rates to slow down. In addition, digestive activity is reduced and urine concentrated to allow a period of uninterrupted sleep.

In sleep the sensory part of the brain is depressed. However, we are still aware of external stimuli which is why we can be woken.

Types of sleep

By monitoring brain activity, scientists have identified two states of sleep. These are referred to as non-rapid eye movement (NREM) sleep and rapid eye movement (REM) sleep. These alternate throughout the night and serve very different roles.

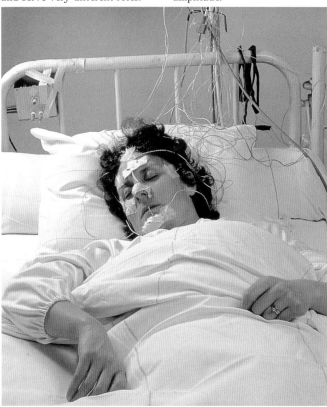

Studies of brain activity during sleep reveal two main stages. During REM sleep the brain is very active, and respiration rates increase.

NREM SLEEP

During the first 45 minutes of sleep, the body passes through four stages of deeper and deeper NREM sleep. This is seen as a decline in the frequency of brain waves, but an increase in their amplitude.

The four stages of NREM sleep are:
■ First stage – the eyes are closed and relaxation begins. Conscious thoughts begin to drift. At this stage arousal is immediate if the body is stimulated
■ Second stage – the EEG becomes more irregular, and arousal becomes more difficult
■ Third stage – as the body slips into this stage the skeletal muscles begin to relax and dreaming is common
■ Fourth and final stage – (slow wave sleep) the body relaxes completely and arousal is difficult. Bedwetting and sleepwalking may occur during this stage.

REM SLEEP

Around an hour after sleep begins the EEG pattern changes, becoming irregular and more frequent, indicating the onset of REM sleep. This change in brain activity is accompanied by an increase in body temperature, heart rate, respiratory rate and blood pressure and a decrease in digestive activity.

The brain pattern seen during this stage of sleep is more typical of the awake state, although the body actually respires more oxygen during this phase of sleep, than when awake.

Typically during this phase, the eyes move rapidly beneath the eyelids, although the rest of the body muscles are inhibited and go limp, resulting in a temporary paralysis designed to prevent us from acting out our dreams. REM sleep makes up around 20 per cent of adult sleep.

DREAMS

Most dreaming occurs during REM sleep. It is hardest to wake somebody during this stage of sleep, although sleepers can wake spontaneously during this time – and will be more likely to remember the details of their dream if they do so.

CHEMICAL MESSENGERS

In addition to changes in brain wave patterns during sleep, there are changes in levels of neurotransmitters (the chemical messengers secreted by the brain). Noradrenaline levels decline and serotonin levels rise. This is because noradrenaline is responsible for maintaining alertness, while serotonin is thought to function as a sleep neurotransmitter.

The role of sleep

Sleep allows the skeletal muscles to relax and our energy levels to be replenished. The amount of sleep the body requires differs among individuals.

The most obvious role of sleep appears to be physical restoration. While we sleep, our muscles relax, allowing them to rest. The body requires more sleep after great physical exertion, or illness.

BRAIN ACTIVITY

Slow wave sleep appears to be the restorative stage of sleep, when most neural mechanisms wind down. Sleep deprivation studies, in which subjects are woken each time they reach a certain stage of sleep, reveal that when continually deprived of REM sleep, subjects become moody and depressed, and exhibit personality disorders.

Many theories exist regarding the function of brain activity during sleep. The most likely theory is that REM sleep gives the brain the opportunity to analyse the day's events, discarding useless information, processing useful information and working through emotional problems in dream imagery.

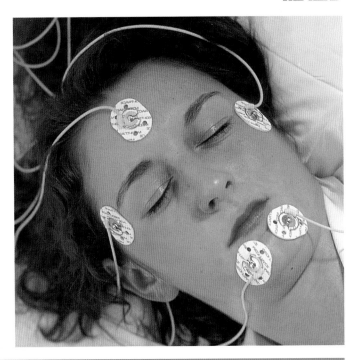

During REM sleep, brain activity increases considerably. This is thought to be a time when the brain assimilates useful information learnt that day.

Sleep requirements

Sleep requirements and patterns change throughout our lives. While a baby can require as much as 16 hours sleep every day, the average adult will only require seven hours.

In old age, the amount of sleep required declines considerably, with people over 60 years requiring shorter spells of sleep, although these tend to be taken more frequently. Elderly people are more likely to take naps during the day.

SLEEP PATTERNS

Sleep patterns also change throughout life: the amount of REM sleep declines from birth, and often disappears completely in people over 60.

This is the reason why many older people sleep more lightly and commonly wake up more frequently in the night. This is because they are not able to attain the profound depths of REM sleep.

Sleep requirements vary from person to person, and with age. Elderly people require shorter, more frequent spells of sleep, and often nap during the day.

Sleep disorders

Although the exact role that sleep fulfils is not entirely known, it is clearly essential to our mental and physical well-being.

Insomnia

Insomniacs suffer from an inability to obtain a sufficient amount or quality of sleep needed to function adequately during the daytime. With prolonged lack of sleep, insomniacs show signs of fatigue, impaired ability to concentrate and carry out everyday tasks and, in some cases, paranoia.

Insomnia can be caused by unfavourable surroundings (noisy neighbours or an uncomfortable bed); physical ailments, such as those causing breathlessness or pain; or an irregular sleep pattern (caused by jet lag or working night shifts). The most common cause of insomnia, however, is psychological disturbance such as anxiety, or depression.

Narcolepsy

Narcolepsy is the complete opposite of insomnia. Sufferers have little control over their sleep patterns, and can lapse into deep sleep spontaneously during waking hours. These episodes of unconsciousness last between 5 and 15 minutes and can occur without warning at any time.

This condition can be very hazardous, for example when the sufferer is taking a bath, or operating machinery. The cause of narcolepsy is not understood, although it seems to arise from an inability to inhibit REM, or dreaming sleep. Most people sleep for some time before falling into a deep sleep, but narcoleptics appear to enter REM as soon as they close their eyes.

Insomnia is a condition in which the sufferer does not have enough quality sleep. The cause is often psychological.

How we dream

Around a fifth of time asleep is spent dreaming, although many people claim not to remember their dreams. The mental activity involved in dreaming is very different from that of waking thought.

Although many people claim not to dream, sleep studies have revealed that the average adult spends around 20 per cent of their sleep in a state of dream activity.

WHAT ARE DREAMS?
Dreams result from a form of mental activity that is very different from waking thought.

A dream is a series of images, thoughts and sensations conjured up by the mind during sleep. Dreams can take the form of pleasant fantasies, everyday scenarios or terrifying nightmares.

DREAM STUDIES
In-depth studies, in which subjects are monitored throughout sleep and woken during dreaming phases and questioned about their dreams, reveal much about the nature of dream activity.

It appears that most dreams are perceptual rather than conceptual – meaning that things are seen and heard rather than thought. In other words, in our dreams we often appear to be an onlooker witnessing events as opposed to conducting them and reflecting upon them.

SENSORY EXPERIENCE
In terms of the senses, visual experience is present in almost all dreams and auditory (hearing) experience in around 40 to 50 per cent of dreams. In comparison the remaining senses – touch, taste and smell – feature in only a small percentage of dreams.

EMOTIONS
The overriding feature of all dreams tends to be a single and strong emotion such as fear, anger or joy, rather than the integrated range of subtle emotions experienced during the waking state.

Most dreams are composed of interrupted stories, partly made up from memories and fragmented scenes. Dreams can range from the very mundane to the truly bizarre.

During sleep the brain conjures up the scenarios we know as dreams. They tend to be composed of vivid visual images and, often, strong emotions.

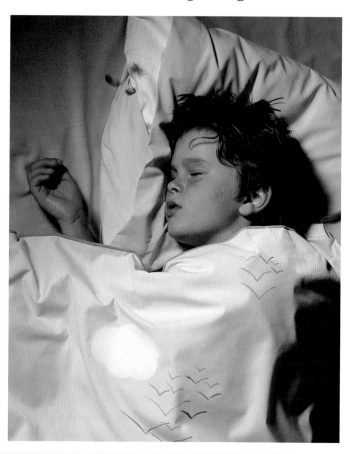

When do dreams occur?

Research in recent years has revealed that two clearly distinguishable states of sleep exist: non-rapid eye movement sleep (NREM) and rapid eye movement sleep (REM).

NREM sleep makes up the greater part of our sleeping time, and is associated with a relatively low pulse and blood pressure and little activation of the autonomic nervous system.

Very few dreams are reported during this state of sleep, and tend to be more like thoughts rather than vivid images.

REM sleep occurs cyclically during the sleep period and is characterized by increased conscious brain activity, the eyes moving rapidly from side to side

Sleep studies reveal that most dreams occur during the REM stage of sleep. This stage is associated with rapid movement of the eyes beneath the lids.

beneath the eyelids, and frequent reports of dreams.

Typically a person will have four or five periods of REM sleep during the night, although usually only a single dream may be remembered the following morning if at all.

REM sleep occurs at intervals of about 90 minutes and makes up around 20 per cent of the night's sleep. Evidence from dream studies suggests that dream periods last for around 5 to 20 minutes.

Sleepwalking

When we sleep, our muscles become very relaxed, with the result that the body becomes temporarily paralysed. This is designed to prevent the body from acting out our dreams.

In some people this mechanism does not work quite so effectively and sleepers can become active while dreaming, sometimes to the point of walking in a semi-conscious state. This phenomenon is known as somnambulism, or sleepwalking.

Sleepwalkers can often perform tasks and even hold conversations. Very often they will have no recollection at all of what happened during the night.

Brain activity while dreaming

When we dream, the limbic system (the part of the brain associated with emotions, senses and long-term memory) is active, while the forebrain (associated with short-term memory and intelligence) is inactive. This may explain the nature of our dreams.

Recent studies using positron emission tomography (PET) scanning, which can be used to measure blood flow to the brain, indicate that different parts of the brain are active when we dream and when we are awake.

PREFRONTAL CORTEX

During the normal waking state the prefrontal cortex – the front part of the brain – is the most active (indicated by increased blood flow to this area on a PET scan). This part of the brain is responsible for our conscious thought, intelligence, reasoning and short-term memory.

LIMBIC SYSTEM

Studies show that during REM sleep the prefrontal cortex of the brain is completely inactive, while the limbic system – the part of the brain that controls emotions, senses and long-term memory – is most active.

It appears that this could account for the heightened emotions experienced during REM dreams, as well as the retrieval of long-term memories (often our dreams can transport us back to events that occurred some time ago).

The fact that short-term memory is de-activated may also account for the bizarre content of dreams – the scene changes, fragmented narratives, and people's identities which seemingly melt into one another.

It may also account for the fact that many people cannot remember their dreams once they awake.

VISUAL IMAGES

PET studies have also revealed that the primary visual cortex – the part of the brain used to see when we are awake – is inactive during sleep. Instead a different visual area, called the extrastriate, is active.

The extrastriate is the visual area that is responsible for the recognition of complex objects like faces and emotions. This could explain the vivid visual images typical of most dreams.

The content of our dreams often seems utterly disconnected from the real world. We dream about situations and scenarios that could never occur in real life.

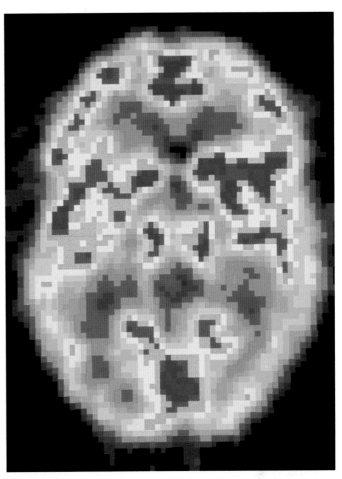

PET studies reveal that different parts of the brain are active during dreaming and when awake. This may explain the strange nature of many dreams.

Role of dreams

Throughout history the role of dreams has attracted many different theories. Ancient cultures placed much importance on dreams, believing that they were spiritual in origin and could even predict the future.

Subconscious expression
The psychologist Sigmund Freud believed that dreams represented a 'road to the subconscious' and were an expression of repressed (usually sexual) desires.

Today, many psychoanalysts use the recounting of dreams as a part of clinical treatment.

It has long been thought that dreams are an expression of the subconscious. Analysis of dreams is often used as a technique by psychoanalysts.

Dreams may express important wishes or fears of the dreamer, and the analysis of dreams can provide great insight into a person's mental functioning.

Dreams and brain function
A more recent theory suggests that dreams are directly linked to the long-term memory system. Research has been carried out in which subjects who were deprived of REM sleep found it more difficult to learn new information, which appears to support this theory.

Also some studies show that REM sleep increases when we are trying to learn a new or difficult task. This suggests that information in the short-term memory is transferred to the long-term memory as we dream.

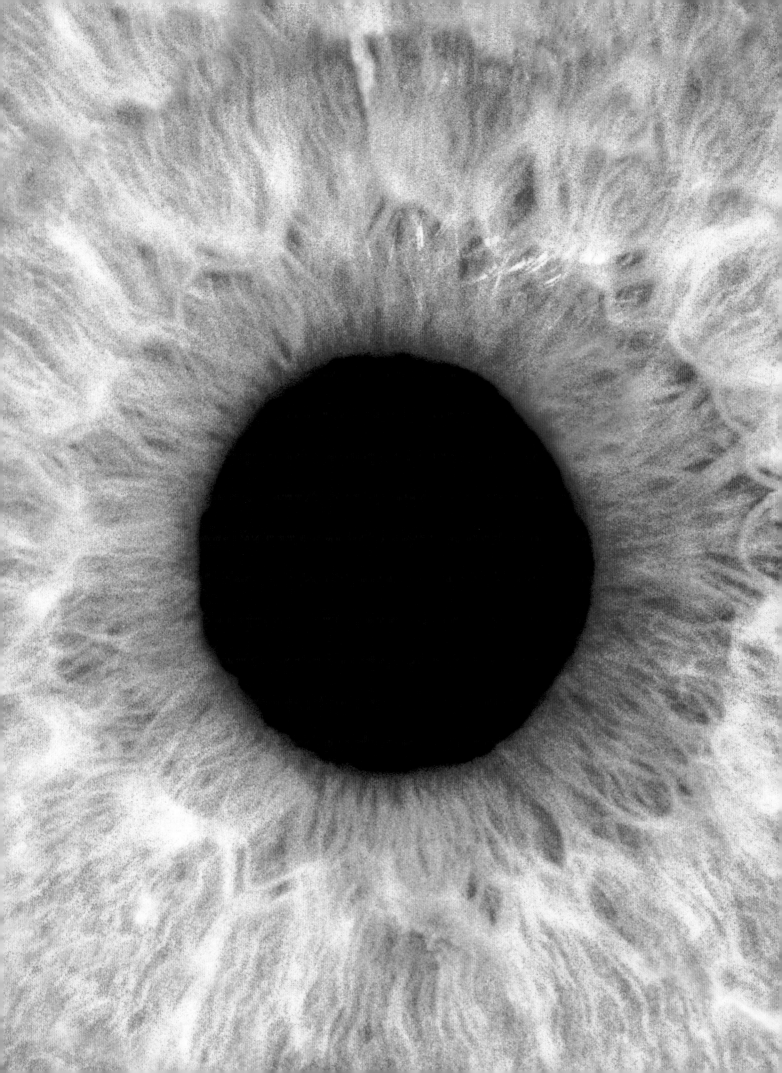

Chapter Two

The Senses

Our bodies are constantly bombarded by stimuli from the outside world. To enable us to cope with this onslaught, a complex and interactive system of receptors contained in sensory organs transmits the data it receives to the brain via cranial nerves. The cranial nerves interpret the data, providing us with the information we need to interact successfully with our environment.

From the inner workings of the eye to the mechanisms that prompt us to sneeze, enable us to smell or maintain our sense of equilibrium, this section explores how our senses operate. From them, we receive essential information on the world around us.

LEFT: Magnified many times, this view of the human eye shows the opening of the pupil and the dense molecular structure of the surrounding iris.

Nose and nasal cavity

'Nose' commonly implies just the external structure, but anatomically it also includes the nasal cavity. The nose is the organ of smell and, as the opening of the respiratory tract, it serves to warm and filter air.

The external nose is a pyramid-shaped structure in the centre of the face, with the tip of the nose forming the apex of the pyramid. The underlying nasal cavity is a relatively large space and is the very first part of the respiratory tract (air passage).

The nasal cavity lies above the oral cavity (mouth) and is separated by a horizontal plate of bone called the hard palate. Both cavities open into the pharynx, a muscular, tube-like passageway.

EXTERNAL STRUCTURE

The external nose is made up of bone in its upper part and cartilage and fibrous tissue in its lower part. The upper part of the skeleton of the nose is mainly made up of a pair of plate-like bones called the nasal bones. These join, by their upper edges, with the frontal bone (forehead). Joining the outer edge of each nasal bone is the frontal process of the maxilla – a projection from the cheekbone between the nasal bone and the inner wall of the orbit (eye socket).

The bridge of the nose consists almost entirely of the two nasal bones, and adjoins the forehead between the two orbits. Because of their location and their relative fragility, the nasal bones are vulnerable to fracturing.

The lower half of the external nose is made up of plates of cartilage on each side. These join each other, and the cartilages of the other side along the midline of the nose.

Lateral view

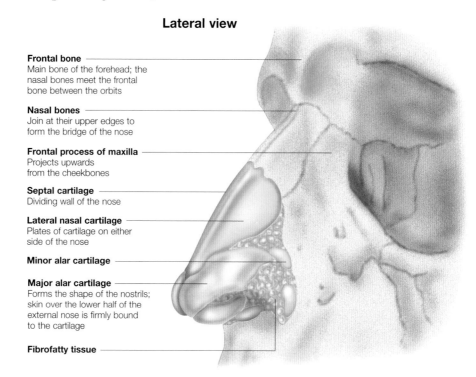

Frontal bone
Main bone of the forehead; the nasal bones meet the frontal bone between the orbits

Nasal bones
Join at their upper edges to form the bridge of the nose

Frontal process of maxilla
Projects upwards from the cheekbones

Septal cartilage
Dividing wall of the nose

Lateral nasal cartilage
Plates of cartilage on either side of the nose

Minor alar cartilage

Major alar cartilage
Forms the shape of the nostrils; skin over the lower half of the external nose is firmly bound to the cartilage

Fibrofatty tissue

Inferior view

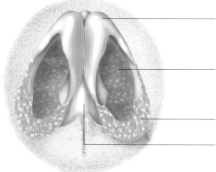

Cartilage
Lower structure of the nose is made up of plates of cartilage, a dense connective tissue

Nostril
One of two external openings of the nose; also known as the naris (*plural* nares)

Fibrofatty tissue

Septal cartilage
Separates the two nostrils; lined with mucous membrane

Inside the nostrils

Nosebleeds are common, particularly in children. This is because the blood-rich lining of the nose ruptures easily.

The part of the nasal cavity immediately above the nostril is somewhat flared and is called the vestibule. It is lined with hair-bearing skin. Elsewhere, the nasal cavity has a more delicate inner lining – the mucous membrane.

The lining of the nasal cavity receives a generous blood flow due to the presence of a rich network of blood vessels. This ensures that inhaled air is adequately warmed and moistened in the nasal cavity before it reaches the lungs.

The delicate lining of the nasal cavity is prone to damage. The most common manifestation of this is bleeding from the membrane, known as a nosebleed (epistaxis).

The lining also contains an abundance of cells, which, when inflamed or infected, tend to secrete an excess of viscous fluid (with a cold, for example).

The roof of the nasal cavity has a lining – the olfactory epithelium – which is different from that of the rest of the nasal cavity. This contains specialized cells that are receptors for the sense of smell.

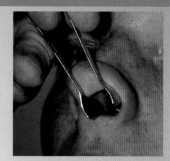

The vestibule is lined with skin that grows hairs. These hairs filter dust and other particles from air entering the nose.

Inside the nasal cavity

The nasal cavity runs from the nostrils to the pharynx, and is divided in two by the septum. The roof forms part of the floor of the cranial cavity.

The nasal cavity is partitioned into two halves by a vertical plate called the nasal septum, which is part bone and part cartilage. Each half of the nasal cavity is open in front at the nostril, and opens into the pharynx at the back through an opening called the choana.

NASAL CAVITY ROOF
The roof of the nasal cavity is arched from front to back. The central part of this roof is the cribriform plate of the ethmoid bone, a strip of bone perforated with a number of holes. This forms part of the floor of the cranial cavity, which contains the brain.

Running through the sieve-like cribriform plate from the nasal cavity to the brain is the olfactory nerve, which transmits the sensation of smell.

These anatomical features explain why head injuries involving fractures to the roof of the nasal cavity sometimes result in a leakage of cerebrospinal fluid (the clear fluid surrounding the brain) into the nose. If the head injury causes significant damage to the olfactory nerve, there may be a resultant loss of the ability to perceive smells. This is a condition called anosmia.

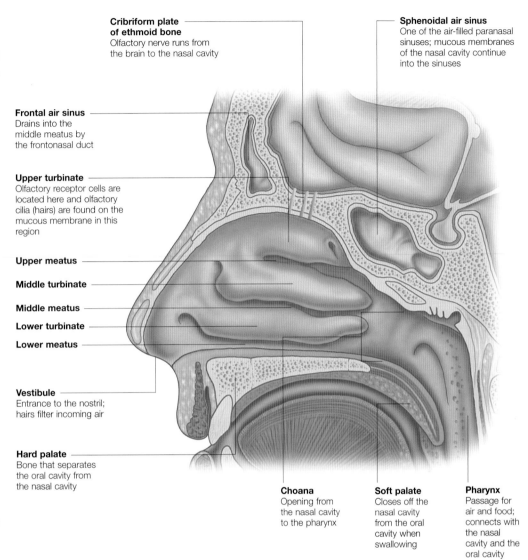

Cribriform plate of ethmoid bone
Olfactory nerve runs from the brain to the nasal cavity

Sphenoidal air sinus
One of the air-filled paranasal sinuses; mucous membranes of the nasal cavity continue into the sinuses

Frontal air sinus
Drains into the middle meatus by the frontonasal duct

Upper turbinate
Olfactory receptor cells are located here and olfactory cilia (hairs) are found on the mucous membrane in this region

Upper meatus

Middle turbinate

Middle meatus

Lower turbinate

Lower meatus

Vestibule
Entrance to the nostril; hairs filter incoming air

Hard palate
Bone that separates the oral cavity from the nasal cavity

Choana
Opening from the nasal cavity to the pharynx

Soft palate
Closes off the nasal cavity from the oral cavity when swallowing

Pharynx
Passage for air and food; connects with the nasal cavity and the oral cavity

Side walls of the nasal cavity

The lateral (side) wall of the nasal cavity is made up of several bones, some of which partially overlap. However, the complexity of the bony architecture of the lateral wall of the nose is not readily apparent, as it is covered by the mucous membrane that lines the cavity.

When viewed from the inside, the side wall of the nasal cavity shows three horizontal, overhanging, inward projections; an upper, a middle and a lower. These projections are called turbinates, and each one is produced by an underlying curled plate of bone.

The facial bones of the skull that form the lateral walls of the nose are reduced in density by air filled cavities - the paranasal sinuses.

Below each turbinate is a space called a meatus. Thus there is an upper meatus, a middle meatus and a lower meatus, each situated below the corresponding turbinate.

Within the bones adjoining the nasal cavity are air-filled spaces – the paranasal sinuses. These communicate with the nasal cavity through tiny openings in the side wall of the nose, situated in one or other meatus. The mucosal lining of the nasal cavity continues through them to line the interior of the paranasal sinuses.

The nasolacrimal duct is an opening in the lower meatus. This is a tube that runs down from the lacrimal sac (within the orbit), allowing tears to drain into the nasal cavity.

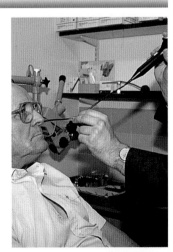

Direct inspection of the nasal cavity is done using a fibre-optic endoscope. It is introduced into the cavity through the nostril.

Pterygopalatine fossa

The pterygopalatine fossa is a funnel-shaped space
between the bones of the head. It contains important nerves and
blood vessels that supply the eye, mouth, nose and face.

The pterygopalatine fossa is an
anatomical area that is difficult
to find when the skull is intact,
and furthermore disappears when
the skull bones are separated.
The easiest way of locating the
fossa is via the pterygomaxillary
fissure, which is a narrow
triangular gap between the
pterygoid plates of the sphenoid
bone and the back of the upper
jaw (maxilla). This leads to the
lateral part of the fossa.

LOCATION OF THE FOSSA

The fossa is a small funnel-
shaped space that tapers
downwards and lies below the
back of the orbit. It is located
behind the maxilla and its back
wall is formed by the pterygoid
plates and the greater wing of
the sphenoid bone. The palatine
bone forms its midline and its
floor. It is a very important
distribution centre as it
communicates with all of the
important regions of the head
including the mouth, nose, eye
and face, infratemporal fossa
and also with the brain.

The main components of the
pterygopalatine fossa are the
maxillary artery and nerve
(branch of the trigeminal nerve)
and the pterygopalatine
ganglion. These enter and exit
the region through the sphenopalatine foramen (hole).

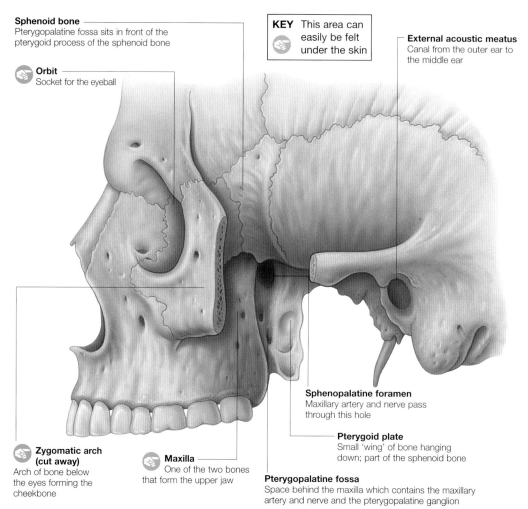

Sphenoid bone
Pterygopalatine fossa sits in front of the
pterygoid process of the sphenoid bone

Orbit
Socket for the eyeball

KEY This area can
easily be felt
under the skin

External acoustic meatus
Canal from the outer ear to
the middle ear

Sphenopalatine foramen
Maxillary artery and nerve pass
through this hole

Pterygoid plate
Small 'wing' of bone hanging
down; part of the sphenoid bone

**Zygomatic arch
(cut away)**
Arch of bone below
the eyes forming the
cheekbone

Maxilla
One of the two bones
that form the upper jaw

Pterygopalatine fossa
Space behind the maxilla which contains the maxillary
artery and nerve and the pterygopalatine ganglion

Maxillary artery

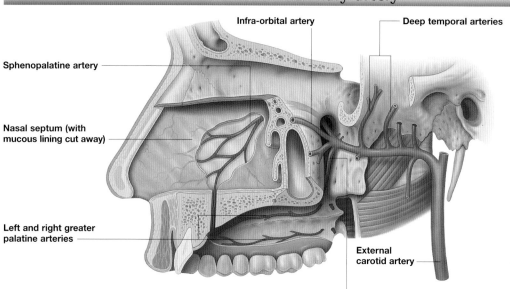

Infra-orbital artery

Deep temporal arteries

Sphenopalatine artery

**Nasal septum (with
mucous lining cut away)**

**Left and right greater
palatine arteries**

**External
carotid artery**

Descending palatine artery

The maxillary artery is one of
the terminal branches of the
large external carotid artery.
Within the infratemporal fossa,
the artery divides into three
parts and the pterygopalatine
section is conventionally
described as the third part.

DIVISIONS OF THE
MAXILLARY ARTERY

The maxillary artery enters the
pterygopalatine fossa via the
pterygomaxillary fissure. Within
the fossa, the artery divides into
numerous branches, which
eventually supply oxygenated
blood to all of the maxillary
(upper) teeth, the hard and soft
palate, the nasal cavity, the
paranasal sinuses, the skin of the
lower eyelid, the nose and the
upper lip.

Maxillary nerve

The maxillary nerve
enters the pterygopalatine
fossa before dividing into
branches which supply
sensation to large areas of
the face.

The maxillary nerve leaves the
cranial part of the skull to enter
directly into the pterygopalatine
fossa by the foramen rotundum.
On entering the fossa, the nerve
contains only fibres for common
sensation including touch, pain
and temperature. It divides
within the fossa to supply these
sensations to the nose, palate,
tonsils and gums, skin of the
cheeks, upper lip and upper
molar teeth.

BRANCHES OF THE MAXILLARY NERVE

The names of the main branches
are the zygomaticotemporal and
zygomaticofacial, greater and
lesser palatine nerves, nasal
nerves and posterior superior
alveolar nerve. The main trunk
of the maxillary nerve leaves the
fossa through the inferior orbital
fissure which is in the floor of
the orbit.

As it leaves the
pterygopalatine fossa, the
maxillary nerve becomes the
infra-orbital nerve. It travels in
the floor of the orbit to emerge
through a foramen in the
maxilla, below the eye.

Branches from the infra-
orbital nerve include the anterior
superior alveolar nerve which
supplies the front upper teeth.

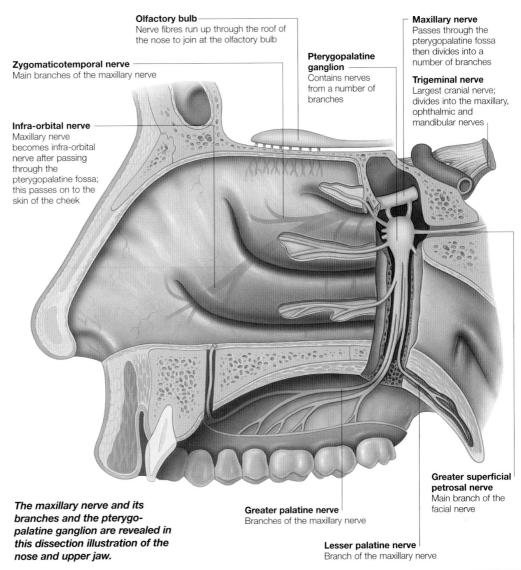

Olfactory bulb
Nerve fibres run up through the roof of
the nose to join at the olfactory bulb

Maxillary nerve
Passes through the
pterygopalatine fossa
then divides into a
number of branches

Pterygopalatine ganglion
Contains nerves
from a number of
branches

Zygomaticotemporal nerve
Main branches of the maxillary nerve

Trigeminal nerve
Largest cranial nerve;
divides into the maxillary,
ophthalmic and
mandibular nerves

Infra-orbital nerve
Maxillary nerve
becomes infra-orbital
nerve after passing
through the
pterygopalatine fossa;
this passes on to the
skin of the cheek

Greater superficial petrosal nerve
Main branch of the
facial nerve

Greater palatine nerve
Branches of the maxillary nerve

Lesser palatine nerve
Branch of the maxillary nerve

*The maxillary nerve and its
branches and the pterygo-
palatine ganglion are revealed in
this dissection illustration of the
nose and upper jaw.*

'Hay fever' ganglion

The pterygopalatine ganglion is a
point of convergence for different
nerves which communicate and
unite within the pterygo-palatine
fossa. Beyond common sensation,
the fossa is also a relay station
for nerve fibres which control
glandular secretions. These are
known as secretomotor fibres and
are not natural components of
the maxillary nerve as the
trigeminal nerve is not capable of
such functions.

SECRETOMOTOR FIBRES

The main source of secretomotor
fibres is the facial nerve. The
main branch of the facial nerve
carrying fibres into the

*The pterygopalatine ganglion
contains nerve fibres which
control secretions of tears and
mucus. It is these fibres that are
stimulated in hay fever.*

pterygopalatine fossa is called
the greater superficial petrosal
nerve. Just before entering the
fossa it unites with a nerve from
the sympathetic nervous system
called the deep petrosal nerve.
Fibres of this nerve are involved
in vasoconstriction – reducing
the blood flow to an area.

The combined nerve is known
as the nerve of the pterygoid
canal. Within the ganglion,
secretomotor fibres and
vasoconstrictor fibres unite with
sensory fibres of branches of the
maxillary nerve. In this way,
secretomotor fibres are
distributed to the lacrimal gland,
which forms tears, and to the
mucus-producing glands within
the nose, palate and sinuses. For
these reasons the pterygopalatine
ganglion is often referred to by
anatomists as the 'hay fever'
ganglion.

Paranasal sinuses

The term 'paranasal' means 'by the side of the nose'. The paranasal sinuses are air-filled cavities in the bones around the nasal cavity.

The paranasal sinuses are paired structures, and the two members of each pair are related to opposite halves of the nasal cavity.

The four pairs of paranasal sinuses are named according to the bones in which they are situated. These four pairs are:
■ Maxillary sinuses
■ Ethmoidal sinuses
■ Frontal sinuses
■ Sphenoidal sinuses.

Each member of a pair of paranasal sinuses opens into its half of the nasal cavity through a tiny opening called an ostium on the side of the nasal cavity.

The paranasal sinuses are very small, or even absent, at the time of birth, and remain small until puberty. Thereafter, the sinuses enlarge fairly rapidly; this enlargement accounts partially for the distinctive change in the size and shape of the face that occurs during adolescence.

FRONTAL SINUSES

The frontal sinuses are situated within the frontal bone (the bone of the forehead). Each is variable in size, corresponding to an area just above the inner part of the eyebrow. The frontal sinuses are situated above the opening into the nasal cavity in the middle meatus. Drainage of mucous secretions is efficient and aided by gravity.

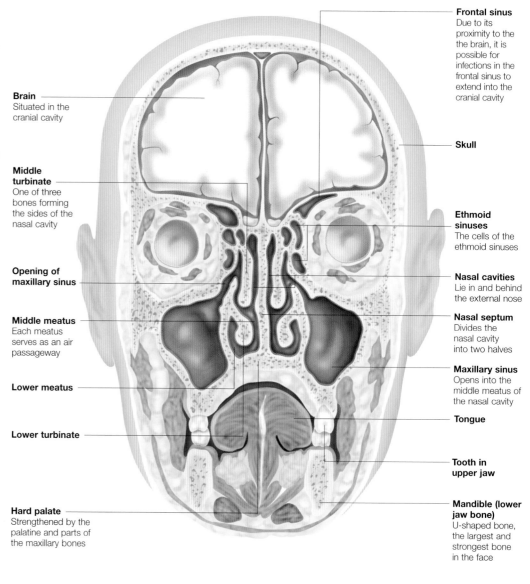

Brain
Situated in the cranial cavity

Middle turbinate
One of three bones forming the sides of the nasal cavity

Opening of maxillary sinus

Middle meatus
Each meatus serves as an air passageway

Lower meatus

Lower turbinate

Hard palate
Strengthened by the palatine and parts of the maxillary bones

Frontal sinus
Due to its proximity to the the brain, it is possible for infections in the frontal sinus to extend into the cranial cavity

Skull

Ethmoid sinuses
The cells of the ethmoid sinuses

Nasal cavities
Lie in and behind the external nose

Nasal septum
Divides the nasal cavity into two halves

Maxillary sinus
Opens into the middle meatus of the nasal cavity

Tongue

Tooth in upper jaw

Mandible (lower jaw bone)
U-shaped bone, the largest and strongest bone in the face

Functions of the paranasal sinuses

An important function of the paranasal sinuses is to help give the voice a warm, rich tone, as the sinuses act as resonators for sound. Patients with chronic sinusitis often have a noticeable lack of resonance in their voice. This strongly suggests that healthy sinuses have a role in modulating the quality of voice.

The paranasal sinuses are also believed to act as thermal insulators by preventing cold, inhaled air from cooling the surrounding structures. Another function of the paranasal sinuses is to lighten the weight of the skull.

They are also responsible for the production of mucus that flows into the nasal cavity.

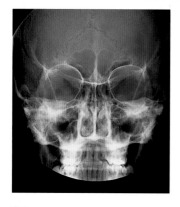

The maxillary sinuses, under the eye sockets, can be seen on this X-ray. They are one of the four pairs of the paranasal sinuses.

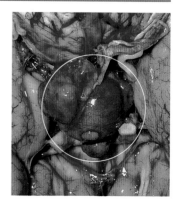

A pituitary gland tumour (circled) will often be surgically removed by approaching the gland through the sphenoidal sinus.

Inside the sinuses

The efficiency of mucous drainage from each of the pairs of sinuses depends on their location. Effective drainage lessens the risk of sinus infection.

SPHENOIDAL SINUSES
The sphenoidal sinuses are behind the roof of the nasal cavity, within the sphenoid bone. The two sphenoidal sinuses lie side by side, separated by a thin, vertical, bony partition. Each sphenoidal sinus opens into the uppermost part of the side wall of the nasal cavity (immediately above the upper turbinate) and also drains fairly efficiently into the nasal cavity.

ETHMOIDAL SINUSES
Each ethmoidal sinus is situated between the thin, inner wall of the orbit (eye socket) and the side wall of the nasal cavity. Unlike the other paranasal sinuses, these sinuses are made up of multiple communicating cavities called ethmoid air cells. These cells are subdivided into front, middle and back groups. The front and middle groups of air cells open into the middle meatus, while the back group opens into the upper meatus. The drainage into the nasal cavity is moderately efficient.

MAXILLARY SINUSES
The largest of the pairs of sinuses are the maxillary sinuses, situated within the maxillae (cheekbones). Infections and inflammation are more common here than in any of the other paranasal sinuses. This is because the drainage of mucous secretions from this sinus to the nasal cavity is not very efficient.

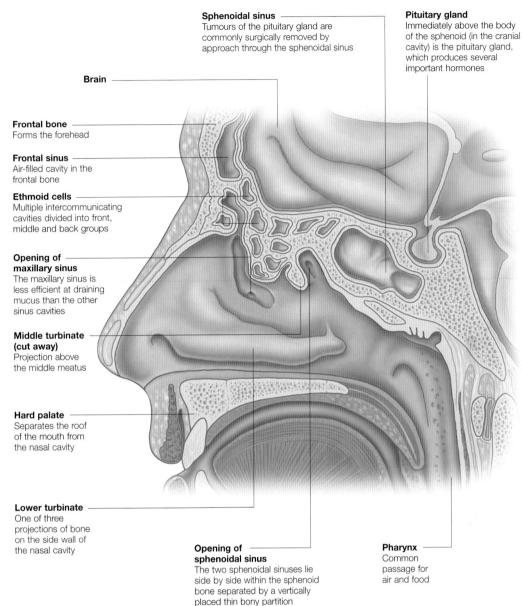

Sphenoidal sinus
Tumours of the pituitary gland are commonly surgically removed by approach through the sphenoidal sinus

Pituitary gland
Immediately above the body of the sphenoid (in the cranial cavity) is the pituitary gland, which produces several important hormones

Brain

Frontal bone
Forms the forehead

Frontal sinus
Air-filled cavity in the frontal bone

Ethmoid cells
Multiple intercommunicating cavities divided into front, middle and back groups

Opening of maxillary sinus
The maxillary sinus is less efficient at draining mucus than the other sinus cavities

Middle turbinate (cut away)
Projection above the middle meatus

Hard palate
Separates the roof of the mouth from the nasal cavity

Lower turbinate
One of three projections of bone on the side wall of the nasal cavity

Opening of sphenoidal sinus
The two sphenoidal sinuses lie side by side within the sphenoid bone separated by a vertically placed thin bony partition

Pharynx
Common passage for air and food

Sinus problems

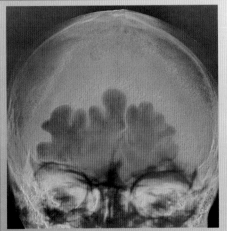

The paranasal sinuses have an inner lining of mucous membrane that is similar to the inner lining of the nasal cavity. As in the nasal cavity, the lining of each paranasal sinus contains many cells that constantly secrete fluid.

Other cells in the lining possess hair-like projections (cilia) on their surface. These projections, by their constant movement, help to propel the secretions into the nasal cavity through the ostia (openings).

Inflammation of the sinuses causes the mucous membrane lining to swell, resulting in

A false-colour X-ray of a person suffering from sinusitis caused by infection reveals the mucus-filled paranasal sinus in the frontal bone. The space is normally air-filled, but is here enlarged and inflamed.

the ostia becoming blocked. This in turn prevents the mucous secretions from draining into the nasal cavity as normal.

Because the lining of the nasal cavity continues through to the paranasal sinuses via the ostia, the paranasal sinuses may be regarded as extensions of the nasal cavity. This arrangement, however, may allow infections of the nasal cavity to spread to the paranasal sinuses.

Sinusitis – inflammation of the paranasal sinuses – is almost always preceded by an infection of the nasal cavity or throat. Symptoms include pain, purulent (pus-containing) discharge and nasal obstruction. The infection can sometimes spread to the meninges of the brain (meningitis), in which case it can be life-threatening.

How we smell

The nostrils carry air towards specialized cells located just
below the front of the skull. These cells are able to detect thousands
of different types of odours at very low concentrations.

Our sense of smell is in many ways similar to our sense of taste. This is because both taste and smell rely on the ability of specialized cells to detect and respond to the presence of many different chemicals.

The olfactory (smell) receptors present in the nose 'transduce' (convert) these chemical signals into electrical signals which travel along nerve fibres to the brain.

OLFACTORY RECEPTORS

Odours are carried into the nose when we inhale, and dissolve in the mucus-coated interior of the nasal cavity. This mucus acts as a solvent, 'capturing' the gaseous odour molecules. It is continuously renewed, ensuring that odour molecules inhaled in each breath have full access to the olfactory receptor cells.

A small patch of mucous membrane located on the roof of the nasal sinuses, just under the base of the brain, contains around 40 million olfactory receptor cells. These are specialized nerve cells which are responsive to odours at concentrations of a few parts per trillion. The tip of each olfactory cell contains up to 20 'hairs', known as cilia, which float in the nasal mucus; these greatly increase the surface area of the cell, thereby enhancing its ability to detect chemicals.

When odour molecules bind to receptor proteins on an olfactory cell they initiate a series of

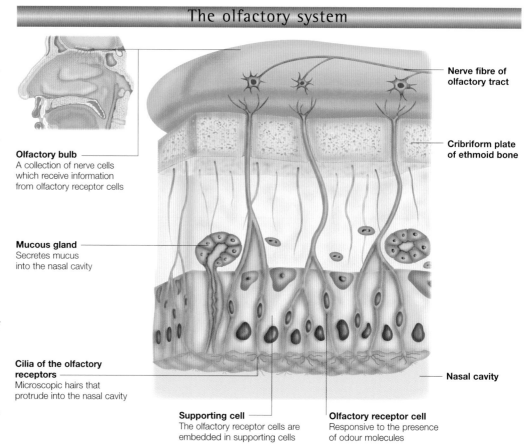

The olfactory system

Nerve fibre of olfactory tract

Cribriform plate of ethmoid bone

Olfactory bulb
A collection of nerve cells which receive information from olfactory receptor cells

Mucous gland
Secretes mucus into the nasal cavity

Cilia of the olfactory receptors
Microscopic hairs that protrude into the nasal cavity

Nasal cavity

Supporting cell
The olfactory receptor cells are embedded in supporting cells

Olfactory receptor cell
Responsive to the presence of odour molecules

nerve impulses. These impulses travel along the cell's axon (a nerve fibre which projects from the nerve cell body), which projects through the cribriform plate, the thin layer of skull immediately above the olfactory epithelia. The olfactory cells in turn communicate with other nerve cells, located in the olfactory bulb, which carry information, via the olfactory nerves (also known as cranial nerve I), to the rest of the brain.

Odour molecules dissolve in mucus secreted into the nasal cavity. Specialized receptors respond to odour molecules by sending nervous impulses to the brain via a structure called the olfactory bulb.

Dimensions of odour

An experienced wine taster can distinguish between numerous odours. Even an untrained nose is thought to be able to detect 20,000 different smells.

Receptors in the retina at the back of the eye are responsive to three colours (red, blue and green). Taste receptors respond to seven modalities. In contrast, there are thought to be hundreds (if not thousands – scientists are not entirely sure) of different types of olfactory receptor.

However, since most of us can differentiate between around 20,000 different odours, it seems unlikely that there is an individual receptor dedicated to

each odour molecule. Rather, it is thought that an odour molecule activates many different types of receptor with varying degrees of success; some receptors are very responsive to a specific odour, whereas other respond only weakly. This pattern of activity is interpreted by the brain to represent a specific smell.

When an odour molecule binds to an olfactory receptor, a complex cascade of chemical reactions is initiated inside the olfactory cell. This has the effect of amplifying the original signal; thus the brain can become aware of odours at remarkably low concentrations.

Memory, emotions and smells

The way the brain interprets smell is different from the way it interprets other senses (for example, vision) – some branches of the olfactory nerves project directly to the areas of the brain which control emotions and memory, without first travelling to the cortex, the region responsible for the development of conscious experience.

In contrast, visual input is first relayed to the visual cortex, an area involved in the conscious perception of vision, before being relayed to the emotion and memory areas.

EFFECT ON MEMORY

The neuroanatomy of the olfactory pathway means that smells can have a very profound effect on our memory. Re-exposure to an odour that was first smelt during childhood, for example, can bring back a flood of memories of that period.

This PET scan of the brain shows olfactory (smell) activity. Areas of low activity are purple; highly active areas are yellow.

Smells first experienced during childhood can evoke strong and intense memories when they are re-encountered in later life.

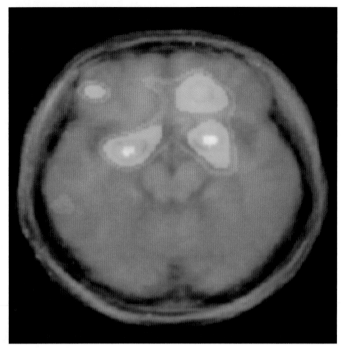

The role of pheromones

Some animals release special types of chemicals, called pheromones, into the air, water or ground in order to influence the behaviour or physiology of other members of their species. There is currently much debate as to the degree that humans use pheromones to unconsciously communicate with each other.

Research suggests that humans do respond to pheromones to some degree. For example, one study showed

It has been suggested that humans unconsciously react to pheromones released by potential sexual partners, but no firm evidence exists to prove this.

that some mothers are able to discriminate between a T-shirt worn by their child and one worn by another child of the same age.

MENSTRUAL SYNCHRONY

Over the past 30 years, a research group in the US has provided much evidence to suggest that the menstrual cycles of female flat-mates tend to converge with time.

A recent study, in which underarm body odour was collected on cotton pads from female donors and then wiped under the noses of recipient women, demonstrated that this is because women are responsive to each others' pheromones.

How good is our sense of smell?

Compared to other animals, humans have very poor smell. To take an extreme example, a dog has 25 times as many olfactory cells as a human, with 30 per cent of its cortex devoted to smell, compared to only five per cent in humans. This explains why trained sniffer dogs can detect odours at concentrations 10,000 times weaker than we can smell.

The evolution of a bipedal gait, which resulted in the nose being raised far from the ground, may have reduced Homo sapiens' reliance on olfaction.

Our sense of smell seems to have been blunted during the process of evolution. One possible explanation for this is that the development of a bipedal gait led to the nose being raised further from the ground; thus there was less advantage, evolutionarily speaking, in having a large area of the cortex

devoted to detecting odours.

The development of higher cognitive functions, such as language – which require considerable cortical processing power – may have also have contributed to the reduced reliance on olfaction.

Losing smell

Anosmia ('without smell') is a term used to describe the sudden loss of the sensation of smell. It often occurs after a blow to the head injures the olfactory nerves, but may also be the result of a nasal infection affecting the olfactory receptors.

Disorders of the brain can also affect the sense of smell. For example, some epileptics may experience an 'olfactory aura' before they have a seizure. Other disorders include olfactory hallucinations in which the affected individual experiences a specific odour, which is usually unpleasant.

How we sneeze

Sneezing is a defence mechanism, designed to protect the respiratory tract from irritant materials. The explosive exhalation that occurs during sneezing serves to clear the upper airway.

The nose is the major route for air to enter the respiratory apparatus. It acts as a very effective air filter, preventing dust and large airborne particles from entering the lungs, and allowing incoming air to adjust to body temperature before passing further down into the lungs.

DEFENCE MECHANISM

Sneezing, a sudden, forceful, involuntary burst of air out through the nose and mouth, is one of the body's many defence mechanisms.

It is designed primarily to protect the respiratory system from irritant particles which could otherwise pass further into the breathing apparatus, causing significant harm. Sneezing also serves to dislodge accumulated particles in the nose, preventing congestion of the nose's filtering system.

EXPLOSIVE REACTION

A sneeze is a blast of air that is forced out of the lungs under pressure through the airway to the mouth and nose. Most of the compressed air of a sneeze escapes through the mouth, but a percentage is directed by the soft palate to flush out the nose.

The speed of the outgoing blast of air in a sneeze can reach up to 160 km/hr (100 miles per hour) – equivalent to the wind-speed of a major typhoon.

A sneeze may carry as many as 5,000 droplets, which may contain infectious material, and be propelled as far as 3.7 metres from the nose.

The sneezing mechanism

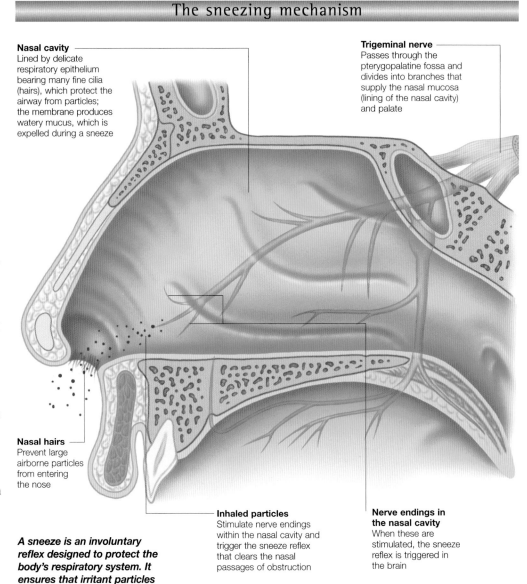

Nasal cavity
Lined by delicate respiratory epithelium bearing many fine cilia (hairs), which protect the airway from particles; the membrane produces watery mucus, which is expelled during a sneeze

Trigeminal nerve
Passes through the pterygopalatine fossa and divides into branches that supply the nasal mucosa (lining of the nasal cavity) and palate

Nasal hairs
Prevent large airborne particles from entering the nose

Inhaled particles
Stimulate nerve endings within the nasal cavity and trigger the sneeze reflex that clears the nasal passages of obstruction

Nerve endings in the nasal cavity
When these are stimulated, the sneeze reflex is triggered in the brain

A sneeze is an involuntary reflex designed to protect the body's respiratory system. It ensures that irritant particles are expelled from the nose.

Common triggers of sneezing

While sneezing can be a classic symptom of the nasal congestion that accompanies a common cold, there are many stimuli which trigger this reflex reaction. The most common include:
■ Inhalation of fine particles in the atmosphere, such as dust, hair, smoke and aerosol sprays

Sneezing is triggered by various stimuli. For example, hay fever sufferers react to the allergen pollen, causing them to sneeze frequently.

■ Allergy to mould, triggered by the inhalation of airborne spores
■ Inhalation of skin and scalp cells (dander), either human or animal
■ Hay fever (allergy to pollen) and house dust mite
■ Upper respiratory tract infections
■ Nasal polyps
■ Looking at bright light, particularly the sun
■ Changes in environmental temperature
■ Inhaled cocaine withdrawal.

What happens when we sneeze

Sneezing occurs as a result of a reflex reaction, triggered by the sensory nerve endings in the lining of the nasal cavity. The voluntary part of the brain is bypassed during this reaction, causing an automatic and involuntary response.

Sneezing occurs when the sensory nerve endings within the mucous membranes lining the nasal cavity are stimulated by triggers such as inhaled dust. This gives rise to the ticklish sensation which often precedes a sneeze. A reflex reaction follows, whereby the secretory cells of the mucous membrane are stimulated to produce watery mucus (sneezing cannot take place through a dry nose).

NERVE IMPULSE

Simultaneously, the sensory nerve fibres within the mucous membrane transmit nerve impulses to the respiratory centre of the brain (the medulla oblongata – located at the base of the brain).

The brain relays these nerve impulses to the respiratory muscles, triggering them to contract. This causes the body to inhale, close the airways and squeeze the chest, and then exhale rapidly.

The air in the lungs 'explodes' upward and outward, expelling the excess secretion along with its trapped particles through the nose and mouth.

The voluntary part of the brain is not involved in this automatic reaction, which is why we have no control over sneezing.

The rapid exhalation brought about by sneezing causes watery mucus to be expelled forcefully. A sneeze cannot be consciously controlled.

Sneezing as a reaction to light

About 25 per cent of people sneeze when exposed to bright lights, such as the sun. This phenomenon has been recognized for at least 40 years, and is often referred to as the 'photic sneeze reflex'.

It is not known exactly why this happens, although it may reflect a crossing of reflex pathways in the brain. In any reflex, a sensory nerve signal directed towards the brain communicates directly with an outgoing neural response pathway bypassing the

Many people sneeze as a reaction to looking at a bright light, especially the sun. The reason for this reaction is unknown.

conscious part of the brain.

Normally, reflex pathways take different and separate routes through the nervous system. In the case of the photic sneeze reflex, it is possible that neural signals cross over between the normal reflex of the eye in response to light and the sneezing reflex. In this situation exposure to bright light simultaneously triggers constriction of the pupil and a sneeze. There is no apparent benefit from 'sun-sneezing', and it is probably a vestigial (redundant) evolutionary trait.

Other unexplained triggers of sneezing include combing hair, plucking eyebrows, and rubbing the inner corner of the eye.

Oral cavity

Also known as the mouth, the oral cavity extends from the lips to the fauces, the opening leading to the pharynx.

The roof of the mouth, viewed from below, shows two distinct structures: the dental arch and the palate. The dental arch is the curved part of the maxilla bone at the front and sides of the roof, and the palate is a horizontal plate of tissue that separates the mouth from the nose.

The front two-thirds of the palate are bony and hard, and are formed by the maxillary bone. The hard palate is covered with a mucous membrane, beneath which run arteries, veins and nerves. These nourish and provide sensation to the palate and the overlying mucous glands, which often form fibrous ridges called rugae. The mucus secreted by these glands lubricates food to facilitate swallowing.

SOFT PALATE

The rear third of the palate is composed of glandula mucosa, muscle and tendon. Forming much of the soft palate are the tensor and levator palati muscles. These muscles close off the nasal cavity from the mouth during swallowing by respectively tensing and elevating the soft palate. They also act with other muscles to open the auditory (Eustachian) tube, which equalizes pressure on either side of the eardrum.

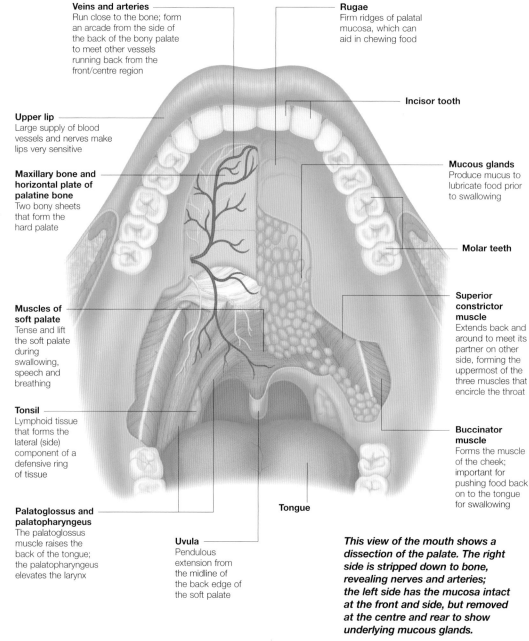

Veins and arteries
Run close to the bone; form an arcade from the side of the back of the bony palate to meet other vessels running back from the front/centre region

Upper lip
Large supply of blood vessels and nerves make lips very sensitive

Maxillary bone and horizontal plate of palatine bone
Two bony sheets that form the hard palate

Muscles of soft palate
Tense and lift the soft palate during swallowing, speech and breathing

Tonsil
Lymphoid tissue that forms the lateral (side) component of a defensive ring of tissue

Palatoglossus and palatopharyngeus
The palatoglossus muscle raises the back of the tongue; the palatopharyngeus elevates the larynx

Uvula
Pendulous extension from the midline of the back edge of the soft palate

Rugae
Firm ridges of palatal mucosa, which can aid in chewing food

Incisor tooth

Mucous glands
Produce mucus to lubricate food prior to swallowing

Molar teeth

Superior constrictor muscle
Extends back and around to meet its partner on other side, forming the uppermost of the three muscles that encircle the throat

Buccinator muscle
Forms the muscle of the cheek; important for pushing food back on to the tongue for swallowing

Tongue

This view of the mouth shows a dissection of the palate. The right side is stripped down to bone, revealing nerves and arteries; the left side has the mucosa intact at the front and side, but removed at the centre and rear to show underlying mucous glands.

Cleft palate

The term cleft palate refers to the condition in which the structures that form the palate do not fuse together properly. This results in a gap along the middle of the roof of the mouth and means that there is no plate to separate the nasal cavity from the oral cavity. If the condition affects the front of mouth, the top lip may be divided too, a deformity known as 'harelip'.

Although it can vary in severity and extent, any significant failure in palate development can lead to major problems with speech and swallowing. An infant with a cleft palate may also have significant problems when suckling from its mother's breast.

The defect can be largely repaired, often with very good cosmetic results, by surgery.

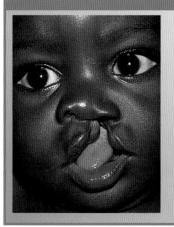

This child has a cleft palate and a harelip. These congenital conditions can prevent the baby from feeding properly.

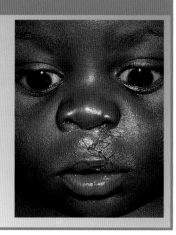

An operation may correct both conditions. Only a small scar on the outer lip will be visible afterwards.

Floor of the mouth

The floor of the mouth acts as the foundation for a network of muscles and glands that are essential to its function.

The tongue is situated over the mylohyoid muscle, which forms the muscular floor of the mouth. It is the hyoglossus muscle that anchors the tongue to the hyoid bone and provides extra strength, while the genioglossus muscle stops the tongue from moving back into the throat.

The temporalis muscles are muscles of mastication (chewing). The lingula is a small bony projection of the mandible. The mandibular nerve passes below this, through the mandibular foramen and runs within the body of the mandible to supply the lower teeth and lower lip with sensation.

SALIVARY GLANDS

There are a pair each of the submandibular and sublingual salivary glands on either side of the oral floor and, together with the paired parotid glands, they make up the six salivary glands. Saliva flows along the submandibular gland duct on the mylohyoid muscle, and emerges in the front of the oral cavity on either side of the tongue, behind the lower front teeth.

Saliva from the sublingual glands either runs into the submandibular duct, or flows out through openings in the mucosa to the side of the tongue.

The lingual nerve provides taste and sensation to the front two-thirds of the tongue.

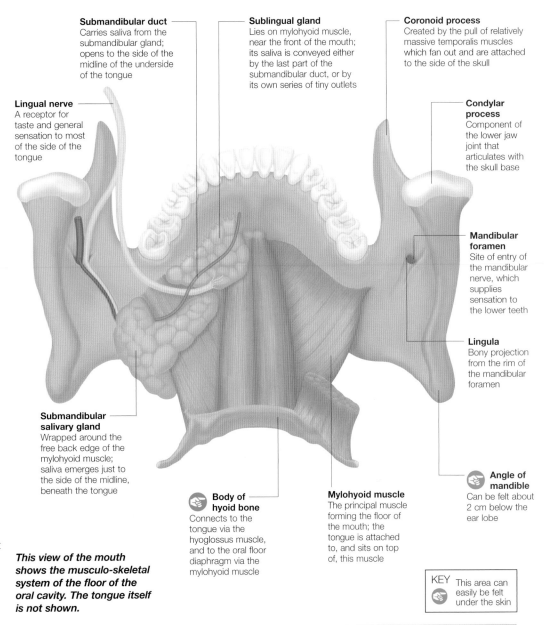

Submandibular duct
Carries saliva from the submandibular gland; opens to the side of the midline of the underside of the tongue

Sublingual gland
Lies on mylohyoid muscle, near the front of the mouth; its saliva is conveyed either by the last part of the submandibular duct, or by its own series of tiny outlets

Coronoid process
Created by the pull of relatively massive temporalis muscles which fan out and are attached to the side of the skull

Lingual nerve
A receptor for taste and general sensation to most of the side of the tongue

Condylar process
Component of the lower jaw joint that articulates with the skull base

Mandibular foramen
Site of entry of the mandibular nerve, which supplies sensation to the lower teeth

Lingula
Bony projection from the rim of the mandibular foramen

Submandibular salivary gland
Wrapped around the free back edge of the mylohyoid muscle; saliva emerges just to the side of the midline, beneath the tongue

Body of hyoid bone
Connects to the tongue via the hyoglossus muscle, and to the oral floor diaphragm via the mylohyoid muscle

Mylohyoid muscle
The principal muscle forming the floor of the mouth; the tongue is attached to, and sits on top of, this muscle

Angle of mandible
Can be felt about 2 cm below the ear lobe

This view of the mouth shows the musculo-skeletal system of the floor of the oral cavity. The tongue itself is not shown.

KEY This area can easily be felt under the skin

Lips and cheeks

The pinkish part normally thought of as the lips is called the 'free red margin'. The lips actually extend to just beneath the nose and above the chin.

Strictly speaking, the lips and the cheeks form a portion of the oral cavity known as the vestibule. However, both work in close association with the dental arches, tongue and palate in activities such as speaking and eating.

The beginning of the digestive tract, the opening of the mouth, is surrounded by the lips. They are extremely sensitive and mobile, being abundantly supplied with nerves, blood and lymph vessels, and consist mainly of muscle fibres and elastic connective tissues. These tissues are covered with a thin, translucent outer layer which allows the small capillaries to show through, giving the lips

their typical reddish colour.

Both the lips and the cheeks are involved in holding food in place so that the teeth can chew effectively. The lips also contain special nerves that help to identify various food textures.

The inner surface of the cheeks are lined with a mucous membrane made up of epithelial cells. These surface cells are rapidly worn away by abrasion with the teeth and are replaced by rapidly dividing cells underneath.

The mucus produced by the mucous membrane of the cheeks helps to lubricate the cheeks against the teeth and coats food, making it easier to swallow.

The outer surface of the lips consists of skin which contains hair follicles and sweat glands. The red part has a translucent membrane.

Tongue

The tongue is basically a mass of muscle, whose complex movement
is essential for speech, mastication and swallowing. Its upper surface
is lined with specialized tissue that contains taste buds.

The dorsal (upper) surface of the
tongue is covered with an
epithelium specialized for the
sense of taste. The anterior two-
thirds of the tongue at rest lies
within the lower dental arcade.
The posterior third slopes back
and down to form part of the
front wall of the oropharynx. Its
musculature and movements are
described in some detail overleaf.

DORSAL SURFACE
The tongue's upper surface is
characterized by filiform
papillae, tiny protuberances
which give the surface a rough
feel. The filiform papillae have
tufts of keratin which, when
elongated, may give the surface
a 'hairy' appearance and feel.
These 'hairs' can be stained by
food, medicine and nicotine.
Scattered among them are the
larger fungiform papillae. Larger
still are the 8–12 circumvallate
papillae, which form an inverse
V at the junction of the anterior
two-thirds and posterior third.
These papillae are the major site
of taste buds, although they do
occur in other papillae and are
scattered over the tongue
surface, the cheek mucosa and
the pharynx.

The posterior third of the
dorsal surface has a cobbled
appearance due to the presence
of 40–100 nodules of lymphoid
tissue, which together form the
lingual tonsil.

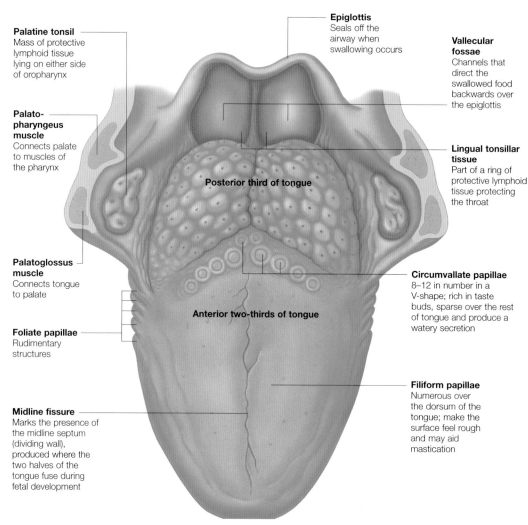

Palatine tonsil
Mass of protective
lymphoid tissue
lying on either side
of oropharynx

Palato-pharyngeus muscle
Connects palate
to muscles of
the pharynx

Palatoglossus muscle
Connects tongue
to palate

Foliate papillae
Rudimentary
structures

Midline fissure
Marks the presence of
the midline septum
(dividing wall),
produced where the
two halves of the
tongue fuse during
fetal development

Epiglottis
Seals off the
airway when
swallowing occurs

Vallecular fossae
Channels that
direct the
swallowed food
backwards over
the epiglottis

Lingual tonsillar tissue
Part of a ring of
protective lymphoid
tissue protecting
the throat

Posterior third of tongue

Anterior two-thirds of tongue

Circumvallate papillae
8–12 in number in a
V-shape; rich in taste
buds, sparse over the rest
of tongue and produce a
watery secretion

Filiform papillae
Numerous over
the dorsum of the
tongue; make the
surface feel rough
and may aid
mastication

Surface of the tongue

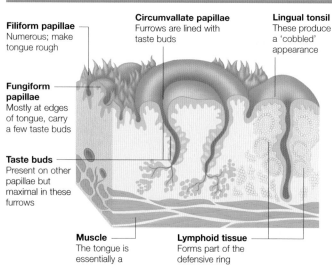

Filiform papillae
Numerous; make
tongue rough

Fungiform papillae
Mostly at edges
of tongue, carry
a few taste buds

Taste buds
Present on other
papillae but
maximal in these
furrows

Circumvallate papillae
Furrows are lined with
taste buds

Lingual tonsil
These produce
a 'cobbled'
appearance

Muscle
The tongue is
essentially a
muscular organ

Lymphoid tissue
Forms part of the
defensive ring
against infection

The taste buds are nests of cells
sensitive to flavoured substances
in solution. It is traditional to
describe tastes as either salt,
sweet, bitter or sour, but the
central processing of taste data
by the brain is complex. It seems
that when a nerve fibre is
carrying data from a taste bud, it
is responding to several or all
four of these basic taste

*This illustration shows the
structure in cross section
through the tongue at the
junction of the posterior third
with the anterior two-thirds
where the taste buds are
concentrated in the furrows
around the circumvallate
papillae.*

sensations with differing
sensitivities. Furthermore, the
sense of taste is interrelated with
the sense of smell, so food
becomes relatively tasteless with
a heavy cold.

The tongue also carries nerve
endings for the 'common'
sensations of touch, pressure,
and pain.

Since the earliest days of
medicine, the tongue has been
used as a barometer of general
health. Hippocrates, in the fifth
century BC, correlated the dry,
heavily coated, fissured tongue
with fever and dehydration, and
he gave a poor prognosis to
patients with a red ulcerated
tongue and mouth due to
prolonged dysentery.

Muscles of the tongue

The muscles within the tongue (intrinsic muscles) comprise three groups of fibre bundles running the length, breadth and depth of the organ.

The intrinsic muscles of the tongue alter the shape of the tongue to facilitate speech, mastication (chewing) and swallowing. The other muscles attached to the tongue (extrinsic muscles), move the organ as a whole. The names of the extrinsic muscles denote their attachments and the general direction of movement promoted.

Protrusion of the tongue (sticking it out), elevation of its sides and depression of its centre are functions of the intrinsic muscles. They also, together with an intact palate, the lips and the teeth, allow the formation of specific sounds in speech.

SWALLOWING
When food has been chewed and mixed with lubricating saliva, it is forced up and back between the hard palate and the upper surface of the tongue by contraction of the styloglossus muscles which pull the tongue up and back. The palatoglossi then contract, squeezing the food bolus into the oral part of the pharynx. The levator palati muscles lift the soft palate to seal off the nasal passage, while the larynx and laryngopharynx are pulled up sealing the airway against the back of the epiglottis while the bolus passes over it.

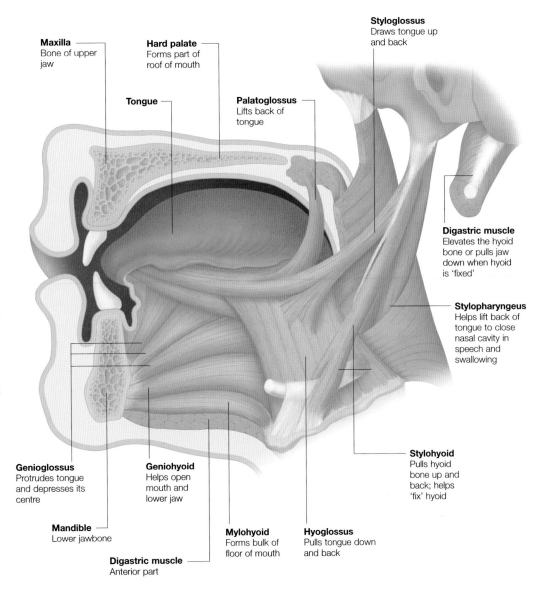

Maxilla
Bone of upper jaw

Hard palate
Forms part of roof of mouth

Tongue

Palatoglossus
Lifts back of tongue

Styloglossus
Draws tongue up and back

Digastric muscle
Elevates the hyoid bone or pulls jaw down when hyoid is 'fixed'

Stylopharyngeus
Helps lift back of tongue to close nasal cavity in speech and swallowing

Stylohyoid
Pulls hyoid bone up and back; helps 'fix' hyoid

Hyoglossus
Pulls tongue down and back

Mylohyoid
Forms bulk of floor of mouth

Digastric muscle
Anterior part

Mandible
Lower jawbone

Geniohyoid
Helps open mouth and lower jaw

Genioglossus
Protrudes tongue and depresses its centre

Lesions on the tongue

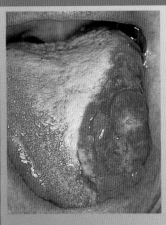

This photograph shows an advanced carcinoma occupying most of the anterior two-thirds of the left side of the patient's tongue.

The commonest mouth ulcers are apthous ulcers, which may be solitary or multiple. Major apthae are greater than 1 cm in diameter, last from weeks to months, and heal with scarring. Minor apthae are less than 1 cm, last for 10–14 days, and heal without leaving scars. Both types are painful, and occur on movable mucosa such as the tongue, lips and soft palate. Treatment is symptomatic and includes the use of oral rinses. Their aetiology (origin) is obscure, but may involve deficiencies in vitamin B12, iron, and folic acid, local trauma and stress.

Recurrent herpetic ulcers tend to arise in crops, and are due to the virus that causes cold sores on lips. In the mouth, unlike the apthae, they develop on mucosa that is bound to hard immovable surfaces such as the hard palate, gingivae and the mucosa covering the jaws. They are very painful, but usually last for less than 10 days. Relief may be obtained, and their course shortened, by using an antiviral, but it must be applied early on.

Cancer of the tongue, the commonest oral malignancy, may develop as an outgrowth or as a chronic ulcer. When treated early, the outlook is favourable, but when extensive disease is present even major surgery and/or radiotherapy do little to improve the poor prognosis. There is currently an extensive campaign to alert the population to the need for regular dental examinations to detect oral cancer early.

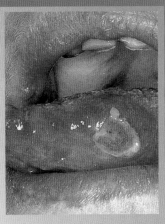

A major aphthous ulcer is easily visible here. Several smaller apthae are also present on the underside of the left of the tongue.

Salivary glands

The salivary glands produce about three-quarters of a litre of saliva a day. Saliva plays a major role in lubricating and protecting the mouth and teeth, as well as aiding swallowing and mastication.

There are three pairs of major salivary glands, which produce about 90 per cent of saliva; the remaining 10 per cent is produced by minor salivary glands located in the cheeks, lips, tongue and palate. The major role of saliva is lubrication, allowing mastication, swallowing and speech. It also has a protective function, keeping the mouth and gums moist and limiting bacterial activity.

The cells producing saliva are located in clusters at the end of a branching series of ducts. Two different types of saliva are produced by two distinctive cell types, called mucous and serous cells. The secretory products of mucous cells form a viscous mucin-rich product; the serous cells produce a watery fluid containing the enzyme amylase.

PAROTID GLAND

The largest of salivary glands are the parotid glands, which secrete serous. Each parotid is superficial, lying just beneath the skin, situated between the mandible (lower jaw) and the ear.

Several important structures pass through the parotid gland. The deepest of these is the external carotid artery; the most superficial is the facial nerve, which supplies the muscles of facial expression.

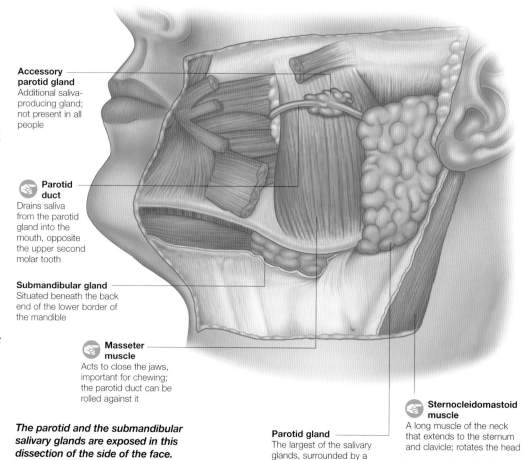

Accessory parotid gland
Additional saliva-producing gland; not present in all people

Parotid duct
Drains saliva from the parotid gland into the mouth, opposite the upper second molar tooth

Submandibular gland
Situated beneath the back end of the lower border of the mandible

Masseter muscle
Acts to close the jaws, important for chewing; the parotid duct can be rolled against it

Parotid gland
The largest of the salivary glands, surrounded by a tough, unyielding, fibrous capsule known as the parotid capsule

Sternocleidomastoid muscle
A long muscle of the neck that extends to the sternum and clavicle; rotates the head

KEY This area can easily be felt under the skin

The parotid and the submandibular salivary glands are exposed in this dissection of the side of the face. The deeper surface of the parotid gland lies on the inner surface of the mandible and close to the wall of the pharynx.

Parotid enlargement

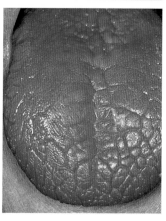

The lump beside this woman's ear is evidence of a tumour of the parotid gland. If benign, this may be the only symptom.

Slow-growing, benign tumours of the parotid gland may have no symptoms apart from an increase in gland size. However, rapidly growing malignant tumours may cause damage to the facial nerve within the gland, causing paralysis of facial muscles on one side – a condition similar to Bell's palsy.

If the majority of muscles are involved, the affected side of the face remains expressionless. The patient has difficulty with speaking and whistling, and cannot prevent food and saliva from leaking out of the corner of the mouth.

If the nerves supplying the muscle closing the eyelid – the orbicularis oris muscle – are also

involved, the ability to blink and spread a film of tears over the cornea is lost. This may result in ulceration of the cornea.

Sjögren's syndrome is a condition in which the parotid glands become enlarged as a result of infiltration by certain blood cells, called lymphocytes, causing destruction and loss of the serous secreting salivary cells. The reduction in saliva production results in a dry mouth (xerostomia). As a result of the lack of saliva cleansing the mouth there is severe gingivitis (inflammation and bleeding of the gums) and periodontitis (inflammation of the tooth-supporting structures), and considerable tooth decay.

Sjögren's syndrome causes wasting of the salivary glands. Xerostomia (dry mouth) can be alleviated with mouthwashes.

Submandibular and sublingual glands

The two smaller pairs of salivary glands are the submandibular and the sublingual glands situated in the floor of the mouth.

The submandibular gland is situated beneath the lower border of the mandible towards the angle of the jaw. It is a mixed salivary gland containing serous cells (about 60 per cent) and mucous cells (about 40 per cent). About the size of a walnut, the gland has two parts: a large, superficial part and a smaller, deep part tucked behind the mylohyoid muscle which forms the floor of the mouth. The saliva produced by the submandibular gland is carried in the submandibular duct, which opens in the sublingual papilla (protuberance) underneath the tongue.

SUBLINGUAL GLANDS

The sublingual gland is the smallest of the three major salivary glands and is almond-shaped. It is composed of about 60 per cent mucous cells and 40 per cent serous cell and lies under the tongue in the sublingual fossa. The two sublingual glands almost meet in the midline, and lie on the mylohyoid muscle.

Behind, the sublingual gland sits close to the deep part of the submandibular gland. Unlike the other glands, the sublingual gland does not have a single major collecting duct, but many smaller ones opening separately into the floor of the mouth or into the submandibular duct.

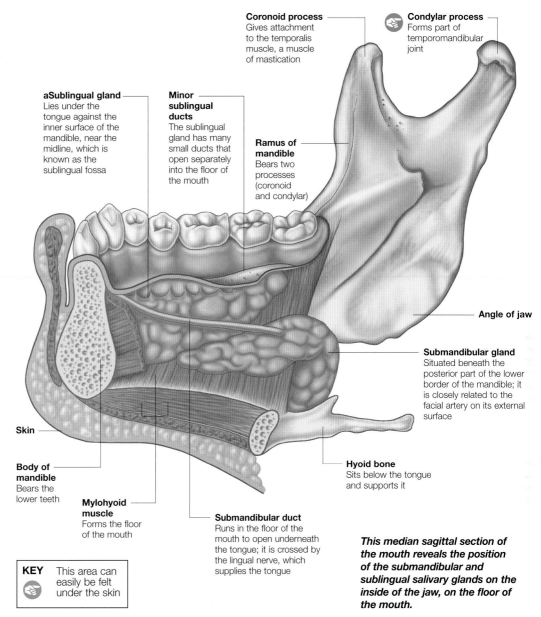

Coronoid process
Gives attachment to the temporalis muscle, a muscle of mastication

Condylar process
Forms part of temporomandibular joint

aSublingual gland
Lies under the tongue against the inner surface of the mandible, near the midline, which is known as the sublingual fossa

Minor sublingual ducts
The sublingual gland has many small ducts that open separately into the floor of the mouth

Ramus of mandible
Bears two processes (coronoid and condylar)

Angle of jaw

Submandibular gland
Situated beneath the posterior part of the lower border of the mandible; it is closely related to the facial artery on its external surface

Skin

Body of mandible
Bears the lower teeth

Mylohyoid muscle
Forms the floor of the mouth

Submandibular duct
Runs in the floor of the mouth to open underneath the tongue; it is crossed by the lingual nerve, which supplies the tongue

Hyoid bone
Sits below the tongue and supports it

KEY This area can easily be felt under the skin

This median sagittal section of the mouth reveals the position of the submandibular and sublingual salivary glands on the inside of the jaw, on the floor of the mouth.

Salivary duct blockages

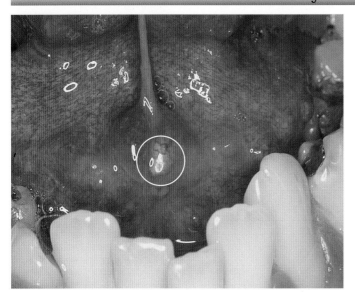

A small calcified stone blocking a salivary duct is visible as a yellowish mass (circled) in the centre of the floor of the mouth.

The submandibular duct is prone to blockage by the development of small calcified stones (calculi or sialoliths). This is partly due to the following factors:
■ Saliva is saturated with calcium and phosphate ions from mineralized calcium phosphate
■ The duct is somewhat twisted, leading to some stagnation of the saliva
■ Submandibular saliva is semi-viscous, and the saliva pools behind the lower incisor teeth near the opening of the duct.

Stones in the salivary duct obstruct the flow of saliva, particularly the increased flow of saliva at mealtimes. A stone also predisposes the mouth to infection. The stone can be readily palpated and is evident on X-ray. Surgical removal is a simple procedure.

The submandibular gland, like the parotid, is subject to the development of tumours. To ensure total removal of all malignant material it is sometimes necessary to remove adjacent nerves that may be affected by the tumour. Removal of the hypoglossal nerve results in loss of movement of one half of the tongue on the affected side, and can lead to atrophy (wasting) of that side of the tongue.

How taste buds work

We have approximately 10,000 taste buds, located mainly on the surface of the tongue and the soft tissues of the mouth. Their sensitivity and distribution means that we are able to discern between food flavours to savour and those to reject.

CHEMICAL SENSE

Taste is, together with smell, a chemical sense. It is reliant on the binding of chemicals from food to receptors located in specific cells, the taste buds, which then transmit via nerves to the brain for interpretation as 'tastes'.

The tongue is, of course, the main organ of taste, as food that the body takes in must pass through the mouth. The tongue's upper surface is covered in numerous small projections called papillae, and it is around these structures that most of the taste buds are clustered. However, a few are found elsewhere in the mouth, such as on the pharynx, the soft palate and the epiglottis.

PAPILLAE

There are three major types of papillae (the word papilla literally means a nipple-shaped protuberance). In increasing order of size, these are filiform (cone-like), fungiform (mushroom-shaped) and circumvallate (circular). In humans, most taste buds are found in these last two. Fungiform papillae are distributed all over the tongue, with a higher number along the sides and the tip. Circumvallate papillae are the largest – there are between 7 and 12 towards the rear of the tongue, arranged in a shallow inverted 'V' form. Taste buds are found in the sides of the circumvallate papillae and on the upper surfaces of the fungiform papillae.

CELLULAR STRUCTURE

Each taste bud is made up of 40 to 100 epithelial cells, which make up the epithelium, the layer that covers the entire external surface of the body and its hollow structures. In the taste buds, there are three types of these: supporting, receptor and basal cells. Receptor cells are also called the gustatory or taste cells, which give rise to taste sensations. The supporting cells form the major part of the taste bud and separate the receptor cells from each other. Taste bud cells are replaced continually – the typical lifespan is about 10 days.

Parts of the tongue

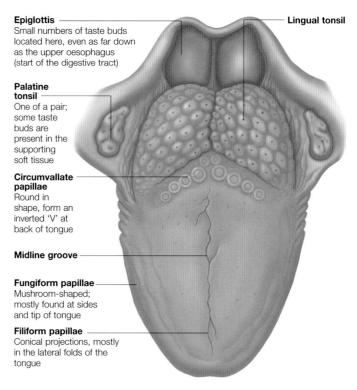

Epiglottis
Small numbers of taste buds located here, even as far down as the upper oesophagus (start of the digestive tract)

Lingual tonsil

Palatine tonsil
One of a pair; some taste buds are present in the supporting soft tissue

Circumvallate papillae
Round in shape, form an inverted 'V' at back of tongue

Midline groove

Fungiform papillae
Mushroom-shaped; mostly found at sides and tip of tongue

Filiform papillae
Conical projections, mostly in the lateral folds of the tongue

Gustatory pathway

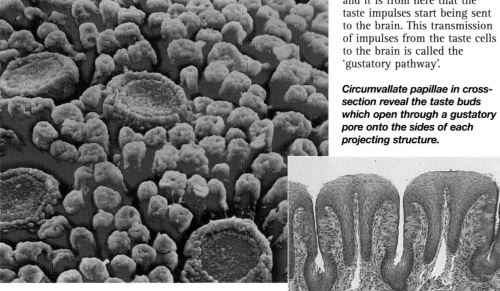

A coloured electron micrograph of the tongue shows fungiform papillae (pink) with taste buds on the surface, surrounded by filiform papillae (blue), whose texture helps manipulate food.

From each gustatory cell fine, sensitive, gustatory hairs project through the layers of epithelial cells to the surface, where they are washed in the saliva in which the substance to be tasted has dissolved. The hairs are sometimes referred to as receptor membranes, in recognition of their role in initially transmitting taste.

Sensory nerve cells form coils around the gustatory cells, and it is from here that the taste impulses start being sent to the brain. This transmission of impulses from the taste cells to the brain is called the 'gustatory pathway'.

Circumvallate papillae in cross-section reveal the taste buds which open through a gustatory pore onto the sides of each projecting structure.

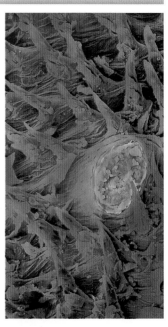

The gustatory pore leads through to a taste bud under the surface. This one is surrounded by lingual papillae, which have a sensory and tactile function.

The tasting mechanism

Once food is dissolved by saliva in the mouth, the taste buds on the tongue's surface
are stimulated. Gustatory cells then convert the chemical reaction into nerve impulses.
When the information reaches the brain, the taste information can be analysed.

When a food chemical binds to a gustatory cell, nerve impulses are sent to the thalamus, the part of the brain that receives sensory information. The thalamus processes the impulses and categorises similar functions together. Subsequently, the thalamus transmits them to the part of the brain associated with the sense of taste – the gustatory or taste cortex. The thalamus is unable to discern to any great extent whether the taste experience is good or bad. This is the job of the more sensitive gustatory cortex.

GUSTATORY CORTEX
The gustatory cortex identifies the food as good or bad and judges whether to continue eating or not. For a substance to be tasted it must be dissolved in saliva and come into contact with the gustatory hairs. From there, nerve impulses are set up to transmit impulses to the brain.

A branch of the facial nerve transmits impulses from the taste buds in the front two-thirds of the tongue, and the lingual branch of the glossopharyngeal nerve serves the rear third of the tongue. It would appear that there is a two-way flow of information to the brain regarding taste and the need to eat certain foods to satisfy the body's requirements.

The gustatory cells in the different regions of the tongue have different thresholds at which they are activated. In the bitter region of the tongue the cells can detect substances such as poisons in very small concentrations. This explains how the apparent disadvantage of its location is overcome and how its 'protective' nature works. The sour receptors are less sensitive and the sweet and salty receptors are the least sensitive of all. The taste receptors react rapidly to a new sensation, usually within three to five seconds.

What is often called taste depends to a great extent on our sense of smell. Taste is about 80 per cent smell, which explains why when we have a heavy cold food never tastes very good. The mouth also contains other receptors which can accentuate the taste sensation. Spicy foods can add to the pleasure of eating by exciting the pain receptors in the mouth.

Taste bud structure

Pore

Gustatory hair (microvilli)
Sensory receptor of taste cell, bathed by saliva

Gustatory cell
Also known as a taste or receptor cell

Supporting cells
Insulate taste cells from each other and from tongue epithelium

Epithelial cells
Form the epithelium (outer layer) of the tongue

Nerve fibres
Transmit impulses to the thalamus region of the brain

Papilla cross-section

Lingual papillae
No taste function, but detect food and provide an abrasive surface

Taste buds
Clustered in groups at the base of papillae

Furrow
The furrow base opens into the Glands of Ebner

Glands of Ebner
Serum secreting glands at the base of the furrow

Which part of the tongue tastes what?

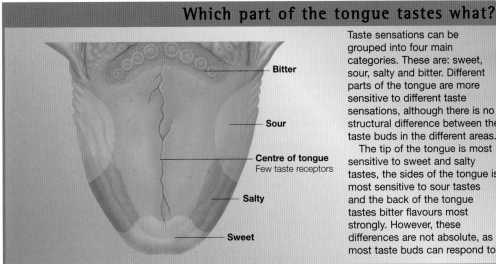

Bitter

Sour

Centre of tongue
Few taste receptors

Salty

Sweet

Taste sensations can be grouped into four main categories. These are: sweet, sour, salty and bitter. Different parts of the tongue are more sensitive to different taste sensations, although there is no structural difference between the taste buds in the different areas.

The tip of the tongue is most sensitive to sweet and salty tastes, the sides of the tongue is most sensitive to sour tastes and the back of the tongue tastes bitter flavours most strongly. However, these differences are not absolute, as most taste buds can respond to two or three – and sometimes all four – taste sensations. Certain substances seem to change in flavour as they move through the mouth: saccharin, for example, tastes sweet at first, but goes on to develop a bitter aftertaste.

Many natural poisons and spoiled foods have a bitter flavour. It is perhaps likely, therefore, that bitterness receptors are located at the back of the tongue as a protective measure. In other words, the back of the tongue screens for 'bad' food and rejects it.

How we speak

All spoken languages are constructed from a number of separate speech sounds, or phonemes. In English, these phonemes all result from the expulsion of air from the lungs.

All the speech sounds produced when speaking the majority of languages are the direct result of expelling air from the lungs. In the first instance, air travels from the lungs, via the trachea, into the larynx (the voice box).

The larynx acts as a valve, sealing off the lungs from harmful irritants during coughing, for example. The opening in the larynx is called the glottis, and this is covered by two flaps of retractable tissue called the vocal folds (the term 'vocal cords' is incorrect because they are not cords at all).

VOCAL FOLDS

As air rushes through the glottis, the vocal folds resonate, producing a buzzing sound. The pitch of this buzz is determined by the tension and position of the vocal folds. However, not all speech sounds rely on the 'voicing' produced by the vocal folds; for example, the sound 'sssss' lacks voicing, whereas the sound 'zzzzzz' requires the vocal folds to vibrate.

EXPULSION OF AIR

The vibrating air then moves up through the pharynx (throat) before leaving the head by travelling either over the tongue and through the mouth, or behind the soft palate and through the nose.

Organs of speech

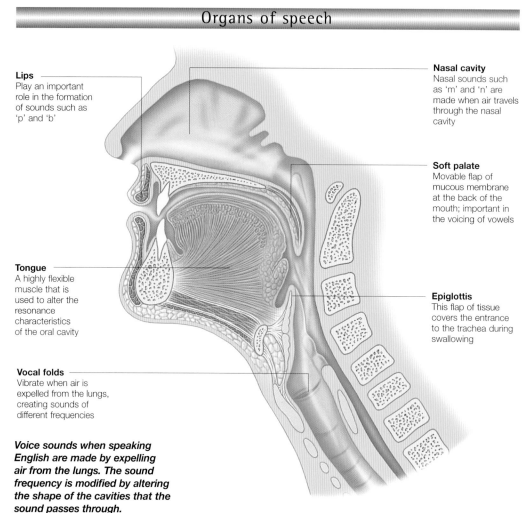

Lips
Play an important role in the formation of sounds such as 'p' and 'b'

Nasal cavity
Nasal sounds such as 'm' and 'n' are made when air travels through the nasal cavity

Soft palate
Movable flap of mucous membrane at the back of the mouth; important in the voicing of vowels

Tongue
A highly flexible muscle that is used to alter the resonance characteristics of the oral cavity

Epiglottis
This flap of tissue covers the entrance to the trachea during swallowing

Vocal folds
Vibrate when air is expelled from the lungs, creating sounds of different frequencies

Voice sounds when speaking English are made by expelling air from the lungs. The sound frequency is modified by altering the shape of the cavities that the sound passes through.

Looking at the vocal folds

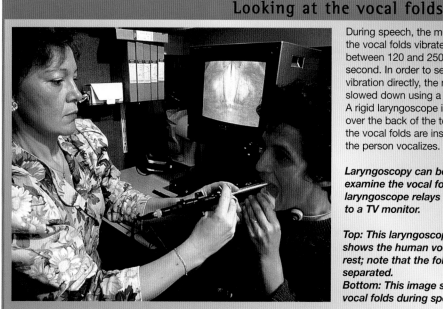

During speech, the mucosa over the vocal folds vibrates at between 120 and 250 times per second. In order to see this vibration directly, the movement is slowed down using a strobe light. A rigid laryngoscope is introduced over the back of the tongue and the vocal folds are inspected while the person vocalizes.

Laryngoscopy can be used to examine the vocal folds. The laryngoscope relays the image to a TV monitor.

Top: This laryngoscope image shows the human vocal folds at rest; note that the folds appear separated.
Bottom: This image shows the vocal folds during speech.

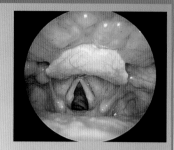

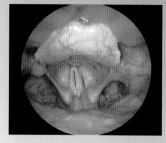

Voice sounds

Each of the chambers that air from the lungs passes through is of a different size and shape; the wavelength of the sound is altered as it travels through these chambers, resulting in a modified sound emitted through the mouth or nose.

VOWELS

Vowel sounds are produced when air is able to travel freely from the larynx to the outside. These vowel sounds are generated by altering the dimensions of the chambers the sound has to pass through.

For example, when you repeat the vowel sounds in 'bet' and 'but' alternately you should be able to feel the body of your tongue move backwards and forwards. This motion alters the resonance characteristics of the mouth cavity, altering the sound produced.

The lips are also important (note the difference in the lip position when pronouncing the vowel sounds in 'loot' and 'look') in determining the final sound, as is the soft palate (the fleshy flap at the back of the roof of the mouth). If the soft palate is opened, air will be able to flow out through the nose, as well as through the mouth, producing a 'nasal twang'.

CONSONANTS

In contrast to vowels, consonants are produced when a barrier is put in the way of the passing air. When the sound 'sssss' is made, the tip of the tongue is brought up just behind the teeth; this narrows the passage that the air can flow through, producing a hissing sound. Sounds like this are called 'fricatives' because they are created by the friction of moving air. Other fricative sounds include 'sh', 'th' and 'f', which are all produced by creating turbulence in the airflow.

Other consonant sounds are made by stopping the flow of air

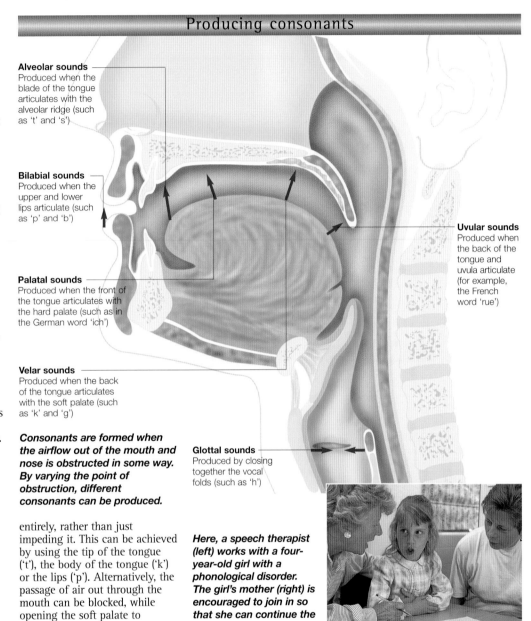

Producing consonants

Alveolar sounds
Produced when the blade of the tongue articulates with the alveolar ridge (such as 't' and 's')

Bilabial sounds
Produced when the upper and lower lips articulate (such as 'p' and 'b')

Palatal sounds
Produced when the front of the tongue articulates with the hard palate (such as in the German word 'ich')

Velar sounds
Produced when the back of the tongue articulates with the soft palate (such as 'k' and 'g')

Uvular sounds
Produced when the back of the tongue and uvula articulate (for example, the French word 'rue')

Glottal sounds
Produced by closing together the vocal folds (such as 'h')

Consonants are formed when the airflow out of the mouth and nose is obstructed in some way. By varying the point of obstruction, different consonants can be produced.

entirely, rather than just impeding it. This can be achieved by using the tip of the tongue ('t'), the body of the tongue ('k') or the lips ('p'). Alternatively, the passage of air out through the mouth can be blocked, while opening the soft palate to produce sounds like 'm' and 'n'.

Here, a speech therapist (left) works with a four-year-old girl with a phonological disorder. The girl's mother (right) is encouraged to join in so that she can continue the practice back at home.

Other speech sounds

South African singer Miriam Makeba is a native speaker of Xhosa. Many of her popular songs contain a large number of 'click consonants'.

Every word in the English language is pronounced based on a set of 40 distinct sounds, called phonemes. However, not every language uses the same set of sounds. Indeed, it has been estimated that the number of phonemes utilized by all the world's languages is in the thousands.

While English speech sounds are made by expelling air from the lungs, other languages often use different techniques:
■ Click sounds are sharp suction sounds made using the tongue or lips (what we might write as 'tut tut', for example) and are widely used in non-European languages
■ Glottalic sounds are made by using the glottis (the space between the vocal folds) to create turbulent air movement. These sounds can be created by moving air inwards (implosive sounds) or outwards (ejective sounds).

Eyeball

The eyes are the specialized organs of sight, designed to respond to light.

Our eyes allow us to receive information from our surroundings by detecting patterns of light. This information is sent to our brain, which processes it so that it can be perceived as images.

Each eyeball is embedded in protective fatty tissue within a bony cavity (the orbit). The orbit has a large opening at the front to allow light to enter, and smaller openings at the back, allowing the optic nerve to pass to the brain, and blood vessels and nerves to enter the orbit.

CHAMBERS

The eyeball is divided into three internal chambers. The two aqueous chambers at the front of the eye are the anterior and posterior chambers, and are separated by the iris. These chambers are filled with clear, watery aqueous humour, which is secreted into the posterior chamber by a layer of cells covering the ciliary body.

This fluid passes into the anterior chamber through the pupil, then into the bloodstream via a number of small channels found where the base of the iris meets the margin of the cornea.

The largest of the chambers is the vitreous body, which lies behind the aqueous chambers, and is separated from them by the lens and the suspensory ligaments (zonular fibres), which connect the lens to the ciliary body. The vitreous body is filled with clear, jelly-like vitreous humour.

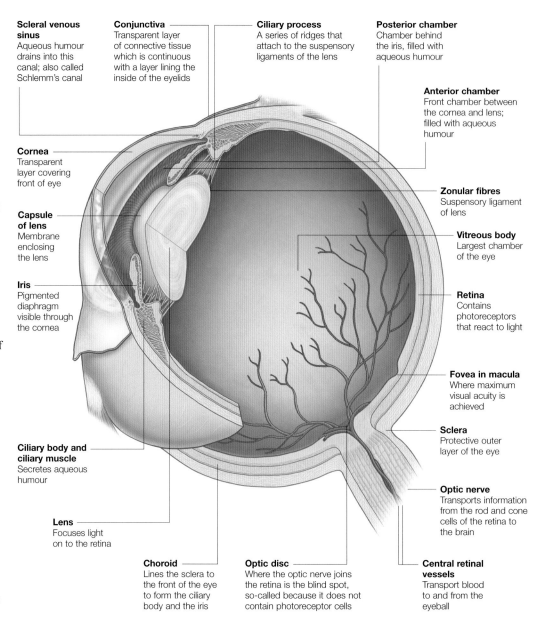

Scleral venous sinus
Aqueous humour drains into this canal; also called Schlemm's canal

Conjunctiva
Transparent layer of connective tissue which is continuous with a layer lining the inside of the eyelids

Ciliary process
A series of ridges that attach to the suspensory ligaments of the lens

Posterior chamber
Chamber behind the iris, filled with aqueous humour

Anterior chamber
Front chamber between the cornea and lens; filled with aqueous humour

Cornea
Transparent layer covering front of eye

Capsule of lens
Membrane enclosing the lens

Iris
Pigmented diaphragm visible through the cornea

Ciliary body and ciliary muscle
Secretes aqueous humour

Lens
Focuses light on to the retina

Zonular fibres
Suspensory ligament of lens

Vitreous body
Largest chamber of the eye

Retina
Contains photoreceptors that react to light

Fovea in macula
Where maximum visual acuity is achieved

Sclera
Protective outer layer of the eye

Optic nerve
Transports information from the rod and cone cells of the retina to the brain

Central retinal vessels
Transport blood to and from the eyeball

Choroid
Lines the sclera to the front of the eye to form the ciliary body and the iris

Optic disc
Where the optic nerve joins the retina is the blind spot, so-called because it does not contain photoreceptor cells

Damage to the eye

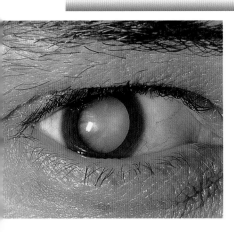

Disease or trauma can cause the transparent structures (the cornea or the lens) to become opaque to varying degrees. Increased opacity of the lens (cataract) is a common condition, especially after middle age. It is usually treated by removal of the lens, and insertion of a lens implant.

Disease and trauma can also lead to retinal damage. The retina can become detached and the detached region will degenerate.

A fairly common reason for retinal damage is glaucoma. This occurs when there is a blockage of the drainage of the aqueous fluid in the anterior chamber, which causes pressure inside the eye to rise (raised intra-ocular pressure). This pressure results in damage to the nerve cells of the retina. Intra-ocular pressure is routinely measured in eye tests for early detection of glaucoma.

A mature cataract can be seen on this man's eye. It was caused by gradual denaturation of the proteins that make up the lens.

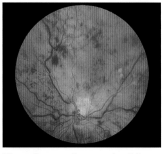

The retina can be examined for evidence of disease. In this case, the arrangement of blood vessels is indicative of diabetes.

Layers of the eye

The eyeball is covered by three different layers, each of which has a special function.

The outer layer of the eyeball is called the sclera, and is a tough, fibrous, protective layer. At the front of the eye, the sclera is visible as the 'white of the eye'. This is covered by the conjunctiva, a transparent layer of connective tissue. The transparent cornea covers the front of the eyeball, allowing light to enter the eye.

UVEA
The intermediate layer, the uvea, contains many blood vessels, nerves and pigmented cells. The uvea is divided into three main regions: the choroid, the ciliary body and the iris. The choroid extends from where the optic nerve meets the eyeball to the front of the eye, where it forms both the ciliary body and the iris.

RETINA
The innermost layer of the eye is the retina, a layer of nerve tissue containing photosensitive (light-sensitive) cells called photoreceptors. It lines all but the most anterior (frontal) part of the vitreous body. There are two types of photoreceptor cells: rods cells detect light intensity and are concentrated towards the periphery of the retina. Cone cells detect colour, and are most concentrated at the fovea at the most posterior part of the eyeball.

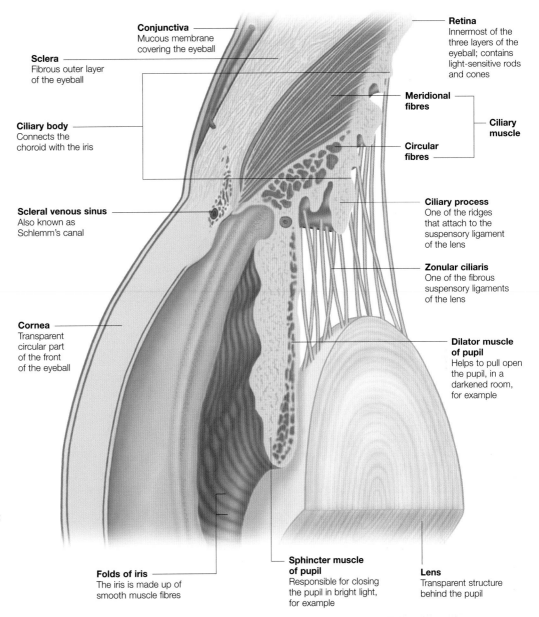

Conjunctiva
Mucous membrane covering the eyeball

Sclera
Fibrous outer layer of the eyeball

Ciliary body
Connects the choroid with the iris

Scleral venous sinus
Also known as Schlemm's canal

Cornea
Transparent circular part of the front of the eyeball

Folds of iris
The iris is made up of smooth muscle fibres

Retina
Innermost of the three layers of the eyeball; contains light-sensitive rods and cones

Meridional fibres

Ciliary muscle

Circular fibres

Ciliary process
One of the ridges that attach to the suspensory ligament of the lens

Zonular ciliaris
One of the fibrous suspensory ligaments of the lens

Dilator muscle of pupil
Helps to pull open the pupil, in a darkened room, for example

Sphincter muscle of pupil
Responsible for closing the pupil in bright light, for example

Lens
Transparent structure behind the pupil

Impaired eyesight

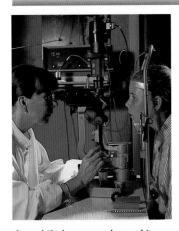

An ophthalmoscope is used to examine the interior of the eye. Serious conditions, such as glaucoma, can be detected in the early, symptomless stages using this viewing technique.

Sight can be impaired in a number of ways. The most common are refractive errors. In a normal eye (emmetropia), light rays are refracted by the cornea and the lens to focus on the retina. When looking at a distant object, the lens is relatively flat, stretched by the ciliary body pulling on the suspensory ligaments.

When looking at a near object, the ciliary muscle fibres in the ciliary body contract. This makes the circle of the ciliary body smaller, leading to the relaxation of the suspensory ligaments. This allows the lens to become more globular, therefore more refractive to focus the near image on to the retina, a process called accommodation.

In short-sightedness (myopia), the image is focused in front of the retina because the eye is too long or the curvature of the lens makes it too refractive, while in long-sightedness (hypermetropia) the opposite is true. In astigmatism, the curvature of the eye is not even, resulting in uneven focus which cannot be compensated for by lens accommodation. Refractive errors are corrected by placing appropriate prescription lenses (glasses or contact lenses) in front of the eye.

In a normal eye (top), light rays converge at the back of the retina. In a myopic eye (bottom), the eye is too long, and light rays fall short of the retina, giving unclear distance vision.

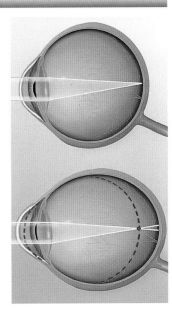

Muscles, blood vessels and nerves of the eye

The rotational movements of the eye are controlled by six rope-like extra-ocular muscles.

The muscles of the eye can be divided into three groups: the muscles inside the eyeball, the muscles of the eyelids and the extra-ocular muscles, which rotate the eyeball within its orbit.

The six extra-ocular muscles are rope-like, attaching directly to the sclera. Four of the muscles are rectus (straight) muscles – superior, inferior, lateral (the temple side of the eye) and medial (nasal side). Each rectus muscle arises from connective tissue, the common tendinous ring (annulus) at the back of the orbit that passes forward to insert just behind the junction of the sclera and cornea.

OBLIQUE MUSCLES

The two extra-ocular muscles are the oblique muscles. The superior oblique arises from bone near the back of the orbit, and extends to the front of the orbit. There, its tendon loops through the trochlea, a 'pulley' made of fibres and cartilage, and turns back to insert into the sclera.

The inferior oblique arises from the floor of the orbit, passing backwards and laterally under the eyeball to insert towards the back of the eye.

LEFT EYE (SIDE VIEW)

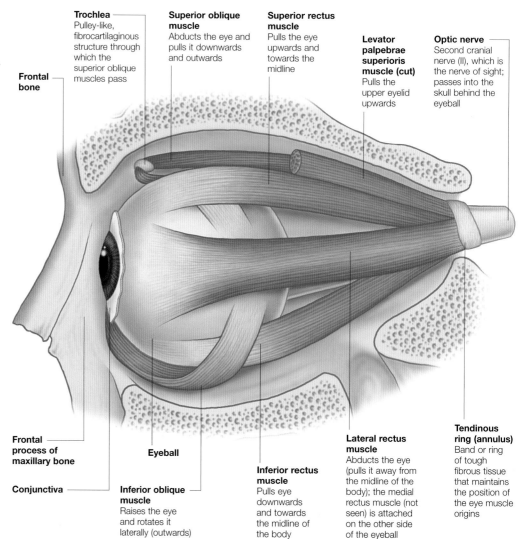

Trochlea
Pulley-like, fibrocartilaginous structure through which the superior oblique muscles pass

Superior oblique muscle
Abducts the eye and pulls it downwards and outwards

Superior rectus muscle
Pulls the eye upwards and towards the midline

Levator palpebrae superioris muscle (cut)
Pulls the upper eyelid upwards

Optic nerve
Second cranial nerve (II), which is the nerve of sight; passes into the skull behind the eyeball

Frontal bone

Frontal process of maxillary bone

Conjunctiva

Eyeball

Inferior oblique muscle
Raises the eye and rotates it laterally (outwards)

Inferior rectus muscle
Pulls eye downwards and towards the midline of the body

Lateral rectus muscle
Abducts the eye (pulls it away from the midline of the body); the medial rectus muscle (not seen) is attached on the other side of the eyeball

Tendinous ring (annulus)
Band or ring of tough fibrous tissue that maintains the position of the eye muscle origins

Eye defects

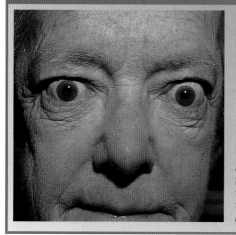

Exophthalmos (bulging eyeballs) can result from weakness of the extraocular muscles. In this patient, this weakness was triggered by thyrotoxicosis, an overproduction of hormones in the thyroid gland.

A squint (strabismus) is the consequence of defects in the function of eye muscles. While it will affect the quality of vision, it may also be indicative of serious disease.

The two main types of squint are non-paralytic strabismus and paralytic strabismus. Non-paralytic occurs when muscle functions are intact. It may be the result of incomplete development of eye movement reflexes (correctable by eye patches and/or surgery), or of serious sight defect (such as severe long or short sightedness, cataract or retinoblastoma).

In paralytic strabismus, one or more of the extraocular muscles is non-functional, usually as the result of a congenital abnormality, or is acquired later in life. It is more common in adults, and is often the result of disease (such as multiple sclerosis, meningitis or brain tumour) or trauma. Determining which eye movements are affected will show which muscles are affected and give clues to potential sites of damage in the eye or brain.

The staring, bulging eyes seen in individuals suffering from hyperthyroidism is the result of effects on eye muscles. With this condition, the levator palpebrae superioris muscle, which opens the upper eyelid, is over-stimulated, resulting in a 'wide-eyed' stare.

Nerves and blood vessels of the eye

The eye muscles are served by a series of nerves and blood vessels that help to make sight our dominant sense.

Nerves of the eye enter and leave the orbit through its openings posteriorly (at the back). Cranial nerve (CN) II – the optic nerve, which carries the visual signals from the retina to the brain – passes from the orbit to the cranial cavity through the optic canal. The other nerves – including branches of the ophthalmic nerve, the sensory nerve of the eye – enter the orbit through the orbital fissure.

Another nerve important for the eye is the facial nerve (CN VII). This supplies orbicularis oculi (a muscle of facial expression), causes blinking and also controls secretion from the lacrimal gland, which keeps the eye moist. They secrete fluid (tears) continuously, which is spread over the surface of the cornea by blinking. Irritation of the cornea can cause an increase in tear production.

ARTERIES OF THE EYE

The main artery of the eye is the ophthalmic artery, which is a branch of the internal carotid artery. The ophthalmic artery enters the orbit within the sheath of the optic nerve, and then branches to the extraocular muscles, the eyeball, the lacrimal gland and surrounding tissues.

The retinal artery remains within the optic nerve stalk until it reaches the optic disc, where it sends out branches supplying the retina. Veins drain the orbit to the cavernous sinus in the cranial cavity and to the facial vein, thus forming a connection between the blood vessels of the face and brain.

LEFT EYE (FROM ABOVE)

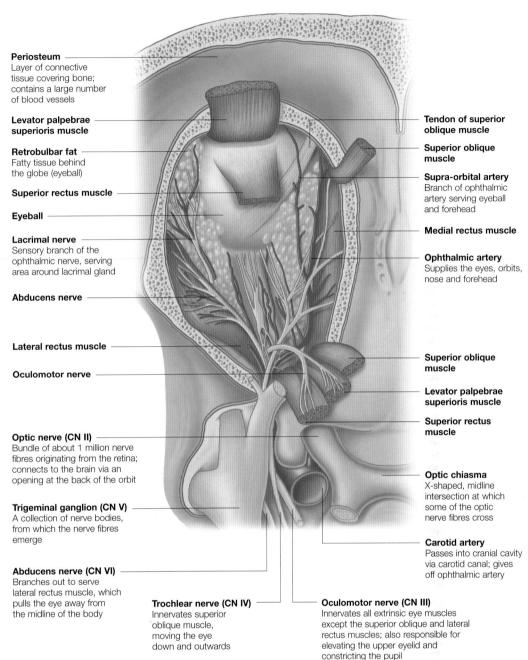

Periosteum
Layer of connective tissue covering bone; contains a large number of blood vessels

Levator palpebrae superioris muscle

Retrobulbar fat
Fatty tissue behind the globe (eyeball)

Superior rectus muscle

Eyeball

Lacrimal nerve
Sensory branch of the ophthalmic nerve, serving area around lacrimal gland

Abducens nerve

Lateral rectus muscle

Oculomotor nerve

Optic nerve (CN II)
Bundle of about 1 million nerve fibres originating from the retina; connects to the brain via an opening at the back of the orbit

Trigeminal ganglion (CN V)
A collection of nerve bodies, from which the nerve fibres emerge

Abducens nerve (CN VI)
Branches out to serve lateral rectus muscle, which pulls the eye away from the midline of the body

Tendon of superior oblique muscle

Superior oblique muscle

Supra-orbital artery
Branch of ophthalmic artery serving eyeball and forehead

Medial rectus muscle

Ophthalmic artery
Supplies the eyes, orbits, nose and forehead

Superior oblique muscle

Levator palpebrae superioris muscle

Superior rectus muscle

Optic chiasma
X-shaped, midline intersection at which some of the optic nerve fibres cross

Carotid artery
Passes into cranial cavity via carotid canal; gives off ophthalmic artery

Trochlear nerve (CN IV)
Innervates superior oblique muscle, moving the eye down and outwards

Oculomotor nerve (CN III)
Innervates all extrinsic eye muscles except the superior oblique and lateral rectus muscles; also responsible for elevating the upper eyelid and constricting the pupil

Movement of the eye

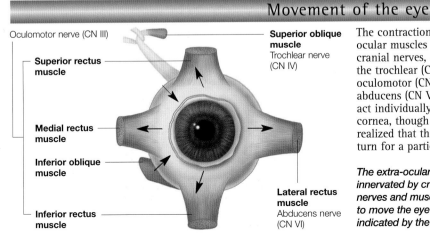

Oculomotor nerve (CN III)

Superior rectus muscle

Medial rectus muscle

Inferior oblique muscle

Inferior rectus muscle

Superior oblique muscle
Trochlear nerve (CN IV)

Lateral rectus muscle
Abducens nerve (CN VI)

The contraction of the extra-ocular muscles is controlled by cranial nerves, and specifically the trochlear (CN IV), oculomotor (CN III) and abducens (CN VI). The muscles act individually to turn the cornea, though it should be realized that the direction of turn for a particular muscle

The extra-ocular muscles are innervated by cranial nerves. The nerves and muscles serve to move the eye in the directions indicated by the arrows.

differs between right and left eyes; for example, in the right eye, lateral rectus will turn the cornea to the right, while in the left eye it would turn it to the left. Since eye movements normally occur in parallel, different muscles in each eye act together to turn the eyes.

For example, to look left, lateral rectus will turn the left eye and medial rectus the right eye. The eye movements of a single eye are usually the result of more than one of these muscles acting together.

How the eye focuses

Sight is the principal human sense, and we rely on our relatively small eyes for all our visual information. Despite their size, we can focus on a distant star or a speck of dust, and see in bright sunlight or near-darkness.

The human eye works like a camera. Light rays from an object pass through an aperture (the pupil) and are focused by a lens on to the retina, a light-sensitive layer at the back of the eye. The optical quality and versatility of the eye are much better than any camera.

The retina – the eye's equivalent of camera film – is a light-sensitive membrane composed of layers of nerve fibres and a pigmented light-sensitive membrane. It contains two kinds of light-sensitive cells: cones and rods.

CONES AND RODS

Cones are sensitive to either red, green or blue light, and their signals enable the brain to interpret a colour image. They also give the eye acute vision.

Rods are extremely sensitive to low light but cannot differentiate between colours, which is why objects appear to lose their colour at night. The rods and cones are linked to the brain by nerve cells which all pass out of the back of the eye via the optic nerve.

To see objects clearly, the muscles of the eye must pull on the lens and focus light on the retina. If this process is faulty, or the lens or eye is the wrong shape, the image will appear blurred, and spectacles, or even surgery, will be needed.

Tendons of the rectus muscles
Connective tissue joining the eye and the rectus muscles, which control the movement of the eye

Vitreous body
Chamber filled with jelly-like vitreous humour

Lens
Transparent crystalline structure that fine-tunes the focusing of images on to the retina

Pupil
Opening in the iris through which light enters

Cornea
Round transparent window in the front of the eye. Refracts light entering the eye on to the lens

Fovea
Shallow depression in the retina where light is most accurately focused

Optic nerve
Bundle of nerves, about 25 mm long, that send signals from the retina to the brain

Aqueous humour
Fluid in front of the lens

Sclera
Outer fibrous coating that gives the eye its shape

Iris
Muscular ring in front of the lens. Controls the amount of light entering the eye

Choroid
Middle layer of the eye wall. Supplies blood and oxygen to the retina

Retina
Innermost wall of the eye consisting of layers of nerve fibres and a light-receptive membrane. Where light entering the eye is focused

Ciliary body
Connects the lens – via suspensory ligaments – to the choroid. Contains muscles which control the shape of the lens

Suspensory ligaments
Ligaments between the lens and ciliary muscle. They pull on the lens, thus changing its shape, when the ciliary muscles contract

Muscles of the eye

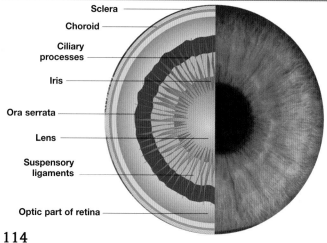

Sclera
Choroid
Ciliary processes
Iris
Ora serrata
Lens
Suspensory ligaments
Optic part of retina

The iris is a muscular, ring-shaped structure with a hole in the middle, which is called the pupil. The iris contains a distinctive coloured pigment. The muscles of the iris are used to make the pupil larger or smaller, thus allowing more or less light into the eye according to the conditions that the person is trying to see in.

The iris muscles are to be found in the ciliary body, which is the part of the eye connecting the choroid (middle layer of the eye wall) with the iris. The ciliary body consists of three parts:

- The ciliary ring, adjoining the choroid
- The ciliary processes, 70 radial ridges around the ciliary body
- The ciliary muscle, which controls lens curvature

This composite picture shows (left) the structure of the eyeball from the inside, with the lens in the centre; and (right) the outer appearance of the eye, where the lens itself is covered by the cornea.

Focusing on the retina

Light entering the eye passes through the cornea and the aqueous humour, both of which cause refraction (bending) of the light rays inwards.

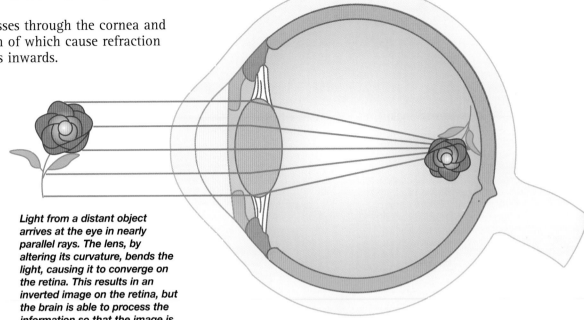

The cornea refracts most of the incoming light, and it is the task of the lens to 'fine tune' the focusing of the rays so that the image falls accurately onto the retina. The lens is a crystalline structure, made up of several layers. It is attached to the muscular ciliary body by suspensory ligaments. Movements of the ciliary muscle alter the shape of the lens, according to whether the eye needs to focus on a distant or nearby object. The diagrams below (viewing the eye from inside and from the side respectively) demonstrate how the shape of the lens is adjusted as necessary.

Light from a distant object arrives at the eye in nearly parallel rays. The lens, by altering its curvature, bends the light, causing it to converge on the retina. This results in an inverted image on the retina, but the brain is able to process the information so that the image is 'seen' the right way up.

Looking at close objects

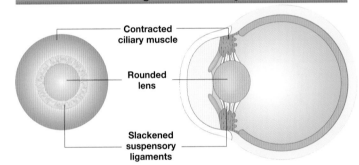

Contracted ciliary muscle

Rounded lens

Slackened suspensory ligaments

Light rays from a nearby object are more diverged, needing greater refraction. The ciliary muscle contracts, reducing the tension on the suspensory ligaments, and the lens gets more rounded. As the light rays pass through the rounded lens they are sharply converged on the back of the eye.

Looking at distant objects

Relaxed ciliary muscle

Flattened lens

Taut suspensory ligaments

Light rays from a distant object are more parallel when they reach the eye, so require less refraction by the lens. The ciliary muscle relaxes and the tension on the suspensory ligament pulls the edges of the lens outwards, thus making it thinner and flatter. The rays are focused on the back of the eye.

Common eye defects

Two common defects of the eye are short-sightedness (myopia) and long-sightedness (hypermetropia).

Short-sightedness is the inability to focus on distant objects. It is usually the result of the eyeball being slightly too long, which means that the sharpest image from a distant object is formed in front of the retina.

Long-sightedness, on the other hand, occurs when the eyeball is too short, with the result that the focus point of light from a nearby object lies behind the retina.

Short-sightedness is corrected by wearing spectacles (or contact lenses) that place a diverging (concave) lens in front of the eye; long-sightedness is corrected by

using spectacles with a converging (convex) lens.

Another common eye defect is **far-sightedness** (presbyopia), which is an inability of the eye to focus on nearby objects as a result of the lens losing its elasticity. The defect occurs naturally as people get older – often in early middle age – and is corrected by the use of converging lenses. This is often the first time that a person needs spectacles to correct a problem with their vision.

Astigmatism is the result of the eyeball being slightly misshapen, causing the image of an object to become distorted. This can be corrected by wearing spectacles with cylindrical lenses which cancel out the distortion caused by the eye itself.

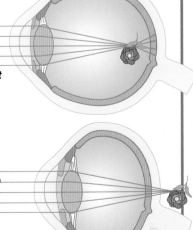

SHORT-SIGHTEDNESS
Parallel light rays are brought to a focus in front of the retina, resulting in distant objects being perceived as blurred. A concave lens diverges the light rays falling on the lens, correcting the vision.

LONG-SIGHTEDNESS
Light rays from an object focus beyond the retina when the muscles that control the focusing of the lens are relaxed. Greater degrees of hypermetropia result in a blurring of near vision.

How the retina works

The retina, located at the back of the eye, contains specialized cells called photoreceptors which are sensitive to light of different colours. These allow us to see in both the light and the dark.

The eye has adapted through evolution to be extremely sensitive to light. However, the bulk of the eye tissue is not responsive to light. Rather, the muscles which surround the eyeball, as well as the iris, cornea and lens, all act to focus light on to the retina, a relatively small area at the back of the eyeball which contains photoreceptors.

STRUCTURE OF THE RETINA

At its simplest level, the retina consists of four layers of cells:
■ At the back of the retina is the outer, pigmented layer – these epithelial cells absorb light (but do not 'detect' it), so preventing it from scattering in the eye
■ Next are aligned a layer of photoreceptors, which are able to convert light energy into electrical energy
■ The electrical potentials the photoreceptors generate are transmitted to the 'bipolar cells'
■ The bipolar cells in turn communicate with 'ganglion cells'; the axons (nerve fibres) of the latter converge and make a right-angled turn before leaving the eye through the optic nerve, which carries information on the visual scene to the brain.

Thus light has first to travel through the ganglion and bipolar cells before it reaches the light-sensitive photoreceptors at the back of the retina. This apparently 'back-to-front' arrangement does not hinder photoreceptors from detecting light.

Structure of the retina

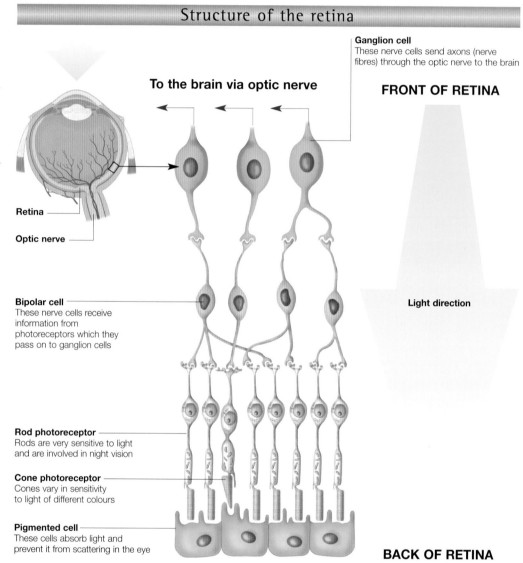

Ganglion cell
These nerve cells send axons (nerve fibres) through the optic nerve to the brain

To the brain via optic nerve

FRONT OF RETINA

Retina

Optic nerve

Light direction

Bipolar cell
These nerve cells receive information from photoreceptors which they pass on to ganglion cells

Rod photoreceptor
Rods are very sensitive to light and are involved in night vision

Cone photoreceptor
Cones vary in sensitivity to light of different colours

Pigmented cell
These cells absorb light and prevent it from scattering in the eye

BACK OF RETINA

Visual acuity

Demonstration of the blind spot

Close your left eye and focus on the square. The circle should disappear when the page is held about six inches away because its image falls on the optic nerve.

This electron micrograph shows the fovea, a crater-like depression in the retina. This area has the greatest visual acuity of the whole of the retina.

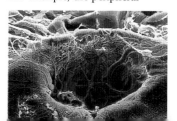

There are two main types of photoreceptors: rods, which operate in dim light and provide low acuity vision in scales of grey; and cones, which operate in bright light and provide high acuity, colour vision.

DISTRIBUTION

The relative distribution of the two types of photoreceptor varies throughout the retina. For example, the peripheral regions contain mainly rods, with relatively few cones. In contrast, at the centre of the retina, directly behind the middle of the lens, is a region the size of a pin head called the fovea which contains entirely cones.

The fovea is the only part of the retina which has cones at sufficient density to provide us with highly detailed colour vision. This is why only one thousandth of our visual field can be in hard focus at any one time; we have to move our eyes continuously in order to comprehend a rapidly changing visual scene, for example when driving along.

Rods and cones

There are two types of photoreceptors: rods, which are sensitive to low levels of light, and cones, which are responsive to light of different colours.

Rods are the most numerous of the two types of photoreceptor; it has been estimated that there are 120 million rod cells compared to only six million cones. Furthermore, rods are about 300 times more sensitive to light than cones.

NIGHT VISION

This sensitivity, coupled with their relative abundance, makes rods ideal for seeing in the dark when light levels are low.

An electron micrograph of a group of rod cells (green). Rod cells are very sensitive to light and so are mainly used for vision in the dark.

However, rods provide the brain with only low acuity vision in scales of grey. This is because a rod cell makes connections with more than one bipolar cell, which, in turn, sends electrical impulses to the brain via many ganglion cells. Thus a ganglion cell – which leaves the eye through the optic nerve – provides the brain with information gathered from a large number of rod cells. This explains why vision seems to be made of lots of large grey dots when a person is out at night.

DAY VISION

In contrast to rods, cones operate primarily in strong light and provide the brain with high acuity, colour information on the visual scene. This is aided by the fact that each individual cone cell has a 'direct line' to the brain; one cone cell is in contact with only one bipolar cell, which in turn communicates with only one ganglion cell. Thus a neuron in the brain can receive information on the activity of a single cone photoreceptor.

Rods and cones have a similar shape. The main difference between them is the photopigment that they contain.

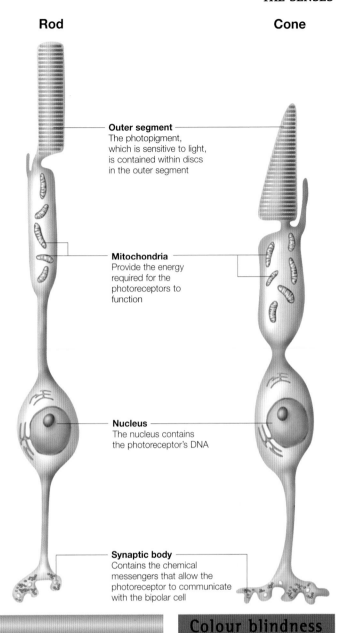

Rod **Cone**

Outer segment
The photopigment, which is sensitive to light, is contained within discs in the outer segment

Mitochondria
Provide the energy required for the photoreceptors to function

Nucleus
The nucleus contains the photoreceptor's DNA

Synaptic body
Contains the chemical messengers that allow the photoreceptor to communicate with the bipolar cell

Colour vision

We are able to see in colour because there are three different types of cone, each of which is sensitive to light of different wavelengths (colours).

The three cone types each contain a different photopigment; a photopigment

is a molecule which is responsive to light of specific wavelengths and which can change the electrical excitability of the photoreceptor cell.

The three cones are called blue, green and red cones. It should be pointed out that these

names do not necessarily correspond to the colour of light that activates them best. For example, green cones are the best of the three groups of cells at responding to green light, though they are activated the most by yellow light.

COLOUR DIFFERENTIATION

We are able to distinguish between different colours because light of a specific wavelength will activate blue, green and red cones to different degrees. The cones send impulses to the brain at a rate proportional to the degree they are activated – the brain interprets the ratio of the nerve impulses arising from the three types of cone as representing a specific colour.

There are three types of cone photoreceptors, each being responsive to a different range of colours.

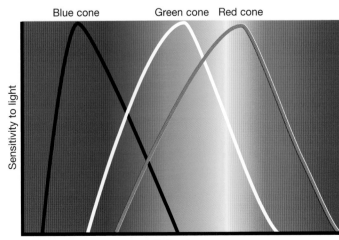

Blue cone   Green cone   Red cone

Sensitivity to light

Colour type

Colour blindness

Red-green colour blindness is a relatively common inherited condition that affects one in 12 men and one in 100 women. Affected individuals have a deficiency in either red or green cones, which makes it impossible for them to differentiate between red and green and between orange and yellow.

Colour-blind people will be unable to see the 'L' in this colour blindness test because the red and green dots appear the same.

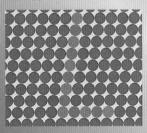

Eyelids and lacrimal apparatus

The eyelids are thin folds of skin that can close over the eye to protect it from injury and excessive light. The lacrimal apparatus is responsible for producing and draining lacrimal fluid.

Each lid is strengthened by a band of dense elastic connective tissue called a tarsal plate. These give the eyelids a curvature that matches that of the eye.

EYELID STRUCTURE
The tarsal plate of the upper eyelid is larger than that of the lower. The inner and outer ends of both tarsal plates are attached to the underlying bone by tiny ligaments. Between the front surface of the tarsal glands and the overlying skin, lie fibres of the orbicularis oculi muscle.

The eyelashes project from the free edge of the eyelids. The follicles of the eyelashes, from which the hairs emerge, have nerve endings which can sense any movement of the lashes.

The tarsal plates contain glands, called meibomian glands, that secrete an oily liquid which prevents the eyelids sticking together. There are also other tiny ciliary glands associated with the eyelash follicles.

EYELID MOVEMENT
The eye closes due to movement of the upper lid. The orbicularis oculi muscle contracts to close the eye, while the upper lid is opened by the levator palpebrae superioris muscle.

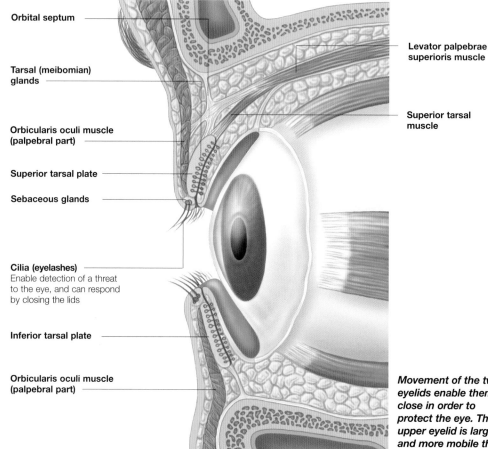
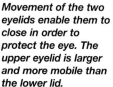

Orbital septum

Tarsal (meibomian) glands

Orbicularis oculi muscle (palpebral part)

Superior tarsal plate

Sebaceous glands

Cilia (eyelashes)
Enable detection of a threat to the eye, and can respond by closing the lids

Inferior tarsal plate

Orbicularis oculi muscle (palpebral part)

Levator palpebrae superioris muscle

Superior tarsal muscle

Movement of the two eyelids enable them to close in order to protect the eye. The upper eyelid is larger and more mobile than the lower lid.

The conjunctiva

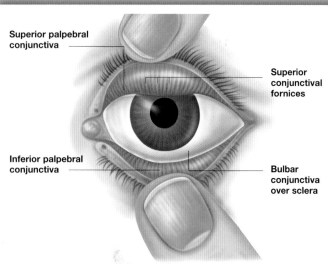

Superior palpebral conjunctiva

Inferior palpebral conjunctiva

Superior conjunctival fornices

Bulbar conjunctiva over sclera

The conjunctiva is a very thin membrane that lines and lubricates the surface of the eyeball and inner surfaces of the eyelid. It can be thought of as having two parts:
■ The bulbar conjunctiva – this covers the white of the eye, the sclera. The conjunctiva here is thin and transparent, and is separated from the sclera by loose connective tissue. It does not cover the cornea, which lies over the iris and pupil, but attaches to its periphery.

The conjunctiva is a membrane covering the white of the eye and lining the inside of the eyelids. The conjunctiva over the eyeball is transparent.

■ The palpebral conjunctiva – this lines the inside of the upper and lower eyelids. Deep recesses, known as conjunctival fornices, are formed where the bulbar and palpebral parts of the conjunctiva meet.

INFECTION/IRRITATION
The bulbar conjunctiva, overlying the eyeball, is normally colourless. However, its tiny blood vessels can become dilated and congested as a response to local irritants such as smoke or dust. The conjunctiva may also be the site of viral or bacterial infections, where redness may be associated with a 'gritty' feeling and even pus formation.

Lacrimal apparatus

The eyes are protected and lubricated by lacrimal fluid, our tears. The lacrimal system produces this fluid and drains the excess to the nasal cavity.

The eyes are kept moist by the continuous production of small amounts of lacrimal fluid by the lacrimal glands. This fluid also contains lysozyme, an antibacterial substance.

Most of the fluid, about 1 ml for each eye per day, is lost through evaporation. What is left is drained by the nasolacrimal ducts to the back of the nose.

LACRIMAL GLAND
The lacrimal gland, which produces the thin, watery lacrimal fluid, lies above the outer side of the eye within a recess in the bony eye socket. It is about 2 cm long and roughly the shape of an almond.

The gland is divided into two parts; an upper orbital part and a lower palpebral part. There are also extra, accessory lacrimal glands which lie predominantly within the upper lid.

LACRIMAL DUCTS
The lacrimal glands each have up to 12 tiny ducts. These lacrimal ducts carry the secretions away from gland and release them into the conjunctival sac through openings under the upper lid in the superior fornix.

CANALICULI
After travelling across the eye during blinking, the lacrimal fluid collects in the lacrimal lake

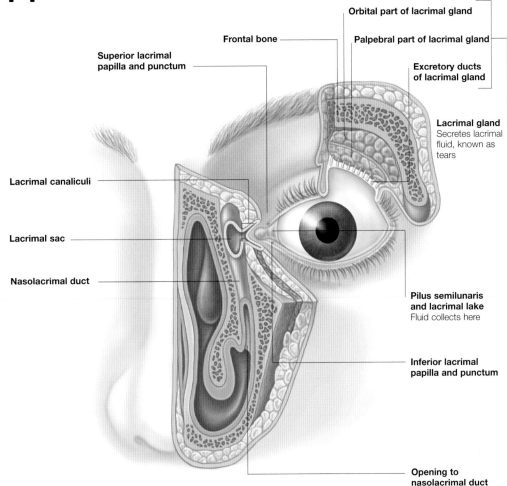

Superior lacrimal papilla and punctum

Frontal bone

Orbital part of lacrimal gland

Palpebral part of lacrimal gland

Excretory ducts of lacrimal gland

Lacrimal gland
Secretes lacrimal fluid, known as tears

Lacrimal canaliculi

Lacrimal sac

Nasolacrimal duct

Pilus semilunaris and lacrimal lake
Fluid collects here

Inferior lacrimal papilla and punctum

Opening to nasolacrimal duct

at the innermost corner.

The upper and lower eyelid both have a raised papilla at their inner ends, which has a tiny opening, the lacrimal punctum.

Excess lacrimal fluid enters these openings to be carried away by the underlying passages, the lacrimal canaliculi.

LACRIMAL SAC AND NASOLACRIMAL DUCT
Lacrimal fluid passes from the canaliculi into a collecting area known as the lacrimal sac.

From here the lacrimal fluid is carried down to the back of the nasal cavity by the nasolacrimal duct. The fluid leaving the nasolacrimal duct normally

The lacrimal apparatus is the system that produces fluid and drains it from the eye. The lacrimal gland secretes the fluid, which drains via the puncta.

evaporates within the nasal cavity, helping to humidify the air there.

Disorders of the lacrimal glands

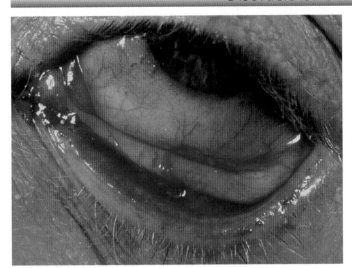

In normal circumstances the eye is slightly moist at all times. Disorders of the lacrimal system may lead to the eye being excessively wet or dry.

WATERY EYE
A watery eye may be due to an excessive flow of fluid, for example where there is physical irritation. The drainage system cannot cope and tears may spill over the front of the lower lid. A watery eye may also be the

Sjögren's syndrome affects the lacrimal glands. The condition causes an under-secretion of lacrimal fluid and results in dry, red eyes.

result of blockage of the exit channels such as may occur if there is an infection in the lacrimal sac or nasolacrimal duct. These infections are usually passed up from the nasal cavity and may become chronic.

DRY EYE
When the lacrimal glands do not produce enough secretions, the eye becomes dry. This can lead to irritation, with ulceration of the cornea in extreme cases.

Under-secretion of the lacrimal gland may be due to medicines taken, disruption of the nerve supply or hardening of the gland in a disease such as Sjögren's syndrome.

How tears are produced

The eyes are highly complex and delicate structures. The constant secretion of fluid from the lacrimal glands above each eye lubricates the eyes and protects them from foreign bodies and infection.

The eye is the organ of visual perception – our dominant sense. It relays vital information to the brain about our surroundings, and plays a crucial role in communication.

MOVEMENT OF THE EYES

In order for us to have a large field of vision, the eyeball is designed to move around within the orbit by fine muscular control.

To facilitate this movement, each eye produces lacrimal secretions. These secretions moisten the conjunctiva (the membrane that covers the eye) and lubricate the eye, allowing it to move more efficiently within the eye socket.

Lacrimal fluid is produced, distributed and carried away by the lacrimal apparatus.

ANATOMY OF THE LACRIMAL APPARATUS

The lacrimal apparatus consists of the lacrimal gland (where lacrimal secretions are produced) and the ducts that drain the excess secretions into the nasal cavity.

Each lacrimal gland lies within the orbit (the eye socket) just above the outer aspect of the eye and is the size and shape of an almond.

These specialized glands are responsible for the constant production of lacrimal secretion, the salty solution otherwise known as tears. Tears are secreted into the eye via several tiny openings in the lacrimal gland.

The eyeball is designed to move within the eye socket. The secretion of lacrimal fluid moistens and lubricates the eye, facilitating this movement.

Tears are produced and secreted by the lacrimal gland, located just above the eye. Excess fluid drains through the tear ducts into the nasal cavity.

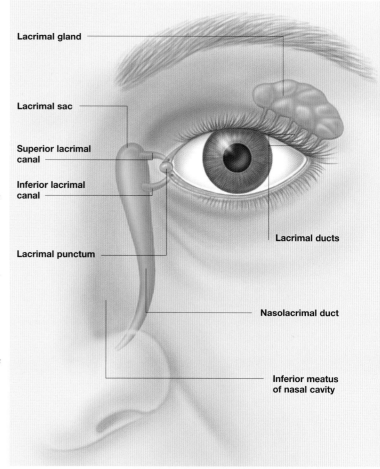

Lacrimal gland

Lacrimal sac

Superior lacrimal canal

Inferior lacrimal canal

Lacrimal punctum

Lacrimal ducts

Nasolacrimal duct

Inferior meatus of nasal cavity

Production and drainage of lacrimal fluid

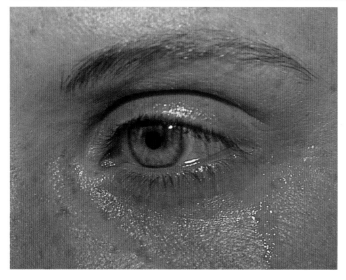

As the eye blinks (around every two to ten seconds) lacrimal fluid is spread downwards, across the eyeball.

Normal amounts of tears washing down over the front of the eyeball are prevented from spilling out onto the cheeks by the oil that the lid glands deposit on the margins of the lids.

Most of the fluid produced by the lacrimal glands evaporates from the surface of the eye, but some collects in the inner corner of the eye.

Lacrimal fluid washes across the eyeball before draining into the lacrimal canal. From here the fluid is drained into the nasal cavity via the nasolacrimal duct.

FLUID DRAINAGE

The excess fluid that collects in the inner corner of the eye enters the lacrimal canals via two tiny openings called the lacrimal puncta. These appear as tiny red dots on the inner margin of each eyelid – adjacent to the bridge of the nose.

From the lacrimal canals, fluid drains into the lacrimal sac, and thence into the nasolacrimal duct, which is an extension of the lacrimal sac. The fluid is then released into the nasal cavity.

It is estimated that the average person produces around 0.75 to 1.1 ml of lacrimal fluid every day, which serves to keep the eyes moist and free from infection.

Tear production

Teardrops are produced when the lacrimal glands are stimulated to secrete increased amounts of fluid. This may be in response to irritants or emotional upset.

When lacrimal secretion increases substantially, the excess spills over the eyelids and drips from the corners of the eyes, forming characteristic drops of fluid that run down the cheeks as teardrops. An excess of lacrimal fluid also fills the nasal cavities, causing congestion and the characteristic sniffling that accompanies tears.

REFLEX REACTION
One trigger of this excess secretion of lacrimal fluid is the presence of foreign bodies such as grit in the eye. When a foreign body enters the eye the lacrimal glands are stimulated to produce increased amounts of lacrimal fluid. This excess fluid irrigates the eye and flushes out the foreign body. In this way, the eye is protected from damage or infection.

In the case of potentially harmful irritants such as noxious chemicals, increased production of lacrimal fluid serves to dilute and wash away the irritating substance.

An example of this mechanism is seen when we chop an onion. The onion releases pungent chemicals into the atmosphere which dissolve in the moisture present on the surface of the eyes, releasing an acid which stings the eyes.

The lacrimal glands are

The pungent chemicals released from a chopped onion stimulate increased production of lacrimal fluid. The excess fluid washes the potentially harmful irritant from the eyes.

stimulated to produce an increased amount of lacrimal fluid, which spills across the eyes, diluting the irritant, and flushing it out of the eyes. As there is an excess of lacrimal fluid it drips from the eyes, giving the appearance of crying.

TRIGGER MECHANISM
Tears are secreted as part of a reflex response (automatic and involuntary) to a variety of stimuli. Examples include irritants to the eye and lining of the nose, as well as hot or peppery foods coming in to contact with the mouth and tongue. Tear flow also occurs in association with vomiting, coughing and yawning.

In each case the reaction is controlled autonomously by a region of the brain known as the hypothalamus.

When receptors in the eye are stimulated, nerve impulses are transmitted to the hypothalamus via the facial nerves supplying the lacrimal glands, and a reflex arc occurs whereby the lacrimal glands are stimulated to produce more tears.

Crying

The production of tears as a result of emotional upset is known as crying or psychical weeping, and is a different kind of response. Research has shown that even if the nerves which cause reflex production of tears to take place are severed, the emotional response of crying can still occur.

EMOTIONAL UPSET
The significance of emotionally induced tears is poorly understood. The discovery that lacrimal secretions contain encephalins (natural opiates) and the fact that only humans shed emotional tears suggest that crying may play a role in reducing stress. This is perhaps why crying is accompanied by a feeling of relief.

Crying is also an effective way of communicating our pain or anguish to others.

Little is understood about the mechanism that triggers crying in response to emotional upset. However, it is certain that tears play a vital role in communication.

Composition of tears

Lacrimal fluid contains mucus, antibodies, and lysozyme, an enzyme that destroys bacteria. The function of this fluid is to cleanse and protect the eye surface as well as moistening and lubricating it.

Some medical conditions such as keratoconjunctivitis sicca (dry-eye) result in impaired lacrimal function, and the eyes become dry and infected. In such cases, patients require artificial saline solution (such as that used by contact lens wearers) to help lubricate the eye and ward off infection.

As we grow older the lacrimal glands become less active, and the eyes tend to become less moist. Consequently the eyes are more prone to infection and irritation during the later years of life.

Antibodies present in tears help to ward off infection. As the lacrimal glands become less active with age the eyes are more prone to infection.

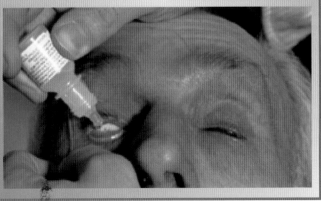

The ear

The ears are vital sensory organs of hearing and balance. Each ear is divided into three parts – outer, middle and inner ear – each of which is designed to respond to sound or movement in a different way.

The ear can be divided anatomically into three different parts: the external, middle and inner ear. The external and middle ear are important in the gathering and transmitting of sound waves. The inner ear is the organ of hearing and is also vital in enabling us to maintain our balance.

TRANSMITTING INFORMATION

The external ear consists of the visible auricle or pinna (earlobe) and the canal that passes into the head – the external auditory meatus. At the inner end of the meatus is the tympanic membrane, or eardrum, which marks the border between the external and middle ear.

The middle ear is connected to the back of the throat via the auditory tube. Within the middle ear are three tiny bones called the ossicles. These bones are linked together in such a way that movements of the eardrum are transmitted via the footplate of the stapes to the oval window (the opening in between the middle and inner ear).

The inner ear contains the main organ of hearing, the cochlea, and the vestibular system that controls balance. Information from both these parts of the ear passes to specific areas within the brainstem via the vestibulocochlear nerve.

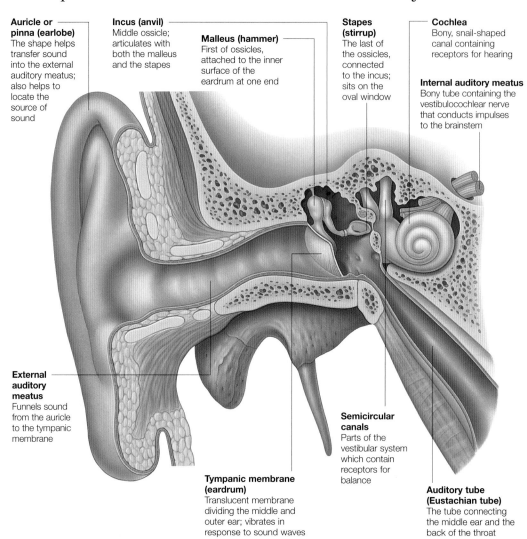

Auricle or pinna (earlobe)
The shape helps transfer sound into the external auditory meatus; also helps to locate the source of sound

Incus (anvil)
Middle ossicle; articulates with both the malleus and the stapes

Malleus (hammer)
First of ossicles, attached to the inner surface of the eardrum at one end

Stapes (stirrup)
The last of the ossicles, connected to the incus; sits on the oval window

Cochlea
Bony, snail-shaped canal containing receptors for hearing

Internal auditory meatus
Bony tube containing the vestibulocochlear nerve that conducts impulses to the brainstem

External auditory meatus
Funnels sound from the auricle to the tympanic membrane

Tympanic membrane (eardrum)
Translucent membrane dividing the middle and outer ear; vibrates in response to sound waves

Semicircular canals
Parts of the vestibular system which contain receptors for balance

Auditory tube (Eustachian tube)
The tube connecting the middle ear and the back of the throat

Viewing the ear through an auriscope

The tympanic membrane (eardrum) can be examined using an instrument called an auriscope, which is inserted into the external auditory meatus. The auriscope illuminates the eardrum, which appears pearly-grey in colour. A reflected cone of light can be seen originating at a small central depression, called the umbo, and radiating downwards and forwards. The umbo marks the site of attachment of the malleus on the other side of the membrane.

The majority of the membrane

Before inserting an auriscope, the earlobe is pulled up and out. This is necessary as the external auditory meatus does not follow a straight course.

is thickened and taut (pars tensa) but there is a small area at the top which is not so fibrous, called the pars flaccida. The membrane is well innervated (served by nerves) and inflammation due to infection may be extremely painful.

A view of a healthy tympanic membrane (eardrum) as seen through an auriscope. The eardrum appears translucent.

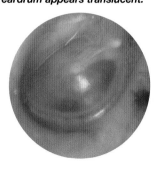

External ear

The pinna is the skin and cartilage that make up the external ear. It serves to channel sound into the middle ear.

The pinna, or auricle, collects sound from the environment and channels it into the external auditory meatus. It consists of a thin sheet of elastic cartilage and a lower portion called the lobule, consisting mainly of fatty tissue, with a tight covering of skin.

The auricle is attached to the head by a series of ligaments and muscles, and the external ear has a complex sensory nerve supply involving three of the cranial nerves.

PROTECTING THE EAR

The external auditory meatus is a tube extending from the lobule to the tympanic membrane and is about 2.5 cm long in adults. The outer third of the tube is made of cartilage (similar to that in the auricle), but the inner two thirds is bony (part of the temporal bone).

In the skin covering the cartilaginous part of the meatus, there are coarse hairs and ceruminous glands that secrete cerumen (earwax). Usually wax dries up and falls out of the ear, but it can build up and interfere with hearing. The combination of wax and hairs prevents dust and foreign objects from entering the ear.

The boundary between the outer and middle ear is the tympanic membrane, or eardrum. This is a translucent membrane which can be viewed using an auriscope. The tympanic membrane can sometimes be perforated due to middle ear infection or high-pressure sound waves.

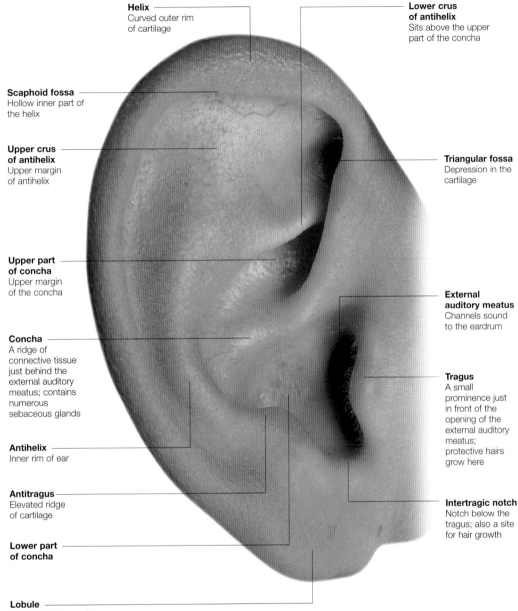

Helix
Curved outer rim of cartilage

Lower crus of antihelix
Sits above the upper part of the concha

Scaphoid fossa
Hollow inner part of the helix

Upper crus of antihelix
Upper margin of antihelix

Triangular fossa
Depression in the cartilage

Upper part of concha
Upper margin of the concha

External auditory meatus
Channels sound to the eardrum

Concha
A ridge of connective tissue just behind the external auditory meatus; contains numerous sebaceous glands

Tragus
A small prominence just in front of the opening of the external auditory meatus; protective hairs grow here

Antihelix
Inner rim of ear

Antitragus
Elevated ridge of cartilage

Intertragic notch
Notch below the tragus; also a site for hair growth

Lower part of concha

Lobule
Soft fatty tissue at the bottom of the earlobe, often the site for ear-piercing; contains no cartilage

Ear deformities

Children with protruding ears, or 'bat' ears, are often self-conscious of their ears and suffer teasing. For this reason, some people choose to have their ears pinned back surgically. Children need to be over five years old to have the operation as before this age the ear cartilage has not yet hardened.

In some cases, protruding ears are caused by excess cartilage around the ear canal, pushing the ear out from the side of the head. To correct this, a surgeon cuts into the back of the ear and creates a fold in the cartilage. The ear is then stitched and allowed to heal.

Cauliflower ears are the result of repeated blows to the earlobe. They represent damage to the cartilage that makes up the earlobe. The problem arises because the cartilage has no blood supply of its own, but instead relies on the vessels within the skin that covers it. When the earlobe is struck, the cartilage may split into several layers. Blood collects within the split cartilage from the surrounding damaged blood vessels. The resultant scaring causes the ear to lose its normal shape.

Bat ears are a common minor cosmetic abnormality. The extent to which the ears protrude is variable.

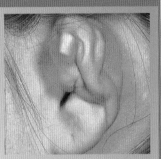

Repeated trauma to the pinna causes cauliflower ear. Scar tissue forms as a result of the damage, deforming the shape.

Inside the ear

The middle ear is an air-filled cavity that contains the eardrum and three small bones that help transmit sound to the inner ear. It is also connected to the throat via the auditory tube.

The middle ear is an air-filled, box-shaped cavity within the temporal bone of the skull. It contains small bones or ossicles – the malleus, incus and stapes – that span the space between the tympanic membrane (the eardrum) and the medial wall of the cavity.

Two small muscles are also present: the tensor tympani, attached to the handle of the malleus; and the stapedius, attached to the stapes. Both help to modulate the movements of the ossicles. The medial wall divides the middle ear from the inner ear and contains two membrane-covered openings; the oval and round windows.

AUDITORY TUBE

The middle ear is connected to the throat by the auditory (Eustachian) tube. This tube is a possible route of infection into the middle ear. If left untreated, infections can spread into the mastoid air cells that lie just behind the middle ear cavity, and may breach the roof of the temporal bone and infect the membranous covering of the brain (the meninges).

Just below the floor of the middle ear cavity is the bulb of the internal jugular vein, and just in front is the internal carotid artery.

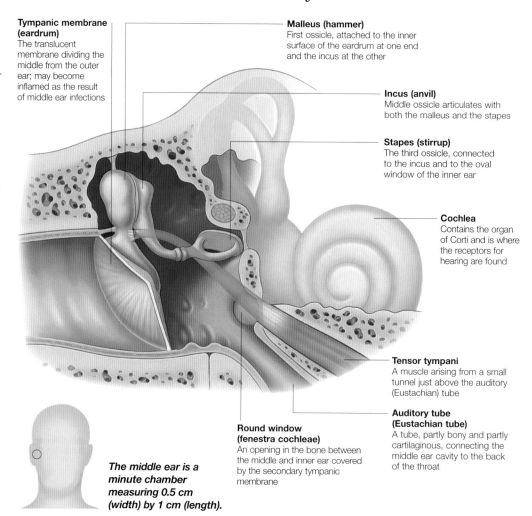

Tympanic membrane (eardrum)
The translucent membrane dividing the middle from the outer ear; may become inflamed as the result of middle ear infections

Malleus (hammer)
First ossicle, attached to the inner surface of the eardrum at one end and the incus at the other

Incus (anvil)
Middle ossicle articulates with both the malleus and the stapes

Stapes (stirrup)
The third ossicle, connected to the incus and to the oval window of the inner ear

Cochlea
Contains the organ of Corti and is where the receptors for hearing are found

Tensor tympani
A muscle arising from a small tunnel just above the auditory (Eustachian) tube

Auditory tube (Eustachian tube)
A tube, partly bony and partly cartilaginous, connecting the middle ear cavity to the back of the throat

Round window (fenestra cochleae)
An opening in the bone between the middle and inner ear covered by the secondary tympanic membrane

The middle ear is a minute chamber measuring 0.5 cm (width) by 1 cm (length).

The ossicles

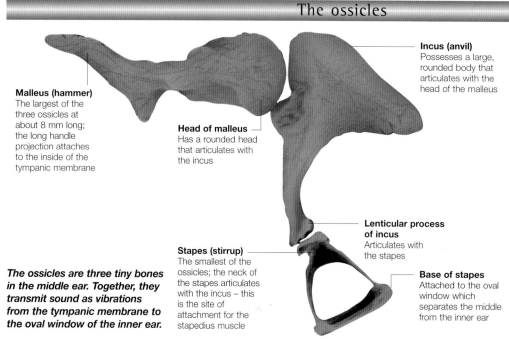

Malleus (hammer)
The largest of the three ossicles at about 8 mm long; the long handle projection attaches to the inside of the tympanic membrane

Head of malleus
Has a rounded head that articulates with the incus

Incus (anvil)
Possesses a large, rounded body that articulates with the head of the malleus

Lenticular process of incus
Articulates with the stapes

Stapes (stirrup)
The smallest of the ossicles; the neck of the stapes articulates with the incus – this is the site of attachment for the stapedius muscle

Base of stapes
Attached to the oval window which separates the middle from the inner ear

The ossicles are three tiny bones in the middle ear. Together, they transmit sound as vibrations from the tympanic membrane to the oval window of the inner ear.

The ossicles are arranged so that vibrations in the tympanic membrane are transmitted across the middle ear to the oval window and to the inner ear. All three bones are held in place by ligaments; in addition, there are two muscles that modulate movement.

Stapedius, the smallest skeletal muscle in the body, arises from a bony projection called the pyramid and attaches to the neck of the stapes. Contraction of this muscle helps to damp down loud sounds.

The other muscle, the tensor tympani, has a similar damping effect but it acts by increasing the tension in the tympanic membrane. People with damage to the facial nerve may suffer from hyperacusis, an abnormal sensitivity to sound.

The inner ear

This part of the ear contains the organs of balance and hearing. It contains the labyrinth that helps us to orientate ourselves, and the cochlea, the organ of hearing.

The inner ear, also known as the labyrinth because of its contorted shape, contains the organ of balance (the vestibule) and of hearing (the cochlea). It can be divided into an outer bony labyrinth and an inner membranous labyrinth. The bony labyrinth is filled with perilymph and the membranous labyrinth contains a fluid called endolymph, with a different chemical composition.

ORIENTATION

The membranous labyrinth contains the utricle and saccule – two linked, sac-like structures within the bony vestibule. They help detect orientation within the environment.

Related to these are the semicircular ducts lying within the bony semicircular canals. Where they are connected to the utricle, the semicircular canals enlarge to form ampullae, containing sensory receptors. Changes in the movement of the fluid in the ducts provides information about acceleration and deceleration of the head.

The cochlea is a bony spiral canal, wound around a central pillar – the modiolus. Within the cochlea are hair cells, the hearing receptors, that react to vibrations in the endolymph caused by the movement of the stapes on the oval window. They lie within the organ of Corti.

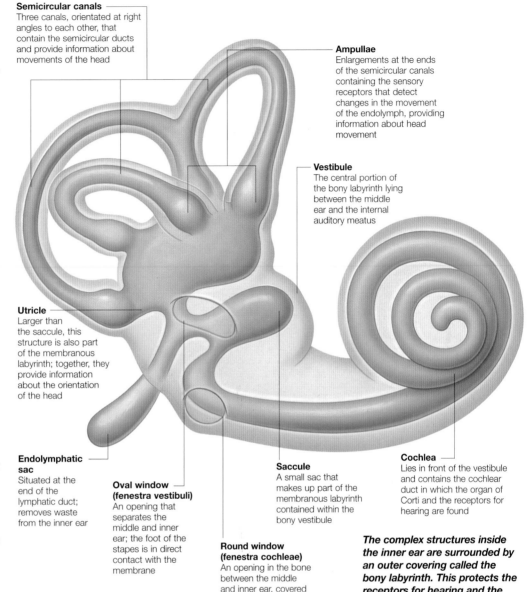

Semicircular canals
Three canals, orientated at right angles to each other, that contain the semicircular ducts and provide information about movements of the head

Ampullae
Enlargements at the ends of the semicircular canals containing the sensory receptors that detect changes in the movement of the endolymph, providing information about head movement

Vestibule
The central portion of the bony labyrinth lying between the middle ear and the internal auditory meatus

Utricle
Larger than the saccule, this structure is also part of the membranous labyrinth; together, they provide information about the orientation of the head

Endolymphatic sac
Situated at the end of the lymphatic duct; removes waste from the inner ear

Oval window (fenestra vestibuli)
An opening that separates the middle and inner ear; the foot of the stapes is in direct contact with the membrane

Saccule
A small sac that makes up part of the membranous labyrinth contained within the bony vestibule

Round window (fenestra cochleae)
An opening in the bone between the middle and inner ear, covered by the secondary tympanic membrane

Cochlea
Lies in front of the vestibule and contains the cochlear duct in which the organ of Corti and the receptors for hearing are found

The complex structures inside the inner ear are surrounded by an outer covering called the bony labyrinth. This protects the receptors for hearing and the delicate organs that detect movements of the head.

Hearing aids and cochlear implants

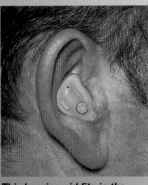

This hearing aid fits in the outer ear. It improves hearing for partially deaf patients by amplifying sounds and transmitting them into the ear.

There are two main forms of deafness. In conductive deafness, meaning a problem conducting sound, the problem is in the outer or middle ear. Sensorineural deafness is when the problem lies with the cochlea or the pathway between the cochlea and the brain.

Conductive hearing loss can often be improved surgically, but can also be helped by using electrical hearing aids. These consist essentially of an amplifier, a microphone, a receiver and a battery, and can either be worn behind the earlobe or, with miniaturization, within the ear itself.

In those patients with sensorineural loss, hearing aids need to be more complex. One approach is the cochlear implant or artificial ear. This requires a tiny microphone to be placed in the ear. This converts sound into electrical impulses which are then fed via electrodes to the cochlea, where they trigger nerve impulses in the vestibulocochlear nerve. This approach is used with patients in whom the hair cells of the organ of Corti have been destroyed. The hearing produced remains relatively crude but provides the patient with some perception of rhythm and intensity of sounds.

The transmitter of a cochlear implant is attached to the scalp, underneath which is a receiver. Electrodes run from the receiver into the cochlea.

How the ear controls our balance

The ear not only facilitates hearing, but is also responsible for maintaining balance when everyday tasks are performed, from climbing the stairs to rollerblading. The intricately designed structures of balance are in the inner ear.

Olympic skiers can maintain their balance at speeds in excess of 50 mph. This is possible because of the structures of the ear.

The ear is made up of three parts. The outer, visible part of the ear (the pinna and auditory canal) gathers and focuses the sound waves. In the middle ear, the eardrum vibrates, and the three ossicles (small bones) transmit these vibrations to the inner ear. The inner ear performs two functions: the cochlea receives the sound waves and helps to transmit them to the brain, where they are interpreted as sound, and the non-auditory or vestibular labyrinth detects changes in the body's position.

BONY LABYRINTH

The part of the inner ear concerned with balance is the bony labyrinth. Within this are the vestibule, the semicircular canals and the membranous labyrinth. The membranous labyrinth is surrounded by a fluid called perilymph. Another fluid, endolymph, is contained in the membranous labyrinth. These fluids do not just fill space; they are a vital part of the whole equilibrium system.

The individual parts of the bony labyrinth are sensitive to movement, rotation and orientation of the head.

Structure of the ear

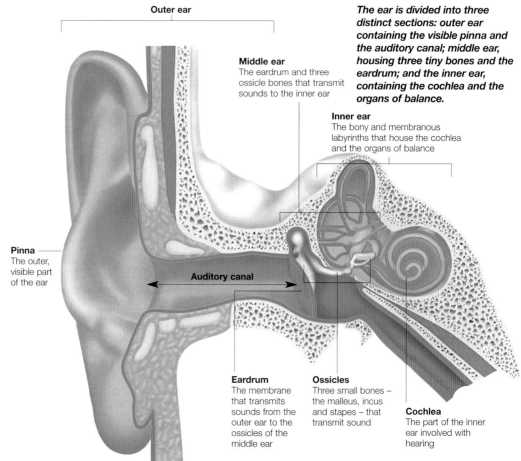

Outer ear

Pinna
The outer, visible part of the ear

Auditory canal

Middle ear
The eardrum and three ossicle bones that transmit sounds to the inner ear

The ear is divided into three distinct sections: outer ear containing the visible pinna and the auditory canal; middle ear, housing three tiny bones and the eardrum; and the inner ear, containing the cochlea and the organs of balance.

Inner ear
The bony and membranous labyrinths that house the cochlea and the organs of balance

Eardrum
The membrane that transmits sounds from the outer ear to the ossicles of the middle ear

Ossicles
Three small bones – the malleus, incus and stapes – that transmit sound

Cochlea
The part of the inner ear involved with hearing

An Olympic slalom skier travels at high speeds and at acute angles, but he is still aware of his body's position thanks to the in-built sense of equilibrium provided by the inner ear.

Loss of balance

When you are stationary, the fluid in the canals and chambers of the ear are in equilibrium. When the head is moved, the fluid moves in the opposite direction, and the brain senses the change in position. The size of this change is different in each ear (depending on which way you turn), but the system remains in equilibrium. However, if the vestibular system of one ear is damaged, the activity of the other ear causes a false sense of turning (vertigo) towards the unaffected side.

If the vestibular function of both ears is damaged, posture and gait can be seriously affected, causing vertigo and disorientation. If our environment changes, as it does when we fly or go to sea, the vestibular system may also react, resulting in air or sea sickness. A similar effect results from drinking too much alcohol.

Recently, space scientists have been studying the effects of weightlessness on the vestibular system. Some astronauts have had minor vestibular symptoms upon return. No disturbances have been permanent.

Parts of the ear that control balance

The tubes and chambers of the bony labyrinth protect the membranous tubes and chambers of the membranous labyrinth, with their fluids and sensors.

SEMICIRCULAR CANALS

The semicircular canals are three bony tubes in each ear that lie roughly at right-angles to each other. Because of their position and structure they are able to detect movement in three-dimensional space and are the parts sensitive to rotation.

Each canal has an expanded end called the ampulla, and is filled with endolymph. There are receptor cells located in the ampulla of each canal which have fine hairs that project up into the endolymph. When we move, these projecting hairs are displaced by the movement of the endolymph. This stimulates the vestibular nerve which sends signals to the cerebellum of the brain.

When we move, a reflex called nystagmus (the back and forth movement of the eyes), helps to prevent dizziness. The eyes move slowly against the direction of the rotation, which allows us to concentrate on a fixed point.

MEMBRANOUS LABYRINTH

The vestibule contains two membranous sacs called the utricle and the saccule. These are known as the otolith organs, and they respond to our orientation. On the inner surface of each sac is a 2mm-wide patch of sensory cells – a macula – which monitors the position of the head.

The utricle maculae lie horizontally and provide information when our heads move from side to side. Less is known about the saccular maculae, but because they are arranged vertically they probably respond to backward and forward tilting of the head. Together, they allow the detection of all the possible positions of the head.

The sensory organs (particularly in the utricle) play an important role in controlling the muscles of the legs, trunk and neck to keep the body and head in an upright position.

The inner ear

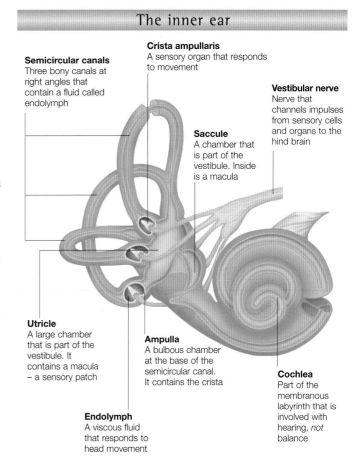

Semicircular canals
Three bony canals at right angles that contain a fluid called endolymph

Crista ampullaris
A sensory organ that responds to movement

Vestibular nerve
Nerve that channels impulses from sensory cells and organs to the hind brain

Saccule
A chamber that is part of the vestibule. Inside is a macula

Utricle
A large chamber that is part of the vestibule. It contains a macula – a sensory patch

Ampulla
A bulbous chamber at the base of the semicircular canal. It contains the crista

Cochlea
Part of the membranous labyrinth that is involved with hearing, *not* balance

Endolymph
A viscous fluid that responds to head movement

How the maculae work

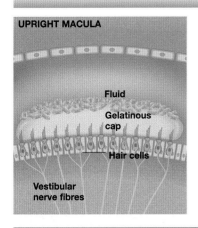

UPRIGHT MACULA

Fluid

Gelatinous cap

Hair cells

Vestibular nerve fibres

Each macula consists of a layer of tissue known as the neuroepitheliam. In this layer are sensory cells called hair cells that send continuous nerve impulses to the brain.

The hair cells are covered by a gelatinous cap that contains small granular particles that weigh against

The macula within the utricle is a horizontal gelatinous cap with tiny hairs embedded in it.

the hairs. When the hair bundles are deflected – because of a tilt of the head, for example – the hair cells are stimulated to alter the rate of nerve impulses being sent.

Hair cells near the centre are rounded, and those on the periphery are cylindrical. This may increase sensitivity to a slight tilting of the head.

When the head tilts, endolymph fluid and gravity pull the cap down, stimulating the hair cells.

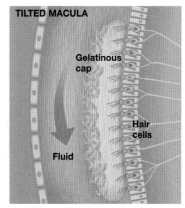

TILTED MACULA

Gelatinous cap

Hair cells

Fluid

What happens to the crista within the ampullae

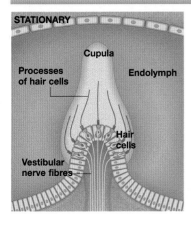

STATIONARY

Cupula

Processes of hair cells

Endolymph

Hair cells

Vestibular nerve fibres

The crista is a cone-shaped sensory structure within the ampulla – the swollen base of each semicircular canal. There are six cristae in each ear. Each crista is surrounded by a fluid called endolymph.

Each crista responds to changes in the rate of movement of the head,

Hair processes in the jelly-like cupula are connected to hair cells and nerve fibres. When the head is stationary, the cupula does not move.

passing information along the vestibular nerve to the brain.

Sensitive hair cells are embedded in a gelatinous cone called the cupula. Any kind of head movement causes fluid to swirl past the cupula, bending it and activating the hair cells.

As the head moves, endolymph fluid displaces the cupula, stimulating the hair processes. These send signals to the brain, which registers movement.

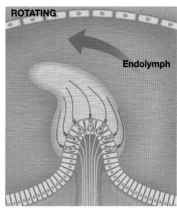

ROTATING

Endolymph

How the brain processes sound

Sound hitting the inner ear is converted into neuronal (nerve) signals. This is a complex and subtle process, which enables the brain to interpret and understand a wide range of sound.

The cochlea – the organ of hearing located in the inner ear – is a coiled bony structure containing a fluid-filled system of cavities.

The central cavity, or cochlear duct, contains the specific structure for hearing, called the spiral organ of Corti. Located on the basilar membrane, this spiral organ contains the thousands of sensory hair cells that convert mechanical movement (caused by sound vibrations resonating through the fluid) into electrical nerve impulses which are then transmitted to the brain.

PATHWAYS TO BRAIN

The neuronal pathways of the auditory system are composed of sequences of neurones arranged in series and parallel. The impulses begin in the organ of Corti and ultimately reach the auditory areas of the cerebral cortex known as the transverse temporal gyri of Heschl.

TRANSIT STATIONS

As neuronal activity is transmitted towards the brain it goes through several 'transit stations'. Certain of these transit stations respond in particular ways to various aspects of the auditory signal thus giving the brain more context to the sound. For example, some cochlear neurones have a sharp burst of activity at the start of a sound, called a primary-like response pattern; this informs the auditory cortex of the start of a sound sequence.

The neurones, transit stations and various brain auditory centres are found on both sides of the body. The auditory centres in the brain receive sound from the opposite ear.

Spiral organ of Corti
Contains vibration-sensitive hair cells that transmit signals via auditory nerve

This cross-section of the cochlea shows how vibrations are transmitted across membranous divisions between the chambers to the organ of Corti hair cells.

Pathway of signals to the brain

Auditory cortex
Area of temporal lobe of cerebral cortex that receives sound signal

Medial geniculate body
Transit station for impulses from auditory nerve

Auditory cortex

Inner ear

Cochlear nuclei in brainstem
Where neurones in the auditory nerve first synapse (meet)

Auditory nerve
Transmits signal from hair cells towards brain

Midbrain

Neurones
These neurones connect the cochlear nuclei to the medial geniculate bodies

Medulla

The nerve signals from the cochlear hair cells travel via the auditory nerve and spinal cord to the auditory cortex.

Interpreting pitch of sound

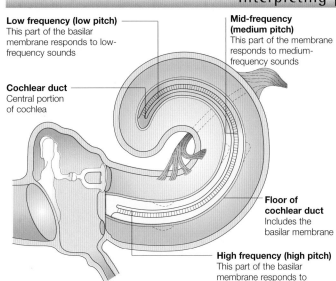

Low frequency (low pitch)
This part of the basilar membrane responds to low-frequency sounds

Mid-frequency (medium pitch)
This part of the membrane responds to medium-frequency sounds

Cochlear duct
Central portion of cochlea

Floor of cochlear duct
Includes the basilar membrane

High frequency (high pitch)
This part of the basilar membrane responds to high-frequency sounds

The stimulation of groups of hair cells at specific locations along the basilar membrane allows the brain to differentiate sounds of different frequencies or pitch.

The hair cells in the spiral organ of Corti are able to convey different tones by responding to different frequencies at different locations along the basilar membrane, thereby contributing to the sound-filtering process.

THE BASILAR MEMBRANE

Cells at the base of the basilar membrane respond more readily to high-frequency sound waves, while those at the tip are more sensitive to low-frequency sounds. This is equivalent to how a grand piano emits sounds, with one end producing high notes and the other low ones.

However, there are additional subtleties used to transduce the different tones.

Imagine a tuning fork that emits the note 'A' is struck. The sound waves reaching the cochlea will all resonate at a frequency of 440 cycles per second (Hertz). This triggers the basilar membrane to vibrate at 440 times a second. However, there is a particular section of the basilar membrane, which is constructed in such a way that it will vibrate with the largest amplitude at 440 times a second. This will then set the neurones from that region signalling at 440 times a second.

How the brain interprets sound signals

Once nerve impulses are transmitted to the auditory cortex, several areas of the brain are responsible for interpreting the signals.

There is still much to be learned about how the brain interprets sound and language. We know that there are several areas of the temporal lobe on both sides of the brain responsible for interpreting different aspects of sound. We also know that these areas receive a lot of additional contextual information from the various staging posts as the basic neuronal signals make their way to the auditory cortex.

IDENTIFYING SOUNDS

The brain identifies sounds by recognizing essential features of each sound – such as volume, pitch, duration, and intervals between sounds. From those elements, the brain creates a unique acoustic 'picture' of each sound, in much the same way that a colour television can reproduce the whole spectrum of colours on a screen using dots of just three colours.

The auditory cortex also has to separate many different sounds arriving at the same time, filtering and analysing them to produce meaningful information. Of course, the brain uses the context in which sound is received to make certain assumptions about what it it will hear. For instance if the visual cortex tells it that a young girl is speaking it will expect speech of a certain pitch to arrive.

AUDITORY ASSOCIATION CORTEX

The auditory association cortex is used to process complex sounds whereby many sound waves arrive at the same time. This is particularly important in language recognition, and damage to this area results in a person detecting sounds without being able to distinguish between them.

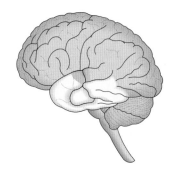

The auditory cortex (pink) recognizes and analyses sounds. The association cortex (yellow) acts to distinguish more complex features of the sounds.

The visual cortex influences the context in which sound is interpreted. When using a phone we have no visual clues and rely solely on auditory input.

Locating sound

Listening to sound from behind

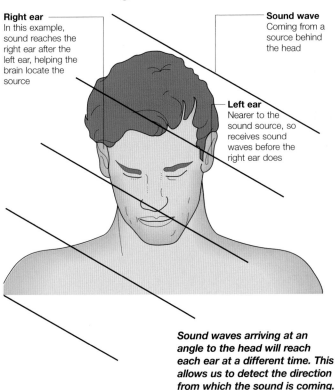

Right ear
In this example, sound reaches the right ear after the left ear, helping the brain locate the source

Sound wave
Coming from a source behind the head

Left ear
Nearer to the sound source, so receives sound waves before the right ear does

Sound waves arriving at an angle to the head will reach each ear at a different time. This allows us to detect the direction from which the sound is coming.

The brain is very accurate at integrating information to locate sound.

The two main ways it does this are by picking up the minute difference in timing and intensity of the sound reaching the two ears.

A sound wave will reach the ear that is closer to the source of the sound a fraction of a second before it reaches the other ear. The brain can interpret the time difference to distinguish the direction of the sound.

In addition to this, if sound is coming from the side, the head causes a 'sound shadow' screening one ear, such that it receives less sound than the other ear. Often, our response is to turn our head in the direction of the sound, evening up the sound to both ears.

However, even with only one ear we would still be able to locate a sound source. This is because small details of the sound we hear, caused by waves deflecting off the irregular surface of the pinna, vary with the angle at which the sound approaches the ear. As we develop, we learn that particular sound differences are associated with particular directions, and from this can detect the direction of a sound source.

From an early age, babies learn to recognize typical details of sounds coming from different directions. They can then use the information as a reference.

Chapter Three

The Neck

The adult neck is an anatomically complex part of the body, densely populated with vital structures. The seven bony sections of the spinal column located here are known as the cervical vertebrae. Each is cushioned between cartilaginous disks that offer a large degree of flexibility, enabling the head to rotate and move up and down.

Although a relatively small section of the skeleton, the neck plays a vital role in supporting the head's weight and protecting nerve pathways travelling between the brain and the body. Furthermore, it houses one of the largest endocrine glands, the thyroid gland, which plays a crucial role in the healthy functioning of the body's metabolism. This chapter examines the structure and function of each of the neck's component parts, highlighting the importance of this intricate part of the body.

LEFT: The neck and shoulder contain a complex mass of muscles and nerves, and are very susceptible to strain and damage.

Inside the neck

The neck is one of the most anatomically complex areas of the body. Many vital structures, including the spinal cord and thyroid gland, are closely packed together within layers of connecting tissue and muscle.

The neck is defined as the region lying between the bottom of the lower jaw and the top of the clavicle (collar bone). Within this relatively small area there are numerous vital structures that are closely packed together between layers of connective tissue. The outermost layer of the neck is the skin. The skin of the neck contains sensory nerve endings from the second, third and fourth cervical nerves. A number of natural stress lines can be seen on the skin that run around the neck horizontally. When the skin is cut during surgical procedures, the incisions are made along, rather than across, these stress lines in order to minimize scarring.

EXTERNAL JUGULAR VEIN

Just beneath the skin is the thin layer of subcutaneous fat and connective tissue called the superficial fascia. Embedded in this layer are blood vessels, such as the external jugular vein and its tributaries. These veins drain blood from the face, scalp and neck. Closely associated with the external jugular vein are the superficial lymph nodes.

One other important structure that can be found in this layer, at the front of the neck, is the very thin platysma muscle that helps to depress the lower jaw.

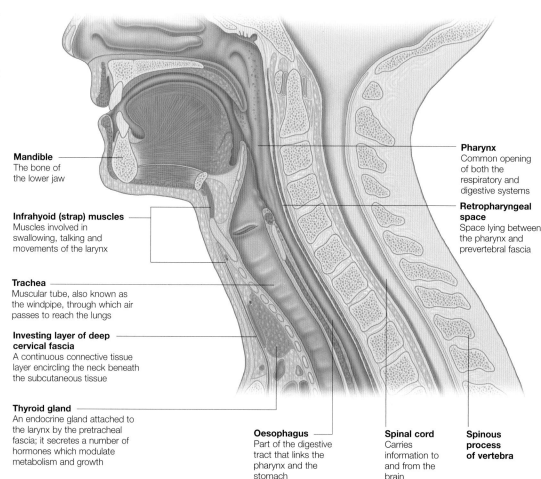

Mandible
The bone of the lower jaw

Infrahyoid (strap) muscles
Muscles involved in swallowing, talking and movements of the larynx

Trachea
Muscular tube, also known as the windpipe, through which air passes to reach the lungs

Investing layer of deep cervical fascia
A continuous connective tissue layer encircling the neck beneath the subcutaneous tissue

Thyroid gland
An endocrine gland attached to the larynx by the pretracheal fascia; it secretes a number of hormones which modulate metabolism and growth

Oesophagus
Part of the digestive tract that links the pharynx and the stomach

Spinal cord
Carries information to and from the brain

Spinous process of vertebra

Pharynx
Common opening of both the respiratory and digestive systems

Retropharyngeal space
Space lying between the pharynx and prevertebral fascia

Lymph nodes

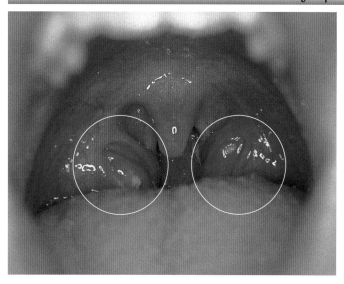

The tonsils (circled) consist of lymphoid tissue, and are concerned with fighting infection. They are vulnerable to infection and inflammation.

Within the neck are lymph nodes, which are essential for protecting the body from infection. Lymph nodes are embedded in connective tissue in various parts of the body, particularly the regions of the groin, the armpits and the neck. They are also present in lymph organs, such as the spleen and the tonsils.

The function of lymph nodes is to filter lymph fluid, which circulates specialized white blood cells (lymphocytes) throughout the body. Lymphocytes, located in the lymph nodes, are also essential to the body's defences, as they produce antibodies which play an important part in fighting infections.

Disorders of the lymphatic system are considered serious, and can include:
- Lymphoedema – results from a blockage of lymphatic drainage, causing gross swelling. May be caused by some parasites, damage to the lymphatic system itself or hereditary abnormalities, such as Milroy's disease
- Lymphangitis – acute inflammation of the lymphatic vessels caused by infection with streptococcal bacteria.

Cross-section of the neck

Deeper layers of the neck reveal interconnected sheets of tissue. These bind to and protect a variety of structures.

Moving deeper into the neck, the connective tissue of the deep cervical fascia is arranged into a number of fibrous sheets. These fasciae surround different groups of muscles, blood vessels and nerves, allowing them to move relative to each other with minimal friction.

The first of these is the investing fascia. This encircles the neck and is anchored to the spinous processes of the cervical vertebrae. It encloses the large sternocleidomastoid muscle at the front and side of the neck, and the trapezius muscle at the back, both of which are important in the movements of the head and neck.

LARYNX AND TRACHEA

The thin pretracheal fascia binds the thyroid gland to the larynx and trachea at the front of the neck. It is anchored to the cricoid cartilage, allowing movement during swallowing.

The pretracheal fascia is continuous with a sheet of tissue (the carotid sheath) which provides protection for the carotid artery, the internal jugular vein and the vagus nerve. Behind the trachea is the oesophagus (gullet), and behind the larynx is the pharynx, the muscular tube connecting the mouth and the oesophagus.

The last deep connective tissue layer is the prevertebral fascia, enclosing the remaining muscles of the neck, the vertebral column and the spinal cord, positioned in the centre of the neck for maximum protection.

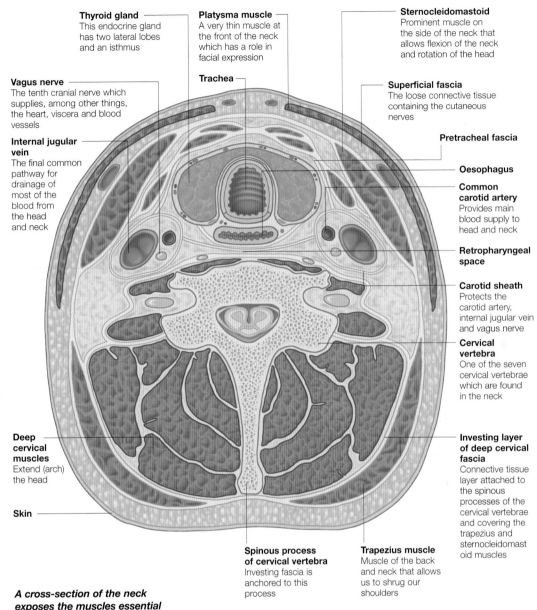

Thyroid gland — This endocrine gland has two lateral lobes and an isthmus

Platysma muscle — A very thin muscle at the front of the neck which has a role in facial expression

Sternocleidomastoid — Prominent muscle on the side of the neck that allows flexion of the neck and rotation of the head

Vagus nerve — The tenth cranial nerve which supplies, among other things, the heart, viscera and blood vessels

Trachea

Superficial fascia — The loose connective tissue containing the cutaneous nerves

Internal jugular vein — The final common pathway for drainage of most of the blood from the head and neck

Pretracheal fascia

Oesophagus

Common carotid artery — Provides main blood supply to head and neck

Retropharyngeal space

Carotid sheath — Protects the carotid artery, internal jugular vein and vagus nerve

Cervical vertebra — One of the seven cervical vertebrae which are found in the neck

Deep cervical muscles — Extend (arch) the head

Skin

Investing layer of deep cervical fascia — Connective tissue layer attached to the spinous processes of the cervical vertebrae and covering the trapezius and sternocleidomastoid muscles

Spinous process of cervical vertebra — Investing fascia is anchored to this process

Trapezius muscle — Muscle of the back and neck that allows us to shrug our shoulders

A cross-section of the neck exposes the muscles essential for movement and facial expression. Among these are connecting tissues and vessels.

Retropharyngeal space

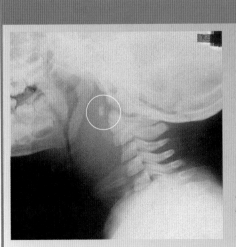

Between the different fascial layers there are areas of much looser connective tissue and spaces, such as the retropharyngeal space. This space lies between the back of the pharynx and the prevertebral fascia and extends from the base of the skull to the chest cavity.

These spaces are clinically important in terms of the spread of infection. An infection of the mouth or teeth may enter the retropharyngeal space and, due to the

The retropharyngeal space can be a site of infection. This X-ray shows an abscess in the retropharyngeal space (circled). This may have been caused by infection spreading from the tonsils.

sheet–like structure of the fascia, the infection may, rarely, spread into the chest.

Retropharyngeal abscesses arise due to spread of infection from areas such as the mouth and tonsils and are most common in children. They can also occur as the result of damage to the posterior (rear) wall of the pharynx that may be caused by objects such as lollipop-sticks.

Patients tend to be feverish and complain of pain when swallowing. Severe swelling can interfere with breathing. The abscess can be identified on a soft tissue X-ray of the side of the neck. Treatment involves making sure the patient has a clear airway, draining the abscess surgically and treating with antibiotics.

Vertebral column

The vertebral column gives our bodies flexibility and keeps us upright. It also protects the delicate spinal cord.

The vertebral column forms the part of the skeleton commonly known as the backbone or spine. The spine supports the skull and gives attachment to the pelvic girdle, supporting the lower limbs. As well as its obvious role in posture and locomotion, the vertebral column surrounds and protects the spinal cord. Like all bones, its marrow is a source of blood cells, and it acts as a reservoir for calcium ions.

The spine exhibits four curvatures when viewed from the side. The cervical and lumbar curvatures are convex anteriorly (forward). The thoracic and sacral curvatures are convex posteriorly (backwards). The cervical curvature develops in infancy as the baby learns to hold its head upright; similarly, the lumbar curvature forms as the baby learns to walk.

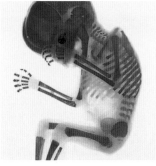

In the fetus, the backbone has a single curvature in the thoracic and sacral regions. Other curvatures develop as the baby sits, stands and walks.

Frontal aspect

Lateral (side) view

Atlas
First cervical vertebra; named after the Greek mythological figure Atlas, who held up the heavens in the same way that this vertebra holds up and articulates with the skull

Axis
Second cervical vertebra; assists in the rotation of the head from side to side

Vertebral body
Disc-shaped unit, common to all vertebrae, that bears the weight of the spinal column; also known as the centrum

Spinous process
Single fin-like posterior protrusion extending from each vertebra; provides attachment point for muscles and ligaments

Transverse processes
Paired 'wings' extending laterally from each side of the vertebrae; provide attachment points for muscles

Fused vertebrae
The sacrum and coccyx are formed from five and four fused bones respectively

Intervertebral discs

Between the bodies of individual vertebrae are the intervertebral discs. Each disc is made of connective tissue with a soft central component (the jelly-like nucleus pulposus), and a tougher surrounding tissue known as the annulus fibrosus.

The intervertebral discs account for about 25 per cent of the total length of the vertebral column. The discs are elastic, and are compressed during the course of the day, meaning that we go to bed a few centimetres shorter than when we wake. The discs also allow for a wide range of different movements, and as the spinal column's natural shock absorbers, protect the vertebrae from excessive pressure.

A disc may be displaced (prolapse) backwards, irritating the spinal nerves and giving rise to the pain associated with prolapsed (also known as 'slipped') discs. Rest is often sufficient to relieve pain, but sometimes surgery is required.

With age, the discs become progressively thinner, partly accounting for the loss of height in older people.

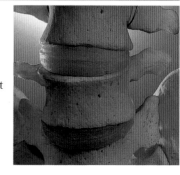

The cushion-like nature of the intervertebral discs is evident in this photograph. They are vital in ensuring the spine's flexibility.

Vertebral connections

The spinal column is divided into five main sections. Each section has a specific function and together they maintain the stability of the skeleton as a whole.

The vertebral column consists of 33 bones, known as the vertebrae. There are seven cervical vertebrae, twelve thoracic, five lumbar, five sacral and four coccygeal. Whereas the cervical, thoracic and lumbar vertebrae are separate bones, the sacral and coccygeal vertebrae are fused (inflexible). The vertebrae show differences along the column, and these will be featured on other sheets.

STRUCTURE

Each vertebra conforms to the same basic plan, consisting of a body in front and a neural arch at the back, which surrounds and protects the spinal cord. From the neural arch arise the transverse processes and spines, which give attachment to muscles and ligaments. Adjacent vertebrae articulate at joints, allowing movement. Movements between adjacent vertebrae are relatively small but, when taken over the whole length of the vertebral column, give the trunk considerable mobility.

The nerves leaving and entering the spinal cord do so through gaps, the intervertebral foramina, between adjacent vertebrae.

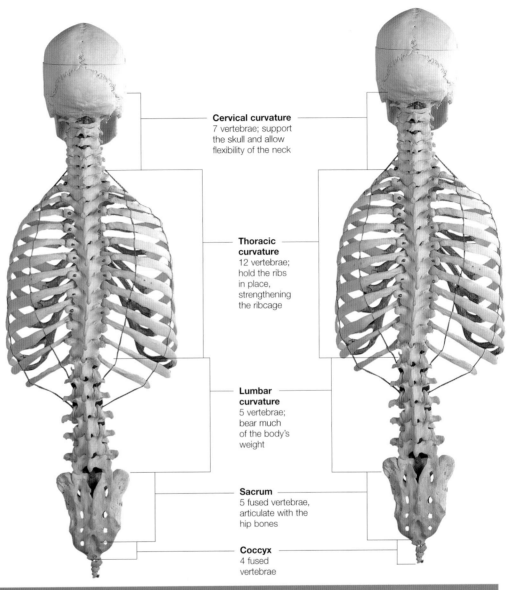

Cervical curvature
7 vertebrae; support the skull and allow flexibility of the neck

Thoracic curvature
12 vertebrae; hold the ribs in place, strengthening the ribcage

Lumbar curvature
5 vertebrae; bear much of the body's weight

Sacrum
5 fused vertebrae, articulate with the hip bones

Coccyx
4 fused vertebrae

Disorders of the spinal column

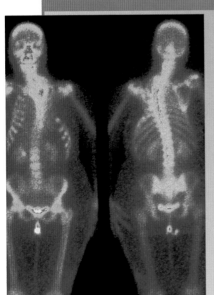

The vertebrae and their joints are affected by conditions common to other bones, such as osteoporosis, osteoarthritis and rheumatoid arthritis. The vertebrae are also commonly affected by fractures, which are considered extremely serious because of the presence of the spinal cord.

Spina bifida (meaning literally 'split spine') is a congenital condition in which the vertebral arches are incomplete and fail to unite at the midline. This causes the spinal cord and its covering to protrude through the vertebrae.

Scoliosis is a congenital abnormal curvature of the thoracic spine. Treatment involves surgery or spinal immobilization with a brace.

The condition results from a defect in fetal development, and is usually apparent on an ultrasound scan after week 16, but it may be unrecognized in mild cases, usually affecting the sacral and lumbar regions. The whole of the spinal column (and skull) may be affected in the most

severe cases. There may also be associated abnormalities of the brain, such as hydrocephalus.

In spina bifida the spinal cord and surrounding meninges project through the skin. Paralysis of the lower limbs is a common symptom.

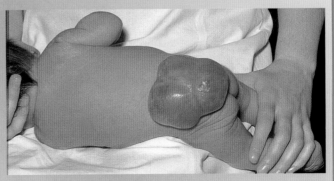

Cervical vertebrae

There are seven cervical vertebrae, which together make up the skeletal structure of the neck. These vertebrae protect the spinal cord, support the skull and allow a range of movement.

Of the seven cervical vertebrae, the lower five appear similar, although the seventh has some distinctive features. The first cervical vertebra (atlas) and the second cervical vertebra (axis) show specializations related to the articulation of the vertebral column with the skull.

TYPICAL CERVICAL VERTEBRA

The third to the sixth cervical vertebrae are comprised of two main components, a body towards the front and a vertebral arch at the rear. These surround the vertebral foramen (hole) which, as part of the vertebral column, forms the vertebral canal. The body is small compared with vertebrae in other regions, and is nearly cylindrical.

The vertebral arch can be subdivided in to two main elements. The pedicles, by which it is attached to the body, contain notches that allow the passage of spinal nerves. The laminae are thin plates of bone which are directed backwards and fuse in the midline, forming a bifid (divided) spine.

Associated with each vertebral arch are a pair of transverse processes. These are sites for muscle attachment, allowing movement and containing foramina through which blood vessels run.

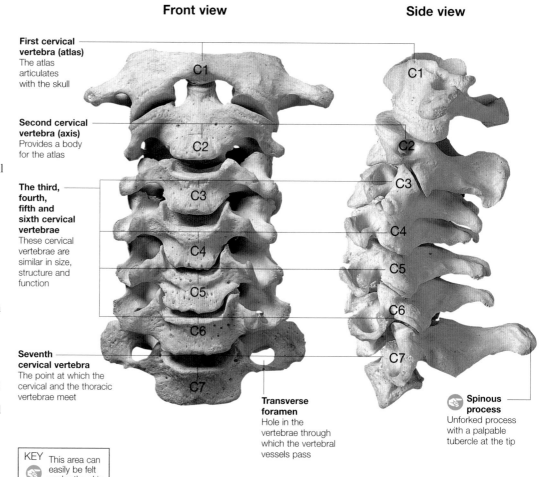

Front view

Side view

First cervical vertebra (atlas)
The atlas articulates with the skull

Second cervical vertebra (axis)
Provides a body for the atlas

The third, fourth, fifth and sixth cervical vertebrae
These cervical vertebrae are similar in size, structure and function

Seventh cervical vertebra
The point at which the cervical and the thoracic vertebrae meet

C1
C2
C3
C4
O5
C6
C7

C1
C2
C3
C4
C5
C6
C7

Transverse foramen
Hole in the vertebrae through which the vertebral vessels pass

Spinous process
Unforked process with a palpable tubercle at the tip

KEY This area can easily be felt under the skin

Damage to the cervical vertebrae

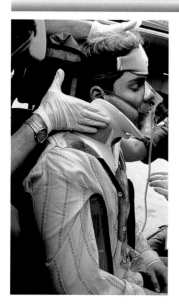

Dislocations of the cervical vertebrae are very serious, as any instability in this region may potentially lead to transection (severing) of the soft spinal cord by the sharp surface of the cervical vertebrae.

The muscles of the diaphragm, essential for breathing, are supplied chiefly by the fourth cervical spinal nerves, and transection above this level is fatal. Thus, where the site of the transection is concerned, distances of a few millimetres may mean the difference between life and death. For this

After trauma, movement of the neck can be limited with a brace or collar. This reduces the risk of damage to the spinal cord.

reason, it is essential at the scene of a traumatic accident that if injury to the cervical vertebrae cannot be excluded, the patient should not be moved until proper equipment and expertise is at hand.

The base of the dens (a toothlike projection of the axis which articulates with the atlas) may be fractured following severe head injuries and this must be checked using X-rays, as displacement of a fragment of bone may not occur immediately. If displacement does occur, it is usually fatal. Other dislocations may occur where the neck is most mobile – between the fourth and fifth or fifth and sixth vertebrae.

The synovial joints associated

with the cervical vertebrae, like similar joints elsewhere in the body, are susceptible to diseases such as rheumatoid arthritis. This is an auto-immune disease causing inflammation of the surface and lining of joints and leads to erosion of bones and deformities of the joint.

This may be extremely dangerous if the joint between the dens of the axis and the arch of the atlas is affected, as this can result in subluxation (dislocation of the joint such that the bones are still in contact but not aligned) of the dens. The dens may then compress the underlying medulla oblongata of the hindbrain, which contains vital brain centres, with disastrous consequences.

Examining the cervical vertebrae

The first, second and seventh cervical vertebrae differ structurally from the others, in relation to their unique functions.

FIRST CERVICAL VERTEBRA

The first cervical vertebra, the atlas, is the vertebra that articulates with the skull. Unlike the other vertebrae, it does not have a body, this being incorporated into the second cervical vertebra as the dens. It also has no spine. Instead, the atlas takes the form of a thin ring of bone with anterior and posterior arches, the surface of which show grooves related to the vertebral arteries before they enter the skull through the foramen magnum.

SECOND CERVICAL VERTEBRA

The second cervical vertebra, the axis, can be distinguished from other cervical vertebrae by the presence of a tooth-like process called the dens (odontoid process). The dens articulates with the facet on the bottom surface of the anterior (front) arch of the atlas. Rotation of the head occurs at this joint.

The body of the axis resembles the bodies of the other cervical vertebrae.

SEVENTH CERVICAL VERTEBRA

This vertebra has the largest spine of any cervical vertebra and, as the first to be readily palpated, it has been termed the vertebra prominens.

The transverse processes are also larger than those of the other cervical vertebrae and the oval foramen transversarium transmits an accessory vertebral vein.

First cervical vertebra (atlas)

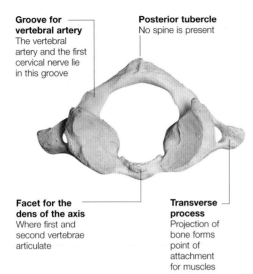

Groove for vertebral artery
The vertebral artery and the first cervical nerve lie in this groove

Posterior tubercle
No spine is present

Facet for the dens of the axis
Where first and second vertebrae articulate

Transverse process
Projection of bone forms point of attachment for muscles

Second cervical vertebra (axis)

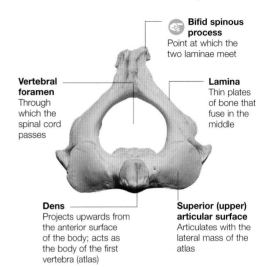

Bifid spinous process
Point at which the two laminae meet

Vertebral foramen
Through which the spinal cord passes

Lamina
Thin plates of bone that fuse in the middle

Dens
Projects upwards from the anterior surface of the body; acts as the body of the first vertebra (atlas)

Superior (upper) articular surface
Articulates with the lateral mass of the atlas

Fifth (typical) cervical vertebra

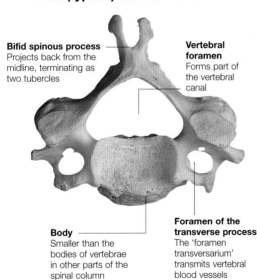

Bifid spinous process
Projects back from the midline, terminating as two tubercles

Vertebral foramen
Forms part of the vertebral canal

Body
Smaller than the bodies of vertebrae in other parts of the spinal column

Foramen of the transverse process
The 'foramen transversarium' transmits vertebral blood vessels

Seventh cervical vertebra

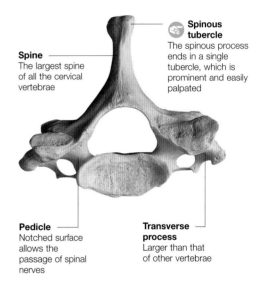

Spine
The largest spine of all the cervical vertebrae

Spinous tubercle
The spinous process ends in a single tubercle, which is prominent and easily palpated

Pedicle
Notched surface allows the passage of spinal nerves

Transverse process
Larger than that of other vertebrae

Abnormalities in the neck

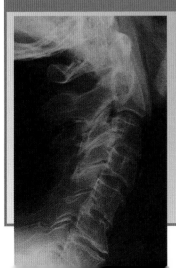

Bony outgrowths, or osteophytes, associated with the vertebra in the region of the spinal nerve pathway may compress the nerves. This may cause symptoms such as pain and numbness in the region supplied by that particular nerve. As the lower cervical spinal nerves contribute to the brachial plexus, this may give rise to symptoms experienced in the arm.

Osteophytes usually occur as a result of osteoarthritis. They cause pain and restricted movement of the neck.

In about 0.5 per cent of people, a small cervical rib may be found associated with the transverse process of (usually) the seventh cervical vertebra. Although usually symptomless, it can sometimes interfere with the blood flow in the subclavian artery. This artery supplies the arm and the condition may give rise to pain as a result of inadequate blood supply.

Cervical ribs occur when the transverse process of a cervical vertebra continues to ossify towards the thoracic rib.

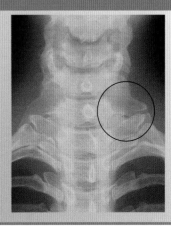

Cervical ligaments

The cervical ligaments play an essential role in securing and stabilizing the vertebrae of the neck. They consist of elastic fibres which enable smooth flexion of the head and neck.

The ligaments of the neck have a vital role in binding the cervical vertebrae strongly together while still allowing the neck a wide range of movements. Some of the cervical ligaments are the continuation of ligaments found further down the spine but other, more specialized ligaments are needed at the point where the spine articulates with the skull.

LIGAMENT FUNCTION

The first two cervical vertebrae, the atlas and the axis, are adapted to the role of supporting and allowing movement of the skull on the spine. The axis, below the atlas, has a spike of bone – the dens – projecting up into the front of the vertebral canal. If the dens moves within the vertebral canal, it could damage the nervous tissue and result in death.

A group of ligaments bind the dens to the atlas and the skull:
■ The cruciform ligament – this cross-shaped ligament has a very thick central band passing from one side of the atlas to the other, holding the dens forward and away from the spinal cord; an upper and a lower band attach the axis to the skull
■ The apical ligament – a small ligament that passes from the tip of the dens to the skull
■ The alar ligaments – two strong bands that pass up and outward from the dens to the skull.

Posterior view of cervical ligaments

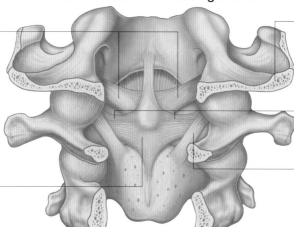

Alar ligaments
Two strong ligaments which pass from the dens of the first cervical vertebra to the base of the skull

Atlas (C1)
First cervical vertebra; Dens of the axis fits into the vertebral canal running through the atlas

Cruciform ligament
Strong cross-shaped ligament – secures the position of the dens, keeping it in front of the spinal cord; also attaches the axis to the skull

Occipital bone (cut)
Bone at base of the skull; alar and cruciform ligaments attach to this bone

Transverse ligament of atlas
Part of cruciform ligament

Tectorial membrane
Accessory part of tectorial membrane

Axis (C2)
Second cervical vertebra; the dens projects upwards into the vertebral canal

Posterior view of dens and deep cervical ligaments

Apical ligament of dens
Thin ligament that attaches the dens to the skull

Dens
Projection of bone from the axis into the spinal canal above; this articular surface is the point of attachment for the transverse ligament of atlas

Alar ligament
Attaches dens to the skull; sits under the cruciform ligament

Atlas (C1)
First cervical vertebra

Axis (C2)
Second cervical vertebra

Together, the cervical ligaments restrict the movement of the dens, guarding particularly against backward movement into the vertebral canal.

Rupture of the cruciform ligament

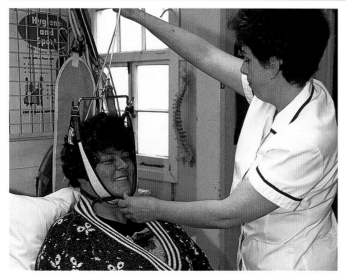

Rarely, the cruciform ligament may be ruptured traumatically or weakened by disease. Traumatic rupture may be caused by an injury which involves sudden flexion (bending forwards) of the neck – as might happen in a head-on collision in a car accident. It has been estimated that a force of about 85 kg is needed to rupture the cruciform ligament. The cruciform ligament is also believed to rupture during hanging.

Neck traction is used to hold the joints of the neck in the correct position following injury. The traction immobilizes and aligns the damaged joints.

Weakening of the cruciform ligament may occur if a patient suffers from a general connective tissue disease or if there is inflammation due to infection.

When the cruciform ligament is ruptured or weakened by disease, the joint between the atlas and the axis may become partially or completely dislocated, enabling the dens to move backwards into the vertebral canal. If this happens, the dens may then cause serious damage to the spinal cord, resulting in paralysis, or damage to the medulla (the lowest part of the brain, concerned with vital functions such as breathing), causing death.

Longitudinal ligaments

The vertebrae of the spine are attached by ligaments running longitudinally. These ligaments support the spine, and its elastic fibres enable a range of movement.

The vertebrae of the spine are linked and supported by two ligaments which pass from skull to sacrum, the anterior and the posterior longitudinal ligaments.

LONGITUDINAL LIGAMENTS

The anterior longitudinal ligament is a broad, strong, fibrous band attached to the front of the intervertebral discs and to the vertebrae. It connects the vertebrae and attaches to the skull as the anterior atlanto-occipital membrane.

The posterior longitudinal ligament is narrower and weaker and connects the backs of the vertebral bodies up to the atlas where it becomes the tectorial membrane attached to the skull.

There are many other short ligaments joining each vertebra to the one above and below. The ligamenta flava, for example, contains elastic tissue and helps preserve the curvatures of the spine and assists with straightening the spine after bending forwards. The ligamenta flava attaches to the skull as the posterior atlanto-occipital membrane.

THE NUCHAL LIGAMENT

The nuchal ligament is an extremely strong, thick band lying in the midline of the back of the neck. It attaches to the spines of the cervical vertebrae and is used as an attachment for muscles such as the trapezius. It is also elastic and helps support the weight of the head; it also stretches while the head and neck is flexed.

Right lateral view of cervical ligaments

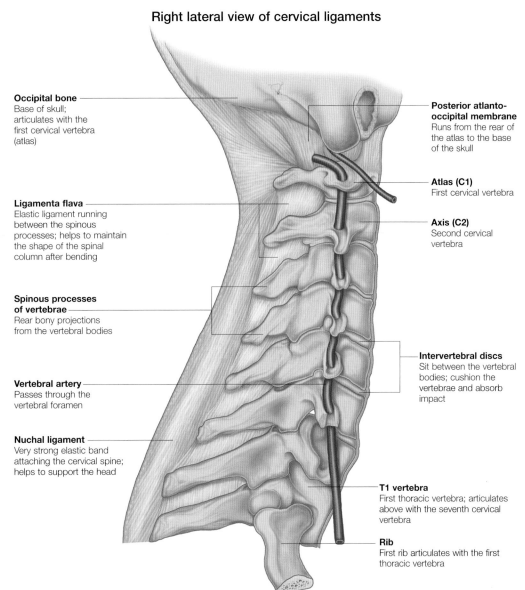

Occipital bone
Base of skull; articulates with the first cervical vertebra (atlas)

Ligamenta flava
Elastic ligament running between the spinous processes; helps to maintain the shape of the spinal column after bending

Spinous processes of vertebrae
Rear bony projections from the vertebral bodies

Vertebral artery
Passes through the vertebral foramen

Nuchal ligament
Very strong elastic band attaching the cervical spine; helps to support the head

Posterior atlanto-occipital membrane
Runs from the rear of the atlas to the base of the skull

Atlas (C1)
First cervical vertebra

Axis (C2)
Second cervical vertebra

Intervertebral discs
Sit between the vertebral bodies; cushion the vertebrae and absorb impact

T1 vertebra
First thoracic vertebra; articulates above with the seventh cervical vertebra

Rib
First rib articulates with the first thoracic vertebra

Structure of the ligaments

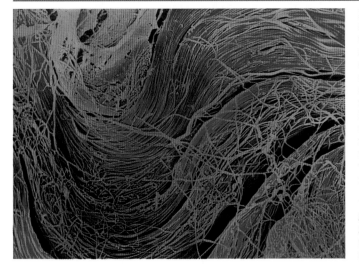

This coloured micrograph shows bundles of wavy collagen fibres. Elastic fibres and collagen within ligaments enable the tissue to stretch under tension.

Ligaments bind bones together at joints and are found throughout the body. They consist of dense, regular connective tissue which contains closely packed bundles of collagen and elastic fibres together with fibroblasts (specialized cells which produce this tissue). The fibres all run in the same direction and allow this strong, flexible tissue to withstand considerable pulling forces in that direction.

Being 'wavy', collagen fibres have a certain amount of give in them which, together with the elastic fibres, allows the ligament to stretch a little under tension and then return to its previous length. This helps a joint to retain its original shape and stability before, during and after movement.

Some ligaments, such as the ligamenta nuchae in the back of the neck, have an unusually high number of elastic fibres which enables them to stretch further than other ligaments without tearing. This allows a greater degree of movement between the bones to which they are attached.

Muscles of the neck

The muscles running up the front of the neck are divided into the suprahyoid and infrahyoid muscle groups. They attach to the hyoid bone and act to raise and lower it and the larynx during swallowing.

Two groups of muscles run longitudinally within the front of the neck from the mandible (jaw) to the sternum (breastbone). These muscles are concerned with movements of the jaw, hyoid bone and larynx and are of particular importance during swallowing. The hyoid bone divides these two groups into the suprahyoid (above) and the infrahyoid (below).

THE SUPRAHYOID MUSCLES

This is a group of paired muscles, lying between the jaw and the hyoid bone. The digastric muscle has two spindle-shaped bellies connected by a tendon in the middle. The anterior belly is attached to the mandible near the mid-line, while the posterior belly arises from the base of the skull. The connecting tendon slides freely through a fibrous 'sling' which is attached to the hyoid bone.

The stylohyoid is a small muscle which passes from the styloid process, a bony projection of the base of the skull, forward and downward to the hyoid bone. Arising from the back of the mandible, the mylohyoid muscles of each side unite in the mid-line to form the floor of the mouth. Posteriorly, they attach to the hyoid bone.

The geniohyoid is a narrow muscle that runs along the floor of the mouth from the back of the mandible in the mid-line to the hyoid bone below.

Infrahyoid and suprahyoid muscles

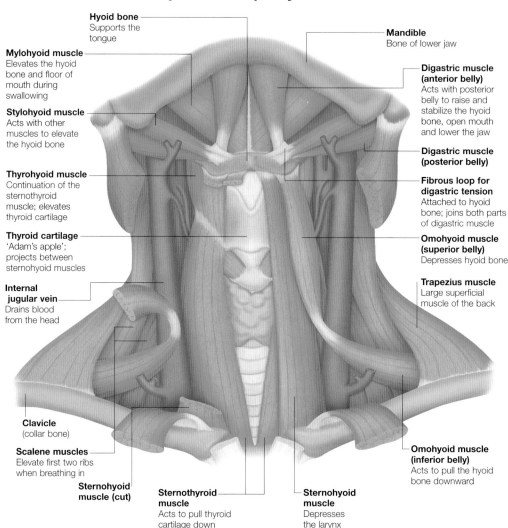

Hyoid bone
Supports the tongue

Mylohyoid muscle
Elevates the hyoid bone and floor of mouth during swallowing

Stylohyoid muscle
Acts with other muscles to elevate the hyoid bone

Thyrohyoid muscle
Continuation of the sternothyroid muscle; elevates thyroid cartilage

Thyroid cartilage
'Adam's apple'; projects between sternohyoid muscles

Internal jugular vein
Drains blood from the head

Clavicle
(collar bone)

Scalene muscles
Elevate first two ribs when breathing in

Sternohyoid muscle (cut)

Sternothyroid muscle
Acts to pull thyroid cartilage down

Sternohyoid muscle
Depresses the larynx

Mandible
Bone of lower jaw

Digastric muscle (anterior belly)
Acts with posterior belly to raise and stabilize the hyoid bone, open mouth and lower the jaw

Digastric muscle (posterior belly)

Fibrous loop for digastric tension
Attached to hyoid bone; joins both parts of digastric muscle

Omohyoid muscle (superior belly)
Depresses hyoid bone

Trapezius muscle
Large superficial muscle of the back

Omohyoid muscle (inferior belly)
Acts to pull the hyoid bone downward

Infrahyoid muscles and platysma

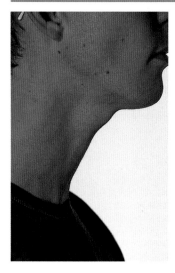

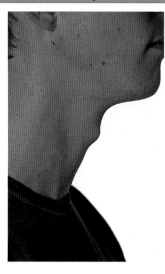

The infrahyoid group of muscles lies between the hyoid bone and the sternum, and comprises the sternohyoid and omohyoid, lying next to each other in the same plane and, more deeply, the thyrohyoid and sternothyroid muscles. Their flat shape leads to the common name of 'strap muscles'. After swallowing, the infrahyoid muscles act to return the hyoid bone and larynx to their previous positions.

The larynx and the hyoid bone are raised during swallowing by the suprahyoid muscles (far left). The infrahyoid muscles then return these structures to their original positions (left).

The platysma is a thin, flat sheet of muscle which lies just under the skin of the front of the neck in the subcutaneous connective tissue layer. It extends from the layer of deep fascia overlying the muscles of the upper chest to the mandible, with some fibres passing up to the corners of the mouth. At the lower end, the two sides are separate but they gradually converge as they ascend until they overlap beneath the chin.

Due to its action on the chin and mouth, the platysma plays a small role in facial expression. It is also the muscle which allows us to draw down the lower lip to show the teeth.

Action of the neck muscles

The suprahyoid and infrahyoid groups of muscles have opposing actions on the larynx and hyoid bone. This enables us to swallow.

The mylohyoid, geniohyoid and the anterior belly of the digastric muscle act together to pull the hyoid and the larynx forward and up during swallowing. They also enable the mouth to be opened against resistance.

The stylohyoid and the posterior belly of the digastric muscle together lift and pull back the hyoid bone and the larynx. The suprahyoid muscles can be tested by asking the patient to open their mouth widely against resistance.

OPPOSING ACTION

The infrahyoid group of muscles act together to pull the hyoid and the larynx back down to their normal positions, as at the end of the act of swallowing. When contracted, the infrahyoid muscles lower and fix the hyoid bone so that the suprahyoid muscles can pull against it to open the mouth.

The infrahyoid muscles can be tested by asking a patient to open their mouth against resistance while the doctor lightly holds the hyoid bone. The hyoid should move down as it is lowered and fixed by the muscles below. If there is weakness of the infrahyoid muscles the hyoid bone will rise up due to the unopposed action of the muscles above.

Action of the infrahyoid and suprahyoid muscles

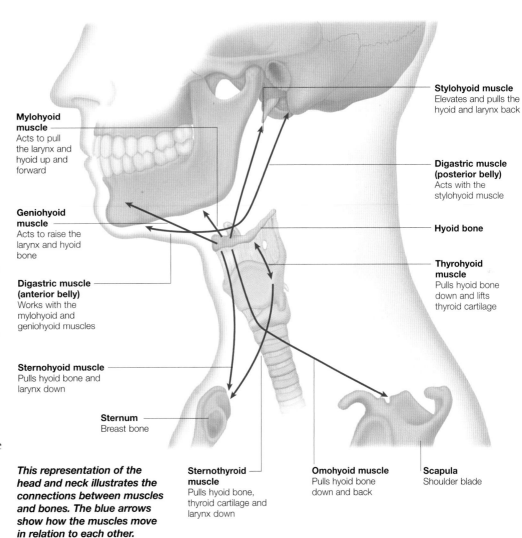

Mylohyoid muscle
Acts to pull the larynx and hyoid up and forward

Geniohyoid muscle
Acts to raise the larynx and hyoid bone

Digastric muscle (anterior belly)
Works with the mylohyoid and geniohyoid muscles

Sternohyoid muscle
Pulls hyoid bone and larynx down

Sternum
Breast bone

Stylohyoid muscle
Elevates and pulls the hyoid and larynx back

Digastric muscle (posterior belly)
Acts with the stylohyoid muscle

Hyoid bone

Thyrohyoid muscle
Pulls hyoid bone down and lifts thyroid cartilage

Sternothyroid muscle
Pulls hyoid bone, thyroid cartilage and larynx down

Omohyoid muscle
Pulls hyoid bone down and back

Scapula
Shoulder blade

This representation of the head and neck illustrates the connections between muscles and bones. The blue arrows show how the muscles move in relation to each other.

The hyoid bone

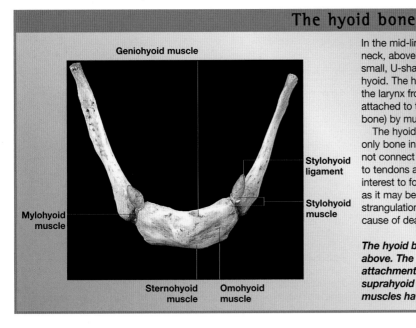

Geniohyoid muscle

Mylohyoid muscle

Sternohyoid muscle

Omohyoid muscle

Stylohyoid ligament

Stylohyoid muscle

In the mid-line of the front of the neck, above the larynx, lies a small, U-shaped bone called the hyoid. The hyoid bone supports the larynx from above and is itself attached to the mandible (jaw bone) by muscles and tendons.

The hyoid is unique as it is the only bone in the body which does not connect to another bone, only to tendons and muscles. It is of interest to forensic pathologists as it may be fractured during strangulation, a vital clue as to the cause of death.

The hyoid bone as seen from above. The points of muscle attachment for some of the suprahyoid and infrahyoid muscles have been marked.

Swallowing involves a rapid series of muscular movements in the back of the mouth, the pharynx and the neck. During the initial stages of swallowing, the suprahyoid muscles act to pull the hyoid bone upward and forward towards the mandible.

The hyoid bone is attached to the larynx by muscle and ligament and so the larynx is also pulled upward and forward. This can be seen in life by observing the 'Adam's apple' rising in the neck.

This movement of the larynx widens the pharynx (the gullet) behind it, thereby allowing food to pass. It also importantly helps to close off the respiratory passage, thus preventing the inhalation of any food.

Neck flexor muscles

The flexor muscles of the neck work to keep the head stable and upright on the spine. These muscles also enable flexion of the neck and head and raise the first two ribs during inspiration.

The centre of gravity of the head lies in front of the spine, and so there needs to be constant activity in the muscles and ligaments of the back of the neck to keep the head from falling forward. Much of the forward and lateral flexion of the head and neck is achieved by the co-ordinated action of the neck flexor muscles: the scalenes, the prevertebral muscles and the powerful sternocleidomastoid.

SCALENE MUSCLES

Three muscles run from the transverse processes either side of the cervical vertebrae to attach to the first and second ribs. The scalenus anterior and the scalenus medius both originate from the third to the sixth cervical vertebrae (C3 to C6) and attach to the first rib. The scalenus posterior muscle may be absent or may be part of the scalenus medius muscle. It passes down to the second rib.

PREVERTEBRAL MUSCLES

The prevertebral muscles lie in front of the cervical vertebrae and extend from the skull down to the junction of the neck and chest. Rectus capitis anterior and lateralis muscles are short muscles that run from the skull to the first cervical vertebra. The longus capitis muscles are longer ribbon-like muscles that lie in line with the tendons of origin of the scalenus anterior muscle. The longus colli connects each of the vertebrae to each other so they move as a unit.

Front view of scalene and prevertebral muscles

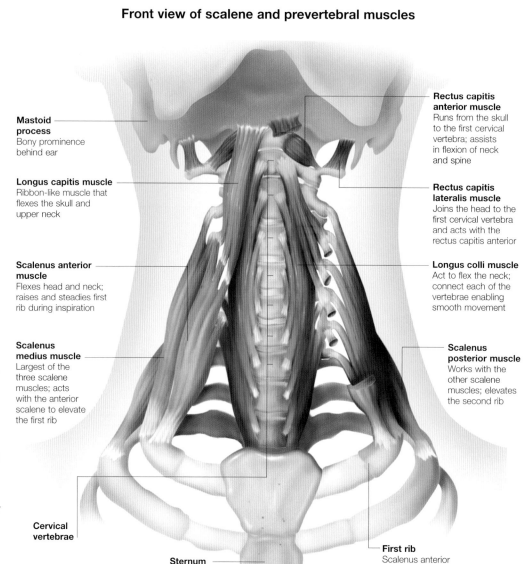

Mastoid process
Bony prominence behind ear

Longus capitis muscle
Ribbon-like muscle that flexes the skull and upper neck

Scalenus anterior muscle
Flexes head and neck; raises and steadies first rib during inspiration

Scalenus medius muscle
Largest of the three scalene muscles; acts with the anterior scalene to elevate the first rib

Cervical vertebrae

Sternum
(breastbone)

Rectus capitis anterior muscle
Runs from the skull to the first cervical vertebra; assists in flexion of neck and spine

Rectus capitis lateralis muscle
Joins the head to the first cervical vertebra and acts with the rectus capitis anterior

Longus colli muscle
Act to flex the neck; connect each of the vertebrae enabling smooth movement

Scalenus posterior muscle
Works with the other scalene muscles; elevates the second rib

First rib
Scalenus anterior and medius muscles attach here

Clinical aspects

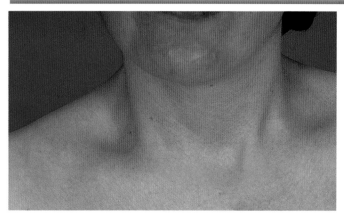

The sternocleidomastoid muscle is involved in the clinical condition known as torticollis or 'wry neck'. In this condition, shortening or abnormal contraction of the muscle on one side fixes the head in a characteristic position, facing upwards and to the opposite side.

This can occasionally occur

Torticollis occurs when the sternocleidomastoid muscle on one side of the neck contracts, pulling the head and neck into an abnormal position.

in newborn babies where it is known as 'congenital torticollis' and may be due to the growth of a fibrous tissue tumour within the muscle of the fetus, or caused by a tear to the sternocleidomastoid muscle during a difficult delivery. The damaged muscle can become replaced by scar tissue which contracts and thus shortens the muscle. Repeated contractions can occur within the sternocleidomastoid of adults in the clinical condition known as 'spasmodic torticollis'.

Action of the neck flexor muscles

The sternocleidomastoid muscles are powerful flexors of the head and cervical spine. They work in conjunction with the scalene and prevertebral muscles.

The sternocleidomastoid muscles are the major head flexor muscles. These powerful muscles can be seen very prominently under the skin on either side of the front of the neck. They run from the mastoid process (a prominence on the base of the skull) down and forwards to the sternum (breastbone) and clavicle (collarbone). At this lower end, each splits into two segments; one part attaches to the front of the upper sternum, while the second, deeper part, attaches to the clavicle.

ACTION OF THE STERNOCLEIDOMASTOID

When the sternocleidomastoid contracts on one side of the neck only, it causes the face to be turned to the opposite side and inclined slightly upwards. If other muscles hold the neck vertical at the same time, simple rotation of the head results. The muscle can be felt under the skin when the head is rotated.

While the sternocleidomastoid and other muscles contract, the muscles at the back of the neck relax, allowing the head to be pulled forward. This movement is known as protraction – when the head is moved forward in relation to the body while keeping it vertical and maintaining a horizontal gaze. An example is the action of the head when someone tries to look over their shoulder.

Side view of sternocleidomastoid muscle

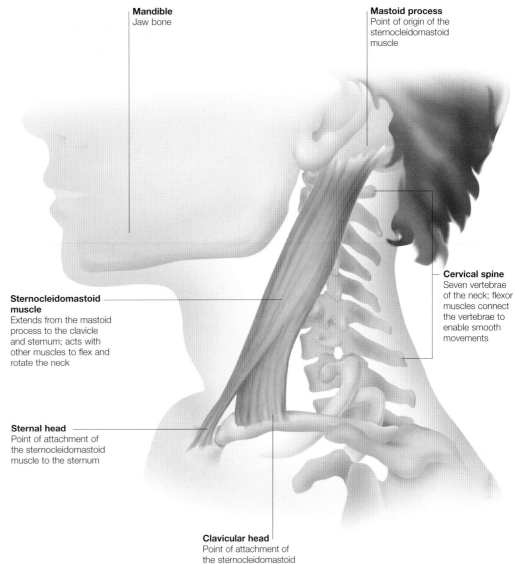

Mandible
Jaw bone

Mastoid process
Point of origin of the sternocleidomastoid muscle

Sternocleidomastoid muscle
Extends from the mastoid process to the clavicle and sternum; acts with other muscles to flex and rotate the neck

Sternal head
Point of attachment of the sternocleidomastoid muscle to the sternum

Clavicular head
Point of attachment of the sternocleidomastoid muscle to the clavicle

Cervical spine
Seven vertebrae of the neck; flexor muscles connect the vertebrae to enable smooth movements

Like the scalene muscles, the sternocleidomastoid also acts as an accessory muscle of respiration, raising the first rib.

Scalene and prevertebral muscles

The anterior, medius and posterior scalene muscles, working with the sternocleidomastoid and deeper muscles of the neck, flex the cervical spine laterally (sideways) and also lift the individual rib to which each is attached and steady it during breathing. When the scalene muscles on both sides contract together they contribute to forward flexion of the neck.

By their action of lifting the first two ribs, they also have the role of 'accessory muscles of respiration'. They are called into play when extra help is needed in expanding the chest to fill the lungs, as might happen after exercise or during an asthma attack.

The sternocleidomastoid muscle can be seen running from behind the ear to the breastbone when the head and neck is rotated to one side. Other neck muscles are also used in this action.

PREVERTEBRAL MUSCLES

The rectus capitis muscles assist in flexion and lateral flexion of the head on the spine. Longus capitis flexes the upper neck and the head on the neck, while longus colli, with no attachment to the skull, acts upon the neck alone.

Brainstem

The brainstem lies at the junction of the brain and spinal cord.
It helps to regulate breathing and blood circulation as well as having
an effect upon a person's level of consciousness.

The brainstem is made up of three distinct parts: the midbrain, the pons, and the medulla oblongata. The midbrain connects with the higher brain above; the medulla is continuous with the spinal cord below.

BRAINSTEM APPEARANCE
The three parts of the brainstem can be viewed from underneath:
■ The medulla oblongata – a bulge at the top of the spinal column. Pyramids, or columns, lie at either side of the midline. Nerve fibres within these columns carry messages from the cerebral cortex to the body. Raised areas known as the olives lie either side of the pyramids
■ The pons – contains a system of nerve fibres which originate in the nerve cell bodies deep within the substance of the pons
■ The midbrain – appears as two large columns, the cerebral crura, separated in the midline by a depression.

CRANIAL NERVES
Also present in the brainstem are some of the cranial nerves which supply much of the head. These nerves carry fibres which are associated with the cranial nerve nuclei, collections of grey matter, that lie inside the brainstem.

The brainstem connects the cerebral hemispheres with the spinal cord. There are three parts to the brain stem; the pons, medulla and the midbrain.

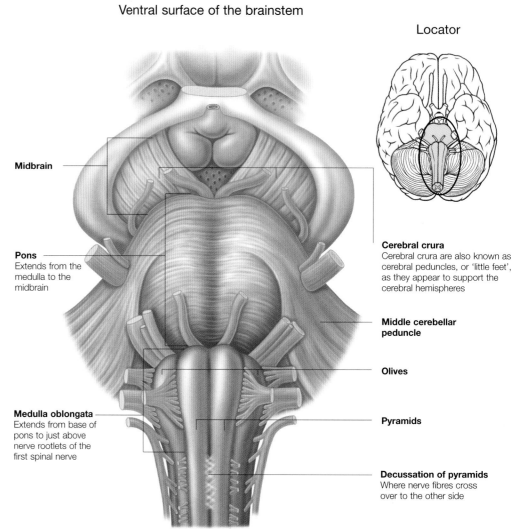

Ventral surface of the brainstem

Locator

Midbrain

Pons
Extends from the medulla to the midbrain

Medulla oblongata
Extends from base of pons to just above nerve rootlets of the first spinal nerve

Cerebral crura
Cerebral crura are also known as cerebral peduncles, or 'little feet', as they appear to support the cerebral hemispheres

Middle cerebellar peduncle

Olives

Pyramids

Decussation of pyramids
Where nerve fibres cross over to the other side

Relationships of the brainstem

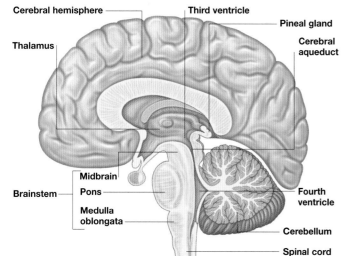

Cerebral hemisphere

Thalamus

Third ventricle

Pineal gland

Cerebral aqueduct

Midbrain
Pons

Brainstem

Medulla oblongata

Fourth ventricle

Cerebellum

Spinal cord

A sagittal section through the brain shows the position of the brainstem in relation to the other parts of the brain and spinal cord:

■ The medulla oblongata – arises as a widening of the spinal cord at the level of the foramen magnum, the large hole in the bottom of the skull. The central canal of the spinal cord widens into the fourth ventricle allowing cerebrospinal fluid

A sagittal view of the brain and spinal cord shows the location of the various structures. The brainstem is located in front of the cerebellum.

(CSF) to circulate between brain and spinal cord.

■ The pons – lies above the medulla, at the level of the cerebellum with which it makes many connections. Above the pons lies the midbrain, encircling the cerebral aqueduct which connects the fourth ventricle to the third ventricle.

■ The midbrain – is the shortest part of the the brainstem and lies under the thalamus, the central core of the brain, which is surrounded by the cerebral hemispheres. It thus lies below the thalamus and hypothalamus, and the tiny pineal gland.

Internal structure of the brainstem

The brainstem contains many areas of neural tissue which have a variety of functions vital to life and health. Responses to visual and auditory stimuli that influence head movement are also controlled here.

Cross sections through the brainstem reveal its internal structure, the arrangement of white and grey matter, which differs according to the level at which the section is taken.

MEDULLA
The features of a section through the medulla are:
■ The inferior olivary nucleus – a bag-like collection of grey matter which lies just under the olives. Other nuclei lying within the medulla include some belonging to the cranial nerves, such as the hypoglossal and the vagus nerves
■ The vestibular nuclear complex – an area that receives information from the ear and is concerned with balance and equilibrium
■ The reticular formation – a complex network of neurones, which is seen here and throughout the brainstem. It has a number of functions vital to life such as the control of respiration and circulation. The reticular formation is present in the midbrain as are several of the cranial nerve nuclei.

MIDBRAIN
A section through the midbrain shows the presence of:
■ The cerebral aqueduct – the channel which connects the

Cross sections of the brainstem

Midbrain (A)

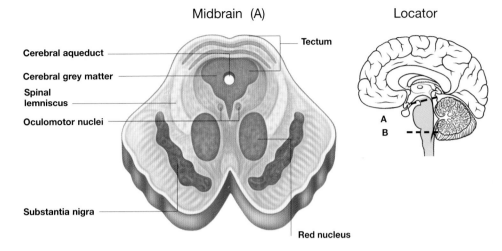

Cerebral aqueduct
Cerebral grey matter
Spinal lemniscus
Oculomotor nuclei
Substantia nigra
Tectum
Red nucleus

Locator

A
B

Medulla (B)

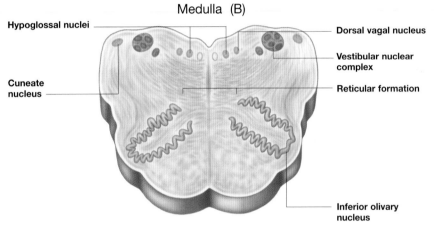

Hypoglossal nuclei
Cuneate nucleus
Dorsal vagal nucleus
Vestibular nuclear complex
Reticular formation
Inferior olivary nucleus

third and fourth ventricles.
Above the aqueduct lies an area called the tectum, while below it lie the large cerebral peduncles
■ The cerebral peduncles – within these lie two structures on each side; the red nucleus and the substantia nigra.
The red nucleus is involved in control of movement, while

damage to the substantia nigra is associated with Parkinson's disease.

PONS
The pons (not illustrated) is divided into upper and lower parts:
■ The lower part – mostly made up of transverse nerve fibres, running across from the nuclei

The numerous nuclei and tracts that are within the brainstem can be seen in these cross sections. They are involved in most functions of the brain.

of the pons to the cerebellum
■ The upper portion – contains a number of cranial nerve nuclei. The pons also contains part of the reticular formation.

Brainstem death

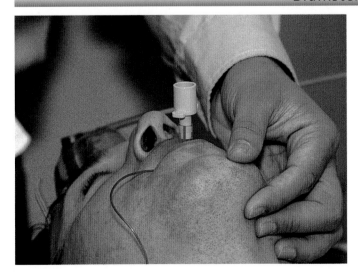

Strict criteria exist when testing for brain stem death. One of these is the patient's ability to breathe independently when disconnected from a ventilator.

It is possible in some cases for life-support machines in intensive care units to maintain breathing and blood circulation in a patient who has suffered brainstem death.

CERTIFYING DEATH
In such cases doctors will certify death using a legally prescribed set of tests and observations. Many of these tests are designed to show death of the brainstem, that part of the brain which

controls the vital functions of consciousness, breathing and circulation.

BRAIN STEM TESTING
Assessing the function of the brainstem includes looking for the following responses:
■ The ability to breathe without the help of a machine
■ Constriction of the pupil in response to light
■ Blinking of the eye when the cornea is touched
■ Eye movement when the ears are flushed with ice-water
■ Coughing or gagging when the airway is stimulated.
The responses are absent if the brainstem is non-functioning.

Brachial plexus

Lying within the root of the neck and extending into the axilla, the brachial plexus is a complicated network of nerves from which arise the major nerves supplying the upper limb.

At the level of each vertebra of the spine there emerges a 'spinal nerve' which divides into dorsal and ventral parts, called 'rami'.

The brachial plexus is formed by the joining and intermixing of the ventral rami at the level of the fifth to the eighth cervical vertebrae and most of the ventral rami from the level of the first thoracic vertebra. These ventral rami are known as the 'roots' of the brachial plexus.

STRUCTURE

The roots of the brachial plexus join to form three 'trunks': superior, middle and inferior.

As the complexity of the brachial plexus increases, each of the three trunks then divides into an anterior (front) and a posterior (back) 'division'. In general, the nerve fibres within the anterior divisions are those which will go on to supply the anterior structures of the upper limb, while the fibres of the posterior divisions will supply posterior upper limb structures.

From the six divisions, three 'cords' are formed, which are named for their positions in relation to the axillary artery to which they lie adjacent: lateral, medial and posterior.

The final part of the brachial plexus consists of the branches given off by the three cords, although other branches also arise at higher levels.

Origins of the brachial plexus

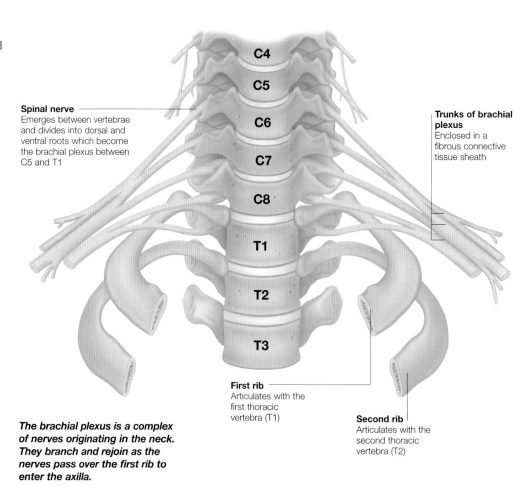

Spinal nerve
Emerges between vertebrae and divides into dorsal and ventral roots which become the brachial plexus between C5 and T1

C4
C5
C6
C7
C8
T1
T2
T3

Trunks of brachial plexus
Enclosed in a fibrous connective tissue sheath

First rib
Articulates with the first thoracic vertebra (T1)

Second rib
Articulates with the second thoracic vertebra (T2)

The brachial plexus is a complex of nerves originating in the neck. They branch and rejoin as the nerves pass over the first rib to enter the axilla.

Parts of the brachial plexus

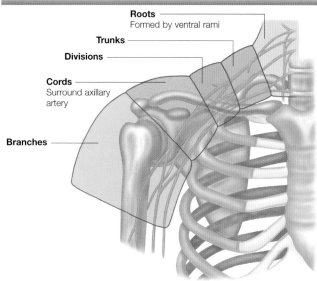

Roots
Formed by ventral rami

Trunks

Divisions

Cords
Surround axillary artery

Branches

Anatomically, the brachial plexus is divided into sections which, starting from the spine, are known as roots, trunks, divisions, cords and branches.

ORIENTATION

■ Roots – the ventral rami of C5 to T1, lying within the neck to either side of the spinal column
■ Three trunks – lie above the clavicle
■ Divisions – arise from the trunks and pass behind the clavicle, entering the axilla

The coloured blocks on this illustration indicate the anatomical levels of each different section of the brachial plexus.

■ Three cords – lie alongside the second part of the axillary artery within the axilla and inside the protective covering of the connective tissue of the axillary sheath
■ The terminal branches of the brachial plexus leave the axilla as they pass into the upper limb.

INJURIES TO THE BRACHIAL PLEXUS

If the brachial plexus is injured, the effect upon function of the upper limb will vary, according to the level of the plexus at which the damage occurs. The nearer the injury is to the spine, the more generalized the resulting damage will be.

Dermatomes

A dermatome is an area of skin which receives its sensory nerve supply from a single spinal nerve (and therefore a single segment of the spinal cord); however, that nerve supply may actually be taken to the skin in two or more cutaneous branches.

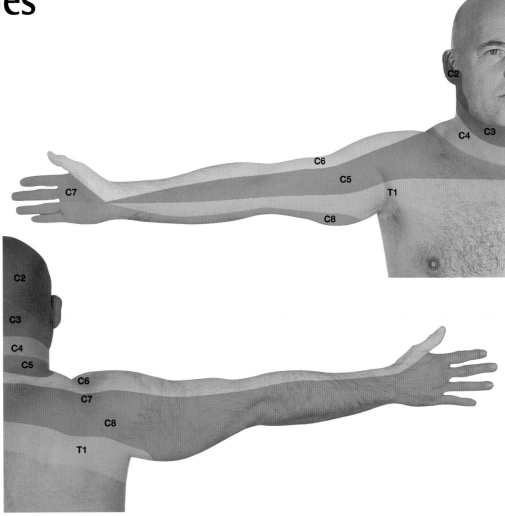

The nerve supply to the skin, and so the dermatomes, of the upper limb comes via the brachial plexus from the ventral rami of the spinal nerves of C5 to T1.

The segmental pattern of the dermatomes can most readily be seen by visualizing the arm held lifted to the side with the thumb uppermost. This pattern has arisen because, during fetal development, the limbs begin as 'buds' from the side of a segmentally arranged embryo which becomes elongated, stretching the dermatome bands as they grow outwards.

ARRANGEMENT OF DERMATOMES

The exact arrangement of the dermatomes of the upper limb may vary but, in general:
■ C5 supplies a band of skin along the length of the front of the arm
■ C6 supplies the lateral surface down to and including the thumb
■ C7 supplies a band along the back of the arm which extends to the first two fingers
■ C8 supplies the last two fingers and a strip of skin along the back of the arm
■ T1 supplies a strip along the front of the arm but not extending to the hand.

This segmental pattern is continued down the length of the body, and is most striking over the thorax and abdomen.

The arrangement of the dermatomes of the upper limb is variable between individuals but follows this basic pattern. They are numbered according to the spinal nerve of origin.

Clinical importance of dermatomes

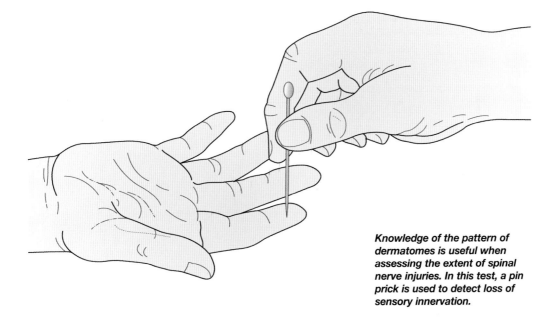

Knowledge of the pattern of dermatomes is useful when assessing the extent of spinal nerve injuries. In this test, a pin prick is used to detect loss of sensory innervation.

Awareness of the pattern of dermatomes over the body is of importance medically because it allows a doctor to test, with a pin prick, whether a particular segment of the spinal cord is working properly. This is of importance if a patient has a suspected spinal injury.

However, if there is a problem with just one spinal nerve or segment there may not be a loss of sensation in the dermatome it supplies as there is actually a good deal of overlap between dermatomes.

This means that even if there is damage to one spinal nerve or segment the skin it supplies may still receive a sensory supply from the adjacent nerves. There are, however, areas of skin over the upper limb which have no such overlap and so will be numb if the spinal nerve is damaged.

Pharynx

The pharynx, situated at the back of the throat, is a passage both for food to the alimentary system and air to the lungs. It can be divided into three major parts, and the entrance is guarded by the tonsils.

The pharynx, a fibromuscular, 15 cm long tube at the back of the throat, is a passage for food and air. The constrictor muscles of the pharynx allow food to be squeezed into the oesophagus.

NASOPHARYNX

The uppermost part of the pharynx, lying above the soft palate, is the nasopharynx. The most prominent feature on each side is the tubal elevation, the end of the auditory (Eustachian) tube that enables air pressure to be equalized between the nasopharynx and the middle ear cavity. Lymphoid (adenoid) tissue is found on the back wall.

OROPHARYNX

The oropharynx lies at the back of the throat. Its roof is the undersurface of the soft palate; the floor is the back of the tongue. The palatine tonsil lies in the side wall, and is bounded by the palatoglossal fold in front and the palatopharyngeal fold behind.

LARYNGOPHARYNX

The laryngopharynx extends from the upper border of the epiglottic cartilage (which covers the opening of the airway during swallowing) to the lower border of the cricoid cartilage, where it continues into the oesophagus. The inlet of the airway lies in the front section.

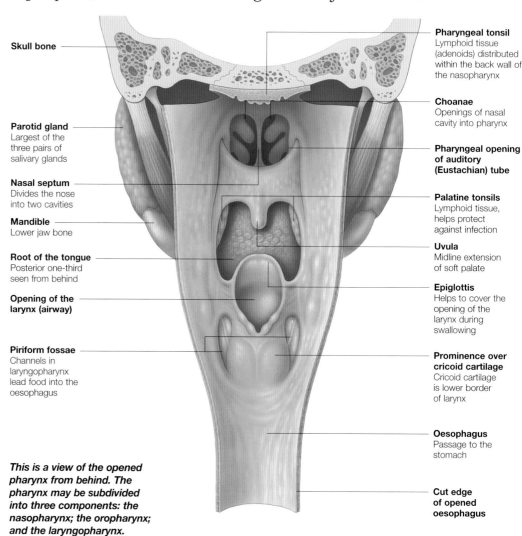

Skull bone

Parotid gland
Largest of the three pairs of salivary glands

Nasal septum
Divides the nose into two cavities

Mandible
Lower jaw bone

Root of the tongue
Posterior one-third seen from behind

Opening of the larynx (airway)

Piriform fossae
Channels in laryngopharynx lead food into the oesophagus

Pharyngeal tonsil
Lymphoid tissue (adenoids) distributed within the back wall of the nasopharynx

Choanae
Openings of nasal cavity into pharynx

Pharyngeal opening of auditory (Eustachian) tube

Palatine tonsils
Lymphoid tissue, helps protect against infection

Uvula
Midline extension of soft palate

Epiglottis
Helps to cover the opening of the larynx during swallowing

Prominence over cricoid cartilage
Cricoid cartilage is lower border of larynx

Oesophagus
Passage to the stomach

Cut edge of opened oesophagus

This is a view of the opened pharynx from behind. The pharynx may be subdivided into three components: the nasopharynx; the oropharynx; and the laryngopharynx.

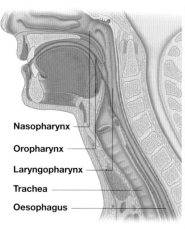

Nasopharynx

Oropharynx

Laryngopharynx

Trachea

Oesophagus

The relative positions of the three major parts of the larynx, the trachea and the oesophagus are evident in this sagittal section.

Infection and inflammation

As the auditory (Eustachian) tube in the nasopharynx is connected with the middle ear cavity, upper respiratory tract infections can spread to cause inflammation of the middle ear (otitis media). Infection may cause blockage of the auditory tube and the middle ear cavity, which contains the three ear ossicles. The cavity may fill with fluid, causing difficulties with hearing. This condition is commonly called glue ear.

The palatine tonsil is a common site of bacterial infection. More seriously, an abscess can develop in and around the tonsil. This can cause swelling, fever and pain, especially during swallowing

and speech. There may be trismus (difficulty opening the mouth) caused by spasm of adjacent muscles. If the abscess is not drained, it can lead to blockage of the airway. This condition was formerly known as 'Vincent's angina', the term 'angina' indicating the constricting nature of the condition.

Inflammation of the epiglottic cartilage is more common in children under the age of two. This is potentially dangerous due to obstruction of the airway. It classically produces stridor (wheezing noises with each intake of breath). This is a sign of impending obstruction and needs prompt medical attention.

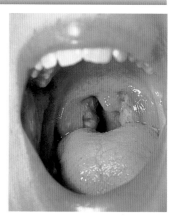

The palatine tonsils are masses of lymphoid tissue on either side of the back of the mouth. The tonsil on the right is infected.

Muscles of the pharynx

There are six pairs of muscles which make up the pharynx. These muscles can be divided into two groups.

One group of pharyngeal muscles comprises three pairs of constrictor muscles that run across the pharynx: the superior, middle and inferior constrictors. These constrict the pharynx, squeezing food downward into the oesophagus.

The other group comprises three pairs of muscles running from above down into the pharynx: the salpingo-pharyngeus, stylopharyngeus and palatopharyngeus. These raise the pharynx during swallowing, elevating the larynx and protecting the airway.

The constrictor muscles overlap each other from below upwards (like three stacked plastic cups inside each other). Important structures enter the pharynx in the intervals between these muscles. The constrictor muscle fibres sweep backwards into a longitudinally running fibrous band in the midline, the pharyngeal raphé, which is attached to the base of the skull.

NERVE SUPPLY

Most of the pharynx derives its sensory nerve supply from the glossopharyngeal (ninth cranial) nerve. Stimulation of the oropharynx at the back of the throat triggers the swallowing and gagging reflexes. The pharynx muscles are supplied chiefly by the 11th cranial (accessory) nerve.

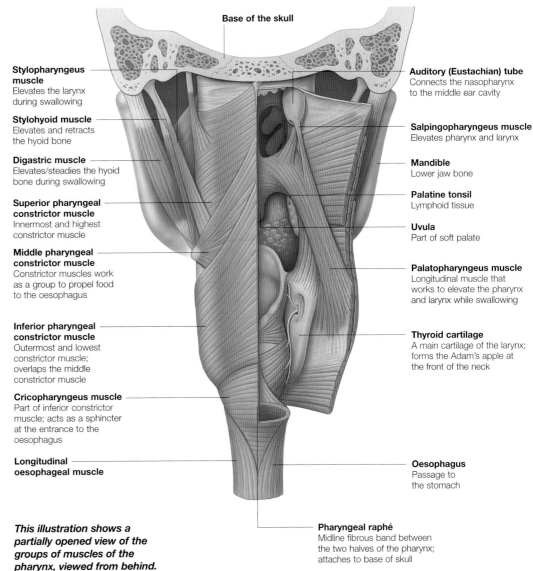

Base of the skull

Stylopharyngeus muscle
Elevates the larynx during swallowing

Stylohyoid muscle
Elevates and retracts the hyoid bone

Digastric muscle
Elevates/steadies the hyoid bone during swallowing

Superior pharyngeal constrictor muscle
Innermost and highest constrictor muscle

Middle pharyngeal constrictor muscle
Constrictor muscles work as a group to propel food to the oesophagus

Inferior pharyngeal constrictor muscle
Outermost and lowest constrictor muscle; overlaps the middle constrictor muscle

Cricopharyngeus muscle
Part of inferior constrictor muscle; acts as a sphincter at the entrance to the oesophagus

Longitudinal oesophageal muscle

Auditory (Eustachian) tube
Connects the nasopharynx to the middle ear cavity

Salpingopharyngeus muscle
Elevates pharynx and larynx

Mandible
Lower jaw bone

Palatine tonsil
Lymphoid tissue

Uvula
Part of soft palate

Palatopharyngeus muscle
Longitudinal muscle that works to elevate the pharynx and larynx while swallowing

Thyroid cartilage
A main cartilage of the larynx; forms the Adam's apple at the front of the neck

Oesophagus
Passage to the stomach

Pharyngeal raphé
Midline fibrous band between the two halves of the pharynx; attaches to base of skull

This illustration shows a partially opened view of the groups of muscles of the pharynx, viewed from behind.

Inhaled objects

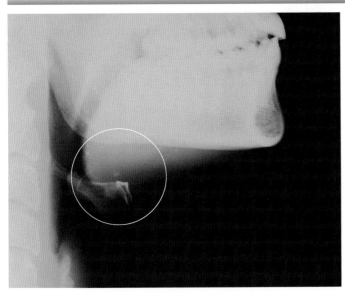

If not removed, a lodged fishbone (circled) may become a site of inflammation, and even rupture the pharyngeal wall.

The epiglottis helps to prevent entry of foreign bodies into the airway. However, it is possible to inhale food, stomach contents or a foreign body. This is more likely to happen if the patient is unconscious, under anaesthesia, drunk or suffering from diseases affecting the nerves or muscles of the pharynx, when there may be loss of the protective cough reflex. Blockage of the airway in such situations is life-threatening.

Between the inlet of the larynx and the side walls of the laryngopharynx lies a groove called the piriform fossa. During swallowing, particles of food may become stuck in the piriform fossa. This is particularly true of small fishbones. Certain fishbones, depending on size and degree of calcification, are more readily identified on X-rays than others.

Denture wearers may be more prone to swallowing foreign bodies as they have diminished palatal sensation due to shielding of the mucosa by the denture, and the tendency to chew food inefficiently. Such foreign bodies lodged in the pharynx can be retrieved by the use of a small, flexible viewing instrument (a naso-endoscope), through which tools can be passed under local anaesthesia.

Larynx

The larynx is situated in the neck below and in front of the pharynx. It is the inlet protecting the lungs, and contains the vocal cords. In men, part of the larynx is visible as the Adam's apple.

The larynx is composed of five cartilages (three single and one paired), connected by membranes, ligaments and muscles. In adult men, the larynx lies opposite the third to sixth cervical vertebrae (slightly higher in women and children), between the base of the tongue and the trachea.

The larynx serves as an inlet to the airways, taking air from the nose and mouth to the trachea. Because air and food share a common pathway, the primary function of the larynx is to prevent food and liquid from entering the airway. This is achieved by three 'sphincters' and by elevation. The larynx has also evolved as an organ of phonation – the act of producing sounds – allowing vocalization.

LARYNGEAL CARTILAGES

The laryngeal prominence (thyroid cartilage protrusion, or Adam's apple) is readily visible in most men. Its greater protrusion in men compared to women is due to the influence of the hormone testosterone.

The thyroid cartilage has two rear extensions, a superior and an inferior 'horn'. The cricoid cartilage, the only complete ring of cartilage in the airway, is partly overlapped above by the thyroid cartilage. Above it sit a pair of mobile, pyramid-shaped arytenoid cartilages.

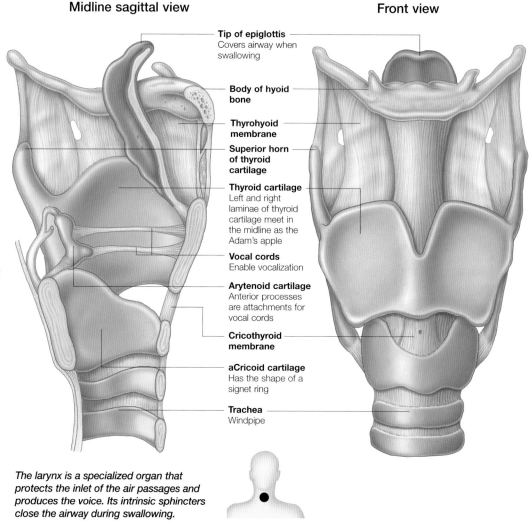

Midline sagittal view

Front view

Tip of epiglottis
Covers airway when swallowing

Body of hyoid bone

Thyrohyoid membrane

Superior horn of thyroid cartilage

Thyroid cartilage
Left and right laminae of thyroid cartilage meet in the midline as the Adam's apple

Vocal cords
Enable vocalization

Arytenoid cartilage
Anterior processes are attachments for vocal cords

Cricothyroid membrane

aCricoid cartilage
Has the shape of a signet ring

Trachea
Windpipe

The larynx is a specialized organ that protects the inlet of the air passages and produces the voice. Its intrinsic sphincters close the airway during swallowing.

Inside the larynx

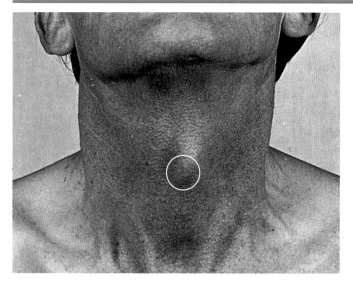

The vocal cords run from the arytenoids to the inner surface of the thyroid cartilage. They can be seen by placing a mirror, or fibre-optic laryngoscope, into the pharynx. They appear white in colour.

The inside of the larynx is lined with mucosa. The mucosa extends from the free edge of the aryepiglottic folds (containing the aryepiglottic muscles) at the laryngeal inlet down over soft tissue masses (vestibular ligaments) that protrude medially, forming two

The area where an emergency cricothyroidostomy would be performed (circled) is shown, just below the Adam's apple.

mucosal folds (false cords or vestibular folds). It then extends laterally, into a recess (sinus) on either side before covering the vocal ligaments, to the vocal cords proper. The glottis is the gap between the vocal folds in front and the arytenoid cartilages behind.

If the glottis becomes obstructed by an inhaled foreign object, breathing is impeded. An emergency procedure may be necessary in such instances, whereby a small hole is made in the cricothyroid membrane to allow air into the larynx below the obstruction. This is called a cricothyroidostomy or laryngostomy, and it can be a life-saving procedure.

Muscles of the larynx

The muscles of the larynx act to close the laryngeal inlet while swallowing and move the vocal cords to enable vocalization.

During swallowing, the epiglottis, along with the rest of the larynx, is raised. As the front surface hits the rear part of the tongue, it flips backwards over the laryngeal inlet.

ARYEPIGLOTTIC FOLDS

The aryepiglottic folds of tissue are the free upper margins of the membranes that run between the epiglottis and the arytenoid cartilages. They contain a pair of transverse aryepiglottic and oblique aryepiglottic muscles. These arise from the muscular process of the opposite arytenoid cartilage, and attach to the sides of the epiglottis. They act like a 'purse-string', closing the laryngeal inlet. The lower ends of each of the quadrangular membranes form the vestibular folds, or 'false' vocal cords.

MUCOUS GLANDS

The quadrangular membranes are covered by a mucosa, and a submucosa rich in mucous glands. These are connected to the inner walls of the thyroid and cricoid cartilages. They keep the vocal cords moist, as the vocal folds have no submucosa themselves, and therefore rely upon these secretions from above. A groove, the piriform fossae, slopes backwards and serves to channel liquids toward the oesophagus and away from the larynx.

Rear view

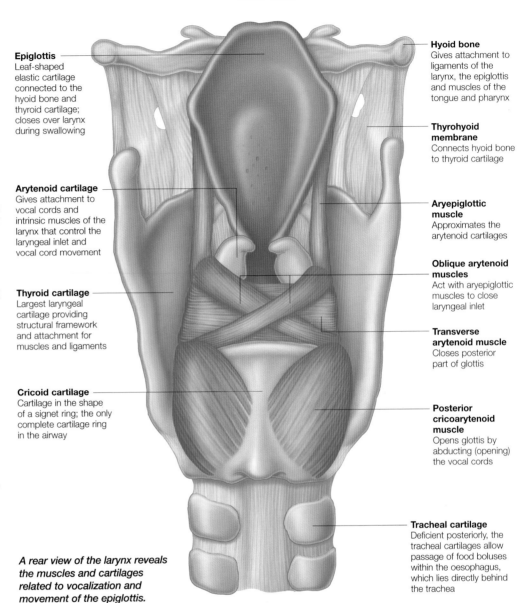

Epiglottis
Leaf-shaped elastic cartilage connected to the hyoid bone and thyroid cartilage; closes over larynx during swallowing

Arytenoid cartilage
Gives attachment to vocal cords and intrinsic muscles of the larynx that control the laryngeal inlet and vocal cord movement

Thyroid cartilage
Largest laryngeal cartilage providing structural framework and attachment for muscles and ligaments

Cricoid cartilage
Cartilage in the shape of a signet ring; the only complete cartilage ring in the airway

Hyoid bone
Gives attachment to ligaments of the larynx, the epiglottis and muscles of the tongue and pharynx

Thyrohyoid membrane
Connects hyoid bone to thyroid cartilage

Aryepiglottic muscle
Approximates the arytenoid cartilages

Oblique arytenoid muscles
Act with aryepiglottic muscles to close laryngeal inlet

Transverse arytenoid muscle
Closes posterior part of glottis

Posterior cricoarytenoid muscle
Opens glottis by abducting (opening) the vocal cords

Tracheal cartilage
Deficient posteriorly, the tracheal cartilages allow passage of food boluses within the oesophagus, which lies directly behind the trachea

A rear view of the larynx reveals the muscles and cartilages related to vocalization and movement of the epiglottis.

Action of the vocal cords

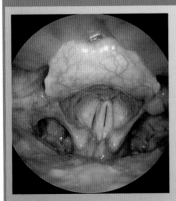

This laryngoscopic image shows the vocal cords during speech. When the cords are in close proximity to each other, air causes them to vibrate, producing vocal noises.

The size of the gap between the vocal cords, called the glottis, varies. Forced respiration requires a wide glottis, while speaking requires a narrow slit. Complete closure of the glottis is used during straining (for example, defecation or childbirth). The size of the glottis is controlled by movements of the arytenoid cartilages, since the vocal cords are attached to these.

These cartilages can be pulled apart (abducted), pulled together (adducted), and rotated by the crico-arytenoid and arytenoid groups of muscles supplied by the recurrent laryngeal nerve. The most important of these muscles is the posterior crico-arytenoid, which opens the glottis (abducts

the vocal folds to allow air into the airways). Paralysis of one posterior crico-arytenoid muscle may not produce any symptoms, as the other cord compensates. Paralysis of both muscles will produce acute breathing difficulties.

Speech is made by passing air over the vocal cords, and tension in these cords controls the pitch of the voice. Tilting the cricoid arch backwards causes an increase in tension, while bringing the thyroid and arytenoid cartilages together causes a relaxation. Fine tuning is accomplished via the vocalis muscle. The sounds that the larynx can produce are complex, and need to be learned. Accents are so acquired, and are often difficult to lose once mastered.

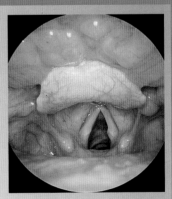

The glottis is wider at rest, as the vocal cords sit apart from one another. The epiglottis, seen as the pale flap above the glottis and vocal cords, closes the larynx while swallowing.

Oesophagus

The oesophagus is the tubular connection between the pharynx in the neck and the stomach. It is used solely as a passage for food, and plays no part in digestion and absorption.

The oesophagus is about 25 cm long in adults and is the muscular tube for the passage of food from mouth to stomach.

SHAPE OF OESOPHAGUS

As it is soft and somewhat flexible, the contour and path of the oesophagus is not straight; rather it curves around, and is indented by firmer structures such as the arch of the aorta and the left main bronchus.

PASSAGE OF FOOD

When no food is passing through the oesophagus, its inner lining lies in folds which fill the lumen, or central space. As a bolus (lump) of food is swallowed and passes down, it distends this lining and the oesophageal walls. Food is carried down the oesophagus by waves of contractions in a process known as peristalsis.

OESOPHAGEAL STRUCTURE

In cross-section, the oesophagus has four layers:
■ Mucosa – the innermost layer lined by stratified squamous epithelium; it is resistant to abrasion by food
■ Submucosa – composed of loose connective tissue; it also contains glands which secrete mucus to aid the passage of food
■ Muscle layer – striated muscle (under voluntary control) lines the upper oesophagus; smooth muscle the lower part; and a combination in the mid-region.
■ Adventitia – a covering layer of fibrous connective tissue.

The oesophagus in cross-section is a multilayered structure. The layers are similar to those found in the rest of the gastro-intestinal tract.

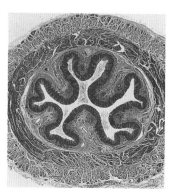

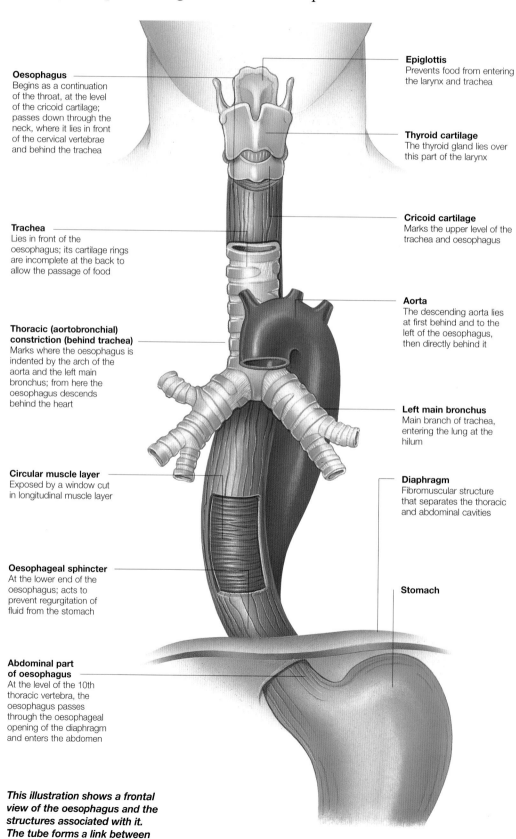

Oesophagus
Begins as a continuation of the throat, at the level of the cricoid cartilage; passes down through the neck, where it lies in front of the cervical vertebrae and behind the trachea

Trachea
Lies in front of the oesophagus; its cartilage rings are incomplete at the back to allow the passage of food

Thoracic (aortobronchial) constriction (behind trachea)
Marks where the oesophagus is indented by the arch of the aorta and the left main bronchus; from here the oesophagus descends behind the heart

Circular muscle layer
Exposed by a window cut in longitudinal muscle layer

Oesophageal sphincter
At the lower end of the oesophagus; acts to prevent regurgitation of fluid from the stomach

Abdominal part of oesophagus
At the level of the 10th thoracic vertebra, the oesophagus passes through the oesophageal opening of the diaphragm and enters the abdomen

Epiglottis
Prevents food from entering the larynx and trachea

Thyroid cartilage
The thyroid gland lies over this part of the larynx

Cricoid cartilage
Marks the upper level of the trachea and oesophagus

Aorta
The descending aorta lies at first behind and to the left of the oesophagus, then directly behind it

Left main bronchus
Main branch of trachea, entering the lung at the hilum

Diaphragm
Fibromuscular structure that separates the thoracic and abdominal cavities

Stomach

This illustration shows a frontal view of the oesophagus and the structures associated with it. The tube forms a link between the mouth and the stomach.

Blood vessels and nerves

The arterial supply of the oesophagus derives from branches of the aorta and subclavian artery. As with much of the body, the veins which drain blood from the oesophagus tend to run alongside the arteries.

A network of small veins surrounds and drains blood from the oesophagus.

UPPER OESOPHAGUS
The veins from the upper third of the oesophagus drain into the inferior thyroid veins. Blood from the middle third of the oesophagus is drained into the azygos venous system.

LOWER OESOPHAGUS
Blood from the lower third of the oesophagus may enter the left gastric vein – part of the portal venous system which drains blood via the liver. This is clinically important because increased pressure in the portal system, may cause blood to travel back up into the oesophageal veins. As a result, the veins in the lower oesophagus become distended (varices form) and may rupture.

The oesophagus is surrounded by a network of veins. These drain either into the SVC via its tributaries or the portal vein via the left gastric vein.

Veins of the oesophagus

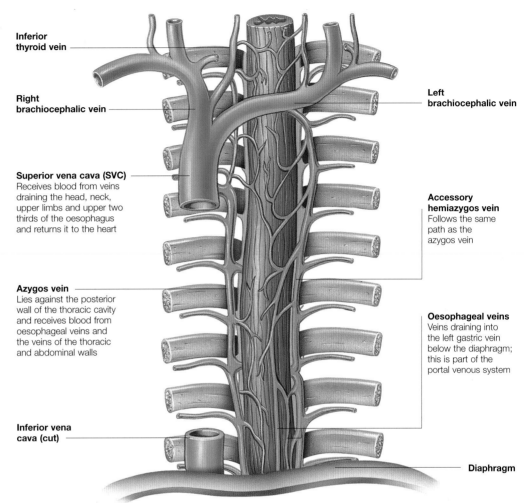

Inferior thyroid vein

Right brachiocephalic vein

Left brachiocephalic vein

Superior vena cava (SVC)
Receives blood from veins draining the head, neck, upper limbs and upper two thirds of the oesophagus and returns it to the heart

Accessory hemiazygos vein
Follows the same path as the azygos vein

Azygos vein
Lies against the posterior wall of the thoracic cavity and receives blood from oesophageal veins and the veins of the thoracic and abdominal walls

Oesophageal veins
Veins draining into the left gastric vein below the diaphragm; this is part of the portal venous system

Inferior vena cava (cut)

Diaphragm

Nerves of the oesophagus

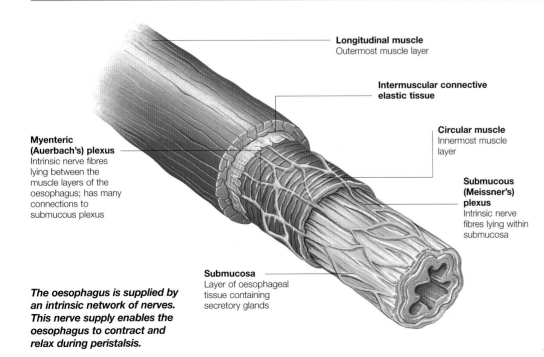

Longitudinal muscle
Outermost muscle layer

Intermuscular connective elastic tissue

Circular muscle
Innermost muscle layer

Myenteric (Auerbach's) plexus
Intrinsic nerve fibres lying between the muscle layers of the oesophagus; has many connections to submucous plexus

Submucous (Meissner's) plexus
Intrinsic nerve fibres lying within submucosa

Submucosa
Layer of oesophageal tissue containing secretory glands

The oesophagus is supplied by an intrinsic network of nerves. This nerve supply enables the oesophagus to contract and relax during peristalsis.

In common with the rest of the gastro-intestinal tract, the oesophagus has its own intrinsic nerve supply, which allows it to contract and relax during the process of peristalsis without any external nervous stimulation.

This intrinsic nerve supply derives from two main nerve plexuses within the walls known as the submucous (Meissner's) plexus and the myenteric (Auerbach's) plexus. These connect with each other, and together regulate the glandular secretion and movements of the oesophagus.

EXTERNAL CONTROL
The functioning of the intrinsic system can be modified by the autonomic nervous system, which regulates the body's internal environment. External nerve fibres come from the sympathetic trunk and from the vagus (10th cranial) nerve.

Thyroid and parathyroid glands

The thyroid and parathyroid glands are situated in the neck. Together, they produce important hormones responsible for regulating growth, metabolism and calcium levels in the blood.

The thyroid gland is an endocrine gland situated in the neck, lying to the front and side of the larynx and trachea. It is similar to a bow-tie in shape, and it produces two iodine-dependent hormones: tri-iodothyronine and thyroxine. These are responsible for controlling metabolism through promotion of metabolic enzyme production.

In addition, the gland secretes calcitonin which is involved in the regulation of calcium levels in the blood. In children, growth is dependent upon this gland, through its stimulation of the metabolism of carbohydrates, proteins and fats.

PYRAMIDAL LOBE

The gland has two conical-shaped lobes connected by an 'isthmus' (a band of tissue connecting the lobes), which usually lies in front of the second and third tracheal cartilage rings. The entire gland is surrounded by a thin connective tissue capsule and a layer of deep cervical fascia. There is often a small third lobe, the pyramidal lobe, which extends upwards from near the isthmus and lies over the cricothyroid membrane and median cricothyroid ligament.

ANTERIOR (FRONT) VIEW

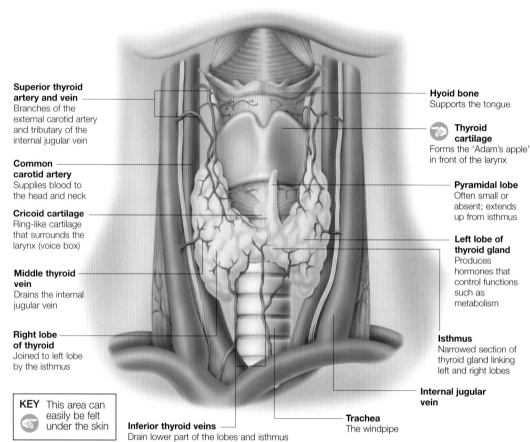

Superior thyroid artery and vein
Branches of the external carotid artery and tributary of the internal jugular vein

Common carotid artery
Supplies blood to the head and neck

Cricoid cartilage
Ring-like cartilage that surrounds the larynx (voice box)

Middle thyroid vein
Drains the internal jugular vein

Right lobe of thyroid
Joined to left lobe by the isthmus

Hyoid bone
Supports the tongue

Thyroid cartilage
Forms the 'Adam's apple' in front of the larynx

Pyramidal lobe
Often small or absent; extends up from isthmus

Left lobe of thyroid gland
Produces hormones that control functions such as metabolism

Isthmus
Narrowed section of thyroid gland linking left and right lobes

Internal jugular vein

Trachea
The windpipe

Inferior thyroid veins
Drain lower part of the lobes and isthmus

KEY This area can easily be felt under the skin

Surface anatomy

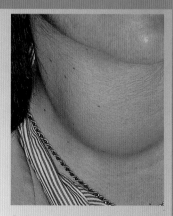

The thyroid gland is held in position by an investing layer of connective tissue, which also attaches to the oblique line on the thyroid cartilage. The gland is not normally palpable, since it is mostly covered by the strap muscles of the neck.

However, should the gland enlarge, it may produce a noticeable swelling which moves upwards upon swallowing, because of its fascial attachments. This is a useful differential to other swellings which might appear in the neck, such as that caused by an enlargement of the pre-tracheal lymph nodes, which are not bound to the same extent.

The top margins of the thyroid lobes are limited by the attachment of the pre-tracheal strap muscles. Therefore, any enlargement of the gland must be downwards into the lower neck. This enlargement of the gland is referred to as a 'goitre'.

Enlargement of the gland can produce difficulty breathing (dyspnoea) due to compression of the trachea, and difficulty and pain on swallowing (dysphagia) due to compression of the oesophagus.

Enlargement may be associated with either under-production of thyroid hormones (hypothyroid goitre), or their over-production (hyperthyroid goitre).

Swelling of the thyroid gland and neck can be caused by a lack of iodine, the presence of tumours or thyrotoxicosis (over-production of hormones).

Goitres – caused by the enlargement of the thyroid gland – can vary in size from a small lump to a large swelling.

Posterior view of thyroid

The posterior view of the thyroid reveals the small parathyroid glands, embedded within the lobes. A rich network of vessels supply the glands.

The thyroid gland is well supplied by blood vessels. The upper pole receives arterial blood from the superior thyroid artery, a branch of the external carotid artery. The lower pole is supplied by the inferior thyroid artery, a branch of the thyrocervical trunk. The lobes of the gland are directly related to the common carotid arteries. The hormones are distributed to the bloodstream via a network (plexus) of veins in and around the gland, which ultimately drain into the internal jugular and brachiocephalic veins.

RELATION TO NERVES
In addition to the blood vessels, the gland is closely related to nerves. Posteriorly, the most important relation is the pair of recurrent laryngeal nerves from the vagus nerve. These ascend in the groove between the oesophagus and trachea, heading towards the larynx where they supply motor nerves to all laryngeal muscles (except the cricothyroid), and sensory nerves to the sub-glottic larynx. Hence, an enlarged thyroid may compress these nerves, causing a hoarseness in the voice.

POSTERIOR (BACK) VIEW

External carotid artery
Gives rise to the superior thyroid artery

Inferior pharyngeal constrictor muscle

Common carotid artery
Supplies blood to the head and neck

Right lobe of thyroid gland

Inferior thyroid artery
Supplies lower pole of lobe

Inferior parathyroid gland

Thyrocervical trunk
Supplies inferior thyroid artery

Internal jugular vein
Drains blood from the scalp, neck, thyroid and face

Superior pharyngeal nerve
External branch

Vagus nerve
Supplies motor and sensory fibres, especially recurrent laryngeal nerves

Superior parathyroid gland

Left recurrent laryngeal nerve
Supplies laryngeal muscles

Left brachiocephalic vein
Blood from gland drains into here

Arch of aorta
Supplies common carotid arteries

Trachea
Airway

Right recurrent laryngeal nerve
Supplies laryngeal muscles

Parathyroid gland

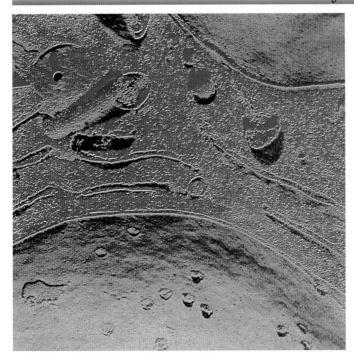

The pea-sized parathyroid glands (superior and inferior) are embedded within the rear tissue of the thyroid gland. They secrete parathormone which, together with calcitonin and vitamin D, controls calcium metabolism.

Disorders of the parathyroids cause problems associated with nerve, muscle and bone, since these tissues utilize calcium. A reduction in parathormone reduces levels of blood calcium, resulting in over-excitability of nerve and muscle. This may cause spasm or convulsions.

Hyperparathyroidism (over-production of parathormone) causes de-calcification of bone, increasing the chance of fractures. It also causes an elevation of calcium excretion

This electron micrograph shows part of a parathyroid gland cell. These glands secrete chemicals vital for calcium metabolism.

by the kidneys, making them prone to stone formation.

Rarely, the parathyroids may be found in a position independent of the thyroid gland. This is because the parathyroids, like the thyroid, develop from tissue in the floor of the embryonic pharynx, and migrate downwards.

Thyroid gland cysts can be found anywhere along its embryonic route. The final location of the parathyroids is normally rear of the thyroid, but may be found anywhere between the hyoid bone and the upper chest cavity behind the sternum.

Such ectopic parathyroids are unlikely to cause problems, whereas an ectopic thyroid gland in the upper chest cavity is likely to block the thoracic inlet, causing difficulty and pain on swallowing (dysphagia), difficulty in breathing (dyspnoea), and swelling of the upper body.

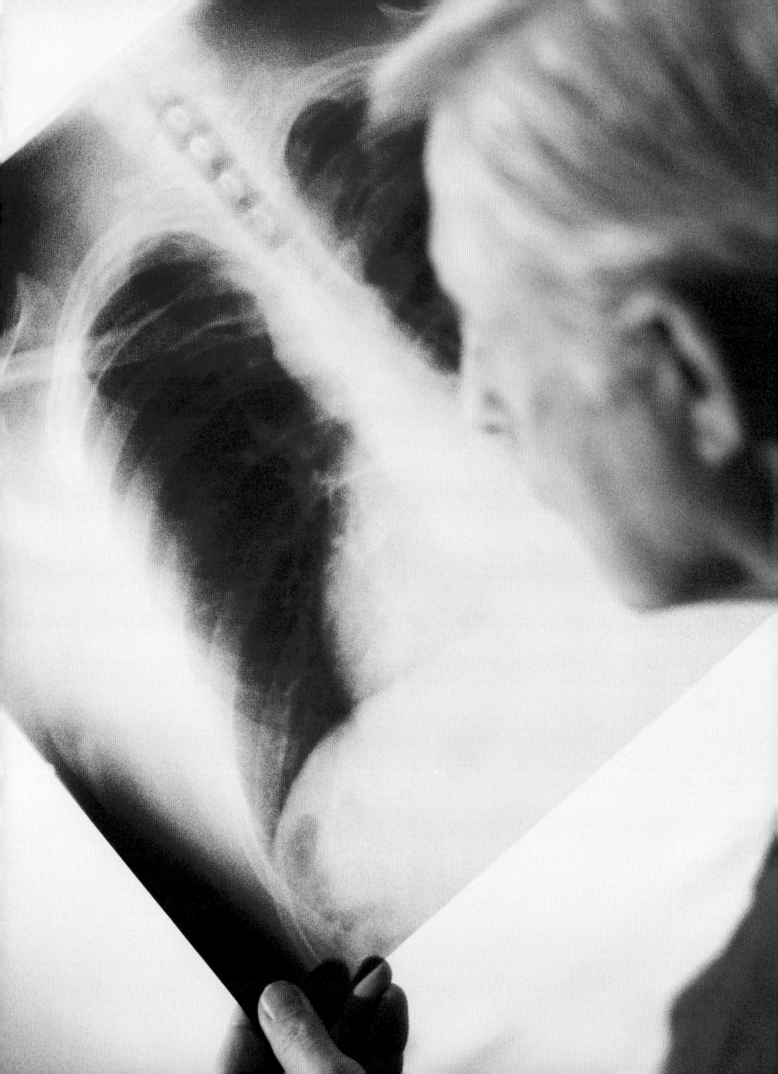

Chapter Four

The Thorax

The skeleton of the thorax, or chest, is a cagelike structure that surrounds and protects some of the body's most vital organs as well as its major blood vessels. This section examines the inner workings of the heart and lungs situated within the thoracic cavity, and the role of the spinal cord running through the thoracic vertebrae.

The dual function of the ribs is explored too, revealing not just their protective role, but also examining their structure and the network of muscles attached to them, which facilitates movement in the back and shoulders.

LEFT: A surgeon examines an x-ray of a human thorax prior to invasive surgery.

Thoracic vertebrae

The 12 thoracic vertebrae are the bones of the spinal column
to which the ribs are attached. The thoracic vertebrae sit between the cervical
vertebrae of the neck and the lumbar vertebrae of the lower back.

Each thoracic vertebrae has two components, a cylindrical body in front and a vertebral arch behind. The body and vertebral arch enclose a hole, called the vertebral foramen, which is rounded. When all the vertebrae are articulated together, the space formed by the linked vertebral foramina forms the vertebral canal. This houses the spinal cord surrounded by three protective layers, called the meninges.

BONY PROCESSES
The part of the vertebral arch that attaches to the body on each side is called the pedicle and the arch is completed behind by two laminae that meet in the midline to form the spinous process. These processes project downwards (like the tiles of a roof), that of the eighth being the longest and most vertical. At the junction of the pedicles and laminae are the projecting transverse processes. These decrease in size from the top down.

MUSCLE ATTACHMENTS
Muscles and ligaments are attached to the spines and transverse processes. The thoracic vertebrae articulate with each other at the intervertebral joints. Between the vertebral bodies are intervertebral discs acting as shock absorbers.

Each vertebra has four surfaces (facets), which form moveable synovial joints with the adjacent vertebrae – one pair of facets articulates with the vertebra above, the other pair with the vertebra below. All of these joints are strengthened by ligaments.

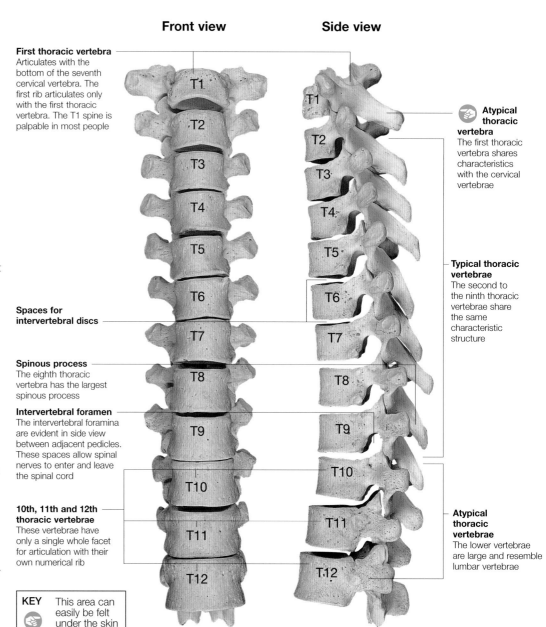

Front view

Side view

First thoracic vertebra
Articulates with the bottom of the seventh cervical vertebra. The first rib articulates only with the first thoracic vertebra. The T1 spine is palpable in most people

Spaces for intervertebral discs

Spinous process
The eighth thoracic vertebra has the largest spinous process

Intervertebral foramen
The intervertebral foramina are evident in side view between adjacent pedicles. These spaces allow spinal nerves to enter and leave the spinal cord

10th, 11th and 12th thoracic vertebrae
These vertebrae have only a single whole facet for articulation with their own numerical rib

Atypical thoracic vertebra
The first thoracic vertebra shares characteristics with the cervical vertebrae

Typical thoracic vertebrae
The second to the ninth thoracic vertebrae share the same characteristic structure

Atypical thoracic vertebrae
The lower vertebrae are large and resemble lumbar vertebrae

KEY This area can easily be felt under the skin

Vertebral deterioration

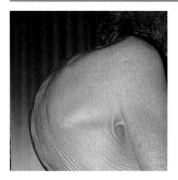

Osteoporosis is a condition which, if left untreated, results in a gradual loss of bone tissue. This particularly affects women following the menopause and may be related to the loss of particular sex hormones.

Osteoporosis results in the

Kyphosis is an abnormal curvature of the spine, giving rise to the characteristic 'hunchback' appearance.

bones becoming weaker and liable to fracture. In the case of the thoracic vertebrae, which help to support the weight of the body, osteoporosis may result in a compression fracture of the vertebral bodies. This causes 'wedging' and can result in kyphosis (curvature of the spine).

Tuberculosis is an infectious disease caused by the bacterium *Mycobacterium tuberculosis.* Although more commonly

affecting the lungs, it can affect any organ or tissue, including bone. Where bone is affected (as seen in Pott's disease), the symptoms may involve pain plus the constitutional symptoms of tuberculosis, including fever and weight loss. Pus (abscess) may form in and around the bone and intervertebral disc, resulting in destruction of much of the vertebra with deformity of posture.

Examining the thoracic vertebrae

The thoracic vertebrae can be distinguished easily from the typical cervical vertebrae.

The thoracic vertebrae differ from the cervical vertebrae in several ways:
■ An absence of the transverse process foramen (the foramen transversarium, through which nerves and blood vessels pass in the cervical vertebrae)
■ A single, rather than bifid (two-part), spine
■ The vertebral canal, through which the spinal cord runs, is smaller and more circular.
■ The most distinguishing feature of the thoracic vertebrae, however, is the presence of facets enabling the ribs to articulate with the spine. Each typical thoracic vertebra has six facets for rib articulation – three on each side.

The head of the rib lies in the region of the intervertebral disc, at the back, and has two hemi-(half) facets that articulate with its own numbered vertebra (upper border) and the vertebra immediately above (lower border).

ATYPICAL VERTEBRAE
The exceptions to the above rule are the first, 10th, 11th and 12th thoracic vertebrae. In the first thoracic vertebra, the facet on the upper border is a whole facet (rather than a half), as the first rib articulates only with its own vertebra.

Each of the 10th, 11th and 12th vertebrae has only one single whole facet to articulate with its own numerical rib. The 11th and 12th vertebrae have no articulation with the tubercle of the corresponding rib (and therefore no articular facet). The last two ribs are called 'floating ribs' as they have no connections to the ribs above.

Fifth (typical) thoracic vertebra (front view)

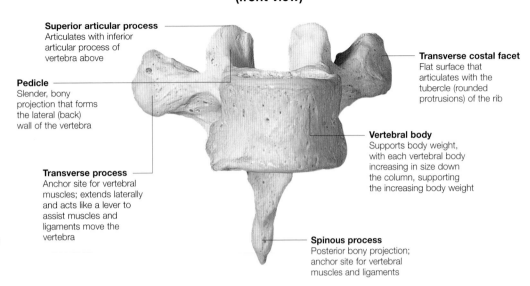

Superior articular process
Articulates with inferior articular process of vertebra above

Pedicle
Slender, bony projection that forms the lateral (back) wall of the vertebra

Transverse process
Anchor site for vertebral muscles; extends laterally and acts like a lever to assist muscles and ligaments move the vertebra

Transverse costal facet
Flat surface that articulates with the tubercle (rounded protrusions) of the rib

Vertebral body
Supports body weight, with each vertebral body increasing in size down the column, supporting the increasing body weight

Spinous process
Posterior bony projection; anchor site for vertebral muscles and ligaments

First (atypical) thoracic vertebra (side view)

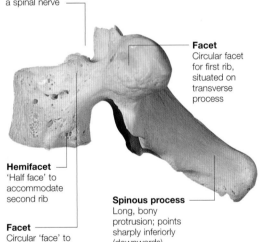

Superior intervertebral notch
Forms foramen with inferior notch below, providing passage for a spinal nerve

Facet
Circular facet for first rib, situated on transverse process

Hemifacet
'Half face' to accommodate second rib

Facet
Circular 'face' to accommodate head of first rib

Spinous process
Long, bony protrusion; points sharply inferiorly (downwards)

12th (atypical) thoracic vertebra (side view)

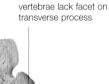

Body of vertebra
Structure of the lower thoracic vertebrae begins to resemble that of the lumbar vertebrae; only one round facet is present each side

Transverse process
11th and 12th thoracic vertebrae lack facet on transverse process

Inferior intervertebral notch
Forms the intervertebral foramen, though which a spinal nerve passes

Spinous process
At the base of the thoracic vertebrae, the spinous processes are small and rounded, resembling those of the lumbar vertebrae

Bone cancers

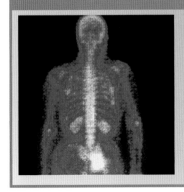

Cancers of bone may be primary, originating in bone, or secondary, when due to spread from another site. The most common cancers which spread to involve bone originate in organs such as the breast, lung, kidney, thyroid and prostate glands. When involving the vertebral column, this causes pain, but may present with varying degrees of weakness or paralysis in the legs due to compression of the spinal cord. Cancer can be seen on X-rays or by utilizing techniques such as radio-isotope bone scans.

Coloured gamma-camera scans (scintigrams) show the 'hot spots' (bright areas) of tumours and cancers spreading to bone.

This scintigram reveals secondary cancer spreading to the thoracic and lumbar vertebral regions.

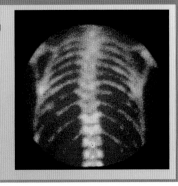

Lumbar vertebrae

The five lumbar vertebrae of the lower back are the strongest vertebrae of the spinal column.

The individual lumbar vertebrae are the largest and strongest in the vertebral column. This is important, as the lower the position of the bones of the spinal column, the more body weight they must bear. The arrangement of the lumbar vertebral joints is designed to allow maximum flexion (allowing us to touch our toes), and some lateral flexion (allowing us to reach sideways), but little rotation (this occurs at the thoracic level).

BASIC STRUCTURE

As with the cervical and thoracic vertebrae, each lumbar vertebra has the same basic plan, consisting of a cylindrical body in the front and a vertebral arch behind which enclose a space, called the vertebral foramen.

Each vertebral arch comprises a number of processes. There are two laterally projecting transverse processes, a centrally positioned spinous process and two pairs of articular facets, one pair above and one pair below. The transverse processes and spines are shorter and thicker than those of other vertebrae and are well adapted for the attachment of the large back muscles and strong ligaments.

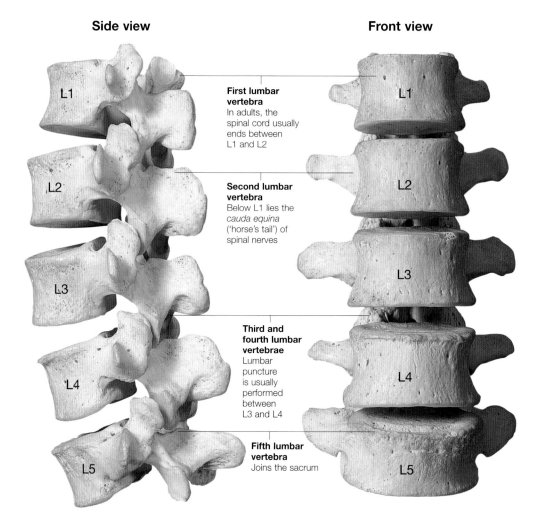

Side view

Front view

First lumbar vertebra
In adults, the spinal cord usually ends between L1 and L2

Second lumbar vertebra
Below L1 lies the *cauda equina* ('horse's tail') of spinal nerves

Third and fourth lumbar vertebrae
Lumbar puncture is usually performed between L3 and L4

Fifth lumbar vertebra
Joins the sacrum

The front of the lumbar vertebrae form a convex curve when viewed from the side, known as lumbar lordosis. This increases strength and helps to absorb shock.

The five lumbar vertebrae are subject to greater vertical compression forces than the rest of the spine. For this reason, these vertebrae are large and strong.

Typical lumbar vertebrae

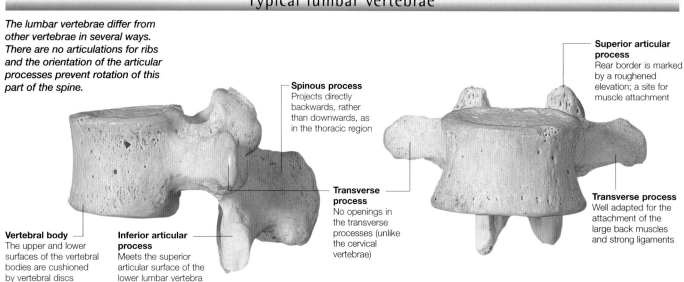

The lumbar vertebrae differ from other vertebrae in several ways. There are no articulations for ribs and the orientation of the articular processes prevent rotation of this part of the spine.

Spinous process
Projects directly backwards, rather than downwards, as in the thoracic region

Superior articular process
Rear border is marked by a roughened elevation; a site for muscle attachment

Transverse process
No openings in the transverse processes (unlike the cervical vertebrae)

Transverse process
Well adapted for the attachment of the large back muscles and strong ligaments

Vertebral body
The upper and lower surfaces of the vertebral bodies are cushioned by vertebral discs

Inferior articular process
Meets the superior articular surface of the lower lumbar vertebra

Lumbar ligaments

The intervertebral discs and connecting ligaments support the bones of the spine. They act as shock absorbers, reducing wear on the vertebrae.

The intervertebral discs link the bones of adjacent vertebrae, prevent dislocation of the vertebral column and also act as shock absorbers between the vertebrae. Intervertebral discs contribute about one-fifth of the length of the vertebral column, and are thickest in the lumbar region where the vertical compression forces are greatest.

STRENGTH AND STABILITY
To reinforce stability, the vertebral bodies are strengthened by tough, longitudinally running ligaments, consisting of fibrous tissue, in the front and rear. These ligaments are firmly attached to the intervertebral disc and adjacent edges of the vertebral body, but loosely attached to the rest of the body.

Movement between vertebrae is the result of the action of muscles attached to the processes of the vertebral arches. The joints associated with the articular processes are synovial joints, allowing the adjacent surfaces to glide smoothly over each other.

Each synovial joint is surrounded by a loose joint capsule. The joints of the vertebral arches are strengthened by various ligaments. The ligamenta flava join the laminae of the adjacent vertebra and contain elastic tissue.

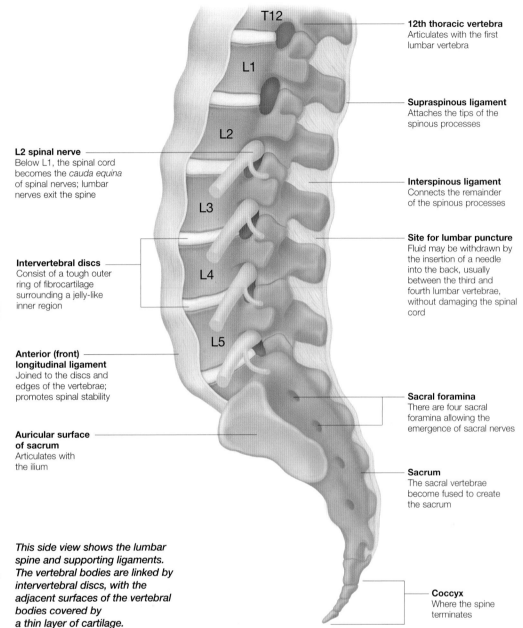

L2 spinal nerve
Below L1, the spinal cord becomes the *cauda equina* of spinal nerves; lumbar nerves exit the spine

Intervertebral discs
Consist of a tough outer ring of fibrocartilage surrounding a jelly-like inner region

Anterior (front) longitudinal ligament
Joined to the discs and edges of the vertebrae; promotes spinal stability

Auricular surface of sacrum
Articulates with the ilium

12th thoracic vertebra
Articulates with the first lumbar vertebra

Supraspinous ligament
Attaches the tips of the spinous processes

Interspinous ligament
Connects the remainder of the spinous processes

Site for lumbar puncture
Fluid may be withdrawn by the insertion of a needle into the back, usually between the third and fourth lumbar vertebrae, without damaging the spinal cord

Sacral foramina
There are four sacral foramina allowing the emergence of sacral nerves

Sacrum
The sacral vertebrae become fused to create the sacrum

Coccyx
Where the spine terminates

This side view shows the lumbar spine and supporting ligaments. The vertebral bodies are linked by intervertebral discs, with the adjacent surfaces of the vertebral bodies covered by a thin layer of cartilage.

Disorders of the lumbar vertebrae

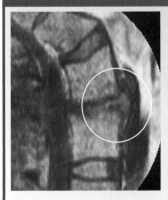

This MR scan shows an intervertebral disc (circled) in the lumbar spine protruding from between the vertebrae. This can cause severe pain.

Each intervertebral disc consists of a tough outer ring of fibrocartilage (the annulus fibrosus), surrounding a jelly-like inner region (the nucleus pulposus). The nucleus pulposus is under constant pressure in the upright position. Degeneration of the disc occurs with age and can allow the nucleus pulposus to protrude backwards through a split in the annulus fibrosus (a so-called slipped disc).

The whole disc does not slip out of place, but it can press on a spinal nerve root. This causes acute pain, known as 'sciatica', which radiates down the back of the thigh and calf, sometimes into the foot, along the sciatic nerve

(the main nerve supply to the leg).

Rarely, the herniated disc may press on the spinal cord itself, causing paralysis of the legs and disturbance of bladder function. Either of these occurrences is an emergency and usually results in surgery to remove the offending part of the herniated disc.

The most common cause of chronic back pain (spondylosis) is degenerative disease of the intervertebral discs and of the facet joints. Bone underlying the damaged cartilage develops ragged projections (osteophytes) that restrict joint movement, causing stiffness and secondary muscle spasm, and may press on nerve roots, causing pain.

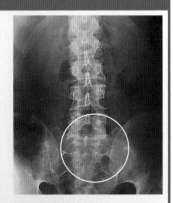

Degeneration of the intervertebral discs (circled) can lead to compression of one vertebra against another. This can be caused by ageing.

Sacrum and coccyx

The sacrum and coccyx form the tail end of the spinal column.
Both are formed from fused vertebrae, allowing attachment for weight-
bearing ligaments and muscles, and helping to protect pelvic organs.

The sacrum is a bony mass composed of five sacral vertebrae which fuse between puberty and the age of 30 years. It performs several functions: it attaches the vertebral column to the pelvic girdle, supporting the body's weight and transmitting it to the legs; it protects pelvic organs, such as the uterus and bladder; and it allows attachment of muscles that move the thigh.

The sacrum is shaped like an upside-down triangle, the five fused vertebral bodies diminishing in size from the wide base above (formed by the first sacral vertebra and the sacral alae, or 'wings') towards the apex below, where the coccyx is attached.

Centrally, horizontal bony ridges indicate the junctions between individual vertebrae; these are the remnants of intervertebral discs. On either side, sacral foramina (holes running through the bone) allow the passage of the ventral sacral motor nerve roots.

THE COCCYX

The coccyx, attached to the base of the sacrum, is the remains of the tail seen in our primate relatives. It consists of a small, pyramid-shaped bone formed from four fused vertebrae, and allows the attachment of ligaments and muscles, forming the anal sphincter.

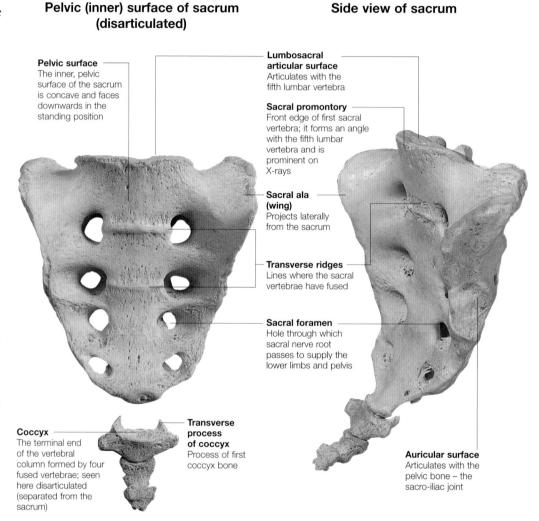

Pelvic (inner) surface of sacrum (disarticulated)

Pelvic surface
The inner, pelvic surface of the sacrum is concave and faces downwards in the standing position

Coccyx
The terminal end of the vertebral column formed by four fused vertebrae; seen here disarticulated (separated from the sacrum)

Transverse process of coccyx
Process of first coccyx bone

Side view of sacrum

Lumbosacral articular surface
Articulates with the fifth lumbar vertebra

Sacral promontory
Front edge of first sacral vertebra; it forms an angle with the fifth lumbar vertebra and is prominent on X-rays

Sacral ala (wing)
Projects laterally from the sacrum

Transverse ridges
Lines where the sacral vertebrae have fused

Sacral foramen
Hole through which sacral nerve root passes to supply the lower limbs and pelvis

Auricular surface
Articulates with the pelvic bone – the sacro-iliac joint

Sacro-iliac joint

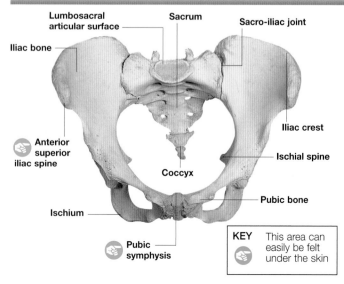

Lumbosacral articular surface

Sacrum

Sacro-iliac joint

Iliac bone

Anterior superior iliac spine

Coccyx

Ischium

Pubic symphysis

Iliac crest

Ischial spine

Pubic bone

KEY This area can easily be felt under the skin

On either side, the sacrum articulates with the pelvic bones at the sacro-iliac joints. The sacral joint surface is known as the auricular surface since it is vaguely ear-shaped.

The sacral joint surface is covered with hyaline cartilage (a type of cartilage typically found in free-moving joints) whereas the joint of the ilium is covered in tough fibrocartilage. The sacro-iliac joint is therefore a mixture of these two types.

The relationship between the sacrum and the pelvis can be seen clearly in this anterior (front) view. The sacrum displays a convex (outwards) curvature, and terminates at the coccyx.

In early life, the sacro-iliac joint is fairly mobile, but becomes progressively less so with age, although its mobility may still be significant. During delivery of a baby, it moves to enlarge the pelvic outlet.

There are sex differences between the male and female sacrum, often sufficient to allow identification of sex. The female sacrum is shorter and wider, allowing a larger pelvic cavity for the passage of an infant during childbirth. The diameter of the pelvic outlet also expands during birth due to the mobility of the coccyx, which moves backward for the baby's passage. The degree of curvature is greater in males than females.

Spinal nerve roots

The genitals, buttocks and lower limbs are supplied by nerve roots that emerge from the lumbar and sacral spine.

The sensory and motor nerve supply to and from the pelvis and legs is derived from a network of nerve roots called the sacral plexus. This lies on the rear wall of the pelvic cavity in front of the piriformis muscle. Contributions to the sacral plexus come from the lumbosacral trunk, representing the 4th and 5th lumbar nerve roots and the sacral nerve roots.

At the sacral plexus these nerve roots exchange nerve fibres and re-form into major nerves. These include the superior and inferior gluteal nerves, supplying the buttocks, and the sciatic nerve, which supplies the muscles of the leg. The parasympathetic splanchnic nerves (S1, S2, S3) regulate urination and defecation by controlling the internal sphincters, and also erection by dilating penile arterioles.

SACRAL FORAMINA

The convex outer sacral surface has a ridge called the median crest in the midline, where the spinous processes fuse. The four posterior sacral foramina transmit the dorsal nerve roots. Nerves pass down the sacrum through the sacral canal.

A normal defect in the fusion of the fifth sacral vertebra posteriorly causes the canal to open out at the sacral hiatus. This is useful to doctors, who can anaesthetize the lower spinal nerves by passing a needle through the open space.

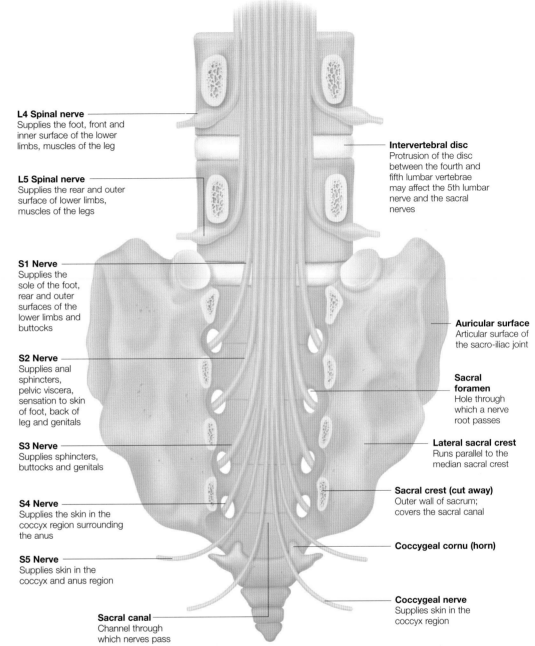

L4 Spinal nerve
Supplies the foot, front and inner surface of the lower limbs, muscles of the leg

L5 Spinal nerve
Supplies the rear and outer surface of lower limbs, muscles of the legs

S1 Nerve
Supplies the sole of the foot, rear and outer surfaces of the lower limbs and buttocks

S2 Nerve
Supplies anal sphincters, pelvic viscera, sensation to skin of foot, back of leg and genitals

S3 Nerve
Supplies sphincters, buttocks and genitals

S4 Nerve
Supplies the skin in the coccyx region surrounding the anus

S5 Nerve
Supplies skin in the coccyx and anus region

Sacral canal
Channel through which nerves pass

Intervertebral disc
Protrusion of the disc between the fourth and fifth lumbar vertebrae may affect the 5th lumbar nerve and the sacral nerves

Auricular surface
Articular surface of the sacro-iliac joint

Sacral foramen
Hole through which a nerve root passes

Lateral sacral crest
Runs parallel to the median sacral crest

Sacral crest (cut away)
Outer wall of sacrum; covers the sacral canal

Coccygeal cornu (horn)

Coccygeal nerve
Supplies skin in the coccyx region

Clinical aspects of the sacrum and coccyx

Tumours and infections of the sacrum are rare. Fractures only occur with severe trauma since the sacrum is very strong.

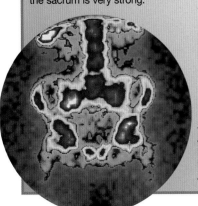

Sacro-iliitis, inflammation of the sacro-iliac joint, is revealed on this false-coloured scintigram of the pelvis. The inflammation appears as red and white areas.

Coccydynia (literally 'pain at the coccyx') is a painful syndrome affecting the base of the spine, rectum, buttocks and lower back. Typically, the pain is worsened or precipitated by sitting down and relieved by standing or lying on the side.

Causes include a fracture or trauma of the coccyx, fibrositis, disc disease, local infection, or idiopathy (no cause identifiable).

Treatment is difficult and consists of steroid and anaesthetic injections around the coccyx.

Sacro-iliitis is inflammation of the sacro-iliac joints. This is most commonly seen in a group of diseases known as the spondylo-arthropathies (literally 'diseases of the spinal joints').

The most common is ankylosing spondylitis. This affects men in their 20s to 40s

This superior (above) MR image reveals a stress fracture in the right sacral ala (circled). Such fractures often result from falling on to hard surfaces.

and causes back pain and stiffness and, in severe cases, a fixed, bent spine. On X-ray, there is irregular erosion, narrowing, and thickening at the sacro-iliac joint and vertebral joints. Other causes of sacro-iliitis are Crohn's disease, Reiter's syndrome and arthritis associated with ulcerative colitis.

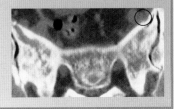

Spinal cord

The spinal cord is the communication pathway between the brain and the body. It allows signals to pass down to control body function and up to inform the brain of what is happening in the body.

The spinal cord is a slightly flattened cylindrical structure of 42–45 cm length in adults, with an average diameter of about 2.5 cm. It begins as a continuation of the medulla oblongata, the lowest part of the brainstem, at the level of the foramen magnum, the largest opening in the base of the skull. It then runs down the length of the neck and back in the vertebral canal, protected by the bony vertebrae which make up the vertebral column.

DEVELOPMENT
Up to the third month of development in the womb, the spinal cord runs the entire length of the vertebral column. Later on, however, the vertebral column outgrows the cord, which by birth ends at the level of the third lumbar vertebra. This more rapid growth of the vertebral column continues so that in the adult, the spinal cord ends at about the level of the disc between the first and second lumbar vertebrae.

ANATOMY OF THE CORD
The cord is enlarged in the region of the neck and lower back. The lower end of the cord tapers off into a cone-shaped region – the conus medullaris. From this, the filum terminale – a thin strand of modified pia mater (one of the membranes that surround the brain and spinal cord) – continues downwards to be attached to the back of the coccyx, anchoring the spinal cord.

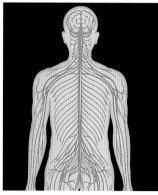

The 31 pairs of spinal nerves that branch off the spinal cord transfer impulses between the brain and all parts of the body.

Posterior view of spinal cord

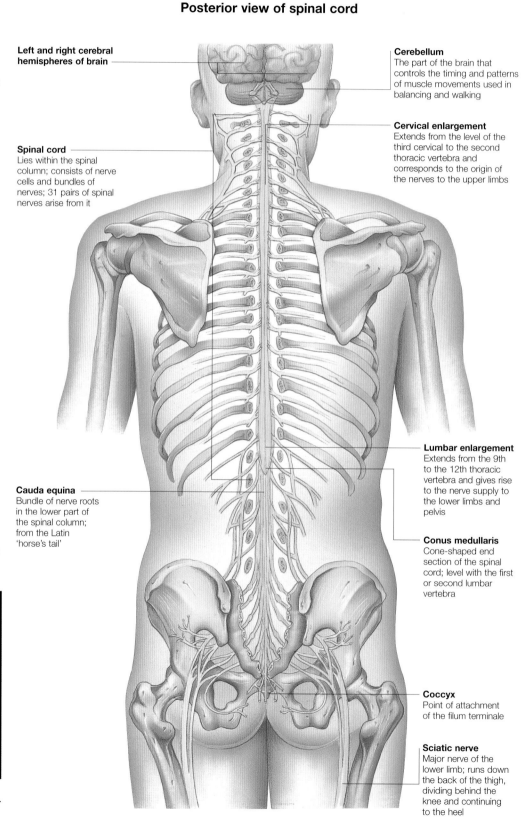

Left and right cerebral hemispheres of brain

Spinal cord
Lies within the spinal column; consists of nerve cells and bundles of nerves; 31 pairs of spinal nerves arise from it

Cauda equina
Bundle of nerve roots in the lower part of the spinal column; from the Latin 'horse's tail'

Cerebellum
The part of the brain that controls the timing and patterns of muscle movements used in balancing and walking

Cervical enlargement
Extends from the level of the third cervical to the second thoracic vertebra and corresponds to the origin of the nerves to the upper limbs

Lumbar enlargement
Extends from the 9th to the 12th thoracic vertebra and gives rise to the nerve supply to the lower limbs and pelvis

Conus medullaris
Cone-shaped end section of the spinal cord; level with the first or second lumbar vertebra

Coccyx
Point of attachment of the filum terminale

Sciatic nerve
Major nerve of the lower limb; runs down the back of the thigh, dividing behind the knee and continuing to the heel

Cross-sections through the spinal cord

The appearance of the spinal cord varies at different levels, according to the amount of muscle supplied by the nerves that emanate from it.

The spinal cord is made up of an inner core of grey matter, which consists mainly of nerve cells and their supporting cells (neuroglia), surrounded by white matter, made up primarily of myelinated nerve fibres – nerves with an insulating sheath of the fatty substance myelin.

In cross-section, the grey matter typically has the shape of a letter H or a butterfly, with two anterior columns or horns, two posterior columns and a thin grey commissure connecting the grey matter in the two halves. There is a small central canal containing cerebrospinal fluid which at its uppermost limit runs into the fourth ventricle in the region of the lower brainstem and cerebellum.

There is some variation in the appearance of a cross-section of the spinal cord at different levels. The amount of grey matter corresponds to the bulk of muscle whose nerve supply comes off at that level.

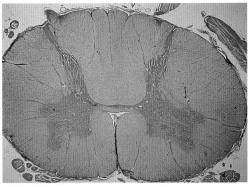

Cervical: the cord is relatively large and has an oval shape. Grey matter (dark red) is prominent, corresponding to the cervical enlargement supplying the upper limbs.

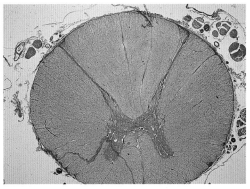

Thoracic: the cord is almost circular and has a smaller diameter. There is an intermediate amount of white matter. The grey matter is not as prominent here.

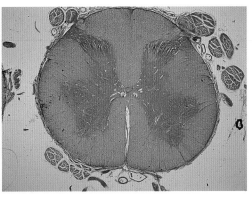

Lumbar: the cord has a larger diameter, corresponding to the increased amount of grey matter in the lumbar enlargement supplying the lower limbs. The white matter is less prominent.

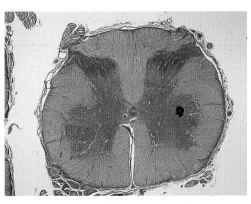

Sacral: in the region of the conus medullaris, the grey matter takes the form of two oval-shaped masses which occupy most of the cord with very little white matter.

Tracts in the spinal cord

Locations of the spinal cord tracts.
Blue: Ascending pathways
Red: Descending pathways
Purple: Fibres that pass in both directions.

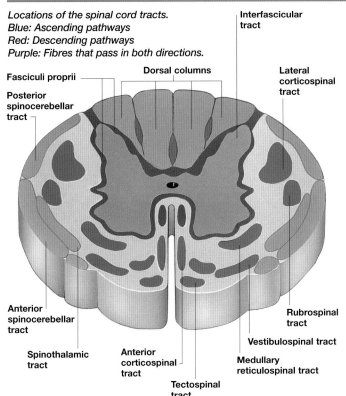

Fasciculi proprii
Posterior spinocerebellar tract
Dorsal columns
Interfascicular tract
Lateral corticospinal tract
Anterior spinocerebellar tract
Spinothalamic tract
Anterior corticospinal tract
Tectospinal tract
Medullary reticulospinal tract
Vestibulospinal tract
Rubrospinal tract

A tract is a collection of nerve axons which all have the same origin, destination and function.

ASCENDING TRACTS
These carry sensory information from the body up to the brain:
1 The dorsal columns carry information to the medulla in the brain about fine touch and pressure from receptors in the skin. They also allow position sense (proprioception) from receptors in the joints, tendons and muscles
2 The anterior and lateral spinothalamic tracts carry information about poorly localized touch, deep pressure sensation, pain and temperature
3 The anterior and posterior spinocerebellar tracts carry information about touch and pressure to the cerebellum to enable it to contribute to the control of voluntary movement.

DESCENDING TRACTS
These carry signals from the brain to the body. They are particularly involved with the control of movement.

The pyramidal or corticospinal tract has its origin in the nerve cells of the cerebral cortex involved with initiating voluntary movement. The tract passes down into the spinal cord with impulses passing out through the ventral spinal nerve roots to skeletal muscles.

EXTRAPYRAMIDAL TRACTS
1 The tectospinal tract begins in the midbrain and passes down in the anterior white column. It contributes to the control of balance and co-ordination
2 The rubrospinal tract originates in the red nucleus of the midbrain and descends in the lateral columns to help control posture and muscle tone
3 The reticulospinal tract begins in the brainstem reticular formation and descends in the anterior and lateral columns. It is involved with muscle tone
4 The vestibulospinal tract has its origin in the vestibular nuclei of the medulla and travels down in the anterior and lateral columns. It also contributes to the control of muscle tone.

Spinal nerves

There are 31 pairs of spinal nerves, arranged on each side of the spinal cord along its length. The pairs are grouped by region: eight cervical, twelve thoracic, five lumbar, five sacral and one coccygeal.

Each spinal nerve has two roots. The anterior, or ventral root, contains the axons of motor nerves which send impulses to control muscle movement. The posterior or dorsal root contains the axons of sensory nerves which send sensory information from the body into the spinal cord on its way to the brain.

SEGMENTS

Each root is formed by a series of small rootlets which attach it to the cord. The portion of the spinal cord which provides the rootlets for one dorsal root is referred to as a segment. In the lumbar and cervical regions, the rootlets are bunched closely, with the cord segments being only about 1 cm long. However, in the thoracic region they are more spread out, with segments more than 2 cm long.

NERVE FORMATION

The ventral and dorsal roots come together to form a single spinal nerve within the intervertebral foramina – small openings between the vertebrae through which the spinal nerves pass.

Just before the point of fusion with the ventral root, there is an enlargement of each dorsal root. This enlargement is known as the dorsal root ganglion – this is a collection of cell bodies of sensory nerves.

Anterior view

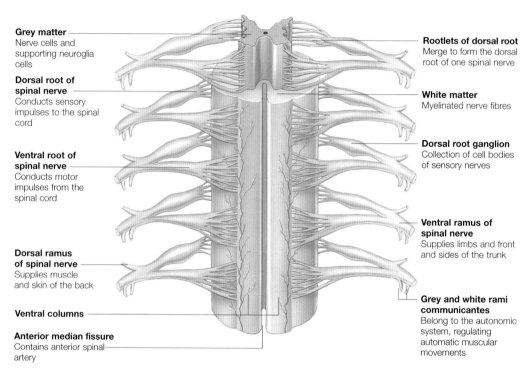

Grey matter
Nerve cells and supporting neuroglia cells

Dorsal root of spinal nerve
Conducts sensory impulses to the spinal cord

Ventral root of spinal nerve
Conducts motor impulses from the spinal cord

Dorsal ramus of spinal nerve
Supplies muscle and skin of the back

Ventral columns

Anterior median fissure
Contains anterior spinal artery

Rootlets of dorsal root
Merge to form the dorsal root of one spinal nerve

White matter
Myelinated nerve fibres

Dorsal root ganglion
Collection of cell bodies of sensory nerves

Ventral ramus of spinal nerve
Supplies limbs and front and sides of the trunk

Grey and white rami communicantes
Belong to the autonomic system, regulating automatic muscular movements

RAMI

Shortly after passing through its intervertebral foramen, each spinal nerve divides into several branches, or rami, including:
■ Ventral ramus: supplies the limbs and front and sides of the trunk
■ Dorsal ramus: supplies the deep muscles and skin of the back
■ Rami communicantes: part of the autonomic nervous system.

CAUDA EQUINA

Because the spinal cord is shorter than the vertebral column, the lower spinal nerve roots exit and travel downwards at quite an oblique angle. The lumbosacral nerve roots are bunched together and pass downwards almost vertically. This gives rise to the name *cauda equina* – Latin for horse's tail – which these lower nerve roots resemble.

Lumbar puncture

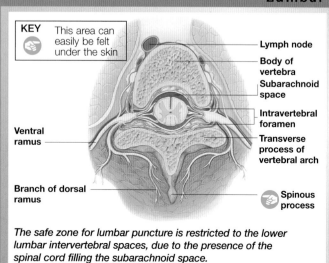

KEY This area can easily be felt under the skin

Ventral ramus

Branch of dorsal ramus

Lymph node

Body of vertebra

Subarachnoid space

Intravertebral foramen

Transverse process of vertebral arch

Spinous process

The safe zone for lumbar puncture is restricted to the lower lumbar intervertebral spaces, due to the presence of the spinal cord filling the subarachnoid space.

Although the spinal cord ends at about the level of the disc between the first and second lumbar vertebrae, the subarachnoid space continues to the second sacral vertebra. Below the end of the cord there is a sac of meninges containing the *cauda equina*, the filum terminale and cerebrospinal fluid (CSF).

A fine needle can be passed between two spinous processes at this level to allow a small amount of CSF to be taken for laboratory examination. This is a common clinical procedure known as lumbar puncture or spinal tap, and can be performed safely, usually between the third and fourth lumbar vertebrae.

A lumbar puncture (spinal tap) is carried out to obtain CSF or to allow drugs to be given directly into the spine. The needle is passed into the subarachnoid space.

Blood supply of the spinal cord

The spinal cord is supplied by a complex arrangement of arteries. This blood supply is vital for the normal functioning of the nervous system.

The anterior spinal arteries originate from the two vertebral arteries at the base of the brain and join together to form a single artery which runs down the front of the spinal cord in the anterior median fissure. Segmental branches of this artery supplies the anterior two-thirds of the spinal cord.

The posterior spinal arteries also arise from the vertebral arteries and split into two descending branches which run either side of the cord, one behind and one in front of the attachment of the dorsal roots. These vessels supply the posterior third of the cord.

There is additional supply from radicular arteries which originate from the deep cervical arteries in the neck, the intercostal arteries in the chest and the lumbar arteries in the lower back. These vessels enter through the intervertebral foramina alongside the spinal nerves.

Usually, one of the anterior radicular arteries is larger than the others and is referred to as the artery of Adamkiewicz. It most commonly arises on the left-hand side from a branch of the descending aorta in the upper lumbar or lower thoracic region. This branch may be the main blood supply to the lower two-thirds of the spinal cord, and injury to it following trauma or during surgery may produce serious neurological damage.

This cast of the aorta (red) and branches shows the rich supply of blood to the spinal column. All spinal nerve roots have associated arteries. The vessels show the segmental arrangement of the vertebrae.

Membranes that protect the spinal cord

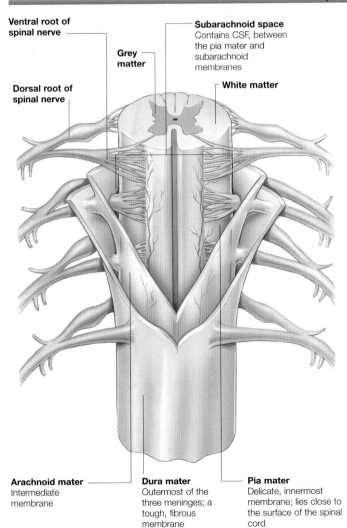

Ventral root of spinal nerve

Grey matter

Dorsal root of spinal nerve

Subarachnoid space
Contains CSF, between the pia mater and subarachnoid membranes

White matter

Arachnoid mater
Intermediate membrane

Dura mater
Outermost of the three meninges; a tough, fibrous membrane

Pia mater
Delicate, innermost membrane; lies close to the surface of the spinal cord

The bones of the vertebral column provide the major protection for the spinal cord, just as the skull does for the brain. However, like the brain, the cord has additional protection from three membranes, which continue down through the foramen magnum from inside the skull.

The dura mater is the tough, fibrous, outer membrane. The extradural or epidural space separates the dura from the bone of the vertebral bodies and contains fatty tissue and a plexus of veins.

The middle membrane is the arachnoid mater, which is much thinner and more delicate, with an arrangement of connective tissue fibres resembling a spider's web. There is a potential subdural space between the dura and the arachnoid, normally containing only a very thin film of fluid.

PIA MATER
The innermost membrane is the fine pia mater, which is closely applied to the surface of the spinal cord. It is transparent and richly supplied with fine blood vessels, which carry oxygen and nutrients to the cord. Between

Like the brain, the spinal cord is surrounded and protected by three membranes. These are the meninges – dura mater, arachnoid mater and pia mater.

the arachnoid and the pia is the subarachnoid space, which contains cerebrospinal fluid (CSF), which cushions the spinal cord, as well as helping to remove chemical waste products produced by nerve activity and metabolism. CSF is formed by the choroid plexuses inside the cerebral ventricles and circulates around the brain and spinal cord.

About 21 triangular extensions of the pia – the denticulate ligaments – pass outwards between the anterior and posterior nerve roots to join with the arachnoid and inner surface of the dura. The spinal cord is suspended by these in its dural sheath.

The end of the spinal cord is shown on this myelogram (a specialized radiograph). Strands of the pia mater anchor the cord to the coccyx.

Vessels and nerves of the thoracic wall

The thoracic wall (the ribcage and surrounding muscles and soft tissues) has a plentiful blood supply carried by the intercostal arteries and veins, which run along in the spaces between the ribs.

The intercostal arteries and veins form connections, or anastomoses, to create a network of blood vessels which encircles the thoracic wall and supplies all its structures. Each intercostal space has within it a posterior intercostal artery, which originates near the spine, and two anterior intercostal arteries, which originate at the front, next to the sternum (breastbone).

POSTERIOR ARTERIES
The first two posterior intercostal arteries arise from the subclavian arteries. The remaining posterior arteries arise directly from the aorta (the large central artery of the body) at the level of each rib. Each posterior intercostal artery gives off:
■ A dorsal branch – which travels backwards to supply the spine, back muscles and overlying skin
■ A collateral artery – a small branch which travels along the upper surface of the rib below.

ANTERIOR ARTERIES
The anterior intercostal arteries arise from the internal thoracic arteries, which run vertically down on either side of the sternum (breast bone). These arteries run along the underside of each rib together with the intercostal vein and nerve, and send a branch along the upper surface of the rib below.

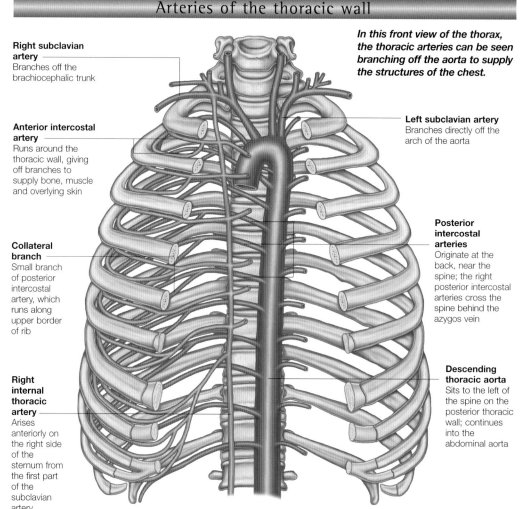

Arteries of the thoracic wall

In this front view of the thorax, the thoracic arteries can be seen branching off the aorta to supply the structures of the chest.

Right subclavian artery
Branches off the brachiocephalic trunk

Anterior intercostal artery
Runs around the thoracic wall, giving off branches to supply bone, muscle and overlying skin

Collateral branch
Small branch of posterior intercostal artery, which runs along upper border of rib

Right internal thoracic artery
Arises anteriorly on the right side of the sternum from the first part of the subclavian artery

Left subclavian artery
Branches directly off the arch of the aorta

Posterior intercostal arteries
Originate at the back, near the spine; the right posterior intercostal arteries cross the spine behind the azygos vein

Descending thoracic aorta
Sits to the left of the spine on the posterior thoracic wall; continues into the abdominal aorta

Veins of the thoracic wall

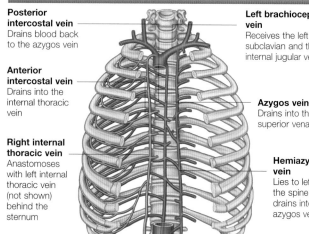

Posterior intercostal vein
Drains blood back to the azygos vein

Anterior intercostal vein
Drains into the internal thoracic vein

Right internal thoracic vein
Anastomoses with left internal thoracic vein (not shown) behind the sternum

Left brachiocephalic vein
Receives the left subclavian and the left internal jugular veins

Azygos vein
Drains into the superior vena cava

Hemiazygos vein
Lies to left of the spine and drains into the azygos vein

The intercostal veins run alongside the intercostal arteries in the spaces between the ribs. There are 11 posterior intercostal veins and one subcostal vein (lying beneath the 12th rib) on each side of the sternum which, like the arteries, anastomose with the corresponding anterior intercostal vessels to form a

Anterior view of the thoracic cage showing veins of the thoracic wall. Intercostal veins accompany the intercostal arteries and nerves and lie uppermost in the costal grooves.

network around the ribcage.
■ **Posterior veins**
These drain blood back to the azygos venous system, which lies in front of the spine at the back of the thoracic wall. This blood is then returned to the heart via the superior vena cava, the main central vein of the upper thorax.
■ **Anterior veins**
Like the arteries in the same position, the anterior veins drain in to the internal thoracic veins, which run vertically alongside the internal thoracic arteries in the front of the thoracic wall.

Nerves of the thoracic wall

There are 12 pairs of nerves which arise from the thoracic part of the spinal cord corresponding to the thoracic vertebrae. These run alongside the blood vessels to supply the muscles and skin of the thoracic wall.

The spinal nerves of the 12 thoracic vertebrae give off anterior branches which become the intercostal nerves (except the 12th, which becomes the subcostal nerve). Each intercostal nerve travels together with its corresponding artery and vein protected in the costal groove along the lower edge of each rib. The posterior branches of the thoracic spinal nerves supply skin and muscles of the back.

NERVE BRANCHES
A typical intercostal nerve has the following branches:
■ **Collateral branches**
These run along the top of the rib below and help supply the intercostal muscles.
■ **Lateral cutaneous branches**
These pierce the intercostal muscles to emerge on their outer surface. They divide into both anterior and posterior branches to supply a band of skin of the thoracic wall.
■ **Anterior cutaneous branches**
These terminal branches of the intercostal nerve pierce the muscles and emerge alongside the sternum to supply the skin of the front of the thorax.
■ **Muscular branches**
These supply the intercostal muscles as well as the various other muscles of respiration within the thoracic wall.

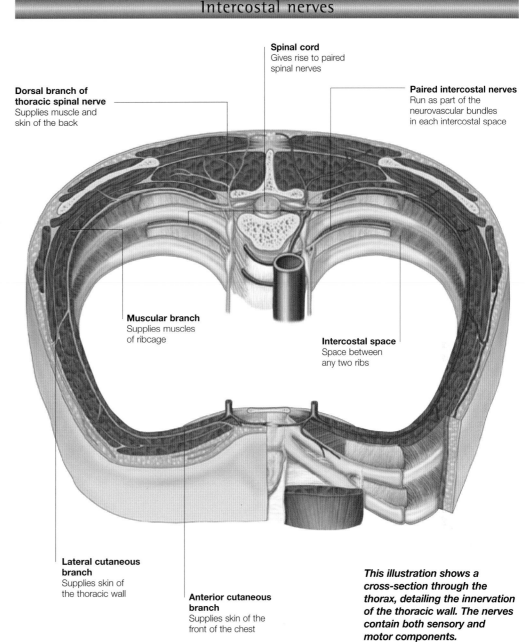

Intercostal nerves

Spinal cord
Gives rise to paired spinal nerves

Dorsal branch of thoracic spinal nerve
Supplies muscle and skin of the back

Paired intercostal nerves
Run as part of the neurovascular bundles in each intercostal space

Muscular branch
Supplies muscles of ribcage

Intercostal space
Space between any two ribs

Lateral cutaneous branch
Supplies skin of the thoracic wall

Anterior cutaneous branch
Supplies skin of the front of the chest

This illustration shows a cross-section through the thorax, detailing the innervation of the thoracic wall. The nerves contain both sensory and motor components.

Intercostal space

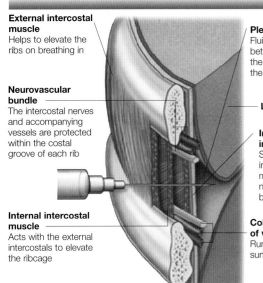

External intercostal muscle
Helps to elevate the ribs on breathing in

Neurovascular bundle
The intercostal nerves and accompanying vessels are protected within the costal groove of each rib

Internal intercostal muscle
Acts with the external intercostals to elevate the ribcage

Pleural cavity
Fluid-filled space between the lining of the thoracic wall and the lining of the lung

Lung

Innermost intercostal muscle
Separated from the internal intercostal muscle by the neurovascular bundles

Collateral branches of vessels and nerve
Run along the upper surface of the rib

The external and internal intercostal muscles span the outer side of the intercostal space between two ribs. Beneath these lie the 'neurovascular bundles' – the intercostal blood vessels and nerves lying within the costal grooves on the lower edges of the inner side of each rib, which act to shield the nerves and blood vessels from trauma. As well as the main intercostal vein, artery and nerve which lie, in that order,

Needle aspiration of pleural fluid is carried out by passing a needle between the ribs, avoiding the neurovascular bundles.

under each rib, there are corresponding collateral vessels and nerves which lie along the upper surface of the rib below. Beneath the neurovascular bundles lie the innermost intercostal and subcostal muscles, which span the inner surface of the intercostal space.

THORACOCENTESIS
In certain circumstances it may be necessary to pass a needle between the ribs in order to aspirate fluid, blood or pus from where it has collected around the lung (thoracocentesis). The needle must be inserted through the thoracic wall, avoiding the neurovascular bundles.

Muscles of the back

The muscles of the back give us our upright posture and allow flexibility and mobility of the spine. The superficial back muscles also act with other muscles to move the shoulders and upper arms.

The deep muscles of the back are concerned with support and movement of the spine, while the superficial muscles act to move the arm and shoulder.

SUPERFICIAL MUSCLES

The trapezius is a large, fan-shaped muscle whose top edge forms the visible slope from neck to shoulder. It attaches to the skull and helps to hold up and rotate the head and enables us to brace the shoulders back. The latissimus dorsi, the largest and most powerful back muscle, is attached to the spine from above the lower edge of the trapezius and runs down to the back of the pelvis. The latissimus dorsi allows a lifted arm to be pulled back into line with the trunk, even against great force.

Smaller muscles also contribute to this superficial muscle layer. Levator scapulae, rhomboid major and rhomboid minor run between the spine and the scapula and act to move the scapula up and inwards.

The 'rotator cuff' is a group of muscles that run between the scapula and head of the humerus (bone of the upper arm) at the shoulder joint. Together, they hold the head of the humerus tightly into the shoulder joint. Serratus posterior runs from the vertebrae to the ribs and moves the ribcage up during breathing.

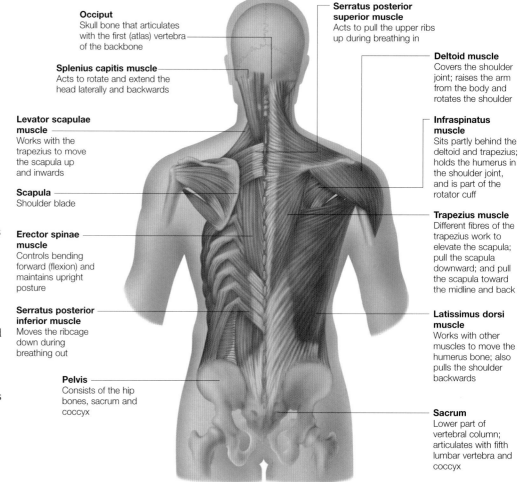

Occiput
Skull bone that articulates with the first (atlas) vertebra of the backbone

Splenius capitis muscle
Acts to rotate and extend the head laterally and backwards

Levator scapulae muscle
Works with the trapezius to move the scapula up and inwards

Scapula
Shoulder blade

Erector spinae muscle
Controls bending forward (flexion) and maintains upright posture

Serratus posterior inferior muscle
Moves the ribcage down during breathing out

Pelvis
Consists of the hip bones, sacrum and coccyx

Serratus posterior superior muscle
Acts to pull the upper ribs up during breathing in

Deltoid muscle
Covers the shoulder joint; raises the arm from the body and rotates the shoulder

Infraspinatus muscle
Sits partly behind the deltoid and trapezius; holds the humerus in the shoulder joint, and is part of the rotator cuff

Trapezius muscle
Different fibres of the trapezius work to elevate the scapula; pull the scapula downward; and pull the scapula toward the midline and back

Latissimus dorsi muscle
Works with other muscles to move the humerus bone; also pulls the shoulder backwards

Sacrum
Lower part of vertebral column; articulates with fifth lumbar vertebra and coccyx

Movements of the spine

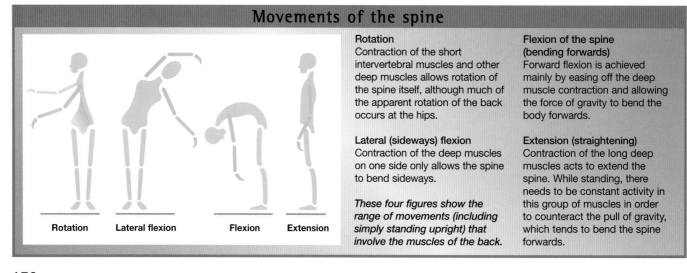

Rotation **Lateral flexion** **Flexion** **Extension**

Rotation
Contraction of the short intervertebral muscles and other deep muscles allows rotation of the spine itself, although much of the apparent rotation of the back occurs at the hips.

Lateral (sideways) flexion
Contraction of the deep muscles on one side only allows the spine to bend sideways.

These four figures show the range of movements (including simply standing upright) that involve the muscles of the back.

Flexion of the spine (bending forwards)
Forward flexion is achieved mainly by easing off the deep muscle contraction and allowing the force of gravity to bend the body forwards.

Extension (straightening)
Contraction of the long deep muscles acts to extend the spine. While standing, there needs to be constant activity in this group of muscles in order to counteract the pull of gravity, which tends to bend the spine forwards.

170

Deep muscles of the back

The deep muscles of the back attach to underlying bones of the spine, pelvis and ribs. They act together to allow smooth movements of the spine.

Muscles need to be attached to bone in order to give them the leverage they need to perform their functions. The bony attachments of the deep muscles of the back include the vertebrae, the ribs, the base of the skull and the pelvis.

DEEP MUSCLE LAYERS
The deep muscles of the back are built up in layers; the most deeply located muscles are very short, running from each vertebra obliquely to the one above. Over these lie muscles which are longer and run vertically between several vertebrae and the ribs. More superficially, the muscles become longer and some are attached to the pelvic bones and the occiput (back of the base of the skull) as well as to the vertebrae.

There are numerous muscles in these layers. Although each muscle is individually named according to its position, in practice they act in varying combinations rather than individually. Together, they form the large group of deep muscles which lie on either side of the spine and act in conjunction to maintain the spine in an S-shaped curve, enabling the fluid movements of the spine.

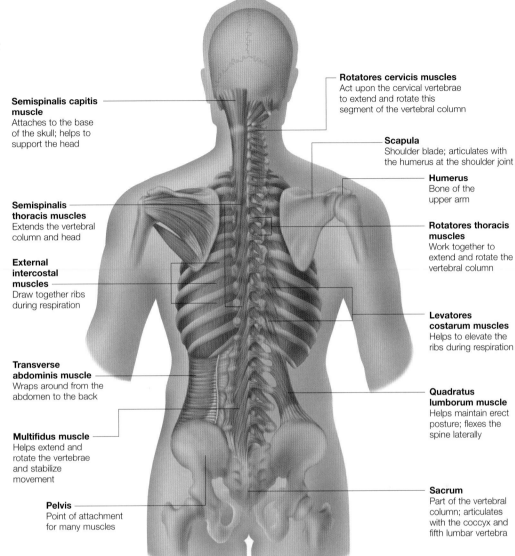

Semispinalis capitis muscle
Attaches to the base of the skull; helps to support the head

Semispinalis thoracis muscles
Extends the vertebral column and head

External intercostal muscles
Draw together ribs during respiration

Transverse abdominis muscle
Wraps around from the abdomen to the back

Multifidus muscle
Helps extend and rotate the vertebrae and stabilize movement

Pelvis
Point of attachment for many muscles

Rotatores cervicis muscles
Act upon the cervical vertebrae to extend and rotate this segment of the vertebral column

Scapula
Shoulder blade; articulates with the humerus at the shoulder joint

Humerus
Bone of the upper arm

Rotatores thoracis muscles
Work together to extend and rotate the vertebral column

Levatores costarum muscles
Helps to elevate the ribs during respiration

Quadratus lumborum muscle
Helps maintain erect posture; flexes the spine laterally

Sacrum
Part of the vertebral column; articulates with the coccyx and fifth lumbar vertebra

Supporting the head and neck

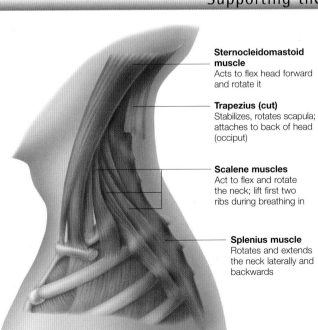

Sternocleidomastoid muscle
Acts to flex head forward and rotate it

Trapezius (cut)
Stabilizes, rotates scapula; attaches to back of head (occiput)

Scalene muscles
Act to flex and rotate the neck; lift first two ribs during breathing in

Splenius muscle
Rotates and extends the neck laterally and backwards

The deep muscles at the top of the spine, attached to the skull, also act to keep the neck extended and the head upright. The centre of gravity of the head is in front of the spine and so constant contraction of the muscles at the back of the neck is needed to prevent the head falling forward – hence the term 'nodding off' when the head nods forward as these muscles relax while falling asleep.

The sternocleidomastoid muscles are large muscles on either side of the neck. They act as the major muscles of flexion of the head and can be braced to support the head when it is

A lateral view of the muscles of the neck reveals some of the major muscles responsible for supporting and flexing the head and neck.

elevated – as happens when rising from a lying position. The sternocleidomastoid muscles are aided in these tasks by several other deep muscles.

The splenius muscles are broad sheets of muscle fibres that wrap around and over the deeper muscles of the neck. The splenius muscles originate in the cervical vertebrae (in the neck) and insert in the occipital bone at the back of the skull. When the muscles of one side of the neck are used alone, the neck and head is extended laterally or rotated; when used together and in collaboration with other muscles, the splenius muscles help to extend the neck backwards.

Extension of the neck is also aided by the action of the trapezius muscle which attaches to the occipital bone, thoracic vertebrae and scapula.

Pectoral girdle

The pectoral, or shoulder, girdle is the bony structure that articulates with and supports the upper limbs. It consists of the clavicles at the front of the chest and the scapulae that lie flat against the back.

The upper limb is connected to the skeleton by the pectoral or shoulder girdle, made up of the clavicle (collar bone) and the scapula (shoulder blade). The pectoral girdle has only one joint with the central skeleton, at the inner end of the clavicle where it articulates with the sternum (breastbone). The stability of the pectoral girdle is provided by muscles and ligaments attached to the skull, ribs, sternum and vertebrae.

THE CLAVICLE

The clavicle is an S-shaped bone that lies horizontally at the upper border of the chest. The front and upper surfaces of the clavicle are mostly smooth, while the under-surfaces are roughened and grooved by the attachments of muscles and ligaments.

The medial (inner) end of the clavicle has a large oval facet for connecting with the sternum at the sternoclavicular joint. A smaller facet lies at the other end where the clavicle articulates with the acromion (a bony prominence of the scapula) at the acromioclavicular joint.

The clavicle acts as a strut to brace the upper limb away from the body, thereby allowing a wide range of free movement. Along with the scapula and its muscular connections, it also transmits the force of impacts on the upper limb to the skeleton.

Pectoral girdle from above

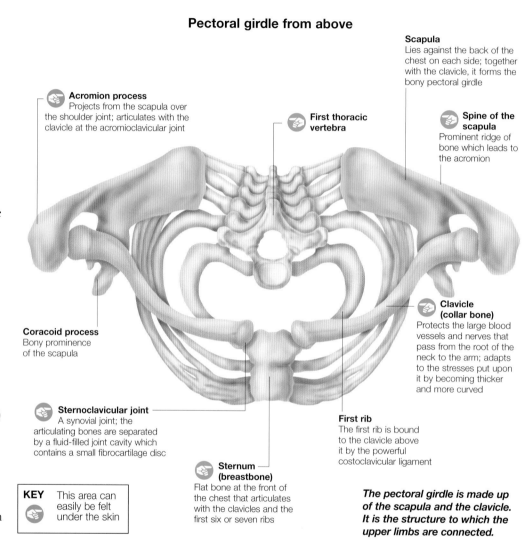

Acromion process
Projects from the scapula over the shoulder joint; articulates with the clavicle at the acromioclavicular joint

First thoracic vertebra

Scapula
Lies against the back of the chest on each side; together with the clavicle, it forms the bony pectoral girdle

Spine of the scapula
Prominent ridge of bone which leads to the acromion

Coracoid process
Bony prominence of the scapula

Clavicle (collar bone)
Protects the large blood vessels and nerves that pass from the root of the neck to the arm; adapts to the stresses put upon it by becoming thicker and more curved

Sternoclavicular joint
A synovial joint; the articulating bones are separated by a fluid-filled joint cavity which contains a small fibrocartilage disc

First rib
The first rib is bound to the clavicle above it by the powerful costoclavicular ligament

Sternum (breastbone)
Flat bone at the front of the chest that articulates with the clavicles and the first six or seven ribs

KEY This area can easily be felt under the skin

The pectoral girdle is made up of the scapula and the clavicle. It is the structure to which the upper limbs are connected.

Joints of the clavicle

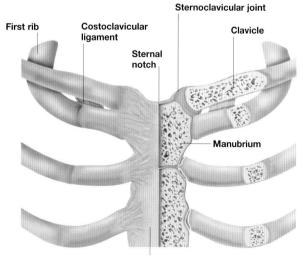

First rib

Costoclavicular ligament

Sternal notch

Sternoclavicular joint

Clavicle

Manubrium

Body of sternum

The sternoclavicular joint is the only bony connection between the pectoral girdle and the rest of the skeleton. It can be felt under the skin, as the sternal end of the clavicle is fairly large and extends above the top of the manubrium (the top of the sternum), both sides together forming the familiar 'sternal notch' at the base of the neck.

The cavity is divided into two by an articular disc made of fibrocartilage, which improves the fit of the bones and keeps the

A tough fibrous capsule (sheath), together with strong surrounding ligaments, holds the sternoclavicular joint firmly in place.

joint stable. The joint is further stabilized by the costoclavicular ligament, which anchors its underside to the first rib.

Only a small degree of movement is possible at the sternoclavicular joint; the outer end of the clavicle can move upward, as when shrugging the shoulders, or forward when the arm reaches out to pick up something in front of the body.

The acromioclavicular joint is formed between the outer end of the clavicle and the acromion of the scapula. The acromio-clavicular joint rotates the scapula on the clavicle under the influence of muscles which attach the scapula to the rest of the skeleton.

Scapula

The scapula is a flat, triangular-shaped bone which lies against the back of the chest. With the clavicle, it forms the bony pectoral girdle.

The scapula, or shoulder blade, lies against the back of the chest on each side overlying the second to seventh ribs. As a rough triangle, the scapula has three borders: medial (inner), lateral (outer) and superior, with three angles between them.

SURFACES

The scapula has two surfaces: anterior (front) and posterior (back). The anterior or costal (rib) surface lies against the ribs at the back of the chest and is concave, having a large hollow called the subscapular fossa that provides a large surface area for the attachment of muscles.

The posterior surface is divided by a prominent spine. The supraspinous fossa is the small area above the spine, while the infraspinous fossa lies below. These hollows also provide sites of attachment for muscles of the same name.

BONY PROCESSES

The spine of the scapula is a thick projecting ridge, continuous with the bony outcrop called the acromion. This is a flattened prominence that forms the tip of the shoulder. The lateral angle, the thickest part of the scapula, contains the glenoid cavity, the depression into which the head of the humerus fits at the shoulder joint. The coracoid process – an important site of attachment of muscles and ligaments – is also palpable in this area.

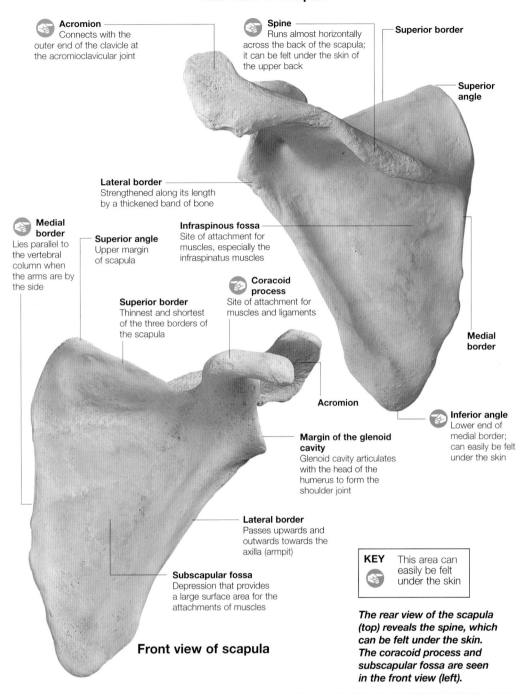

Rear view of scapula

Acromion
Connects with the outer end of the clavicle at the acromioclavicular joint

Spine
Runs almost horizontally across the back of the scapula; it can be felt under the skin of the upper back

Superior border

Superior angle

Lateral border
Strengthened along its length by a thickened band of bone

Infraspinous fossa
Site of attachment for muscles, especially the infraspinatus muscles

Medial border

Medial border
Lies parallel to the vertebral column when the arms are by the side

Superior angle
Upper margin of scapula

Coracoid process
Site of attachment for muscles and ligaments

Superior border
Thinnest and shortest of the three borders of the scapula

Inferior angle
Lower end of medial border; can easily be felt under the skin

Acromion

Margin of the glenoid cavity
Glenoid cavity articulates with the head of the humerus to form the shoulder joint

Lateral border
Passes upwards and outwards towards the axilla (armpit)

Subscapular fossa
Depression that provides a large surface area for the attachments of muscles

Front view of scapula

KEY This area can easily be felt under the skin

The rear view of the scapula (top) reveals the spine, which can be felt under the skin. The coracoid process and subscapular fossa are seen in the front view (left).

Winged scapula

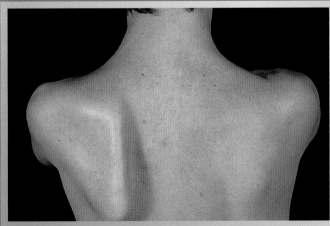

As it has no bony connections with the spine or the ribs, the scapula is held tightly against the posterior wall of the chest by the action of muscles, mainly the serratus anterior muscle.

Serratus anterior is supplied by the long thoracic nerve which descends from the axilla (armpit) on the surface of the muscle under

The position of this patient's left scapula is the result of damage to the long thoracic nerve. This nerve supplies the serratus anterior muscle that holds the scapula to the chest.

the skin, where it is vulnerable to injury. If this nerve is damaged, for instance by a penetrating wound, the muscle will be paralysed and the contraction within it that holds the scapula flat against the ribs will cease.

In this situation, the medial border and inferior angle of the scapula become more prominent and move away from the midline, the scapula jutting out like a wing. This gives rise to the name of 'winged scapula' for this condition, which is most obvious when the arm is pushed against a door or wall.

Muscles of the pectoral girdle

The pectoral girdle consists of the scapulae and clavicles, and is responsible for attaching the upper limbs to the central skeleton. The pectoral girdle muscles hold the scapulae and clavicles in place.

The pectoral girdle is defined as the structure which attaches the upper limbs to the axial skeleton, namely the clavicles and scapulae, to which the muscles of the pectoral girdle attach; however, there are a few muscles which connect the upper limb directly to the central skeleton, and cause indirect movements of the pectoral girdle. This group of muscles lie superficially on the trunk; pectoralis major at the front and latissimus dorsi at the back.

PECTORALIS MAJOR

Pectoralis major arises via two heads, one from the sternum (breastbone) and adjacent rib (costal) cartilages, and another from the middle third of the clavicle. Its tendon twists anti-clockwise as it courses towards the outer lip of the bicipital groove, on the upper end of the humerus. This twisting gives the clavicular head a greater mechanical advantage during flexion (bending) of the arm.

Pectoralis major derives its nerve supply and blood supply from a wide source. The sternocostal head of pectoralis major is a powerful adductor of the arm (pulling the limb towards the body), and is hence well-developed in climbers and weight-lifters. If the arm is kept fixed, this muscle can elevate the ribs as an accessory muscle of inspiration.

Muscles of the pectoral girdle from the front

Superficial

Deep

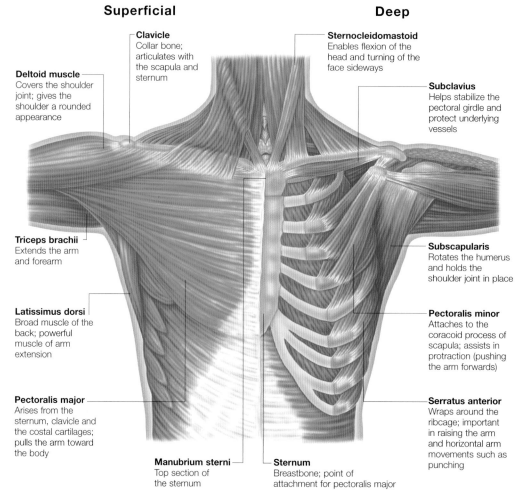

Clavicle
Collar bone; articulates with the scapula and sternum

Deltoid muscle
Covers the shoulder joint; gives the shoulder a rounded appearance

Triceps brachii
Extends the arm and forearm

Latissimus dorsi
Broad muscle of the back; powerful muscle of arm extension

Pectoralis major
Arises from the sternum, clavicle and the costal cartilages; pulls the arm toward the body

Manubrium sterni
Top section of the sternum

Sternocleidomastoid
Enables flexion of the head and turning of the face sideways

Subclavius
Helps stabilize the pectoral girdle and protect underlying vessels

Subscapularis
Rotates the humerus and holds the shoulder joint in place

Pectoralis minor
Attaches to the coracoid process of scapula; assists in protraction (pushing the arm forwards)

Serratus anterior
Wraps around the ribcage; important in raising the arm and horizontal arm movements such as punching

Sternum
Breastbone; point of attachment for pectoralis major

Beneath the pectoralis major

Deep to (lying beneath) the pectoralis major are subclavius and pectoralis minor. Subclavius is a rather insignificant muscle that probably helps to stabilize the clavicle during movements of the pectoral girdle. Following a clavicular fracture, the subclavius, together with deltoid

If the arm is kept fixed, pectoralis major can elevate the ribs, and so act as an accessory muscle of inspiration. Exhausted sprinters, with hands on knees, exploit this feature.

and gravity, acts to pull downward on the outer fragment, while the medial fragment is pulled upwards by the unrestrained action of sternocleidomastoid. The separation of fragments may pose a particular threat to the subclavian vessels close by.

Sternocleidomastoid, which arises from the medial third of the clavicle and manubrium sternum, is mainly involved with movements of the head and neck.

Pectoralis minor arises from the second, third, fourth and

fifth ribs, and attaches to the coracoid process of the scapula. It assists in pulling the scapula against and around the trunk wall. This action (protraction) is necessary to 'throw' a punch.

The main muscle which performs protraction is serratus anterior. This muscle wraps itself around the wall of the rib cage, to attach to the scapula's inner edge. The lower four digitations converge on the inferior angle of the scapula, and are involved in assisting trapezius during scapular rotations.

Pectoral girdle from the back

The large trapezius and latissimus dorsi are superficial muscles of the back which attach to and influence the movement of the pectoral girdle.

Latissimus dorsi arises from the lower thoracic and the lumbar and sacral vertebrae. It also arises from the thoracolumbar fascia and posterior part of the iliac crest, with a few fibres attaching to the lower four ribs. From this broad base (latissimus means 'broadest' in Latin), it converges onto the floor of the bicipital groove at the upper end of the humerus.

This muscle assists pectoralis major in pulling the arm towards the body (adduction). Since the muscle wraps itself around the lower ribs, it assists during forceful expiration (breathing out), for example during coughing.

TRAPEZIUS

Partly overlapping the latissimus dorsi is the lower part of the trapezius muscle. The trapezius also has a broad origin from the base of the skull (occipital protuberance) to the spines of the twelve thoracic vertebrae. The lower fibres attach to the spine of the scapula; intermediate fibres to the acromion process; and upper fibres to the outer third of the clavicle. The upper part serves to shrug the shoulders; the middle and lower parts serve to laterally rotate the scapula.

Muscles of the pectoral girdle from the back

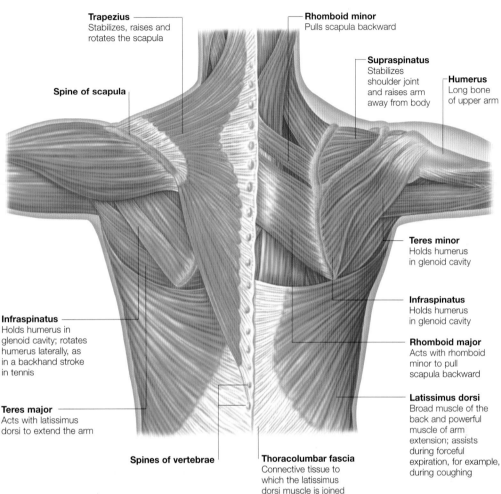

Superficial

Deep

Trapezius
Stabilizes, raises and rotates the scapula

Spine of scapula

Infraspinatus
Holds humerus in glenoid cavity; rotates humerus laterally, as in a backhand stroke in tennis

Teres major
Acts with latissimus dorsi to extend the arm

Spines of vertebrae

Rhomboid minor
Pulls scapula backward

Supraspinatus
Stabilizes shoulder joint and raises arm away from body

Humerus
Long bone of upper arm

Teres minor
Holds humerus in glenoid cavity

Infraspinatus
Holds humerus in glenoid cavity

Rhomboid major
Acts with rhomboid minor to pull scapula backward

Latissimus dorsi
Broad muscle of the back and powerful muscle of arm extension; assists during forceful expiration, for example, during coughing

Thoracolumbar fascia
Connective tissue to which the latissimus dorsi muscle is joined

The rhomboids act together to retract the scapula to enable powerful protraction. This action can be seen before a boxer throws a punch.

Deep dissection

The scapula is pulled backwards (retracted) by the action of the rhomboids (rhomboid major and minor). They attach the inner edge of the scapula to the vertebral column. These muscles allow the shoulder to be 'braced'; as seen before a punch is thrown, or before a forceful push, since this maximizes the force in protraction (forwards).

Since these muscles lie deep to trapezius, they are difficult to see and feel. However, should they become paralysed on one side, the scapula on that side would be displaced further away from the midline.

Rhomboid major and minor, together with the levator scapulae, also act to rotate the scapula medially, and hence counteract the actions of the trapezius and serratus anterior.

MOVEMENT OF SCAPULA

Movements of the scapula are essential in providing the widest range of motion at the shoulder joint. Although there is no anatomical joint between the scapula and the trunk, clinicians often refer to a scapulothoracic 'joint', since there is a great deal of movement between the two.

These movements are also transferred to the sterno-clavicular joint through the clavicle. The sternoclavicular joint is the sole connecting link between the pectoral girdle and the trunk.

Ribcage

The ribcage protects the vital organs of the thorax, as well as providing sites for the attachment of muscles of the back, chest and shoulders. It is also light enough to move during breathing.

The ribcage is supported at the back by the 12 thoracic vertebrae of the spinal column and is formed by the 12 paired ribs, the costal cartilages and the bony sternum, or breastbone, at the front.

THE RIBS

Each of the 12 pairs of ribs is attached posteriorly (at the back) to the corresponding numbered thoracic vertebra. The ribs then curve down and around the chest towards the anterior (front) surface of the body.

The 12 ribs can be divided into two groups according to their anterior (front) site of attachment:

■ **True (vertebrosternal) ribs**
The first seven pairs of ribs attach anteriorly directly to the sternum via individual costal cartilages.

■ **False ribs**
These do not attach directly to the sternum. Rib pairs eight to 10 (vertebrochondral ribs) attach indirectly to the sternum via fused costal cartilages. Rib pairs 11 and 12 do not have attachments to bone or cartilage and so are known as 'vertebral' or 'floating' ribs. Their anterior ends lie buried within the musculature of the lateral abdominal wall.

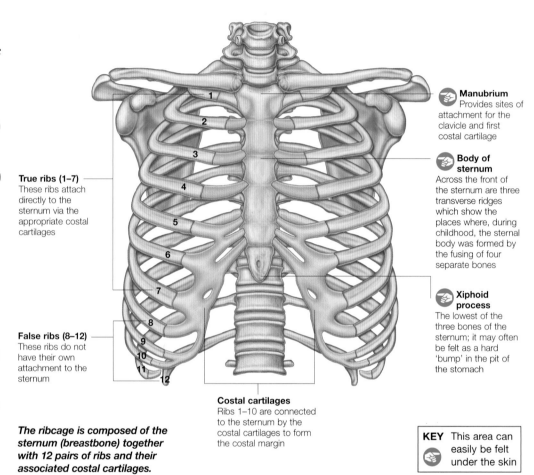

True ribs (1–7)
These ribs attach directly to the sternum via the appropriate costal cartilages

False ribs (8–12)
These ribs do not have their own attachment to the sternum

Manubrium
Provides sites of attachment for the clavicle and first costal cartilage

Body of sternum
Across the front of the sternum are three transverse ridges which show the places where, during childhood, the sternal body was formed by the fusing of four separate bones

Xiphoid process
The lowest of the three bones of the sternum; it may often be felt as a hard 'bump' in the pit of the stomach

Costal cartilages
Ribs 1–10 are connected to the sternum by the costal cartilages to form the costal margin

KEY This area can easily be felt under the skin

The ribcage is composed of the sternum (breastbone) together with 12 pairs of ribs and their associated costal cartilages.

Rib structure

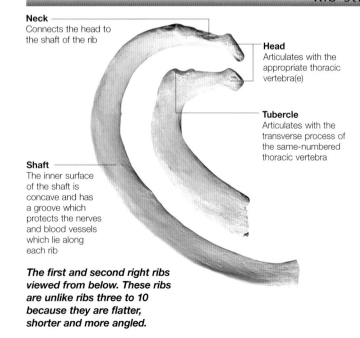

Neck
Connects the head to the shaft of the rib

Head
Articulates with the appropriate thoracic vertebra(e)

Tubercle
Articulates with transverse process of the same-numbered thoracic vertebra

Shaft
The inner surface of the shaft is concave and has a groove which protects the nerves and blood vessels which lie along each rib

The first and second right ribs viewed from below. These ribs are unlike ribs three to 10 because they are flatter, shorter and more angled.

While they all vary slightly in their structure, ribs three to 10 are similar enough to be described as 'typical ribs'. They consist of the following parts:

■ **Head.** This connects with the thoracic vertebra with the same numeric value and the one immediately above that (for example the fourth rib attaches to both the third and fourth thoracic vertebrae)

■ **Neck.** This narrowed length of rib connects the head to the shaft or body

■ **Tubercle.** This raised, roughened area lies at the junction of neck and shaft and bears a facet for articulation with the transverse process of the thoracic vertebra

■ **Shaft.** The rib continues as a flattened, curved bone which bends around at the 'costal angle' to encircle the thorax.

DISSIMILAR RIBS

■ **First rib.** This is the widest, shortest and most flattened rib; it has only one facet on its head for articulation with the first thoracic vertebra. On its upper surface it has a prominent 'scalene tubercle'

■ **Second rib.** This rib is thinner than the first, its shaft being more like that of a typical rib. Half-way down the shaft it has a second prominent tubercle for the attachment of muscles

■ **11th and 12th ribs (floating ribs).** These have only a single facet on their heads and do not have a point of articulation between their tubercle and the transverse process of the corresponding thoracic vertebrae. The ends of their shafts carry only a cap of cartilage and do not connect with any of the other ribs.

The sternum

The sternum (breastbone) is a long, flat bone which lies vertically at the centre of the anterior (front) surface of the ribcage.

The sternum has three parts:
- **The manubrium.** This bone forms the upper part of the sternum and is in the shape of a rough triangle with a prominent, and easily palpable, notch in the centre of its superior surface, the 'suprasternal notch'
- **The body.** The manubrium and the body of the sternum lie in slightly different planes, angled so that their junction, the manubriosternal joint, projects forwards forming the 'sternal angle of Louis'. The body of the sternum is longer than the manubrium, forming the greater length of the breastbone
- **The xiphoid process.** This is a small pointed bone which projects downwards and slightly backwards from the lower end of the body of the sternum. In young people it may be cartilaginous, but it usually becomes completely ossified (changed to bone) by 40–50 years of age.

The sternum (breastbone) consists of three parts: the manubrium, the body and the xiphoid process.

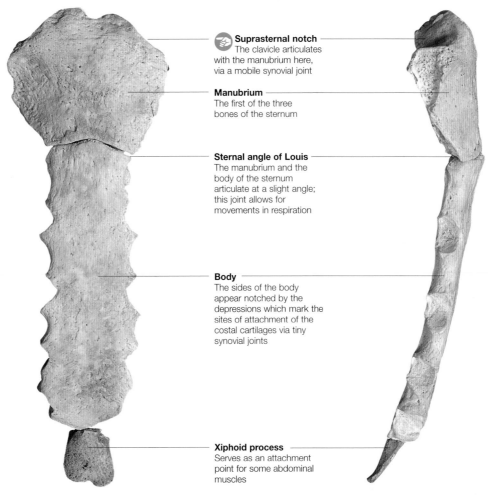

Suprasternal notch
The clavicle articulates with the manubrium here, via a mobile synovial joint

Manubrium
The first of the three bones of the sternum

Sternal angle of Louis
The manubrium and the body of the sternum articulate at a slight angle; this joint allows for movements in respiration

Body
The sides of the body appear notched by the depressions which mark the sites of attachment of the costal cartilages via tiny synovial joints

Xiphoid process
Serves as an attachment point for some abdominal muscles

The costal cartilages

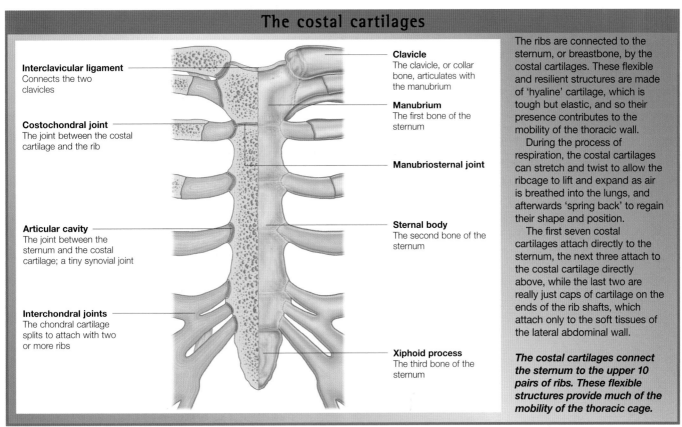

Interclavicular ligament
Connects the two clavicles

Costochondral joint
The joint between the costal cartilage and the rib

Articular cavity
The joint between the sternum and the costal cartilage; a tiny synovial joint

Interchondral joints
The chondral cartilage splits to attach with two or more ribs

Clavicle
The clavicle, or collar bone, articulates with the manubrium

Manubrium
The first bone of the sternum

Manubriosternal joint

Sternal body
The second bone of the sternum

Xiphoid process
The third bone of the sternum

The ribs are connected to the sternum, or breastbone, by the costal cartilages. These flexible and resilient structures are made of 'hyaline' cartilage, which is tough but elastic, and so their presence contributes to the mobility of the thoracic wall.

During the process of respiration, the costal cartilages can stretch and twist to allow the ribcage to lift and expand as air is breathed into the lungs, and afterwards 'spring back' to regain their shape and position.

The first seven costal cartilages attach directly to the sternum, the next three attach to the costal cartilage directly above, while the last two are really just caps of cartilage on the ends of the rib shafts, which attach only to the soft tissues of the lateral abdominal wall.

The costal cartilages connect the sternum to the upper 10 pairs of ribs. These flexible structures provide much of the mobility of the thoracic cage.

Muscles and movements of the ribcage

The bony skeleton of the ribcage is sheathed in several layers
of muscle which include many of the powerful muscles of the upper
limb and back, as well as those which act upon the ribcage alone.

The integral muscles of the ribcage are concerned with respiration (breathing). They attach only to the ribcage and the thoracic spine. They form the structure of the thoracic wall, enclosing and protecting the vital internal organs of the thorax.

INTERCOSTAL MUSCLES

The intercostal muscles fill the 11 intercostal spaces between the ribs. They lie in three layers, the external intercostals lying superficially, then the internal intercostals, with the innermost intercostals at the deepest level.

■ **External intercostal muscles**
The fibres of each external intercostal muscle run downwards and forwards to the rib below and their contraction acts to lift the ribs during inspiration (breathing in).

■ **Internal intercostal muscles**
The internal intercostal muscles lie just deep to the external intercostals and at right angles to them; that is, their fibres run downwards and backwards from the upper to the lower rib. Like the external intercostal muscles, they act to assist in inspiration.

■ **Innermost intercostal muscles**
These lie deep to the internal intercostal muscles, their fibres running in the same direction. They are separated from the internal intercostals by connective tissue containing the nerves and blood vessels.

Internal view of the chest wall

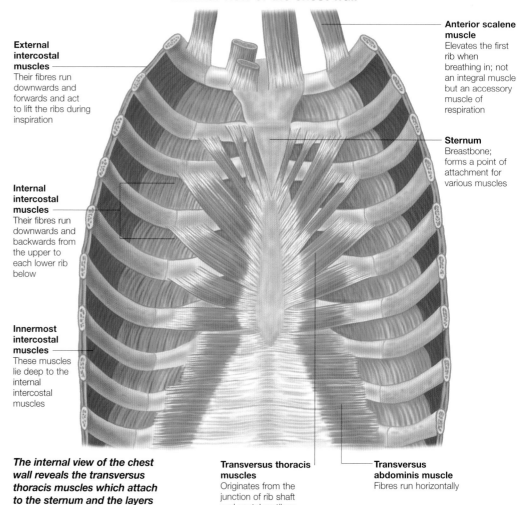

External intercostal muscles
Their fibres run downwards and forwards and act to lift the ribs during inspiration

Internal intercostal muscles
Their fibres run downwards and backwards from the upper to each lower rib below

Innermost intercostal muscles
These muscles lie deep to the internal intercostal muscles

Anterior scalene muscle
Elevates the first rib when breathing in; not an integral muscle but an accessory muscle of respiration

Sternum
Breastbone; forms a point of attachment for various muscles

The internal view of the chest wall reveals the transversus thoracis muscles which attach to the sternum and the layers of intercostal muscles.

Transversus thoracis muscles
Originates from the junction of rib shaft and costal cartilage

Transversus abdominis muscle
Fibres run horizontally

Integral muscles of the ribcage

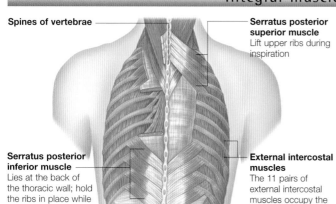

Spines of vertebrae

Serratus posterior superior muscle
Lift upper ribs during inspiration

Serratus posterior inferior muscle
Lies at the back of the thoracic wall; hold the ribs in place while breathing

External intercostal muscles
The 11 pairs of external intercostal muscles occupy the intercostal spaces

The integral muscles that form the structure of the ribcage include:

■ **Intercostal muscles**
These lie in three layers filling each intercostal space between the ribs.

■ **Subcostal muscles**
These are small muscles which run down on the inner surface of the posterior thoracic wall

The view of the back of the thoracic wall reveals the external intercostal muscles and serratus inferior and superior muscles.

between the lower ribs. Their fibres run in the same direction as those of the internal intercostal muscles and act to help in elevation of the ribs.

■ **Transversus thoracis muscles**
Small muscles which may vary in size and shape lying on the inside of the front of the chest.

■ **Serratus posterior**
These lie at the back of the thoracic wall in two parts: superior, which lifts the upper ribs during inspiration; and inferior, which holds the ribs in place during breathing.

Movements of ribcage

During the action of breathing, the chest cavity expands and contracts, causing air to enter and leave. Expansion of the chest cavity is achieved by contraction of the diaphragm and movements of the ribcage.

Movements of the ribcage during quiet respiration are due to the action of the respiratory muscles, the most important being the intercostals. Contraction of the intercostal muscles causes the ribcage to expand both sideways and from front to back.

EXPANDING THE CHEST
The lower ribs, especially, are lifted up and out to the sides in what had been described as a 'bucket handle' movement that increases the width of the ribcage. When the upper ribs are elevated, the sternum (breastbone) is also pulled up and rotated slightly so that its lower end moves forward. This action increases the depth of the ribcage and is described as a 'pump handle' movement.

Together, these movements act to increase the volume of the ribcage which, in turn, expands the underlying lungs and draws air in. When the intercostal muscles relax at the end of inspiration (breathing in), the ribcage descends again under the influence of gravity and the natural elasticity of the lungs.

Ribcage when breathing in

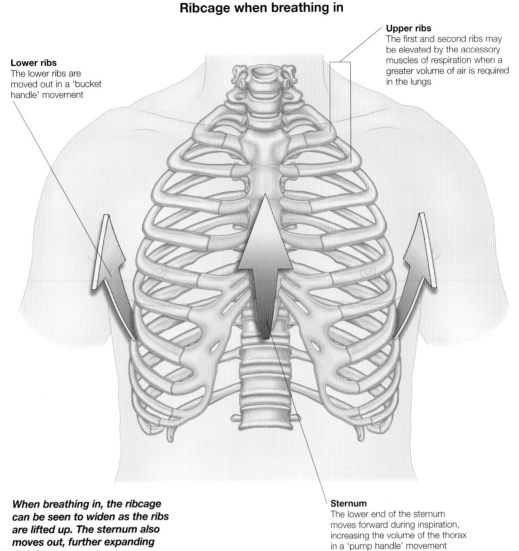

Lower ribs
The lower ribs are moved out in a 'bucket handle' movement

Upper ribs
The first and second ribs may be elevated by the accessory muscles of respiration when a greater volume of air is required in the lungs

Sternum
The lower end of the sternum moves forward during inspiration, increasing the volume of the thorax in a 'pump handle' movement

When breathing in, the ribcage can be seen to widen as the ribs are lifted up. The sternum also moves out, further expanding the chest.

Accessory muscles of respiration

Sternocleidomastoid muscle
This strong muscle, which turns the head, is also used in deep inspiration

Scalene muscles
Elevate the ribs during inspiration

Pectoralis minor muscle
Draws ribcage upwards and outwards when the shoulder is fixed

Pectoralis major muscle
Large muscle which can aid pulling the ribcage upward when the shoulder is in a fixed position

Rectus abdominis muscle
This strong segmented muscle aids in forced expiration, such as coughing

The accessory muscles of respiration are brought into play when a large volume of air is require, for example, after exercising.

There are times when a much greater volume of air must enter the lungs or when, due to lung disease, there is more resistance to the entry of air. At these times, the accessory muscles of respiration are brought into play.

These are muscles with attachments both to the ribcage and to other parts of the upper skeleton, their normal function being to move the head, neck or upper limbs. If the origin of these powerful muscles is fixed at the other end, their contraction will lead to forcible movement and expansion of the ribcage.

This can be seen in the posture of an athlete after a race when the head is held back and the hands are held on hips or knees to brace the arms. This allows the muscles of the neck and shoulder girdle to expand the chest powerfully.

Looking inside the chest

A plain chest X-ray covers some of the most important organs
in the body, notably the heart and the lungs, and can
reveal important information on a wide range of medical conditions.

The chest X-ray is one of the hospital investigations most frequently requested by doctors, and it is a powerful diagnostic tool. It can provide a great deal of information about chest and heart disease and, in certain circumstances, problems in the abdominal cavity. It is a relatively inexpensive and readily available procedure; few other tests offer such value for money. Often the chest X-ray provides vital information that will lead to a correct diagnosis, allowing the appropriate treatment.

The chest X-ray is analysed in a systematic way: the soft tissues, bones, lungs, hila (central inner parts of the lung), the heart and its major vessels, the diaphragm and the airways are analysed separately. These may all yield subtle or obvious clues which could be crucial for diagnosis and treatment.

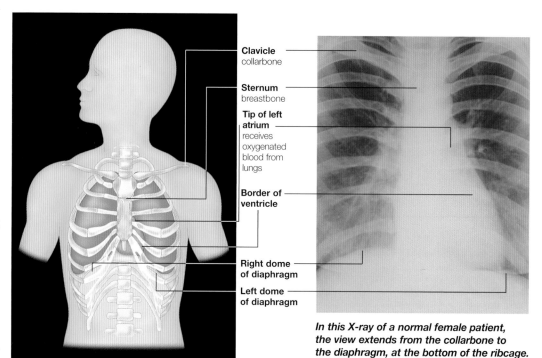

Clavicle
collarbone

Sternum
breastbone

Tip of left atrium
receives oxygenated blood from lungs

Border of ventricle

Right dome of diaphragm

Left dome of diaphragm

In this X-ray of a normal female patient, the view extends from the collarbone to the diaphragm, at the bottom of the ribcage. The left side of the heart is clearly visible.

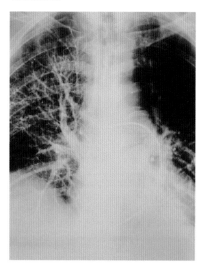

Pulmonary embolism
A pulmonary embolism is an obstruction of the pulmonary artery (which carries blood from the heart to the lungs, where it receives oxygen). This is usually caused by thrombosis – a blood clot that has travelled from its original site of formation, usually in the leg. Large pulmonary emboli can cause heart failure or sudden death. In this X-ray, a blood clot is blocking branches of the pulmonary artery in the left upper zone.

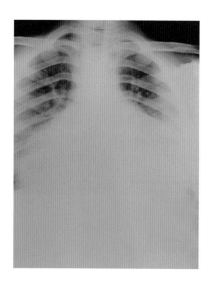

Pericardial effusion
In pericardial effusion, excess fluid accumulates in the pericardium – the membrane around the heart – often due to inflammation (pericarditis). The incoming fluid puts pressure on the heart, reducing the amount of blood that can flow through it – a serious condition known as cardiac tamponade. The globular and enlarged heart can be seen to the right of this X-ray.

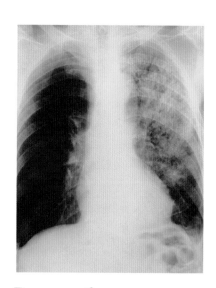

Pneumothorax
This is caused by air – usually from the lung – leaking into the pleural cavity (the space between the lung and the chest wall) causing the lung to collapse. This can occur for no apparent reason (spontaneous pneumothorax), due to chest injuries (traumatic pneumothorax) or when a hole in the lung prevents breathing out (tension pneumothorax). In this X-ray, there is air between the liver (on the right) and diaphragm.

Examining lung damage

A range of different lung problems can be identified and examined by studying a plain chest X-ray. They are a vital diagnostic tool.

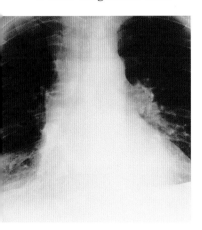

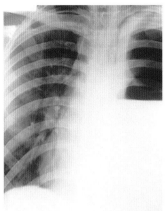

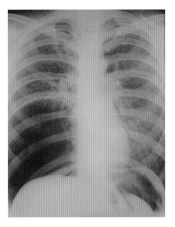

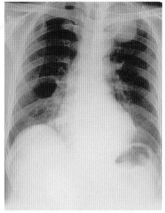

Asbestosis

The X-ray above shows the results of long-term exposure to asbestos fibres. The covering of the lungs (pleura) has thickened by chronic irritation caused by the inhalation of tiny fibres and developed plaques made up of deposited calcium compounds. Asbestosis is common among former boiler-makers, shipyard workers, those in textile manufacturing, in the construction industry and people making brake linings. People who smoke, and have also been exposed to asbestos, have a 90 times greater risk of contracting lung cancer.

Pleural effusion

The lungs and inside of the chest are covered with pleura – lubricated membranes that slide painlessly past each other during breathing. In pleural effusion, fluid, pus or blood may collect between the two layers. Lung cancer, infection or heart failure are possible causes. Pleural effusion causes fluid to accumulate around the lung. This causes breathlessness and chest pain when taking a breath.The level of fluid in the left lung is clearly visible.

Pneumonia

Pneumonia has been caused in this instance by Pseudomonas bacteria. The tiny air sacs (alveoli) of the lungs fill with fluid, converting into a solid mass in a process called consolidation. When the chest is tapped, these areas produce a dull sound. In this image, as a consequence of infection in the left lung, air has been replaced by fluid, causing the widespread inflammation shown.

Bronchial carcinoma

Bronchial carcinoma (lung cancer) at the apex of the left lung is visible in this X-ray. Such a tumour can spread through the upper ribs giving rise to clinical symptoms, such as weakness and pain in the arm. The well defined, circular region at top right of the picture (the apex of the left lung) shows cancer cells filling the air spaces.

How chest X-rays are taken

The standard chest X-ray is posteroanterior (PA), which means that the beam enters from behind, with the film cassette in front of the patient. However, many emergency chest X-rays are anteroposterior (AP), the beam entering the front and hitting the film behind. An AP film can be taken with the patient either sitting, or lying in bed, but the results can be quite different in each case.

This is one of several technical factors which must be taken into account by the clinician or radiologist reviewing the film. Other considerations include how deeply the breath can be held, patient orientation and the energy of the X-ray beam. Variations in any of these can

make a chest film of the same patient look markedly different. It is up to the radiographer to position the patient, film and X-ray tube in order to optimize final film quality.

Lateral chest X-rays, with the beam entering the side of the chest, may also be taken. These are usually carried out in non-urgent situations, when the exact position of an abnormal shadow needs to be located, or all of the lung fields need checking for a cancer in patients with symptoms such as haemoptysis (coughing up blood). This is because the standard (PA) view does not reveal the lower quarter of the lung fields situated behind the diaphragm.

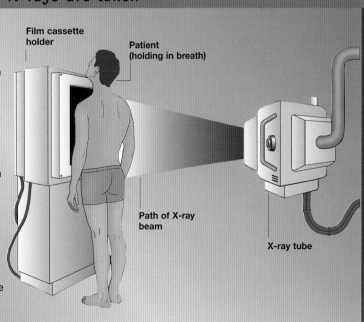

Film cassette holder

Patient (holding in breath)

Path of X-ray beam

X-ray tube

Female breast

The breast undergoes structural changes throughout the life of a woman. The most obvious changes occur during pregnancy as the breast prepares for its function as the source of milk for the baby.

Men and women both have breast tissue, but the breast is normally a well-developed structure only in women. The two female breasts are roughly hemispherical and are composed of fat and glandular tissue which overlie the muscle layer of the front of the chest wall on either side of the sternum (breastbone).

BREAST STRUCTURE

The base of the breast is roughly circular in shape and extends from the level of the second rib above to the sixth rib below. In addition, there may be an extension of breast tissue towards the axilla (armpit), known as the 'axillary tail'.

Breast size varies greatly between women; this is mainly due to the amount of fatty tissue present, as there is generally the same amount of glandular tissue in every breast.

The mammary glands consist of 15 to 20 lobules – clusters of secretory tissue from which milk is produced. Milk is carried to the surface of the breast from each lobule by a tube known as a 'lactiferous duct', which has its opening at the nipple.

The nipple is a protruding structure surrounded by a circular, pigmented area, called the areola. The skin of the nipple is very thin and delicate and has no hair follicles or sweat glands.

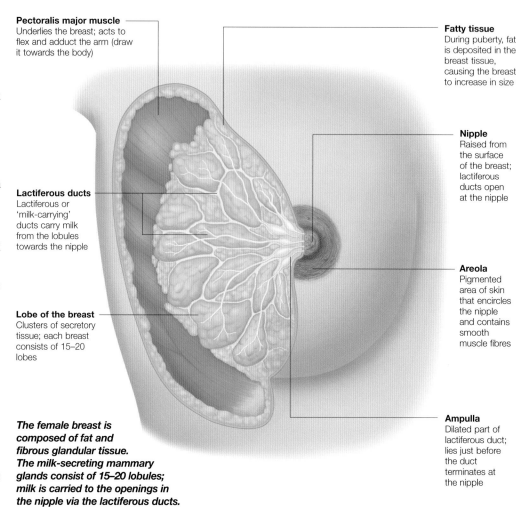

Pectoralis major muscle
Underlies the breast; acts to flex and adduct the arm (draw it towards the body)

Lactiferous ducts
Lactiferous or 'milk-carrying' ducts carry milk from the lobules towards the nipple

Lobe of the breast
Clusters of secretory tissue; each breast consists of 15–20 lobes

Fatty tissue
During puberty, fat is deposited in the breast tissue, causing the breast to increase in size

Nipple
Raised from the surface of the breast; lactiferous ducts open at the nipple

Areola
Pigmented area of skin that encircles the nipple and contains smooth muscle fibres

Ampulla
Dilated part of lactiferous duct; lies just before the duct terminates at the nipple

The female breast is composed of fat and fibrous glandular tissue. The milk-secreting mammary glands consist of 15–20 lobules; milk is carried to the openings in the nipple via the lactiferous ducts.

Blood vessels of the breast

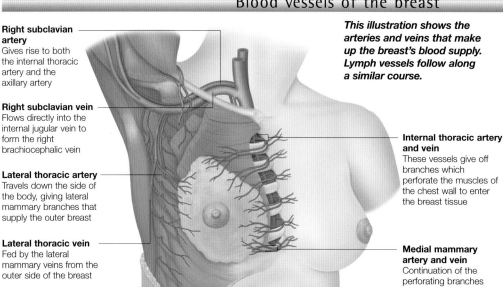

Right subclavian artery
Gives rise to both the internal thoracic artery and the axillary artery

Right subclavian vein
Flows directly into the internal jugular vein to form the right brachiocephalic vein

Lateral thoracic artery
Travels down the side of the body, giving lateral mammary branches that supply the outer breast

Lateral thoracic vein
Fed by the lateral mammary veins from the outer side of the breast

This illustration shows the arteries and veins that make up the breast's blood supply. Lymph vessels follow along a similar course.

Internal thoracic artery and vein
These vessels give off branches which perforate the muscles of the chest wall to enter the breast tissue

Medial mammary artery and vein
Continuation of the perforating branches supplying breast tissue near the sternum (breastbone)

The blood supply to the breast comes from a number of sources; these include the internal thoracic artery, which runs down the length of the front of the chest, and the lateral thoracic artery, which supplies the outer part of the breast and some of the posterior intercostal arteries.

A network of superficial veins underlies the skin of the breast, especially in the region of the areola, and these veins may become very prominent during pregnancy.

The blood collected in these veins drains in various directions, following a similar pattern to the arterial supply, travelling via the internal thoracic veins, the lateral thoracic veins and the posterior intercostal veins to the large veins that return blood to the heart.

Lymphatic drainage of the breast

Lymph, the fluid which leaks out of blood vessels into the spaces between cells, is returned to the blood circulation by the lymphatic system. Lymph passes through a series of lymph nodes, which act as filters to remove bacteria, cells and other particles.

Tiny lymphatic vessels arise from the tissue spaces and converge to form larger vessels which carry the (usually) clear lymph away from the tissues and into the venous system.

Lymph drains from the nipple, areola and mammary gland lobules into a network of small lymphatic vessels, the 'subareolar lymphatic plexus'. From this plexus the lymph may be carried in several different directions.

PATTERN OF DRAINAGE

About 75 per cent of the lymph from the subareolar plexus drains to the lymph nodes of the armpit, mostly from the outer quadrants of the breast. The lymph passes through a series of nodes in the region of the armpit draining into the subclavian lymph trunk, and ultimately into the right lymphatic trunk, which returns the lymph to the veins above the heart.

Most of the remaining lymph, mainly from the inner quadrants of the breast, is carried to the 'parasternal' lymph nodes, which lie towards the mid-line of the front of the chest. A small percentage of lymphatic vessels from the breast take another route and travel to the posterior intercostal nodes.

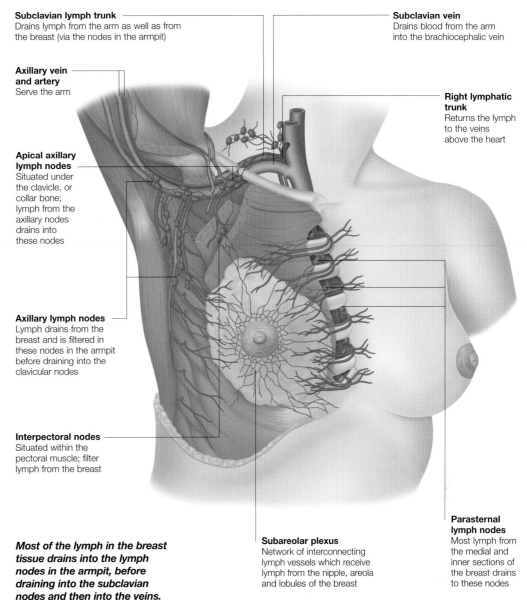

Subclavian lymph trunk
Drains lymph from the arm as well as from the breast (via the nodes in the armpit)

Axillary vein and artery
Serve the arm

Apical axillary lymph nodes
Situated under the clavicle, or collar bone; lymph from the axillary nodes drains into these nodes

Axillary lymph nodes
Lymph drains from the breast and is filtered in these nodes in the armpit before draining into the clavicular nodes

Interpectoral nodes
Situated within the pectoral muscle; filter lymph from the breast

Subclavian vein
Drains blood from the arm into the brachiocephalic vein

Right lymphatic trunk
Returns the lymph to the veins above the heart

Parasternal lymph nodes
Most lymph from the medial and inner sections of the breast drains to these nodes

Subareolar plexus
Network of interconnecting lymph vessels which receive lymph from the nipple, areola and lobules of the breast

Most of the lymph in the breast tissue drains into the lymph nodes in the armpit, before draining into the subclavian nodes and then into the veins.

Lymphatic drainage and breast cancer

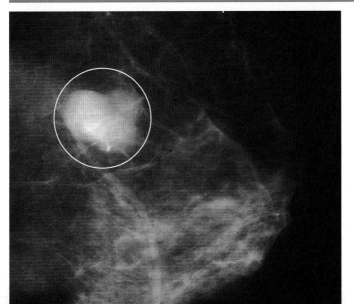

This mammogram shows a malignant tumour in the breast. The tumour is apparent as the dense area (circled) within the breast tissue.

Lymph fluid often contains particles such as cells which it has cleared from the tissue spaces. If the lymph has come from an area which contains a growing cancer, then it may contain cells which have broken off from that tumour. These cells will be filtered out by the lymph nodes where they may lodge and grow to form a secondary tumour, or 'metastasis'.

Knowledge of the pattern of lymph drainage of each area of the body, and especially of an area as prone to cancer as the breast, is therefore important to doctors. If a breast lump is found it is important for the doctor to check the associated lymph nodes for secondary spread of cancer cells.

MAMMOGRAPHY

As well as examination of the breast by the doctor or the woman herself, mammography (X-ray examination of the breasts), can be used to check for breast cancer. Mammograms help to detect the presence of cancer of the breast at an early, and therefore more easily treatable, stage.

Diaphragm

The diaphragm is a sheet of muscle that separates the thorax from the abdominal cavity. It is essential for breathing as its contraction expands the chest cavity, allowing air to enter.

The diaphragm is the main muscle involved in respiration and has several apertures for the passage of important structures which must pass between thorax and abdomen. It is made up of peripheral muscle fibres inserting into a central sheet of tendon which, unlike most tendons, does not have any attachment to bone.

MUSCLE OF THE DIAPHRAGM

The muscle tissue of the diaphragm arises from three areas of the chest wall, merges to form a continuous sheet and converges on the central tendon, which acts as a site of muscular attachment.

The three areas of origin of the diaphragm give rise to three separately named parts: the sternal part, the costal part and the lumbar or vertebral part, which arises from the crus and arcuate ligaments.

CENTRAL TENDON

Muscle fibres of the diaphragm insert into the central tendon, which has a three-leaved shape. The central part lies just beneath, and is depressed by, the heart. It is attached by ligaments to the pericardium, the membrane surrounding the heart. The two lateral leaves lie towards the back and help form the right and left domes (cupolae) of the diaphragm.

Abdominal surface of the diaphragm

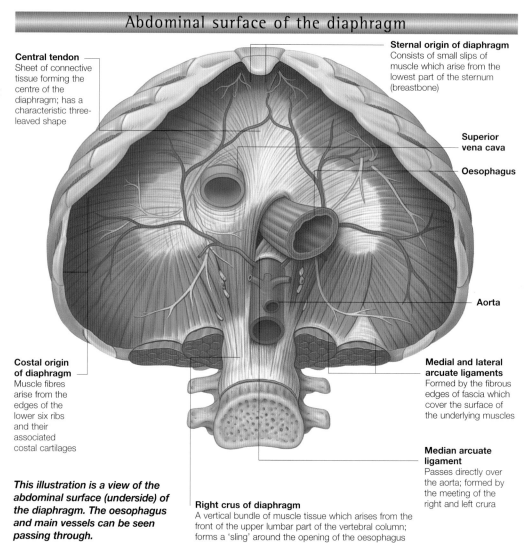

Central tendon
Sheet of connective tissue forming the centre of the diaphragm; has a characteristic three-leaved shape

Sternal origin of diaphragm
Consists of small slips of muscle which arise from the lowest part of the sternum (breastbone)

Superior vena cava

Oesophagus

Aorta

Costal origin of diaphragm
Muscle fibres arise from the edges of the lower six ribs and their associated costal cartilages

Medial and lateral arcuate ligaments
Formed by the fibrous edges of fascia which cover the surface of the underlying muscles

Median arcuate ligament
Passes directly over the aorta; formed by the meeting of the right and left crura

This illustration is a view of the abdominal surface (underside) of the diaphragm. The oesophagus and main vessels can be seen passing through.

Right crus of diaphragm
A vertical bundle of muscle tissue which arises from the front of the upper lumbar part of the vertebral column; forms a 'sling' around the opening of the oesophagus

Nerve supply of the diaphragm

The phrenic nerves carry both motor and sensory fibres to the diaphragm. Each nerve supplies its corresponding side of the diaphragm.

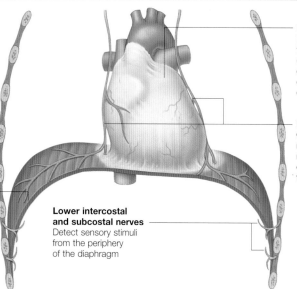

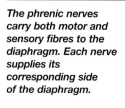

Heart
Enclosed in the fibrous pericardium (sac), the heart lies on the central part of the diaphragm, depressing it slightly

Phrenic nerves
Originate in the neck region; the phrenic nerves approach the diaphragm along the fibrous pericardium

Muscular diaphragm
Movement of the diaphragm is controlled by phrenic nerves which spread out over the surface of the muscle

Lower intercostal and subcostal nerves
Detect sensory stimuli from the periphery of the diaphragm

The motor nerve supply of the diaphragm (which causes the muscle of the diaphragm to contract) comes entirely from a nerve on each side called the 'phrenic nerve'. These nerves originate from each side of the spinal cord in the neck, at the level of the third, fourth and fifth cervical vertebrae.

SENSORY NERVE SUPPLY

The phrenic nerves also provide a sensory nerve supply, detecting pain and giving information on position, to the greater, central part of the diaphragm. The periphery of the diaphragm receives its sensory nerve supply from the lower intercostal nerves and the subcostal nerves.

Thoracic surface of the diaphragm

The upper aspect of the diaphragm is convex and forms the floor of the thoracic (chest) cavity. It is perforated by major vessels and structures which must pass through the muscle sheet in order to reach the abdomen.

The central part of the surface of the diaphragm is covered by the pericardium, the membrane which surrounds the heart. To either side, the upper surface of the diaphragm is lined with the diaphragmatic part of the parietal pleura (the thin membrane which lines the chest cavity). This is continuous around the edges of the diaphragm with the costal pleura, which covers the inside of the chest wall.

DIAPHRAGMATIC APERTURES

Although the diaphragm acts to separate the chest and abdominal cavities, certain structures do pass through 'diaphragmatic apertures'. The three largest of these are:

■ The caval aperture

This is an opening in the central tendon of the diaphragm which allows passage of the inferior vena cava, the main vein of the abdomen and lower limbs. As the opening is in the central tendon rather than the muscle of the diaphragm it will not close when the diaphragm contracts during inspiration; in fact the opening widens and blood flow increases. The opening also contains branches of the right phrenic nerve and lymphatic vessels.

Diaphragm from above

Diaphragmatic pleura
Cut away here to show the underlying diaphragm

Right leaflet of central tendon
Non-muscular (fibrous) part of the diaphragm

BACK

Aorta lying in the aortic aperture
Not strictly an aperture as the aorta, thoracic duct and azygos vein actually lie against the vertebral column behind the diaphragm

Caval aperture
Opening in the central tendon, just to the right of the midline, through which the vena cava passes

Pericardium
Sac containing the heart

Left phrenic nerve
Seen here travelling along the border of the pericardium

FRONT

Inferior vena cava
Main vein of the abdomen and lower limbs

Oesophagus
Lies in the oesophageal aperture, just to the left of the midline within the right crus of the diaphragm

This cross-section of the chest shows the diaphragm from above. The diaphragm provides a division between the thoracic and abdominal cavities.

■ The oesophageal aperture

This allows the passage of the oesophagus (gullet) through the diaphragm to reach the stomach. The muscle fibres of the right crus act as a sphincter, closing off the oesophageal opening when the diaphragm contracts during inspiration. As well as the oesophagus, the aperture also gives passage to nerves (vagus), arteries and lymphatic vessels.

■ The aortic aperture

This opening lies behind the diaphragm rather than within it. As the aorta does not actually pierce the diaphragm, the flow of blood within it is not affected by diaphragmatic contractions while breathing. The aorta emerges under the median arcuate ligament, in front of the vertebral column. The aortic aperture also transmits the thoracic duct (major lymphatic channel) and the azygos vein.

Position and function of the diaphragm

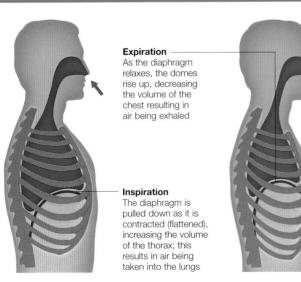

Expiration
As the diaphragm relaxes, the domes rise up, decreasing the volume of the chest resulting in air being exhaled

Inspiration
The diaphragm is pulled down as it is contracted (flattened), increasing the volume of the thorax; this results in air being taken into the lungs

The diaphragm lies across the body separating the thoracic cavity from the abdominal cavity. It curves upwards into two domes, right and left, separated by a central depression where the heart rests. The right dome is normally higher than the left because of the liver beneath.

The periphery of the diaphragm is at a constant level as it is attached to the thoracic wall, but the heights of the domes vary depending on the degree to which the muscle of

The diaphragm is the main muscle of respiration. By working in conjunction with the changing shape of the ribcage, air is inhaled and exhaled.

the diaphragm is contracted. The right dome can reach as high as the fifth rib, the left slightly lower.

ACTIONS OF THE DIAPHRAGM

Contraction of the muscle fibres cause the domes to be pulled down, which expands the thoracic cavity above, and air enters. Relaxation of the diaphragmatic muscle allows the domes to rise and air is exhaled.

Contraction of the diaphragm also causes the abdominal cavity below to become smaller, putting the contents under greater pressure. Its contraction is thus used to help expel abdominal contents, such as in defecation.

Mediastinum

The mediastinum is the name given to the core of vital structures that runs down through the centre of the thoracic (chest) cavity. It contains the heart and is flanked on either side by the lungs.

The mediastinum takes up the whole of the central space of the thoracic cavity, extending from the root of the neck to the diaphragm, and from the sternum and costal cartilages back to the bones of the spine.

It is covered on each side by the thin, lubricated membranes of the mediastinal pleura, and the structures within the mediastinum are held only loosely together by fatty connective tissue. This arrangement allows for the movements brought about within the mediastinum by changes in posture of the body or by the changes in pressure and volume of the thoracic cavity during breathing.

CONTENTS

Contained within the mediastinum are the heart and major blood vessels, the thymus gland, the trachea, the oesophagus and some important nerves, including the vagus and phrenic nerves.

The loose connective tissue of the mediastinum also contains many important lymph nodes, which receive lymph from the lungs and other mediastinal structures.

The mediastinum is a moveable partition extending from the neck to the diaphragm. Its size and shape varies on inspiration and expiration.

Major vessels of the mediastinum

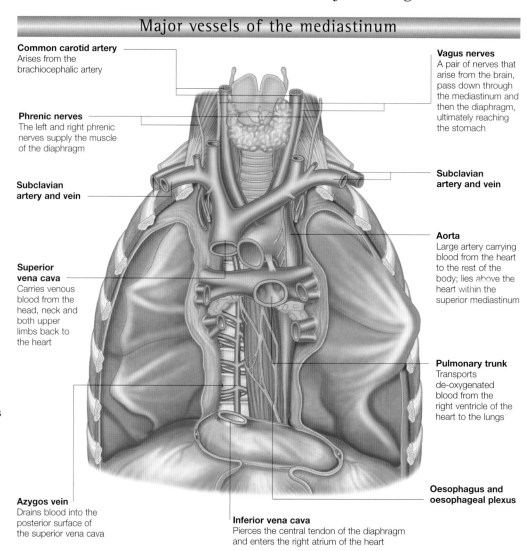

Common carotid artery
Arises from the brachiocephalic artery

Phrenic nerves
The left and right phrenic nerves supply the muscle of the diaphragm

Subclavian artery and vein

Superior vena cava
Carries venous blood from the head, neck and both upper limbs back to the heart

Azygos vein
Drains blood into the posterior surface of the superior vena cava

Vagus nerves
A pair of nerves that arise from the brain, pass down through the mediastinum and then the diaphragm, ultimately reaching the stomach

Subclavian artery and vein

Aorta
Large artery carrying blood from the heart to the rest of the body; lies above the heart within the superior mediastinum

Pulmonary trunk
Transports de-oxygenated blood from the right ventricle of the heart to the lungs

Oesophagus and oesophageal plexus

Inferior vena cava
Pierces the central tendon of the diaphragm and enters the right atrium of the heart

Divisions of the mediastinum

Sagittal section of the mediastinum

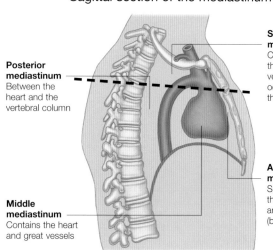

Posterior mediastinum
Between the heart and the vertebral column

Middle mediastinum
Contains the heart and great vessels

Superior mediastinum
Contains the thymus, major vessels, trachea, oesophagus and thoracic duct

Anterior mediastinum
Space between the pericardium and the sternum (breastbone)

For descriptive purposes, the mediastinum is divided into two main parts; the superior (upper) mediastinum and the inferior (lower) mediastinum. The inferior mediastinum is divided still further into anterior, middle and posterior divisions.

SUPERIOR MEDIASTINUM

The superior division of the mediastinum extends down as far as an imaginary line that passes from the sternal angle

This illustration shows the location of the superior mediastinum and, below the dotted line, the three divisions of the inferior mediastinum: anterior, middle and posterior.

anteriorly (at the front) to the level of the intervertebral disc between the fourth and fifth thoracic vertebrae.

INFERIOR MEDIASTINUM

From this imaginary line down to the diaphragm lies the inferior mediastinum, which is divided into three sections. From front to back, these are:

■ Anterior mediastinum – the area in front of the heart and behind the sternum
■ Middle mediastinum – containing the heart and great vessels
■ Posterior mediastinum – the area behind the heart and in front of the vertebral column.

Thymus

The thymus gland forms a vital part of the body's immune system, being the site of development of important specialized immune cells called T-lymphocytes. It lies in the anterior mediastinum.

The thymus is a pink, flattened, often bilobed gland which lies within the anterior mediastinum and extends up to the superior mediastinum, in front of the large blood vessels and the trachea, just behind the manubrium (the upper part of the breast bone).

During childhood, when the gland is at its largest, it may also extend upwards into the root of the neck and downwards into the anterior mediastinum in front of the heart.

The thymus is surrounded by a fibrous capsule from which extensions enter the thymus tissue itself to divide the gland into lobules.

BLOOD SUPPLY

There is a generous blood supply from the internal thoracic arteries via their anterior intercostal and anterior mediastinal branches. Venous drainage is via the thymic veins, which drain into the brachiocephalic vein close to the heart.

GROWTH OF THE THYMUS

The thymus gland is at its largest during infancy and childhood. Occasionally, in some newborn babies, it may be

Location of the thymus

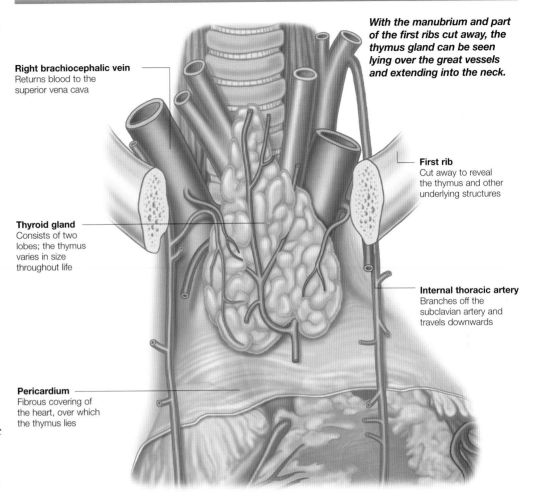

With the manubrium and part of the first ribs cut away, the thymus gland can be seen lying over the great vessels and extending into the neck.

Right brachiocephalic vein
Returns blood to the superior vena cava

Thyroid gland
Consists of two lobes; the thymus varies in size throughout life

Pericardium
Fibrous covering of the heart, over which the thymus lies

First rib
Cut away to reveal the thymus and other underlying structures

Internal thoracic artery
Branches off the subclavian artery and travels downwards

so large that it projects up into the neck, compressing the trachea and causing breathing difficulties.

Similar to other lymphoid tissue throughout the body, the thymus becomes relatively less significant in size as a child grows into adulthood. At birth the thymus may weigh 10–15 g, but this increases to only 30–40 g at puberty, despite the fact that the body weight as a whole has increased tenfold.

CHANGES THROUGH LIFE

After puberty, the functional tissue of the thymus gradually becomes increasingly smaller and is replaced by fat until, in old age, the gland may be very difficult to find at all.

The mediastinum on X-ray

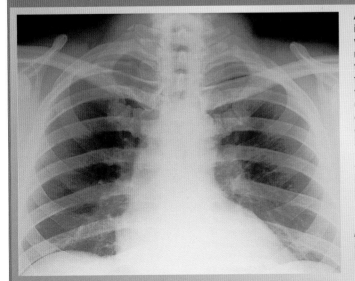

A chest X-ray is a very common investigation undertaken for a wide variety of reasons. The normal X-ray shadow cast by the mediastinum is produced by the individual shadows of the various structures superimposed upon one another. On examination of the radiograph, a doctor will be looking carefully at the outline of the mediastinum, as any variations from normal may give important information about the health of the patient.

This chest X-ray shows a person with normal healthy lungs and heart. The white central portion outlines the area of the mediastinum.

Mediastinal widening
Mediastinal widening, an increase in the width of the mediastinum that is visible on a chest X-ray, may be due to disease or injury of any of the mediastinal structures. It is sometimes seen after trauma, such as from a head-on vehicle collision, when there has been bleeding into the mediastinum from tearing of the large blood vessels within it.

Other causes of mediastinal widening may include malignant lymphoma (cancer of the lymphatic tissues) when there may be enlargement of the mediastinal lymph nodes, and enlargement of the heart due to cardiac failure.

Lungs

The paired lungs are cone-shaped organs of respiration which occupy the thoracic cavity, lying to either side of the heart, great blood vessels and other structures of the central mediastinum.

The right and left lungs are separate entities, each enclosed within a bag of membranes, the right and left pleural sacs. Each lung lies free within the thoracic cavity, attached to the mediastinum only by a root made up of the main bronchus and large blood vessels.

The lung tissue is soft and spongy and has great elasticity. In children the lungs are pink in colour, but they usually become darkener and mottled later in life as they are exposed to dust which is taken in by the defence cells of the lining of the airways.

Each lung has:
■ An apex, which projects up into the base of the neck behind the clavicle (collarbone)
■ A base, with a concave surface, which rests on the superior surface of the diaphragm
■ A concave mediastinal surface, which lies against various structures of the mediastinum.

LOBES AND FISSURES

The lungs are divided into sections known as lobes by deep fissures. The right lung has three lobes while the left lung, which is slightly smaller (due to the position of the heart), has two. Each lobe is independent of the others, receiving air via its own lobar bronchus and blood from lobar arteries.

The fissures are deep, extending right through the structure of the lung, and are lined by the pleural membrane.

Anterior view of the lungs

Right lung

Left lung

Horizontal fissure of right lung
Lies anteriorly behind the fourth costal cartilage; extends back through the lung tissue to meet the oblique fissure about halfway along its length

Trachea
Divides into the two bronchi

Hilum
Area on the centre of the inner surface where the structures which form the root enter and leave the lung

UPPER LOBE

UPPER LOBE

MIDDLE LOBE

LOWER LOBE

LOWER LOBE

Cardiac notch
A 'cut-out' in the left lung which accommodates the heart

The right lung consists of three lobes but the left lung has only two. This is to accommodate the heart which lies in the left chest.

Oblique fissure of right lung
Lies between the middle and lower lobes of the right lung

Oblique fissure of left lung
Runs down and forward through the lung tissue from a point about 6 cm below the apex posteriorly

Bronchopulmonary segments

Right lung

Left lung

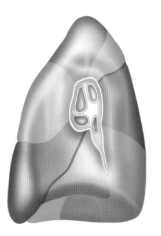
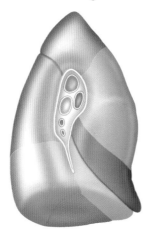

The lobes of the lung are further subdivided into units known as bronchopulmonary segments, which are separated from each other by a layer of connective tissue. Each segment is roughly pyramidal in shape, with its base on the pleural surface and its apex at the root of the lung, where the bronchus and major vessels enter and leave.

The lobes of each lung are further divided into several distinct parts, known as the bronchopulmonary segments. Medial (inward-looking) views of the right and left lungs are illustrated here.

CLINICAL SIGNIFICANCE

Knowledge of the layout and structure of the bronchopulmonary segments is of particular importance to thoracic surgeons who may need to remove a lung tumour or abscess.

Like the whole lobe itself, each segment has its own blood supply and receives air from a segmental bronchus, thus making it independent from the others. For this reason, one or more segments may be surgically removed – because of disease or trauma, for instance – without adversely affecting the others.

The pleura

The lungs are covered by a thin membrane known as the pleura. The pleura lines both the outer surface of the lung and the inner surface of the thoracic cage.

The layer of pleura covering the lung is called the visceral pleura, while that lining the thoracic cage is the parietal pleura.

VISCERAL PLEURA
This thin membrane covers the lung surface, dipping down into the fissures between the lobes of the lung.

PARIETAL PLEURA
This is continuous with the visceral pleura at the hilum of the lung. Here, the membrane reflects back and lines all the inner surfaces of the thoracic cavity. The parietal pleura is one continuous membrane divided into areas that are named after the surfaces they cover:
■ Costal pleura – lines the inside of the ribcage, the back of the sternum and the sides of the vertebral bodies of the spine
■ Mediastinal pleura – covers the mediastinum, the central area of the thoracic cavity
■ Diaphragmatic pleura – lines the upper surface of the diaphragm, except where it is covered by the pericardium
■ Cervical pleura – covers the tip of the lung as it projects up into the base of the neck.

Position of the lungs and pleura

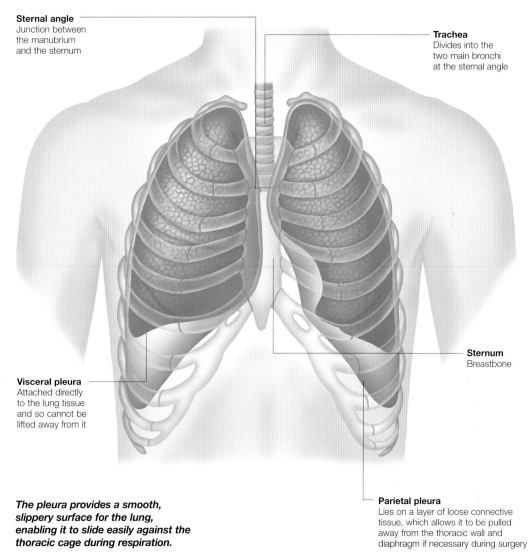

Sternal angle
Junction between the manubrium and the sternum

Trachea
Divides into the two main bronchi at the sternal angle

Visceral pleura
Attached directly to the lung tissue and so cannot be lifted away from it

Sternum
Breastbone

Parietal pleura
Lies on a layer of loose connective tissue, which allows it to be pulled away from the thoracic wall and diaphragm if necessary during surgery

The pleura provides a smooth, slippery surface for the lung, enabling it to slide easily against the thoracic cage during respiration.

Pleural cavity and recesses of the pleura

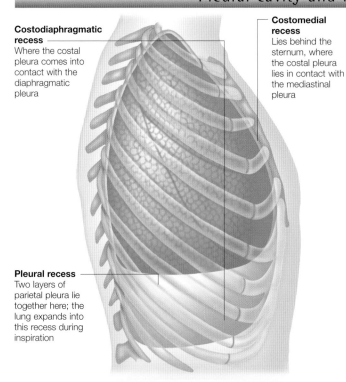

Costodiaphragmatic recess
Where the costal pleura comes into contact with the diaphragmatic pleura

Costomedial recess
Lies behind the sternum, where the costal pleura lies in contact with the mediastinal pleura

Pleural recess
Two layers of parietal pleura lie together here; the lung expands into this recess during inspiration

This view from the right side of the chest shows how the lung does not entirely fill the pleural sac within which it lies.

The pleural cavity, which lies between the visceral and parietal layers of pleura, is a narrow area filled with a small amount of pleural fluid. The fluid lubricates the movement of the lung within the thoracic cavity and also acts to provide a tight seal, holding the lung against the thoracic wall and diaphragm by surface tension. It is this seal that forces the elastic tissue of the lung to expand when the diaphragm contracts and the ribcage lifts during inspiration.

PLEURAL RECESSES
During quiet breathing, the lungs do not completely fill the pleural sacs within which they lie. There is room for expansion in the pleural recesses, areas where the sacs are empty and where parietal pleura comes into contact with itself rather than visceral pleura overlying lung tissue. The lungs only expand fully into these recesses during deep inspiration, when lung volume is at a maximum.

PLEURAL EFFUSION
At the base of the thoracic cavity, the lowermost parts of the costal pleura come into contact with the diaphragmatic pleura in the costodiaphragmatic recess. This recess is of importance clinically because it provides a potential space that may become filled with fluid – a so-called pleural effusion – in certain medical conditions, such as heart failure.

The costomediastinal recess is smaller and of less clinical importance.

How the lungs work

The lungs, which take up most of the upper chest cavity, have a surface area equivalent to a tennis court. They work tirelessly to sustain life, supplying the body with oxygen and filtering harmful carbon dioxide from the blood.

The lungs are a pair of large, cone-shaped, spongy organs that remove waste carbon dioxide from the body and exchange it for a fresh supply of oxygen. Air is drawn into the lungs by expanding the chest cavity and then expelled by allowing the cavity to collapse or by forcing the air out.

The lungs occupy most of the chest cavity. The upper part of the cavity is bounded by the ribs and the intercostal muscles. The base of the cavity is bounded by the diaphragm – a flat sheet of tissue that forms a wall between the chest and the abdomen.

INSIDE THE LUNGS

Inside the lungs is a dense, branching latticework of tubes that get progressively smaller. The largest tubes are the two bronchi, which connect with the base of the trachea. Inside the lungs, the bronchi divide into smaller branches known as bronchioles, finally ending in clusters of tiny air sacs, or alveoli (singular: alveolus) In total, the lungs contain over 2,400 km of airway, and the surface area inside is about 260 m² – equivalent to an area the size of a tennis court.

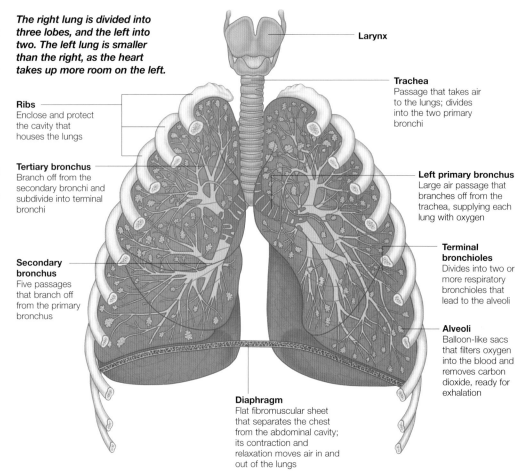

The right lung is divided into three lobes, and the left into two. The left lung is smaller than the right, as the heart takes up more room on the left.

Ribs
Enclose and protect the cavity that houses the lungs

Tertiary bronchus
Branch off from the secondary bronchi and subdivide into terminal bronchi

Secondary bronchus
Five passages that branch off from the primary bronchus

Larynx

Trachea
Passage that takes air to the lungs; divides into the two primary bronchi

Left primary bronchus
Large air passage that branches off from the trachea, supplying each lung with oxygen

Terminal bronchioles
Divides into two or more respiratory bronchioles that lead to the alveoli

Alveoli
Balloon-like sacs that filters oxygen into the blood and removes carbon dioxide, ready for exhalation

Diaphragm
Flat fibromuscular sheet that separates the chest from the abdominal cavity; its contraction and relaxation moves air in and out of the lungs

Breathing in and out

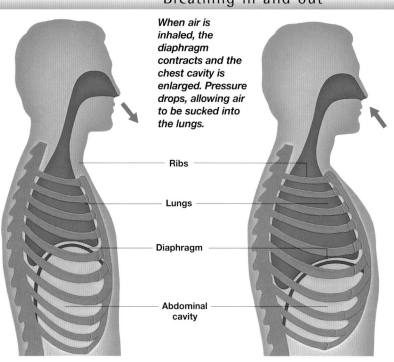

As air is expelled, the diaphragm relaxes and moves up. This causes pressure to rise in the lower chest cavity; exhalation equalizes the pressure.

At rest, a person will inhale and exhale about 500 ml of air, 13–17 times a minute. The lungs will expand and contract between 15 and 85 times within this time, depending on the body's activity.

When air is inhaled, the diaphragm contracts and the chest cavity is enlarged. Pressure drops, allowing air to be sucked into the lungs.

Ribs

Lungs

Diaphragm

Abdominal cavity

The lungs have a natural tendency to collapse. Inside the chest cavity, they are held open by surface tension created by a fluid produced by the inner pleural membrane. To draw air into the lungs, the chest cavity is expanded. The muscles of the diaphragm contract, causing it to become flatter. At the same time, the intercostal muscles of the ribs contract, lifting the ribs upwards and outwards. The pressure in the chest cavity is thus reduced and the lungs expand, drawing in air via the mouth or nose.

When the intercostal muscles relax, the ribs fall downwards and inwards and the lungs collapse, forcing the air out. At the same time, the diaphragm relaxes and is pulled up into the chest cavity. In order to force more air out of the chest cavity the abdominal muscles can be used to push the diaphragm further into the chest cavity.

Controlling breathing

An adult's lung capacity is about 55 litres, but during normal breathing only 500 ml of air is exchanged. This movement is associated with inside and outside pressure.

Although breathing can be controlled voluntarily, respiratory movements are generally a series of reflex actions. There are controlled by the respiratory centre in the hindbrain (the part of the brain that regulates basic bodily systems). This has two regions: an inspiratory centre and an expiratory centre.

Nerve impulses from the inspiratory centre cause the contraction of the intercostal muscles (which move the ribs) and the diaphragm. This is the beginning of an intake of air into the lungs. As the lungs expand,

This resin cast of the pulmonary arteries and bronchi clearly shows the network of vessels that supply blood and air to the lungs.

stretch receptors in the walls of the lungs send back signals that begin to inhibit the signals from the inspiratory centre.

At the same time, impulses from the inspiratory centre activate the expiratory centre, which sends back inhibitory signals. The result is the relaxation of the intercostal muscles and the diaphragm, which ceases the intake of air and begins exhalation. The whole process is now ready to start again.

Breathing is also controlled and regulated by the level of carbon dioxide (CO_2) in the blood. Excess CO_2 causes the blood to become more acid. This is detected by the brain, and the inspiratory centre starts to produce deeper breathing until the CO_2 level is reduced.

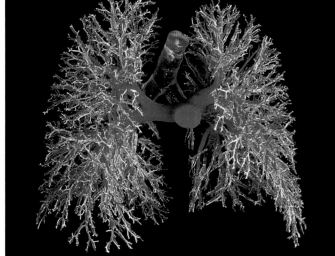

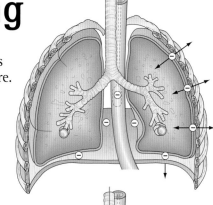

LUNGS AT REST:
At rest, the muscles associated with respiration are relaxed. The air in the trachea and bronchi is at atmospheric pressure (the standard pressure exerted on the environment) and there is no airflow. Recoil of lung and chest wall are equal but opposite.

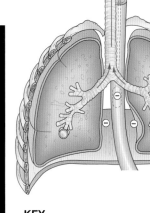

BREATHING IN:
During inspiration, muscles contract and the chest expands. Pressure in the alveolar becomes less than that outside the lungs, and air flows into the airways.

BREATHING OUT:
During expiration, muscle relaxation causes the lungs to contract. This increases the pressure in the alveolar to a point where it is greater than the pressure at the airway opening. Air now flows out of the lungs.

KEY

⊕ Pressure within the alveolar greater than external air pressure

⊖ Pressure within the alveolar lesser than external air pressure

➤ Air flow to and from the lungs

➤ Internal or external forces, resulting in an increase or decrease of pressure

➤ Muscular contraction or relaxation

Gas exchange

The exchange of gases takes place in the alveoli, of which there are about 300 million. When fully expanded, the lungs can contain between 4 and 6 litres of air, but a much smaller amount is normally breathed in and out. When engaged in a quiet activity, a person breathes in and out about 15 times a minute, moving around 500 ml of air with each breath. During strenuous activity, however, the breathing rate may increase to up to 80 breaths a minute, and the volume of air moved may increase to between three and five litres.

Air that is breathed in contains about 21 per cent

oxygen. Inside the alveoli, some of this oxygen dissolves in the surface moisture and passes through the thin lining into the blood, where most of it is picked up by the haemoglobin of the red blood cells. At the same time, carbon dioxide, most of which is carried in the blood plasma, passes into the lungs, where it is released as a gas ready to be breathed out. Exhaled air contains about 16 per cent oxygen.

The spaces seen in this meshwork of lung tissue are alveoli, which perform the lung's primary role of gas exchange.

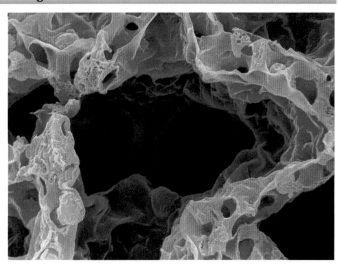

Vessels of the lung

The primary function of the lungs is to reoxygenate the blood used by the tissues of the body and to remove accumulated waste carbon dioxide. This is effected via the pulmonary blood circulation.

Blood from the body returns to the right side of the heart and from there passes directly to the lungs via the pulmonary arteries.

Having been oxygenated by its passage through the lungs, the blood returns to the left side of the heart in the pulmonary veins. The oxygen-rich blood is then pumped around the body. Collectively, the pulmonary arteries and veins and their branches are referred to as the pulmonary circulation.

PULMONARY VESSELS

A large artery known as the pulmonary trunk arises from the heart's right ventricle carrying dark-red deoxygenated blood from the body into the lungs.

The pulmonary trunk divides into two smaller branches, the right and left pulmonary arteries, which run horizontally and enter the lungs at the hilum, alongside the bronchi (main airways).

Within the lungs the arteries divide to supply each lobe of their respective lung; two on the left and three on the right. The lobar arteries divide further to give the segmental arteries, which supply the bronchopulmonary segments (structural units of the lung). Each segmental artery ends in a network of capillaries.

Oxygenated blood returns to the left side of the heart through a system of pulmonary veins running alongside the arteries.

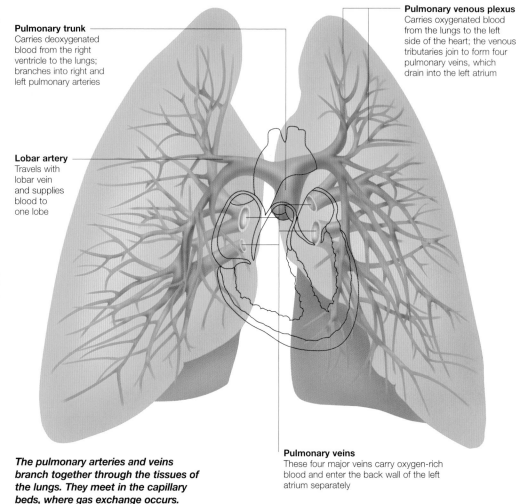

Pulmonary trunk
Carries deoxygenated blood from the right ventricle to the lungs; branches into right and left pulmonary arteries

Lobar artery
Travels with lobar vein and supplies blood to one lobe

Pulmonary venous plexus
Carries oxygenated blood from the lungs to the left side of the heart; the venous tributaries join to form four pulmonary veins, which drain into the left atrium

Pulmonary veins
These four major veins carry oxygen-rich blood and enter the back wall of the left atrium separately

The pulmonary arteries and veins branch together through the tissues of the lungs. They meet in the capillary beds, where gas exchange occurs.

Alveolar capillary plexus

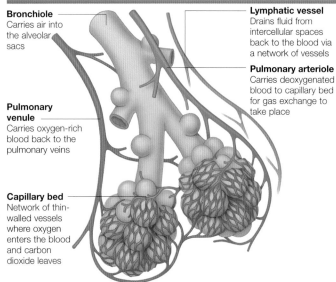

Bronchiole
Carries air into the alveolar sacs

Pulmonary venule
Carries oxygen-rich blood back to the pulmonary veins

Capillary bed
Network of thin-walled vessels where oxygen enters the blood and carbon dioxide leaves

Lymphatic vessel
Drains fluid from intercellular spaces back to the blood via a network of vessels

Pulmonary arteriole
Carries deoxygenated blood to capillary bed for gas exchange to take place

Within the lung, repeated division of the pulmonary arteries ultimately results in a network (plexus) of tiny blood vessels (capillaries), around each of the millions of alveolar sacs. The walls of the capillaries are extremely thin, which allows the blood within them to come into close contact with the walls of the alveoli, through which gas exchange takes place.

As oxygen enters and carbon dioxide leaves the pulmonary blood it changes from dark to light red. The newly oxygenated blood is collected into venules which drain each capillary plexus, these venules ultimately joining to form the pulmonary veins, which complete the pulmonary circulation by returning the blood to the heart.

INTRINSIC BLOOD SUPPLY

The tissues of the smallest airways can absorb oxygen from the air they contain, but this is not true for the larger airways, the supporting connective tissue of the lung and the pleura covering the lung. These structures receive their blood supply directly from two small bronchial arteries which arise from the thoracic aorta.

Each alveolus is surrounded by a capillary plexus. Deoxygenated blood is oxygenated by gaseous exchange, which occurs through the walls of the alveoli.

Lymphatics of the lung

Lymphatic drainage of the lung originates in two main networks, or plexuses: the superficial (subpleural) plexus and the deep lymphatic plexus. These communicate freely with each other.

Lymph is a fluid which is collected from the spaces between cells and carried in lymphatic vessels back to the venous circulation. On its way, the lymph must pass through a series of lymph nodes, which act as filters to remove particulate matter and any invading micro-organisms.

SUPERFICIAL PLEXUS

This network of fine lymphatic vessels extends over the surface of the lung, just beneath the visceral pleura (covering of the lung). The superficial plexus drains lymph from the lung towards the bronchi and trachea, where the main groups of lymph nodes are found.

Lymph from the superficial plexus arrives first at the bronchopulmonary group of lymph nodes, which lie at the hilum of the lung.

DEEP PLEXUS

The lymphatic vessels of the deep plexus originate in the connective tissue surrounding the small airways, bronchioles and bronchi (the alveoli have no lymphatic vessels). There are also small lymphatic vessels within the lining of the larger airways.

These lymphatic vessels join and run back along the route of the bronchi and pulmonary blood vessels, passing through

Trachea
Airway through which air moves between the lungs and the atmosphere via the nose and mouth

Right subclavian lymphatic trunk
Drains lymph from upper limb and joins vessels draining right lung

Interlobar fissure
Separates adjacent lobes of the lung

Tracheo-bronchial (carinal) lymph nodes
Drain lymph from the nodes in the hilum

Brachiocephalic vein
Drains into left subclavian vein

Thoracic duct
Drains lymph from the left chest and most of the body below the diaphragm into the origin of the left brachiocephalic vein

Paratracheal node
Drains lymph from lung via carinal and hilar nodes

Deep lymphatic vessels
Drain lymph from within the substance of the lung

Intrapulmonary lymph node
Filters lymph in vessels of the deep lymphatic plexus

Bronchopulmonary lymph nodes
Lymph nodes at the hilum of the lung

intrapulmonary nodes which lie within the lung. From these nodes lymph passes through vessels which drain towards the hilum into the broncho-pulmonary lymph nodes.

The bronchopulmonary nodes at the hilum of the lung therefore receive lymph from both superficial and deep lymphatic plexuses.

LYMPH NODES OF THE BRONCHI AND TRACHEA

From the bronchopulmonary nodes, lymph drains to the tracheobronchial (carinal) lymph nodes. From here, lymph passes up through the paratracheal nodes lying alongside the trachea, into the paired bronchomediastinal lymph trunks, which return the fluid to the venous system in the neck.

Mottled lungs

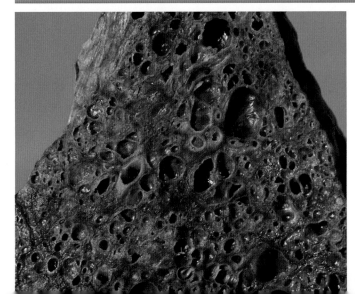

Living in an urban environment where atmospheric pollution is high, or smoking cigarettes, will lead to the inhalation of air containing many particles of dust and carbon.

Specialized cells in the lungs, called phagocytes, are able to protect the delicate lung tissue from these potential irritants by ingesting them. The process of ingestion is known as phagocytosis.

This post-mortem lung specimen shows the black discoloration of a smoker's lung. The air sacs are also abnormally enlarged due to emphysema.

Phagocytes containing particles can be carried away in the lymphatics and become lodged in the superficial plexus just under the surface of the lung. This gives a darkly mottled, 'honeycomb' appearance to the lung surface. This dark staining may also be apparent in the various groups of lymph nodes throughout the lung.

An understanding of the layout of the lymphatic vessels and lymph nodes of the lung can be important clinically in the assessment of lung cancer, which can spread via the lymphatic system.

Respiratory airways

The airways form a network along which air travels to, from and within the lungs. The airways branch repeatedly, each branch narrowing until the end terminals – the alveoli – are reached.

As a breath is taken, air enters through the nose and mouth, then passes down through the larynx to enter the trachea (windpipe). The air is carried down into the chest by the trachea, which then divides into smaller tubes – the bronchi – which take the air into the lungs.

The bronchi divide to form progressively smaller tubes that reach all areas of the lung. These tubes terminate in the alveolar sacs, which form the substance of the lung. It is in these thin-walled sacs that gas exchange with the blood occurs.

TRACHEA

The trachea extends down from the cricoid cartilage just below the larynx in the neck to enter the chest. At the level of the sternal angle it ends by dividing into two branches, the right and left main bronchi.

The trachea is composed of strong fibroelastic tissue, within which are embedded a series of incomplete rings of hyaline cartilage, the tracheal cartilages. In adults the trachea is quite wide (approximately 2.5 cm), but it is much narrower in infants – about the width of a pencil.

The posterior (back) surface of the trachea has no cartilaginous support and instead consists of fibrous tissue and trachealis muscle fibres. This posterior wall lies in contact with the oesophagus, which is directly behind the trachea.

The major airways

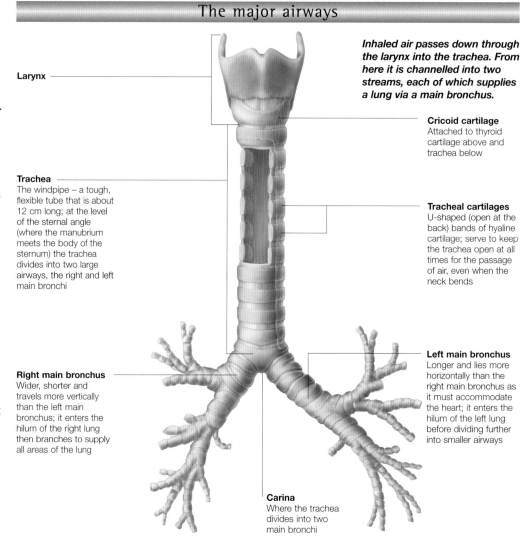

Inhaled air passes down through the larynx into the trachea. From here it is channelled into two streams, each of which supplies a lung via a main bronchus.

Larynx

Trachea
The windpipe – a tough, flexible tube that is about 12 cm long; at the level of the sternal angle (where the manubrium meets the body of the sternum) the trachea divides into two large airways, the right and left main bronchi

Right main bronchus
Wider, shorter and travels more vertically than the left main bronchus; it enters the hilum of the right lung then branches to supply all areas of the lung

Cricoid cartilage
Attached to thyroid cartilage above and trachea below

Tracheal cartilages
U-shaped (open at the back) bands of hyaline cartilage; serve to keep the trachea open at all times for the passage of air, even when the neck bends

Left main bronchus
Longer and lies more horizontally than the right main bronchus as it must accommodate the heart; it enters the hilum of the left lung before dividing further into smaller airways

Carina
Where the trachea divides into two main bronchi

Cross-section through the trachea

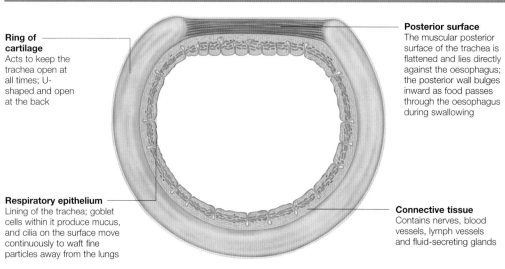

Ring of cartilage
Acts to keep the trachea open at all times; U-shaped and open at the back

Respiratory epithelium
Lining of the trachea; goblet cells within it produce mucus, and cilia on the surface move continuously to waft fine particles away from the lungs

Posterior surface
The muscular posterior surface of the trachea is flattened and lies directly against the oesophagus; the posterior wall bulges inward as food passes through the oesophagus during swallowing

Connective tissue
Contains nerves, blood vessels, lymph vessels and fluid-secreting glands

The trachea in cross-section is an incomplete ring. The epithelium (cellular lining) of the trachea contains goblet cells, which secrete mucus onto the surface, and tiny, brush-like cilia (hairs) which together help to catch dust particles and move them back up towards the larynx and away from the lung.

Between the epithelium and the rings of cartilage lies a layer of connective tissue containing small blood vessels, nerves, lymphatic vessels and glands that produce watery mucus which is secreted into the trachea. There are also many elastic fibres, which help to give the trachea its flexibility.

Smaller airways and alveoli

On entering the lung the main bronchus divides again and again, forming the 'bronchial tree', which takes air to all parts of the lung.

The first division of the main bronchus gives rise to the lobar bronchi, three on the right and two on the left, which each supply one lobe of the lung. Each of these lobar bronchi divides to form the smaller bronchi, which supply each of the independent bronchopulmonary segments.

BRONCHI STRUCTURE
The bronchi have a similar structure to the trachea, being very elastic and flexible, having cartilage in their walls and being lined with respiratory epithelium. There are also numerous muscle fibres, which allow for changes in diameter of these tubes.

BRONCHIOLES
Within the bronchopulmonary segments the bronchi continue to divide, perhaps as many as 25 times, before they terminate in the blind-ended alveolar sacs.

At each division the tubes become smaller, although the total cross-sectional area increases. When the air tubes have an internal diameter of less than 1 mm they become known as bronchioles.

Bronchioles differ from bronchi in that they have no cartilage in their walls nor any mucus-secreting cells in their lining. They do, however, still have muscle fibres in their walls.

Bronchioles and alveoli

Terminal bronchiole
Leads into the distal bronchial tree

Respiratory bronchiole
Has a diameter of less than 1 mm; each bronchiole gives rise to 2–11 alveolar ducts, each of which gives rise to five or six alveolar sacs

Elastic fibres
Allow expansion of the alveoli walls on filling with air

Smooth muscle
Expands and contracts the lumen of the bronchioles and supports the wall in place of cartilage

Alveolar sac
Thin-walled, basic structure for gas exchange

Opening of alveolar duct
Where the ducts (branches off the bronchioles), open into the alveolar sacs

Further divisions lead to the formation of terminal bronchioles, which in turn divide to form a series of respiratory bronchioles, the smallest and finest air passages. Respiratory bronchioles are so named because they have a few alveoli (air sacs) opening directly into them. Most of the alveoli, however, arise in clusters from alveolar ducts, which are formed from division of the respiratory bronchioles.

The branching of the bronchial tree ends in numerous terminal bronchioles from which arise the respiratory bronchioles, alveolar ducts and alveoli. The clusters of alveoli provide a large surface area for gaseous exchange.

Alveoli

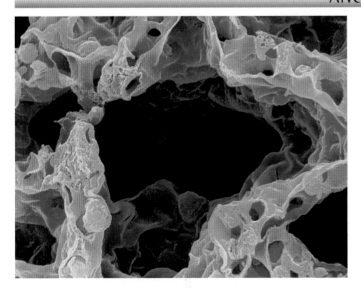

The alveoli, tiny hollow sacs with extremely thin walls, are the sites of gaseous exchange within the lungs. It is through the alveolar walls that oxygen diffuses from the air into the pulmonary bloodstream, and waste carbon dioxide diffuses out.

There are many millions of alveoli in the human lung which together give a huge surface area (about 140 m^2) for this exchange to take place.

Highly magnified image of alveolar sacs. It is across their thin walls that oxygen diffuses into the blood and waste gases are removed. Each adult lung contains about 300 million alveoli.

The alveoli lie in clusters like bunches of grapes around the alveolar ducts, each having a narrowed opening into a duct. They also have small holes, or pores, through which they connect with neighbouring alveoli. The alveolar walls are lined by flattened epithelial cells, and are supported by a framework of elastic and collagen fibres.

Two other types of cell are found in the alveoli: macrophages (defence cells), which engulf any foreign particles that get down the respiratory tract; and cells which produce surfactant, an important substance that lowers the surface tension in the fluid lining the alveoli, preventing their collapse.

Heart

The adult heart is about the size of a clenched fist and lies within the mediastinum in the thoracic cavity. It rests on the central tendon of the diaphragm and is flanked on either side by the lungs.

Surrounding the heart is a protective sac of connective tissue called the pericardium.

The heart is hollow and is composed almost entirely of muscle. The typical weight of the normal heart is only about 250 to 350 grams yet it has incredible power and stamina, beating over 70 times every minute to pump blood around the body.

SURFACES OF THE HEART
Roughly the shape of a pyramid on its side, the heart is said to have a base, three surfaces and an apex:

■ The base of the heart lies posteriorly (at the back) and is formed mainly by the left atrium, the chamber of the heart that receives oxygenated blood from the lungs

■ The inferior or diaphragmatic surface lies on the underside and is formed by the left and right ventricles separated by the posterior interventricular groove. The right and left ventricles are the large chambers which pump blood around the lungs and the body respectively

■ The anterior, or sternocostal, surface lies at the front of the heart just behind the sternum and the ribs and is formed mainly by the right ventricle

■ The left, or pulmonary, surface is formed mainly by the large left ventricle, which lies in a concavity of the left lung.

Position of the heart

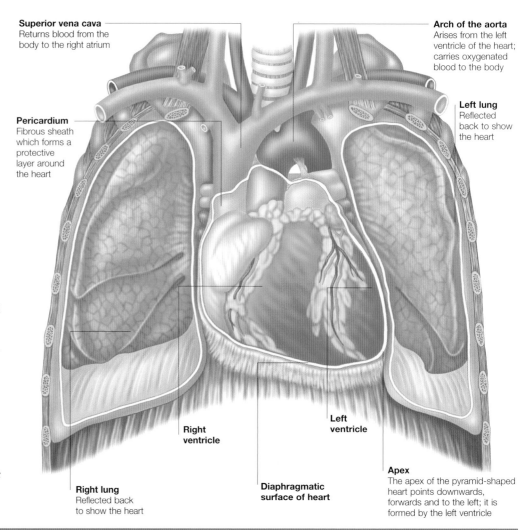

Superior vena cava
Returns blood from the body to the right atrium

Pericardium
Fibrous sheath which forms a protective layer around the heart

Arch of the aorta
Arises from the left ventricle of the heart; carries oxygenated blood to the body

Left lung
Reflected back to show the heart

Right ventricle

Right lung
Reflected back to show the heart

Diaphragmatic surface of heart

Left ventricle

Apex
The apex of the pyramid-shaped heart points downwards, forwards and to the left; it is formed by the left ventricle

Position of the heart

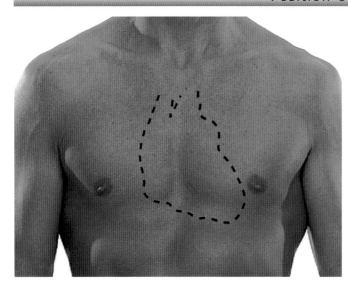

The heart lies behind the body of the sternum (breastbone), extending from the second rib above to the fifth intercostal space below. About two-thirds of the heart lies to the left of the midline of the chest, with the remaining third to the right.

BORDERS
The heart has four borders. The right border is formed by the right atrium and is slightly convex. The left border is formed mainly by the left ventricle and

The heart fills the central part of the thorax, with the apex extending to the left. The heart's shape and position vary as it beats and also with respiration.

slopes upwards and inwards to merge with the superior border, which is formed by the atria and great blood vessels. The inferior border, which lies nearly horizontally, is formed mainly by the right ventricle.

APEX OF THE HEART
The apex of the heart normally lies behind the fifth intercostal space a hand's breadth from the midline. The pulsations of the heart can usually be felt, and often observed, at this point.

As it is attached only to other soft tissues the heart is quite mobile within the thoracic cavity and can change position as the diaphragm, on which it lies, contracts and relaxes.

The pericardium

The heart is enclosed within a protective triple-walled bag of connective tissue called the pericardium. The pericardium is composed of two parts, the fibrous pericardium and the serous pericardium.

FIBROUS PERICARDIUM

The fibrous pericardium forms the outer part of the bag and is composed of tough fibrous connective tissue. It has three main functions:

■ **Protection.** The fibrous pericardium is strong enough to provide some protection from trauma for such a vital structure as the heart

■ **Attachment.** There are fibrous attachments between this part of the pericardium and both the sternum and the diaphragm. In addition, the fibrous pericardium fuses with the strong walls of the arteries which pass through it from the heart. These attachments help to anchor the heart to its surrounding structures

■ **Prevention of overfilling of the heart.** Because the fibrous pericardium is non-elastic, it does not allow the heart to expand with blood beyond a certain safe limit.

SEROUS PERICARDIUM

The serous pericardium covers and surrounds the heart in the same way as the pleura does the lungs. This part of the pericardium is a thin membrane which has two parts that are continuous with each other, called the parietal and the visceral layers.

The pericardial sac with the heart removed

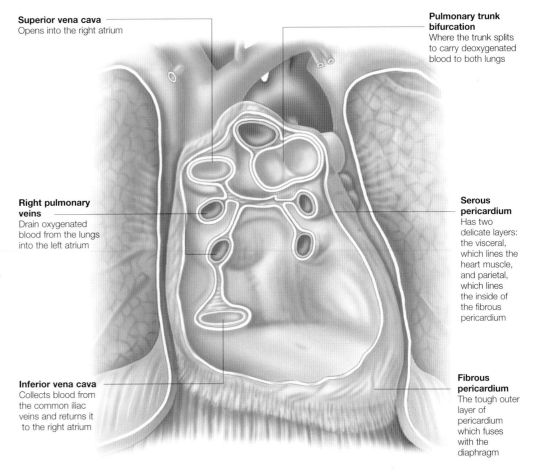

Superior vena cava
Opens into the right atrium

Pulmonary trunk bifurcation
Where the trunk splits to carry deoxygenated blood to both lungs

Right pulmonary veins
Drain oxygenated blood from the lungs into the left atrium

Serous pericardium
Has two delicate layers: the visceral, which lines the heart muscle, and parietal, which lines the inside of the fibrous pericardium

Inferior vena cava
Collects blood from the common iliac veins and returns it to the right atrium

Fibrous pericardium
The tough outer layer of pericardium which fuses with the diaphragm

The parietal pericardium lines the inner surface of the fibrous pericardium and reflects back onto the surface of the heart at the roots of the large blood vessels to form the visceral pericardium.

Between the two layers of serous pericardium lies a slit-like cavity, the pericardial cavity, which is filled with a very small amount of fluid. The presence of this fine fluid layer, together with the slipperiness of the layers of serous pericardium, allows the chambers of the heart to move freely within the pericardium as the heart beats.

If the pericardial cavity becomes filled with an abnormally large amount of fluid, as may happen in infection or inflammation, the heart becomes compressed within the confines of the fibrous pericardium and is unable to function properly. In extreme cases, when it is known as 'cardiac tamponade', this is life-threatening.

Layers of the heart wall

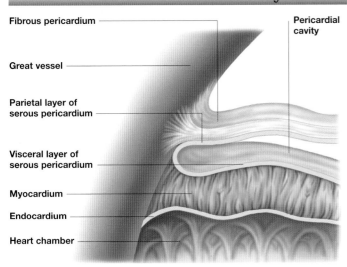

Fibrous pericardium

Great vessel

Parietal layer of serous pericardium

Visceral layer of serous pericardium

Myocardium

Endocardium

Heart chamber

Pericardial cavity

Inside the pericardial cavity, the heart wall is made up of three layers: the epicardium, the myocardium and the endocardium.

■ The epicardium is the visceral layer of the serous pericardium, which covers the outer surface of the heart and is attached firmly to it

■ The myocardium makes up the bulk of the heart wall and is

A section taken through the heart at the junction of a typical great vessel reveals the different layers of pericardium and heart wall.

composed of specialized cardiac muscle fibres. This type of muscle occurs only in the heart and is adapted for the special role it plays there. The muscle fibres of the myocardium are supported and held together by interlocking fibres of connective tissue

■ The endocardium is a smooth, delicate membrane, formed by a very thin layer of cells, and lines the inner surface of the heart chambers and valves. The blood vessels entering and leaving the heart are lined by a similar layer – the endothelium – which is a continuation of the endocardium.

Chambers of the heart

The heart is divided into four chambers: two thin-walled atria, which receive venous blood, and two larger, thick-walled ventricles, which pump blood into the arterial system.

The heart is divided into left and right sides, each having an atrium and ventricle.

THE VENTRICLES

The two ventricles make up the bulk of the muscle of the heart, the left being larger and more powerful than the right. The right ventricle lies in front, forming much of the anterior surface of the heart, while the left lies behind and below, comprising the greater part of the inferior surface. The apex of the heart is formed by the tip of the left ventricle.

The right ventricle receives blood from the right atrium, back flow being prevented by the tricuspid valve. Blood is then pumped by contraction of the ventricular muscle up through the pulmonary valve into the pulmonary trunk and from there into the lungs.

The left ventricle receives blood from the left atrium through the left atrioventricular orifice, which bears the mitral valve. Powerful contractions of the left ventricle then pump the blood up through the aortic valve into the aorta, the main artery of the body.

This illustration shows the internal structure of the heart when opened along a plane connecting the root of the aorta and the apex of the heart.

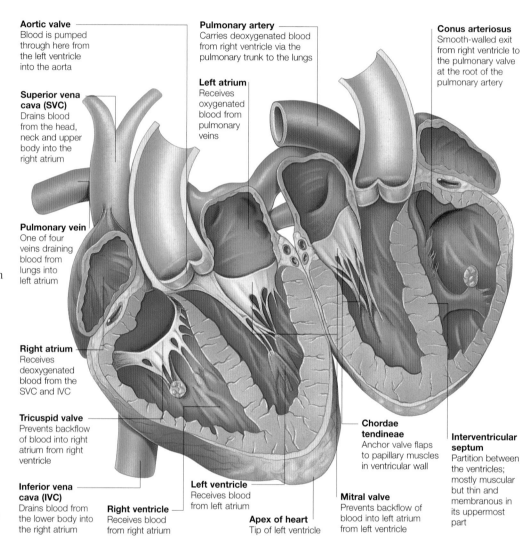

Aortic valve
Blood is pumped through here from the left ventricle into the aorta

Superior vena cava (SVC)
Drains blood from the head, neck and upper body into the right atrium

Pulmonary vein
One of four veins draining blood from lungs into left atrium

Right atrium
Receives deoxygenated blood from the SVC and IVC

Tricuspid valve
Prevents backflow of blood into right atrium from right ventricle

Inferior vena cava (IVC)
Drains blood from the lower body into the right atrium

Right ventricle
Receives blood from right atrium

Pulmonary artery
Carries deoxygenated blood from right ventricle via the pulmonary trunk to the lungs

Left atrium
Receives oxygenated blood from pulmonary veins

Left ventricle
Receives blood from left atrium

Apex of heart
Tip of left ventricle

Conus arteriosus
Smooth-walled exit from right ventricle to the pulmonary valve at the root of the pulmonary artery

Chordae tendineae
Anchor valve flaps to papillary muscles in ventricular wall

Mitral valve
Prevents backflow of blood into left atrium from left ventricle

Interventricular septum
Partition between the ventricles; mostly muscular but thin and membranous in its uppermost part

Architecture of the ventricular walls

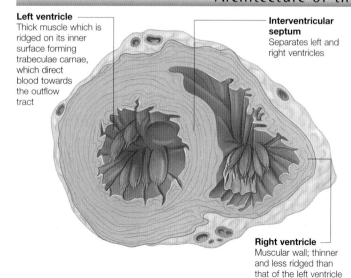

Left ventricle
Thick muscle which is ridged on its inner surface forming trabeculae carnae, which direct blood towards the outflow tract

Interventricular septum
Separates left and right ventricles

Right ventricle
Muscular wall; thinner and less ridged than that of the left ventricle

The muscular walls of the left ventricle are twice as thick as those of the right, and form a rough circle in cross-section. The right ventricle forms a crescent in cross-section as it is deformed by the more muscular left ventricle.

The difference in muscle thickness between the chambers reflects the pressure required to empty the relevant chamber when the muscle contracts.

Arising from the walls of both ventricles are the papillary

A cross-section of the heart through the ventricles shows the difference in thickness of the muscular walls of the left and right ventricles.

muscles, which taper to a point and bear tendinous chords (chordae tendineae) that attach to the tricuspid and mitral valves to stabilize them during pumping.

The inner surfaces of the ventricular walls, especially where blood enters, are roughened by irregular ridges of muscle, the trabeculae carnae, which give way to smoother walls near the outflow tracts through which blood is pumped out. There is only a small area of smooth wall in the left ventricle, just before the aortic valve. The right ventricle has a larger, funnel-shaped area of smooth wall below the pulmonary valve known as the conus arteriosus, or infundibulum.

The atria

The atria are the two smaller, thin-walled chambers of the heart. They sit above the ventricles separated by the atriovbentricular valves.

All the venous blood from the body is delivered to the right atrium by the two great veins, the superior and inferior vena cavae (SVC and IVC respectively). The coronary sinus, the vessel which collects venous blood from the heart tissues, also drains into the right atrium.

The interior has a smooth-walled posterior part and a rough-walled anterior section. These two areas are separated by a ridge of tissue known as the crista terminalis.

The roughened anterior wall is thicker than the posterior part, being composed of the pectinate muscles, which give a comb-like appearance to the inner surface. The fossa ovalis is a depression on the wall adjoining the left atrium.

The pectinate muscles extend into a small, ear-like outpouching of the right atrium called the auricle. This conical chamber wraps around the outside of the main artery from the heart – the aorta – and acts to increase the capacity of the right atrium.

OPENINGS INTO THE RIGHT ATRIUM

The SVC, which receives blood from the upper half of the body, opens into the upper part of the smooth area of the right atrium.

The IVC, which receives blood from the lower half of the body, enters the lower part of the right atrium. The SVC has no valve to

prevent backflow of blood; the IVC only bears a rudimentary non-functional valve.

The opening of the coronary sinus lies between the IVC opening and the opening that allows blood through into the right ventricle (the right atrioventricular orifice).

THE LEFT ATRIUM

The left atrium is smaller than the right, and forms the main part of the base of the heart. It is roughly cuboid in shape and has smooth walls, except for the lining of the left auricle, which is roughened by muscle ridges. The four pulmonary veins,

which bring oxygenated blood back from the lungs, open into the posterior part of the left atrium. There are no valves in these orifices.

In the wall adjoining the right atrium lies the oval fossa, which corresponds to the oval fossa on the right side.

Right atrium of the heart

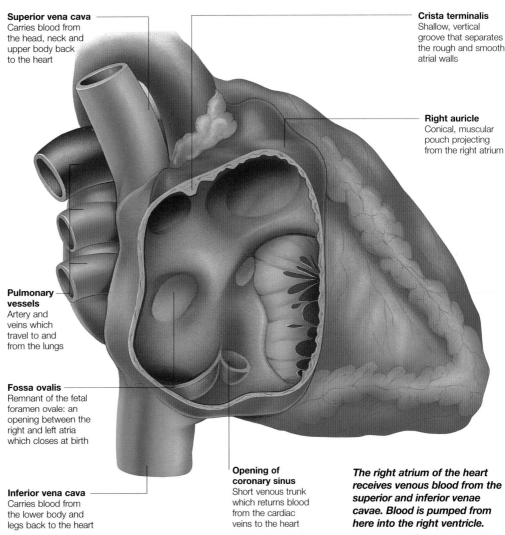

Superior vena cava
Carries blood from the head, neck and upper body back to the heart

Pulmonary vessels
Artery and veins which travel to and from the lungs

Fossa ovalis
Remnant of the fetal foramen ovale: an opening between the right and left atria which closes at birth

Inferior vena cava
Carries blood from the lower body and legs back to the heart

Crista terminalis
Shallow, vertical groove that separates the rough and smooth atrial walls

Right auricle
Conical, muscular pouch projecting from the right atrium

Opening of coronary sinus
Short venous trunk which returns blood from the cardiac veins to the heart

The right atrium of the heart receives venous blood from the superior and inferior venae cavae. Blood is pumped from here into the right ventricle.

The fetal heart

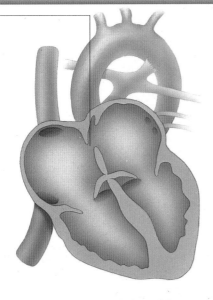

Foramen ovale
Gap between the two atria through which blood can flow before birth

In the fetus, the route of bloodflow through the heart is different from that after birth. Instead of being passed into the right ventricle to be pumped into the lungs, blood passes from the right atrium directly into the left atrium to be pumped around the body.

The blood passes through the foramen ovale, a hole in the wall which divides the two atria and has a flap-like valve to prevent backflow.

The circulation within the fetal heart is different from that of the postnatal heart. Blood flows directly between the atria via the foramen ovale.

CHANGES AFTER BIRTH

After birth the foramen ovale closes and blood is pumped into the lungs from the right ventricle. The site of the foramen ovale is marked by a depression in the wall (septum) between the atria, known as the fossa ovalis.

ATRIAL SEPTAL DEFECTS

In about 15 to 25 per cent of adults a small opening may still be present at this site although it may be found only incidentally during investigations of the heart and does not usually cause problems. Openings of this kind are known as atrial septal defects, or 'holes in the heart'.

Valves of the heart

The heart is a powerful muscular pump through which blood flows in a forward direction only. Backflow is prevented by the four heart valves, which have a vital role in maintaining the circulation.

Each of the two sides of the heart has two valves. On the right side of the heart, the tricuspid valve lies between the atrium and the ventricle, and the pulmonary valve lies at the junction of the ventricle and the pulmonary trunk. On the left side, the mitral valve separates the atrium and ventricle while the aortic valve lies between the ventricle and the aorta.

THE TRICUSPID AND MITRAL VALVES

The tricuspid and mitral valves are also known as the atrioventricular valves as they lie between the atria and the ventricles on each side. They are composed of tough connective tissue covered with endocardium, the thin layer of cells which lines the entire heart. The upper surface of the valves is smooth whereas the lower surface carries the attachments of the chordae tendineae.

The tricuspid valve has three cusps, or flaps. In contrast, the mitral valve has only two and is consequently also known as the bicuspid valve; the name 'mitral' comes from its supposed likeness to a bishop's mitre.

THE HEARTBEAT

During its contraction, the normal heart makes a two-component sound (often described as 'lub-dup') which can be heard using a stethoscope. The first of these sounds comes from the closure of the atrioventricular valves while the second is due to the closure of the pulmonary and aortic valves.

Diastolic heart with the atria removed

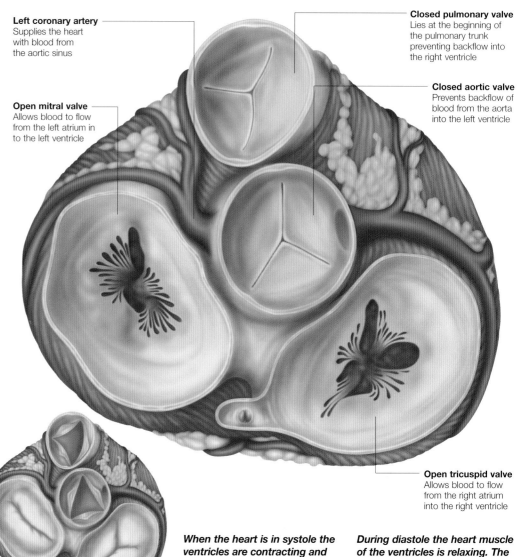

Left coronary artery
Supplies the heart with blood from the aortic sinus

Open mitral valve
Allows blood to flow from the left atrium in to the left ventricle

Closed pulmonary valve
Lies at the beginning of the pulmonary trunk preventing backflow into the right ventricle

Closed aortic valve
Prevents backflow of blood from the aorta into the left ventricle

Open tricuspid valve
Allows blood to flow from the right atrium into the right ventricle

When the heart is in systole the ventricles are contracting and the aortic and pulmonary valves open, allowing blood to be pumped out of the heart.

During diastole the heart muscle of the ventricles is relaxing. The tricuspid and mitral valves are open allowing blood to flow from the atria to fill the ventricles.

The chordae tendineae

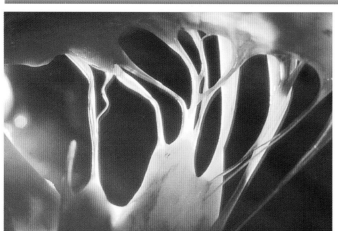

Attached to the edges and lower surfaces of the tricuspid and mitral valves are many thin tendinous chords of collagen – the chordae tendineae – which pass down to the papillary muscles which project into the cavity of the ventricle from the muscular walls.

The chordae tendineae of the mitral valve attach the cusps of the valve to papillary muscles, which in turn are attached to the wall of the ventricle.

ACTION OF THE CHORDS

These chords act like guy ropes to anchor the valves and prevent the cusps of the valves from giving way or being blown inside out like an umbrella under the high pressure of blood during contraction of the ventricles. Chords attached to neighbouring cusps also act to keep those cusps held tightly together during the ventricular contraction so that no blood can leak between them when they are closed.

Aortic and pulmonary valves

The pulmonary and aortic valves are also known as the semilunar valves. They guard the route of exit of blood from the heart, preventing backflow of blood into the ventricles as they relax after a contraction.

Each of these two valves is composed of three semilunar pocket-like cusps, which have a core of connective tissue covered by a lining of endothelium. This lining ensures a smooth surface for the passage of blood.

AORTIC VALVE

The aortic valve lies between the left ventricle and the aorta, the main artery that carries oxygenated blood to the body. It is stronger and more robust than the pulmonary valve as it has to cope with the higher pressures of the systemic circulation (to the body).

Above each cusp of the valve, formed by bulges of the aortic wall, lie the aortic sinuses. From two of these sinuses arise the right and left coronary arteries, which carry blood to the muscle and coverings of the heart itself.

PULMONARY VALVE

The pulmonary valve separates the ventricle from the pulmonary trunk, the large artery that carries blood from the heart towards the lungs. Just above each cusp of the valve the pulmonary trunk bulges slightly to form the pulmonary sinuses, blood-filled spaces that prevent the cusps from sticking to the arterial wall behind them when they open.

View of the left ventricle opened up

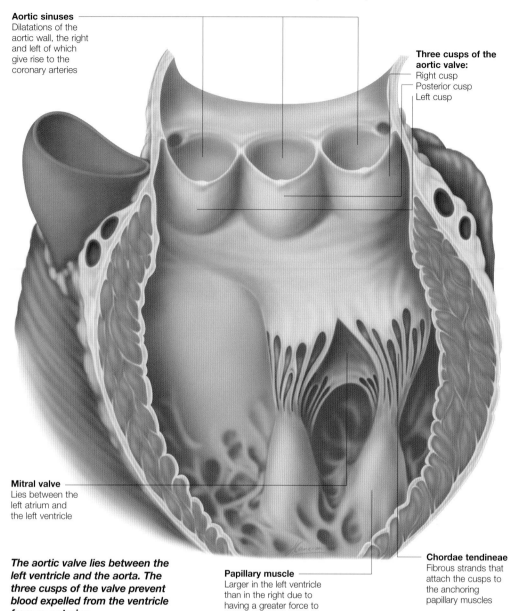

Aortic sinuses
Dilatations of the aortic wall, the right and left of which give rise to the coronary arteries

Three cusps of the aortic valve:
Right cusp
Posterior cusp
Left cusp

Mitral valve
Lies between the left atrium and the left ventricle

Papillary muscle
Larger in the left ventricle than in the right due to having a greater force to overcome

Chordae tendineae
Fibrous strands that attach the cusps to the anchoring papillary muscles

The aortic valve lies between the left ventricle and the aorta. The three cusps of the valve prevent blood expelled from the ventricle from re-entering.

Action of the valves

Open valve

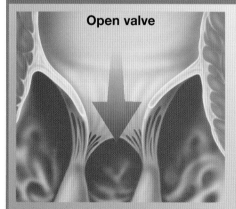

When an atrioventricular valve is open, the papillary muscles are relaxed and the cusps flap downwards. Blood from the atrium can flow into the ventricle.

Closed valve

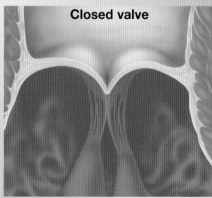

As the ventricle fills with blood the atrioventricular valve snaps shut, tensing the chordae tendineae. When the ventricle contracts the blood is thus pushed forwards.

When the atria contract, blood passes through the open and relaxed tricuspid and mitral valves into the ventricles.

As the ventricles contract in turn, the sudden rising pressure of blood within each ventricle causes the valves to close, so preventing backflow of that blood into the atria. The pull of the chordae tendineae steadies the valves and enables them to withstand the pressure of the blood within the ventricle.

As the atrioventricular valves are now closed the blood must travel up and out through the semilunar valves into the pulmonary trunk and the aorta. The semilunar valves are forced open by the high-pressure flow of blood from the ventricles, but snap shut again as soon as the ventricles stop contracting and start to relax.

Vessels of the heart

Blood is delivered to the heart by two large veins – the superior and inferior venae cavae – and pumped out into the aorta. The venae cavae and aorta are collectively known as the great vessels.

THE VENAE CAVAE

The superior vena cava is the large vein that drains blood from the upper body to the right atrium of the heart. It is formed by the union of the right and left brachiocephalic veins which, in turn, have been formed by smaller veins that receive blood from the head, neck and upper limbs.

The inferior vena cava is the widest vein in the body, but only its last part lies within the thorax as it passes up through the diaphragm to deliver blood to the right atrium.

THE AORTA

The aorta is the largest artery in the body, having an internal diameter of about 2.5 cm in adults. Its relatively thick walls contain elastic connective tissue which allows the vessel to expand slightly, as blood is pumped into it under pressure, and then recoil, thus maintaining blood pressure between heart beats.

The aorta passes upwards initially, then curves around to the left and travels down into the abdomen. It consists of the ascending aorta, the arch of the aorta and the descending (thoracic) aorta. The various sections of the aorta are named for their shape or the positions in which they lie, and each has branches which carry blood to the tissues of the body.

The heart and great vessels

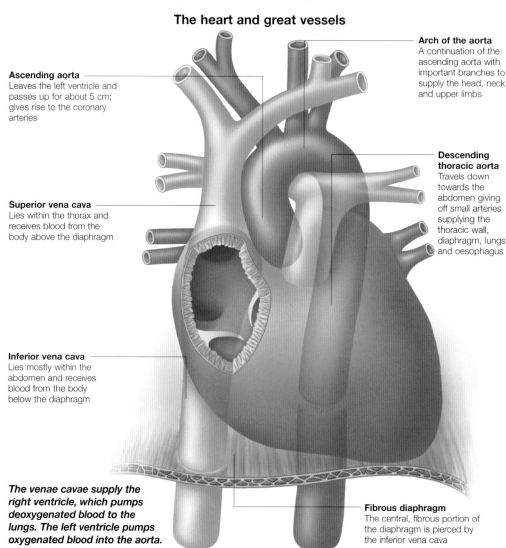

Ascending aorta
Leaves the left ventricle and passes up for about 5 cm; gives rise to the coronary arteries

Superior vena cava
Lies within the thorax and receives blood from the body above the diaphragm

Inferior vena cava
Lies mostly within the abdomen and receives blood from the body below the diaphragm

Arch of the aorta
A continuation of the ascending aorta with important branches to supply the head, neck and upper limbs

Descending thoracic aorta
Travels down towards the abdomen giving off small arteries supplying the thoracic wall, diaphragm, lungs and oesophagus

Fibrous diaphragm
The central, fibrous portion of the diaphragm is pierced by the inferior vena cava

The venae cavae supply the right ventricle, which pumps deoxygenated blood to the lungs. The left ventricle pumps oxygenated blood into the aorta.

How the fetal heart changes after birth

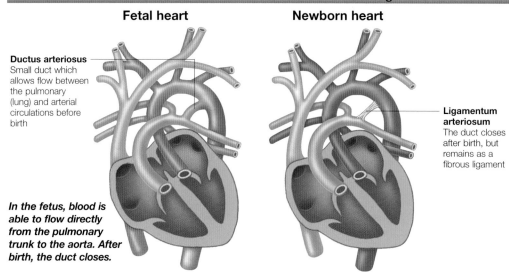

Fetal heart

Ductus arteriosus
Small duct which allows flow between the pulmonary (lung) and arterial circulations before birth

In the fetus, blood is able to flow directly from the pulmonary trunk to the aorta. After birth, the duct closes.

Newborn heart

Ligamentum arteriosum
The duct closes after birth, but remains as a fibrous ligament

In the fetus, there is a connecting blood vessel which allows blood to travel from the pulmonary trunk to the aorta, bypassing the lungs. This vessel, known as the ductus arteriosus, closes after birth. Then, blood from the right ventricle only passes into the pulmonary circulation.

The site of this fetal blood vessel is marked by the ligamentum arteriosum, a fibrous band that passes from the pulmonary trunk to the arch of the aorta. Sometimes, the ductus arteriosus does not close at birth and high pressure aortic blood will enter the relatively low pressure pulmonary system. Surgical closure is then required.

Supplying blood to the heart

The heart muscle itself and the coverings of the heart need their own blood supply, which is provided by the coronary arteries.

There are two coronary arteries: right and left. These arise from the ascending aorta just above the aortic valve and run around the heart just beneath the epicardium, embedded in fat.

■ **The right coronary artery**
This arises within the right aortic sinus, a small outpouching of the arterial wall just behind the aortic valve. It runs down and to the right, along the groove between the right atrium and the right ventricle until it lies along the inferior surface of the heart. Here it terminates in an anastomosis (connecting network) with branches of the left coronary artery. The right coronary artery gives off several branches.

■ **The left coronary artery**
This arises from the coronary sinus above the aortic valve and runs down towards the apex of the heart. The left coronary artery divides early on into two branches.

VENOUS DRAINAGE
The main vein of the heart is the coronary sinus. It receives blood from the cardiac veins and empties into the right atrium. In general, cardiac veins follow the routes of the coronary arteries.

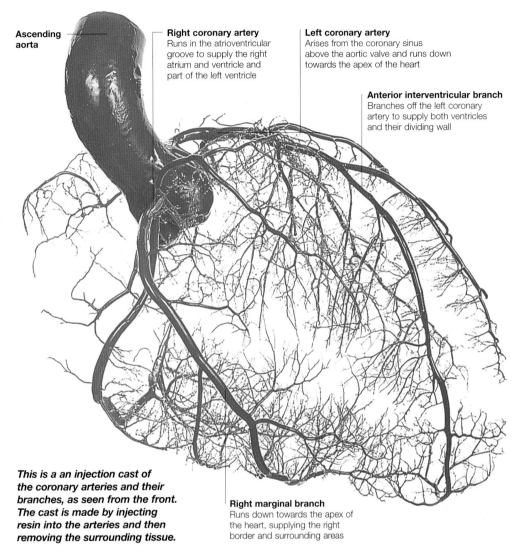

Ascending aorta

Right coronary artery
Runs in the atrioventricular groove to supply the right atrium and ventricle and part of the left ventricle

Left coronary artery
Arises from the coronary sinus above the aortic valve and runs down towards the apex of the heart

Anterior interventricular branch
Branches off the left coronary artery to supply both ventricles and their dividing wall

This is a an injection cast of the coronary arteries and their branches, as seen from the front. The cast is made by injecting resin into the arteries and then removing the surrounding tissue.

Right marginal branch
Runs down towards the apex of the heart, supplying the right border and surrounding areas

Variations in the coronary arteries

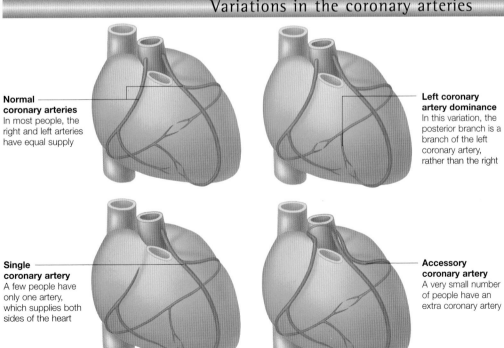

Normal coronary arteries
In most people, the right and left arteries have equal supply

Single coronary artery
A few people have only one artery, which supplies both sides of the heart

Left coronary artery dominance
In this variation, the posterior branch is a branch of the left coronary artery, rather than the right

Accessory coronary artery
A very small number of people have an extra coronary artery

In most people, the right and left coronary arteries are equally responsible for the blood supply to the heart; however, there may be great variation in the branching patterns of the coronary arteries between individuals.

VARIATIONS
However, in about 15 per cent of people, the left coronary artery provides a greater supply as it gives rise to the large posterior interventricular artery (normally a branch of the right coronary artery). Very occasionally, there may be only one coronary artery or sometimes an extra, accessory coronary artery. There are many other possible variations.

The blood supply of the heart can vary; either the left or right coronary artery can be dominant, an artery can be missing or an extra one present.

Conducting system of the heart

When the body is at rest, the heart beats at a rate of about 70 to 80 beats per minute. Within its muscular walls, a conducting system sets the pace and ensures that the muscle contracts in a co-ordinated way.

SINO-ATRIAL NODE

The sino-atrial (SA) node is a collection of cells within the wall of the right atrium.

Each contraction of the cells of the SA node generates an electrical impulse, which is passed to the other muscle cells of the right and left atria and then to the atrio-ventricular (AV) node.

ATRIOVENTRICULAR NODE

The cells of the AV node will initiate contractions of their own, and pass on impulses at a slower rate, if not stimulated by the SA node. Impulses from the AV node are passed to the ventricles through the next stage of conducting tissue.

ATRIOVENTRICULAR BUNDLE

The AV bundle passes from the atria to the ventricles through an insulating layer of fibrous tissue. It then divides into two parts, the right and left bundle branches, which supply the right and left ventricles respectively.

The intrinsic conduction system of the heart carries a wave of nerve impulses, which create synchronized contraction of the heart muscle.

The intrinsic conduction system of the heart

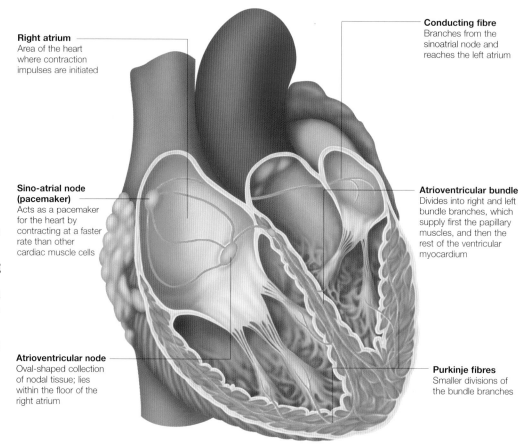

Right atrium
Area of the heart where contraction impulses are initiated

Conducting fibre
Branches from the sinoatrial node and reaches the left atrium

Sino-atrial node (pacemaker)
Acts as a pacemaker for the heart by contracting at a faster rate than other cardiac muscle cells

Atrioventricular bundle
Divides into right and left bundle branches, which supply first the papillary muscles, and then the rest of the ventricular myocardium

Atrioventricular node
Oval-shaped collection of nodal tissue; lies within the floor of the right atrium

Purkinje fibres
Smaller divisions of the bundle branches

Nerve supply of the heart

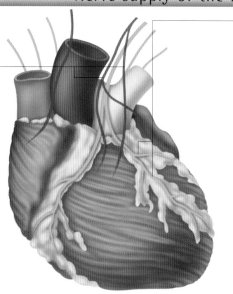

Parasympathetic nerves
These nerves have a 'braking' effect and slow the heart

Sympathetic nerves
Impulses from these nerves increase both the rate and force of the heartbeat

The nerve supply to the heart is from the autonomic nervous system. This is the system of nerves that regulates the internal organs of the body without our conscious control.

The heart beats regularly without external stimulation; its nerve supply affects its rate and force of contraction.

EXTERNAL NERVE SUPPLY

The autonomic nerve supply is carried to the heart from the cardiac plexuses, networks of nervous tissue that lie just behind the ascending aorta above the heart.

Autonomic nerves are divided into two groups: sympathetic fibres, which arise from the cervical and upper thoracic parts of the sympathetic trunks (which lie alongside the spine); and parasympathetic fibres, which come via the vagus (10th cranial) nerves.

The cardiac cycle

The cardiac cycle is the series of changes within the heart which causes blood to be pumped around the body. It is divided into a period when heart muscle contracts, known as systole and a period when it is relaxed, known as diastole.

VENTRICULAR FILLING

During diastole the tricuspid and mitral valves are open. Blood from the great veins fills the atria and then passes through these open valves to fill the relaxing ventricles.

ATRIAL CONTRACTION

As diastole ends and systole begins, the SA node sparks off a contraction of the atrial muscle which forces more blood into the ventricles.

VENTRICULAR CONTRACTION

The wave of contraction reaches the ventricles via the AV bundles and the Purkinje fibres. The tricuspid and mitral valves snap shut as pressure increases. The blood pushes against the closed pulmonary and aortic valves and causes them to open.

As the wave of contraction dies away, the ventricles relax. The cycle begins again with the next SA node impulse about a second later.

The movements of the heart cause the circulation of blood. The sequence of contraction is repeated, in normal adults, about 70 to 90 times a minute.

Events of the cardiac cycle

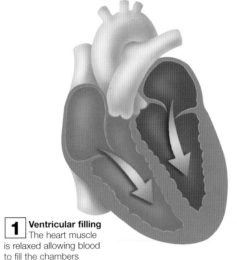

1 **Ventricular filling**
The heart muscle is relaxed allowing blood to fill the chambers

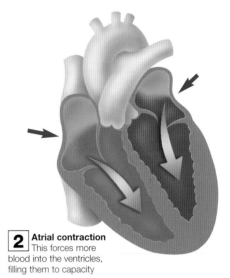

2 **Atrial contraction**
This forces more blood into the ventricles, filling them to capacity

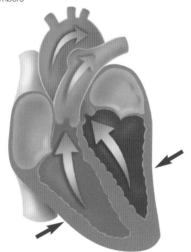

3 **Ventricular contraction**
Pulmonary and aortic valves open letting blood pass up and out into the pulmonary trunk and the aorta

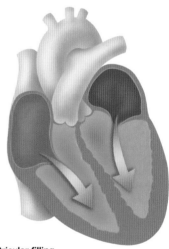

4 **Ventricular filling**
As the wave of contraction dies away the ventricles relax and allow blood to enter again

Fibrous skeleton of the heart

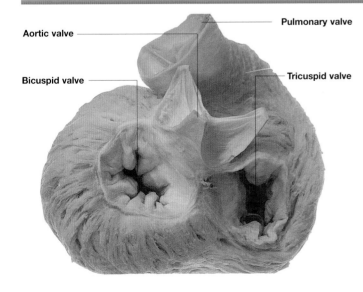

Aortic valve

Bicuspid valve

Pulmonary valve

Tricuspid valve

Like the muscles of the bony skeleton, when the muscle of the heart wall contracts it needs something to pull against to be effective. There are no bones or solid structures in the heart, but it does have a skeleton of tough fibrous connective tissue, which performs a similar function in that it offers a site for attachment of the cardiac muscle fibres.

SUPPORT

The fibrous skeleton of the heart also helps to support the heart

The skeleton of the heart is made of tough fibrous connective tissue. It forms the rigid framework that the heart pulls against when it contracts.

valves and prevent them from becoming pulled out of shape by the great pressure of blood during systole. It also provides a base for the cusps of those valves.

INSULATION

Another vital function of the fibrous skeleton is to separate and insulate the myocardium (the middle of the three layers forming the wall of the heart) of the atria from that of the ventricles so that the wave of contraction during systole can only pass via the AV bundle.

This ensures that the ventricles will contract slightly after the atria contract, which gives time for them to be filled as a result of atrial contraction.

How the heart pumps

The adult heart beats over 100,000 times and pumps about 8,000 litres of blood in a 24-hour period. Although the heart is a muscle, it does not tire like other muscles, and it never rests.

The heart is a powerful muscle that performs two vital tasks. It propels blood enriched with oxygen to all parts of the body, and it also pumps deoxygenated (used) blood to the lungs, where it will be re-oxygenated.

The heart is divided in two by a strong muscular wall called the septum. Each half is further divided into two chambers – the left and right upper chambers are known as the atria (singular: atrium), and the two lower chambers are the ventricles. Each of these four chambers plays a specific role in circulating blood around the heart and then back out into the body or to the lungs.

THE CARDIAC MUSCLE

The wall of the heart is composed of three layers: the epicardium (outer layer), the myocardium (middle layer) and the endocardium (inner layer). The myocardial layer is responsible for the contraction of the heart. Muscle fibres are arranged to give a 'wringing' movement that effectively squeezes blood out of the heart.

The thickness of the myocardial layer varies according to the pressure generated within the various chambers in the heart. The right ventricular myocardial layer is moderately thick, as blood is only pumped through the pulmonary artery to the lungs. The myocardium of the left ventricle is much thicker because greater pressure is needed to pump blood into all parts of the body. The myocardial layer of the atria is relatively thin.

Inner structure of the heart

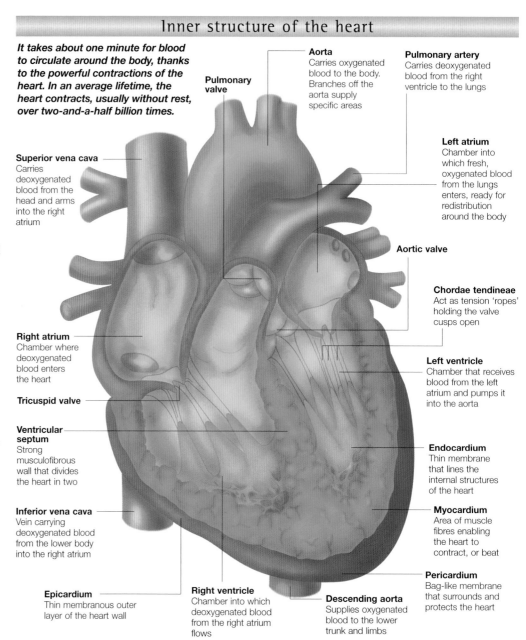

It takes about one minute for blood to circulate around the body, thanks to the powerful contractions of the heart. In an average lifetime, the heart contracts, usually without rest, over two-and-a-half billion times.

Aorta
Carries oxygenated blood to the body. Branches off the aorta supply specific areas

Pulmonary artery
Carries deoxygenated blood from the right ventricle to the lungs

Pulmonary valve

Superior vena cava
Carries deoxygenated blood from the head and arms into the right atrium

Left atrium
Chamber into which fresh, oxygenated blood from the lungs enters, ready for redistribution around the body

Aortic valve

Chordae tendineae
Act as tension 'ropes' holding the valve cusps open

Right atrium
Chamber where deoxygenated blood enters the heart

Left ventricle
Chamber that receives blood from the left atrium and pumps it into the aorta

Tricuspid valve

Ventricular septum
Strong musculofibrous wall that divides the heart in two

Endocardium
Thin membrane that lines the internal structures of the heart

Inferior vena cava
Vein carrying deoxygenated blood from the lower body into the right atrium

Myocardium
Area of muscle fibres enabling the heart to contract, or beat

Pericardium
Bag-like membrane that surrounds and protects the heart

Epicardium
Thin membranous outer layer of the heart wall

Right ventricle
Chamber into which deoxygenated blood from the right atrium flows

Descending aorta
Supplies oxygenated blood to the lower trunk and limbs

Controlling the flow of blood around the heart

Blood flow around the four chambers of the heart is controlled by four valves. The atrioventricular valves (the tricuspid valve and the mitral, or bicuspid, valve) lie between the atria and the ventricles. The two semilunar valves lie at the openings of the pulmonary artery and the aorta. The pulmonary artery takes deoxygenated blood to the

lungs; the aorta carries blood to the body's organs and tissues.

The heart valves ensure that blood only flows in one direction. As pressure reaches a critical point, the valves open, allowing blood to pass through. When the heart is relaxed between contractions the aortic and pulmonary valves stay tightly sealed, but the atrioventricular valves stay open.

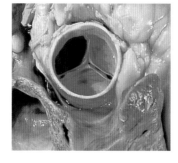

Oxygenated blood enters the left atrium of the heart, is pushed into the left ventricle and is finally pumped to the arteries of the body via the aorta. The aortic valve (pictured) has three semilunar cusps. The purpose of the valve is to prevent the back-flow of blood into the left ventricle, maintaining the directional flow of blood through the heart.

The heartbeat cycle

There are three phases to every beat of our heart. As the cardiac muscle contracts, blood is moved around the internal chambers in strict rotation. At the same time, blood is pumped out to the body's organs and tissue, or transported back to the lungs, where it will be re-oxygenated, ready for re-use.

1 **DIASTOLE**

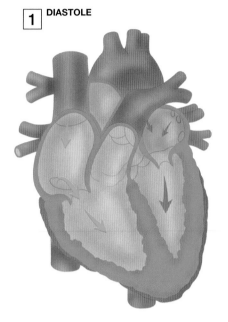

2 **ATRIAL SYSTOLE**

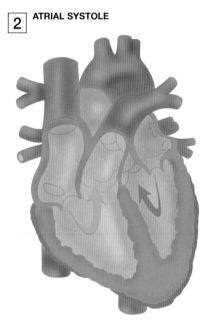

3 **VENTRICULAR SYSTOLE**

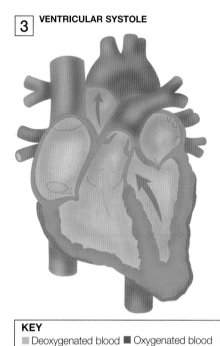

KEY
■ Deoxygenated blood ■ Oxygenated blood

During the first phase (diastole), deoxygenated blood enters the right atrium and oxygenated blood enters the left atrium. As these chambers reach capacity, blood flows into the ventricles.

In the atrial systole phase, the heart muscles (myocardium) surrounding the atria contract, which in turn causes the two atrial chambers to empty and blood remaining in the atria is forced into the ventricles.

The ventricles contract during the third phase. Semilunar valves open, and blood is pumped out to the body via the aorta, or carried to the lungs by the pulmonary artery. The whole cycle can now start again.

Deoxygenated blood enters the right side of the heart and is pumped out to the lungs to be re-oxygenated. This fresh blood then flows to the left side of the heart, from where it will be circulated throughout the body. This nutrient-rich blood flow is known as systemic circulation.

The atria and the ventricles hold blood as it is pumped into the heart. In the first phase of a heartbeat, deoxygenated blood flows into the right atrium and oxygenated blood enters the left atrium, causing both chambers to expand. As pressure builds up, the atria contract, forcing blood into the two ventricles. This is the second phase of the heartbeat cycle.

In the final phase, pressure begins to build in the ventricles as they fill with blood. At a critical point, blood is ejected from the heart via the aorta. The oxygenated blood is then distributed to the organs and tissue, and deoxygenated blood is pumped to the lungs.

Cardiac muscle action

The rhythmic contractions of the heart are caused by the 'wringing' motion of the cardiac muscle. Heart muscle is unique in that it is inherently contractile – it undergoes rhythmical contractions even if the heart is removed from the body briefly.

The process that causes automatic contraction of the heart muscle is called self-excitation. Many cardiac muscle fibres display this capability, but it is especially true of fibres found in the heart's specialised conducting system, which

controls the heartbeat cycle. The phases of the heartbeat cycle can be heard heart using a stethoscope. This technique is called auscultation of the heart.

In a normal person, the heart contracts about 72 times a minute and two heart sounds are distinguishable in each beat. These are described as 'lub dup'. The first sound, lub, is caused by closure of the mitral and tricuspid valves, and the second, dup, is caused by the closure of the aortic and pulmonary valves.

Blood flow to and from the heart

Blood flows into the right atrium via the superior vena cava, the large vein that drains used blood from the head, neck, arms and parts of the chest. Blood also flows from the inferior vena cava – which drains blood from the rest of the body – and the coronary sinus, which drains blood from the heart itself.

Oxygenated blood leaves the heart via the aorta, which branches into the arterial system. It is the means by which nutrients and oxygen are carried to the cells. The arteries divide into arterioles and ultimately into capillaries. It is in these microscopic capillaries that fluid, nutrients and waste are exchanged between the

tissue and the blood. The blood now containing waste products, drains into venules and veins, and ultimately into the great veins which drain into the heart.

For the heart to function, it must be constantly fed with a supply of oxygen and nutrients. These nutrients are carried to the heart by the blood via the coronary arterial system. Two coronary arteries (left and right) branch over the surface of the heart. Branches of these systems diffuse over the surface of the heart so the muscle fibre can receive oxygen and nutrients. The coronary veins carry used blood and waste products away from the heart's surface into the coronary sinus.

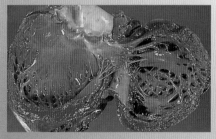

To function effectively, the heart (seen here in cross-section) needs a substantial supply of blood. The brain is the only organ in the body that requires a larger supply.

How the heart beats

The heart contains specialized tissue that generates an intrinsic rhythmic beat. The brain controls the heart rate by sending nervous impulses that alter this inherent rhythm.

A remarkable feature of the heart is that as long as it is bathed with a solution containing vital nutrients, it will continue to beat for long periods when removed from the body. This is because the heartbeat originates from within the heart itself, rather than resulting from electrical impulses from the brain.

Specialized pacemaker tissue and an electrical conducting system are responsible for generating the electrical 'spark' underlying the heartbeat and transmitting it in an orderly sequence across the upper and lower heart chambers.

SINOATRIAL NODE

The primary pacemaker of the heart is called the sinoatrial (SA) node, which is a small area of tissue located in the right atrium (upper chamber) of the heart. The SA node is about 20 mm long by 5 mm wide. Specialized electrical properties of the cells in the SA node allow it to generate regular 'sparks' of electricity that initiate each heartbeat.

Heart muscle cells are connected to one another in such a way that electrical events pass rapidly from cell to cell. Thus, when the cells of the SA node generate electrical impulses, the emerging wave of electrical excitation spreads very rapidly across both atria. This leads to a synchronized contraction of the atria, which pushes blood into the lower chambers of the heart – the ventricles.

The heart can continue to contract long after it has been excised from the body. However, it must be bathed with the appropriate nutrients.

Spread of electrical activity across the heart

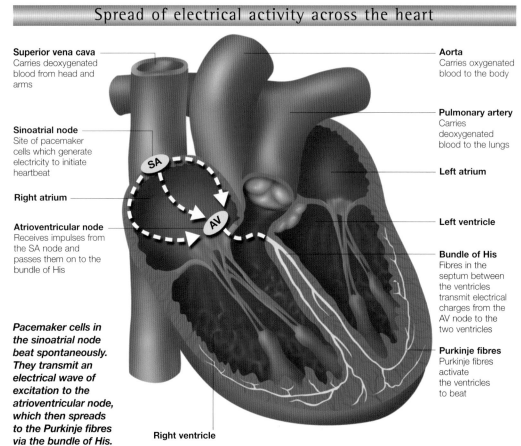

Superior vena cava
Carries deoxygenated blood from head and arms

Sinoatrial node
Site of pacemaker cells which generate electricity to initiate heartbeat

Right atrium

Atrioventricular node
Receives impulses from the SA node and passes them on to the bundle of His

Aorta
Carries oxygenated blood to the body

Pulmonary artery
Carries deoxygenated blood to the lungs

Left atrium

Left ventricle

Bundle of His
Fibres in the septum between the ventricles transmit electrical charges from the AV node to the two ventricles

Purkinje fibres
Purkinje fibres activate the ventricles to beat

Right ventricle

Pacemaker cells in the sinoatrial node beat spontaneously. They transmit an electrical wave of excitation to the atrioventricular node, which then spreads to the Purkinje fibres via the bundle of His.

Atrioventricular node

In order for an impulse to reach the ventricles, it must pass through a specialized region called the atrioventricular (AV) node. The AV node functions as a kind of electrical junction box. Electrical conduction is normally slower through the AV node than through other areas. As a consequence, the impulse is delayed in its passage through the AV node for approximately 0.1 of a second, at resting heart rates. This allows time for the ventricles to receive the blood being pumped into them during contraction of the upper chambers.

From the AV node, the impulse enters a cluster of fibres between the ventricles called the bundle of His. This divides into two branches, spreading into a network of conducting fibres called Purkinje fibres that rapidly distribute electrical excitation throughout the ventricles. Once excited, the ventricles contract, pumping blood into the circulation. Thus, a sequence of events starting with a spontaneous electrical impulse in the SA node ends with ventricular contraction.

When the heart stops beating, it can sometimes be started again by delivering a large electrical shock across the chest wall.

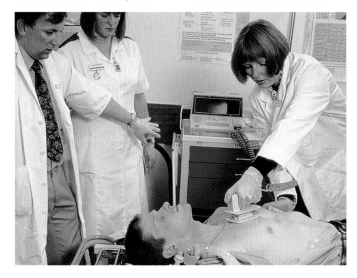

How the heart rate is controlled

The brain is able to modulate the heart rate via parasympathetic and sympathetic nerve fibres. These adapt the strength and timing of the heartbeat during rest and exercise or intense emotion. The electrical activity of the heart can be studied using an ECG.

Although the heartbeat arises in the sinoatrial nodal tissue of the heart, it can be modulated by the brain via a series of nerve fibres. These nerve fibres are subdivided anatomically and functionally into two groups:

■ Parasympathetic nerves – these decrease heart rate.

■ Sympathetic nerves – these increase the rate and strength of the heart's beating.

PARASYMPATHETIC CONTROL

In the absence of any influence from the nervous system, the inherent rate of SA node impulse generation is approximately 100 beats/minute in humans. This is somewhat higher than the normal resting heart rate which is near 70 beats/minute. The reason for this is that parasympathetic activity (via the vagus nerve) slows the rate of automatic sinoatrial impulse generation.

At rest, therefore, the heart is considered to be under 'vagal tone'; this allows the brain to increase the heart rate by reducing the activity of the vagus nerve.

SYMPATHETIC CONTROL

During increased demands on the circulation, such as occurs during exercise, sympathetic fibres release noradrenaline, which speeds up the rate of sinoatrial nodal impulse generation. In addition, sympathetic activity increases the speed of electrical conduction through the AV node allowing the ventricles to be excited and therefore beat more frequently.

As well as exercise, intense emotional states (for example, fear) can increase the heart rate via increased sympathetic activity. This can be off-set by 'beta-blockers', drugs which block the excitatory effects of noradrenaline and circulating adrenaline.

The brain is able to alter the heart's rate and strength of beating by sending nervous impulses along sympathetic and parasympathetic nerve fibres.

Nervous control of the heart

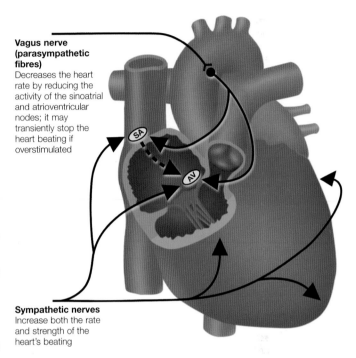

Vagus nerve (parasympathetic fibres)
Decreases the heart rate by reducing the activity of the sinoatrial and atrioventricular nodes; it may transiently stop the heart beating if overstimulated

Sympathetic nerves
Increase both the rate and strength of the heart's beating

Recording the heart's electrical activity

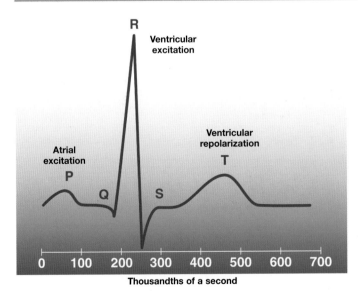

Thousandths of a second

The sequence of electrical events that occur during a single heartbeat can be detected on the body surface via an electrocardiogram (ECG).

SEQUENCE OF EVENTS DURING AN ECG TRACE

During each heartbeat, the first perceptible event on the ECG is the combined electrical excitation of atrial tissue. This corresponds to the 'P wave' of the ECG. A period then follows during which atrial contraction occurs and excitation travels through the AV node.

The excitation of ventricular tissue which follows gives rise to the 'QRS' complex of the ECG. Ventricular tissue remains uniformly excited for 0.2–0.3 seconds, then recovers during electrical recovery

This ECG shows a recording from a healthy person's heart. ECG equipment can detect the sequence of electrical events that occur during one heartbeat.

(repolarization), which corresponds to the 'T wave' of the ECG.

For the cardiologist, ECG recordings provide valuable information in the diagnosis of a range of pathological conditions involving abnormal impulse generation and propagation, as these produce characteristic alterations to the ECG profile.

Abnormal heart rhythms are grouped together under the general terms 'arrhythmias' or 'dysrhythmias' and their treatment involves a range of pharmacological and non-pharmacological strategies aimed towards restoring the normal timing and rate of impulse generation and conduction.

ECG recordings are made by attaching electrodes to the skin. The leads are connected to a monitor which displays voltage changes on the ECG.

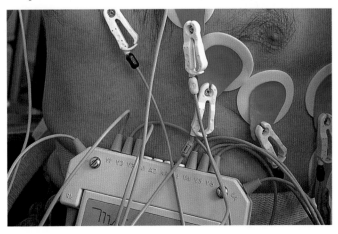

Chapter Five

The Upper Limbs

Many everyday tasks are dependent on the workings of our upper limbs, from the strength needed for climbing, pulling and lifting to the dexterity and fine motor skills required for threading a needle or manipulating a pen. Here we explore how the shoulder joints, arms and hands are designed to enable a range of tasks and movements.

As well as examining the structure of the upper limbs, this chapter also reveals the importance of the hand. Not just as a manipulative tool, it also has a rich supply of nerve endings, making it our main source of tactile communication with our surroundings.

LEFT: As a socket joint, the shoulder is one of the most flexible joints in the body.

Shoulder joint

The glenohumeral, or shoulder joint, is a ball-and-socket joint at the point of articulation of the humerus and the scapula. The construction of this joint allows the arm a wide range of movement.

The glenohumeral, or shoulder joint, is the point of articulation between the glenoid cavity of the scapula (shoulder blade) and the head of the humerus (bone of the upper arm). It is a ball-and-socket synovial (fluid-filled) joint constructed to allow the upper limb a very wide range of movement.

ARTICULAR SURFACE

To permit a wide range of movement, the head of the humerus provides a large articular surface. The glenoid cavity of the scapula, deepened by a ring of tough fibrocartilage (the glenoid labrum), offers only a shallow socket. The resulting ball-and-socket is so shallow that the joint needs to be held firmly together by the surrounding muscles and ligaments.

A thin layer of smooth articular (or hyaline) cartilage allows the bones to slip over each other with minimum friction.

JOINT CAPSULE

The shoulder joint is surrounded by a loose capsule of fibrous tissue. This capsule is lined by the synovial membrane which covers all the inner surfaces of the joint except those covered with articular cartilage.

The cells of this synovial membrane secrete synovial fluid, a viscous liquid which lubricates and nourishes the joint.

Shoulder joint viewed from the front

Acromion
Projects from the scapula

Coracoid process
Site of attachment for ligaments which stabilize the shoulder

Clavicle (collar bone)
Forms part of the pectoral girdle; articulates with the scapula at the acromioclavicular joint

Head of humerus
Articulates with the glenoid cavity

Anatomical neck of humerus
Fibrous tissue of joint capsule attaches to the anatomical neck of the humerus

Glenoid cavity of scapula
Shallow socket of the glenohumeral joint

Humerus
Long bone of the upper arm; articulates with the radius and ulna at the elbow and the scapula at the shoulder joint

Scapula (shoulder blade)
Triangular bone which lies flat against the back of the chest

Subscapular fossa
Site of attachment for subscapularis muscle

Inferior angle of scapula
Lower border of scapula

KEY This area can easily be felt under the skin

Bursae of the shoulder joint

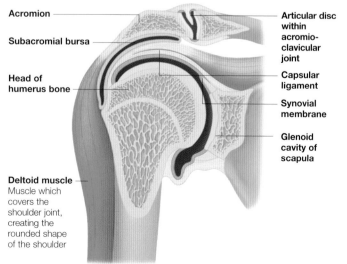

Acromion

Subacromial bursa

Head of humerus bone

Deltoid muscle
Muscle which covers the shoulder joint, creating the rounded shape of the shoulder

Articular disc within acromio-clavicular joint

Capsular ligament

Synovial membrane

Glenoid cavity of scapula

A coronal section through the glenohumeral joint shows the position of the bursa of the shoulder joint.

A bursa (*plural*: bursae) is a flattened fibrous sac lined with synovial membrane which contains a small amount of viscous synovial fluid. Bursae act to reduce the friction between structures which necessarily rub against each other during normal movement. Bursae are located at various points around the body where ligaments, muscle and tendons rub against bone.

Bursae may become abnormally enlarged at a point of unusual pressure, such as

occurs with a bunion at the base of the big toe where a shoe rubs.

The shoulder joint has several important bursae:

■ **Subscapular bursa**
This protects the tendon of the subscapularis muscle as it passes over the neck of the scapula. It usually has an opening which leads into the joint cavity and so may actually be thought of as an outpouching of that cavity.

■ **Subacromial bursa**
This lies above the glenohumeral joint beneath the acromion and the coraco-acromial ligament. It allows free movement of the muscles which pass beneath it. It is usually a true bursa, with no connection to the joint cavity of the shoulder.

Ligaments of the shoulder joint

The ligaments of the shoulder joint, along with the surrounding muscles, are crucial for the stability of this shallow ball-and-socket joint.

The ligaments around any joint contribute to its stability by holding the bones firmly together. In the shoulder joint, the main stabilizers are the surrounding muscles, but ligaments also play a role.

STABILIZING LIGAMENTS
The fibrous joint capsule has ligaments within it which help to strengthen the joint:
■ The glenohumeral ligaments are three weak, fibrous bands which reinforce the front of the capsule
■ The coracohumeral ligament is a strong, broad band which strengthens the upper aspect of the capsule. Although not actually part of the glenohumeral joint itself, the coraco-acromial ligament is important as it spans the gap between the acromion and the coracoid process of the scapula. The arch of bone and ligament is so strong that even if the humerus is forcibly pushed up, it will not break; the clavicle or the humerus will give way first
■ The transverse humeral ligament runs from the greater to the lesser tuberosity of the humerus, creating a tunnel for the passage of the biceps brachii tendon in its synovial sheath.

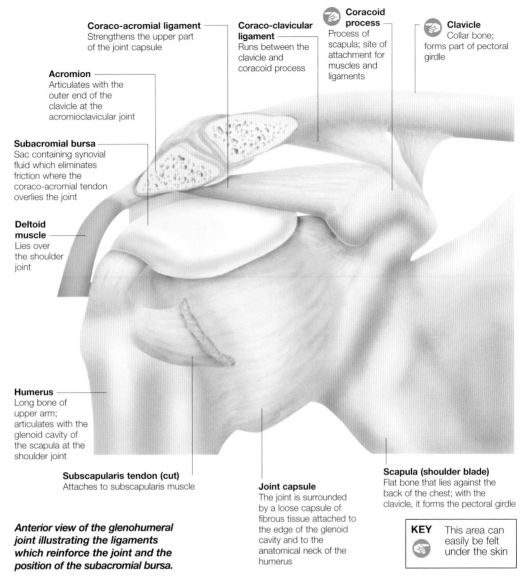

Coraco-acromial ligament
Strengthens the upper part of the joint capsule

Coraco-clavicular ligament
Runs between the clavicle and coracoid process

Coracoid process
Process of scapula; site of attachment for muscles and ligaments

Clavicle
Collar bone; forms part of pectoral girdle

Acromion
Articulates with the outer end of the clavicle at the acromioclavicular joint

Subacromial bursa
Sac containing synovial fluid which eliminates friction where the coraco-acromial tendon overlies the joint

Deltoid muscle
Lies over the shoulder joint

Humerus
Long bone of upper arm; articulates with the glenoid cavity of the scapula at the shoulder joint

Subscapularis tendon (cut)
Attaches to subscapularis muscle

Joint capsule
The joint is surrounded by a loose capsule of fibrous tissue attached to the edge of the glenoid cavity and to the anatomical neck of the humerus

Scapula (shoulder blade)
Flat bone that lies against the back of the chest; with the clavicle, it forms the pectoral girdle

KEY This area can easily be felt under the skin

Anterior view of the glenohumeral joint illustrating the ligaments which reinforce the joint and the position of the subacromial bursa.

Instability of the shoulder joint

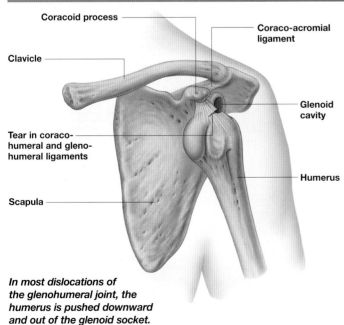

Coracoid process

Coraco-acromial ligament

Clavicle

Glenoid cavity

Tear in coraco-humeral and gleno-humeral ligaments

Humerus

Scapula

In most dislocations of the glenohumeral joint, the humerus is pushed downward and out of the glenoid socket.

A wide range of movement is one of the characteristics of the shoulder joint, but this is achieved at the expense of stability. The features which allow such free movement cause the joint to be relatively unstable; the fibrous capsule is lax, the ligaments are weak and loose and the socket of this ball-and-socket joint is shallow.

The shoulder joint is stabilized primarily by the action of the short muscles that surround it, holding the head of the humerus into the glenoid cavity, the so-called rotator cuff.

DISLOCATED SHOULDER
Of all the joints in the body, the shoulder is the most likely to dislocate. Dislocation of the shoulder joint most commonly occurs in a downward direction as the top and sides of the joint

are supported by the rotator cuff and the coraco-acromial ligament, leaving the under side supported only by the lax fibrous capsule.

Dislocation of the shoulder may occur in athletes and is usually the result of a sudden force being applied along the length of the humerus when the arm is lifted high to one side. The head of the humerus is pushed down over the lip of the glenoid cavity and, under the influence of the muscles which act upon it, usually comes to rest just beneath the coracoid process of the scapula.

After dislocation, the arm cannot be used until the bones are returned to the correct alignment. Once a dislocation has occurred, damage to the joint may leave the shoulder susceptible to future dislocations.

Movements of the shoulder joint

The shoulder joint is a ball-and-socket joint which allows 360° of movement to give maximum flexibility. In addition to enabling these movements, the muscles of the pectoral girdle add stability.

The movements of the shoulder joint take place around three axes: a horizontal axis through the centre of the glenoid fossa; axis perpendicular to this (front-back) through the humeral head; and a third axis running vertically through the shaft of the humerus. These give the axes of flexion and extension, adduction (movement towards the body) and abduction (movement away from the body), and medial (internal) and lateral (external) rotation respectively. A combination of these movements can allow a circular motion of the limb called circumduction.

MUSCLES OF SHOULDER MOVEMENT

Many of the muscles involved in these movements are attached to the pectoral girdle (the clavicles and scapulae). The scapula has muscles attached to its rear and front surfaces and the coracoid process, a bony projection. Some muscles arise directly from the trunk (pectoralis major and latissimus dorsi). Other muscles influence the movement of the humerus even though they are not attached to it directly (such as trapezius). They do this by moving the scapula, and hence the shoulder joint.

Front view of muscles of the shoulder

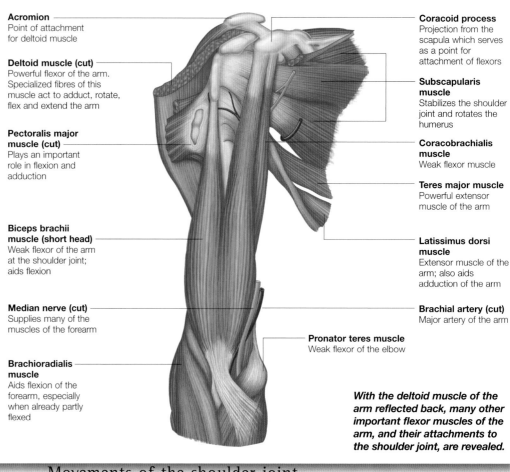

Acromion
Point of attachment for deltoid muscle

Deltoid muscle (cut)
Powerful flexor of the arm. Specialized fibres of this muscle act to adduct, rotate, flex and extend the arm

Pectoralis major muscle (cut)
Plays an important role in flexion and adduction

Biceps brachii muscle (short head)
Weak flexor of the arm at the shoulder joint; aids flexion

Median nerve (cut)
Supplies many of the muscles of the forearm

Brachioradialis muscle
Aids flexion of the forearm, especially when already partly flexed

Coracoid process
Projection from the scapula which serves as a point for attachment of flexors

Subscapularis muscle
Stabilizes the shoulder joint and rotates the humerus

Coracobrachialis muscle
Weak flexor muscle

Teres major muscle
Powerful extensor muscle of the arm

Latissimus dorsi muscle
Extensor muscle of the arm; also aids adduction of the arm

Brachial artery (cut)
Major artery of the arm

Pronator teres muscle
Weak flexor of the elbow

With the deltoid muscle of the arm reflected back, many other important flexor muscles of the arm, and their attachments to the shoulder joint, are revealed.

Movements of the shoulder joint

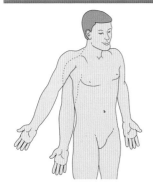

Adduction (towards the body) of the arm is brought about by pectoralis major and latissimus dorsi muscles; abduction (away from the body) by supraspinatus and the deltoid.

Muscles which bring about lateral rotation are infraspinatus, teres minor and the posterior fibres of the deltoid muscle. Medial rotators form the 'rotator cuff' group of muscles.

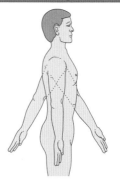

Flexion (forward movement) is due to biceps, coracobrachialis, deltoid and pectoralis major. Extension (backward movement) is due to rear fibres of deltoid, latissimus dorsi and teres major.

Circumduction is a combination of these movements. It is dependent on the clavicle holding the shoulder joint in the glenoid cavity and contractions of different muscle groups.

Rotation of the arm and 'rotator cuff'

The rotator cuff muscles include subscapularis, supraspinatus, infraspinatus and teres minor. These muscles act to strengthen and increase the stability of the shoulder joint. They also act individually to move the humerus and upper arm.

The pectoralis major, anterior fibres of deltoid, teres major and latissimus dorsi muscles also cause medial rotation of the humerus. The most powerful medial rotator, however, is subscapularis. This muscle occupies the entire front surface of the scapula, and attaches to the joint capsule around the lesser tuberosity of the humerus.

ROTATOR CUFF

Subscapularis is one of a set of four short muscles, collectively called the 'rotator cuff', which attach to and strengthen the joint capsule. In addition, they pull the humerus into the socket of the joint (glenoid fossa), increasing contact of the bony elements. This is the most important factor contributing to the stability of the joint.

The other muscles of the group are supraspinatus, infraspinatus and teres minor. These latter three muscles attach to the three facets on the greater tuberosity of the humerus. Infraspinatus and teres minor are lateral rotators of the shoulder joint, together with the posterior fibres of the deltoid.

Injury to the rotator cuff muscles is disabling, because the stability of the humerus in the joint is lost. The other muscles of the arm lose the ability to move the humerus correctly, resulting in dislocation of the joint.

Muscles of shoulder movement (front)
Muscles of shoulder movement (back)

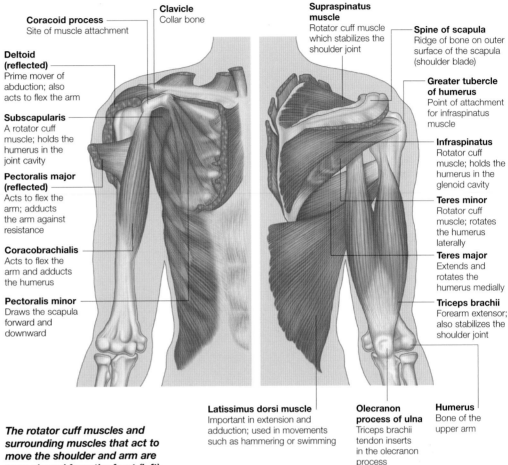

Coracoid process
Site of muscle attachment

Deltoid (reflected)
Prime mover of abduction; also acts to flex the arm

Subscapularis
A rotator cuff muscle; holds the humerus in the joint cavity

Pectoralis major (reflected)
Acts to flex the arm; adducts the arm against resistance

Coracobrachialis
Acts to flex the arm and adducts the humerus

Pectoralis minor
Draws the scapula forward and downward

Clavicle
Collar bone

Supraspinatus muscle
Rotator cuff muscle which stabilizes the shoulder joint

Spine of scapula
Ridge of bone on outer surface of the scapula (shoulder blade)

Greater tubercle of humerus
Point of attachment for infraspinatus muscle

Infraspinatus
Rotator cuff muscle; holds the humerus in the glenoid cavity

Teres minor
Rotator cuff muscle; rotates the humerus laterally

Teres major
Extends and rotates the humerus medially

Triceps brachii
Forearm extensor; also stabilizes the shoulder joint

Latissimus dorsi muscle
Important in extension and adduction; used in movements such as hammering or swimming

Olecranon process of ulna
Triceps brachii tendon inserts in the olecranon process

Humerus
Bone of the upper arm

The rotator cuff muscles and surrounding muscles that act to move the shoulder and arm are seen viewed from the front (left) and behind (right).

Abduction of the arm

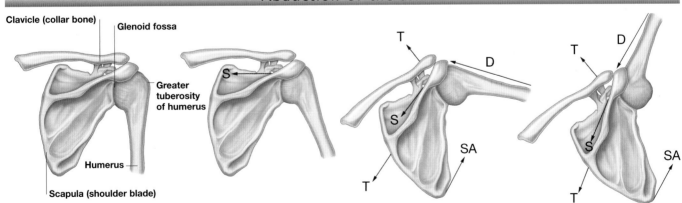

Clavicle (collar bone)
Glenoid fossa
Greater tuberosity of humerus
Humerus
Scapula (shoulder blade)

Abduction (movement away from the body) is a weak action performed by supraspinatus (S) and the acromial (middle) part of deltoid (D) muscle.

KEY	S	= Supraspinatus
---	D	= Deltoid
	T	= Trapezius
	SA	= Serratus anterior

From the resting position, the deltoid muscle can pull the humerus upwards but not outwards. Thankfully, the supraspinatus muscle is in a much better mechanical position to initiate abduction. Once the motion has started, the deltoid muscle takes over and continues the movement.

The next major obstacle to full abduction is the bony contact of the greater tuberosity of the humerus and the acromion process of the scapula. This bony contact would prevent raising of the arm higher than horizontal (for example, arms held straight outwards at shoulder height).

We are able to put raise our hands above our heads by rotating the scapula using the trapezius muscle (T). The scapula rotates so that the glenoid fossa points upwards, taking the acromion process with it. The humerus also rotates so that articular contact of the joint is maintained.

215

Axilla

The axilla, or armpit, is a roughly pyramidal space where the upper arm joins the thorax. It contains a number of important structures, such as blood vessels and nerves passing to and from the upper limb.

Vessels, nerves and lymphatics serving the upper limb all pass through the axilla. The structures lie embedded in fatty connective tissue, which occupies the axillary space.

THE AXILLARY ARTERY

The axillary artery and its branches supply oxygenated blood to the upper limb.

As it passes through the axilla this artery gives off several branches which supply the surrounding structures of the shoulder and pectoral regions.

THE AXILLARY VEIN

The axillary vein runs through the axilla on the medial side of the axillary artery.

The pattern of veins and venous drainage is variable but the axillary vein, in general, receives blood from tributary veins which correspond to the branches of the axillary artery.

NERVES IN THE AXILLA

The nerves which lie in the axilla are part of a complex network known as the 'brachial plexus'.

LYMPHATICS

Within the fatty connective tissue of the axilla lie a series of groups of lymph nodes which are connected by lymphatic vessels. Lymph nodes are scattered throughout the fat of the axilla.

Front view of the shoulder showing the structures of the axilla

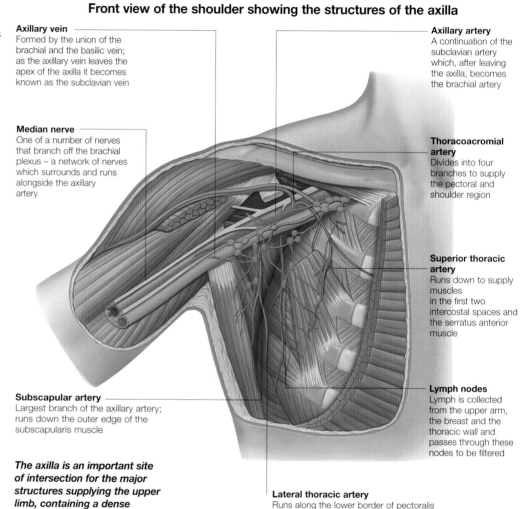

Axillary vein
Formed by the union of the brachial and the basilic vein; as the axillary vein leaves the apex of the axilla it becomes known as the subclavian vein

Median nerve
One of a number of nerves that branch off the brachial plexus – a network of nerves which surrounds and runs alongside the axillary artery

Subscapular artery
Largest branch of the axillary artery; runs down the outer edge of the subscapularis muscle

Axillary artery
A continuation of the subclavian artery which, after leaving the axilla, becomes the brachial artery

Thoracoacromial artery
Divides into four branches to supply the pectoral and shoulder region

Superior thoracic artery
Runs down to supply muscles in the first two intercostal spaces and the serratus anterior muscle

Lymph nodes
Lymph is collected from the upper arm, the breast and the thoracic wall and passes through these nodes to be filtered

Lateral thoracic artery
Runs along the lower border of pectoralis minor and gives off branches to supply the outer side of the mammary gland

The axilla is an important site of intersection for the major structures supplying the upper limb, containing a dense network of blood vessels, nerves and lymphatics.

Passage of vessels and nerves

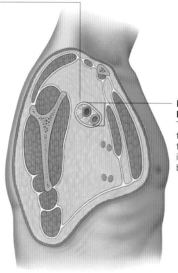

Axillary sheath
Formed from deep fascia in the neck, the sheath extends to enclose important structures of the axilla

Neurovascular bundle
The vein lies on the inner side of the artery, which is surrounded by nerves

Throughout the body it is a common pattern for arteries, veins and nerves to travel together in 'neurovascular bundles' and, indeed, the blood vessels and nerves of the upper limb pass through the axilla together in such an arrangement.

THE AXILLARY SHEATH

Within the axilla (which is vulnerable to trauma, especially from below), these important structures are protected and

This section of the axilla shows muscles, vessels, bones and lymph nodes. The nerves and vessels can be seen travelling in the axillary sheath.

enclosed by a tube of strong connective tissue, the 'axillary sheath'. Within the axillary sheath the vein lies on the medial (inner) side of the artery, with the nerves (parts of the brachial plexus) surrounding the artery.

The artery, vein and nerves supplying the upper limb all lie within the protection of the axillary sheath.

Because of the situation of the sheath, it is an important site for administration of general anaesthesia. If the lower end of the sheath is occluded (blocked) with finger pressure, the anaesthetic injected into the proximal sheath causes brachial plexus nerve block.

The clavipectoral fascia

This is a sheet of strong connective tissue, which is attached at its upper border to the coracoid process of the scapula and the clavicle.

The clavipectoral fascia descends to enclose the subclavius muscle and the pectoralis minor muscle and then joins with the overlying axillary fascia in the base of the axilla.

The part of the clavipectoral fascia that lies above the pectoralis minor muscle is known as the 'costocoracoid membrane' and is pierced by the nerve which supplies the overlying pectoralis minor muscle.

Below the pectoralis minor muscle, the fascia becomes the 'suspensory ligament of the axilla', which attaches to the skin of the armpit and is responsible for pulling that skin up when the arm is raised.

The clavipectoral fascia is continuous with the brachial fascia, which envelops the arm like a sleeve.

The fascia is pierced by a number of veins, arteries and nerves. These are the cephalic vein, the thoracoacromial artery (a branch of the axillary artery) and the lateral pectoral nerve.

The clavipectoral fascia protects the contents of the axilla. It does this by filling the gap between the clavicle and the pectoralis minor muscle.

Front view of the axilla showing the clavipectoral fascia

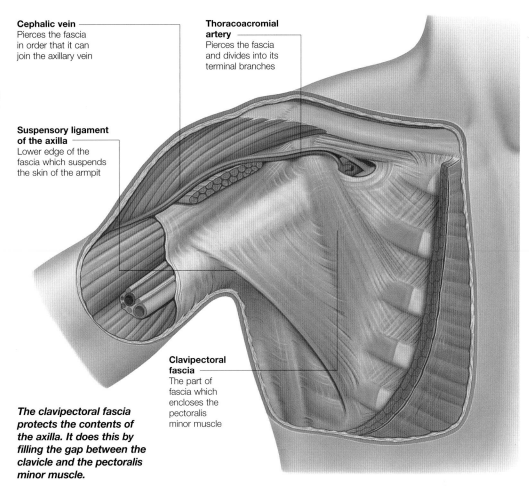

Cephalic vein
Pierces the fascia in order that it can join the axillary vein

Thoracoacromial artery
Pierces the fascia and divides into its terminal branches

Suspensory ligament of the axilla
Lower edge of the fascia which suspends the skin of the armpit

Clavipectoral fascia
The part of fascia which encloses the pectoralis minor muscle

Borders of the axilla

The shape of the axilla varies according to the position of the arm. When the arm is raised the axilla is a wide-based pyramid, and when the arm is lowered it is a narrow, compressed space.

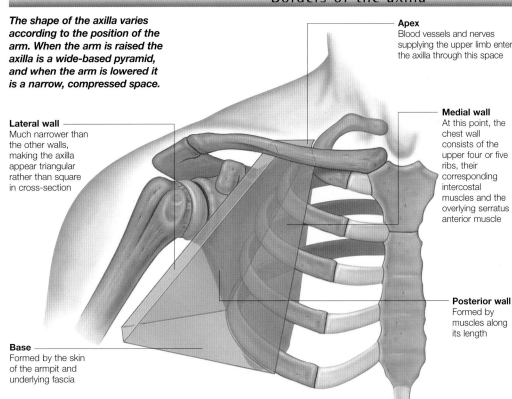

Lateral wall
Much narrower than the other walls, making the axilla appear triangular rather than square in cross-section

Base
Formed by the skin of the armpit and underlying fascia

Apex
Blood vessels and nerves supplying the upper limb enter the axilla through this space

Medial wall
At this point, the chest wall consists of the upper four or five ribs, their corresponding intercostal muscles and the overlying serratus anterior muscle

Posterior wall
Formed by muscles along its length

The axilla is said to have an apex, a base, and four walls:

■ The apex, or top point of the axilla, is a space between the clavicle (collarbone) at the front, the first rib on the medial (inner) side and the top of the scapula (shoulder blade) behind
■ The base is formed by the skin of the armpit and underlying axillary fascia, a layer of strong connective tissue
■ The anterior wall is formed by the pectoral and subclavius muscles and by the clavipectoral fascia
■ The posterior wall is formed by the subscapularis muscle lying on the scapula, and, lower down, by the latissimus dorsi and teres major muscles
■ The medial (inner) wall of the axilla is formed by the chest wall
■ The lateral (outer) wall is formed by the muscles attached to the bone of the upper arm, the humerus. These muscles are the coracobrachialis and biceps brachii.

Structure of the humerus

The humerus, a typical 'long bone', is found in the upper arm. It has a long shaft with expanded ends that connect with the scapula at the shoulder joint and the radius and ulna at the elbow.

At the top of the humerus (the proximal end) lies the smooth, hemispherical head that fits into the glenoid cavity of the scapula at the shoulder joint. Behind the head is a shallow constriction known as the 'anatomical neck' of the humerus, which separates the head from two bony prominences, the greater and the lesser tuberosities. These are sites for muscle attachment and are separated by the intertubercular (or bicipital) groove.

THE SHAFT

At the upper end of the shaft is the slightly narrowed 'surgical neck' of the humerus – a common site for fractures. The relatively smooth shaft has two distinctive features. About half way down the shaft, on the lateral (outer) side, lies the deltoid tuberosity, a raised site of attachment of the deltoid muscle. The second feature is the radial (or spiral) groove which runs across the back of the middle part of the shaft. This depression marks the path of the radial nerve and the profunda brachii artery.

Ridges at each side of the lower shaft pass down to end in the prominent medial (inner) and lateral epicondyles. There are two main parts to the articular surface: the trochlea, which articulates with the ulna; and the capitulum, which articulates with the radius.

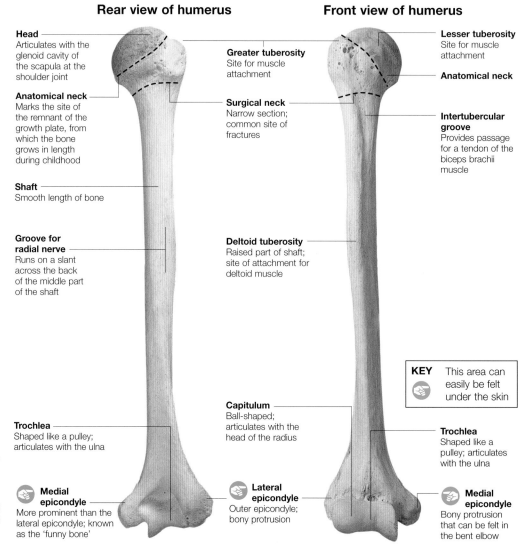

Rear view of humerus

Head
Articulates with the glenoid cavity of the scapula at the shoulder joint

Anatomical neck
Marks the site of the remnant of the growth plate, from which the bone grows in length during childhood

Shaft
Smooth length of bone

Groove for radial nerve
Runs on a slant across the back of the middle part of the shaft

Trochlea
Shaped like a pulley; articulates with the ulna

Medial epicondyle
More prominent than the lateral epicondyle; known as the 'funny bone'

Greater tuberosity
Site for muscle attachment

Surgical neck
Narrow section; common site of fractures

Deltoid tuberosity
Raised part of shaft; site of attachment for deltoid muscle

Capitulum
Ball-shaped; articulates with the head of the radius

Lateral epicondyle
Outer epicondyle; bony protrusion

Front view of humerus

Lesser tuberosity
Site for muscle attachment

Anatomical neck

Intertubercular groove
Provides passage for a tendon of the biceps brachii muscle

KEY This area can easily be felt under the skin

Trochlea
Shaped like a pulley; articulates with the ulna

Medial epicondyle
Bony protrusion that can be felt in the bent elbow

Fractures of the humerus

Most fractures of the upper end of the humerus occur at the surgical neck, and often happen as a result of a fall onto an outstretched hand. Fractures of the shaft of the humerus may cause damage to the radial nerve as it lies along the bone in the radial groove, resulting in the condition 'wrist drop', where the muscles at the back of the forearm which are innervated by the radial nerve become paralysed.

In children, fractures of the humerus are often supracondylar (at the lower end just above the elbow joint) and occur when the child falls on an outstretched hand with the elbow slightly bent. In these cases, there may be damage to the nearby nerves and arteries.

This X-ray shows a fracture of the upper part of the humerus. This may occur after falling onto an outstretched hand.

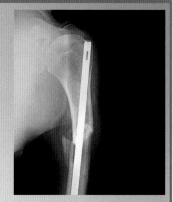

Some fractures of the humerus may need to be stabilized with a metal pin. This holds the broken ends of bone together.

Inside the humerus

The structure of the humerus is typical of the long bones. The bone is divided into the diaphysis (shaft) and the epiphysis (head) at either end.

Long bones are elongated in shape and longer than they are wide. Most of the bones of the limbs are long bones, even the small bones of the fingers, and as such they have many features in common with the humerus.

The humerus consists of a diaphysis, or shaft, with an epiphysis (expanded head) at each end. The diaphysis is of tubular construction with an outer layer of dense, thick bone surrounding a central medulla (inner region) containing fat cells. The epiphyses of the humerus are, at the upper end, the head and at the lower end the condylar region. These are composed of a thin layer of compact bone covering cancellous (spongy) bone which makes up the greater volume.

BONE SURFACE

The surface of the humerus (and all long bones) is covered by a thick membrane, the periosteum. The articular surfaces at the joints are the only parts of the bone not covered by the periosteum. These surfaces are covered by tough articular (or hyaline) cartilage which is smooth, allowing the bones to glide over each other.

The outer compact bone receives its blood supply from the arteries of the periosteum, and will die if that periosteum is stripped off, while the inner parts of the bone are supplied by occasional nutrient arteries which pierce the compact bone.

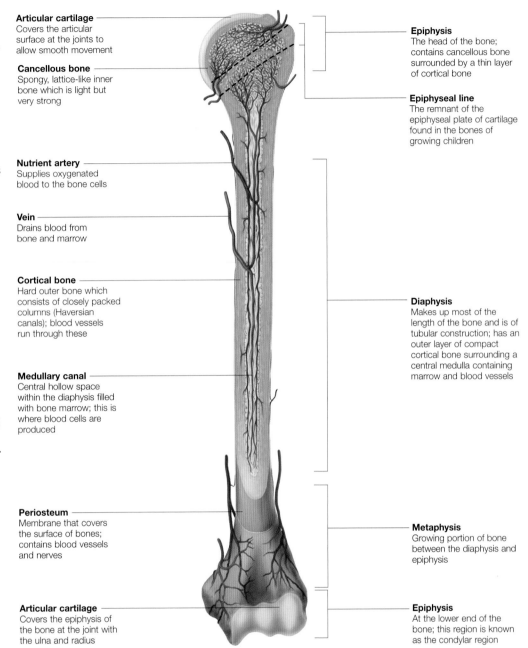

Articular cartilage
Covers the articular surface at the joints to allow smooth movement

Cancellous bone
Spongy, lattice-like inner bone which is light but very strong

Nutrient artery
Supplies oxygenated blood to the bone cells

Vein
Drains blood from bone and marrow

Cortical bone
Hard outer bone which consists of closely packed columns (Haversian canals); blood vessels run through these

Medullary canal
Central hollow space within the diaphysis filled with bone marrow; this is where blood cells are produced

Periosteum
Membrane that covers the surface of bones; contains blood vessels and nerves

Articular cartilage
Covers the epiphysis of the bone at the joint with the ulna and radius

Epiphysis
The head of the bone; contains cancellous bone surrounded by a thin layer of cortical bone

Epiphyseal line
The remnant of the epiphyseal plate of cartilage found in the bones of growing children

Diaphysis
Makes up most of the length of the bone and is of tubular construction; has an outer layer of compact cortical bone surrounding a central medulla containing marrow and blood vessels

Metaphysis
Growing portion of bone between the diaphysis and epiphysis

Epiphysis
At the lower end of the bone; this region is known as the condylar region

Types of bone tissue found in the body

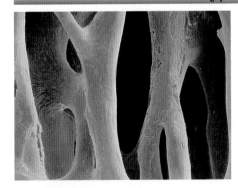

Cancellous (or spongy) bone. This tissue, seen on this electron micrograph, fills the interior of bones. Cancellous bone is lattice-like in structure and has a low density.

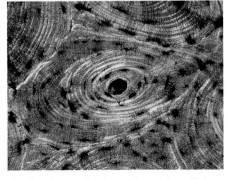

Cortical bone. This is made up of parallel columns called Haversian canals. These consist of concentric layers around channels containing blood vessels and nerves.

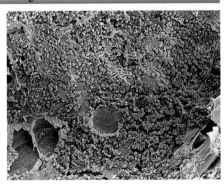

Bone marrow. This is found within the spaces of the cancellous bone in the centre of long bones. Bone marrow contains stem cells which produce several types of blood cells.

Ulna and radius

The ulna and the radius are the long bones of the forearm. They articulate with the humerus and the wrist bones and are uniquely adapted to enable rotation of the hand and forearm.

The ulna and radius are the two parallel long bones of the forearm and lie between the elbow and wrist joints. The ulna lies on the same side as the little finger (medially), while the radius lies on the same side as the thumb (laterally).

The radio-ulnar joints allow the ulna and radius to rotate around each other in the movements peculiar to the forearm known as 'pronation' (rotating the forearm so that the palm faces down), and 'supination' (rotating the forearm so that the palm faces up).

THE ULNA

The ulna is longer than the radius and is the main stabilizing bone of the forearm. It has a long shaft with two expanded ends. The upper end of the ulna has two prominent projections, the olecranon and the coronoid process, which are separated by the deep trochlear notch, which articulates with the trochlea of the humerus.

On the lateral (outer) side of the coronoid process, there is a small, rounded recess (the radial notch), which is the site of articulation of the upper end of the ulna with the neighbouring head of the radius. The head of the ulna is separated from the wrist joint by an articular disc and does not play much part in the wrist joint itself.

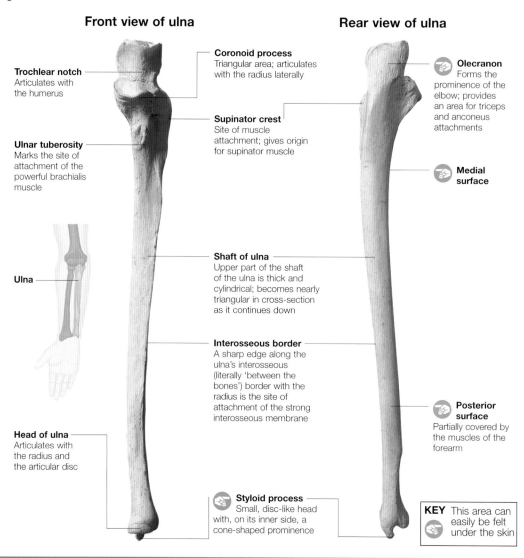

Front view of ulna

Trochlear notch
Articulates with the humerus

Ulnar tuberosity
Marks the site of attachment of the powerful brachialis muscle

Ulna

Head of ulna
Articulates with the radius and the articular disc

Coronoid process
Triangular area; articulates with the radius laterally

Supinator crest
Site of muscle attachment; gives origin for supinator muscle

Shaft of ulna
Upper part of the shaft of the ulna is thick and cylindrical; becomes nearly triangular in cross-section as it continues down

Interosseous border
A sharp edge along the ulna's interosseous (literally 'between the bones') border with the radius is the site of attachment of the strong interosseous membrane

Styloid process
Small, disc-like head with, on its inner side, a cone-shaped prominence

Rear view of ulna

Olecranon
Forms the prominence of the elbow; provides an area for triceps and anconeus attachments

Medial surface

Posterior surface
Partially covered by the muscles of the forearm

KEY This area can easily be felt under the skin

The interosseous membrane

Cross-section through the bones of the forearm

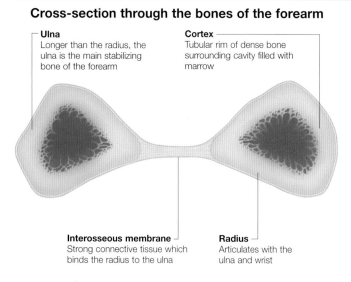

Ulna
Longer than the radius, the ulna is the main stabilizing bone of the forearm

Cortex
Tubular rim of dense bone surrounding cavity filled with marrow

Interosseous membrane
Strong connective tissue which binds the radius to the ulna

Radius
Articulates with the ulna and wrist

The radius and the ulna are connected by a thin sheet of tough, fibrous connective tissue which ties them tightly together, called the interosseous membrane. This membrane is broad enough to enable a good deal of movement between the bones during the actions of supination and pronation (turning the palm up and then down). The interosseous membrane is also strong enough to provide sites for the attachment of some of the deep muscles of the forearm.

The interosseous membrane, composed of tough connective tissue, binds the radius and ulna together. It also divides the forearm into two compartments.

The interosseous membrane has an important role to play in the transmission of force through the forearm. If force were to be applied to the wrist – as in breaking a fall onto an outstretched hand – that force would be received first by the end of the radius as it makes up the greater part of the wrist joint.

The tough fibres of the interosseous membrane lie in such a direction that the force is then transmitted effectively to the ulna, which makes up the greater part of the elbow joint. The transmission of this force via the interosseous membrane to the ulna allows the impact to be further absorbed by the bone of the upper arm, the humerus.

The radius

The radius is the shorter of the two bones of the forearm and articulates with the wrist. It is joined firmly to the ulna by a tough layer of connective tissue.

Like the ulna, the radius has a long shaft with upper and lower expanded ends. While the ulna is the forearm bone which contributes most to the elbow, the radius plays a major part in the wrist joint.

HEAD OF THE RADIUS

The disc-like head of the radius is concave above, where it articulates with the capitulum of the humerus in the elbow joint. The cartilage that covers this concavity continues down over the head, especially on the side nearest the ulna, to allow the smooth articulation of the head of the radius with the radial notch at the upper end of the ulna.

THE SHAFT

The shaft of the radius becomes progressively thicker as it continues down to the wrist. It also has a sharp edge for attachment of the interosseous membrane. On the inner side, next to the ulna, there is a concavity (the ulnar notch), which is the site for articulation with the head of the ulna.

Extending from the opposite side is the radial styloid process, a blunt cone which projects a little further down than the ulnar styloid process. At the back of the end of the radius, and easily felt at the back of the wrist, is the dorsal tubercle.

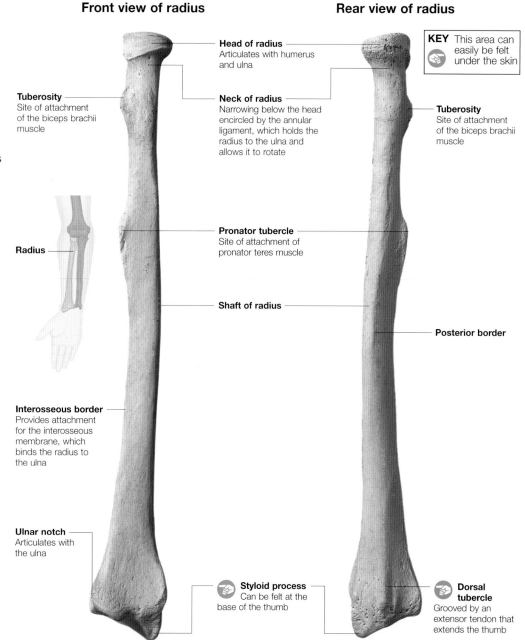

Front view of radius

Rear view of radius

KEY This area can easily be felt under the skin

Head of radius
Articulates with humerus and ulna

Tuberosity
Site of attachment of the biceps brachii muscle

Neck of radius
Narrowing below the head encircled by the annular ligament, which holds the radius to the ulna and allows it to rotate

Tuberosity
Site of attachment of the biceps brachii muscle

Radius

Pronator tubercle
Site of attachment of pronator teres muscle

Shaft of radius

Posterior border

Interosseous border
Provides attachment for the interosseous membrane, which binds the radius to the ulna

Ulnar notch
Articulates with the ulna

Styloid process
Can be felt at the base of the thumb

Dorsal tubercle
Grooved by an extensor tendon that extends the thumb

Colles' fracture

In adults over the age of 50, and especially in women who may have bones which have become thinned due to osteoporosis, the most commonly occurring fracture of the bones of the forearm is the Colles' fracture.

This is a fracture of the lower end of the radius, near the wrist joint, and usually happens as a result of a fall where the outstretched hand hits the ground first. The broken end of the radius, together with the wrist and hand to which it is attached, is displaced dorsally (in the direction away from the palm) leading to the characteristic 'dinner fork' deformity of the lower forearm and wrist.

In 40 per cent of cases of Colles' fracture, the styloid process of the ulna – a cone-shaped process at the lower end of the ulna – is also broken off. Because of the rich blood supply to the end of the radius, healing of the fracture is usually straightforward.

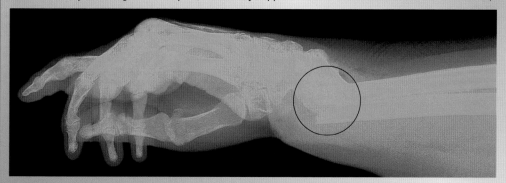

A Colles' fracture results from a fall onto an outstretched hand. It is characterized by a break in the lower end of the radius (circled), above the wrist joint.

Elbow

The elbow is the fluid-filled joint where the humerus of the upper arm and the radius and ulna of the forearm articulate. The joint structure only allows hinge-like movement but is extremely stable.

The elbow is a synovial (fluid-filled) joint between the lower end of the humerus and the upper ends of the ulna and radius. It is the best example of a 'hinge' joint, where the only movements permitted are flexion (bending) and extension (straightening). Its structure gives the joint great stability and in adults dislocation rarely occurs.

STRUCTURE OF THE ELBOW

At the elbow, the pulley-shaped trochlea of the lower end of the humerus articulates with the deep trochlear notch of the ulna, while its hemispherical capitulum articulates with the head of the radius. All the opposing joint surfaces are covered by smooth articular cartilage (hyaline) to reduce friction between the bony surfaces during movement.

The whole joint is surrounded by a fibrous capsule which extends down from the articular surfaces of the humerus to the upper end of the ulna. The capsule is loose at the back of the elbow to allow flexion and extension. The capsule is lined with synovial membrane which secretes thick synovial fluid filling the joint cavity. This fluid nourishes the joint and acts as a lubricant. The joint cavity is continuous with that of the superior radioulnar joint below.

View of right elbow from front

View of right elbow from behind

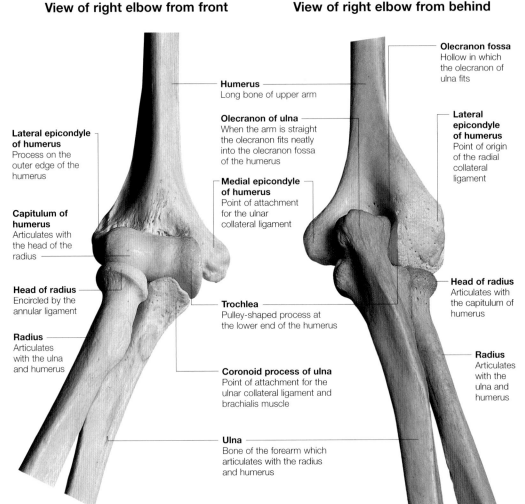

Humerus
Long bone of upper arm

Olecranon of ulna
When the arm is straight the olecranon fits neatly into the olecranon fossa of the humerus

Medial epicondyle of humerus
Point of attachment for the ulnar collateral ligament

Lateral epicondyle of humerus
Process on the outer edge of the humerus

Capitulum of humerus
Articulates with the head of the radius

Head of radius
Encircled by the annular ligament

Radius
Articulates with the ulna and humerus

Trochlea
Pulley-shaped process at the lower end of the humerus

Coronoid process of ulna
Point of attachment for the ulnar collateral ligament and brachialis muscle

Ulna
Bone of the forearm which articulates with the radius and humerus

Olecranon fossa
Hollow in which the olecranon of ulna fits

Lateral epicondyle of humerus
Point of origin of the radial collateral ligament

Head of radius
Articulates with the capitulum of humerus

Radius
Articulates with the ulna and humerus

Stability and movement of the elbow

Bending the elbow joint

Humerus

Flexion

Coronoid fossa

Radius

Lateral epicondyle

Axis

Capitulum

Ulna

Extension

The elbow performs only two movements, bending and straightening, so the structure is very stable. The main stability of the elbow comes from the size and depth of the trochlear notch of the ulna, which effectively grips the lower end of the humerus like a wrench.

The depth of this bony notch is increased by the presence of a band of the medial collateral ligament. Because of the shape of this joint, and the presence of the strong collateral ligaments on each side, the elbow can move only as a hinge.

The elbow joint is capable of two movements: flexion (bending) and extension (straightening), as indicated by the purple arrow.

FLEXION AND EXTENSION

Flexion (bending) of the elbow is achieved by contraction of the powerful muscles at the front of the upper arm such as brachialis and the well-known biceps brachii. The movement is limited at its fullest extent by the coming together of the forearm and the upper arm.

Extension of the elbow is mainly achieved by contraction of the triceps muscle at the back of the upper arm, assisted by gravity. At full extension, with the arm straight, the olecranon of the ulna fits neatly into the olecranon fossa (hollow) of the lower end of the back of the humerus. This fitting together of the two bones prevents over-extension of the elbow, and so adds to its stability.

Ligaments of the elbow

The elbow is supported and strengthened at each side by the strong collateral ligaments. These are thickenings of the joint capsule.

The radial collateral ligament is a fan-shaped ligament that originates from the lateral epicondyle – a bony prominence on the outer side of the lower end of the humerus – and runs down to blend with the annular ligament, which encircles the head of the radius. It is not attached to the radius itself so does not restrict movement of the radius during pronation (when the forearm is rotated so that the palm faces down) and supination (when the forearm is rotated so the palm faces up).

The ulnar collateral ligament runs between the medial (inner) epicondyle of the humerus and the upper end of the ulna and is in three parts, which form a rough triangle.

CARRYING ANGLE

When the arm is fully extended downwards with the palm facing forwards, the long axis of the forearm is not in line with the long axis of the upper arm, but deviates slightly outwards.

The angle so formed at the elbow is known as the 'carrying angle' and is greater in women than in men (by about 10 degrees), possibly to accommodate the wider hips of the female body. The carrying angle disappears when the forearm is pronated (turned so the palm faces in to the body).

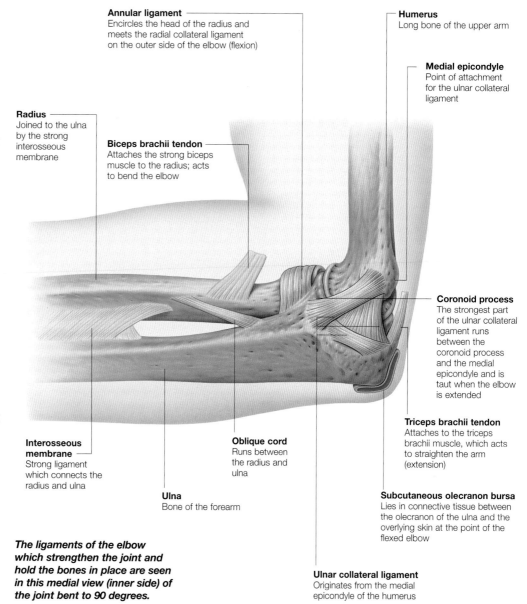

Annular ligament
Encircles the head of the radius and meets the radial collateral ligament on the outer side of the elbow (flexion)

Humerus
Long bone of the upper arm

Medial epicondyle
Point of attachment for the ulnar collateral ligament

Radius
Joined to the ulna by the strong interosseous membrane

Biceps brachii tendon
Attaches the strong biceps muscle to the radius; acts to bend the elbow

Coronoid process
The strongest part of the ulnar collateral ligament runs between the coronoid process and the medial epicondyle and is taut when the elbow is extended

Triceps brachii tendon
Attaches to the triceps brachii muscle, which acts to straighten the arm (extension)

Interosseous membrane
Strong ligament which connects the radius and ulna

Oblique cord
Runs between the radius and ulna

Ulna
Bone of the forearm

Subcutaneous olecranon bursa
Lies in connective tissue between the olecranon of the ulna and the overlying skin at the point of the flexed elbow

Ulnar collateral ligament
Originates from the medial epicondyle of the humerus

The ligaments of the elbow which strengthen the joint and hold the bones in place are seen in this medial view (inner side) of the joint bent to 90 degrees.

Clinical features

Cross-section through elbow joint

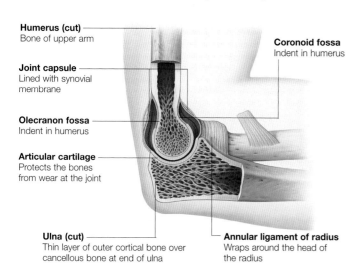

Humerus (cut)
Bone of upper arm

Joint capsule
Lined with synovial membrane

Olecranon fossa
Indent in humerus

Articular cartilage
Protects the bones from wear at the joint

Coronoid fossa
Indent in humerus

Ulna (cut)
Thin layer of outer cortical bone over cancellous bone at end of ulna

Annular ligament of radius
Wraps around the head of the radius

A cross-section through the elbow joint reveals the articular surfaces and the positions of the overlying joint capsules in front of and behind the humerus.

DISLOCATION

The bones which stabilize the elbow joint are incompletely developed in children and dislocations of the elbow, where the ulna and radius move posteriorly (backwards with respect to the humerus) may occur when a child falls onto an outstretched, partially flexed arm. There is often tearing of the ulnar collateral ligament and sometimes a fracture of the upper parts of the radius and ulna associated with this dislocation.

TENNIS ELBOW

Despite the name, tennis elbow is not a condition of the elbow joint itself but is a painful inflammation of the site of attachment of muscles to the lateral epicondyle of the humerus. The muscles concerned are those which extend the wrist and the fingers and so are used when, for instance, hitting a ball with a backhand stroke in tennis.

The condition occurs as a result of acute trauma to the area or normal but excessively repetitive actions involving those muscles. Pain is felt over the lateral epicondyle and down the back of the forearm, especially when the hand is being used.

Forearm movements

The ability to rotate the forearm – movements called pronation
and supination – is of enormous value in adding to the range
of movements of the hand and the versatility of the upper limb.

The terms pronation and supination are peculiar to the movements of the forearm. To pronate the forearm is to rotate it so that the palm faces down (or backwards if the arm is by the side). To supinate the forearm is to rotate it so that the palm is facing upwards (or forwards if the arm is held by the side).

Pronation and supination are achieved by contraction of muscles which rotate the radius around the relatively stationary ulna with the upper and lower radio-ulnar joints acting as pivots.

MUSCLES OF PRONATION

■ **Pronator teres.** This muscle lies in front of the bones of the forearm and acts to pronate the forearm and flex (bend) the elbow. It arises from two heads, which are attached to the coronoid process of the ulna and the medial epicondyle of the humerus (a bony prominence on the inner side of the elbow). It passes down and outwards to the middle of the outer edge of the radius where it can exert maximum leverage on the bone.

■ **Pronator quadratus.** This is a small muscle which connects the front of the lower quarter of the ulna to the radius and acts to help the interosseous membrane tie the two bones together, as well as to pronate the forearm.

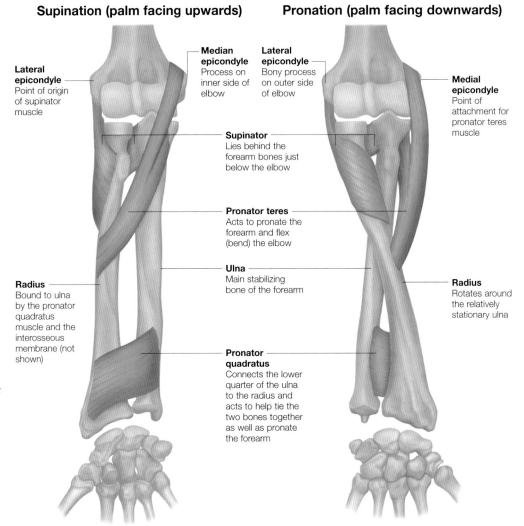

Supination (palm facing upwards)

Pronation (palm facing downwards)

Lateral epicondyle
Point of origin of supinator muscle

Median epicondyle
Process on inner side of elbow

Lateral epicondyle
Bony process on outer side of elbow

Medial epicondyle
Point of attachment for pronator teres muscle

Supinator
Lies behind the forearm bones just below the elbow

Pronator teres
Acts to pronate the forearm and flex (bend) the elbow

Radius
Bound to ulna by the pronator quadratus muscle and the interosseous membrane (not shown)

Ulna
Main stabilizing bone of the forearm

Radius
Rotates around the relatively stationary ulna

Pronator quadratus
Connects the lower quarter of the ulna to the radius and acts to help tie the two bones together as well as pronate the forearm

---

Muscles involved in rotating the forearm

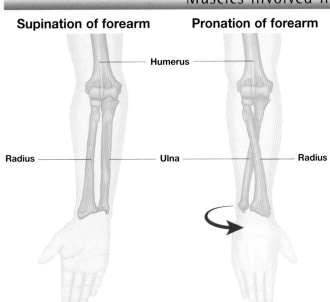

Supination of forearm

Pronation of forearm

Humerus

Radius

Ulna

Radius

The muscles responsible for the actions of pronation and supination generally lie deep to the other muscles of the forearm which are mainly concerned with movements of the wrist and fingers. The muscles which enable supination include:

■ **Supinator.** This is a deep muscle which lies behind the forearm bones just below the elbow. It arises from the outer edge of the elbow and from a raised line on the ulna, the supinator crest. It then passes down and around the outer edge

When the forearm is pronated, so that the palm of the hand faces down, the deeper muscles act to rotate the radius around and over the ulna bone.

of the upper radius to insert into the upper third of the shaft

■ **Biceps brachii.** This powerful muscle acts to supinate the forearm as well as its better known function of flexing the elbow. It is the muscle involved when powerful supination is needed, such as when a right-handed person drives in a screw

■ **Brachioradialis.** This is a muscle which runs from the outer part of the humerus just above the elbow down to the outer side of the lower end of the radius. It acts to pull the forearm back to the position of rest, midway between pronation and supination; if the forearm is fully pronated it will act as a supinator; if fully supinated, it will act as a pronator.

Radius and ulna bone connections

As well as the articulations at the elbow and wrist, the radius and ulna articulate with each other at both their upper and lower ends.

The radio-ulnar joints at both ends of the forearm allow for pronation (the hand turning palm down) and supination (the hand turning palm up) as the radius rotates around the ulna.

UPPER (SUPERIOR) RADIO-ULNAR JOINT

This is a synovial (fluid-filled) joint which acts like a pivot. The head of the radius articulates with the radial notch of the upper end of the ulna and is held in place by a strong annular (ring-like) ligament.

The whole joint is surrounded and supported by the fibrous joint capsule (lined by synovial membrane, which produces the lubricating synovial fluid). During the actions of pronation and supination, the head of the radius freely rotates within the circle formed by the annular ligament and the radial notch.

LOWER (INFERIOR) RADIO-ULNAR JOINT

This is also a pivot joint where the lower end of the radius rotates around the relatively stationary head of the ulna. The rounded head of the ulna articulates with the ulnar notch on the inner side of the lower end of the radius. The bones are bound together by the 'triangular ligament', which is a tough fibrocartilaginous disc separating this joint from the joint cavity of the wrist.

Upper radio-ulnar joint of left forearm (at elbow) in supination

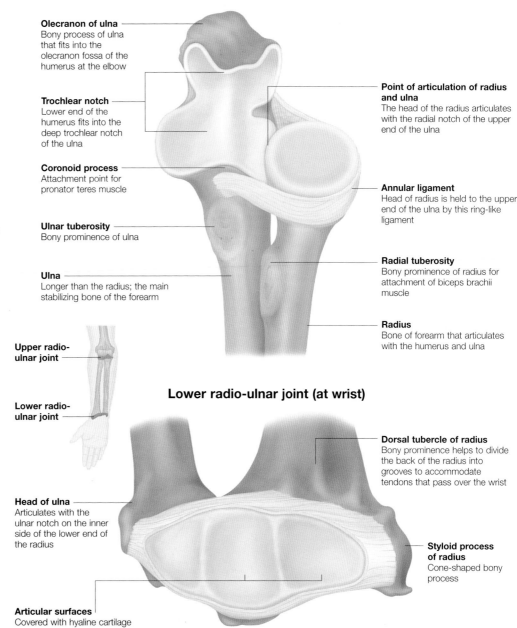

Olecranon of ulna
Bony process of ulna that fits into the olecranon fossa of the humerus at the elbow

Trochlear notch
Lower end of the humerus fits into the deep trochlear notch of the ulna

Coronoid process
Attachment point for pronator teres muscle

Ulnar tuberosity
Bony prominence of ulna

Ulna
Longer than the radius; the main stabilizing bone of the forearm

Upper radio-ulnar joint

Lower radio-ulnar joint

Point of articulation of radius and ulna
The head of the radius articulates with the radial notch of the upper end of the ulna

Annular ligament
Head of radius is held to the upper end of the ulna by this ring-like ligament

Radial tuberosity
Bony prominence of radius for attachment of biceps brachii muscle

Radius
Bone of forearm that articulates with the humerus and ulna

Lower radio-ulnar joint (at wrist)

Head of ulna
Articulates with the ulnar notch on the inner side of the lower end of the radius

Articular surfaces
Covered with hyaline cartilage

Dorsal tubercle of radius
Bony prominence helps to divide the back of the radius into grooves to accommodate tendons that pass over the wrist

Styloid process of radius
Cone-shaped bony process

Dislocation of the radial head (pulled elbow)

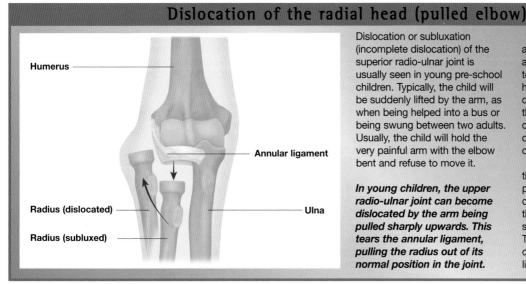

Humerus

Annular ligament

Radius (dislocated)

Radius (subluxed)

Ulna

Dislocation or subluxation (incomplete dislocation) of the superior radio-ulnar joint is usually seen in young pre-school children. Typically, the child will be suddenly lifted by the arm, as when being helped into a bus or being swung between two adults. Usually, the child will hold the very painful arm with the elbow bent and refuse to move it.

In young children, the upper radio-ulnar joint can become dislocated by the arm being pulled sharply upwards. This tears the annular ligament, pulling the radius out of its normal position in the joint.

As a result of this action, the annular ligament which wraps around the head of the radius, is torn by the sudden force and the head of the radius can then 'pop out' of the joint. The upper end of the radius may come to be displaced sideways (as seen in dislocation), or may remain in the correct alignment (subluxation).

Treatment is by manipulation of the bones back into their correct positions in the joint soon after dislocation. This is followed by the child wearing a supportive sling for a two-week period. There is usually complete healing of the supporting annular ligament.

Muscles of the upper arm

The musculature of the upper arm is divided into two distinct compartments. The muscles of the anterior compartment act to flex the arm and the muscles of the posterior compartment extend it.

The muscles of the anterior (front) compartment of the upper arm are all flexors:

■ Biceps brachii. This muscle arises from two heads, which join together to form the body of the muscle. The bulging body then tapers as it runs down to form the strong tendon of insertion.

   When the elbow is straight, biceps acts to flex the forearm. However, when the elbow is already bent the biceps muscle is a powerful supinator of the forearm, rotating the forearm so that the hand is palm up.
■ Brachialis. This arises from the lower half of the anterior surface of the humerus and passes down to cover the front of the elbow joint, its tendon inserting into the coronoid process and tuberosity of the ulna.

   Brachialis is the main flexor muscle of the elbow, whatever the position of the forearm.
■ Coracobrachialis. This muscle arises from the tip of the coracoid process of the scapula and runs down and outwards to insert into the inner surface of the humerus. This muscle helps to flex the upper arm at the shoulder and to pull it back into line with the body (adduction).

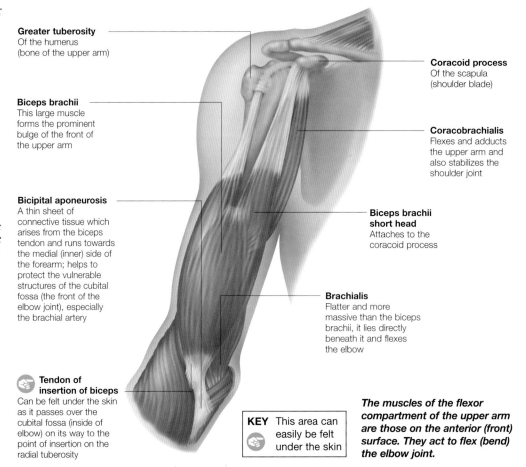

Greater tuberosity
Of the humerus (bone of the upper arm)

Biceps brachii
This large muscle forms the prominent bulge of the front of the upper arm

Bicipital aponeurosis
A thin sheet of connective tissue which arises from the biceps tendon and runs towards the medial (inner) side of the forearm; helps to protect the vulnerable structures of the cubital fossa (the front of the elbow joint), especially the brachial artery

Tendon of insertion of biceps
Can be felt under the skin as it passes over the cubital fossa (inside of elbow) on its way to the point of insertion on the radial tuberosity

Coracoid process
Of the scapula (shoulder blade)

Coracobrachialis
Flexes and adducts the upper arm and also stabilizes the shoulder joint

Biceps brachii short head
Attaches to the coracoid process

Brachialis
Flatter and more massive than the biceps brachii, it lies directly beneath it and flexes the elbow

KEY This area can easily be felt under the skin

The muscles of the flexor compartment of the upper arm are those on the anterior (front) surface. They act to flex (bend) the elbow joint.

The long head of biceps

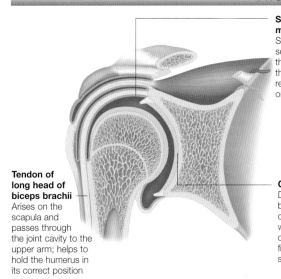

Tendon of long head of biceps brachii
Arises on the scapula and passes through the joint cavity to the upper arm; helps to hold the humerus in its correct position

Synovial membrane
Sheath of fluid-secreting tissue that lubricates the tendon, reducing friction on movement

Glenoid fossa
Depression of bone, lined with cartilage, into which the head of the humerus fits at the shoulder joint

The long head of biceps brachii arises from a point on the scapula just above the glenoid fossa. The rounded tendon crosses the head of the humerus actually within the cavity of the shoulder joint before emerging in the upper arm.

   As it passes out of the shoulder joint cavity, the tendon of the long head runs in the 'bicipital groove' between the lesser and greater tubercles of the humerus, surrounded by a sheath of synovial membrane

This section through the shoulder joint is angled to show the tendon of the long head of biceps. The name biceps means 'having two heads'.

(fluid-secreting connective tissue). The fluid acts to lubricate the tendon, reducing friction on movement. The position of attachment of this head of the biceps muscle allows it to help in stabilizing the shoulder joint, as well as in flexing the arm.

RUPTURE OF THE TENDON
Very occasionally, and usually after inflammation, the tendon of the long head of biceps may rupture as it passes within the bicipital groove. On flexing the elbow, a large 'lump' will appear, which is the bunched-up body and long head of the biceps muscle, with a depression above it showing that the long head is not in its usual position.

Muscles of the posterior compartment

The muscles of the back of the upper arm act to extend the elbow, so straightening the forearm with the upper arm.

The posterior compartment has only one major muscle, the triceps brachii, which is a powerful extensor (straightens the arm). The only other muscle in this compartment is the small, relatively insignificant anconeus.

TRICEPS BRACHII

This is a large, bulky muscle which lies posterior to the humerus and, as its name implies, has three heads:
■ The long head
■ The lateral head
■ The medial head.

The three heads converge in the middle of the upper arm on a wide, flattened tendon which passes down, over a small bursa, to attach to the olecranon process of the ulna.

The main action of triceps is to extend (straighten) the elbow joint. In addition, because of its position, the long head of the triceps muscle helps to stabilize the shoulder joint.

ANCONEUS

The small anconeus muscle lies behind and below the elbow joint and is triangular in shape. As with the triceps, it extends the elbow and also has a function in the stabilization of the elbow joint.

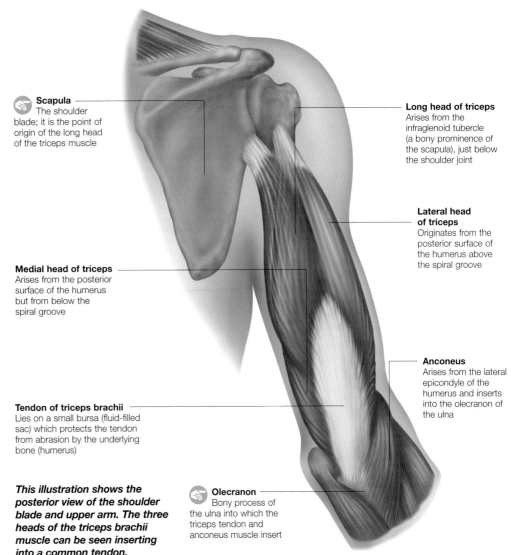

Scapula
The shoulder blade; it is the point of origin of the long head of the triceps muscle

Long head of triceps
Arises from the infraglenoid tubercle (a bony prominence of the scapula), just below the shoulder joint

Lateral head of triceps
Originates from the posterior surface of the humerus above the spiral groove

Medial head of triceps
Arises from the posterior surface of the humerus but from below the spiral groove

Anconeus
Arises from the lateral epicondyle of the humerus and inserts into the olecranon of the ulna

Tendon of triceps brachii
Lies on a small bursa (fluid-filled sac) which protects the tendon from abrasion by the underlying bone (humerus)

Olecranon
Bony process of the ulna into which the triceps tendon and anconeus muscle insert

This illustration shows the posterior view of the shoulder blade and upper arm. The three heads of the triceps brachii muscle can be seen inserting into a common tendon.

Cross-section of the upper arm

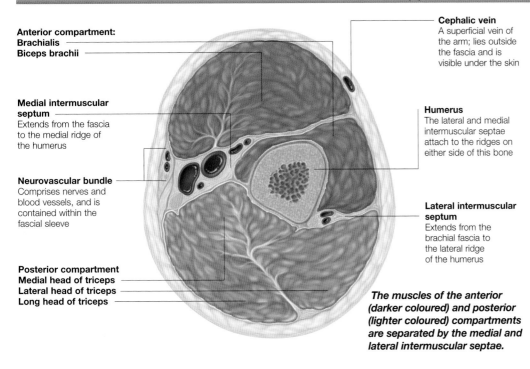

Anterior compartment:
Brachialis
Biceps brachii

Medial intermuscular septum
Extends from the fascia to the medial ridge of the humerus

Neurovascular bundle
Comprises nerves and blood vessels, and is contained within the fascial sleeve

Posterior compartment
Medial head of triceps
Lateral head of triceps
Long head of triceps

Cephalic vein
A superficial vein of the arm; lies outside the fascia and is visible under the skin

Humerus
The lateral and medial intermuscular septae attach to the ridges on either side of this bone

Lateral intermuscular septum
Extends from the brachial fascia to the lateral ridge of the humerus

The muscles of the anterior (darker coloured) and posterior (lighter coloured) compartments are separated by the medial and lateral intermuscular septae.

Viewing the upper arm in cross-section illustrates the fact that the arm is divided into distinct compartments, surrounded and enclosed by sheets of connective tissue known as fascia.

The brachial fascia is like a sleeve under the skin within which the major structures of the arm lie. Further fascial divisions, the lateral and medial intermuscular septae, arise from this brachial fascia and attach to the ridges on either side of the humerus, dividing the upper arm into the anterior and posterior muscular compartments.

NEUROVASCULAR BUNDLES

Also contained within the fascial 'sleeve' are the nerves and blood vessels of the upper arm. Nerves and blood vessels often travel together in the body within 'neurovascular bundles' and this is apparent in the case of the upper arm.

Muscles of the forearm

The flexor muscles of the front compartment of the forearm act to flex the hand, wrist and fingers. They are divided into superficial and deep muscles of the flexor and extensor compartments.

Superficial flexor muscles

This compartment, or section, of the forearm lies in the front of the forearm and contains muscles which flex the wrist and fingers as well as some which pronate the forearm (turn the hand palm down). They are sub-divided into superficial and deep layers according to position.

The superficial group contains five muscles which all originate at the medial epicondyle of the humerus, where their fibres merge to form the 'common flexor tendon':

■ Pronator teres – pronates the forearm and flexes the elbow
■ Flexor carpi radialis – acts to produce flexion and abduction (bending away from the midline of the body) of the wrist

■ Palmaris longus – this small muscle is absent in 14 per cent of people; it acts to flex the wrist
■ Flexor carpi ulnaris – this muscle flexes and adducts the wrist (bends away from the midline of the body); unlike the other muscles of the flexor compartment, this muscle is innervated by the ulnar nerve
■ Flexor digitorum superficialis – this is the largest superficial muscle of the forearm and it acts, as its name suggests, to flex the fingers, or digits.

The five main superficial flexor muscles of the forearm are shown in this illustration. These muscles originate from the humerus bone of the upper arm.

Superficial flexor muscles

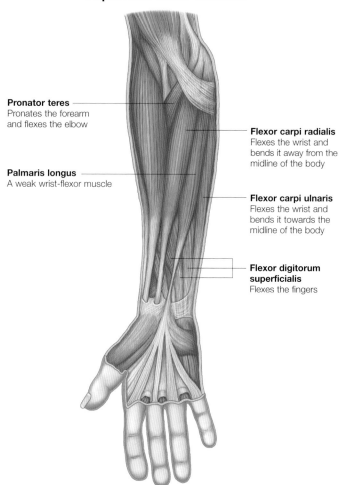

Pronator teres
Pronates the forearm and flexes the elbow

Palmaris longus
A weak wrist-flexor muscle

Flexor carpi radialis
Flexes the wrist and bends it away from the midline of the body

Flexor carpi ulnaris
Flexes the wrist and bends it towards the midline of the body

Flexor digitorum superficialis
Flexes the fingers

Deep flexor muscles

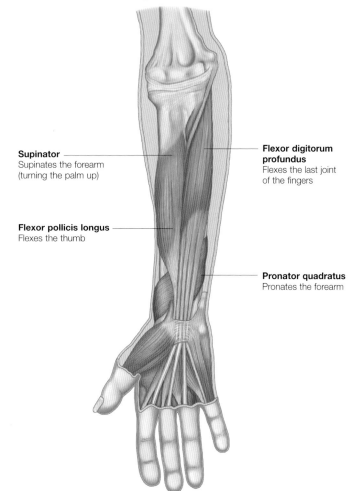

Supinator
Supinates the forearm (turning the palm up)

Flexor pollicis longus
Flexes the thumb

Flexor digitorum profundus
Flexes the last joint of the fingers

Pronator quadratus
Pronates the forearm

The deep flexor muscles lie close to the bones of the forearm (the ulna and radius). These act to flex the hand, wrist and fingers.

Deep flexor muscles

The deep layer of the flexor compartment consists of three muscles:

■ **Flexor digitorum profundus**
This bulky muscle originates from a wide area of the ulna and neighbouring interosseous membrane (a strong sheet of tissue connecting the radius and ulna). It is the only muscle which flexes the last joint of the fingers and so acts with its more superficial counterpart to curl the fingers. Like the flexor digitorum superficialis muscle, this deeper muscle divides into four tendons, which pass through the carpal tunnel within the same synovial sheath. The tendons insert into the bases of the distal (far end) phalanges of the four fingers.

■ **Flexor pollicis longus**
This muscle flexes the thumb. Its long, flat tendon passes through the carpal tunnel within its own synovial sheath and inserts into the base of the distal phalanx of the thumb (which, unlike the fingers, has only two phalanges).

■ **Pronator quadratus**
The deepest muscle of the anterior compartment, the pronator quadratus acts to pronate the forearm and is the only muscle which attaches solely to the radius and ulna. It also assists the interosseous membrane in binding the radius and ulna tightly together.

Flexing the hand

The muscles of the forearm are divided into front and rear compartments. The front flexor muscles bend the wrist and fingers and the rear extensor muscles act to straighten them again.

The muscles of the forearm are roughly divided into two groups, according to their function. These two groups are isolated from each other by the radius and ulna bones and by fascial layers (sheets of connective tissue) to form the 'anterior flexor compartment' and the 'posterior extensor compartment' of the forearm.

OPPOSING ACTIONS
The flexor muscles act to flex (bend) the wrist joint and the fingers, while the extensors act to extend (straighten) the same joints. Within these two groups are both deep and superficial muscles which act together to give the wide range of movements characteristic of the wrist and hand.

FOREARM TENDONS
So that the wrist and hand may move flexibly, the bulk of muscle around the lower end of the upper limb is kept to a minimum. This is achieved by using long tendons from muscles higher up in the forearm to work the wrist, and the fingers.

The muscles concerned are forearm muscles and need to be longer than the forearm will allow to work at maximum efficiency and thus many originate from the lower end of the humerus. The humerus has developed two projections called the medial (inner) and lateral (outer) epicondyles. The flexor muscles are attached to the medial epicondyle while the extensor muscles are attached to the lateral epicondyle.

Cross-section through the forearm

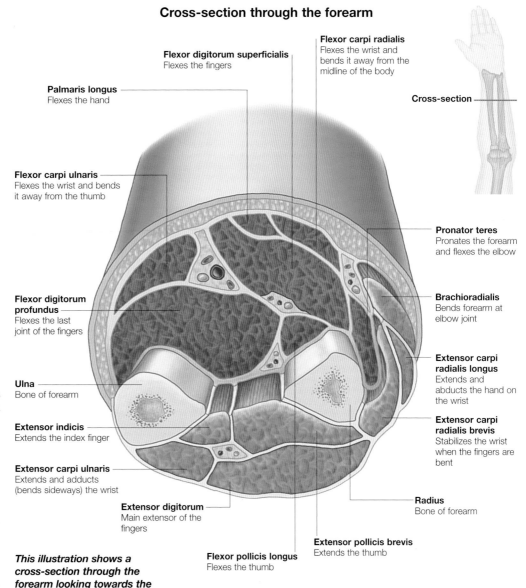

Flexor digitorum superficialis
Flexes the fingers

Flexor carpi radialis
Flexes the wrist and bends it away from the midline of the body

Palmaris longus
Flexes the hand

Cross-section

Flexor carpi ulnaris
Flexes the wrist and bends it away from the thumb

Pronator teres
Pronates the forearm and flexes the elbow

Flexor digitorum profundus
Flexes the last joint of the fingers

Brachioradialis
Bends forearm at elbow joint

Ulna
Bone of forearm

Extensor carpi radialis longus
Extends and abducts the hand on the wrist

Extensor indicis
Extends the index finger

Extensor carpi radialis brevis
Stabilizes the wrist when the fingers are bent

Extensor carpi ulnaris
Extends and adducts (bends sideways) the wrist

Radius
Bone of forearm

Extensor digitorum
Main extensor of the fingers

Extensor pollicis brevis
Extends the thumb

Flexor pollicis longus
Flexes the thumb

This illustration shows a cross-section through the forearm looking towards the hand with the palm upturned.

Compression of the forearm tissues

The strong fascial layers, together with the bones and interosseous membrane which surrounds and encloses the compartments of the forearm, can be of clinical significance after a fracture to the forearm bones, especially if a plaster cast is applied too tightly.

In this situation, the bleeding and swelling of the tissues that accompanies any fracture can cause a rise in pressure within the

It is important that plaster casts applied to the forearm are not overtight. If the dressing is too constricting, the underlying tissues may become damaged.

forearm compartments. The soft veins within the compartment are compressed, preventing blood from leaving the area and increasing the swelling.

As the pressure rises, the walled arteries are compressed, decreasing the blood supply and oxygen available to the nerves and muscles, and the cells of these structures begin to die. Dead tissues are replaced with fibrous scar tissue, which causes the damaged muscle to shorten permanently, leaving a deformity of the hand and sometimes the wrist. This deformity is known as Volkmann's ischaemic contracture.

Extensor muscles

Working together with the flexor muscles, the extensor muscles of the forearm enable a range of actions that allow the great mobility of the wrist, hand, fingers and thumb.

The posterior (rear) extensor compartment of the forearm contains muscles which act to straighten and pull back the wrist and fingers. These are separated from the flexor muscles at the front of the forearm by the radius and ulna bones, the strong membrane between them and the enveloping layer of thin connective tissue, the fascia of the forearm.

EXTENSOR ACTIONS
The extensors act to give the wide range of mobility required by the wrist and hand. The extensor muscles can be divided into three groups according to their functions:
■ Muscles that move the hand or the wrist – straighten the wrist, pull the hand back or allow the hand to bend sideways
■ Muscles that straighten the fingers, excluding the thumb
■ Muscles that act only upon the thumb to extend it or pull it out sideways.

SUPERFICIAL EXTENSORS
■ **Extensor carpi radialis longus**
It acts to extend and abduct the hand on the wrist (bend sideways away from the little finger).

■ **Extensor carpi radialis brevis**
This muscle acts together with the extensor carpi radialis longus to stabilize the wrist joint when the four fingers are being flexed (bent into the palm).

■ **Extensor carpi ulnaris**
This long, slender muscle lies along the inner edge of the forearm and extends and adducts the wrist and is also needed for the action of clenching the fist.

■ **Extensor digitorum**
This muscle is the main extensor of the four fingers and its bulk makes up a good proportion of the back of the forearm.

■ **Extensor digiti minimi**
This runs down alongside the extensor digitorum and acts to help extend the little finger.

■ **Brachioradialis**
Although in the 'extensor compartment' of the forearm, this moderately powerful muscle acts to flex (bend) the forearm at the elbow joint. It returns a pronated or supinated forearm to the working position (thumb up).

The superficial extensor muscles lie close to the surface of the forearm. They are bound together by a strap of tissue called the extensor retinaculum.

Superficial extensor muscles

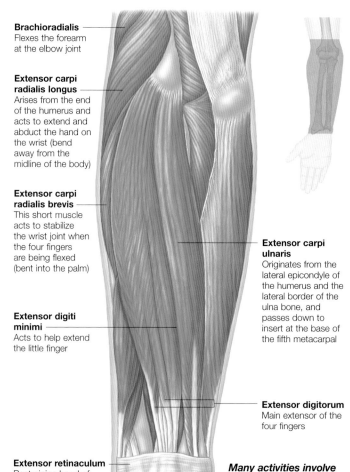

Brachioradialis
Flexes the forearm at the elbow joint

Extensor carpi radialis longus
Arises from the end of the humerus and acts to extend and abduct the hand on the wrist (bend away from the midline of the body)

Extensor carpi radialis brevis
This short muscle acts to stabilize the wrist joint when the four fingers are being flexed (bent into the palm)

Extensor digiti minimi
Acts to help extend the little finger

Extensor retinaculum
Restraining band of tissue which lies across the back of the wrist

Extensor carpi ulnaris
Originates from the lateral epicondyle of the humerus and the lateral border of the ulna bone, and passes down to insert at the base of the fifth metacarpal

Extensor digitorum
Main extensor of the four fingers

Many activities involve the use of the extensor muscles of the forearm. Table-tennis players in particular rely on their versatility of action.

Synovial cyst of the wrist (ganglion)

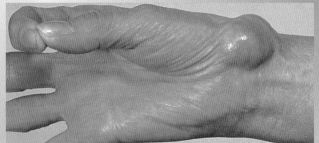

The long tendons of the extensor muscles in the forearm pass over the back of the wrist contained within synovial sheaths – fluid-filled sacs which act to protect and lubricate them as they run over the bones of the wrist.

Occasionally, connected to one of these synovial sheaths, there may develop a thin-walled cyst containing clear, thick fluid,

A ganglion is an abnormal swelling of a tendon sheath. These cysts occur most often at the wrist joint and, although prominent, are harmless.

which appears as a swelling. This non-tender swelling may vary in size and is known as a 'ganglion'. If ganglions do not resolve spontaneously, they can be surgically removed.

Deep extensor muscles

Lying closer to the underlying bones, the deep extensor layer includes muscles which act upon the thumb and the little finger individually.

The muscles of the extensor compartment, including the deep extensor muscles, receive their nerve supply from the radial nerve, which runs down the forearm alongside them.

The deep extensor muscles include:

■ **Extensor indicis**
This muscle can act together with the extensor digitorum, or alone to extend the index finger. The fact that the index finger can be extended alone (as in the action of pointing), without the other fingers following, allows greater versatility to the movements of the hand.

■ **Abductor pollicis longus**
This long muscle abducts the thumb, giving the action of lifting the thumb up and away from the plane of the palm. It can also act to extend the thumb, which is the action of moving the thumb out away from the other fingers sideways in the same plane. It originates from the back of the ulna, radius and adjoining interosseous membrane and runs down to insert at the base of the first metacarpal.

■ **Extensor pollicis brevis**
This short muscle extends the whole thumb, originating from the back of the radius and interosseous membrane and inserting at the base of the proximal phalanx of the thumb (the first bone in the thumb).

■ **Extensor pollicis longus**
This extensor of the thumb is larger than pollicis brevis.

■ **Supinator**
Together with the brachioradialis this muscle forms the floor of the cubital fossa. It is the major muscle involved in the action of supinating the forearm, which it does by rotating the radius. It is supplied by the deep branch of the radial nerve.

Together with the flexor muscles of the front of the forearm, the extensors act to give the wide range of mobility required by the wrist and hand.

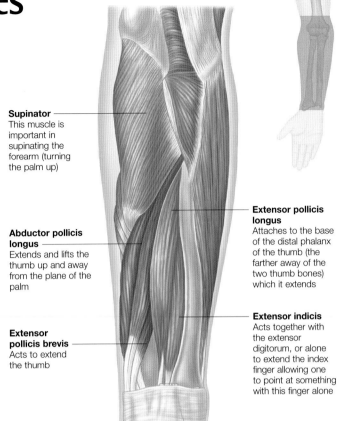

Supinator
This muscle is important in supinating the forearm (turning the palm up)

Abductor pollicis longus
Extends and lifts the thumb up and away from the plane of the palm

Extensor pollicis brevis
Acts to extend the thumb

Extensor pollicis longus
Attaches to the base of the distal phalanx of the thumb (the farther away of the two thumb bones) which it extends

Extensor indicis
Acts together with the extensor digitorum, or alone to extend the index finger allowing one to point at something with this finger alone

---

Attachments of extensor tendons

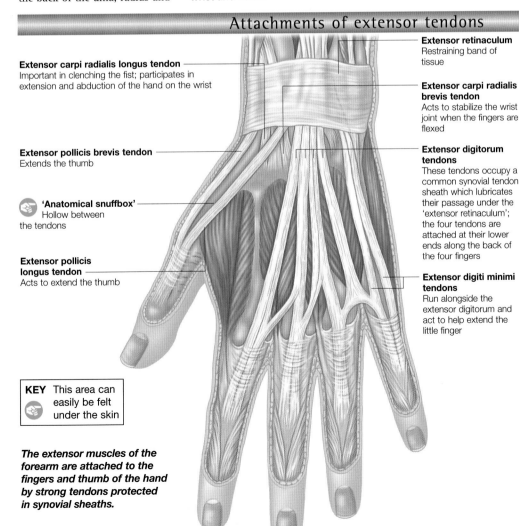

Extensor carpi radialis longus tendon
Important in clenching the fist; participates in extension and abduction of the hand on the wrist

Extensor pollicis brevis tendon
Extends the thumb

'Anatomical snuffbox'
Hollow between the tendons

Extensor pollicis longus tendon
Acts to extend the thumb

Extensor retinaculum
Restraining band of tissue

Extensor carpi radialis brevis tendon
Acts to stabilize the wrist joint when the fingers are flexed

Extensor digitorum tendons
These tendons occupy a common synovial tendon sheath which lubricates their passage under the 'extensor retinaculum'; the four tendons are attached at their lower ends along the back of the four fingers

Extensor digiti minimi tendons
Run alongside the extensor digitorum and act to help extend the little finger

KEY This area can easily be felt under the skin

The extensor muscles of the forearm are attached to the fingers and thumb of the hand by strong tendons protected in synovial sheaths.

Most of the muscles of the posterior extensor compartment of the forearm terminate in long tendons which pass down over the back of the wrist to attach to the bones of the hands and fingers. In this way, the muscles which lie within the upper forearm can bring about extension (straightening and bending back) of the hand and fingers by 'remote control', so allowing the hand itself to be less bulky than it would otherwise have to be.

The site of attachment of each tendon, and whether it is to a hand- or finger-bone, will determine which joint of the hand will be straightened when that muscle contracts.

PROTECTING THE TENDON
As the extensor tendons pass over the back of the wrist they pass under the 'extensor retinaculum', a horizontal restraining band of strong connective tissue which holds them in place against the joint as the hand moves.

To protect the tendons where they may rub against the underlying bones, and to lubricate their passage, the tendons lie within a fluid-filled synovial sheath.

Blood vessels of the arm

The arteries of the arm supply blood to the soft tissues and bones. The main arteries divide to form many smaller vessels which communicate at networks – anastomoses – at the elbow and wrist.

The main blood supply to the arm is provided by the brachial artery, a continuation of the axillary artery, which runs down the inner side of the upper arm. It gives rise to many smaller branches that supply surrounding muscles and the humerus (upper bone of the arm). The largest of these is the profunda brachii artery, which supplies the muscles that straighten the elbow.

The profunda brachii artery and the other, smaller arteries given off by the lower part of the brachial artery run down around the elbow joint. They then form a network of connecting arteries before rejoining the main arteries of the forearm.

FOREARM AND HAND

The brachial artery divides below the elbow joint into the radial and the ulnar arteries. The radial artery runs from the cubital fossa along the length of the radius (bone of the forearm). At the lower end of the radius, it lies under the skin and connective tissue; pulsations can be felt here. The ulnar artery runs towards the base of the ulna (the other bone of the forearm).

The hand has a profuse blood supply from the end branches of the radial and ulnar arteries. Branches of the two arteries join together in the palm to form the deep and the superficial palmar arches, from which small arteries arise to supply the fingers.

The pulse can be felt at several points: in the brachial artery at the elbow; in the ulnar artery on the outside of the wrist; and in the radial artery on the inner side of the wrist (shown here).

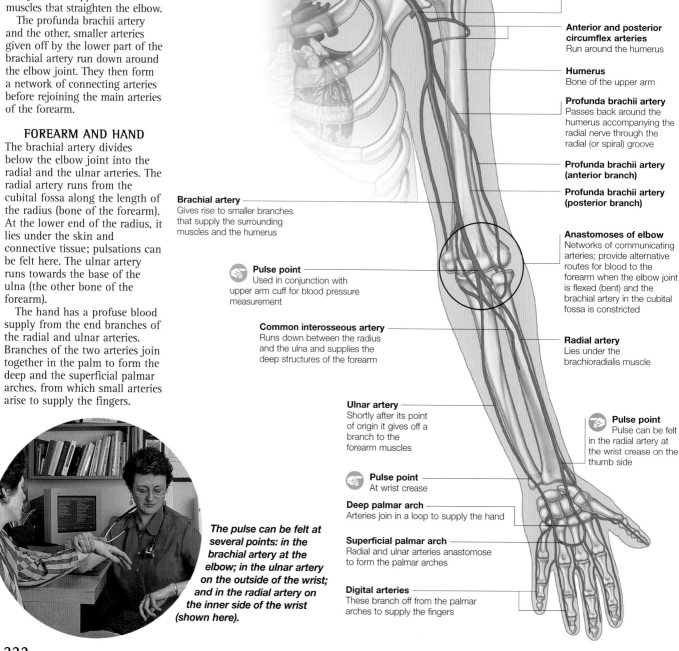

Arteries of the arm

Subclavian artery
Arises from the aortic arch and runs behind the clavicle

Axillary artery
Becomes the brachial artery

Anterior and posterior circumflex arteries
Run around the humerus

Humerus
Bone of the upper arm

Profunda brachii artery
Passes back around the humerus accompanying the radial nerve through the radial (or spiral) groove

Profunda brachii artery (anterior branch)

Profunda brachii artery (posterior branch)

Anastomoses of elbow
Networks of communicating arteries; provide alternative routes for blood to the forearm when the elbow joint is flexed (bent) and the brachial artery in the cubital fossa is constricted

Radial artery
Lies under the brachioradialis muscle

Brachial artery
Gives rise to smaller branches that supply the surrounding muscles and the humerus

Pulse point
Used in conjunction with upper arm cuff for blood pressure measurement

Common interosseous artery
Runs down between the radius and the ulna and supplies the deep structures of the forearm

Ulnar artery
Shortly after its point of origin it gives off a branch to the forearm muscles

Pulse point
At wrist crease

Deep palmar arch
Arteries join in a loop to supply the hand

Superficial palmar arch
Radial and ulnar arteries anastomose to form the palmar arches

Digital arteries
These branch off from the palmar arches to supply the fingers

Pulse point
Pulse can be felt in the radial artery at the wrist crease on the thumb side

232

Veins of the arm

The veins of the upper limb are divided into deep and superficial veins. The superficial veins lie close to the skin's surface and are often easily visible.

Venous drainage of the upper limb is achieved by two interconnecting series of veins, the deep and the superficial systems. Deep veins run alongside the arteries, while superficial veins lie in the subcutaneous tissue. The layout of the veins is very variable but usually resembles the pattern detailed below.

DEEP VEINS

In most cases, the deep veins are paired or double veins (venae comitantes) that lie on either side of the artery they accompany, making frequent anastomoses and forming a network surrounding the artery. The pulsations of blood within the artery alternately compress and release the surrounding veins, helping blood return to the heart.

The radial and ulnar veins arise from the palmar venous arches of the hand and run up the forearm to merge at the elbow, forming the brachial vein. This, in turn, merges with the basilic vein to form the large axillary vein.

SUPERFICIAL VEINS

There are two main superficial veins of the arm, the cephalic and the basilic veins, which originate at the dorsal venous arch of the hand. The cephalic vein runs under the skin along the radial side of the forearm.

The basilic vein runs up the ulnar side of the forearm, crossing the elbow to lie along the border of the biceps muscle. About halfway up the upper arm it turns inwards to become a deep vein.

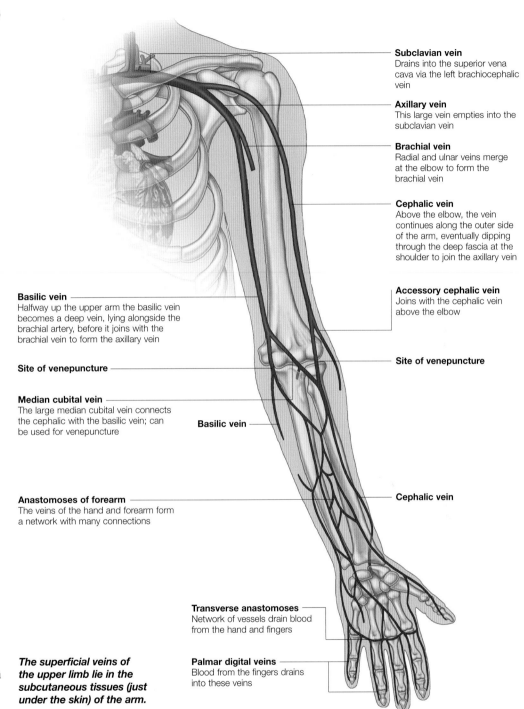

Subclavian vein
Drains into the superior vena cava via the left brachiocephalic vein

Axillary vein
This large vein empties into the subclavian vein

Brachial vein
Radial and ulnar veins merge at the elbow to form the brachial vein

Cephalic vein
Above the elbow, the vein continues along the outer side of the arm, eventually dipping through the deep fascia at the shoulder to join the axillary vein

Accessory cephalic vein
Joins with the cephalic vein above the elbow

Site of venepuncture

Basilic vein

Cephalic vein

Basilic vein
Halfway up the upper arm the basilic vein becomes a deep vein, lying alongside the brachial artery, before it joins with the brachial vein to form the axillary vein

Site of venepuncture

Median cubital vein
The large median cubital vein connects the cephalic with the basilic vein; can be used for venepuncture

Anastomoses of forearm
The veins of the hand and forearm form a network with many connections

Transverse anastomoses
Network of vessels drain blood from the hand and fingers

Palmar digital veins
Blood from the fingers drains into these veins

The superficial veins of the upper limb lie in the subcutaneous tissues (just under the skin) of the arm.

Venepuncture

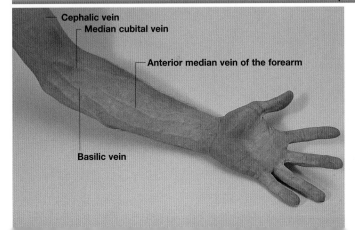

Cephalic vein
Median cubital vein
Anterior median vein of the forearm
Basilic vein

The presence of the large median cubital vein in the cubital fossa makes this an ideal site for the removal of a sample of venous blood for laboratory analysis. This large vein is usually easy to see or feel but may be difficult to find if the patient is obese.

The superficial veins in men's arms are often quite visible. This is because men tend to have less cutaneous fat covering their veins than women.

There are, however, some risks associated with using the median cubital vein to take a blood sample. The biceps tendon and the brachial artery lie deep to the median cubital vein and care must therefore be taken to not penetrate the tissues too deeply.

In some cases, it may be necessary to place a tourniquet around the upper arm to engorge the veins of the forearm making them more prominent.

Nerves of the arm

The nerves of the arm supply the skin and muscles of the forearm and hand. There are four main nerves in the arm: the radial, musculocutaneous, median and ulnar nerves.

The nerve supply to the upper limb is provided by four main nerves and their branches. These receive sensory information from the hand and arm, and also innervate the numerous muscles of the upper limb. The radial and musculocutaneous nerves supply muscles and skin of all parts of the arm, while the median and ulnar nerves only supply structures below the elbow.

RADIAL NERVE

The radial nerve is of great importance as it is the main supplier of innervation to the extensor muscles which straighten the bent elbow, wrist and fingers. It arises as the largest branch of the 'brachial plexus', a network of nerves from the spinal cord in the neck.

Near the lateral epicondyle, the radial nerve divides into its two terminal branches:
■ The superficial terminal branch – sensory nerve supply to the skin over the back of the hand, thumb and adjoining two and a half fingers
■ The deep terminal branch – motor nerve supply to all of the extensor muscles of the forearm.

MUSCULOCUTANEOUS NERVE

The musculocutaneous nerve supplies both muscles and skin in the front of the upper arm. Below the elbow it becomes the lateral cutaneous nerve of the forearm, a sensory nerve which supplies a large area of forearm skin.

Radial nerve damage

The radial nerve is most vulnerable as it passes along the bone through the radial groove at the back of the humerus. Here, it may be damaged during fracture of the shaft of the humerus or may be compressed against the bone, and thus bruised, by a direct blow to the back of the arm.

Damage to the radial nerve may paralyse all of the extensor muscles of the wrist and fingers. This leads to the characteristic clinical sign of 'wrist drop', when the wrist is held flexed due to the unopposed action of the flexor muscles and gravity.

Rear view of the nerves of the arm

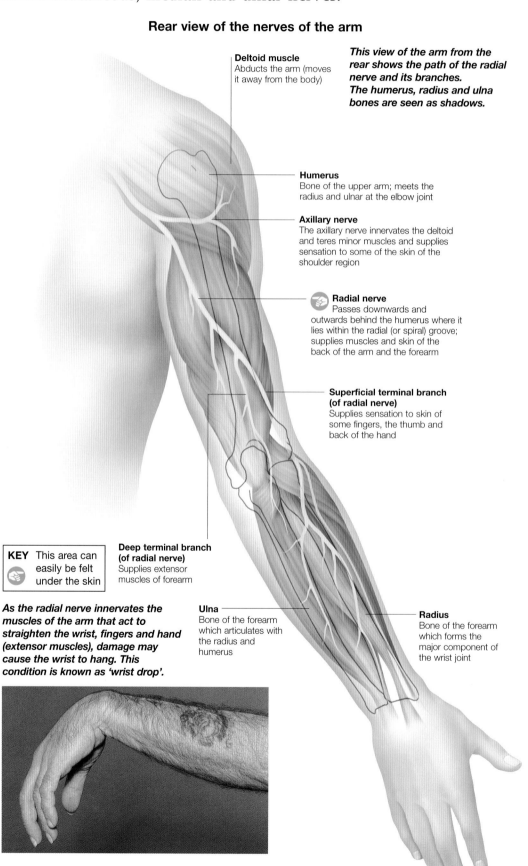

This view of the arm from the rear shows the path of the radial nerve and its branches. The humerus, radius and ulna bones are seen as shadows.

Deltoid muscle
Abducts the arm (moves it away from the body)

Humerus
Bone of the upper arm; meets the radius and ulnar at the elbow joint

Axillary nerve
The axillary nerve innervates the deltoid and teres minor muscles and supplies sensation to some of the skin of the shoulder region

Radial nerve
Passes downwards and outwards behind the humerus where it lies within the radial (or spiral) groove; supplies muscles and skin of the back of the arm and the forearm

Superficial terminal branch (of radial nerve)
Supplies sensation to skin of some fingers, the thumb and back of the hand

KEY This area can easily be felt under the skin

Deep terminal branch (of radial nerve)
Supplies extensor muscles of forearm

Ulna
Bone of the forearm which articulates with the radius and humerus

Radius
Bone of the forearm which forms the major component of the wrist joint

As the radial nerve innervates the muscles of the arm that act to straighten the wrist, fingers and hand (extensor muscles), damage may cause the wrist to hang. This condition is known as 'wrist drop'.

234

Median and ulnar nerves

The median nerve supplies the forearm muscles enabling the actions of flexion and pronation. The ulnar nerve passes behind the elbow – where it may be felt if the 'funny bone' is knocked – to supply some of the small muscles of the hand.

The median nerve of the upper limb arises from the brachial plexus and runs downwards centrally to the elbow. It is the main nerve of the front of the forearm, which contains the muscles of flexion and pronation.

At the wrist, the median nerve passes through the carpal tunnel. The median nerve ends in branches that supply some of the small muscles of the hand, as well as the skin over the thumb and some neighbouring fingers.

ULNAR NERVE

The ulnar nerve passes down along the humerus to the elbow where it loops behind the medial epicondyle, beneath the skin where it can easily be felt. It gives off branches to supply the elbow, two of the muscles of the forearm and several areas of overlying skin before entering the hand. In the hand, the ulnar nerve divides into deep and superficial branches.

MEDIAN NERVE DAMAGE

The median nerve can be damaged by fractures of the lower end of the humerus or compressed by swollen muscle tendons within the carpal tunnel (carpal tunnel syndrome). Median nerve injury can make it difficult to use the 'pincer grip' of the thumb and fingers, as the nerve supplies the small muscles of the thenar eminence (the fleshy prominence below the base of the thumb).

The ulnar nerve is most vulnerable to injury as it passes behind the medial epicondyle of the humerus. The feeling from hitting the 'funny bone' occurs when the nerve is compressed against the underlying bone. Severe damage can lead to sensory loss, paralysis and wasting of the muscles it supplies.

If the ulnar nerve is damaged, the first dorsal interosseus muscle – at the back of the thumb – may atrophy. Wasting of this muscle can be seen circled below.

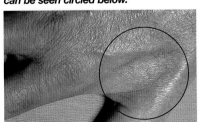

Front view of the nerves of the arm

The paths of the ulnar, median and musculocutaneous nerves can be seen in this dissected illustration of the arm from the front.

KEY This area can easily be felt under the skin

Humerus
Bone of the upper arm

Musculocutaneous nerve
This nerve supplies both muscles and skin in the arm; it is protected by muscles along its course, and is rarely injured

Median nerve
Innervates the flexor muscles of the front of the forearm as well as muscles of the outer wrist and first two fingers; also supplies sensation to the thumb and two-and-a-half fingers on the front of the hand

Ulnar nerve
Innervates the elbow and some flexor muscles of the forearm; lies close to the surface of the elbow and, if knocked, causes a 'funny bone' sensation; it can be palpated just behind the medial epicondyle

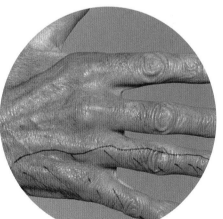

Branch of ulnar nerve
Innervates many of the intrinsic muscles of the hand as well as sensation to one-and-a-half fingers on the front and back of the hand

The marked section of this hand corresponds to the area of skin supplied by the ulnar nerve. The radial and median nerves supply other parts of the hand.

235

Deep fascia of the arm

Between the subcutaneous tissue and muscles of the arm lies
the deep fascia. This thin layer of connective tissue runs around the
upper arm (brachial fascia) and the forearm (antebrachial fascia).

The deep fascia of the arm is a
layer of thin, tough connective
tissue that lies between the
subcutaneous tissue and the
muscles, enveloping the limb like
a sleeve.

BRACHIAL FASCIA
The deep fascia of the upper arm
is also known as the brachial
fascia. It is continuous above
with the pectoral fascia, which
encloses the pectoral muscles of
the front of the chest, and the
axillary fascia, which forms the
floor of the armpit.

Around the elbow the brachial
fascia is attached to the
epicondyles of the humerus
(bony prominences at either side
of the lower end of this bone),
and to the olecranon of the ulna
(one of the forearm bones).

ANTEBRACHIAL FASCIA
The deep fascia of the forearm is
also known as the antebrachial
fascia. It is continuous with the
brachial fascia at the elbow and
runs down to enclose the tissues
of the forearm, forming the
horizontal bands of the extensor
and flexor retinacula at the wrist.

The bicipital aponeurosis is a
membranous band which runs
across the front of the elbow to
merge with the antebrachial
fascia. It arises from the tendon
of the biceps muscle and inserts,
with the deep fascia, into the
subcutaneous border of the ulna.

The deep fascia forms partitions
(septa) which pass down from its
deep surface to attach to the
bones beneath, dividing the arm
into compartments.

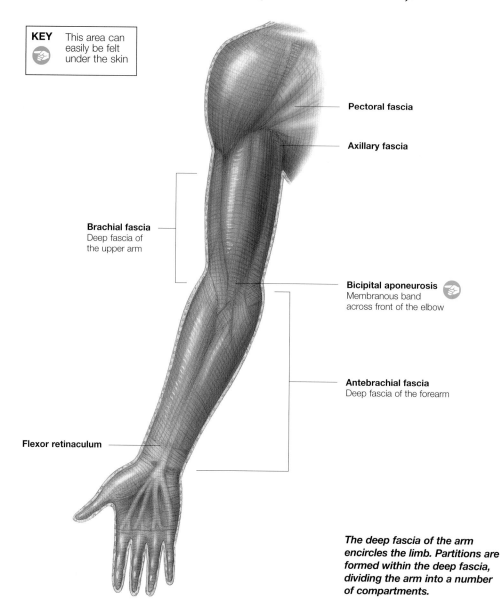

KEY This area can
easily be felt
under the skin

Pectoral fascia

Axillary fascia

Brachial fascia
Deep fascia of
the upper arm

Bicipital aponeurosis
Membranous band
across front of the elbow

Antebrachial fascia
Deep fascia of the forearm

Flexor retinaculum

*The deep fascia of the arm
encircles the limb. Partitions are
formed within the deep fascia,
dividing the arm into a number
of compartments.*

Extensor retinaculum

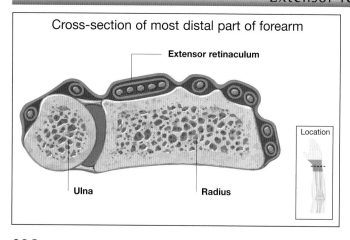

Cross-section of most distal part of forearm

Extensor retinaculum

Ulna

Radius

Location

A retinaculum is a band of
fascia that stretches across a
joint to hold down or 'retain'
tendons against the joint as it
moves. The deep fascia of the
arm gives rise to the extensor
retinaculum of the wrist, which
performs this function for the
long tendons passing over the
back of the wrist to reach the
bones of the hand and fingers.

*The extensor retinaculum is
formed by the continuation of
the deep fascia over the back
of the wrist. It keeps the tendons
against the joint as it moves.*

FOREARM BONES
The extensor retinaculum
stretches over the ends of the
radius and ulna (the two bones
of the forearm), and gives off
fibrous septa which attach to
these bones in several places.

The extensor tendons lie
between the retinaculum and the
bones and are divided by these
attachments into six groups. The
presence of the extensor
retinaculum ensures that the
long tendons of the back of the
hand remain in position as the
wrist moves rather than slipping
sideways or 'bowstringing'.

Deep fascia of the hand

The continuation of the antebrachial fascia over the wrist forms the deep fascia of the hand. Within the hand, this fascia forms the palmar aponeurosis, a central thickened area. Partitions divide the palm into compartments and spaces.

The deep fascia of the hand is the continuation of the antebrachial fascia running down over the wrist to enclose the hand like a glove.

PALMAR APONEUROSIS
In the palm of the hand, the deep fascia is thin at the sides, but thickened centrally to form a tough, fibrous sheet known as the palmar aponeurosis.

The triangular palmar aponeurosis lies subcutaneously, over the long flexor tendons and other soft tissues of the hand. One of its roles is to provide a firm site of attachment for the overlying skin, which helps to improve grip in the hand.

FLEXOR RETINACULUM
The proximal (wrist) end of the palmar aponeurosis is attached to the flexor retinaculum, the horizontal band of fascia that runs across the wrist to hold down the long flexor tendons.

Distally (towards the fingers), four longitudinal bands within the palmar aponeurosis pass up to the fingers, helping to form the fibrous sheaths which enclose the tendons there.

COMPARTMENTS
Medial and lateral septa from the deep fascia divide the muscles of the hand into a number of compartments. These

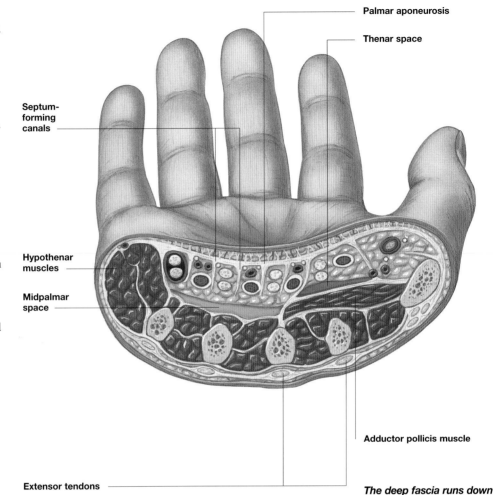

Labels: Palmar aponeurosis · Thenar space · Septum-forming canals · Hypothenar muscles · Midpalmar space · Extensor tendons · Adductor pollicis muscle

The deep fascia runs down around the forearm to enclose the hand. At the centre of the palm it forms a thickened area – the palmar aponeurosis.

compartments comprise:
- ■ The hypothenar compartment – contains the muscles associated with the little finger, nerves and blood vessels. It is enclosed within the hypothenar fascia
- ■ The thenar compartment – this is surrounded by the thenar fascia, and houses some of the muscles that act upon the thumb
- ■ The adductor compartment – this is named for the action of

its muscle, namely the adduction (pulling in) of the thumb
- ■ The central compartment – this lies under the palmar aponeurosis and contains flexor tendons, nerves and blood vessels.

Spaces of the hand

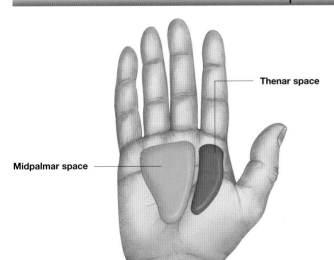

Labels: Thenar space · Midpalmar space

The deep fascia of the hand and, in particular, the strong lateral fibrous septum divide the palm into two potential spaces: the thenar space and the midpalmar space. Despite their names, these areas are not actually empty spaces, but are normally filled with loose connective tissue.

The thenar space is the smaller of the two spaces. It lies on the

side of the hand nearest the thumb, lateral to the fibrous lateral septum of deep fascia. The midpalmar space is larger and lies medial to this septum, within the centre of the palm.

INFECTION
The fascial spaces of the hand may become filled with pus if part of the hand becomes infected. The surrounding partitions of fascia form boundaries that contain the pus within each potential space and determine to what extent the infection can spread.

The palm is divided into two spaces by the deep fascia: the thenar space and the midpalmar space. Each area is made up of loose connective tissue.

Bones of the wrist

The wrist lies between the radius and ulna of the forearm and the bones of the fingers. It is made up of eight marble-sized bones which move together to allow flexibility of the wrist joint and the hand.

The area that we commonly think of as the wrist is the end area of the forearm overlying the lower ends of the radius and ulna bones of the forearm. The wrist, in fact, lies in the base of the hand, and comprises eight bones held together by ligaments. They move in relation to each other, thus allowing the wrist to be flexible.

The carpal bones form two rows of four bones each – the proximal row (nearer the forearm) and the distal row (nearer the fingers). The main joint of the wrist is between the first of these two rows and the lower end of the radius.

THE PROXIMAL ROW
The proximal row of the wrist consists of the following bones:
- **Scaphoid** – a 'boat-shaped' bone which has a large facet for articulation with the lower end of the radius; it articulates with three bones of the distal row
- **Lunate** – this is a moon-shaped bone that articulates with the lower end of the radius
- **Triquetral** – this pyramid-shaped bone articulates with the disc of the inferior radioulnar joint and the pisiform bone
- **Pisiform** – although usually considered part of the proximal row, this small bone plays no part in the wrist joint. It is about the size and shape of a pea and is a 'sesamoid' bone, a bone that lies within a muscle tendon.

The way that the bones of the wrist sit in relation to each other is seen in this image. The top (distal) bones – closest to the fingers – are tinted orange; the bottom (proximal) bones – closest to the forearm are purple.

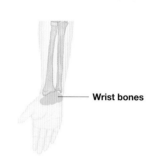

Wrist bones

KEY This area can easily be felt under the skin

Bones of the left wrist viewed from above

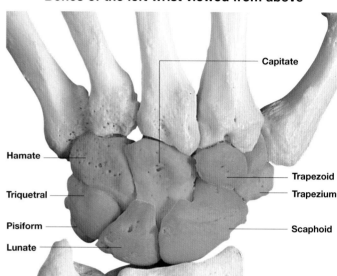

Capitate

Hamate

Triquetral

Pisiform

Lunate

Trapezoid

Trapezium

Scaphoid

Bottom (proximal) row of wrist bones

The proximal row of carpal bones includes two bones that can be easily felt: the pisiform and scaphoid bones.

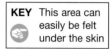

Triquetral
Has a small facet which is the site of articulation with the pisiform bone

Pisiform
Lies within the tendon of the flexor carpi ulnaris muscle

Lunate
Articulates with the lower end of the radius

Scaphoid
Has a narrowed 'waist' which is of importance clinically as it may be the site of a fracture

The distal row

Top (distal) row of wrist bones

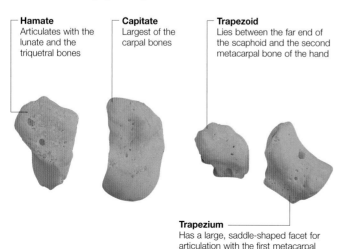

Hamate
Articulates with the lunate and the triquetral bones

Capitate
Largest of the carpal bones

Trapezoid
Lies between the far end of the scaphoid and the second metacarpal bone of the hand

Trapezium
Has a large, saddle-shaped facet for articulation with the first metacarpal

The distal row of the bones of the wrist include the:
- **Trapezium**
This four-sided bone lies between the scaphoid and the first metacarpal (the lowest bone of the thumb). It has a large, saddle-shaped facet for articulation with the first metacarpal and a prominent tubercle (bump) on its palmar side.
- **Trapezoid**
A small, wedge-shaped bone which lies between the far end of the scaphoid and the second

The top (distal) row of carpal bones lies between the bones of the hand and the bottom (proximal) row. Both of these rows are held together by ligaments.

metacarpal, the bone of the hand which runs up to the base of the index finger.
- **Capitate**
The largest of the carpal bones, the capitate is named after its large, rounded head, which lies in the cup-shaped hollow formed by the scaphoid and the lunate bones. It articulates at its far end with the third metacarpal and also the second and fourth metacarpal bones.
- **Hamate**
A triangular bone which is much wider at its far end than at the proximal end. It articulates with the lunate and the triquetral bones. This bone has a hook-like process on its palmar surface, the 'hook of the hamate'.

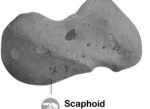

The wrist joint

The bones of the wrist are covered in cartilage and enclosed by a synovial membrane. This secretes a viscous fluid which allows the bones to move in relation to one another with minimum friction.

The wrist, or radiocarpal, joint is a synovial (fluid-filled) joint. On one side lie the lower end of the radius and the articular disc of the inferior radioulnar joint; while on the other are three bones of the first row of carpal (wrist) bones: the scaphoid, lunate and triquetral bones. The fourth bone of this first row, the pisiform, plays no part in the wrist joint.

THE RADIOCARPAL JOINT
The radiocarpal joint is made up of three areas, which are separated by two low ridges:
■ A lateral area (outer, on the same side as the thumb), which is formed by the lateral half of the end of the radius as it articulates with the scaphoid bone
■ A middle area, where the medial (inner) half of the end of the radius articulates with the lunate bone
■ A medial area (inner, on the side of the little finger), where the articular disc which separates the wrist joint from the inferior radioulnar joint articulates with the triquetral bone.
   All the articular surfaces are covered with smooth, articular (hyaline) cartilage for reduction of friction during movement. The joint is lined with synovial membrane which secretes thick, lubricating synovial fluid, and is surrounded by a fibrous

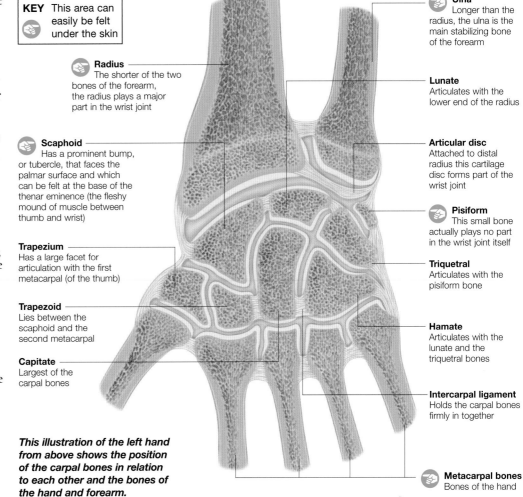

KEY This area can easily be felt under the skin

Radius
The shorter of the two bones of the forearm, the radius plays a major part in the wrist joint

Scaphoid
Has a prominent bump, or tubercle, that faces the palmar surface and which can be felt at the base of the thenar eminence (the fleshy mound of muscle between thumb and wrist)

Trapezium
Has a large facet for articulation with the first metacarpal (of the thumb)

Trapezoid
Lies between the scaphoid and the second metacarpal

Capitate
Largest of the carpal bones

Ulna
Longer than the radius, the ulna is the main stabilizing bone of the forearm

Lunate
Articulates with the lower end of the radius

Articular disc
Attached to distal radius this cartilage disc forms part of the wrist joint

Pisiform
This small bone actually plays no part in the wrist joint itself

Triquetral
Articulates with the pisiform bone

Hamate
Articulates with the lunate and the triquetral bones

Intercarpal ligament
Holds the carpal bones firmly in together

Metacarpal bones
Bones of the hand

This illustration of the left hand from above shows the position of the carpal bones in relation to each other and the bones of the hand and forearm.

capsule that is strengthened by ligaments.
   The articular surfaces as a whole form an ellipsoid shape, with the long axis of the ellipse lying across the width of the wrist. The shape of the articular surfaces of a joint helps to determine the range of movements; an ellipsoid shape

does not allow rotation of that joint. The joint is convex towards the hand.

THE INTERCARPAL JOINTS
As well as the joint between the lower end of the forearm and the first row of carpal bones, there is articulation between the carpal bones themselves.

There is a large, irregular, 'midcarpal' joint which lies between the two rows of carpal bones. It is a synovial joint with a joint cavity which extends into the gaps between the eight bones and allows them to glide over one another, giving the flexibility which is needed in the wrist.

Fracture of the scaphoid

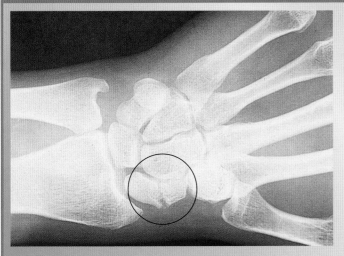

The most frequently fractured carpal bone is the scaphoid bone; fractures of this bone are one of the most common types of wrist injury. The fracture is often the result of a fall onto the outstretched hand and is usually across the narrowed 'waist' of the bone.
   Fracture of the scaphoid is important clinically as the pattern of arteries around the bone is such that after a fracture across

The scaphoid is prone to fracture through the narrow waist in the centre of the bone. This X-ray reveals the fracture (circled) but it is not always visible.

the middle region, part of the bone may be left without a blood supply and may undergo avascular necrosis (tissue death due to lack of a blood supply). This may lead to damage and loss of function of the whole wrist.
   Clinical suspicion of a scaphoid fracture is aroused by the presence of pain when pressure is applied by the doctor to the 'anatomical snuffbox', a depression at the back of the wrist near the base of the thumb; the scaphoid bone lies at the base of this depression. Diagnosis is complicated as there may be no obvious changes on an X-ray of the wrist.

Carpal tunnel

The strong ligaments of the wrist bind together the carpal bones,
allowing stability and flexibility. Within the wrist is a fibrous band
through which important tendons and nerves run – the carpal tunnel.

The eight carpal bones fit
together in the wrist to form the
shape of an arch. The back of
the wrist, the dorsal surface, is
gently convex upwards while the
palmar surface is concave. The
arch is deepened on the palmar
aspect of the wrist by the
presence of the prominent
tubercles of the scaphoid and
trapezium bones on one side and
the hook of the hamate and the
pisiform bone on the other.

STRUCTURE OF THE WRIST

This bony arch is converted into
a tunnel by a tough band of
fibrous tissue, the flexor
retinaculum, which lies across the
palmar surface and is attached on
each side to the bony projections.
This is called the carpal tunnel;
through it run the long tendons
of the muscles which flex (bend)
the fingers. The presence of this
band ensures that these tendons
are held close to the wrist even
when the wrist is bent, so
allowing flexion of the fingers
at every position of the wrist.

*The flexor retinaculum holds the
tendons tightly together,
allowing flexion in all positions.
This is called the carpal tunnel.*

Tendons of the wrist

Radius
Bone of the forearm

Interosseous membrane
Binds the radius and ulna
firmly together

**Flexor carpi radialis
tendon**
Flexes the wrist and bends
it away from the body

Radial artery
Major artery of the forearm

Flexor pollicis longus tendon
Flexes the thumb

Median nerve
Main nerve supplying
the hand

Flexor retinaculum
Tough band of fibrous
tissue which hold the
tendons in place

Metacarpal bones
Bones of the fingers

**Ulnar artery
and nerve**

**Flexor carpi ulnaris
tendon**
Flexes the wrist

**Flexor digitorum
profundus tendons**
Flexes the last joint of
the fingers

**Flexor digitorum
superficialis
tendons**
Flexes the fingers

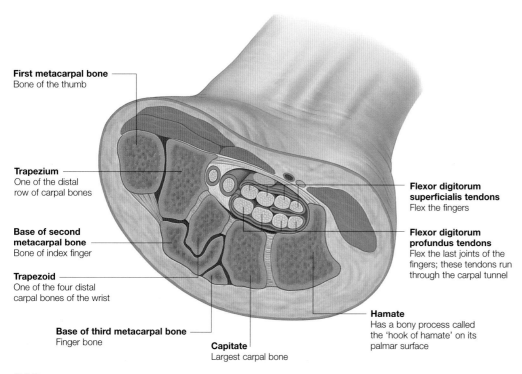

Cross-section of the right wrist

First metacarpal bone
Bone of the thumb

Trapezium
One of the distal
row of carpal bones

**Base of second
metacarpal bone**
Bone of index finger

Trapezoid
One of the four distal
carpal bones of the wrist

Base of third metacarpal bone
Finger bone

Capitate
Largest carpal bone

**Flexor digitorum
superficialis tendons**
Flex the fingers

**Flexor digitorum
profundus tendons**
Flex the last joints of the
fingers; these tendons run
through the carpal tunnel

Hamate
Has a bony process called
the 'hook of hamate' on its
palmar surface

NERVE SUPPLY

As well as the long flexor
tendons, the carpal tunnel also
contains the median nerve, one
of the main nerves supplying the
hand. If there is any swelling
within the confined space of the
carpal tunnel, the median nerve
will be constricted. Such
swelling may occur due to
inflammation of the long flexor
tendons caused by repetitive
strain injuries, or due to general
fluid retention such as is
occasionally seen in pregnancy.

This leads to the condition
known as 'carpal tunnel
syndrome' whereby constriction
of the median nerve causes 'pins
and needles' or a burning pain in
the skin of the side of the hand.
There may also be weakness of
the muscles at the base of the
thumb as these are supplied
from the median nerve as well.

*Compression of the carpal
tunnel (blue) may affect the
median nerve and therefore
the functioning of the fingers.*

Ligaments of the wrist

The ligaments of the wrist joint are thickenings of the joint capsule which help tie the wrist strongly to the lower ends of the radius and ulna.

The wrist joint itself cannot rotate and so rotation of the hand is achieved by pronation and supination of the forearm. The strong ligaments between the carpal bones and the radius are important as they 'carry' the hand round with the forearm during these actions. These ligaments include:

■ **Palmar radiocarpal ligaments**
Run from the radius to the carpal bones on the palm side of the hand. The fibres are directed so that the hand will go with the forearm during supination.
■ **Dorsal radiocarpal ligaments**
Run at the back of the wrist from the radius to the carpal bones and carry the hand back during pronation.

COLLATERAL LIGAMENTS
Strong collateral ligaments run down each side of the wrist to strengthen the joint capsule and add to the stability of the wrist. These limit the movements of the wrist joint when it is bent.
■ **Radial collateral ligament**
Runs between the styloid process of the radius and the scaphoid bone in the wrist.
■ **Ulnar collateral ligament**
Runs between the styloid process of the ulna and the triquetral bone.

Dorsal view of the ligaments of the wrist

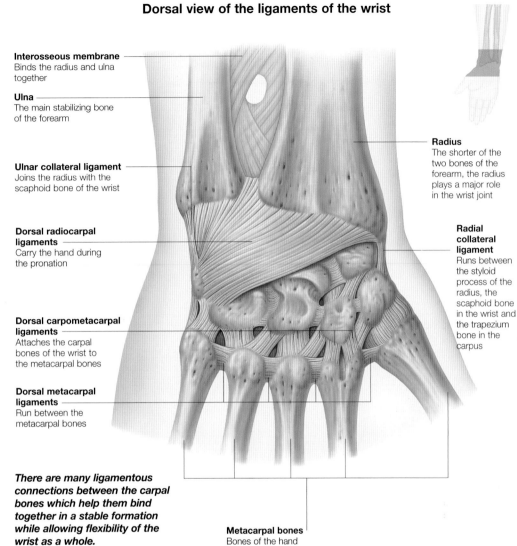

Interosseous membrane
Binds the radius and ulna together

Ulna
The main stabilizing bone of the forearm

Ulnar collateral ligament
Joins the radius with the scaphoid bone of the wrist

Dorsal radiocarpal ligaments
Carry the hand during the pronation

Dorsal carpometacarpal ligaments
Attaches the carpal bones of the wrist to the metacarpal bones

Dorsal metacarpal ligaments
Run between the metacarpal bones

Radius
The shorter of the two bones of the forearm, the radius plays a major role in the wrist joint

Radial collateral ligament
Runs between the styloid process of the radius, the scaphoid bone in the wrist and the trapezium bone in the carpus

There are many ligamentous connections between the carpal bones which help them bind together in a stable formation while allowing flexibility of the wrist as a whole.

Metacarpal bones
Bones of the hand

Movements of the wrist

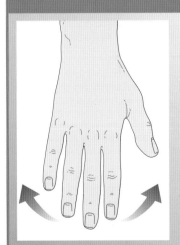

Abduction is the the action of bending the wrist sideways towards the side of the thumb. This is limited to about 15 degrees. Adduction is the action of bending the wrist away from the thumb.

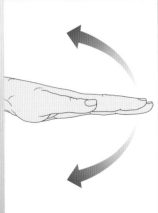

Flexion (bending) of the wrist is limited by the pull of the tendons at the back of the hand. The wrist can normally flex about 80 degrees, whereas extension (bending back) can usually only achieve 60 degrees.

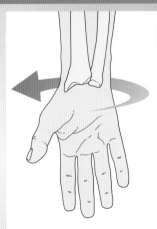

The wrist is rotated by the action of the forearm in pronation (turning the palm face down) and supination (turning the palm face up). This is possible due to the strong ligaments of the wrist.

Movements at the wrist joint are added to by smaller movements between the carpal bones themselves. The movements possible at the wrist joint are flexion and extension (bending forwards and backward), abduction (bending toward the thumb) and adduction (bending away from the thumb). Put together, these actions can bring about circumduction which is a complete circular movement of the hand at the wrist.

The most stable position of the wrist, when the bones are held most firmly together, is when it is in full extension, with the hands bent back. In this position the strong anterior radiocarpal ligaments are stretched taut. This is the position of the wrist that is naturally used when we push a heavy load, or put out the hand to break a fall.

Bones of the hand

The bones of the hand are divided into the metacarpal bones which support the palm and the phalanges or finger bones. The joints of these bones allow the fingers and the thumb great mobility.

The skeleton of the hand is made up of eight carpal bones of the wrist, the five metacarpal bones, which support the palm, and the 14 phalanges, or finger bones.

THE METACARPALS

Five slender bones radiate out from the wrist bones towards the fingers to form the support of the palm of the hand. They are numbered from one to five, starting at the thumb.

Each of the metacarpals is made up of a body, or shaft, and two slightly bulbous ends. The proximal end (near to the wrist) or base articulates with one of the carpal bones. The distal end (away from the wrist), or head, articulates with the first phalanx of the corresponding finger. In a clenched fist, the heads of the metacarpals are the knuckles.

THE THUMB

The first metacarpal, at the base of the thumb, is the shortest and thickest of the five bones and is rotated slightly out of line. It is extremely mobile, allowing a wider range of movement to the thumb than to the fingers, including the action of opposition whereby the thumb can touch the tips of each of the fingers.

The metacarpal bones

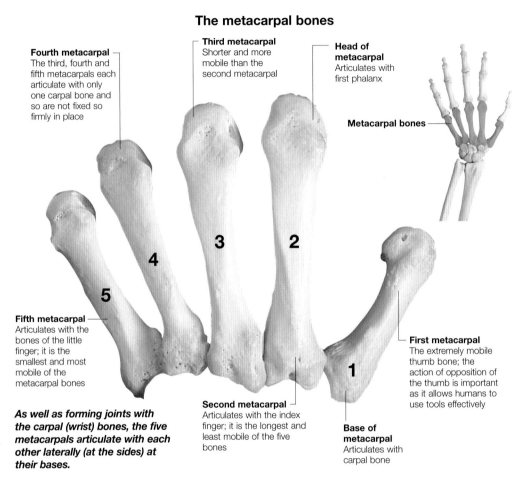

Fourth metacarpal
The third, fourth and fifth metacarpals each articulate with only one carpal bone and so are not fixed so firmly in place

Third metacarpal
Shorter and more mobile than the second metacarpal

Head of metacarpal
Articulates with first phalanx

Metacarpal bones

Fifth metacarpal
Articulates with the bones of the little finger; it is the smallest and most mobile of the metacarpal bones

Second metacarpal
Articulates with the index finger; it is the longest and least mobile of the five bones

First metacarpal
The extremely mobile thumb bone; the action of opposition of the thumb is important as it allows humans to use tools effectively

Base of metacarpal
Articulates with carpal bone

As well as forming joints with the carpal (wrist) bones, the five metacarpals articulate with each other laterally (at the sides) at their bases.

The phalanges

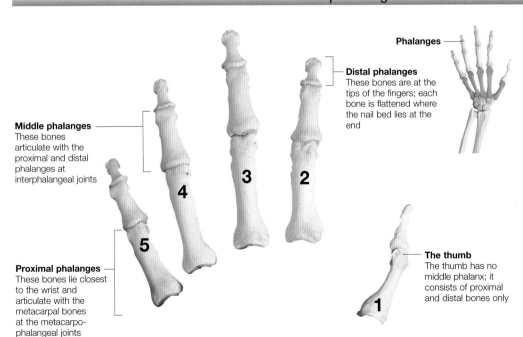

Middle phalanges
These bones articulate with the proximal and distal phalanges at interphalangeal joints

Phalanges

Distal phalanges
These bones are at the tips of the fingers; each bone is flattened where the nail bed lies at the end

Proximal phalanges
These bones lie closest to the wrist and articulate with the metacarpal bones at the metacarpo-phalangeal joints

The thumb
The thumb has no middle phalanx; it consists of proximal and distal bones only

The phalanges (singular: 'phalanx') are the bones of the fingers, or digits. The digits are numbered one to five, with the thumb being number one. The first digit, the thumb, has only two phalanges, the other four digits each have three.

Each phalanx is a miniature long bone with a slender shaft, or body, and two expanded ends. In each digit the first, or proximal, phalanx is the largest while the end, or distal, phalanx is the smallest. The phalanges of the thumb are shorter and thicker than those in the other digits.

Each of the small distal phalanges is characteristically flattened at the tip to form the skeletal support of the nail bed.

Each of the fingers, with the exception of the thumb, consists of three bones. These articulate with each other and with the metacarpal bones.

Finger joints

The joints between the phalanges are surrounded by fibrous capsules, lined with synovial membrane and supported by strong collateral ligaments.

The joints of the metacarpal bones with the carpal bones of the wrist – the carpometacarpal joints – are synovial (fluid-filled). The thumb has a saddle-shaped joint with the trapezium allowing a wide range of movement while the other metacarpals form 'plane' joints where the articulating surfaces are flat; they therefore have a limited range of movements.

The carpometacarpal joints are surrounded by fibrous joint capsules lined with synovial membrane. This secretes the lubricating synovial fluid which fills the joint cavity. In most people there is a single, continuous joint cavity for the second to fifth carpometacarpal joints. The joint of the first metacarpal with the trapezium has its own separate joint cavity.

METACARPOPHALANGEAL JOINTS

The joints between the metacarpals and the proximal phalanges are 'condyloid' synovial joints – this shape allows movement in two planes. The fingers may flex and extend (bend and straighten), or abduct and adduct (move apart and together sideways, spreading the fingers). This adds to the mobility and versatility of the hand, as the fingers can be placed in a wide variety of positions.

INTERPHALANGEAL JOINTS

The joints between each phalanx and the next are simple hinge-shaped joints which allow flexion and extension only.

Synovial membrane
Secretes the lubricating synovial fluid which fills the joint cavity

Metacarpal bone
The five metacarpal bones support the palm; the head of each metacarpal articulates with the first phalanx of the corresponding finger to form a 'knuckle'

Metacarpophalangeal joint
Synovial joint where the metacarpal bone of the hand and the proximal phalanges meet

Joint capsule
Each joint is surrounded by a fibrous capsule which is lined by synovial membrane and supported by a strong collateral ligament on each side

Middle phalanx
Found only in the second to fifth digits and is absent in the thumb

Interphalangeal joint
Hinge-shaped joint connects the individual phalanges; these joints allow flexion and extension alone

Distal phalanges
Bone of the tip of the finger which is flattened under the nail bed

Each finger has two inter-phalangeal joints, where the phalanges articulate with each other. These joints allow flexion and extension.

The hinge-shaped joints of the fingers – the interphalangeal joints – enable flexion and extension. The fingers may be flexed without flexing the hand.

Dislocated fingers

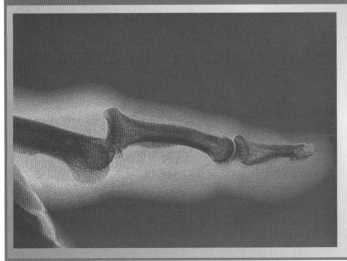

Dislocation is an injury to a joint which results in the bones of that joint becoming misaligned with respect to each other. Dislocation may be accompanied by damage to soft tissues surrounding the joint, the synovial membrane lining the joint cavity, the ligaments, and to muscles, nerves and blood vessels.

Dislocation of the interphalangeal joints is relatively common. Typically the finger is

bent backwards so forcibly that the ligaments surrounding the joint are unable to keep the bones aligned. Another cause of finger dislocation is rheumatoid arthritis where tissues surrounding the joint are softened by inflammation.

The dislocated finger joint may be bruised, swollen and painful on trying to bend it. There may be obvious deformity of the finger, with a corresponding loss of function.

Treatment involves returning the finger bones to their correct alignment, a process known as 'reduction'. Damage to surrounding tissues may require surgical intervention.

This false-colour X-ray shows a dislocation of the finger at the proximal interphalangeal joint. It is treated by manipulating the bones back into position.

Muscles of the hand

The human hand is an exceptionally versatile structure, capable of powerful and delicate movements. These are produced by the actions and interactions of the numerous muscles which act upon it.

Many powerful movements of the hand, which need the contractile strength of a large bulk of muscle tissue, are controlled by the action of muscles in the forearm via tendons, rather than the hand.

Precise and delicate actions are produced by small, or 'intrinsic', muscles. These can be divided into three groups:
- The muscles of the thenar eminence (the bulge of muscle which lies between the base of the thumb and the wrist), which move the thumb
- The muscles of the hypothenar eminence (muscle between the little finger and the wrist), which move the little finger
- The short muscles that run deep in the palm of the hand.

There are two groups of muscles which run longitudinally deep within the hand – the lumbricals and the interossei.

THE LUMBRICALS

There are four lumbrical muscles arising in the palm from the tendons of the flexor digitorum profundus, a powerful muscle of the forearm. The four lumbrical muscles pass around the thumb side of the corresponding digit and insert into the area on the back of the finger which contains the extensor tendons (extensor expansion, or hood).

The lumbrical muscles

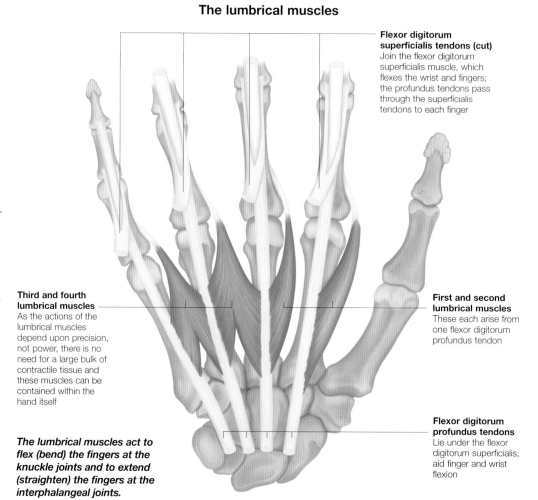

Flexor digitorum superficialis tendons (cut)
Join the flexor digitorum superficialis muscle, which flexes the wrist and fingers; the profundus tendons pass through the superficialis tendons to each finger

Third and fourth lumbrical muscles
As the actions of the lumbrical muscles depend upon precision, not power, there is no need for a large bulk of contractile tissue and these muscles can be contained within the hand itself

First and second lumbrical muscles
These each arise from one flexor digitorum profundus tendon

Flexor digitorum profundus tendons
Lie under the flexor digitorum superficialis; aid finger and wrist flexion

The lumbrical muscles act to flex (bend) the fingers at the knuckle joints and to extend (straighten) the fingers at the interphalangeal joints.

Interosseous muscles

Extensor digitorum tendons
Allow extension of the fingers and wrist (cut ends shown)

Dorsal interosseous muscles
These are the deepest of the intrinsic hand muscles and act in opposition to the palmar interosseous muscles to abduct, or spread, the fingers

Abductor pollicis brevis
Lifts the thumb up and away from the plane of the palm

Carpal bones
The carpal bones of the wrist join the metacarpal bones of the hand

Extensor expansion or hood

Abductor digiti minimi
Lies just under the skin and has its origin in the flexor retinaculum

These muscles are named after their position, interossei meaning 'between the bones'. This is a dorsal view of the interosseous muscles of the right hand.

The interosseous muscles lie in two layers, those near the palm, the 'palmar interossei', and the deeper layer, the 'dorsal interossei'.
- **Palmar interossei**
These small muscles arise from the palmar surface of the metacarpals (excluding the third). The first two pass round the medial side of each digit before inserting into the dorsal (back) surface. Those which pass to digits four and five pass around the lateral (thumb) side. Contraction of these muscles pulls the fingers in together to give the action of adduction.
- **Dorsal interossei**
These larger muscles lie between the metacarpal bones of the hand, deep to the palmar interossei. Each arises from the sides of the metacarpal bones and act to spread the fingers.

Moving the thumb and little finger

The muscles that move the thumb are contained in the thenar eminence, at the base of the thumb; those that move the little finger are found in the hypothenar eminence, between the little finger and the wrist.

The four small muscles of the thenar eminence act together to allow the thumb to move in the manner that is so important to humans. This action is known as 'opposition' and is the action whereby the tip of the thumb is brought around to touch the tip of any of the fingers.

The muscles of the thenar eminence that move the thumb include:

■ **Abductor pollicis brevis**
Abductor pollicis brevis literally means the short muscle which abducts the thumb (lifts the thumb up away from the palm).
■ **Flexor pollicis brevis**
Flexor pollicis brevis (which flexes the thumb) lies near to the centre of the palm.
■ **Opponens pollicis**
Opponens pollicis (the muscle that opposes the thumb) originates in the flexor retinaculum and the trapezium bone of the wrist, and it inserts into the outer border of the first metacarpal bone.

Palmar view of the right hand

The muscles which activate the thumb and little finger are seen in this illustration of the palm of the right hand.

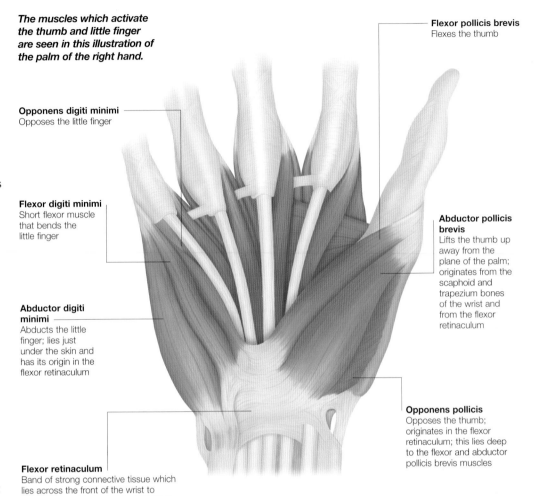

Opponens digiti minimi
Opposes the little finger

Flexor digiti minimi
Short flexor muscle that bends the little finger

Abductor digiti minimi
Abducts the little finger; lies just under the skin and has its origin in the flexor retinaculum

Flexor retinaculum
Band of strong connective tissue which lies across the front of the wrist to prevent 'bowstringing' of long tendons

Flexor pollicis brevis
Flexes the thumb

Abductor pollicis brevis
Lifts the thumb up away from the plane of the palm; originates from the scaphoid and trapezium bones of the wrist and from the flexor retinaculum

Opponens pollicis
Opposes the thumb; originates in the flexor retinaculum; this lies deep to the flexor and abductor pollicis brevis muscles

Section through the hand

Flexor digiti minimi brevis

Abductor digiti minimi

Palmaris brevis

Opponens digiti minimi

Flexor pollicis brevis (deep head)

Adductor pollicis

Flexor pollicis brevis (superficial head)

Abductor pollicis brevis

Third palmar interosseous

Fourth dorsal interosseous

Second palmar interosseous

Third dorsal interosseous

First palmar interosseous

Second dorsal interosseous

First dorsal interosseous

Opponens pollicis

■ **Adductor pollicis**
Adductor pollicis brings the abducted thumb back in line with the palm. This is a deeply placed muscle which has two heads of origin, separated by the radial artery, which join to form a tendon. This tendon often contains a 'sesamoid' bone, a small bone which lies completely within the tendon and makes no connections with other bones.

THE HYPOTHENAR EMINENCE

The muscles of the smaller hypothenar eminence form the swelling which lies between the little finger and the wrist. These muscles act together to move the little finger around towards the thumb in the action of cupping the hand, or when gripping the lid of a jar to twist it off.

This illustration of a cross-section of the hand seen palm-up reveals the position of many of the hand muscles in relation to one another.

■ **Abductor digiti minimi**
This muscle lies just under the skin and has its origin in the flexor retinaculum and the pisiform bone of the wrist. It inserts into the side of the base of the little finger.
■ **Flexor digiti minimi**
This short flexor muscle lies alongside the previous muscle but nearer to the centre of the palm. It originates in the flexor retinaculum and the hamate bone of the wrist, and inserts into the base of the little finger.
■ **Opponens digiti minimi**
This muscle, which opposes the little finger, lies underneath the more superficial muscles of the hypothenar eminence.
■ **Palmaris brevis muscle**
This short muscle has no attachments to bone, but originates in the palmar aponeurosis (connective tissue sheet which lies in the palm) and inserts into the skin overlying the hypothenar eminence. It acts to wrinkle the skin, which is believed to aid grip.

Nerves and blood vessels of the hand

The hand is supplied with numerous arteries and veins. These join to form networks of small, interconnecting blood vessels which ensure a good blood supply to all fingers, even if one artery is damaged.

The hand has a plentiful blood supply from the ulnar and radial arteries. These have many interconnections (anastomoses), maintaining the blood supply even if one artery is damaged.

■ **Superficial palmar arch**
The ulnar artery enters the hand on the same side as the little finger and crosses the palm to join with the radial artery to form the 'superficial palmar arch'. This gives off small digital arteries which supply blood to the little, ring and middle fingers.

■ **Deep palmar arch**
The deep palmar arch is formed by a continuation of the radial artery. This enters the palm from below the base of the thumb and gives off small arteries which supply the thumb and index finger as well as metacarpal branches which anastomose with the digital arteries.

■ **Back of the hand**
An irregular network of small arteries lies over the back of the wrist, which supplies the back of the hand and fingers.

Palmar view of arteries of the left hand

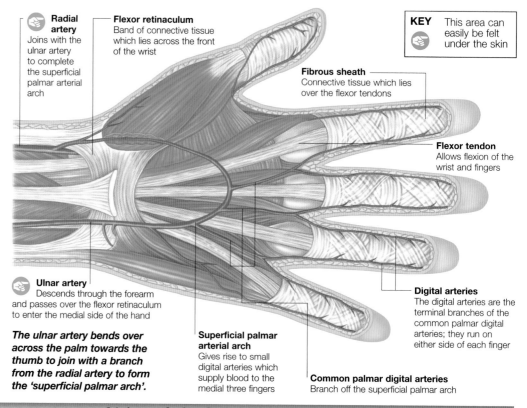

Radial artery
Joins with the ulnar artery to complete the superficial palmar arterial arch

Flexor retinaculum
Band of connective tissue which lies across the front of the wrist

KEY This area can easily be felt under the skin

Fibrous sheath
Connective tissue which lies over the flexor tendons

Flexor tendon
Allows flexion of the wrist and fingers

Ulnar artery
Descends through the forearm and passes over the flexor retinaculum to enter the medial side of the hand

The ulnar artery bends over across the palm towards the thumb to join with a branch from the radial artery to form the 'superficial palmar arch'.

Superficial palmar arterial arch
Gives rise to small digital arteries which supply blood to the medial three fingers

Digital arteries
The digital arteries are the terminal branches of the common palmar digital arteries; they run on either side of each finger

Common palmar digital arteries
Branch off the superficial palmar arch

Veins of the hand

Dorsal view of the veins of the left hand

The small veins of the hand join together to form the larger veins of the arm, which carry the blood back to the heart.

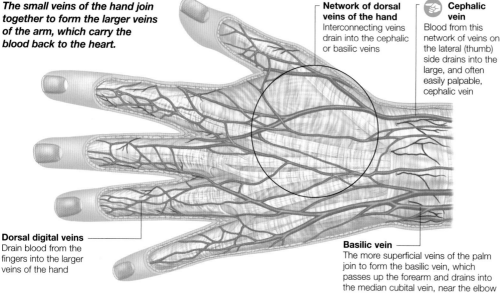

Network of dorsal veins of the hand
Interconnecting veins drain into the cephalic or basilic veins

Cephalic vein
Blood from this network of veins on the lateral (thumb) side drains into the large, and often easily palpable, cephalic vein

Dorsal digital veins
Drain blood from the fingers into the larger veins of the hand

Basilic vein
The more superficial veins of the palm join to form the basilic vein, which passes up the forearm and drains into the median cubital vein, near the elbow

The veins of the back of the fingers join to form the prominent dorsal venous arch. Blood from this network of veins on the lateral (thumb) side drains into the large cephalic vein. On the other side of the back of the hand, the network drains into the large basilic vein.

PALMAR VEINS
The veins of the palm form interconnecting arches which lie deeply alongside the deep palmar arterial arch and, more superficially, accompany the superficial palmar arch. They receive blood from the fingers via small digital veins running down either side of each finger.

The deep veins of the palm run with the radial and ulnar arteries of the forearm. The superficial palmar veins run with their equivalent arteries.

Nerves of the hand

The structures of the hand receive their nerve supply from terminal branches of the three main nerves of the upper limb: the median, ulnar and radial nerves.

The median nerve enters the hand on the palmar side by passing under the flexor retinaculum (a restraining band of connective tissue) within the carpal tunnel.

In the hand the median nerve supplies:
■ The three muscles of the thenar eminence – abductor pollicis brevis, flexor pollicis brevis and opponens pollicis. If the median nerve is damaged there will be loss of innervation to these muscles with corresponding loss of function of the thumb. This will include the inability to perform the important action of opposition of the thumb.
■ The first and second lumbrical muscles.
■ The skin of the palm and palmar surface of the first three-and-a-half digits as well as the dorsal surface (back) of the tips of those fingers. The branch of the median nerve which supplies the skin of the central palm arises before the median nerve enters the carpal tunnel and passes over, not under, the flexor retinaculum so the skin will continue to receive its nerve supply if the median nerve is damaged there.

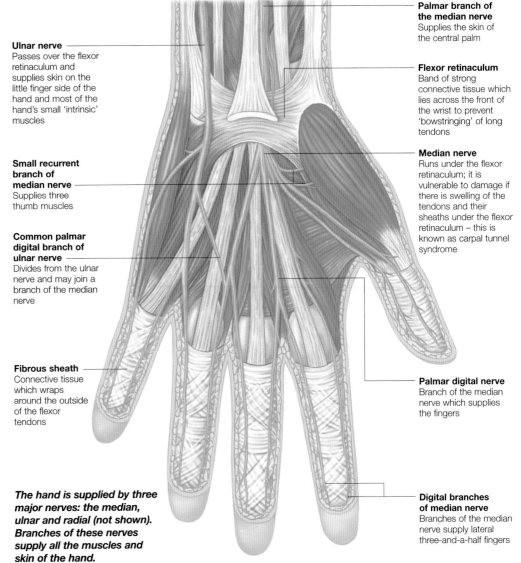

Ulnar nerve
Passes over the flexor retinaculum and supplies skin on the little finger side of the hand and most of the hand's small 'intrinsic' muscles

Small recurrent branch of median nerve
Supplies three thumb muscles

Common palmar digital branch of ulnar nerve
Divides from the ulnar nerve and may join a branch of the median nerve

Fibrous sheath
Connective tissue which wraps around the outside of the flexor tendons

Palmar branch of the median nerve
Supplies the skin of the central palm

Flexor retinaculum
Band of strong connective tissue which lies across the front of the wrist to prevent 'bowstringing' of long tendons

Median nerve
Runs under the flexor retinaculum; it is vulnerable to damage if there is swelling of the tendons and their sheaths under the flexor retinaculum – this is known as carpal tunnel syndrome

Palmar digital nerve
Branch of the median nerve which supplies the fingers

Digital branches of median nerve
Branches of the median nerve supply lateral three-and-a-half fingers

The hand is supplied by three major nerves: the median, ulnar and radial (not shown). Branches of these nerves supply all the muscles and skin of the hand.

Nerve supply of the hand

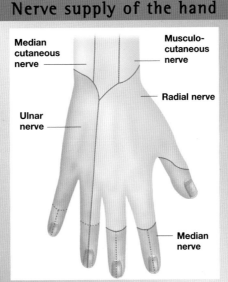

Musculo-cutaneous nerve

Radial nerve

Median nerve

Median cutaneous nerve

Ulnar nerve

Median cutaneous nerve

Ulnar nerve

Musculo-cutaneous nerve

Radial nerve

Median nerve

These two diagrams show the areas of skin innervated by the nerves of the hand. The ulnar nerve supplies the areas shaded purple. The median nerve supplies the pink areas. The radial nerve supplies the yellow areas. The median cutaneous nerve and the musculocutaneous nerve areas are shaded green and blue respectively.

The ulnar nerve
The ulnar nerve enters the medial side of the hand by passing over the flexor retinaculum. In the hand the ulnar nerve supplies:
■ The skin on the medial side of the palm, via its palmar cutaneous branch
■ The skin of the medial half of the back of the hand, little finger and medial half of the ring finger via its dorsal cutaneous branch
■ The skin of the palmar side of the little finger and half the ring finger via its superficial branch
■ The muscles of the hypothenar eminence via its deep branch
■ The adductor pollicis muscle, which acts to pull the thumb back down to the palm
■ The third and fourth lumbrical muscles and all the interosseous muscles.

The radial nerve
The radial nerve only supplies the skin within the hand. It runs down the back of the forearm to the dorsal surface (back) of the hand to supply the skin of the back of the lateral three-and-a-half digits.

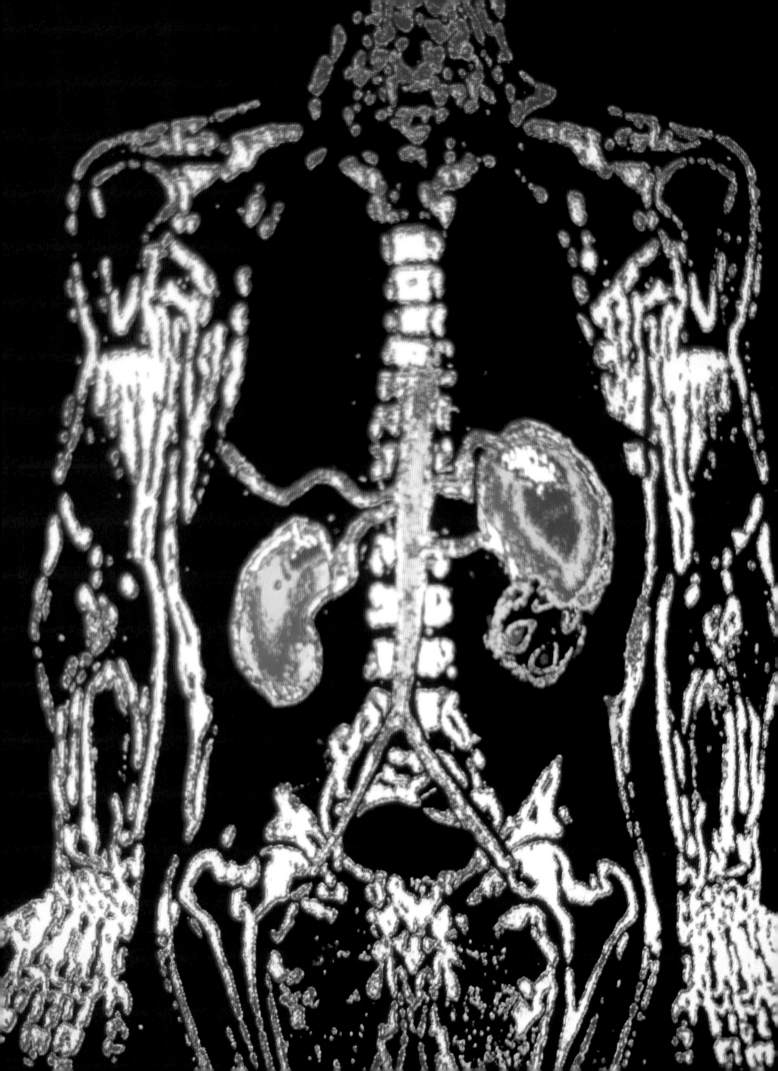

Chapter Six

The Abdomen & Stomach

The largest cavity in the human body, the abdomen contains the majority of the digestive tract and organs. Most of the absorption and digestion of the food we eat occurs here.

This chapter explains the fascinating process of digestion, detailing how each of the foods we eat is broken down and used by the body. We learn the importance of a balanced diet in providing us with the major food groups, and discover how these foods are essential for building and nourishing cells throughout the body, helping to prevent disease, aid growth and maintain energy levels.

LEFT: This Thermograph of the human torso shows the two kidneys.

Overview of the abdomen

The abdomen is the part of the trunk which lies between the thorax (above) and the pelvis (below). The contents of the abdominal cavity are supported by a bony framework and the abdominal wall.

The organs of the upper part of the abdominal cavity – the liver, gall bladder, stomach and spleen – lie under the domes of the diaphragm and are protected by the lower ribs.

The vertebrae and their associated muscles, form the back wall of the abdominal cavity, while the bones of the pelvis support it from beneath.

The abdomen is relatively unprotected by bone. This does, however, allow for mobility of the trunk and enables the abdomen to distend when necessary, such as during pregnancy or after a large meal.

ABDOMINAL CONTENTS

The contents of the abdominal cavity include:
- Much of the gastro-intestinal tract
- Liver
- Pancreas
- Spleen
- Kidneys.

As well as these viscera the abdominal cavity contains all the blood vessels, lymphatics and nerves which supply them, together with a variable amount of fatty tissue.

The abdominal cavity contains the digestive organs, such as the stomach and intestines, known collectively as viscera.

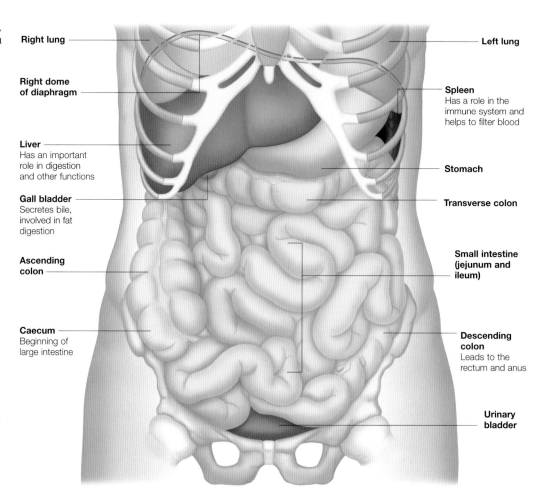

Right lung

Right dome of diaphragm

Liver
Has an important role in digestion and other functions

Gall bladder
Secretes bile, involved in fat digestion

Ascending colon

Caecum
Beginning of large intestine

Left lung

Spleen
Has a role in the immune system and helps to filter blood

Stomach

Transverse colon

Small intestine (jejunum and ileum)

Descending colon
Leads to the rectum and anus

Urinary bladder

The omentum

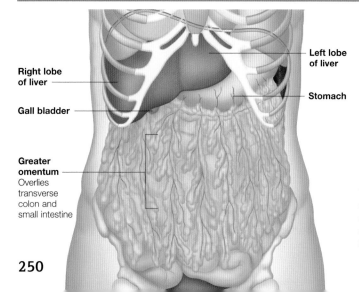

Right lobe of liver

Gall bladder

Greater omentum
Overlies transverse colon and small intestine

Left lobe of liver

Stomach

The majority of the contents of the abdominal cavity are clothed in a thin, lubricating sheet of tissue, known as the peritoneum. Folds of peritoneum attach the abdominal viscera to the walls of the abdominal cavity, and enable these organs to slide easily over one another.

The most noticeable part of the peritoneum is the greater omentum, which hangs down from the lower border of the

The greater omentum is a curtain of fatty tissue that hangs in front of the abdominal organs. It acts to protect and insulate the underlying structures.

stomach and covers the transverse colon and the coils of small bowel like an apron. The greater omentum contains a large amount of fat within it, which gives it a yellowish appearance.

PROTECTIVE ROLE

The greater omentum has been called the 'abdominal policeman' due to its action in wrapping itself around an inflamed organ to prevent spread of infection to other organs.

It also helps to protect the abdominal organs from injury and to insulate the abdomen against heat loss.

Planes and regions of the abdomen

In order to describe the position of organs or the site of abdominal pain, doctors find it useful to divide the abdomen into regions defined by imaginary vertical and horizontal planes. These areas help in the making of a clinical diagnosis.

The abdomen may be divided into nine regions for precise descriptions. These regions are delineated by two horizontal (subcostal and transtubercular) planes and two vertical (midclavicular) planes.

The nine regions are:
- Right hypochondrium
- Epigastrium
- Left hypochondrium
- Right flank (lumbar)
- Umbilical
- Left flank (lumbar)
- Right inguinal (groin)
- Suprapubic
- Left inguinal (groin).

FOUR QUADRANTS
For general clinical purposes it is usually sufficient to divide the abdomen into just four sections delineated by one horizontal (transumbilical) plane and one vertical (median) plane.

The four sections are known simply as the right and left upper quadrants and the right and left lower quadrants.

CLINICAL IMPORTANCE
It is important to know which of the abdominal contents lie in each of the regions. If an abnormality is found, or if the patient has an abdominal pain, these regions can be used to indicate the site in the patient's notes for future reference.

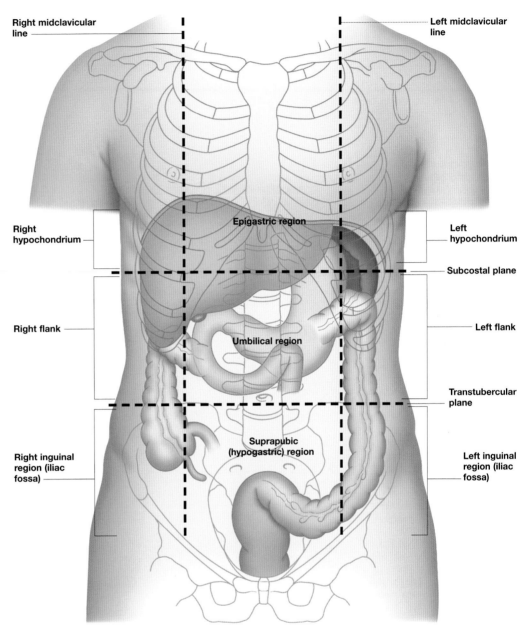

Right midclavicular line

Left midclavicular line

Right hypochondrium

Epigastric region

Left hypochondrium

Subcostal plane

Right flank

Umbilical region

Left flank

Transtubercular plane

Right inguinal region (iliac fossa)

Suprapubic (hypogastric) region

Left inguinal region (iliac fossa)

Surgical incisions

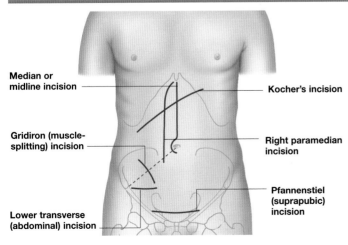

Median or midline incision

Kocher's incision

Gridiron (muscle-splitting) incision

Right paramedian incision

Pfannenstiel (suprapubic) incision

Lower transverse (abdominal) incision

Knowledge of the anatomy of the abdominal wall is important for surgeons performing abdominal operations.

If possible the surgeon will cut along the natural lines of cleavage (Langer's lines) of the skin, which lie parallel to the collagen fibres in the skin and so will close more neatly.

Other anatomical factors will be taken into account such as

Surgeons employ a number of incisions to gain access to the abdominal cavity. Incisions need to allow adequate exposure during the operation.

the layout of nerves supplying the abdominal wall, the direction in which the muscle fibres run and the position of connective tissue sheets (aponeuroses) so that the surgeon can minimize the damage he or she does to the healthy structures of the abdominal wall.

TYPES OF INCISION
These various factors, together with the particular requirements of each clinical case, lead to a variety of incisions being used including the midline (median), the paramedian and the transverse (abdominal) incisions.

251

Abdominal wall

The abdominal cavity lies between the diaphragm and the pelvis.
The abdominal wall at the front and sides of the body consists of
different muscular layers, surrounding and supporting the cavity.

The posterior (rear) abdominal wall is formed by the lower ribs, the spine and accompanying muscles, while the anterolateral (front and side) wall consists entirely of muscle and fibrous sheets (aponeuroses).

Under the skin and subcutaneous fat layer there lie the muscle layers of the abdominal wall. The muscles here lie in three broad sheets: the external oblique, the internal oblique and the transversus abdominis, which give the abdomen support in all directions. In addition, there is a wide band of muscle, the rectus abdominis, which runs vertically from the front of the ribcage down to the front of the pelvis.

EXTERNAL OBLIQUE

The external oblique muscle forms the most superficial layer of abdominal muscles. It is in the form of a broad, thin sheet whose fibres run down and inwards.

The muscle arises from the under-surfaces of the lower ribs. The fibres fan out into the wide sheet of tough connective tissue known as the external oblique aponeurosis. At the lower end the fibres insert into the top of the pubic bones.

The external oblique muscle is part of the anterior abdominal wall. It is the longest and most superficial of the anterolateral flat abdominal muscles.

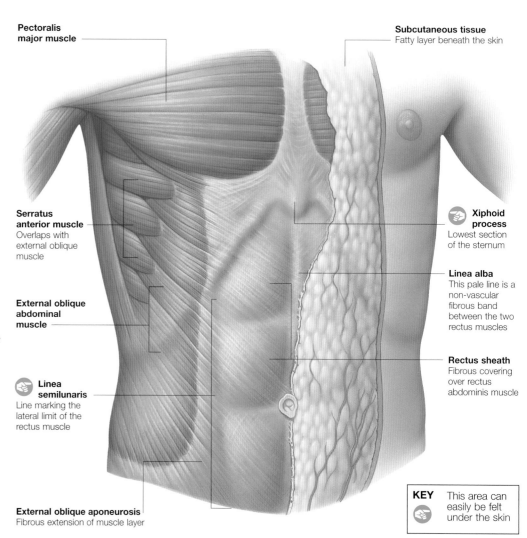

Pectoralis major muscle

Serratus anterior muscle
Overlaps with external oblique muscle

External oblique abdominal muscle

Linea semilunaris
Line marking the lateral limit of the rectus muscle

External oblique aponeurosis
Fibrous extension of muscle layer

Subcutaneous tissue
Fatty layer beneath the skin

Xiphoid process
Lowest section of the sternum

Linea alba
This pale line is a non-vascular fibrous band between the two rectus muscles

Rectus sheath
Fibrous covering over rectus abdominis muscle

KEY This area can easily be felt under the skin

Layers of the abdominal wall

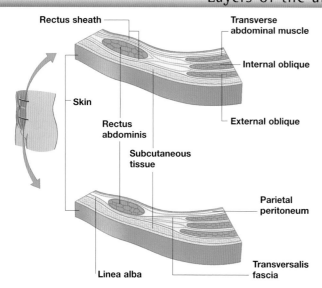

Rectus sheath

Skin

Rectus abdominis

Subcutaneous tissue

Linea alba

Transverse abdominal muscle

Internal oblique

External oblique

Parietal peritoneum

Transversalis fascia

The layers of the abdominal wall include:
- Skin – the natural lines of cleavage of the skin lie horizontally over the greater part of the abdominal wall
- Superficial fatty layer, or Camper's fascia – this may be very thick and, in obese people, may cause the abdominal wall to lie in folds
- Superficial membranous layer, or Scarpa's fascia – this thin layer is continuous with the superficial fascia in adjoining

These transverse sections of the abdominal wall show its layered structure. The fibrous layers from the muscle interweave around the rectus abdominis.

parts of the body
- Three muscle layers: external and internal oblique, and transversus abdominis
- Deep fascia layers lying between, and separating, these muscle layers
- Transversalis fascia – this firm, membranous sheet lines most of the abdominal wall, merging with the tissues lining the underside of the diaphragm above and the pelvis below
- Fat, lying between the transversalis fascia and the peritoneum
- Peritoneum – this is the delicate, lubricating membrane that lines the abdominal cavity and covers the surfaces of many abdominal organs.

Deeper muscles of the abdominal wall

Beneath the large external oblique muscle lie two more layers of sheet-like muscle, the internal oblique and the transversus abdominis. In addition, running vertically down the centre of the abdominal wall is the rectus abdominis.

The internal oblique muscle is a broad, thin sheet which lies deep to the external oblique. Its fibres run upwards and inwards, at approximately 90 degrees to those of the external oblique.

The fibres of the internal oblique originate from the lumbar fascia (a layer of connective tissue on either side of the spine), the iliac crest of the pelvis and the inguinal ligament (formed in the groin).

Like the external oblique, the internal oblique muscle inserts into a tough, broad aponeurosis which splits to enclose the rectus abdominis muscle (rectus sheath).

TRANSVERSUS ABDOMINIS
This is the innermost of the three sheets of muscle which support the abdominal contents. Its fibres run horizontally to insert into an aponeurosis which lies behind the rectus abdominis muscle for much of its length.

RECTUS ABDOMINIS
These two strap-like muscles run vertically down the front of the abdominal wall.

The upper part of each muscle is wider and thinner than the lower part. Lying between the muscles is a thin, tendinous band of tough connective tissue, the linea alba.

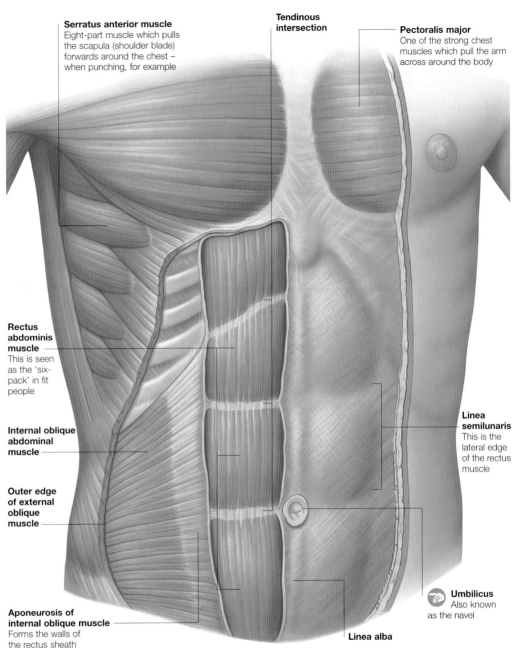

Serratus anterior muscle
Eight-part muscle which pulls the scapula (shoulder blade) forwards around the chest – when punching, for example

Tendinous intersection

Pectoralis major
One of the strong chest muscles which pull the arm across around the body

Rectus abdominis muscle
This is seen as the 'six-pack' in fit people

Internal oblique abdominal muscle

Outer edge of external oblique muscle

Aponeurosis of internal oblique muscle
Forms the walls of the rectus sheath

Linea semilunaris
This is the lateral edge of the rectus muscle

Umbilicus
Also known as the navel

Linea alba

The rectus sheath

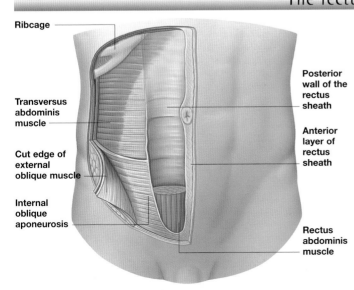

Ribcage

Transversus abdominis muscle

Cut edge of external oblique muscle

Internal oblique aponeurosis

Posterior wall of the rectus sheath

Anterior layer of rectus sheath

Rectus abdominis muscle

The rectus abdominis muscle is enclosed within a sheath of connective tissue formed by the coming together of the aponeuroses of the three muscle sheets of the abdominal wall. (An aponeurosis is a thin but strong sheet of fibrous tissue.)

The upper three-quarters of the rectus sheath differ from the lower quarter due to the way in which the three aponeuroses interweave.

The rectus abdominis muscle extends along the front of the abdomen. It is enclosed by connective tissue known as the rectus sheath.

■ **The upper rectus sheath**
The anterior wall is formed by the external oblique aponeurosis and half of the internal oblique aponeurosis, while the posterior wall is formed by the remaining half of the internal oblique and the transversus aponeurosis.

■ **The lower rectus sheath**
Three aponeuroses lie in front of the rectus abdominis muscle which lies, therefore, directly on the transversalis fascia beneath.

The three aponeuroses meet in the mid-line to form the tough linea alba. As well as the rectus abdominis muscle, the rectus sheath contains blood vessels which lie deep to the muscle.

253

Stomach

The stomach is the expanded part of the digestive tract that receives swallowed food from the oesophagus. Food is stored here before being propelled into the small intestine as digestion continues.

The stomach is a distendable muscular bag lined by mucous membrane. It is fixed at two points: the oesophageal opening at the top and at the beginning of the small intestine below. Between these points it is mobile and can vary in position.

STOMACH LINING

When empty, the stomach lining lies in numerous folds, or rugae, which run from one opening to the other.

The walls of the stomach are similar to other parts of the gut but with some modifications:
■ The gastric epithelium – this is the layer of cells which lines the stomach; it contains many glands that secrete protective mucus, and others that produce enzymes and acid, which begin the process of digestion.
■ The muscle layer – this has an inner oblique layer of muscle as well as the usual longitudinal and circular fibres. This arrangement helps the stomach to churn food thoroughly before propelling it on towards the small intestine.

REGIONS OF THE STOMACH

The stomach is said to have four parts, and two curvatures:
■ The cardia
■ The fundus
■ The body
■ The pyloric region – the outlet area of the stomach
■ The lesser curvature
■ The greater curvature.

Location and structure of the stomach

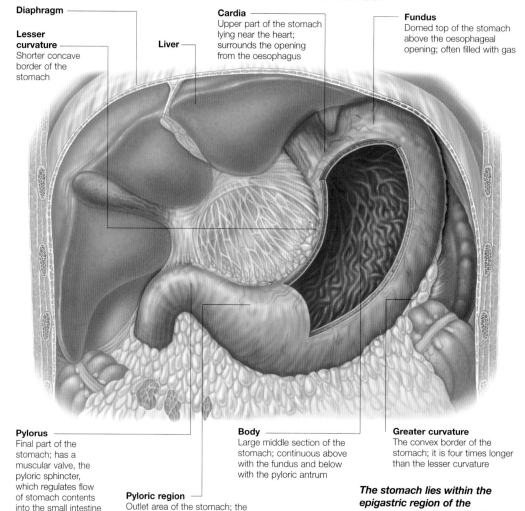

Diaphragm

Lesser curvature
Shorter concave border of the stomach

Liver

Cardia
Upper part of the stomach lying near the heart; surrounds the opening from the oesophagus

Fundus
Domed top of the stomach above the oesophageal opening; often filled with gas

Pylorus
Final part of the stomach; has a muscular valve, the pyloric sphincter, which regulates flow of stomach contents into the small intestine

Pyloric region
Outlet area of the stomach; the funnel-like pyloric antrum leads into the narrow pyloric canal; at the end of the canal lies the pylorus

Body
Large middle section of the stomach; continuous above with the fundus and below with the pyloric antrum

Greater curvature
The convex border of the stomach; it is four times longer than the lesser curvature

The stomach lies within the epigastric region of the abdomen, below the diaphragm. It lies to the right of the spleen and partly under the liver.

The gastro-oesophageal junction

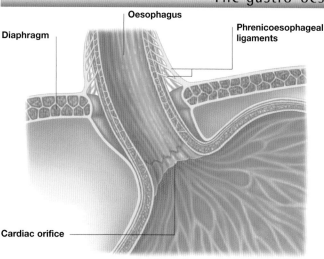

Oesophagus

Diaphragm

Phrenicoesophageal ligaments

Cardiac orifice

At the lower end of the oesophagus the epithelium, or lining layer of cells, changes from multilayered, stratified squamous, to the typical gastric mucosa in a zig-zag junction.

CONNECTIVE LIGAMENTS

The oesophagus and upper part of the stomach are held to the diaphragm by the phrenicoesophageal ligaments.

The muscular tube of the oesophagus becomes continuous with the stomach just below the diaphragm. This is where the oesophageal contents enter the stomach.

These ligaments are extensions of the fascia, a connective tissue that covers the diaphragm's surface.

PHYSIOLOGICAL SPHINCTER

There is no identifiable valve at the top of the stomach to control the passage of food. However, the surrounding muscle fibres of the diaphragm act to keep the tube closed except when a bolus (swallowed mass) of food passes through. This is referred to as the physiological oesophageal sphincter, through which the oesophagus passes.

Blood supply of the stomach

The stomach has a profuse blood supply, which comes from the various branches of the coeliac trunk.

The vessels that supply the stomach are:
■ Left gastric artery – a branch of the coeliac trunk
■ Right gastric artery – usually arises from the hepatic artery (a branch of the coeliac trunk)
■ Right gastroepiploic artery – arises from the gastroduodenal branch of the hepatic artery
■ Left gastroepiploic artery – arises from the splenic artery
■ Short gastric arteries – arise from the splenic artery.

VEINS AND LYMPHATICS
The gastric veins run alongside the various gastric arteries. Blood from the stomach is drained ultimately into the portal venous system, which takes blood through the liver before returning it to the heart.

Lymph collected from the stomach walls drains through lymphatic vessels into the many lymph nodes which lie in groups along the lesser and greater curvature. It is then transported to the coeliac lymph nodes.

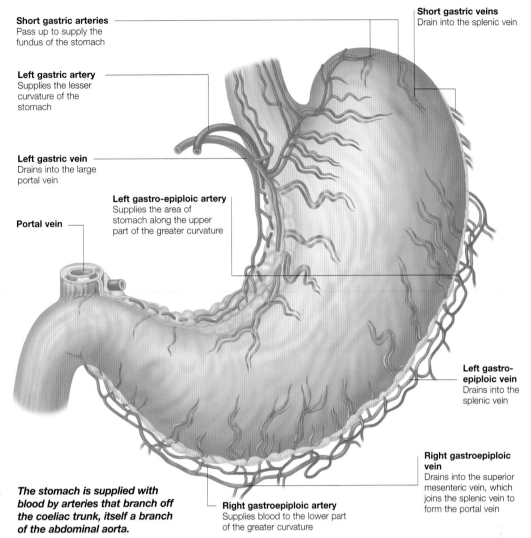

Short gastric arteries
Pass up to supply the fundus of the stomach

Left gastric artery
Supplies the lesser curvature of the stomach

Left gastric vein
Drains into the large portal vein

Portal vein

Left gastro-epiploic artery
Supplies the area of stomach along the upper part of the greater curvature

Short gastric veins
Drain into the splenic vein

Left gastro-epiploic vein
Drains into the splenic vein

Right gastroepiploic vein
Drains into the superior mesenteric vein, which joins the splenic vein to form the portal vein

Right gastroepiploic artery
Supplies blood to the lower part of the greater curvature

The stomach is supplied with blood by arteries that branch off the coeliac trunk, itself a branch of the abdominal aorta.

Shape and position of the stomach

The stomach can expand greatly to accept food. Since it is fixed only at its upper and lower ends, it can vary considerably in its position, size and shape.

Normal

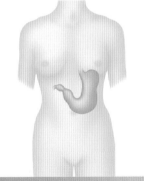

The normal stomach is an elongated pouch whose size and shape will vary with the position of the body and the degree of filling. This can also be affected by other abdominal contents, such as the fetus in pregnancy.

Active

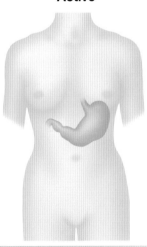

When the stomach is active, and its muscular tone high, it will lie higher and more horizontally. This is common in short, stout people. When there is less muscle activity the stomach may descend into a long 'J' shape.

Stretched

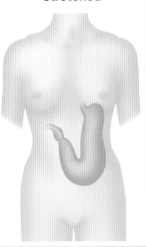

The stomach can hold up to three litres of food and may extend down to below the umbilicus after a particularly large meal. The stomach can become permanently stretched as a result of overeating.

Pregnant

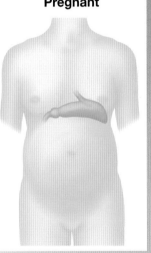

A heavily pregnant uterus will push the stomach up to a more horizontal position and even affect its ability to fill. This explains why pregnant women tend to eat little and often, and are more prone to heartburn.

Small intestine

The small intestine extends from the stomach to the junction with the large intestine. It is made up of three parts, and is the main site in the body where food is digested and absorbed.

The small intestine is the main site of digestion and absorption of food. It is about seven metres in length in adults and extends from the stomach to the junction with the large intestine. It is divided into three parts: the duodenum, the jejunum and the ileum.

THE DUODENUM

The duodenum is the first part of the small intestine and the shortest (about 25 cm in length). It receives the contents of the stomach with each wave of contraction of the stomach walls. In the duodenum the contents are mixed with secretions from the duodenal walls, pancreas and gall bladder.

The duodenum cannot move, but it is fixed in place behind the peritoneum, the sheet of connective tissue that lines the abdominal cavity.

BLOOD SUPPLY OF THE DUODENUM

The duodenum receives arterial blood from various branches off the aorta. These, in turn, give off small branches that provide each part of the duodenum with a rich supply of blood. Venous blood supply mirrors the arterial pattern, returning blood to the hepatic portal venous system.

The duodenum is the first part of the small intestine. It is roughly C-shaped and is made up of four parts.

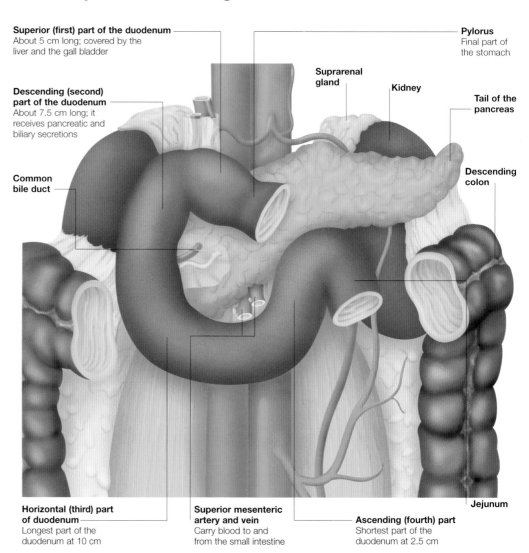

Superior (first) part of the duodenum
About 5 cm long; covered by the liver and the gall bladder

Pylorus
Final part of the stomach

Suprarenal gland

Kidney

Tail of the pancreas

Descending (second) part of the duodenum
About 7.5 cm long; it receives pancreatic and biliary secretions

Descending colon

Common bile duct

Horizontal (third) part of duodenum
Longest part of the duodenum at 10 cm

Superior mesenteric artery and vein
Carry blood to and from the small intestine

Ascending (fourth) part
Shortest part of the duodenum at 2.5 cm

Jejunum

Structure of the duodenum

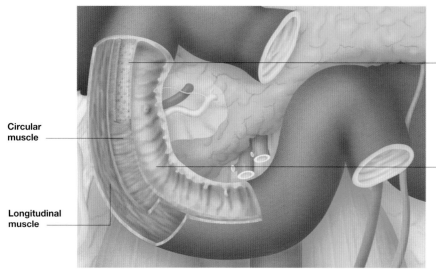

Circular muscle

Longitudinal muscle

Submucosa
Brunner's glands are embedded within, which secrete alkaline-rich mucus

Plicae
Deep folds of mucous tissue

The duodenum's walls have two layers of muscle fibres, one circular and one longitudinal. The mucosa, or lining, of the duodenum is particularly thick. It contains numerous glands, (Brunner's glands), which secrete a thick alkaline fluid that helps to counteract the acidic nature of the contents that have reached the duodenum from the stomach.

The mucosa in the first part of the duodenum is smooth, but thereafter it is thrown into deep, permanent folds of tissue, known as plicae.

There are two layers of muscle fibres in the duodenum. Together, they produce the waves of contraction known as peristalsis.

The jejunum and ileum

The jejunum and the ileum together form the longest part of the small intestine. Unlike the duodenum, they can move within the abdomen.

The jejunum and the ileum comprise the longest part of the small intestine. They are surrounded and supported by a fan-shaped fold of the peritoneum – the mesentery – which allows them to move within the abdominal cavity. The mesentery is 15 cm long.

BLOOD SUPPLY
The jejunum and the ileum receive their arterial blood supply from 15–18 branches of the superior mesenteric artery. These branches anastomose (join) to form arches, called arterial arcades. Straight arteries pass out from the arterial arcades to supply all parts of the small intestine. Venous blood from the jejunum and ileum enters the superior mesenteric vein. This vein lies alongside the superior mesenteric artery and drains into the hepatic portal venous system.

ROLE OF LYMPH IN DIGESTION
Fat is absorbed from the contents of the small intestine into specialized lymphatic vessels, known as lacteals, which are found within the mucosa. The milky lymphatic fluid produced by this absorption enters lymphatic plexuses (networks of lymphatic vessels) within the walls of the intestine. The fluid is then carried to special nodes called mesenteric lymph nodes.

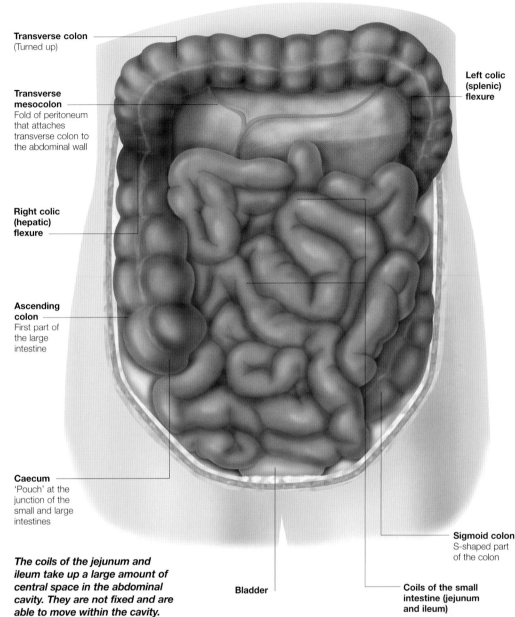

Transverse colon (Turned up)

Transverse mesocolon
Fold of peritoneum that attaches transverse colon to the abdominal wall

Right colic (hepatic) flexure

Ascending colon
First part of the large intestine

Caecum
'Pouch' at the junction of the small and large intestines

Left colic (splenic) flexure

Sigmoid colon
S-shaped part of the colon

Bladder

Coils of the small intestine (jejunum and ileum)

The coils of the jejunum and ileum take up a large amount of central space in the abdominal cavity. They are not fixed and are able to move within the cavity.

Differences between the jejunum and the ileum

Jejunum

Mesentery

Straight arteries

Anastomotic loop of arteries

Ileum

Anastomotic loops of arteries

Mesentery

Lymphoid nodules (Peyer's patches)

Straight arteries

There are many structural differences between the jejunum and the ileum. There are differences in:

■ Plicae – the walls of the jejunum are a deeper red and thicker than those of the ileum; the thickness is due to the presence of numerous plicae. These plicae help the lining of the jejunum to absorb nutrients by increasing its surface area and making the jejunum's contents travel more slowly.

There are a number of structural differences between the jejunum and the ileum. The transition from the jejunum to the ileum occurs gradually.

■ The mesentery – the arterial arcades of the jejunum are made up by a few large loops, which pass only infrequent straight branches out to the intestinal wall, while those of the ileum have many short loops with numerous straight branches.
■ Fat deposits – the jejunum has less fat lying near the root of the mesentery than the ileum, which has a much greater amount of fat which is distributed throughout the mesentery.
■ Lymphoid tissue – the lower ileum has many areas of lymphoid tissue (called Peyer's patches) at its lower end, while the jejunum has only a few solitary lymphoid nodules.

257

How digestion begins

Digestion involves the movement of food through the body's alimentary canal so that nutrients can be absorbed. The first stage is the ingestion of food from the mouth into the alimentary canal.

Digestion is the process by which the complex chemicals in food are broken down into simpler chemicals that can be absorbed into the body. It takes place in the alimentary canal, which is made up of the mouth, oesophagus, stomach, small and large intestine, and rectum.

The alimentary canal – also known as the gastrointestinal tract – contains the structures and associated organs concerned with digestion. It runs from the mouth to the anus, absorbing nutrients and expelling waste material.

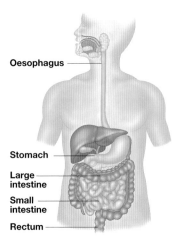

Oesophagus

Stomach

Large intestine

Small intestine

Rectum

CHEWING

Digestion begins in the mouth, where chewing breaks down the food into smaller pieces and mixes it with saliva. The tongue, a muscular organ capable of a variety of movements, has two main functions in this process. Firstly, it moves the food around the mouth, acting with muscles in the neck and jaws, presenting it to the teeth for chewing.

It is also concerned with taste: its surface is covered with thousands of papillae (nipple-shaped protuberances) that increase the surface area coming into contact with food.

SWALLOWING

The first stage of swallowing is under voluntary control. When chewing is complete, the tongue pushes up against the hard palate, and food is forced to the back of the mouth where it is formed into a soft mass (bolus).

The bolus is forced into the pharynx, where swallowing is a reflex action. The tongue prevents the food from re-entering the mouth, and the soft palate moves upwards to close off the nasal cavity. The epiglottis then closes the trachea, and pharynx muscles squeeze the bolus into the oesophagus.

The action of swallowing

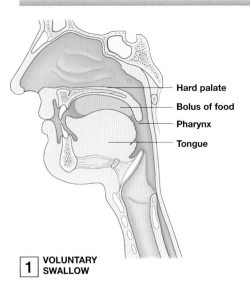

Hard palate

Bolus of food

Pharynx

Tongue

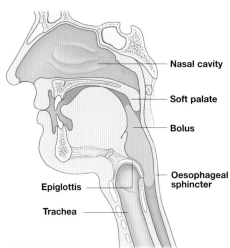

Nasal cavity

Soft palate

Bolus

Oesophageal sphincter

Epiglottis

Trachea

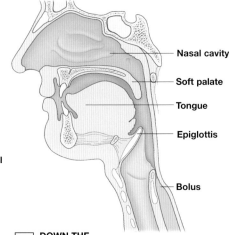

Nasal cavity

Soft palate

Tongue

Epiglottis

Bolus

1 VOLUNTARY SWALLOW

In the voluntary stage of swallowing, the tongue rises up towards the hard palate. This forces the bolus into the pharynx (throat).

2 THROUGH THE THROAT

As the bolus passes through the pharynx, the nasal cavity and the trachea are both closed off. The upper oesophageal sphincter relaxes.

3 DOWN THE OESOPHAGUS

Once the bolus has passed through, the sphincter contracts, forcing the bolus down the oesophagus towards the stomach.

What saliva does

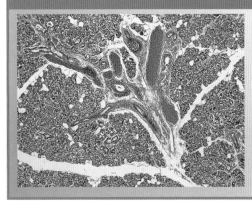

Saliva is a watery secretion produced by the salivary glands. There are three pairs of salivary glands situated in the face and neck, and many smaller ones in the tongue and lining of the mouth.

Saliva contains mucus, which surrounds the chewed pieces of food in the mouth. This lubricates them, assisting their passage down the oesophagus

The structure of a salivary gland situated beneath the base of the tongue is shown in this light micrograph. The mauve regions contain the excretory ducts.

when they are swallowed. It also contains a chemical called lysozyme, which acts as a disinfectant, and an enzyme called ptyalin that begins the process of digesting certain starches by splitting them into disaccharide sugars, such as dextrose and maltose.

Saliva is secreted constantly (about 1.7 litres each day), but the rate of flow can be altered by nervous stimulation. For example, more saliva is produced when we smell food and when food is in the mouth, while nervousness may cause a decrease – the 'dry mouth of fear'.

Moving food down to the stomach

Swallowed food – in the form of a bolus – passes down the
oesophagus and into the stomach. Here, it is temporarily stored
while the process of chemical breakdown begins.

Oesophagus

The oesophagus is an elastic, muscular tube about 25 cm long and lined with a mucous membrane that allows food to pass through easily. Its outer wall contains longitudinal and circular muscles that enable a process known as peristalsis to occur, in which waves of contractions pass down the tube.

Successive muscular contractions (peristalsis) propel the bolus along the oesophagus and into the stomach.

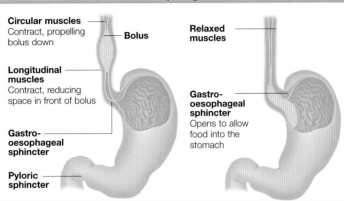

Circular muscles
Contract, propelling bolus down

Bolus

Longitudinal muscles
Contract, reducing space in front of bolus

Gastro-oesophageal sphincter

Pyloric sphincter

Relaxed muscles

Gastro-oesophageal sphincter
Opens to allow food into the stomach

PERISTALSIS

The presence of a bolus of food automatically triggers peristalsis, and the bolus is carried progressively towards the stomach by the contractions.

The contents of the stomach are normally prevented from returning upwards by the muscular walls of the oesophagus. This effectively forms a sphincter at the lower end, although its structure is not noticeably different from the rest of the oesophageal wall.

Stomach

The stomach is a muscular bag situated in the upper part of the abdomen. It consists of four regions: the cardia is the part that immediately adjoins the oesophagus; the fundus is the upper, dome-shaped part; the corpus, or body, is the main central part; and the antrum is the lower third. At its lower end, the stomach is separated from the small intestine by the pyloric sphincter. This opens at intervals to allow some of the contents through to the intestine.

MUSCULAR WALL

The function of the stomach is to act as a reservoir for food and to begin the process of digesting proteins and fats. The walls are made up of muscles that run up and down, transversely and

diagonally across. Rhythmical contractions of these muscles mix the food with gastric juices to form a thick, creamy, acid fluid known as chyme. On average, the stomach contains 1–1.5 litres of chyme, but it can expand to hold much more.

When full, the stomach is shaped like a boxing glove, about 25–30 cm long and with a diameter of 10–12 cm at its widest point. When empty, its walls contract, developing internal folds, or rugae, of the lining mucosa. In this state, it is shaped more like a letter 'J'.

The wall of the stomach is made up of a muscular layer, a layer of connective submucosa and a lining of mucosa containing millions of gastric pits.

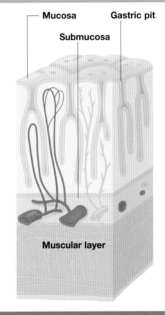

Mucosa

Gastric pit

Submucosa

Muscular layer

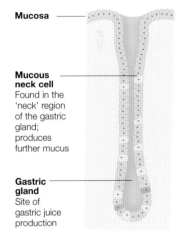

Mucosa

Mucous neck cell
Found in the 'neck' region of the gastric gland; produces further mucus

Gastric gland
Site of gastric juice production

A gastric pit leads into a gastric gland of which there are three kinds. These secrete chemicals that form the gastric juices.

Digestive juices

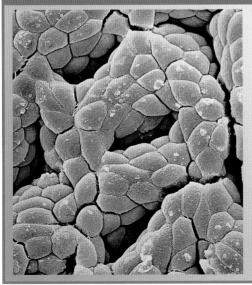

A false-colour micrograph reveals the complex structure of the stomach lining. The cells on the surface (green) secrete mucus; between them are the deeper 'pits' containing the gastric glands.

Embedded in the wall of the stomach are glands which contain a variety of secretory cells that together produce the gastric juices. Hydrochloric acid is produced by parietal cells, which are found mostly in the body and fundus of the stomach. Hydrochloric acid enables the gastric enzyme pepsin to work, and helps to sterilize the food by killing most types of bacteria and other micro-organisms.

Production of hydrochloric acid is stimulated by the hormone gastrin, which is secreted by glands in the antrum, the lower part of the stomach. Gastrin is then absorbed into the body of the stomach and

carried to parietal cells in the blood.
Gastric juices contain three enzymes:
■ Renin: coagulates milk and is more important in infants than in adults
■ Pepsin: begins the digestion of proteins by splitting them into short chain molecules known as peptides
■ Gastric lipase: starts to convert fats to fatty acids and glycerol.
Another secretion, known as 'intrinsic factor', enables the body to absorb vitamin B_{12}, a substance vital for the healthy function of most body tissue.

The contents of the stomach are acidic enough to dissolve a razor blade. To prevent the stomach wall from being digested, it is protected by a layer of alkaline mucus. In addition, the cells of the stomach lining are replaced continuously at the rate of half a million each minute, so that the stomach effectively has a new lining every three days.

How food is absorbed

Digestion begins in the mouth and stomach, but it is in the small intestine that most of the digestive processes take place. This part of the digestive tract is divided into three sections: the duodenum, jejunum and ileum.

The total length of the small intestine is 6.5 m. The duodenum is about 25 cm long, and it is here that material from the stomach is mixed with digestive juices. The jejunum is about 2.5 m long and merges with the ileum, which makes up the rest of the small intestine. The division between them is gradual, but the jejunum has a thicker wall and a larger diameter (about 3.8 cm).

Food moves along the bowel by peristalsis (muscular contraction) and the process of digestion continues throughout the small intestine. The main function of the jejunum and ileum is to absorb the products of digestion into the body.

DIGESTIVE JUICES

The digestive juices of the duodenum contain the alkali sodium bicarbonate, which neutralizes the acid produced in the stomach and provides an alkaline environment that allows the intestinal enzymes to work.

The digestive juices of the duodenum have two sources. Firstly, the glands in the duodenal wall produce the enzymes maltase, sucrase, enterokinase and erepsin.

The second source is the pancreas, which in addition to its endocrine function, produces three digestive enzymes: lipase, amylase and trypsinogen. Together, these enzymes continue the digestion of proteins, sugars and fats.

DIGESTION OF PROTEINS, FATS AND CARBOHYDRATES

Some proteins are broken down into peptides (small chains of amino acids, the building blocks of protein) in the stomach. In the small intestine, enterokinase activates pancreatic trypsin. This continues the process of protein digestion by breaking down both proteins and peptides into amino acids. Duodenal erepsin converts peptides into amino acids.

The digestion of fats is aided by salts present in the greenish mixture called bile, produced by the liver and stored in the gall bladder. Bile enters the duodenum via the bile duct. Bile salts emulsify the fats, producing small globules that present a greater surface area to the enzyme lipase, which converts fats into fatty acids and glycerol.

Any starch not already acted on by the ptyalin in saliva is now converted into the sugar maltose by the pancreatic enzyme amylase. Maltase continues the process, breaking down maltose into glucose. Sucrase converts sucrose into glucose and fructose.

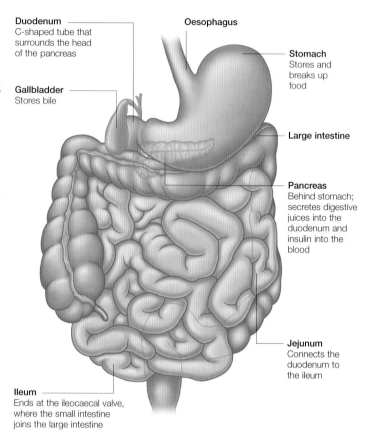

Duodenum
C-shaped tube that surrounds the head of the pancreas

Oesophagus

Gallbladder
Stores bile

Stomach
Stores and breaks up food

Large intestine

Pancreas
Behind stomach; secretes digestive juices into the duodenum and insulin into the blood

Jejunum
Connects the duodenum to the ileum

Ileum
Ends at the ileocaecal valve, where the small intestine joins the large intestine

The small intestine starts with the duodenum, which receives bile from the gall bladder and secretions from the pancreas. The intestine then continues through the jejunum and on to the ileum.

How nutrients are absorbed

The lining of the jejenum and ileum is the main absorptive surface for the products of digestion. The total volume of fluid that is absorbed by the intestines every day amounts to about 9 litres. About 7.5 litres of this is absorbed by the small intestine.

The inner surfaces of the jejunum and ileum are covered with small finger-like projections called villi, which protrude about 1 mm into the centre of the intestinal tube. The purpose of these specially adapted structures is to greatly increase the surface area over which

absorption can take place.

The walls of each villus are formed by long epithelial cells. Inside each villus is a network of small capillaries and a single lacteal – a blind-ending tube connected to the body's lymphatic system.

The epithelial cells absorb the products of digestion together with litres of water and pass the sugars and amino acids into the blood stream. Fatty acids and glycerol are converted by the epithelial cells back into fats, which form a fine whitish emulsion that passes directly into the lacteals.

Villus

Goblet cells
Secrete mucus

Artery

Vein

Intestinal crypts
Lining cells here secrete intestinal juice – a carrier for nutrients

Lymph vessel

A section through the mucosa that lines the wall of the small intestine reveals the structure of the villi.

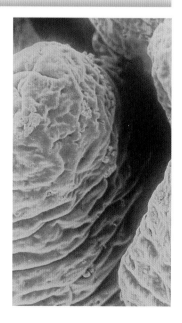

The surface area of the small intestine is increased by villi. In this micrograph, food particles are coloured green.

The liver's role in digestion

Although not actually a part of the digestive tract, the liver is vital for the digestion of food, along with the pancreas and gall bladder. The liver is the body's chemical unit for the processing of the products of digestion.

The products of digestion are processed by the liver. Chemical processing takes place in the liver cells, or hepatocytes, which line blood-filled spaces, or sinusoids, inside the liver.

These cells carry out the liver's several important functions, including a regulatory role in the maintenance of glucose (sugar) in the blood.

After eating, the blood contains a large amount of glucose. Blood pumped from the intestine arrives in the liver through the hepatic portal vein, and cells within the liver remove excess glucose from the blood and store it in the form of glycogen. As glucose is used elsewhere in the body, and the blood sugar level falls, the liver gradually reconverts glycogen back into glucose.

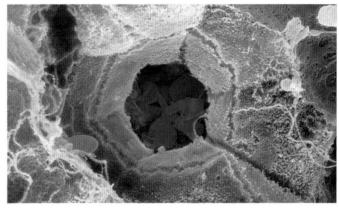

AMINO ACIDS

The amino acids produced during digestion cannot be stored in the body. Some are converted immediately into proteins, a process that occurs in most body cells, but those that are not required are broken down in the liver by a process called deamination. The nitrogen they contain is used to make ammonia. This is immediately converted into urea and transported in the blood to the kidneys to be excreted.

LIVER FUNCTION

The liver also manufactures blood proteins, such as fibrinogen, and stores iron for use in the manufacture of the red blood cell pigment haemoglobin. It also breaks down the haemoglobin of worn-out red blood cells, producing a

This false-colour electron micrograph shows one of the functional units within the liver. Hepatic cells (brown) surround sinusoid channels through which blood cells (red, at centre) flow to a central vein.

substance called bile. This is removed via the bile canaliculi and the bile duct and then stored in the gall bladder.

The liver is very versatile: if certain types of food material are in short supply, the liver can convert certain kinds of food into others. Carbohydrates can be converted into fats, such as cholesterol, for storage, and some amino acids may be converted into carbohydrates or fats.

The liver also deals with toxins ingested by the body (such as alcohol), breaking them down to render them less harmful.

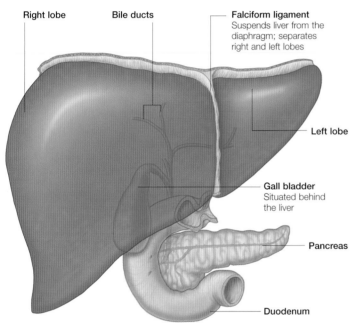

Right lobe   **Bile ducts**

Falciform ligament
Suspends liver from the diaphragm; separates right and left lobes

Left lobe

Gall bladder
Situated behind the liver

Pancreas

Duodenum

The liver and associated structures are situated in the right of the abdominal cavity.

How gallstones form

Gallstones are solid masses that form in the gall bladder and bile duct. Single stones may be very small or they can grow to the size of a hen's egg.

Fifteen per cent of gallstones are formed by the crystallization of bile pigment salts. This is often associated with the excessive destruction of red blood cells.

Eighty per cent of gallstones are composed of cholesterol. An excess of cholesterol in the blood compared to the amount of bile available to suspend it leads to the excess cholesterol crystallizing in the gall bladder.

Gallstones are found in about 30 per cent of the adult population in Europe, and they are more common in females.

A gallstone is visible as the elongated, pitted mass on this coloured X-ray. It has developed because of an upset in the chemical composition of bile.

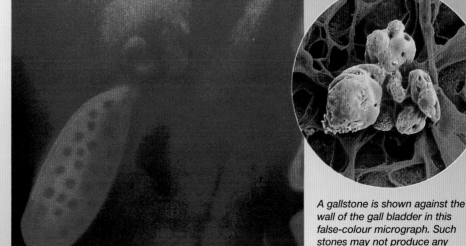

A gallstone is shown against the wall of the gall bladder in this false-colour micrograph. Such stones may not produce any symptoms until they become stuck in the bile duct.

How the body uses carbohydrates

Carbohydrates, also known as saccharides, are used in the body as a fuel source, an energy store and as building blocks for more complex molecules.

Carbohydrates are made entirely from carbon, hydrogen and oxygen atoms. They are classified, according to their size, into three main groups: monosaccharides, disaccharides and polysaccharides.

MONOSACCHARIDES

The most common monosaccharides are fructose, galactose and glucose. Of these, glucose is the most important because the body's cells are unable to metabolize directly any other saccharides; they must first be converted into glucose before they can be broken down to release energy. Thus the level of free glucose in the blood is very important.

Monosaccharides

Glucose

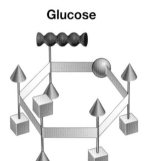

Glucose is the most important carbohydrate in the body. Free glucose is not found in many foods; it is obtained by breaking down complex saccharides.

Galactose

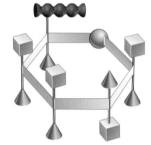

Although galactose has a similar chemical structure to glucose, cells cannot metabolize galactose, so it is converted into glucose in the liver.

Fructose

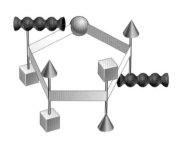

Fructose is found in sweet fruits and fruit juices. When fructose is chemically joined to glucose, the disaccharide sucrose (or 'table sugar') is produced.

Disaccharides – ingestible sugars

Disaccharides consist of two monosaccharide molecules joined together. For example, lactose, a disaccharide found in milk, consists of a glucose and a galactose molecule joined together. Lactose is the only disaccharide made by the body. The other two common disaccharides are sucrose ('table sugar') and maltose ('malt sugar').

Lactose is a disaccharide that is found in milk and dairy products. Some people cannot digest lactose because they lack the enzyme lactase.

Lactose

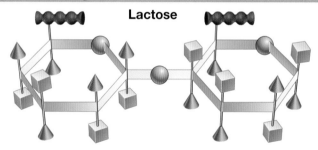

Saccharides are also important components of many complex molecules including cartilage and bone. They are also present in small quantities within cell membranes.

Disaccharides are made up of two monosaccharides joined together. For example, lactose is made by the joining of a molecule of glucose to a molecule of galactose.

Polysaccharides – energy stores

Polysaccharides are long branched chains of monosaccharides joined together. Their large size makes them relatively insoluble in water and also means that they remain trapped inside the cell. This makes them ideal energy stores.

Two polysaccharides, starch and glycogen, are especially important. Both are made of long chains of glucose molecules:

■ Starch is the major long-term store of carbohydrates for plants and so is an important part of the human diet

■ Glycogen, in contrast, is synthesized and used as a carbohydrate store by animals. It is mainly found in skeletal muscle and liver cells. When blood glucose levels fall, glycogen is rapidly converted into glucose.

Potatoes contain a large quantity of starch, a polysaccharide made by a number of glucose molecules joining together.

How the body uses lipids

Lipids are a large group of organic molecules (i.e. they contain carbon) that are insoluble in water, but are soluble in alcohol. There are three main groups of lipids: triglycerides, phospholipids and steroids.

Triglycerides – long-term energy stores

Triglycerides are made up of one glycerol molecule (an alcohol) attached to three long chains of fatty acids. The glycerol backbone is the same in all triglycerides, but the composition of the fatty acid chains varies, creating a large number of different triglycerides.

Fatty acids yield a large amount of energy when they are metabolized inside the cell. This, together with the fact that they are insoluble in water, makes them an excellent energy store. Indeed, a large proportion of the body's long-term energy requirements are catered for by fatty acids.

SATURATED AND UNSATURATED FATS

The carbon atoms of saturated fats have a full complement of hydrogen atoms attached (hence they are said to be 'saturated') and are common in animal fats. In contrast, the carbon atoms in unsaturated fats can bind to additional hydrogen atoms. Exactly how many hydrogen atoms they can bind to determines whether they are mono- or polyunsaturated fats.

Olive oil is rich in monounsaturated fats. In contrast, sunflower oil contains mainly polyunsaturated fats.

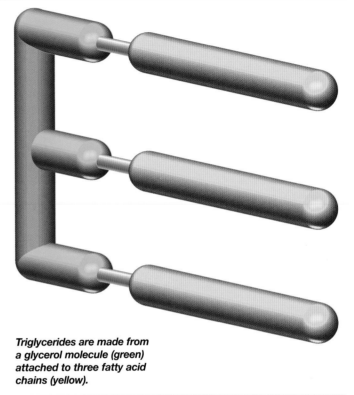

Triglycerides are made from a glycerol molecule (green) attached to three fatty acid chains (yellow).

Phospholipids – building blocks of cell membranes

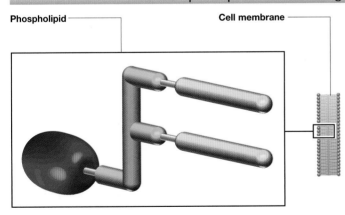

Phospholipid   Cell membrane

Phospholipids are similar to triglycerides, in that they have a glycerol backbone. However, unlike triglycerides, phospholipids have only two fatty acid chains; instead of a third chain, they have a phosphorus-containing head.

The 'tail' of a phospholipid (made up of the two fatty acid chains) has no electrical charge

Phospholipids are made up of a phosphorus-containing head (red), a glycerol backbone (green) and two fatty acid chains (yellow).

and so does not mix with water (water is electrically charged), whereas the phosphorus-containing 'head' attracts water because it is electrically charged.

It is this property that makes phospholipids ideal building blocks for the cell membrane. Cell membranes are made of two layers of phospholipid molecules; the 'water-hating' (hydrophobic) tails point towards each other, whereas the 'water-loving' (hydrophilic) heads point towards water – which is present both inside and outside the cell.

Steroids

Steroids have a very different structure from both triglycerides and phospholipids, though they are classified as lipids since they are fat soluble. Probably the most important steroid in the human body is cholesterol as it is the precursor for many of the steroid hormones which are essential for human development and long-term health. Other steroids, for example the sex hormones, are present in very small amounts, but are still essential.

Anabolic steroids are derivatives of the sex hormone testosterone. One of their physiological effects is to increase muscle mass.

Other lipid-based molecules

Lipids are also major components of three other important groups of molecules.

Fat-soluble vitamins
Fat-soluble vitamins include vitamins A, D, E and K. Since these vitamins can only be absorbed after they have bound to ingested lipids, anything which interferes with fat absorption (such as cystic fibrosis) also prevents fat-soluble vitamins from being absorbed.

Eicosanoids
These include prostaglandins and leukotrienes, which are both involved in inflammation, and thromboxanes, which cause blood vessels to constrict.

Lipoproteins
These chemicals transport fatty acids and cholesterol in the bloodstream. The two main groups are high-density lipoproteins (HDL) and low-density lipoproteins (LDL).

How proteins work

Proteins are vital to the structure, growth and metabolism of humans and all living organisms. Their large molecules are built up from smaller units – amino acids – and often have very complex structures.

Proteins of various kinds play important roles in the human body. Structural proteins include collagen in connective tissue; keratin in skin; actin and myosin in muscles; and tubulin in cells. Proteins in cell membranes act as carriers, transporting chemicals in to and out of cells.

Other proteins include enzymes, which promote chemical reactions within cells, essential hormones and antibodies which play an important role in defence against disease. Blood proteins include haemoglobin, albumin and a series of proteins needed to ensure that the blood clots properly when injury occurs.

AMINO ACID BUILDING BLOCKS

Like all organic chemicals, protein molecules are made up largely of carbon, hydrogen and oxygen atoms. However, proteins also contain nitrogen, and, in many cases, sulphur as well.

The basic units of protein molecules are amino acids. There are only 20 types of amino acids, but they can be joined together in an enormous number of different combinations. Some amino acids can be synthesized in the body; others have to be obtained from proteins in food. In order to maintain health, a human must eat at least 30 grams of protein each day.

An amino acid molecule consists of a core chain of

This model of collagen shows individual molecules as spheres. Collagen is found in connective tissues such as bone and skin.

carbon atoms. There are two different chemical groups at the ends of the molecule: at one end an amino group; at the other end a carboxylic acid group. The amino group of one amino acid can bond chemically with the carboxylic acid group of an adjacent amino acid, releasing a molecule of water in the process – this facilitates the formation of long chains of amino acids.

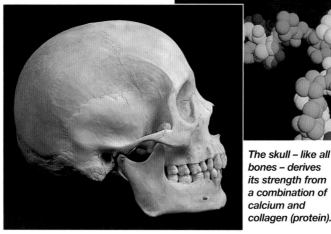

The skull – like all bones – derives its strength from a combination of calcium and collagen (protein).

The structure of an amino acid makes it soluble in water and renders it amphoteric (it can act as either an acid or an alkali in solution). This allows it to resist changes in acidity and alkalinity, and to act as a buffer (a regulator of pH), playing an important role in homeostasis – the maintenance of a constant internal environment.

The template for the manufacture of proteins is DNA. The cell synthesizes amino acid chains (the basis of proteins) on ribosomes (structures in the cell) using a DNA related molecule – this is called RNA.

PRIMARY STRUCTURE

The sequence of amino acids assembled on the ribosome gives the protein its primary structure, like the pattern of beads on a string. This sequence is dictated by the DNA sequence, and forms the 'backbone' of the protein molecule.

Protein denaturation

Under normal circumstances, proteins are relatively stable. Their activity is dependent on their 3-D structure and the bonds holding the molecule together. However,

The white of an egg is made primarily from albumin. When cooked, it changes from clear to opaque white because the protein has denatured.

these bonds are sensitive to factors such as acidity and heat.

When proteins lose their 3-D shape, they are said to be denatured. Often this can be reversed and proteins will regain their normal shape when conditions are restored. However, if the change in pH or temperature is extreme, they become irreversibly denatured.

Amino acids have a common basic structure. Binding of amino groups and acid groups between adjacent molecules allows the formation of very long chains. The side chain (box) varies between different amino acids.

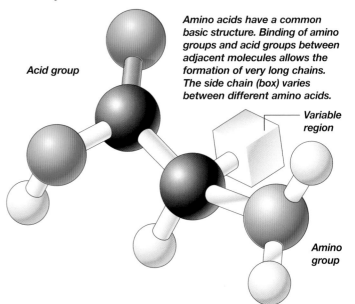

Acid group

Variable region

Amino group

How proteins fold

The sequence of amino acids in a protein determines its final three-dimensional shape. The resulting shapes and folds will confer a protein's particular properties.

α-helix

β-helix

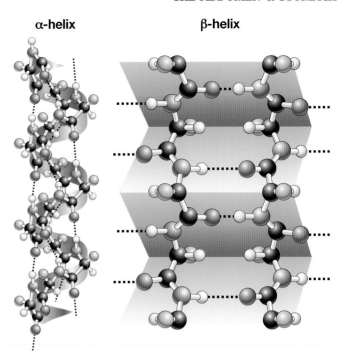

SECONDARY STRUCTURE

When a long polypeptide chain (string of amino acids) is formed, it rarely exists as a simple long chain, but tends to arrange itself in a complex shape or pattern. This is a result of a type of chemical bond called a hydrogen bond, which is relatively weak but strong enough to pull the chain into a particular shape. Hydrogen atoms form these bonds with certain other atoms in the structure and two distinct shapes tend to be created.

The most common pattern is an α-helix, a right-handed spiral. This shape is the result of

Depending on their sequence of amino acids, proteins can form either long helical shapes (α-helix,) or pleated sheets (β-sheet).

hydrogen bonds between approximately every fourth amino acid. The other is the ß-pleated sheet, a flat structure in which hydrogen bonds form between two polypeptide chains running parallel to each other, similar to an accordion. In some proteins both kinds of secondary structure can be seen in different places along the length of the chain.

Tertiary and quaternary structures

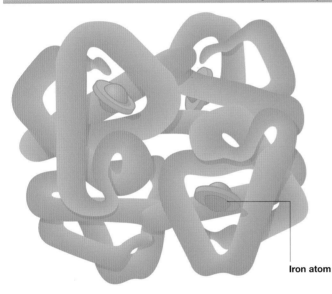

Iron atom

A protein with a very long polypeptide chain tends to have an additional tertiary structure superimposed on the secondary structure. This occurs when spirals, sheets and other bends in the molecule fold in on each other, producing a ball-like or globular shape. The structure is maintained by attractive forces resulting from the proximity of chemical groups to each other, especially those in amino acids that contain sulphur atoms.

Finally, the protein may acquire a quaternary structure

Haemoglobin, pictured, forms a complicated globular shape. This shape allows the binding of iron atoms, which confer its oxygen-binding properties.

in which two or more polypeptide chains that already have a complex tertiary structure become bonded to one another to form an even more complicated molecule. This may be further enhanced by the binding of non-protein groups, for example an iron atom, in the blood protein haemoglobin. This has a quaternary structure formed from four globular polypeptide chains, each of which incorporates an iron-containing 'haem' group.

The overall configuration of a protein with a tertiary or quaternary structure is very specific to the protein and is determined by its primary structure; that is, the sequence of amino acids.

Fibrous and globular proteins

Proteins are classified in two groups according to their general shape. Fibrous or structural proteins resemble the strands of a rope. They are stable and provide the body's tissue with strength and support. Most fibrous proteins have a secondary structure but some also have a

quaternary structure. Collagen, found in all connective tissues, is a triple helix of three polypeptide chains. Other fibrous proteins include keratin, elastin and actin.

Globular, or functional, proteins are more chemically active and play roles in the body's chemical processes. They are soluble in water and have spherical shapes with at least a tertiary structure. Enzymes are globular proteins.

Fingernails and animal horn consist of keratin. This protein provides strength and stability.

Globular proteins (X-shaped structure, right) are found in the outer membranes of cells. They control cellular entry and exit of chemicals.

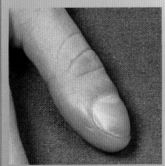

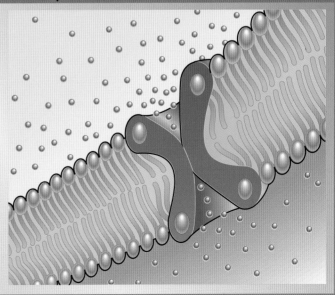

How the body uses vitamins

The body requires all of the thirteen vitamins for growth, maintenance and repair. Long-term deficiencies, often caused by intestinal disorders or alcoholism, can lead to serious disease.

Vitamins are organic (carbon-containing) compounds that are found only in living organisms (plant or animal). They are required in tiny amounts by the body for it to function effectively.

Vitamins act as catalysts: they combine with proteins to create enzymes, which in turn give rise to important chemical reactions throughout the body. Without vitamins, many of these reactions would slow down or stop altogether.

VITAMIN SOURCES

The term 'vitamin' was first used by the chemist Casimir Funk in 1912, who noted that certain diseases appeared to be linked to a lack of specific substances in the diet.

Vitamins were defined as substances absolutely necessary for life which cannot be produced by the body. Later research showed that the exception to this is vitamin D, which can be synthesized by the skin when exposed to sunlight and niacin (B_3), which can be synthesized in tiny amounts by the liver.

The two main sources of vitamins are food and drink.

VITAMIN SUPPLEMENTS

A well-balanced diet should provide all the vitamins that the body requires. However, certain people, such as those on restricted diets, pregnant or breast-feeding women or those with intestinal disorders, may require vitamin supplements to bolster their metabolism.

Vitamin supplements should not be seen as a substitute for a healthy diet; nutritionists believe the naturally occurring balance of micronutrients within food may be of great importance.

The 13 vitamins can be

divided into two main groups:
- Fat-soluble – vitamins A, D, E and K
- Water-soluble – Vitamins C and B complex.

A balanced diet provides a perfect mix of water and fat-soluble vitamins in a form the body can easily use. All 13 vitamins are essential for life.

Fat-soluble vitamins

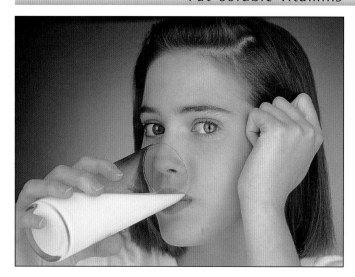

The four fat-soluble vitamins are mainly ingested in meat and dairy products. They can be stored in the body, so do not have to be consumed every day.

VITAMIN A

Vitamin A is stored mainly in the liver. It is vital for the formation and health of the skin, mucous membranes, bones and teeth, and for vision and reproduction. Vitamin A can be obtained from liver, eggs, cream or butter, or derived from beta-carotene (a pigment that occurs in leafy green vegetables and in orange fruits and vegetables).

VITAMIN D

Vitamin D is necessary for the formation of healthy bone and

The calcium found in milk can only be retained by the body in the presence of vitamin D. A deficiency of this vitamin can cause rickets.

for retention of calcium and phosphorus in the body. It is found in eggs, liver and fish oil, and is also made by the body during interaction with sunlight.

VITAMIN E

Vitamin E is found in vegetable oils, wheat germ, liver and green, leafy vegetables, and it is stored mainly in body fat. It is an antioxidant (a substance that is able to neutralize certain harmful molecules) and plays a role in the formation of red blood cells and muscle.

VITAMIN K

Vitamin K is mainly necessary for blood coagulation, by helping to form prothrombin (an enzyme required during blood clotting). Alfalfa and liver are both rich sources of vitamin K, as are leafy green vegetables, eggs and soybean oil.

Vitamin C

Vitamin C, otherwise known as ascorbate, is a water-soluble vitamin that is important in the formation and maintenance of collagen. This is a connective tissue used in the formation of bones, cartilage, muscle and blood vessels. Vitamin C also enhances the absorption of iron from vegetables and plays a role in metabolizing food.

Sources of vitamin C include most fruits (particularly citrus), green peppers, tomatoes, broccoli, potatoes and cabbage. Interestingly, all other meat-eating mammals are able to synthesize vitamin C within their bodies; humans are the only ones to rely entirely on outside sources for vitamin C.

Research has shown that vitamin C acts as an antioxidant and protects the body's cells and tissues against the harmful effects of free radicals (damaging molecules produced in metabolic reactions and by certain factors, such as disease and UV radiation).

Water-soluble vitamins

The body is largely unable to store water-soluble vitamins and consequently they should be consumed every day.

THE B VITAMINS

The large group of B vitamins includes:

■ Thiamine (B_1) – this is essential for the metabolism of carbohydrates and the proper functioning of the nervous system. Whole grain cereals, bread, red meat, eggs and brown rice are all good sources

■ Riboflavin (B_2) – this is required to complete certain metabolic reactions. It is also vital for healthy skin, mucous membranes, cornea and nerve sheaths. Riboflavin is found in

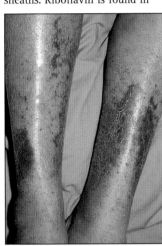

Pregnant women are advised to take folic acid supplements. This B vitamin is essential for the healthy brain and nerve development of the fetus.

meat, dairy products, whole grain products and peas

■ Niacin (B_3) – essential for the metabolism of food and maintaining healthy skin, nerves and gastrointestinal tract. It is found in protein-rich foods such as meat, fish, brewer's yeast, milk, eggs, legumes (pod vegetables), potatoes and peanuts

■ Pyridoxine (B_6) – this is vital to the metabolism of amino acids, glucose and fatty acids and for the production of red blood cells. Pyridoxine is found in many foods, so a deficiency is rare, except in alcoholics. It is present in many foods, including liver, brown rice, fish, and whole grain cereals

■ Cyanocobalamin (B_{12}) – this is a compound that functions in all cells, most significantly in the intestines, nervous system and bone marrow. It is used in the manufacture of healthy blood cells and is essential for

A severe vitamin C deficiency can result in scurvy. This disease causes subcutaneous bleeding, swollen and bleeding gums and if untreated, ultimately death.

maintaining nerve sheaths and synthesizing nucleic acids (the building blocks of DNA). Rich sources include liver, meat, eggs, and milk

■ Folic acid – this interacts with B_{12} to allow the synthesis of nucleic acids and is used in the formation of red blood cells. It is essential for brain and nerve development in the fetus. Folic acid is found in many foods,

notably yeast, liver and green vegetables

■ Pantothenic acid and biotin – these B vitamins are produced by bacteria in the intestines and are important in a number of metabolic reactions. Pantothenic acid is found in abundance in meats, legumes and whole grain cereals, and biotin is found in beef liver, eggs, brewer's yeast, peanuts and mushrooms.

Vitamin deficiencies

Vitamin	Result of deficiency	People at risk
Vitamin A	Dry skin, reduced mucus secretion, poor night vision	Those with cystic fibrosis or liver disease; alcoholics
Vitamin B_1 (thiamine)	Beriberi disease, Wernicke-Korsakoff's syndrome	Alcoholics
Vitamin B_2 (riboflavin)	Skin disorders, anaemia, light sensitivity, cracked lips, sore tongue	Those on poor diets
Vitamin B_3 (niacin)	Pellagra (sore mouth and skin, diarrhoea, dementia)	Alcoholics and transients
Vitamin B_6 (pyridoxine)	Skin disorders, depression, poor co-ordination, insomnia	Alcoholics; women on birth control pills
Vitamin B_{12} (cyanocobalamin)	Pernicious anaemia, brain disorders, mouth irritation	Strict vegetarians; elderly people (absorption decreases with age)
Vitamin B_9 (folic acid)	Folate-deficient anaemia (gastro-intestinal problems, ulcers)	Alcoholics; pregnant women
Vitamin C	Scurvy (skin and tissue haemorrhages, stiff limbs)	Elderly people on restricted diets; babies drinking only cows' milk
Vitamin D	Rickets (due to the body's subsequent inability to absorb calcium)	Babies; elderly people with low sunlight exposure
Vitamin E	None known	None known
Vitamin K	Blood clotting disorders; can affect baby during pregnancy	People with jaundice, liver cirrhosis; those on long-term antibiotics

Alcoholism

Three types of malnutrition affect alcoholics:

■ Primary malnutrition due to a decreased nutrient intake

■ Secondary malnutrition caused by digestion impairment and malabsorption of nutrients

■ Tertiary malnutrition due to an inability to convert nutrients.

In addition, alcohol itself inhibits fat absorption and with it, all fat-soluble vitamins.

Resulting deficiencies

Alcoholics typically become severely deficient in:

■ Vitamin A – even moderate alcoholic disease can cause a severe vitamin A deficiency

■ Vitamin B – alcoholics become deficient in all of the B vitamins, but particularly B_1 (thiamine), leading to Wernicke-Korsakoff's syndrome (causing disorientation, lack of memory and a tendency to invent material to fill memory blanks)

■ Vitamin B_9 (folic acid) – the most common deficiency, this causes anaemia and adversely alters the small intestine.

How the body uses minerals

Minerals are inorganic elements from the earth and make up five per cent of the total body weight. Minerals are essential to the body's function, and are required in small amounts only.

Although vitamins are essential to the functioning of the body, they cannot be assimilated without minerals. Minerals are inorganic elements, which make up 4-5 per cent of the total body weight. They are vital to mental and physical wellbeing and are essential constituents of the bones, teeth, soft tissue, blood, muscle and nerve cells.

Like vitamins, minerals act as catalysts or co-enzymes for a number of biological reactions within the body, including muscle control, the transmission of nerve impulses, hormone production, digestion and the assimilation of nutrients. The body utilizes over 80 minerals for optimum performance.

MINERAL SOURCES

Minerals originate in the earth and, as they are inorganic, they cannot be made by living systems. Plants obtain minerals from the soil, and most of the minerals in our diet come directly from plants or indirectly from animal sources.

Foods with a high mineral content include vegetables, legumes (vegetables containing pods) and milk and milk products, whereas refined foods such as cereals, bread, fats and sugary foods contain hardly any minerals.

TWO GROUPS

As long as a person eats a well-balanced diet, their body will receive all the minerals it needs to function adequately. Minerals can be categorized into two main groups: macrominerals and trace minerals.

Vegetables, legumes and fruits have a high mineral content. Eaten as part of a balanced diet, they enable the body to acquire all the minerals required.

Macrominerals

Macrominerals (from the Greek word 'macro', meaning 'large') are required by the body in larger amounts than other minerals and include:
■ Calcium – required for the development and maintenance of bones and teeth. It also contributes to the formation of cell membranes, and regulates nerve transmission and muscular contraction. About 90 per cent of calcium in the body is stored in the bones, forming a reservoir

which can be reabsorbed by blood and tissue. A deficiency of calcium can lead to bone disorders such as osteoporosis
■ Phosphorus – combines with calcium in the bones and teeth and plays a role in cell metabolism of carbohydrates, lipids and proteins
■ Potassium – the third most abundant mineral in the body. It works with sodium and chloride to maintain fluid distribution and pH balance, and plays a role

Nerve cells (shown here magnified) require magnesium in order to function well. Magnesium can be found in vegetables such as broccoli.

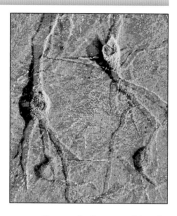

in transmitting nerve impulses, muscle contraction and regulation of heartbeat and blood pressure. Potassium participates in the process by which the blood vessel constricting effects of adrenaline are moderated, thus reducing the rise in blood pressure that occurs during stress. Potassium is also required for protein synthesis, metabolism of carbohydrates and insulin secretion by the pancreas
■ Sodium – helps to maintain fluid balance in the body. Together with potassium, sodium also helps to control muscle contraction and nerve function. Most of the sodium in the diet

comes from salt. Increased levels of sodium can cause the body to lose potassium and retain water, elevating blood pressure
■ Magnesium – plays a role in nerve and muscle function and is required for healthy bones. It helps the body to absorb calcium and protects the atrial lining of the heart from the stress of sudden blood pressure changes. Magnesium deficiency may be related to angina and an increased risk of heart attack. A lack of it has also been linked to pre-menstrual syndrome (PMS).

Sodium is found in the earth's crust. The mineral – seen here in block form in a petri dish – is essential for the body's fluid and electrolyte balance.

Trace minerals

Trace or microminerals are minerals that the body requires in only tiny amounts. Even in very small quantities, trace minerals can have a powerful effect on health. Trace minerals include the following:

■ Zinc – important in growth, appetite, development of the testicles, skin integrity, mental activity, wound healing and proper functioning of the immune system.

Zinc is a cofactor for many enzymes, and is a necessary component in a number of biological reactions, including the metabolism of carbohydrates, lipids and proteins. Zinc also plays an important role in the regulation of bone calcification

■ Copper – this mineral is indispensable to health and plays a number of important roles including: the formation of haemoglobin, absorption and

Iodine (seen in crystal form under the microscope) is an essential trace element. Its benefits have been recognized for centuries.

assimilation of iron, regulation of heart rate and blood pressure, strengthening of blood vessels, bones, tendons and nerves, and the promotion of fertility

■ Fluoride – this mineral is required for healthy teeth and bones. It helps form tooth enamel (which prevents teeth from decay) and also increases bone strength. Fluoride may be added to water supplies as well as toothpaste in order to promote healthy teeth

■ Manganese – this is essential to the formation and maintenance of bone, cartilage and connective tissue. Manganese contributes to the synthesis of proteins and genetic material and helps produce energy from food. Manganese is necessary for normal skeletal development and maintains sex hormone production

■ Chromium – works with insulin to help regulate the body's use of sugar and is essential to the metabolism of fatty acids. Supplemental chromium may be used to treat some cases of adult-onset diabetes, to reduce insulin requirements of some diabetic children, and to relieve the symptoms of hypoglycaemia

■ Selenium – this mineral is thought to stimulate metabolism and, in combination with vitamin E, acts as an antioxidant to protect cells and tissues from damage due to free radicals. Selenium also promotes immune function

■ Iodine – one of the first minerals recognized to be

essential to human health, iodine has been used for hundreds of years to treat goitres (swellings caused by enlargement of the thyroid gland).

Iodine is a constituent of several thyroid hormones and plays a role in metabolism, nerve and muscle function, nail, hair, skin and teeth condition, as well as physical and mental development. Seafoods such as shellfish and saltwater fish are rich sources of iodine. It also occurs in bread and dairy foods

Shellfish and saltwater fish are good sources of iodine. The thyroid gland needs iodine for thyroid hormones, which are essential for growth in children.

■ Iron – this mineral is essential to the formation of haemoglobin, a blood protein that transports oxygen. Iron is also an important component of myoglobin, a protein that provides oxygen to muscles during exertion. Iron deficiency may lead to anaemia.

Mineral supplements

Despite a well-balanced diet, some people may require mineral supplements. Women who suffer from excessive menstrual bleeding, for example, may benefit from taking iron supplements.

However, it is important that any such supplements are discussed with a doctor first. Because minerals are stored in bone and muscle tissue, stores of minerals can build up to toxic

Even if people are getting a balanced diet, mineral supplements may be needed. Amounts taken should be discussed with a doctor first.

levels. Toxicity risks increase when one isolated mineral is ingested without any supportive cofactor nutrients, which help the body to assimilate the mineral.

Toxic levels
Toxic levels only accumulate if massive overdoses persist for a prolonged period of time. Mildly elevated levels of any minerals, such as may occur in the consumption of contaminated water or excessive mineral supplements, can give rise to the development of adverse symptoms such as nausea, diarrhoea, dizziness, headaches and abdominal pain.

The role of fat

Fat has a number of important roles in the body,
including the provision of energy. However, an excess of fat can
lead to problems such as obesity and heart disease.

Fat is a necessary component of a healthy diet and constitutes the body's greatest source of stored energy. Fat also protects the organs, strengthens the joints, aids hormone production, and helps absorption of fat-soluble vitamins.

STORING FAT
Fat (usually in the form of triglycerides) is derived from food and stored in adipose tissue

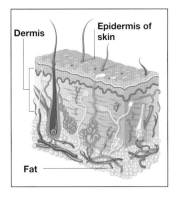

Most of the body's fat is stored in adipose tissue beneath the dermis of the skin. When no carbohydrates are available, this fat is broken down for energy.

throughout the body. Some fat is stored just above the kidneys, but most is located just beneath the dermis of the skin, in the subcutaneous layer, which has a rich supply of blood vessels.

The distribution of this fat is influenced by gender:
■ Men tend to store fat around the chest, abdomen and buttocks (producing an apple-like shape)
■ Women tend to carry fat on the breasts, hips, waist and buttocks (creating a pear-like appearance).

The difference in distribution is due to the sex hormones oestrogen and testosterone.

TYPES OF FAT
There are two types of body fat:
■ White fat – important for

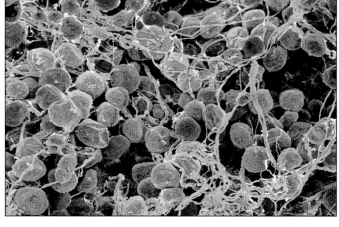

energy metabolism, insulation and also forms a protective layer for the skeleton and organs
■ Brown fat – mostly found in newborns, this is important in the production of heat.

Fat tissue is made up of fat cells known as adipocytes:
■ White fat cells are large cells

Fat that is not used up by metabolic processes is stored in fat cells (shown in brown and yellow). These rounded cells are supported by connective tissue.

with one large fat droplet
■ Brown fat cells are smaller, with many small fat droplets.

The digestion and storage of fat

In order for any food component to be used by the body, it must first be absorbed into the body cells. Fat molecules are too large to pass directly across cell membranes, so fats must first be broken down into their component parts.

ABSORBING FAT
Fats are absorbed as follows:
■ Food containing fat (mostly in the form of triglycerides) passes into the stomach and intestines
■ Bile salts produced by the liver mix with the large fat droplets in a process called emulsification. The bile breaks up the large fat droplets into several smaller droplets known as micelles. This increases the surface area of the fat droplets and speeds up their digestion
■ Meanwhile, the pancreas secretes enzymes known as lipases that attack the surface of each micelle, breaking down the fats into their component parts: glycerol and fatty acids. These can then be absorbed into the cells lining the intestine
■ Once absorbed into the intestinal cells, the fat components are reassembled into microparticles known as

chylomicrons. These have a protein coating to make the fat dissolve more easily in water
■ As the chylomicrons are too large to pass directly through capillary walls into the bloodstream, they pass first into the lymphatic system
■ The lymphatic system eventually drains into the veins

and the chylomicrons pass into the bloodstream.

STORING FAT
Once in the bloodstream, chylomicrons last only a few minutes before they are broken down again. Enzymes known as lipoprotein lipases (found in the walls of blood vessels that

supply fat tissue, muscle tissue and heart muscle) break the fats down into fatty acids.

The activity of these enzymes depends upon the body's levels of insulin (a hormone produced by the pancreas):
■ If insulin is high, the lipases will be highly active and fat will be broken down quite rapidly
■ If levels of insulin are low, the lipases will be inactive.

The resulting fatty acids can then be absorbed from the blood into fat cells, muscle cells and liver cells, where they are made into fat molecules once more and stored as fat droplets. As the body stores more fat, the number of fat cells remains the same, but the size of each fat cell gets bigger.

ABSORBING MOLECULES
It is also possible for fat cells to absorb other food molecules, such as glucose and amino acids, (derived from protein) and to convert them into fat for storage.

Fat from food must be absorbed into the body's cells to release energy. It moves through the intestine and lymphatic system before reaching the blood.

Converting fat into energy

The body is able to produce energy from fats by a process called lipolysis. This involves the break-down of fats by enzymes into glycerol and fatty acids.

The body's prime source of energy is glucose, which is usually obtained by the breakdown of carbohydrates in the diet.

However, during stamina-building exercise such as walking or cycling, the body relies on fat as a rich reserve of stored energy. Fatty acids are also used by the body for energy whenever glucose (in the form of carbohydrates) is not available.

ENERGY FROM FAT

Energy is derived from fat by the process of lipolysis (the breakdown of fats into glycerol and fatty acids). This process is activated by enzymes (lipases) in the fat cell, which are in turn controlled by various hormones, such as glucagon and adrenaline.

The resulting fatty acids are then released into the blood and are carried in the bloodstream to the liver. Once in the liver, the glycerol and fatty acids can be broken down further or converted into glucose by a multi-step process known as gluconeogenesis.

Long-distance swimming and other forms of endurance training utilize the body's fat stores as the most efficient form of muscle fuel.

Excess body fat

Most nutritionists recommend a diet that includes around 35 per cent fat. This should be unsaturated, such as olive oil, rather than saturated, such as fats derived from meat.

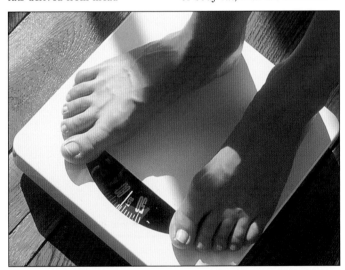

If a person consumes more fat than they metabolize, it is stored in fat reserves, causing the person to gain weight.

Obesity is defined by the level of body fat; men are classified as obese at more than 25 per cent body fat and women 32 per cent. Although fat plays a vital role, excess fat is linked to a number of health problems.

HIGH BLOOD PRESSURE

Obese people tend to have high cholesterol levels, making them more prone to atherosclerosis (in which fatty plaque deposits cause a narrowing of the arteries). This becomes life-threatening when blood vessels become so narrow that vital organs are deprived of blood.

In addition, the narrowing of blood vessels forces the heart to work harder, causing blood pressure to rise. High blood pressure carries many serious

People become obese if their body fat approaches one-third of their total weight. This can cause severe health problems, such as diabetes and heart attacks.

health risks including heart attack, kidney failure and stroke.

DIABETES

Obesity increases the risk of diabetes by disrupting the delicate balance between blood sugar, body fat and insulin.

This is because excess blood sugar is stored in the liver and other vital organs. When these organs are 'full', the excess blood sugar is converted to fat. As the fat cells themselves become full, they take in less blood sugar.

In some people the pancreas produces more and more insulin to regulate this excess sugar, which the body cannot use, and as a result the whole system becomes overwhelmed.

This poor regulation of blood sugar results in diabetes, a disease associated with long-term consequences including heart disease, kidney failure and blindness.

Heat production

When babies are first born, their bodies do not contain much fat to help insulate them and retain body heat. While they do have white fat cells, there is hardly any fat stored in them.

Newborn babies produce heat by breaking down fat molecules into fatty acids within brown cells (heat-producing cells found mainly around the central organs):
■ Instead of the fatty acids leaving the brown fat cells (as is the case with white fat cells), they remain within them
■ They are broken down further

in the mitochondria (the part of the cell that produces energy)
■ This releases energy in the form of heat. (The same process occurs in hibernating animals that have more brown fat reserves than humans).

Once the baby starts to eat more, layers of white fat develop and the brown fat disappears.

Newborn babies do not have enough stored fat to retain body heat effectively. Instead they have special heat-producing brown fat cells.

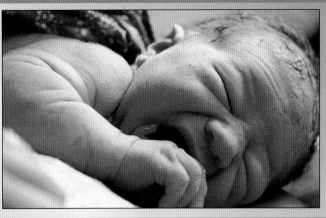

How enzymes work

Enzymes are vital for the body's chemistry. Without them, many of the chemical reactions on which life depends would not occur – such as those in which glucose is broken down to produce energy.

The life of every cell in the human body depends on the production of energy. Yet the chemical reactions that release this energy normally require a temperature in excess of 90 °C. It is enzymes that allow these reactions to occur – and so life to exist – at normal body temperatures.

The vast majority of enzymes are complex proteins – that is, strings of amino acids, which in turn, are made up of carbon, hydrogen, oxygen and nitrogen

Enzymes are utilized in a number of products, such as this pregnancy testing kit. Enzymes embedded in the test stick react with chemicals in urine, causing a colour change.

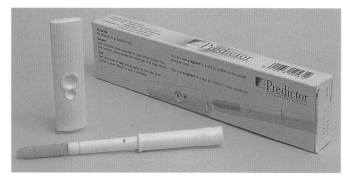

atoms, and in many cases sulphur atoms as well. Like all proteins, they are produced by the cell using DNA as a template. A few of them, however, are made up of RNA (ribonucleic acid, which along with DNA is part of the genetic code), in which case they are known as ribozymes.

WHAT ENZYMES DO

Enzymes act as catalysts to chemical reactions, speeding them up and reducing the amount of energy that they require. They may be either 'catabolic' – involved in breaking complex substances down to their simpler components – or 'anabolic', when they help reactions that build up materials by putting their components together. Other enzymes are involved in helping chemicals cross the membrane that encloses each cell.

An enzyme called sucrase, for example, is catabolic: it helps break sucrose (sugar) down into glucose and fructose, forms in which it can be more easily digested. An enzyme called carbonic anhydrase is anabolic: it helps water to combine with carbon dioxide (CO_2), a by-product of the energy-producing process within the cells, to make

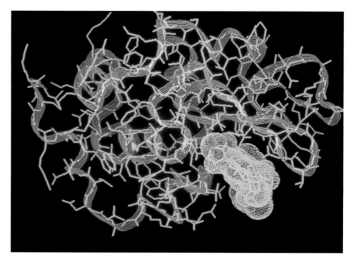

This image is a representation of the enzyme lysozyme, which is present in tears. It is able to break up chains of sugar molecules (yellow), which make up bacterial cell walls.

carbonic acid. In this form, it is transported in the blood to the lungs, to be expelled from the body as CO_2. An enzyme called glucose permease (the names of most enzymes end in '-ase') helps transport glucose across cell membranes so that it can be used to produce energy.

The 'lock and key' model

Every enzyme in the body has a specific task, and various theories have been put forward to explain how each enzyme fulfils this one task and no other. The first such theory is known as the 'lock and key' hypothesis. This theory works on the principle that there is an area or 'active site' on the surface of the enzyme molecule into which a molecule of a chemical, known as 'the substrate', fits and is then held in place by electrical attraction. As the reaction proceeds, the substrate is turned into another chemical (the 'product') that has different electrical properties, so the electrical attraction disappears and the product molecule abandons the site. The whole process repeats itself with new substrate molecules many thousands of times in a fraction of a second.

Unfortunately, the lock and key theory does not completely fit the facts. For one thing, enzyme activity can be inhibited by factors such as a change in temperature or pH, which would not be the case if the fit was a purely physical matter. Also, molecules other than the substrate can lock on to the site.

Another concept, known as the 'induced fit' theory, satisfies these objections. It holds that the active site has elastic properties and expands and contracts as necessary to accommodate the substrate – rather as a glove changes shape to accommodate the hand.

In the 'lock and key' model, a region of the enzyme (purple) is complementary to the substrate (orange). Here the enzyme breaks up the substrate without being chemically involved.

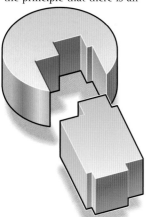

Enzymes and energy

Enzymes reduce the amount of energy required for a chemical reaction. However, being proteins, they are sensitive to changes in their surroundings.

Whenever a chemical reaction takes place, energy is used. For a reaction to take place, bonds between atoms must be broken for new ones to form – the energy required to break these bonds is called 'activation energy.' For instance, many substances will burn, but only after energy has been supplied (such as the heat from a lighted match). Enzymes work by lowering this activation energy, allowing reactions to proceed at lower temperatures, without chemically becoming involved in the reaction.

For a chemical reaction to take place, activation energy must be supplied. Enzymes reduce the amount of activation energy, speeding up the reaction.

WHAT AFFECTS ENZYME ACTIVITY?

The three main factors that affect enzyme activity are temperature, pH and the presence of other chemicals that either occupy the active site or distort its shape. Depending on their action, such chemicals are known as competitive and non-competitive inhibitors.

For each enzyme in the body, there is a temperature range within which it works at maximum efficiency. Outside this range, the bonds that hold together the complex protein structures of which most enzymes are made, start to break down. As a result, the shape of the enzyme's active site changes, making it impossible for the substrate to lock on to the

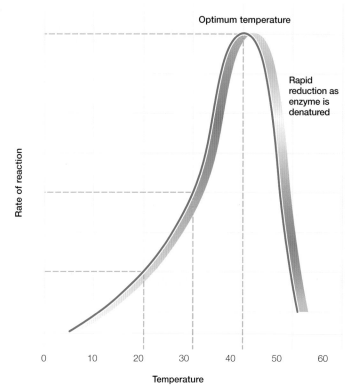

Optimum temperature

Rate of reaction

Rapid reduction as enzyme is denatured

0   10   20   30   40   50   60

Temperature

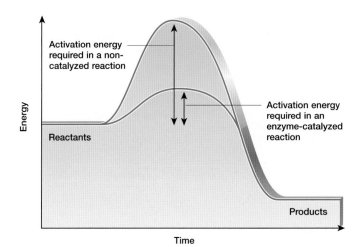

Activation energy required in a non-catalyzed reaction

Activation energy required in an enzyme-catalyzed reaction

Energy

Reactants

Products

Time

enzyme. This is why body systems start to shut down when the temperature is too high (hyperthermia), or too low (hypothermia).

As with temperature, enzymes have an optimum range of pH values within which they work most effectively. Outside this range, the action of an enzyme is inhibited, and at extremes it does not work at all. (pH is a value that indicates the concentration of hydrogen ions in a solution and so how acid or alkaline a solution is; distilled water, for example, has a pH of 7, while bleach has a pH of 12 and orange juice has a pH of 2.)

The optimum pH range varies

The activity of most enzymes rises with temperature until about 40 °C, where it peaks. Beyond that, the protein begins to denature and the enzyme rapidly loses its activity.

from enzyme to enzyme, but this is not normally a problem because it is matched to the enzyme's environment, and a buffering system (which compensates for small changes in pH) helps keep pH in different areas of the body relatively constant. Stomach enzymes, such as pepsin and chymotrypsin, work best at a low, acidic pH, which is brought about by the acidic conditions in the stomach.

Competitive and non-competitive inhibition

Enzyme activity can be inhibited by other chemicals. In competitive inhibition, the active site is taken up by a competing molecule. Competitive inhibition accounts for the ability of various chemicals – including cyanide – to shut down enzyme activity and cause death. Other poisons, such as lead and mercury, have the same effect, but are 'non-competitive inhibitors', since they do not 'compete' for occupation of the active site, but fix themselves onto the enzyme, thereby distorting its shape.

In competitive inhibition (top), two molecules with similar shapes compete for the active site of the enzyme. In non-competitive inhibition (bottom), the inhibitor binds to another site on the enzyme, changing the shape of the active site and preventing the reaction.

This woman and child are receiving oxygen for carbon monoxide (CO) poisoning. CO is a competitive inhibitor for oxygen, reducing the amount available to the body.

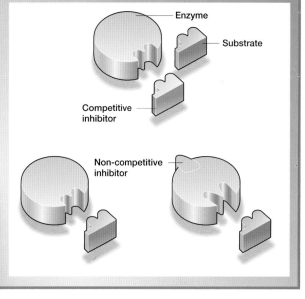

Enzyme

Substrate

Competitive inhibitor

Non-competitive inhibitor

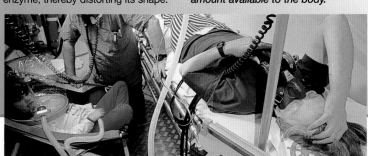

How blood sugar is controlled

Sugar is an important source of energy for the body, and it exists naturally in the blood in the form of glucose. The correct balance of blood-sugar levels is vital to life, and this is regulated by hormones secreted from the pancreas.

Glucose is a simple sugar that is vital for brain function, and an important source of energy for the rest of the body. Glucose is stored in the body in the form of glycogen – which is simply long chains of sugar molecules found in the liver and muscles – and transferred around the body in the blood.

There is a natural level of glucose in the blood, but when we eat, or do not eat enough, this level changes. The size of this change is regulated in the pancreas by hormones (which means 'to spur on').

THE PANCREAS

The pancreas is a long, whitish gland, about 20–25 cm in length, that lies just behind the lower part of the stomach and is connected to the duodenum. It produces enzymes that flow along a duct into the duodenum and assist in the digestion of food. But this is not its only job.

The digestive portion of the pancreas makes up 90 per cent of its total cell mass. About five per cent is made of cells that produce the hormones that regulate the blood sugar level: insulin and glucagon.

These 'endocrine' cells, known as islets, are clustered in groups throughout the pancreas. Unlike most pancreatic products, the hormones do not enter the duct leading to the duodenum; instead, they are delivered directly into the bloodstream.

Blood sugar

The ideal level of blood glucose is between 70 and 110 mg per 100 ml. After a meal, it is normal for the sugar level to rise for a few hours, but it should not go beyond 180 mg. Anyone with a higher glucose level is described as suffering from *hyperglycaemia*. Anyone with a blood glucose level of 70 or lower is described as having *hypoglycaemia*, or being *hypoglycaemic*.

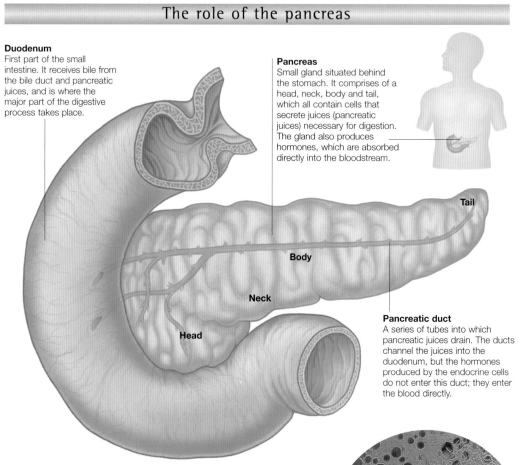

The role of the pancreas

Duodenum
First part of the small intestine. It receives bile from the bile duct and pancreatic juices, and is where the major part of the digestive process takes place.

Pancreas
Small gland situated behind the stomach. It comprises of a head, neck, body and tail, which all contain cells that secrete juices (pancreatic juices) necessary for digestion. The gland also produces hormones, which are absorbed directly into the bloodstream.

Tail

Body

Neck

Head

Pancreatic duct
A series of tubes into which pancreatic juices drain. The ducts channel the juices into the duodenum, but the hormones produced by the endocrine cells do not enter this duct; they enter the blood directly.

INSULIN

There are different types of pancreatic cell, each responsible for producing a different hormone. Insulin is normally secreted by the beta cells of the pancreatic islets. A low level of the hormone is secreted continuously, but if the amount of glucose in the blood increases, the cells are stimulated into producing more insulin. If the blood glucose level falls, insulin production decreases.

Insulin has an effect on a number of body cells, including muscle cells, red blood cells and fat cells. When insulin levels increase, these cells are forced to absorb more glucose from the blood and use it to produce energy. Insulin production is also controlled by another hormone, called somatostatin. This is secreted in response to high levels of other hormones, and its action is to slow down, among other things, the production of insulin.

GLUCAGON

Glucagon is secreted by the alpha cells of the pancreatic islets. These cells are stimulated into action when the level of glucose in the blood becomes too low. The hormone causes glycogen, particularly in the liver, to be converted into glucose and released into the blood. It also induces the liver, muscle and other body cells to make glucose from other chemicals in the body.

Under the microscope, endocrine cells resemble islands, hence their name 'pancreatic islets'. They are also known as 'islets of Langerhans', after the German physician Paul Langerhans. These cells produce the hormones insulin and glucagon.

Abnormal blood-sugar levels

A fine balance in blood-sugar levels is vital for good health. A fall in glucose levels can result in profuse sweating, confusion and even coma. A rise can lead to the disorder diabetes.

Diabetes

The inability of the pancreas to produce enough insulin leads to the disorder known as diabetes mellitus. Lack of insulin means that the body's cells are unable to take up glucose, which then accumulates in the blood. Because the muscles cannot absorb enough sugar, they become weak and start to waste away.

The exact cause of pancreas failure is not known, but diabetes often runs in families. Diabetes usually appears quite slowly in middle-aged people, and those who are overweight are particularly at risk. However, it can also appear in children, when the onset tends to be sudden.

Diabetes may be triggered in susceptible people by a range of factors, such as exposure to cold and wet, overwork and depression, but the most common cause is infection, particularly by viruses.

Symptoms include weakness, loss of weight, increased thirst and increased production of urine. Constipation and dryness of the mouth and skin also occur. Most patients do not develop any complications, but the disorder can affect the heart, blood vessels and nerves, and diabetics are more prone to cataracts of the eyes.

In extreme cases, patients can suffer pulmonary tuberculosis or a diabetic coma.

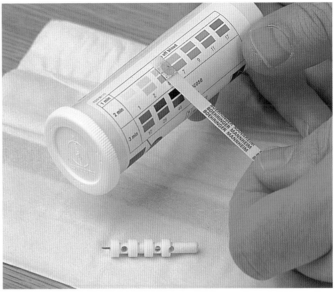

Blood glucose levels can be tested using a simple diagnostic kit. A lancet is used to prick the finger, and blood is placed on a test strip. At the tip of the strip are two reactive patches, which *change colour from yellow to black and white to dark blue depending on the level of sugar in the blood. In this case, the reading is normal. Raised levels can lead to diabetes.*

Controlling diabetes

Diabetes cannot be anticipated, prevented or cured. However, it can be controlled with proper treatment. In older patients, this may involve simply eating regularly and keeping to a reduced-sugar diet prescribed by a doctor. The doctor may also prescribe tablets that increase the effect of the insulin that is present in the blood.

In severe cases, it is necessary to take insulin. This has to be administered in the form of injections once or twice a day, as insulin is destroyed if taken by mouth.

Care must be taken with injections, as excess insulin induces hypoglycaemia, which brings on sweating, unsteadiness and disturbed behaviour. In extreme cases, the patient appears drunk, at which point medical help must be sought, as there is a risk of coma.

Diabetics give themselves an intramuscular insulin injection, which regulates blood-sugar levels. If there is a natural deficiency in insulin, regular replacement therapy is needed to prevent coma and death.

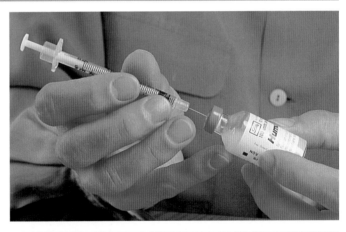

Natural balance of glucose levels

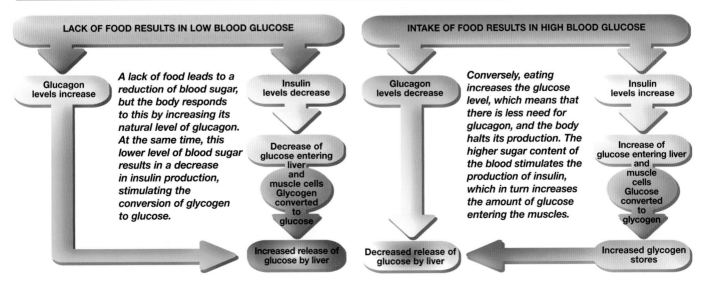

LACK OF FOOD RESULTS IN LOW BLOOD GLUCOSE

Glucagon levels increase

A lack of food leads to a reduction of blood sugar, but the body responds to this by increasing its natural level of glucagon. At the same time, this lower level of blood sugar results in a decrease in insulin production, stimulating the conversion of glycogen to glucose.

Insulin levels decrease

Decrease of glucose entering liver and muscle cells Glycogen converted to glucose

Increased release of glucose by liver

INTAKE OF FOOD RESULTS IN HIGH BLOOD GLUCOSE

Glucagon levels decrease

Conversely, eating increases the glucose level, which means that there is less need for glucagon, and the body halts its production. The higher sugar content of the blood stimulates the production of insulin, which in turn increases the amount of glucose entering the muscles.

Insulin levels increase

Increase of glucose entering liver and muscle cells Glucose converted to glycogen

Decreased release of glucose by liver

Increased glycogen stores

The role of water

Without water, human beings could not survive as it is essential for hydration and certain bodily processes, and is a vital source of minerals. If fluid intake is insufficient, dehydration can occur.

Of all forms of nourishment, water is the most vital. A person can survive for up to several weeks without food, but without water they would perish in a matter of days.

The reason for this is that a very large percentage of the body is actually made up of water. Most body cells consist of approximately 80 per cent water, while plasma (the liquid component of blood) consists of 92 per cent water.

SOURCES OF WATER

It is essential that water intake is sufficient to meet the body's needs. Most foods contain some water; meat ranges from 40–75 per cent water, and vegetables can consist of as much as 95 per cent water. Surprisingly, even dry foods, such as bread and cereal, can contain as much as 30 per cent water.

It is vital, however, that food is supplemented with regular drinks of water in order for the body to gain optimum levels of hydration. As a rough guideline, people should drink around two litres (about eight to 10 glasses) of water per day.

FLUID TYPE

Unfortunately, most people do not drink enough fluid, and those who do often fail to drink the right type of fluid.

For many people, the bulk of fluid intake comes from drinks such as tea, coffee or cola drinks. Despite the fact that these drinks do contain water, they also contain caffeine – a diuretic – which actually serves to deplete the body of valuable water by increasing urine production. Drinks containing alcohol also serve to dehydrate the body.

Water is essential for survival, as it makes up a large proportion of the body. People should ideally drink around two litres of water every day.

Benefits of water

The benefits of drinking water are often overlooked despite the fact that a lack of water can have a serious impact on health. As well as quenching thirst, water is essential to a number of body processes: it helps to maintain body temperature, acts as a lubricant, enables a variety of chemical reactions to take place and serves as a mixing medium.

BODY TEMPERATURE

Water has what is known as a high specific heat, which means, in effect, that it takes a relatively large amount of energy to raise its temperature. Consequently, water within the body helps it to resist great fluctuations in body temperature. In addition, water in the body enables it to be cooled in high temperatures or during physical exertion. This occurs through the evaporation of water in the form of sweat.

PROTECTION

Water acts as a lubricant to prevent friction within the body. For example, tears produced by the lacrimal glands prevent the surface of the eye from rubbing against the eyelid. Water also forms a cushion within joints and between organs to protect them from trauma, as seen, for example, in the cerebrospinal fluid around the brain.

CHEMICAL REACTIONS

The chemical reactions that take place in the body could not occur without water. This is because molecules must be dissolved in water to form ions (electrically charged atoms)

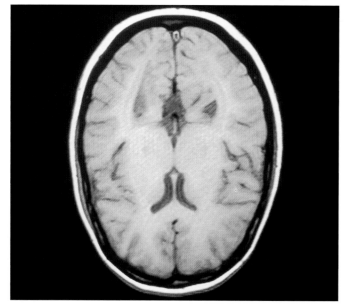

Cerebrospinal fluid is watery fluid that surrounds the brain and spinal cord. It cushions the brain from contact with the skull when the head is shaken.

before they can react. For example, when sodium chloride is dissolved in water, it splits to form separate sodium and chloride ions, which are then free to react with other ions.

In addition, cell membranes rely on water for the movement of enzymes into and out of cells. Enzymes are essential to cell function and so, without sufficient water, these reactions could not occur.

MIXING MEDIUM

Water mixes with other substances to form a solution (for example, when sodium chloride is dissolved in water to form sweat), a suspension (for example, red blood cells in plasma), or a colloid (a liquid that contains non-dissolved materials that do not settle out of the liquid, for example the water and protein within cells).

The ability of water to mix with other substances enables it to act as an effective medium for the transport of nutrients, gases and waste products throughout the body in body fluids such as plasma.

Loss of water

A number of regulatory mechanisms ensure that body fluids are kept within fine limits (homeostasis). This balance may be upset by extreme temperatures or illness, however, when dehydration may occur due to extreme water loss.

THREE ROUTES

Water is lost from the body through three main routes:

■ Perspiration – fluid loss through sweat depends upon a number of factors, such as environmental temperature, humidity and the degree of physical exertion.

The volume of sweat lost by a person at rest in a cool temperature is negligible. When exercising in elevated temperatures, or during a fever, however, the volume of water lost through sweat increases substantially. A person working outdoors in the summer, for example, could lose up to five litres of sweat

■ Urination – the amount of fluid ingested usually exceeds the body's needs; the excess is eliminated by the kidneys in the form of urine. Drugs or medical conditions that cause increased urination can cause dehydration

■ Defecation – only a small portion of body fluids is lost through defecation, since water is reabsorbed in the colon. However, this is not true in the case of diarrhoea. Diarrhoea occurs when increased waves of movement in the intestinal tract

cause faecal matter to pass through the colon too rapidly for water reabsorption to occur. This can cause a great deal of fluid to be lost from the body.

DEHYDRATION

Loss of body fluids can lead to dehydration, which is potentially fatal and must be treated promptly. The young and the old (the sense of thirst tends to dim with age) are most at risk of developing dehydration.

Symptoms of dehydration include raised body temperature, fatigue, nausea, extreme thirst, the passing of small amounts of dark urine, headaches and confusion.

Severe dehydration is defined as fluid loss of more than 1 per cent in body weight. For example, a person weighing 70 kg would have to lose 700 grams of fluid to be considered seriously dehydrated. Affected individuals can experience low blood pressure, loss of consciousness, severe cramping of the arms, legs, stomach and back, convulsions, heart failure, sunken eyes, inelasticity of the skin, and deep, rapid breathing.

Dehydration is treated by replacing lost fluids, as well as the electrolytes (salts) contained within them. In mild cases, this can be achieved by taking oral

In the developing world, education programmes help to prevent dehydration. Mild dehydration can be treated by the use of oral rehydration salts.

rehydration salts (a sachet of powder which is mixed with water to replace both water and salts such as salt and glucose). Severe cases are treated with an intravenous infusion of saline solution to restore the patient's fluid levels rapidly.

Sadly, in the developing world, where uncontaminated water supplies are limited and hospital facilities are not always available, many people die from dehydration every day.

The thirst mechanism

When the thirst centre within the hypothalamus is stimulated, it triggers a sensation of thirst. As water is drunk in response, the body's 'problem' is solved.

The body requires a fairly constant water balance. This means that the water lost, mainly through sweating and urination, must be replaced. Certain circumstances, such as excessive heat or exercise, use of diuretics, or a high-salt diet, can result in a rapid depletion of water.

Maintaining the balance

The body monitors water levels constantly, by checking blood volume and concentration. If the volume becomes low, and/or the concentration high, it registers the need both to conserve water and to increase intake via drinking. This is mainly controlled by the hypothalamus within the brain.

The hypothalamus sends 'instructions' to the kidneys to

The level of water in the body affects blood concentration and volume. The brain will trigger the thirst sensation when these need correction.

decrease the amount of urine passed, while its thirst centre initiates the sensation of thirst. This triggers an impulse to drink. The thirst centre is also activated by a dry mouth, whether caused by diet, drugs or nervousness.

How we vomit

Vomiting (emesis) is a protective reflex that serves to remove toxins from the stomach and small intestine. The unpleasant sensation that often precedes vomiting (nausea) is also a component of this reflex.

Feeling nauseated usually occurs prior to vomiting and serves as an 'early warning signal'. Nausea prevents further ingestion of a toxin and induces a powerful aversive response to stop future toxin ingestion.

However, nausea and vomiting can occur inappropriately due to pregnancy, motion, radiation, cytotoxic chemotherapeutic agents (for example, cisplatin) and anaesthetics (post-operative sickness). Under these conditions, there is no toxin in the stomach, so ejecting its contents from the body is clearly not going to be beneficial to the subject.

NERVOUS INPUT

Mucosal enterochromaffin cells located in the stomach and small intestine respond to the presence of toxins by releasing the neurotransmitter serotonin.

The serotonin molecules in turn activate nearby nerve fibre endings in the vagus nerve. This nerve carries electrical impulses through the abdominal and thoracic cavities, eventually terminating in a region of the brainstem (situated just above the spinal cord) called the nucleus of the tractus solitarius (NTS).

The vomiting reflex is co-ordinated by groups of neurones that are located in the brainstem; the precise location(s) of these neurones, which are often termed 'the vomiting centre', is currently unknown. However, the inputs capable of

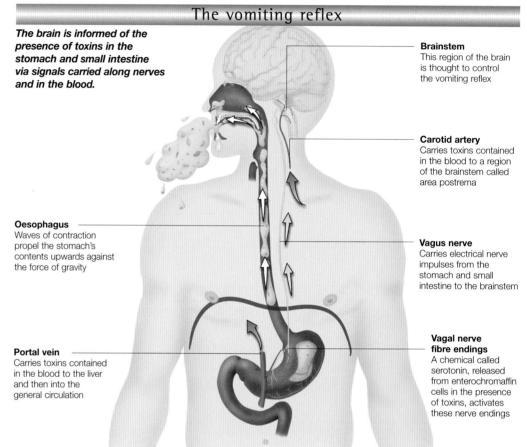

The vomiting reflex

The brain is informed of the presence of toxins in the stomach and small intestine via signals carried along nerves and in the blood.

Brainstem
This region of the brain is thought to control the vomiting reflex

Carotid artery
Carries toxins contained in the blood to a region of the brainstem called area postrema

Oesophagus
Waves of contraction propel the stomach's contents upwards against the force of gravity

Vagus nerve
Carries electrical nerve impulses from the stomach and small intestine to the brainstem

Portal vein
Carries toxins contained in the blood to the liver and then into the general circulation

Vagal nerve fibre endings
A chemical called serotonin, released from enterochromaffin cells in the presence of toxins, activates these nerve endings

activating the vomiting reflex all converge on the NTS, suggesting that NTS neurones either form part of the vomiting centre itself, or are able to modulate its activity in some way.

BLOODSTREAM

Alternatively, toxins absorbed into the bloodstream from the stomach or small intestine may activate a brainstem region adjacent to the NTS, called the

area postrema. These neurones are thought to be able to detect the presence of toxins in the blood. The NTS and area postrema communicate with each other via a series of nerves.

Motion sickness

Scientists believe that motion sickness occurs when information about the body's position supplied by the eyes does not match information supplied by the vestibular (balance) system, which is located in the inner ear.

This is supported by the fact that when the region of the brain that receives input from the balance organs is damaged – following a stroke, for example – the patient no longer suffers from motion sickness.

A rollercoaster ride can often induce nausea owing to the disorientating effects of conflicting sensory inputs.

The vestibular apparatus located in the inner ear controls our sense of balance, carrying information about the body's movements in space.

Ejecting stomach contents

The preliminaries to vomiting are a relaxation of the
stomach muscles, followed by retching – a repeated contraction
of the abdominal muscles and diaphragm.

Approximately 2 to 10 minutes prior to vomiting the stomach relaxes. Next, a giant migrating contraction starts in the mid portion of the small intestine and spreads rapidly (5–10 cm per second) towards the stomach. This contraction forces the contents of the small

Nausea and vomiting are often evoked inappropriately during pregnancy. One cause of this is the regurgitation of stomach acid into the oesophagus.

intestine back into the stomach, confining the ingested toxin to the stomach and preventing it from being absorbed further.

MUSCLE CONTRACTIONS

The abdominal muscles and diaphragm repeatedly contract and relax in a synchronous manner, squeezing the stomach and forcing its contents into and out of the oesophagus (retching). Retching gives the contents of the stomach momentum.

A characteristic body posture is often assumed to assist in the movement of the vomit – bent over with the head forward and a straight back. Then, the abdominal and diaphragm muscles produce a strong, maintained contraction, which produces an intra-abdominal pressure of 200 mm Hg.

Meanwhile, the glottis closes (preventing the stomach contents entering the lungs), the oesophagus shortens, the mouth involuntarily opens and internal pressure propels the stomach contents out of the body.

This illustration demonstrates how various muscular actions combine to expel the stomach contents from the body. This is commonly known as retching.

The mechanics of vomiting

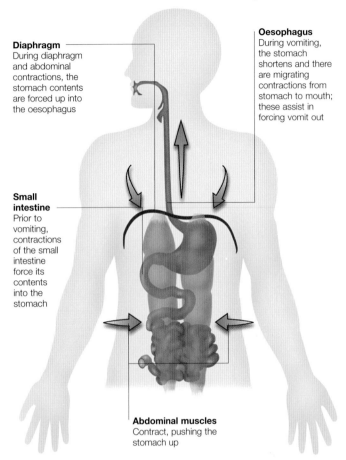

Diaphragm
During diaphragm and abdominal contractions, the stomach contents are forced up into the oesophagus

Oesophagus
During vomiting, the stomach shortens and there are migrating contractions from stomach to mouth; these assist in forcing vomit out

Small intestine
Prior to vomiting, contractions of the small intestine force its contents into the stomach

Abdominal muscles
Contract, pushing the stomach up

Anti-emetic therapy

Clinically, nausea and vomiting are a significant problem. They can be so severe in cancer patients receiving chemotherapy or radiotherapy that some patients will refuse to take their life-saving treatment. Indeed, the experience can be so unpleasant that patients have been known to be sick at the mere thought of returning to the hospital.

Anti-emetic drugs are usually administered by intravenous injection before a cancer patient undergoes chemotherapy.

SEROTONIN ANTAGONISTS

Consequently, drugs such as ondansetron, which block serotonin receptors, have been widely used for the prevention of vomiting caused by chemotherapeutic agents.

These drugs work by binding to the serotonin receptors located on the abdominal vagal fibres, preventing the serotonin released from enterochromaffin cells from activating them.

Serotonin receptor antagonists are not a universal anti-emetic, however. Their clinical efficacy is limited to vomiting induced by radiation and cytotoxic drugs. They are ineffective against nausea and vomiting caused by motion, anaesthetics and other pharmacological agents, such as L-dopa (used to treat Parkinson's disease).

This ampoule contains a single dose of an emetic drug used to control the nausea and vomiting associated with chemotherapy.

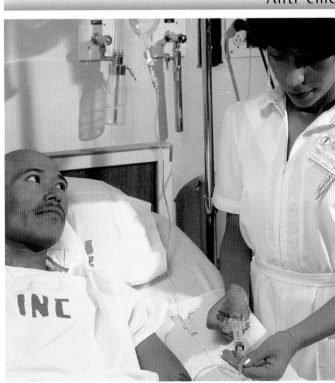

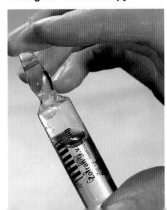

Caecum and appendix

The caecum and appendix lie at the junction of the large and small intestine, an area also known as the ileocaecal region. The caecum, from which the appendix arises, receives food from the small intestine.

The caecum is the first part of the large intestine. Food is passed from the terminal ileum, part of the small intestine, into the large intestine through the ileocaecal valve; the caecum lies below this valve. The caecum is a blind-ending pouch, about 7.5 cm in length and breadth, which continues above as the ascending colon, the next part of the large intestine. The appendix, a long, thin pouch of intestine arises from the caecum.

MUSCLE FIBRES

The muscular 'coat' of the small intestine is continued in the walls of the large intestine, but here it becomes separated into three strips of muscle, the taeniae coli. As food passes through the intestine, the caecum may become distended with faeces or gas. It may then be palpable through the abdominal wall.

BLOOD SUPPLY

The arterial blood supply to the caecum is from the anterior and posterior caecal arteries, which arise from the ileocolic artery. Venous blood returns through a similar layout of veins, ultimately draining into the superior mesenteric vein.

The ileocaecal region is the area surrounding the junction where the small intestine meets the large intestine. It consists of the caecum and the appendix.

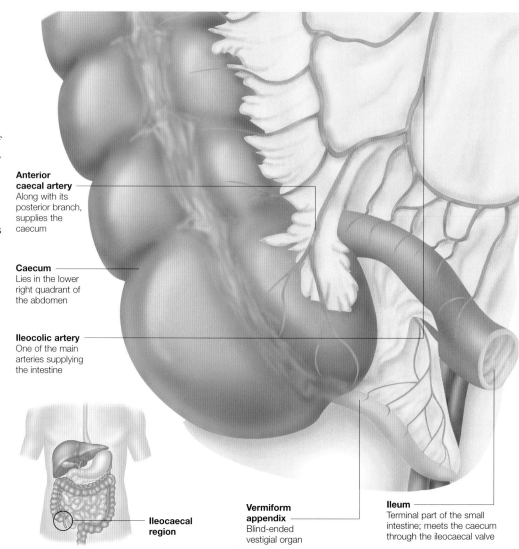

Anterior caecal artery
Along with its posterior branch, supplies the caecum

Caecum
Lies in the lower right quadrant of the abdomen

Ileocolic artery
One of the main arteries supplying the intestine

Ileocaecal region

Vermiform appendix
Blind-ended vestigial organ

Ileum
Terminal part of the small intestine; meets the caecum through the ileocaecal valve

The ileocaecal valve

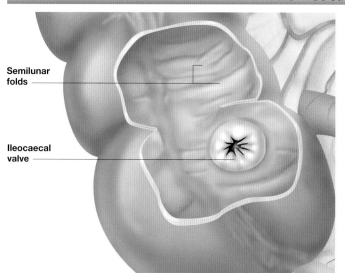

Semilunar folds

Ileocaecal valve

The ileocaecal valve surrounds the orifice through which intestinal contents pass into the caecum. This valve is not believed to be very effective.

The ileocaecal valve surrounds the orifice, or opening, through which the liquefied contents of the terminal ileum, the last part of the small intestine, enter the caecum.

ANATOMICAL STUDIES

Anatomical studies of cadavers (dead bodies) in the past had shown this orifice to be enclosed between folds or ridges in the caecal wall, which were thought to act like a valve.

Now that it is possible to study this area in living people, using endoscopes, it is clear that the opening looks quite different in life. The ileocaecal orifice is in fact raised above the caecal wall and is surrounded by a ring of circular muscle fibres, which help to keep it closed.

BARIUM STUDIES

Although the contents of the caecum do not easily pass back into the ileum when the caecal walls contract, the ileocaecal valve is not very effective. Barium X-ray studies of the large intestine commonly show leakage of contents backwards from the caecum into the terminal ileum through the ileocaecal valve.

The appendix

The appendix is a narrow, muscular outpouching of the caecum. It is usually between six and 10 cm in length, although it may be much longer or shorter. It arises from the back of the caecum, its lower end being free and mobile.

The vermiform (or 'wormlike') appendix is attached to the caecum at the beginning of the large intestine. The walls of the appendix contain lymphoid tissue. The lymphoid tissue of the appendix and that within the walls of the small intestine, protects the body from micro-organisms within the gut.

MUSCLE LAYER
Whereas the longitudinal muscle in the walls of the rest of the large intestine is present only in three strips – the taeniae coli – the appendix has a complete muscle layer. This is because the three taeniae coli converge on the base of the appendix and their fibres join to cover its entire surface.

PERITONEUM
The appendix is enclosed within a covering of peritoneum which forms a fold between the ileum, the caecum and the first part of the appendix. This fold is known as the mesoappendix.

BASE OF THE APPENDIX
The base of the appendix, where it arises from the caecum, is usually in a fixed position. The corresponding area on the surface of the abdomen is known as McBurney's point.

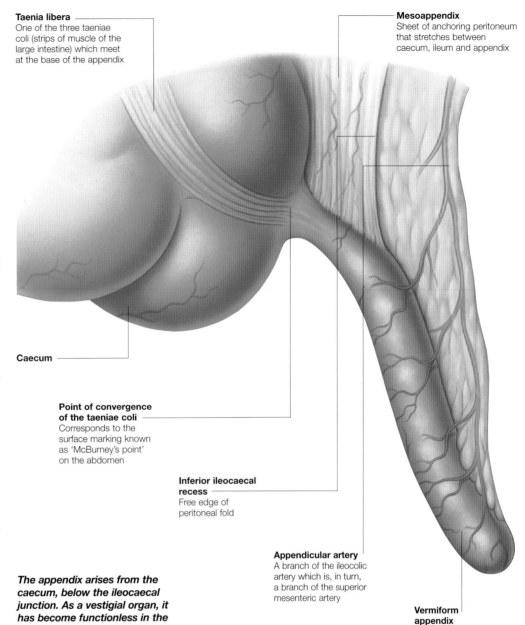

Taenia libera
One of the three taeniae coli (strips of muscle of the large intestine) which meet at the base of the appendix

Mesoappendix
Sheet of anchoring peritoneum that stretches between caecum, ileum and appendix

Caecum

Point of convergence of the taeniae coli
Corresponds to the surface marking known as 'McBurney's point' on the abdomen

Inferior ileocaecal recess
Free edge of peritoneal fold

Appendicular artery
A branch of the ileocolic artery which is, in turn, a branch of the superior mesenteric artery

Vermiform appendix

The appendix arises from the caecum, below the ileocaecal junction. As a vestigial organ, it has become functionless in the course of evolution, and has no role in the process of digestion.

Positions of the appendix

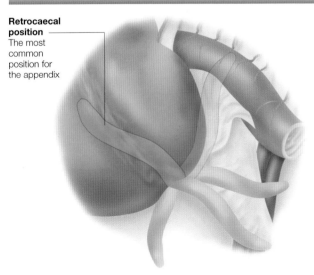

Retrocaecal position
The most common position for the appendix

Although the base of the appendix is usually at a fixed point, the far end is free and may lie in a variety of positions.

COMMON POSITIONS
The most common position is the retrocaecal position, wherein the appendix passes upwards to lie behind the caecum.

In other cases the appendix may lie next to the terminal ileum, or project downwards into the pelvis.

There are various locations for the appendix. The position will determine where pain and tenderness is felt if it becomes inflamed.

UNCOMMON POSITIONS
Uncommonly, the caecum may lie abnormally high or low and therefore the appendix will also be in an unusual position.

APPENDICITIS
The position of the appendix is important in the diagnosis of appendicitis, when the appendix becomes inflamed and swollen. The pain and tenderness this causes may be felt in different places according to where the appendix lies.

The position of the appendix can cause confusion: when an appendix lies in the pelvis the symptoms may be similar to a urinary tract infection.

Colon

The colon forms the main part of the large intestine. Although a continuous tube, the colon has four parts: the ascending colon, the transverse colon, the descending colon and the sigmoid colon.

The colon receives the liquefied contents of the small intestine and reabsorbs the water to form semi-solid waste, which is then expelled through the rectum and anal canal as faeces. There are two sharp bends, or flexures, in the colon known as the right colic (or hepatic) flexure and the left colic (or splenic) flexure.

ASCENDING COLON

The ascending colon runs from the ileocaecal valve up to the right colic flexure, where it becomes the transverse colon. It is about 12 cm long and lies against the posterior (back) abdominal wall, being covered on the front and sides only by the peritoneum, the thin sheet of connective tissue that lines the abdominal organs.

TRANSVERSE COLON

The transverse colon begins at the right colic flexure, under the right lobe of the liver, and runs across the body towards the left colic flexure next to the spleen.

With a length of about 45 cm, the transverse colon is the longest and the most mobile part of the large intestine, as it hangs down suspended within a fold of peritoneum (or mesentery).

THE DESCENDING COLON

The descending colon runs from the left colic flexure down to the brim of the pelvis where it becomes the sigmoid colon. As the left colic flexure is higher than the right, the descending colon is consequently longer than the ascending colon.

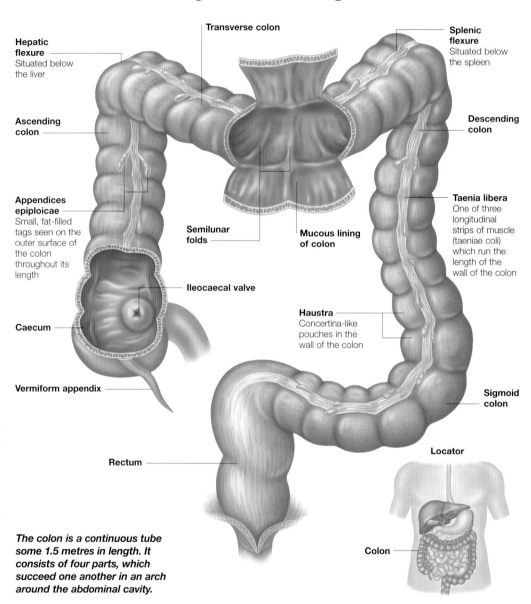

Transverse colon

Hepatic flexure
Situated below the liver

Splenic flexure
Situated below the spleen

Ascending colon

Descending colon

Appendices epiploicae
Small, fat-filled tags seen on the outer surface of the colon throughout its length

Semilunar folds

Mucous lining of colon

Taenia libera
One of three longitudinal strips of muscle (taeniae coli) which run the length of the wall of the colon

Ileocaecal valve

Caecum

Haustra
Concertina-like pouches in the wall of the colon

Vermiform appendix

Sigmoid colon

Locator

Rectum

Colon

The colon is a continuous tube some 1.5 metres in length. It consists of four parts, which succeed one another in an arch around the abdominal cavity.

Sigmoid colon and the lining of the colon

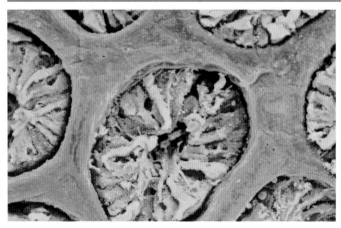

The lining of the colon – the mucosa (green) – contains glands (yellow). Cells in these glands are involved in water absorption and mucus secretion.

The sigmoid – 'S-shaped' – colon is the continuation of the descending colon, starting at the pelvic brim.

CHARACTERISTICS

It is about 40 cm long and, unlike the descending colon, quite mobile as it lies within its mesentery, or fold of peritoneum. At its far end the sigmoid colon leads into the rectum. The function of the sigmoid colon is to store faeces before defecation, and so its size and position vary depending on whether it is full or empty and dietary intake.

LINING OF THE COLON

The lining of the colon has a simple layer of cells with many deep depressions, or crypts, which contain mucus-secreting cells. The mucus is important for lubricating the passage of faeces and protecting the walls from acids and gases produced by the intestinal bacteria.

Blood supply and drainage of the colon

Like the rest of the intestine, each of the parts of the colon is readily supplied with blood from a network of arteries.

Venous blood draining from the colon passes through the hepatic portal system, for treatment by the liver, before re-entering the general circulation.

ARTERIAL SUPPLY OF THE COLON

The arterial supply to the colon comes from the superior and inferior mesenteric branches of the aorta, the large central artery of the abdomen.

The ascending colon and first two thirds of the transverse colon are supplied by the superior mesenteric artery, while the last third of the transverse colon, the descending colon and the sigmoid colon are supplied by the inferior mesenteric.

PATTERN OF THE ARTERIES

As in other parts of the gastro-intestinal tract, there are anastomoses, or connections, between the branches of these two major arteries.

The superior mesenteric artery gives off the ileocolic, right colic and middle colic arteries which anastomose with each other and with the left colic and sigmoid branches of the inferior mesenteric artery.

In this way an 'arcade' of arteries is formed around the wall of the colon, supplying all parts with arterial blood.

Arterial system of the colon

- Transverse colon
- Superior mesenteric artery
- Middle colic artery
- Right colic artery
- Ileocolic artery
- Colic branch of ileocolic artery
- Ascending colon
- Abdominal aorta
- Left colic artery
- Inferior mesenteric artery
- Descending colon
- Sigmoid arteries
- Sigmoid colon

The arteries of the ascending colon and greater part of the transverse colon are supplied by the superior mesenteric artery. The inferior mesenteric artery supplies the descending colon and the left part of the transverse colon.

Venous drainage of the colon

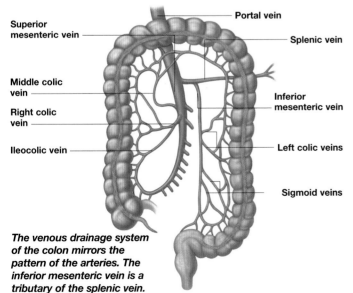

- Superior mesenteric vein
- Middle colic vein
- Right colic vein
- Ileocolic vein
- Portal vein
- Splenic vein
- Inferior mesenteric vein
- Left colic veins
- Sigmoid veins

The venous drainage system of the colon mirrors the pattern of the arteries. The inferior mesenteric vein is a tributary of the splenic vein.

Venous blood from the colon is collected ultimately into the portal vein. In general, blood from the ascending colon and first two-thirds of the transverse colon runs into the superior mesenteric vein, with blood from the remainder of the colon being drained by the inferior mesenteric vein.

The inferior mesenteric vein drains into the splenic vein, which then joins with the superior mesenteric vein to form the portal vein. The portal vein then carries all the venous blood through the liver on its way back to the heart.

LYMPHATIC DRAINAGE

Lymph collected from the walls of the colon travels in lymphatic vessels back alongside the arteries towards the main abdominal lymph-collecting vessel, the cysterna chyli. There are many lymph nodes which filter the fluid before it is returned to the venous system.

Lymph passes through the lymph nodes on the wall of the colon, through the nodes adjacent to the small arteries supplying the colon and then through the superior and inferior mesenteric nodes.

COLON CHARACTERISTICS

Unlike the small intestine, the walls of the colon are puckered into concertina-like pockets, or haustra, which show up quite clearly on direct examination, although this pattern can be absent if there is chronic inflammation such as in colitis.

Rectum and anal canal

The rectum and anal canal together form the last part of the
gastro-intestinal tract. They receive waste matter in the form of
faeces and allow it to be passed out of the body.

The rectum continues on from
the sigmoid colon, which lies at
the level of the third sacral
vertebra. Rectum means 'straight'
but in fact the rectum follows
the curve of the sacrum and
coccyx, which form the back of
the bony pelvis.

The lower end of the rectum
joins to the anal canal with an
80–90 degree change in
direction. This anorectal flexure
prevents faeces passing into the
anal canal until required.

The longitudinal muscle of the
rectum is in two broad bands,
which run down the front and
the back surfaces. There are
three horizontal folds in the wall
of the rectum, known as the
superior (upper), middle and
inferior (lower) transverse folds.
Below the inferior fold the
rectum widens into the ampulla.

THE ANAL CANAL

The anal canal runs from the
anorectal flexure down to the
anus. Except during defecation
the canal is empty and closed.

The lining of the anal canal
changes along its length. The
upper part carries longitudinal
ridges called anal columns which
begin above at the anorectal
junction and end below at the
pectinate line.

At the lower end of the anal
columns are the anal sinuses and
anal valves. The anal sinuses
produce mucus when faeces are
being passed, which acts as a
lubricant. The valves help to
prevent the passage of mucus out
of the anal canal at other times.

Coronal section through the rectum and anal canal

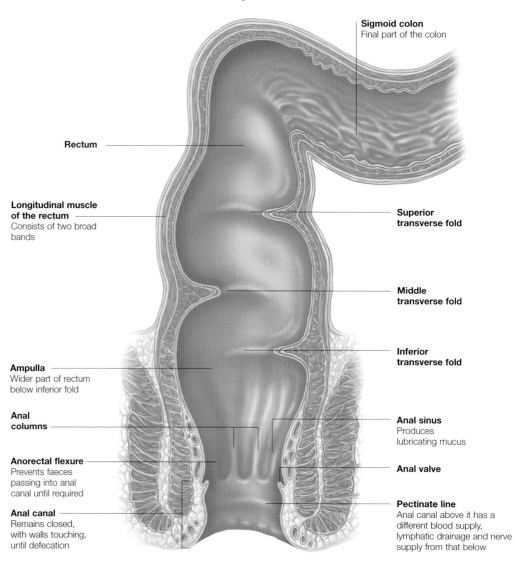

Sigmoid colon
Final part of the colon

Rectum

**Longitudinal muscle
of the rectum**
Consists of two broad
bands

**Superior
transverse fold**

**Middle
transverse fold**

Ampulla
Wider part of rectum
below inferior fold

**Inferior
transverse fold**

**Anal
columns**

Anal sinus
Produces
lubricating mucus

Anorectal flexure
Prevents faeces
passing into anal
canal until required

Anal valve

Anal canal
Remains closed,
with walls touching,
until defecation

Pectinate line
Anal canal above it has a
different blood supply,
lymphatic drainage and nerve
supply from that below

The anal sphincter

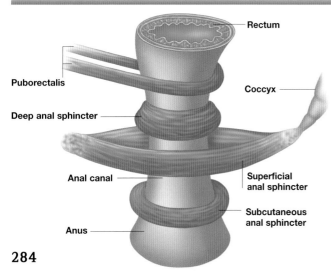

Rectum

Puborectalis

Coccyx

Deep anal sphincter

Anal canal

**Superficial
anal sphincter**

**Subcutaneous
anal sphincter**

Anus

The contents of the intestines are
constantly being moved on to
the next stage without conscious
awareness of it.

However, it is obviously
important that there is control of
such motions at the final stage.
This control is achieved through
the function of the anal
sphincter, which is made up of
several parts:

*The anal sphincter, which
consists of several parts,
controls the release of faeces
from the body. Only the
external anal sphincter is
under voluntary control.*

■ **The internal anal sphincter** A
thickening of the normal circular
muscle layer of the bowel in the
upper two thirds of the anal
canal. It is not under voluntary
control.

■ **The puborectalis muscle**
A sling of muscle which loops
around the anorectal junction
forming an angle and preventing
passage of the contents of the
rectum into the anal canal.

■ **External anal sphincter**
In three parts, deep, superficial
and subcutaneous, this sphincter
is under voluntary control, and
so can be relaxed by an act of
will when convenient.

Vessels of the rectum and anus

The rectum and anal canal have a rich blood supply. A network of veins drains blood from this area.

Beneath the lining of the rectum and anal canal lies a network of small veins, the rectal venous plexus. This is in two parts:
■ The internal rectal venous plexus – lies just under the lining
■ The external rectal venous plexus – lies outside the muscle layer.

These receive blood from the tissues and carry it to the larger veins that drain the area. These larger veins are the superior, middle and inferior rectal veins which drain the corresponding parts of the rectum.

The internal venous plexus of the anal canal drains blood in two directions on either side of the pectinate line region. Above this level blood drains mainly into the superior rectal vein while from below it drains into the inferior rectal vein.

ARTERIAL BLOOD SUPPLY
The rectum receives its blood supply from three sources. The upper part is supplied by the superior rectal artery, the lower portion is supplied by the middle rectal arteries while the anorectal junction receives blood from the inferior rectal arteries.

Within the anal canal the superior rectal artery travels down to provide blood above the pectinate line. The two inferior rectal arteries, branches of the pudendal, supply the anal canal below the pectinate line.

Venous drainage system of the rectum and anus

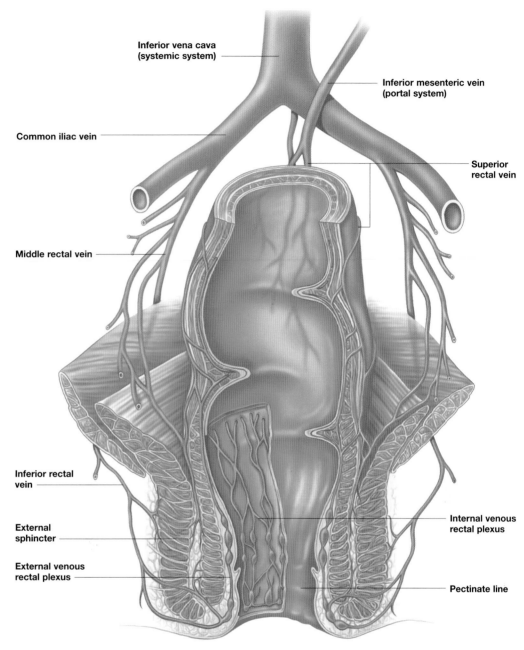

Inferior vena cava (systemic system)

Inferior mesenteric vein (portal system)

Common iliac vein

Superior rectal vein

Middle rectal vein

Inferior rectal vein

Internal venous rectal plexus

External sphincter

External venous rectal plexus

Pectinate line

Nerves of the rectum and anal canal

Nerve supply

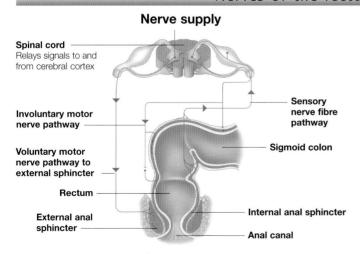

Spinal cord
Relays signals to and from cerebral cortex

Involuntary motor nerve pathway

Voluntary motor nerve pathway to external sphincter

Rectum

External anal sphincter

Sensory nerve fibre pathway

Sigmoid colon

Internal anal sphincter

Anal canal

Like the rest of the gastro-intestinal tract, the walls of the rectum and anal canal have a nerve supply from the body's autonomic nervous system. This system works 'in the background', usually without us being aware of it, to regulate and control the body's internal functions.

These nerves can sense the filling of the rectum and can

When the rectum is full, a defecation reflex is triggered in the spinal cord. Signals are sent to the rectal muscles to start contracting.

then cause the reflex contraction of the rectal walls to push the faeces into the anal canal and the relaxation of the internal anal sphincter.

However, the anal canal, or more specifically the external anal sphincter, also has a nerve supply from the 'voluntary' nervous system.

These nerves, which originate from the second, third and fourth sacral spinal nerves, allow us to contract the sphincter muscle by an act of will and so prevent filling of the anal canal until an appropriate time for defecation.

285

How waste is excreted

Almost all of the useful nutrients in food are absorbed into the body by the small intestine. The function of the large intestine is to excrete the unwanted waste material while retaining any useful chemicals that remain.

The large intestine is about 1.5 m in length and forms an arch that surrounds the folds and twists of the small intestine. It consists of four sections: the caecum, colon, rectum and anal canal. Food passes from the ileum of the small intestine into the upper end of the caecum via the ileocaecal valve. This valve prevents material in the large intestine from returning to the small intestine, even when the large intestine is distended. The caecum is a downward-pointing pouch that ends in the worm-like appendage called the vermiform appendix.

THE COLON

The caecum leads into the ascending colon, a straight section of the colon that travels up to the liver. There it bends and becomes the transverse colon, which crosses the abdomen and bends to become the descending colon and then the sigmoid colon. Altogether the colon measures about 1.3 m in length and is the longest section of the large intestine.

The primary function of the colon is the propulsion of faeces towards the anal canal. This process can be achieved by a relatively short length of intestine, which is why some or even all of the colon can, if necessary, be surgically removed. The colon is long to provide the maximum possible area for the reabsorption of water, dissolved salts and water-soluble vitamins.

The large intestine

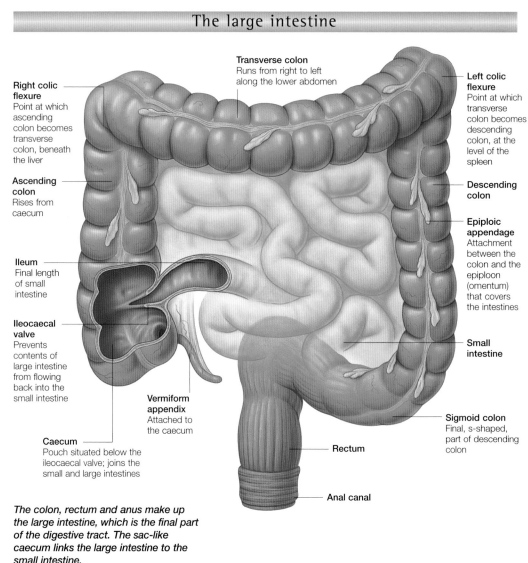

Transverse colon
Runs from right to left along the lower abdomen

Right colic flexure
Point at which ascending colon becomes transverse colon, beneath the liver

Ascending colon
Rises from caecum

Ileum
Final length of small intestine

Ileocaecal valve
Prevents contents of large intestine from flowing back into the small intestine

Vermiform appendix
Attached to the caecum

Caecum
Pouch situated below the ileocaecal valve; joins the small and large intestines

Left colic flexure
Point at which transverse colon becomes descending colon, at the level of the spleen

Descending colon

Epiploic appendage
Attachment between the colon and the epiploon (omentum) that covers the intestines

Small intestine

Sigmoid colon
Final, s-shaped, part of descending colon

Rectum

Anal canal

The colon, rectum and anus make up the large intestine, which is the final part of the digestive tract. The sac-like caecum links the large intestine to the small intestine.

Passage of faeces

Movement of the faeces along the colon is achieved by muscular movements in the colon wall. Three types of muscular movement occur and these serve not only to propel faeces along, but also to keep mixing the material, allowing water to be taken up by the colon wall more easily.

Faecal material passes through the colon far more slowly than through the small intestine. Each day, the large intestine will absorb approximately 1.4 litres of water, and smaller amounts of sodium and chloride.

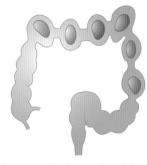

Segmentation is a series of ring-like contractions that churn the faecal material without moving it forwards. Water is thus more easily absorbed.

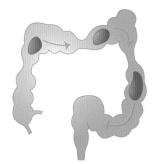

Peristaltic contractions, like those in the small intestine, mix and move the faeces. Muscles behind each segment contract, while those in front relax.

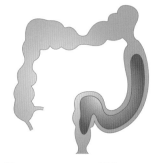

Mass movements, which are stronger than the peristaltic contractions, propel large quantities of faeces. They occur two or three times a day.

Defecation

The process of digestion ends when the waste products are removed from the body. Although an involuntary impulse, defecation can be consciously delayed.

The colon leads to the rectum, which is about 12 cm long and is equipped with muscular walls. These are capable of stretching, allowing the rectum to act as a reservoir for faeces, and are used to expel the faeces into the anal canal. Faeces arriving in the rectum are relatively dry, but the rectum has a glandular lining that secretes mucus, helping to lubricate the faeces and ease their passage through the rectum and anal canal.

The faeces that arrive in the rectum contain undigested food residues, mucus, epithelial cells (from the lining of the digestive tract), bacteria and enough water to enable smooth passage.

LACK OF BULK
If the faeces in the colon lack bulk, due to a lack of fibrous, indigestible material in the diet, the colon may become narrowed and its contractions, with nothing to work against, can become too powerful. This causes an increased pressure on the walls of the colon and can result in the formation of sac-like herniations (pouches) known as diverticula.

Diverticulosis usually occurs in the region of the sigmoid colon. It is associated with left-sided pelvic pain and may have serious consequences. The diverticula may rupture, releasing faecal matter into the abdominal cavity, and this can lead to severe infection.

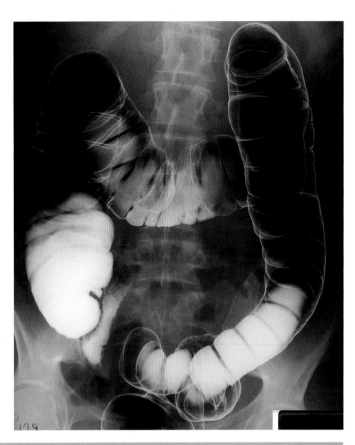

A barium X-ray shows the twists and turns of the colon.
Of the 500 ml of food residue that usually enters the caecum daily, approximately only 150 ml becomes faeces.

Anal canal

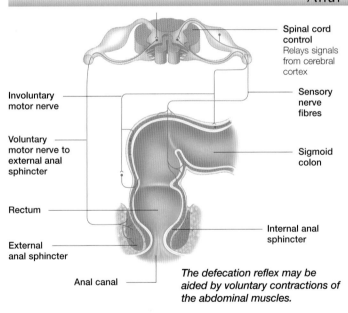

Spinal cord control
Relays signals from cerebral cortex

Involuntary motor nerve

Voluntary motor nerve to external anal sphincter

Rectum

External anal sphincter

Anal canal

Sensory nerve fibres

Sigmoid colon

Internal anal sphincter

The defecation reflex may be aided by voluntary contractions of the abdominal muscles.

The anal canal, or anus, is a short, narrow tube, about 4 cm long surrounded by two rings of muscle, known as the internal and external anal sphincters. The function of the anal canal is to keep the opening to the exterior closed until a person is ready to excrete the faeces.

DEFECATION CONTROL
The process of defecation is controlled by the brain, which most of the time sends signals to the sphincter muscles to keep them contracted.

When the rectum becomes so full that it cannot stretch any further, a defecation reflex is triggered in the spinal cord, and signals are sent to the rectal muscles to start contracting. At the same time, signals are sent to the brain, giving warning that defecation is necessary, but the brain remains in conscious control of the sphincter muscles until such time as it becomes convenient to defecate. When the decision to defecate is made, the brain allows the sphincter muscles to relax and the muscular wall of the rectum propels the faeces through the anal canal.

Involuntary defecation occurs in infants because they have not yet gained control of their external and anal sphincter. It also occurs in those who have damage to the spinal cord. Watery stools (diarrhoea) result when the food residue is rushed through the large intestine. Dehydration can result from inadequate water absorption.

The appendix

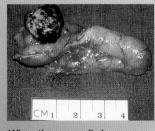

When the appendix becomes inflamed, the result is appendicitis. This condition can be life-threatening.

Neither the caecum nor the appendix has any obvious function in humans and both are probably relics of our ancestry. The appendix, for example, is a narrow, blind-ending piece of gut, up to 10 cm long and 1 cm in diameter, which extends downwards from the caecum. It is found only in humans, certain species of ape and, curiously, in the wombat, an Australian marsupial.

Its origin is probably connected with the fact that a number of herbivorous animals have an organ in the same position that acts as an extra stomach in which the cellulose in their plant food is digested by bacteria. If so, the human appendix is clearly in this respect a vestigial organ, as humans cannot digest cellulose. It does, however, appear to have developed a secondary function; that of acting as an early warning of infection: like the adenoids and tonsils, the appendix contains a large number of lymph glands whose purpose is to counter infection.

However, the appendix can itself become inflamed and if this happens appendicitis results. This can be fatal and the appendix usually has to be removed surgically. It can be removed at any age, but is more likely to be necessary in early life, as by the time a person has reached the age of 40 their appendix has almost completely shrivelled up.

Liver and biliary system

The liver is the largest abdominal organ, weighing about 1.5 kg in adult men. It plays an important role in digestion, and also produces bile, which is secreted into the duodenum.

The liver lies under the diaphragm in the abdominal cavity, on the right side, protected largely by the ribcage.

The tissue of the liver is soft and pliable, and reddish brown in colour. It has a rich blood supply from both the portal vein and the hepatic artery and so will bleed profusely if cut or damaged.

LOBES OF THE LIVER

Although it has four lobes, functionally, the liver is divided into two parts, right and left, each receiving its own separate blood supply. The two smaller lobes, the caudate and the quadrate, can only be seen on the underside of the liver.

PERITONEAL COVERINGS

The greater part of the liver is covered with the peritoneum, a sheet of connective tissue which lines the walls and structures of the abdomen. Folds of the peritoneum form the various ligaments of the liver.

As it lies against the diaphragm, the position of the liver may vary during respiration. It is pushed down on inhalation, and rises again as the breath is exhaled.

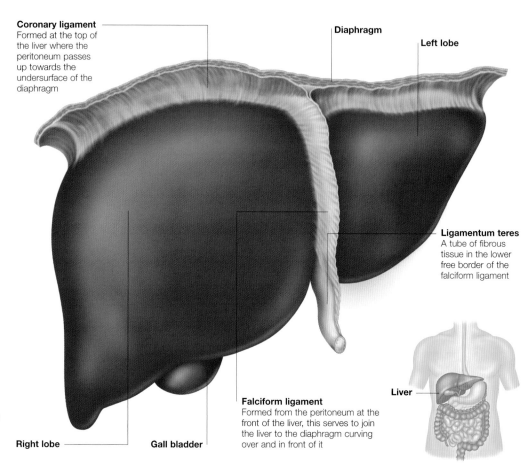

Coronary ligament
Formed at the top of the liver where the peritoneum passes up towards the undersurface of the diaphragm

Diaphragm

Left lobe

Ligamentum teres
A tube of fibrous tissue in the lower free border of the falciform ligament

Liver

Right lobe

Gall bladder

Falciform ligament
Formed from the peritoneum at the front of the liver, this serves to join the liver to the diaphragm curving over and in front of it

Microscopic anatomy of the liver

The sinusoids within each lobule contain tiny specialized cells known as Kupffer cells. These remove debris and worn out blood cells from the blood before it is taken back to the heart.

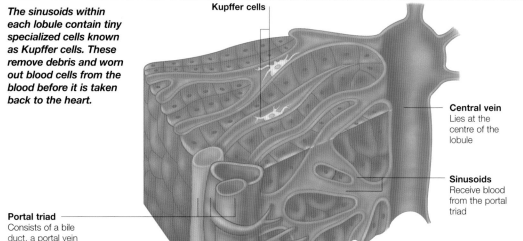

Kupffer cells

Central vein
Lies at the centre of the lobule

Sinusoids
Receive blood from the portal triad

Portal triad
Consists of a bile duct, a portal vein and a portal arteriole

The liver is composed of numerous tiny groups of cells called lobules, which are hexagonal in shape. They have a distinctive structure, with hepatocytes (liver cells) arranged like the spokes of a wheel around a central vein, which is a tributary of the hepatic vein. Blood flows past the hepatocytes and into this central vein through tiny vessels known as sinusoids.

The sinusoids receive blood from the vessels of the portal triads, groupings of three vessels which lie at the six points of the lobule. The portal triad is made up of a small branch of the hepatic artery, a small branch of the portal vein and a small biliary duct which collects the bile made by the liver cells.

Visceral surface of the liver

The underside of the liver is known as the visceral surface as it lies against the abdominal organs, or viscera. The impressions of adjacent organs, the related vessels and the positions of the inferior vena cava and gall bladder can be seen.

The liver lies closely against many other organs in the abdomen. Because the tissue of the liver is soft and pliable, these surrounding structures may leave impressions on its surface. The largest and most obvious impressions are seen on the surfaces of the right and left lobes.

PORTA HEPATIS

The porta hepatis is an area which is similar to the hilum of the lungs, in that major vessels enter and leave the liver together clothed in a sleeve of connective tissue, in this case peritoneum.

Structures which pass through the porta hepatis include the portal vein, the hepatic artery, the bile ducts, lymphatic vessels and nerves.

BLOOD SUPPLY

The liver is unusual in that it receives blood from two sources:

■ The hepatic artery. Conveys 30 per cent of the liver's blood supply. It arises from the common hepatic artery

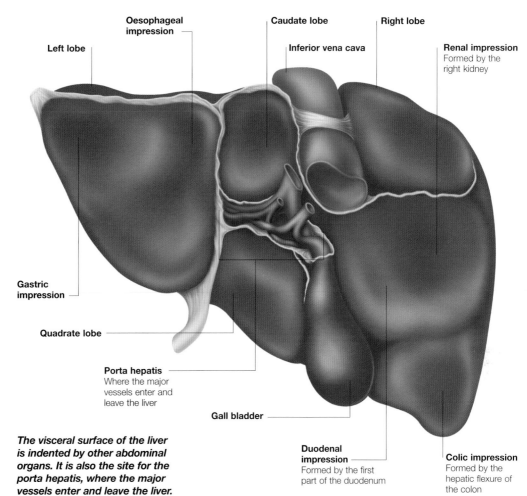

Left lobe

Oesophageal impression

Caudate lobe

Right lobe

Inferior vena cava

Renal impression
Formed by the right kidney

Gastric impression

Quadrate lobe

Porta hepatis
Where the major vessels enter and leave the liver

Gall bladder

Duodenal impression
Formed by the first part of the duodenum

Colic impression
Formed by the hepatic flexure of the colon

The visceral surface of the liver is indented by other abdominal organs. It is also the site for the porta hepatis, where the major vessels enter and leave the liver.

and carries fresh oxygenated blood. On entering the liver it divides into right and left branches. The right branch supplies the right lobe and the left branch supplies the caudate, quadrate and left lobes.

■ The hepatic portal vein. Conveys 70 per cent of the liver's blood supply. This large vein drains blood from the gastro-intestinal tract, from the stomach to the rectum. Portal blood is rich in nutrients which

have been absorbed after digestion in the gut. Like the hepatic artery, it divides into right and left branches with similar distributions. Venous blood from the liver is returned to the heart via the hepatic vein.

The biliary system

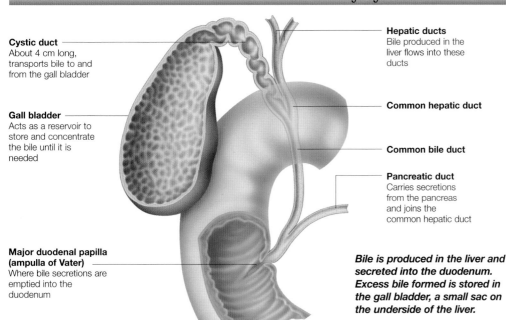

Cystic duct
About 4 cm long, transports bile to and from the gall bladder

Gall bladder
Acts as a reservoir to store and concentrate the bile until it is needed

Major duodenal papilla (ampulla of Vater)
Where bile secretions are emptied into the duodenum

Hepatic ducts
Bile produced in the liver flows into these ducts

Common hepatic duct

Common bile duct

Pancreatic duct
Carries secretions from the pancreas and joins the common hepatic duct

Bile is produced in the liver and secreted into the duodenum. Excess bile formed is stored in the gall bladder, a small sac on the underside of the liver.

Bile is a greenish fluid which aids the digestion of fats within the small intestine. It is secreted by the cells of the liver.

PASSAGE OF BILE

Bile is passed into small bile ducts which merge to form the right and left hepatic ducts. These ducts pass out of the liver through the porta hepatis then unite to form the common hepatic duct.

THE COMMON BILE DUCT

The common hepatic duct is then joined by the cystic duct, forming the common bile duct. This continues down towards the duodenum where, together with the duct carrying secretions of the pancreas, it empties through the major duodenal papilla (or ampulla of Vater).

289

How the liver works

The liver is one of the most complex organs in the body.
It controls more than 500 chemical reactions, and manufactures
and stores substances that are vital to sustaining life.

The liver is the largest of the body's internal organs, weighing about 1.8 kg in a man and 1.3 kg in a woman. It forms a right-angled triangle whose bulk is on the right side of the abdomen although it continues across the midline of the body to lie below the apex of the heart and behind the stomach on the left side. Its top lies beneath the fifth rib, and it reaches down on the right side to just below the 10th rib – this is why doctors push their fingers underneath the ribs on the patient's right side to check whether the liver is enlarged.

LIVER STRUCTURE
Reddish-brown in colour, the liver is not only the largest internal organ but also the most complex. It consists of eight lobes, each one made up of hexagonally shaped areas called lobules, which consist of a central vein surrounded by liver cells.

The whole structure is permeated by a network of veins, arteries and ducts. The ducts are channels that collect bile, which is produced by liver cells, and direct it to the gall bladder, where it is stored. This is a pear-shaped sac, about 8 cm long, that lies underneath and extends just below the ninth rib. When the gall bladder is swollen, it can sometimes be felt just below the ninth rib and a few centimetres to the left of its point.

Cross-section through the liver

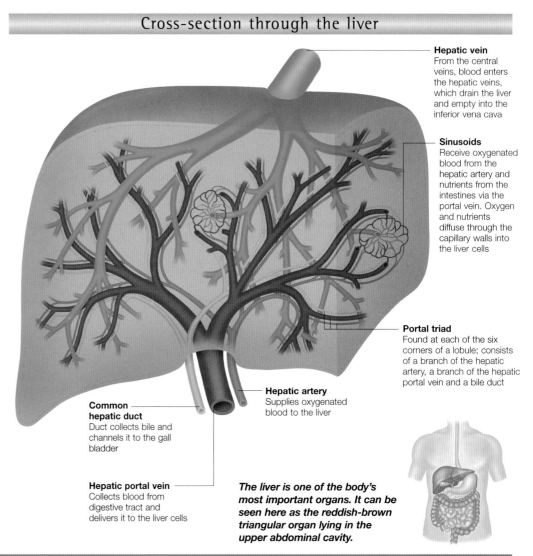

Hepatic vein
From the central veins, blood enters the hepatic veins, which drain the liver and empty into the inferior vena cava

Sinusoids
Receive oxygenated blood from the hepatic artery and nutrients from the intestines via the portal vein. Oxygen and nutrients diffuse through the capillary walls into the liver cells

Portal triad
Found at each of the six corners of a lobule; consists of a branch of the hepatic artery, a branch of the hepatic portal vein and a bile duct

Common hepatic duct
Duct collects bile and channels it to the gall bladder

Hepatic artery
Supplies oxygenated blood to the liver

Hepatic portal vein
Collects blood from digestive tract and delivers it to the liver cells

The liver is one of the body's most important organs. It can be seen here as the reddish-brown triangular organ lying in the upper abdominal cavity.

Processes in the liver

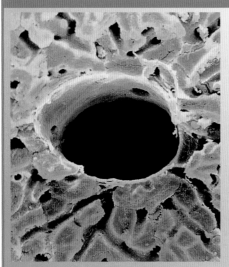

The functions of the liver involve the control of more than 500 chemical reactions, making the liver the most important organ of metabolism – that is, the process whereby chemicals are changed in the body. These include:
■ Storing carbohydrates. The liver breaks down glucose, the form in which carbohydrates are carried in the blood, and stores it as glycogen. The process is reversed when blood glucose levels fall or there is a sudden demand for extra energy
■ Disposal of amino acids. The liver breaks down surplus amino acids, which make up proteins, and turn the ammonia that is produced into urea, a constituent of urine

Blood is detoxified as it flows along the sinusoids towards the centre of the lobule. All the lobules of the liver have a central vein (pictured).

■ Using fat to provide energy. When there is insufficient carbohydrate in the diet to fulfil energy needs, the liver breaks down stored fat into chemicals (ketones), which are used to produce energy and heat
■ Manufacturing cholesterol. Naturally produced cholesterol is essential for the production of bile and hormones such as cortisol and progesterone
■ Storing minerals and vitamins. The liver stores sufficient minerals, such as iron and copper (needed for red blood cells), and vitamins A (which it synthesizes as well as stores), B_{12} and D to meet the body's requirements for a year
■ Processing blood. The liver breaks down old red blood cells, using some of their constituents to make bile pigments. It also manufactures prothrombin and heparin – proteins that affect blood clotting.

Circulation within the liver

The liver has its own circulation system made up of an intricate network of veins and arteries.

The veins and arteries form the liver's own circulation system, known as the 'hepatic portal system' (the Latin word for the liver is *hepaticus*). The purpose of this system is to remove harmful substances from the digestive organs and deal with them before they reach the heart. It also takes some constituents of food out of the digestive tract so that they can be stored for future use.

The portal vein collects blood from the digestive tract and delivers it to the liver cells for processing, while the hepatic artery branches off the aorta to supply nutrients to the liver cells. After it has circulated through the liver's capillaries, the blood is collected by the hepatic veins at the centre of each lobule and passed through the main hepatic vein to the inferior vena cava, to be transported back to the heart.

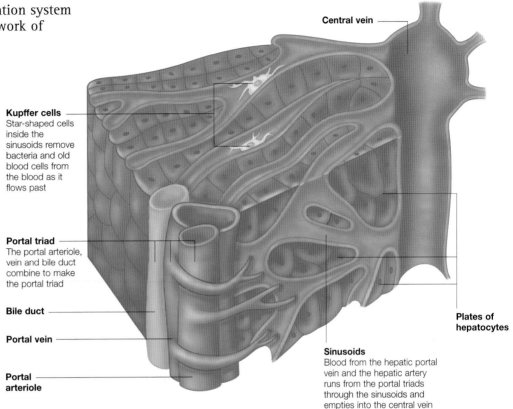

Central vein

Kupffer cells
Star-shaped cells inside the sinusoids remove bacteria and old blood cells from the blood as it flows past

Portal triad
The portal arteriole, vein and bile duct combine to make the portal triad

Bile duct

Portal vein

Portal arteriole

Plates of hepatocytes

Sinusoids
Blood from the hepatic portal vein and the hepatic artery runs from the portal triads through the sinusoids and empties into the central vein

The role of bile

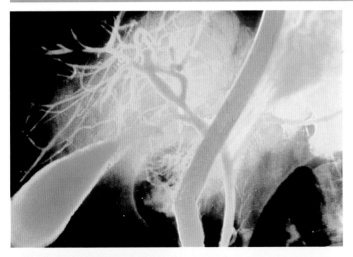

Bile, which is essential for the absorption of vitamins D and E and the breakdown of fats, is produced by liver cells and stored in the gall bladder; it consists of bile salts and bile pigments, which come from the breakdown of red blood cells, cholesterol and lecithin.

The presence of fats in the stomach causes the gall bladder to contract, which squeezes bile through the common bile duct

This false-colour X-ray of the gall bladder and bile ducts has been enhanced by injecting a contrast medium. Bile is stored in the gall bladder.

and into the duodenum. Here, it emulsifies fats, making them easier to digest.

Electron micrograph of red blood cells. Each minute, between 1.2 and 1.7 litres of blood pass through the liver.

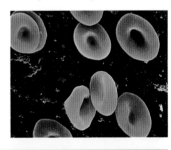

Liver problems

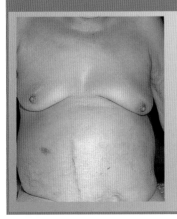

The liver breaks down toxic substances, such as alcohol, into harmless constituents that can be excreted from the body. It also processes chemicals produced naturally in the body, although their constituents are usually recycled. This is the reason why substance abuse can cause serious damage

by overloading the liver.

In extreme cases, cirrhosis can result. This is a serious disorder in which healthy liver tissue is damaged and replaced by fibrous scar tissue, eventually hardening the whole organ. The long-term effect is that the liver's regenerative capability is greatly reduced.

Jaundice – yellowing of the skin due to excessive bilirubin in the blood – is a possible symptom of liver damage caused by the intake of high levels of alcohol.

Cirrhosis is a liver disease that can be caused by alcohol. In this condition, fibrous tissue breaks up the internal structure of the liver (shown here).

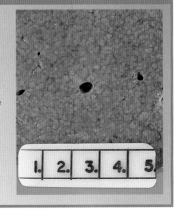

Pancreas and spleen

The pancreas is a large gland that produces both enzymes and hormones. It lies in the upper abdomen behind the stomach, one end in the curve of the duodenum and the other end touching the spleen.

The pancreas secretes enzymes into the duodenum, the first part of the small intestine, to aid the digestion of food. It also produces the hormones insulin and glucagon, which regulate the use of glucose by cells.

Lying across the posterior wall of the abdomen, the pancreas is said to have four parts:
■ The head – which lies within the C-shaped curve of the duodenum. It is attached to the inner side of the duodenum; a small, hook-like projection, the uncinate process, projects towards the midline
■ The neck – which is narrower than the head, due to the large hepatic portal vein behind; it lies over the superior mesenteric blood vessels
■ The body – which is triangular in cross-section and lies in front of the aorta; it passes up and to the left to merge with the tail
■ The tail – which comes to a tapering end within the concavity of the spleen.

BLOOD SUPPLY
The pancreas has a very rich blood supply. The pancreatic head is supplied from two arterial arcades which are formed from the superior and inferior pancreaticoduodenal arteries. The body and tail of the pancreas are supplied with blood by branches of the splenic artery.

Venous blood from the pancreas travels to the liver via the portal venous system, the veins lying in an arrangement which mirrors the arterial supply.

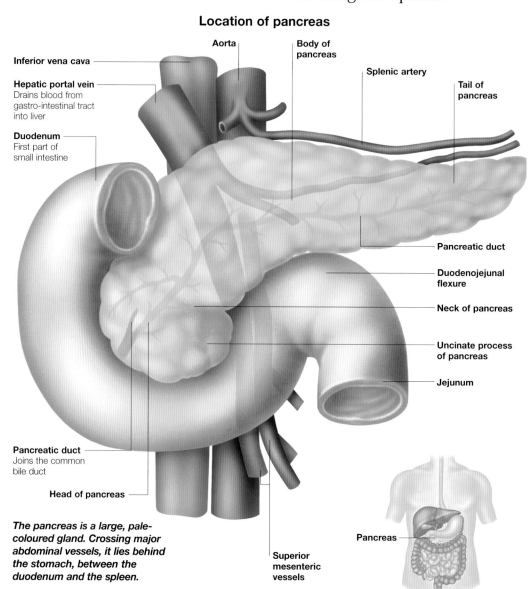

Location of pancreas

Aorta

Body of pancreas

Inferior vena cava

Splenic artery

Hepatic portal vein
Drains blood from gastro-intestinal tract into liver

Tail of pancreas

Duodenum
First part of small intestine

Pancreatic duct

Duodenojejunal flexure

Neck of pancreas

Uncinate process of pancreas

Jejunum

Pancreatic duct
Joins the common bile duct

Head of pancreas

The pancreas is a large, pale-coloured gland. Crossing major abdominal vessels, it lies behind the stomach, between the duodenum and the spleen.

Superior mesenteric vessels

Pancreas

The pancreatic duct and duodenal papilla

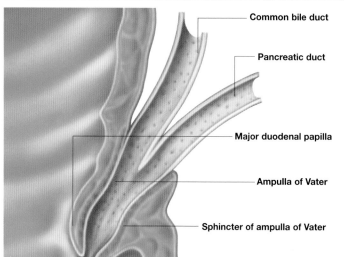

Common bile duct

Pancreatic duct

Major duodenal papilla

Ampulla of Vater

Sphincter of ampulla of Vater

The main pancreatic duct runs the length of the pancreas from tail to head, receiving smaller tributaries as it goes.

At the head of the pancreas it joins with the bile duct to form a short, dilated tube known as the hepatopancreatic ampulla, or ampulla of Vater. This duct opens to discharge its contents into the duodenum at the tip of the major duodenal papilla.

The pancreatic duct and common bile duct join within the head of the pancreas to form the ampulla of Vater. This empties into the duodenum.

MUSCLE FIBRES
Involuntary muscle fibres run around the walls of the two ducts, and around the wall of the combined duct they form, to provide sphincters (specialized rings of muscle that surround an orifice), which regulate the flow of the ducts' contents into the duodenum.

ACCESSORY DUCT
An accessory pancreatic duct may be present in addition to the main duct, and this may have its own, smaller opening into the duodenum, known as the minor duodenal papilla.

The spleen

The spleen is the largest of the lymphatic organs. It is dark purple in colour and lies under the lower ribs on the left side of the upper abdomen.

The dimensions of the spleen can vary greatly, but is usually about the size of a clenched fist. In old age, the spleen naturally atrophies and reduces in size.

The hilum of the spleen contains its blood vessels (the splenic artery and vein) and some lymphatic vessels. The hilum also contains lymph nodes and the tail of the pancreas, all enclosed within the lienorenal ligament – a fold of peritoneum.

SURFACE OF THE SPLEEN

The spleen shows indentations of the organs which surround it. The surface which lies against the diaphragm is curved smoothly, while the visceral surface carries the impressions of the stomach, the left kidney and the splenic flexure of the colon.

SPLEEN COVERINGS

The spleen is surrounded and protected by a thin capsule, which is composed of irregular fibro-elastic connective tissue. Contained within the tissue of the capsule are muscle fibres that allow the spleen to contract periodically. These contractions expel the blood the spleen has filtered back into the circulation.

Outside the capsule the spleen is completely enclosed by the peritoneum, the thin sheet of connective tissue which lines the abdominal cavity and covers the organs within it.

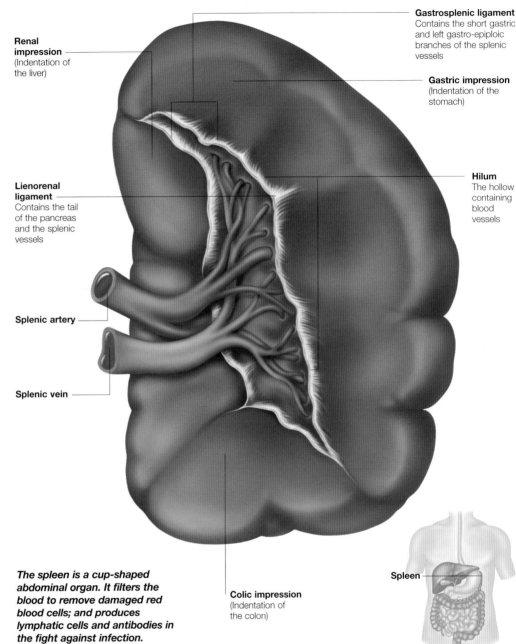

Renal impression
(Indentation of the liver)

Gastrosplenic ligament
Contains the short gastric and left gastro-epiploic branches of the splenic vessels

Gastric impression
(Indentation of the stomach)

Lienorenal ligament
Contains the tail of the pancreas and the splenic vessels

Hilum
The hollow containing blood vessels

Splenic artery

Splenic vein

Colic impression
(Indentation of the colon)

Spleen

The spleen is a cup-shaped abdominal organ. It filters the blood to remove damaged red blood cells; and produces lymphatic cells and antibodies in the fight against infection.

Microanatomy of the spleen

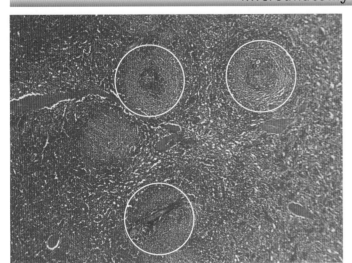

The spleen is enclosed within a capsule, projections of which (the trabeculae) pass down into its substance. The trabeculae support the soft splenic tissue and also carry numerous blood vessels.

By cutting its surface, the tissue of the spleen itself can be seen to be composed of pale areas lying within a red background, the two types of tissue being known as white pulp and red pulp.

A micrograph of the spleen shows areas of white pulp (circled) embedded in the matrix of the red pulp. Each area of white pulp has a central artery.

WHITE PULP

The white pulp is composed mainly of lymphoid cells lying clustered around the small branches of the splenic artery, which bring blood into the spleen.

RED PULP

The red pulp (within which the islands of white pulp lie) consists of connective tissue. This connective tissue contains red blood cells and macrophages, cells which can engulf and destroy other cells.

The function of this tissue is to filter the blood and remove damaged red cells from the bloodstream.

Inguinal region

The inguinal region, commonly known as the groin, is the site of inguinal hernias. The abdominal wall has an area of weakness, which may allow the abdominal contents to protrude through it.

The bilateral areas of weakness in the inguinal region are due to the presence of the inguinal canals, tubes through which pass the spermatic cords in males, and the round ligaments in females.

INGUINAL CANAL
The design of the inguinal canal minimizes the likelihood of herniation (protrusion) of abdominal contents. It passes down towards the midline of the body from its origin, the deep inguinal ring (the entrance to the inguinal canal), to emerge at the superficial inguinal ring (the exit from the canal).

WALLS OF THE CANAL
The inguinal canal has a roof, a floor and two walls:
■ Roof – formed by the arching fibres of the internal oblique and the transversus abdominis muscle
■ Floor – a shallow gutter, formed by the inguinal ligament
■ Anterior wall – formed mainly by the strong aponeurosis of the external oblique muscle, with a contribution from the internal oblique at the outer edges
■ Posterior wall – formed by the transversalis fascia with the medial part of the wall being reinforced by the conjoint tendon.

The inguinal canal passes through the layers of the lower abdominal wall. In adults, it is about 4 cm long, although in babies it is much shorter.

Inguinal region in a male

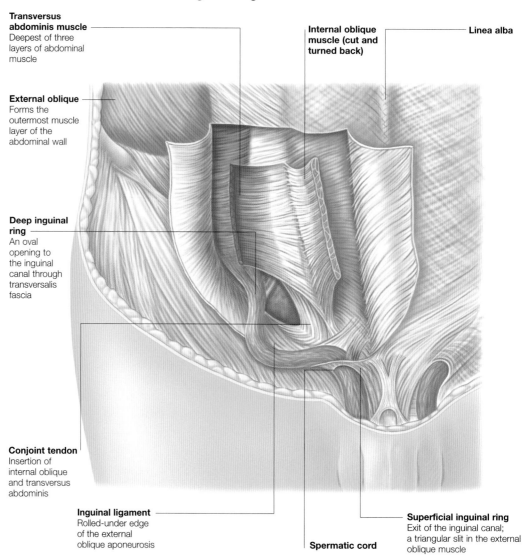

Transversus abdominis muscle
Deepest of three layers of abdominal muscle

External oblique
Forms the outermost muscle layer of the abdominal wall

Deep inguinal ring
An oval opening to the inguinal canal through transversalis fascia

Conjoint tendon
Insertion of internal oblique and transversus abdominis

Inguinal ligament
Rolled-under edge of the external oblique aponeurosis

Spermatic cord

Internal oblique muscle (cut and turned back)

Linea alba

Superficial inguinal ring
Exit of the inguinal canal; a triangular slit in the external oblique muscle

The inguinal ligament

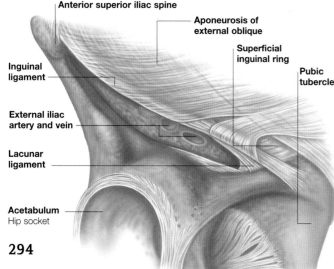

Anterior superior iliac spine

Aponeurosis of external oblique

Superficial inguinal ring

Pubic tubercle

Inguinal ligament

External iliac artery and vein

Lacunar ligament

Acetabulum
Hip socket

The inguinal ligament is a tough, fibrous band that bridges a gap at the front of the pelvis.

The ligament passes across the groin from the anterior superior iliac spine, a prominence of the pelvis above the hip, to the pubic tubercle, a small prominence of the pelvis near the midline.

The inguinal ligament is formed by the inferior, underturned fibres of the aponeurosis of the external oblique muscle. The ligament forms the floor of the inguinal canal.

FORMATION OF THE INGUINAL LIGAMENT
The inguinal ligament is formed from the lower edge of the aponeurosis of the external oblique muscle as it rolls under and reflects back upon itself. The shallow trough produced by this reflection forms the floor of the inguinal canal.

At the medial (inner) end of the inguinal ligament some of its fibres splay out to form the lacunar ligament, knowledge of which is important for the surgeon performing an operation to correct an inguinal hernia.

Behind the inguinal ligament

The inguinal ligament encloses behind it a number of vital structures including blood vessels and nerves serving the lower limb and two groups of lymph nodes (deep and superficial).

There are two major blood vessels which pass behind the inguinal ligament:
■ Femoral artery – the main vessel supplying blood to the lower limb
■ Femoral vein – lies medially to the femoral artery (on the inner side).

FEMORAL NERVE
Lateral to these vessels, on the outer side, lies the femoral nerve which is the largest branch of the lumbar plexus, a network of nerves within the abdomen.

FEMORAL SHEATH
The femoral blood vessels are enclosed within a thin, funnel-shaped sheet of connective tissue, known as the femoral sheath. This allows the femoral vessels to glide harmlessly against the inguinal ligament during movements of the hip.

INGUINAL LYMPH NODES
Within the groin lie two groups of lymph nodes:
■ Superficial inguinal nodes – these lie just under the skin in a horizontal and a vertical group, and drain areas which include the buttocks, external genitalia and superficial layers of the lower limb
■ Deep inguinal nodes – these lie around the femoral artery and vein as they pass under the inguinal ligament, and drain lymph from the lower limb.

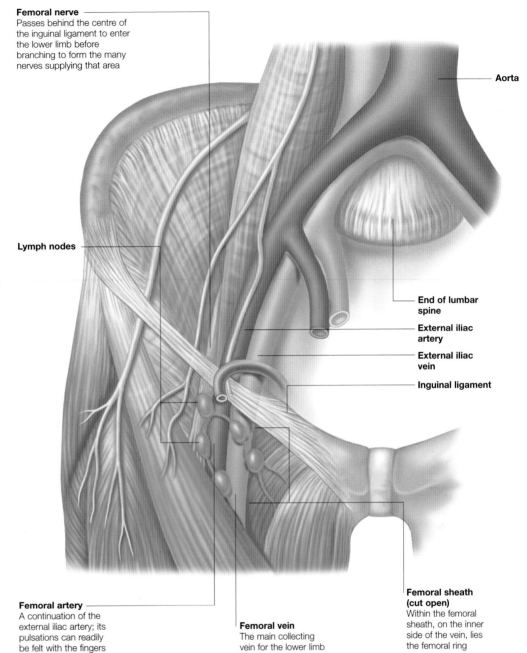

Femoral nerve
Passes behind the centre of the inguinal ligament to enter the lower limb before branching to form the many nerves supplying that area

Aorta

Lymph nodes

End of lumbar spine

External iliac artery

External iliac vein

Inguinal ligament

Femoral sheath (cut open)
Within the femoral sheath, on the inner side of the vein, lies the femoral ring

Femoral artery
A continuation of the external iliac artery; its pulsations can readily be felt with the fingers

Femoral vein
The main collecting vein for the lower limb

The inguinal canal

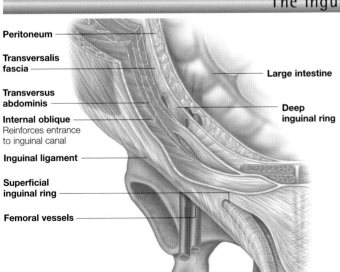

Peritoneum

Transversalis fascia

Transversus abdominis

Internal oblique
Reinforces entrance to inguinal canal

Inguinal ligament

Superficial inguinal ring

Femoral vessels

Large intestine

Deep inguinal ring

The presence of the inguinal canal leaves a potential defect in the otherwise continuous abdominal wall, through which the abdominal contents may herniate (protrude). The risk of this happening is minimized by a number of features:
■ Length – other than in infants, the inguinal canal is a relatively long structure with the entrance and exit some distance apart

The inguinal canal is supported by numerous structures, such as muscles and tendons. These structures prevent abdominal hernias developing

■ Deep ring – the entrance; is reinforced in front by the strong internal oblique muscle
■ Superficial ring – the exit; is reinforced behind by the strong conjoint tendon
■ Raised pressure in the abdomen – the muscle fibres arching over the canal automatically contract (when sneezing and coughing) to close it and compress the contents.
 During defecation and childbirth, the body naturally assumes a squatting position, so that the fronts of the thighs come up to support the inguinal area.

Overview of the urinary tract

The urinary tract consists of the kidneys, ureters, urinary bladder and urethra. Together, these organs are responsible for the production of urine and its expulsion from the body.

The paired kidneys filter the blood to remove waste chemicals and excess fluid, which they excrete as urine. Urine passes down through the narrow ureters to the bladder, which stores it temporarily before it is expelled through the urethra.

■ **Kidneys**
The bean-shaped kidneys lie within the abdomen, against the posterior abdominal wall behind the intestines

■ **Ureters**
From the hilus, or 'stalk' of each of the two kidneys, emerge the right and left ureters. These are narrow tubes which receive the urine produced continuously by the kidney

■ **Bladder**
Urine is received and stored temporarily in the urinary bladder, a collapsible, balloon-like structure which lies within the pelvis

■ **Urethra**
When appropriate, the bladder contracts to expel its contents through the urethra, a thin-walled muscular tube.

The urinary tract consists of the structures involved in the, production, storage and expulsion of urine. It extends from the abdomen into the pelvis.

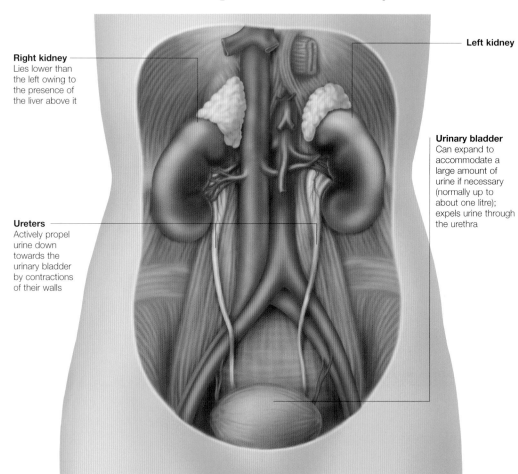

Right kidney
Lies lower than the left owing to the presence of the liver above it

Left kidney

Ureters
Actively propel urine down towards the urinary bladder by contractions of their walls

Urinary bladder
Can expand to accommodate a large amount of urine if necessary (normally up to about one litre); expels urine through the urethra

Rear view of the kidneys

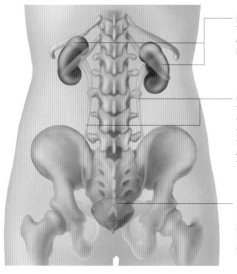

Kidneys
These bean-shaped organs are protected by the lower rib-cage

Ureters
These lengthy tubes extend from the kidneys down to the lower pelvis, and drain urine into the bladder

Bladder
Lies low in the pelvis but rises higher in the abdomen as it fills

The kidneys lie against the posterior (back) wall of the abdomen, their upper poles lying under the 11th and 12th ribs. Because of this posterior position, surgery on the kidneys is usually performed from the back.

POSITION OF KIDNEYS
The right kidney lies approximately 2.5 cm lower than the left. Both kidneys move up and down during respiration and with changes in posture.

The urinary tract extends down the length of the abdomen and pelvis. The kidneys lie behind the lower ribs and the bladder is situated on the pelvic floor.

PROTECTION
The kidneys are protected by the lower ribs, which form a bony cage around them. In addition they are surrounded by a protective layer of fat. The ureters are also well cushioned as they are buried deep within a dense mass of tissue.

PALPATION OF KIDNEYS
Palpation of the lower pole of the right kidney is usually possible by using two hands, one behind the flank and one pressing down from the front. The left kidney, being higher, is not usually palpable unless it is abnormally enlarged or contains a large cyst or tumour.

The adrenal glands

The adrenal glands are situated above the kidneys, but are not part of the urinary tract. Each one consists of two separate parts: a medulla surrounded by a cortex.

Lying on top of the kidneys are the paired adrenal glands, also known as the suprarenal glands. Although they are physically close to the kidneys they play no part in the urinary system: they are endocrine glands which produce hormones vital to the healthy functioning of the body.

SURROUNDING TISSUES

The yellowish adrenal glands lie above the kidneys and under the diaphragm. They are surrounded by a thick layer of fatty tissue and are enclosed by renal fascia although they are separated from the kidneys themselves by fibrous tissue.

This separation allows a kidney to be surgically removed without damaging these delicate and important glands.

GLAND DIFFERENCES

Due to the position of the surrounding structures the soft adrenal glands differ in appearance:

■ **The right adrenal gland**
The right adrenal gland is pyramid-shaped and sits on the upper pole of the right kidney. It lies in contact with the diaphragm, the liver and the inferior vena cava, the main vein of the abdomen

■ **The left adrenal gland**
The left adrenal gland has the shape of a half moon and lies along the upper surface of the

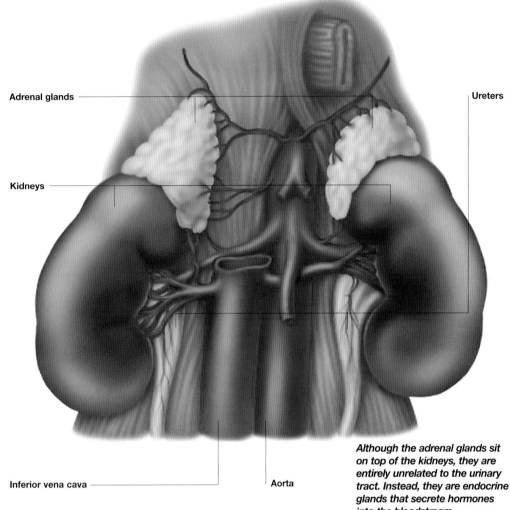

Adrenal glands

Ureters

Kidneys

Inferior vena cava

Aorta

Although the adrenal glands sit on top of the kidneys, they are entirely unrelated to the urinary tract. Instead, they are endocrine glands that secrete hormones into the bloodstream.

left kidney from the pole down to the hilus. It lies in contact with the spleen, the stomach, the pancreas and the diaphragm.

BLOOD SUPPLY

As with other endocrine glands, which secrete their hormones directly into the bloodstream, the adrenal glands have a very rich blood supply. They receive arterial blood from three sources – the superior, middle and inferior adrenal arteries – which arise from the inferior phrenic artery, the aorta and the renal artery respectively.

Near the adrenal glands these arteries branch repeatedly, numerous tiny arteries thereby entering the glands over their entire surface.

A single vein leaves the adrenal glands on each side to drain blood into the inferior vena cava on the right and the renal vein on the left.

Structure of the adrenal glands

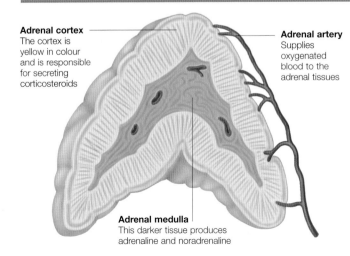

Adrenal cortex
The cortex is yellow in colour and is responsible for secreting corticosteroids

Adrenal artery
Supplies oxygenated blood to the adrenal tissues

Adrenal medulla
This darker tissue produces adrenaline and noradrenaline

Within the protective capsule, each adrenal gland is composed of an outer region, known as a cortex, and an inner region – a medulla. The regions are made up of two separate types of tissue which have differing functions.

ADRENAL CORTEX

The yellow adrenal cortex makes up the bulk of the gland. It makes and secretes a wide

Adrenal glands consist of two types of tissue – cortex and medulla. Each area is responsible for producing different types of hormones.

variety of hormones collectively known as corticosteroids which are vital for control of the metabolism of the body, fluid balance and response to stress. The cortex also produces a very small amount of male sex hormones (androgens).

ADRENAL MEDULLA

The darker adrenal medulla is made up of a 'knot' of nervous tissue surrounded by numerous small blood vessels, and is the site of formation of the hormones adrenaline and noradrenaline. These hormones prepare the body for the 'fight or flight' response to stress.

Kidneys

The kidneys are a pair of solid organs situated at the back of the abdomen. They act as filtering units for blood and maintain the balance and composition of fluids within the body.

The paired kidneys lie within the abdominal cavity against the posterior abdominal wall. Each kidney is about 10 cm in length, reddish brown in colour and has the characteristic shape, after which the 'kidney bean' is named. On the medial, or inward facing, surface lies the hilus of the kidney from which the blood vessels enter and leave. The hilus is also the site of exit for the right and left ureters, via which urine leaves the kidney and is transported to the bladder.

REGIONS OF THE KIDNEY

The kidney has three regions, each of which plays a role in the production or collection of urine:
■ The renal cortex – the most superficial layer; it is quite pale and has a granular appearance
■ The renal medulla – composed of dark reddish tissue, it lies within the cortex in the form of 'pyramids'
■ The renal pelvis – the central, funnel-like area of the kidney which collects the urine and is continuous with the ureters at the hilus.

OUTER LAYERS

Each kidney is covered by a tough, fibrous capsule. Outside the kidney lies a protective layer of fat which is contained within the renal fascia – a dense connective tissue that anchors the kidneys and adrenal glands, to surrounding structures.

The kidneys are responsible for the excretion of waste from the blood. Each has three regions: cortex, medulla and renal pelvis.

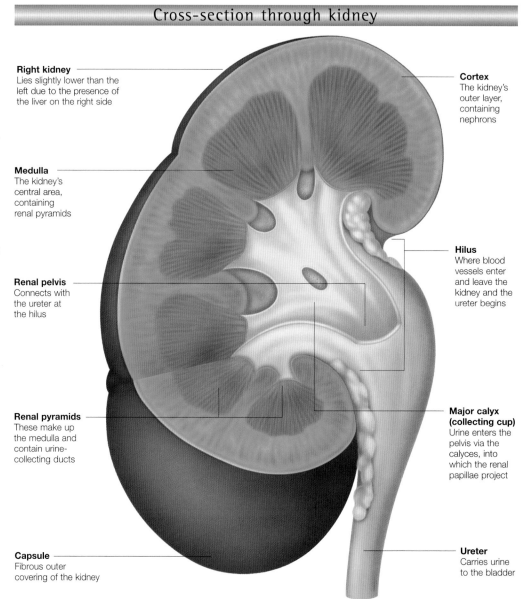

Cross-section through kidney

Right kidney
Lies slightly lower than the left due to the presence of the liver on the right side

Medulla
The kidney's central area, containing renal pyramids

Renal pelvis
Connects with the ureter at the hilus

Renal pyramids
These make up the medulla and contain urine-collecting ducts

Capsule
Fibrous outer covering of the kidney

Cortex
The kidney's outer layer, containing nephrons

Hilus
Where blood vessels enter and leave the kidney and the ureter begins

Major calyx (collecting cup)
Urine enters the pelvis via the calyces, into which the renal papillae project

Ureter
Carries urine to the bladder

Nephrons of the kidney

Glomerulus

Bowman's capsule

Urine-collecting tubule

Distal tubule

Afferent arteriole

Efferent arteriole

Proximal tubule

Loop of Henle

The work of the kidneys is achieved by the action of over a million tiny nephrons, or blood-processing units. Each nephron contains a renal corpuscle within the medulla, from which projects a long loop of renal tubule:
■ **Renal corpuscle**
The renal corpuscle is composed of a clump of tiny arterioles, the

Water and solutes from arterial blood pass across a membrane in the glomerulus. This fluid, or urine, passes into the renal tubule where it is processed.

glomerulus, surrounded by an expanded cup of renal tubule, known as the Bowman's capsule. Fluid filters out of the blood to enter the renal tubule here for processing
■ **Renal tubule**
The renal tubule makes a long journey from its origin at the Bowman's capsule into the cortex and back again as Henle's loop. It then ultimately drains its contents, processed urine, into a collecting tubule which carries it away to the renal pelvis.

Blood supply to the kidneys

The function of the kidneys is to filter blood, for which they receive an exceedingly rich blood supply. As with other parts of the body, the pattern of drainage of venous blood mirrors the pattern of arterial supply.

Arterial blood is carried to the kidneys by the right and left renal arteries, which arise directly from the main artery of the body, the aorta. The right renal artery is longer than the left as the aorta lies slightly to the left of the midline. One in three people have an additional, accessory, renal artery.

RENAL ARTERIES

The renal artery enters the kidney at the hilus and divides into between three and five segmental arteries, each of which further divides into lobar arteries. There are no connections between branches of neighbouring segmental arteries.

The interlobar arteries pass between the renal pyramids and branch to form the arcuate arteries, which run along the junction of cortex and medulla. Numerous interlobular arteries pass into the tissue of the renal cortex to carry blood to the glomeruli of the nephrons, where it is filtered to remove excess fluid and waste products.

VENOUS DRAINAGE

Blood enters the interlobular, arcuate and then interlobar veins before being collected by the renal vein and returned to the inferior vena cava, the main collecting vein of the abdomen.

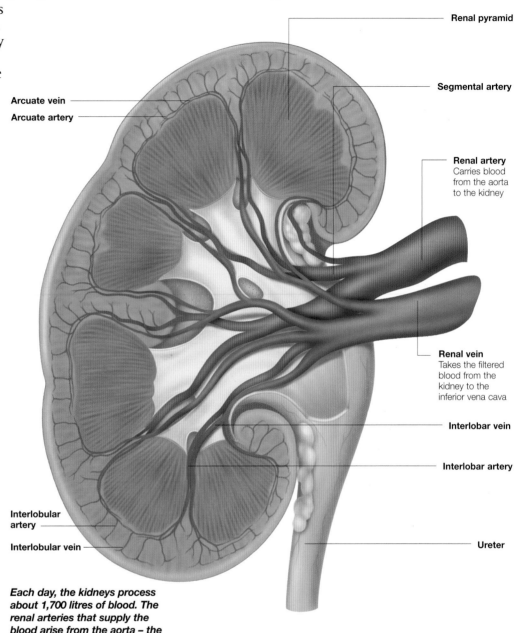

Arcuate vein

Arcuate artery

Renal pyramid

Segmental artery

Renal artery
Carries blood from the aorta to the kidney

Renal vein
Takes the filtered blood from the kidney to the inferior vena cava

Interlobar vein

Interlobar artery

Interlobular artery

Interlobular vein

Ureter

Each day, the kidneys process about 1,700 litres of blood. The renal arteries that supply the blood arise from the aorta – the main blood vessel in the body.

Congenital abnormalities

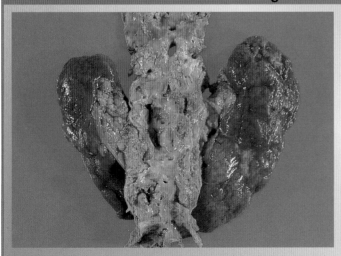

During early fetal life, the kidneys develop close together in the pelvis and then later ascend into their final resting position on the posterior abdominal wall under the diaphragm. Very occasionally, the kidneys and their associated structures do not develop normally, which leads to congenital abnormalities:

■ **Horseshoe kidney**
In about one in 600 children, the

A horsehoe kidney is an abnormality in which the kidneys become fused together during development. Usually, kidney function is not affected.

kidneys become fused together at their lower poles during development. The resulting U-shaped kidney usually lies at a lower level than normal kidneys

■ **Renal agenesis**
Occasionally, a baby is born with only one kidney. However, it is possible to live normally with a single kidney, which enlarges to cope with the increased workload

■ **Duplex ureters**
Some children are born with duplicated ureters. This condition, which is not uncommon, can occur on one or both sides, and may be partial or complete.

How kidneys produce urine

The kidneys are responsible for maintaining the volume and chemical composition of bodily fluids. They do this by filtering impurities from the blood and excreting excess water and metabolic by-products as urine.

The kidneys are the major excretory organs of the body, and are situated towards the back of the abdomen, below the diaphragm. They are responsible for maintaining the constancy of body fluids by filtering toxins, metabolic waste products and excess ions from the blood. The end result of this process is the excretory fluid urine.

At the same time, the kidneys also maintain blood volume (the correct balance of water and salts) and the correct acidity of body fluids. This complex process is called homeostasis.

INSIDE THE KIDNEY

There are three distinct zones within the kidney: the renal cortex (outermost zone), the renal pelvis (inner zone) and the renal medulla (middle zone). The cortex is granular and pale in appearance, and contains a network of arteries, veins and capillaries. The medulla is a darker, striped area divided into conical structures known as renal pyramids. At the apex of each pyramid are papillae, nipple-shaped projections that extend into the renal pelvis via cavities known as calyces.

There are over one million blood processing units within the kidney that are called nephrons. Urine produced by the nephrons drains into the pelvis via calyces. In turn, the pelvis is linked to the ureter, the tubes that channel the urine to the bladder.

Internal structure of the kidney

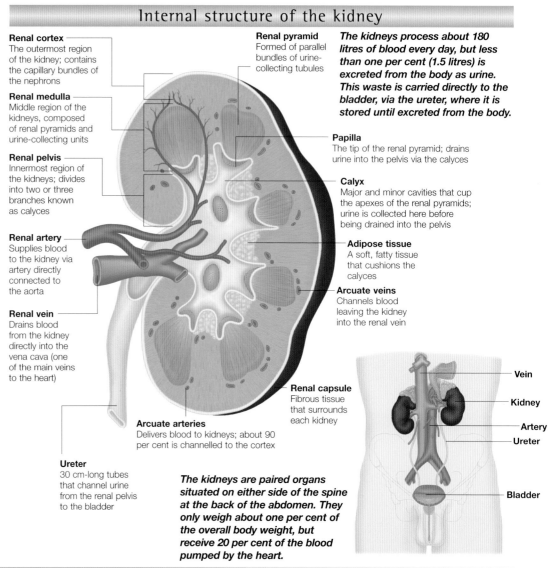

Renal cortex
The outermost region of the kidney; contains the capillary bundles of the nephrons

Renal medulla
Middle region of the kidneys, composed of renal pyramids and urine-collecting units

Renal pelvis
Innermost region of the kidneys; divides into two or three branches known as calyces

Renal artery
Supplies blood to the kidney via artery directly connected to the aorta

Renal vein
Drains blood from the kidney directly into the vena cava (one of the main veins to the heart)

Ureter
30 cm-long tubes that channel urine from the renal pelvis to the bladder

Arcuate arteries
Delivers blood to kidneys; about 90 per cent is channelled to the cortex

Renal pyramid
Formed of parallel bundles of urine-collecting tubules

Papilla
The tip of the renal pyramid; drains urine into the pelvis via the calyces

Calyx
Major and minor cavities that cup the apexes of the renal pyramids; urine is collected here before being drained into the pelvis

Adipose tissue
A soft, fatty tissue that cushions the calyces

Arcuate veins
Channels blood leaving the kidney into the renal vein

Renal capsule
Fibrous tissue that surrounds each kidney

The kidneys process about 180 litres of blood every day, but less than one per cent (1.5 litres) is excreted from the body as urine. This waste is carried directly to the bladder, via the ureter, where it is stored until excreted from the body.

Vein

Kidney

Artery

Ureter

Bladder

The kidneys are paired organs situated on either side of the spine at the back of the abdomen. They only weigh about one per cent of the overall body weight, but receive 20 per cent of the blood pumped by the heart.

Urinary drainage

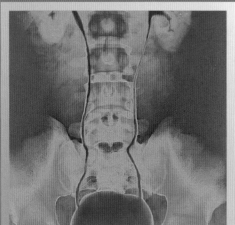

The production of urine is a three-step process: filtration, reabsorption and secretion. Once the required water and essential nutrients have been reabsorbed, the fluid remaining in the tubule is urine, which empties into the collecting ducts and then to the ureters to be excreted from the body via the bladder.

The walls of the ureter are muscular. Regular waves of contraction (peristalsis) move the urine from the renal pelvis towards

This contrast medium X-ray clearly shows the kidneys (green) and the ureters (red – the vessels connecting the ureters to the kidneys). The bladder is the dark red circular mass at the bottom of the X-ray.

the bladder every 10–60 seconds. The ureters pass obliquely through the bladder wall, tending to close the uretic opening except during a peristaltic contraction. This prevents the backflow of urine.

The bladder muscle is controlled by involuntary nerve action. The bladder fills without increasing internal pressure until it is near capacity. When the bladder is full, the pressure within rises dramatically, triggering a spinal nerve reflex which acts to cause the bladder muscle to contract and empty its contents via the urethra. This is the process of micturition (urination). The first urge to micturate is felt when the bladder volume is about 150 ml. This increases to a sense of urgency at 400 ml.

Urine production

Approximately one litre of blood flows into an adult kidney every minute.
There are over one million urine-producing units within the kidney, and from all
of these, one millilitre of urine is produced every minute.

The nephron is the functional, structural unit of the kidney which filters blood and is responsible for urine production. There are over a million nephrons in each kidney, as well as thousands of collecting ducts into which the urine drains.

The nephron is formed from two main units: a glomerulus and its associated renal tubule. The glomerulus is a tight ball of capillaries situated in the renal cortex, and its tubules, through which water and chemicals absorb into the blood, extend down into the medulla.

BOWMAN'S CAPSULE

At one end of the renal tubule, completely encasing the glomerulus, is a closed unit called the Bowman's capsule. Together, the Bowman's capsule and its glomerulus are called a renal corpuscle, and are responsible for filtering waste products into the renal tubule.

The other end of the renal tubule connects to a urine-collecting tubule. The specific nature and function of the cells within the renal tubule are essential to the excretory and homeostatic function of the nephron as a whole.

The nephron and its blood supply

Glomerulus
Tight knot of capillary blood vessels located in the renal cortex; blood is fed into this capillary network and drained out via two arterioles

Bowman's capsule
Cup-shaped end of a nephron which encloses the glomerulus; it is the site of blood filtration into the kidney tubule

Afferent arteriole
Arteriole (blood vessel linking capillaries to arteries) that feeds blood into the glomerulus from the interlobular artery

Interlobular artery
Branch of the renal artery, which delivers blood to the kidney

Arcuate vein
Branch of the renal vein, which empties blood into the heart

Loop of Henle
Hairpin bend in the renal tubule; nutrient reabsorption can also take place here

Proximal convoluted tubule
Location of the first stage of the reabsorption process, in which water and useful chemicals begin to re-enter the blood

Efferent arteriole
Drains blood from the glomerulus into the renal tubules.

Distal convoluted tubule
Another section of the renal tubule concerned with reabsorption; also largely responsible for water regulation and the balance of chemically active solutions

Urine collecting tubule
Drains urine into the ureter, for removal to the bladder

The nephron is the active unit of filtration within the kidneys. It is composed of two main elements: the glomerulus, which filters blood, and the renal tubule, which re-absorbs the useful substances back into the blood and extracts the waste material. The tubule is divided into distinct segments: the proximal tubule, the loop of Henle and the distal tubule.

Glomeruli are tight knots of blood capillaries (seen here in blue) in the kidneys. Each glomerulus forms part of a tiny filtration unit that removes toxic waste from the blood.

EXCRETION OF METABOLIC WASTE PRODUCTS

The waste products of metabolism are eliminated by the kidneys via the nephrons. They also excrete toxins ingested or produced by the body. The principle waste products in urine are urea (from protein metabolism), creatinine (from muscle), uric acid (from metabolism of nucleic acids), bilirubin (from haemoglobin metabolism) and the broken down products of hormones.

The nephron works by a process of secretion followed by reabsorption. Nutrients and waste products flow freely out of the blood in the glomerulus into the Bowman's capsule. These chemicals are accompanied by water and many essential nutrients, which must be reclaimed by the body.

This reabsorption occurs in the remaining parts of the nephron and renal tubules. The waste eventually drains into the collecting ducts to be eliminated from the body.

Most of this reabsorption takes place in a section of the renal tubules called the distal convoluted tubule (see diagram above). The reabsorption and some secretion that takes place here, and in another section known as the loop of Henle, is dependent upon the body's requirements at the time.

Closely associated with the capillary bed of the glomerulus and the renal tubules are the peritubular capillaries. These are another vital element to the reabsorption process. The pressure in these capillaries is much lower than that of the glomerulus and allows water and nutrients to flow freely into them, re-absorbing them back into the blood.

Capillary networks

On entering the kidney, the renal artery divides into several branches, each radiating towards the cortex. In the cortex, the branches subdivide repeatedly into smaller and smaller vessels. The final sub-branch is called an arteriole. Each arteriole supplies blood to one nephron.

The anatomy of the arterial blood supply to the kidney nephrons is unique, in that each nephron is supplied by two, rather than one, capillary beds. The arteriole supplying the nephron is known as the efferent arteriole. It is the tight knotting of the resulting capillaries that forms the glomerulus.

On leaving the capillary tuft, the microvessels join together to form the outgoing arteriole, known as the afferent arteriole. This arteriole then redivides into the peritubular capillaries – a second network of microvessels surrounding the urine collecting tubule further down its length. These capillaries empty into the vessels of the venous system, eventually draining into the renal vein.

The pressure in the glomerulus is high, forcing fluid, nutrients and waste products out of the blood into the nephron capsule. The pressure in the peritubular capillaries is low, allowing fluid reabsorption. Adjustments to the pressure differences between the two capillary beds control the excretion and reabsorption of water and chemicals within the blood.

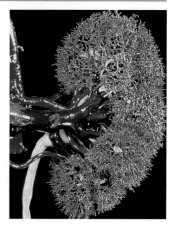

A cast of a normal kidney shows the complex capillary networks within the organ. There are approximately one million arterioles in each kidney.

How the kidneys control blood pressure

The kidneys play a fundamental role in the long-term regulation of blood pressure. The blood pressure must be kept stable so that organs receive an adequate supply of blood and oxygen.

The kidneys are two bean-shaped organs located on either side of the pelvis. They have two main roles:
■ Regulating the salt and water balance in the body
■ Excreting waste substances such as urea, excess salt and other minerals, in the form of urine.

FILTRATION SYSTEM

The kidneys contain millions of microscopic filtering units, called nephrons, which are the working components of the kidneys. Certain substances in the blood (such as glucose) are filtered but reabsorbed back into the bloodstream, while harmful wastes and excess water are excreted in the form of urine.

BLOOD PRESSURE

The kidneys play an extremely important role in the long-term regulation of blood pressure. Blood pressure is defined as the pressure of blood against the walls of the main arteries and is an indication of the efficiency of a person's circulation.

REGULATION

Blood pressure must be regulated in order to provide an adequate supply of blood and oxygen to the organs.
■ Hypotension (low blood pressure) may indicate that there is insufficient blood in the circulation. This can result in vital organs being deprived of oxygen rich blood and result in shock
■ Hypertension (abnormally high blood pressure) means that the heart has to work harder to pump blood against a greater resistance within the arterial circulation, putting great strain on the heart.

Blood is filtered through the kidneys. Some substances are reabsorbed into the blood, while others, such as excess water and waste, are excreted as urine.

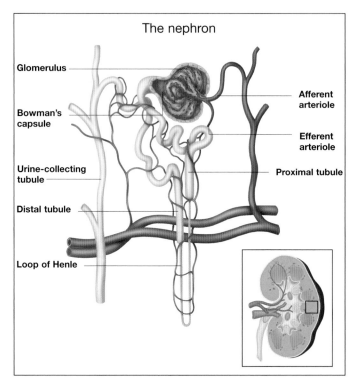

The nephron

- Glomerulus
- Bowman's capsule
- Urine-collecting tubule
- Distal tubule
- Loop of Henle
- Afferent arteriole
- Efferent arteriole
- Proximal tubule

Blood volume

Drinking water is an important way of maintaining blood volume. The kidneys use water levels and salt concentration to control blood pressure.

A number of mechanisms within the body act to ensure that blood pressure is kept within normal limits, on both a short- and long-term basis. The kidneys play an important part in this long-term regulation of blood pressure.

BLOOD VOLUME

The kidneys help to maintain homeostasis (equilibrium) in the circulation by regulating blood volume. Although the volume of blood varies with age and gender, the kidneys usually maintain total circulating volume at around five litres.

Any significant alteration in this level will affect blood pressure:
■ An increase in blood volume leads to an increase in blood pressure. For example, an excessive intake of salt with resulting water retention can lead to a higher blood pressure
■ A decrease in blood volume causes a decrease in blood pressure. Severe blood loss or dehydration are common causes. A sudden drop in blood pressure may indicate internal bleeding.

FEEDBACK SYSTEM

The role of the kidneys is to detect any changes in blood volume or pressure via a feedback system and to react accordingly.
■ When blood volume increases, the kidneys remove more water from the blood, reducing the blood volume and restoring normal blood pressure
■ When blood volume decreases, due to dehydration for example, the kidneys absorb less water, thus restoring blood pressure.

Renal hormones

Blood volume is a direct indicator of blood pressure. The kidneys continually monitor blood absorption and sodium levels to maintain an even pressure.

The kidneys regulate blood pressure by altering the amount of urine passed, thereby regulating blood volume. When blood pressure is low, the kidneys conserve water in the circulation and when it is raised, they ensure that greater volumes of water are passed as urine.

FILTRATION RATE
Within each nephron (functional unit of the kidney) is a bundle of arterioles (blood vessels) called the glomerulus. Water and solutes are 'pushed' out of the blood into the collecting tubules by the higher blood pressure in the glomerulus. An average person filters about 125 mls of filtrate per minute. If the blood pressure is too low, water will remain in the circulation to help boost the blood pressure. If the blood pressure is high, more water is forced into the tubules and passed as urine.

FEED-BACK MECHANISM
The walls of the blood vessels supplying the nephrons contain specialized cells that are able to detect blood pressure. It is these cells that set into motion additional processes needed to rectify abnormal pressure.
■ The blood pressure falls below normal limits and the specialized cells detect this change
■ A hormone called renin is secreted into the blood stream
■ Renin converts a substance called angiotensin into angiotensin I, which then becomes angiotensin II as it passes through the lungs in the blood
■ Angiotensin II stimulates the adrenal glands (located on the top of the kidneys) to produce aldosterone
■ Aldosterone acts directly on the nephrons in the kidneys so that more salt and water are reabsorbed back into the blood circulation. This results in an increase in blood pressure
In addition to this mechanism, angiotensin II constricts blood vessels, thus increasing the pressure within them.

ANTI-DIURETIC HORMONE
The hypothalamus in the brain also has a role to play. When the water concentration in the blood is low, potentially leading to a drop in blood pressure, the hypothalamus secretes anti-diuretic hormone (ADH). This acts on the tubules in the nephrons, making them more permeable so that more water is reabsorbed into the blood.

The kidneys help to control blood pressure using a feedback mechanism. This diagram shows the sequence of events following a change in pressure.

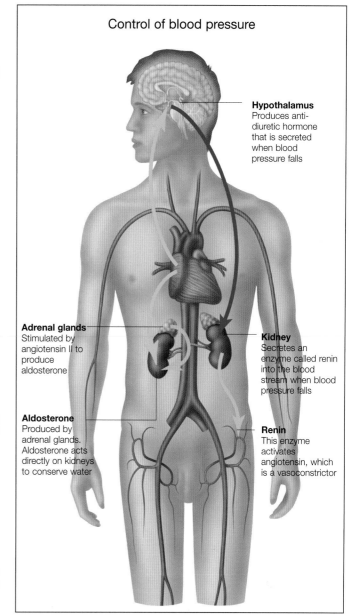

Control of blood pressure

Hypothalamus
Produces anti-diuretic hormone that is secreted when blood pressure falls

Adrenal glands
Stimulated by angiotensin II to produce aldosterone

Aldosterone
Produced by adrenal glands. Aldosterone acts directly on kidneys to conserve water

Kidney
Secretes an enzyme called renin into the blood stream when blood pressure falls

Renin
This enzyme activates angiotensin, which is a vasoconstrictor

The causes of high and low blood pressure

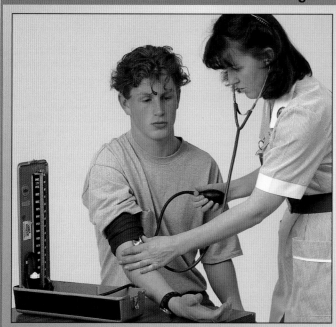

The normal blood pressure of a resting adult is usually about 120/80 mmHg, but this can be influenced by a wide range of factors:
■ Age. Blood pressure naturally increases throughout life. This is because the arteries lose the elasticity that, in younger people, absorbs the force of heart contractions
■ Gender. Men generally experience higher blood pressure than women or children
■ Lifestyle choices. Being overweight, consuming high levels of alcohol or enduring a long period of stress can all contribute to high blood pressure.

Blood pressure is influenced by a number of factors, such as age and stress. Regular monitoring and life-style advice are vital in those at risk.

Hypertension
Abnormally high blood pressure (hypertension) may be caused by a number of factors, but is commonly caused by atherosclerosis, a disease that causes narrowing of the blood vessels.
  When the disease affects the arteries of the kidney (renal arteries), it may cause long-term problems with blood pressure regulation.

Hypotension
Abnormally low blood pressure (hypotension) is usually due to reduced blood volume or increased blood-vessel capacity. This can happen in the case of severe burns or dehydration, which both lower blood volume, or through an infection such as septicaemia which causes a widening of the blood vessels.

Bladder and ureters

The ureters channel urine produced by the kidneys down their length and into the urinary bladder. Urine is stored in the bladder until it is expelled from the body via the urethra.

Urine is continuously produced by the kidneys and is carried down to the urinary bladder by two muscular tubes, the ureters.

THE BLADDER

The bladder stores urine until it is passed out via the urethra. When the bladder is empty, it is pyramidal in shape, its walls thrown into folds, or rugae, which flatten out on filling. The position of the bladder varies:

■ In adults the empty bladder lies low within the pelvis, rising up into the abdomen as it fills

■ In infants the bladder is higher, being within the abdomen even when empty

■ The walls contain many muscle fibres, collectively known as the detrusor muscle, which allow the bladder to contract and expel its contents.

TRIGONE

The trigone is a triangular area of the bladder wall at the base of the structure. The wall here contains muscle fibres which act to prevent urine from ascending the ureters when the bladder contracts. A muscular sphincter around the urethral opening keeps it closed until urine is passed out of the body.

The urinary bladder is flexible enough to expand as it fills. It is made from strong muscle fibres that facilitate the expulsion of urine when necessary.

Coronal section of female bladder and urethra

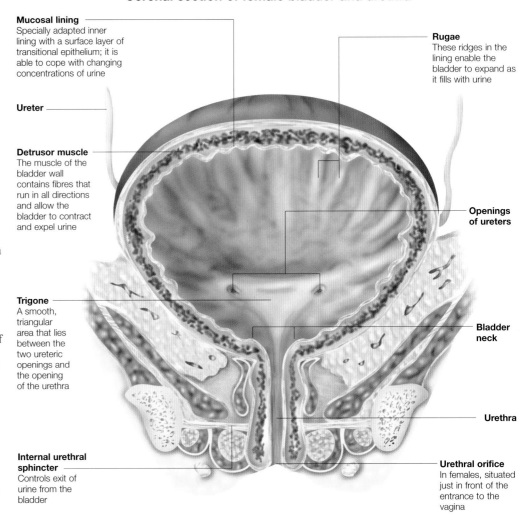

Mucosal lining
Specially adapted inner lining with a surface layer of transitional epithelium; it is able to cope with changing concentrations of urine

Ureter

Detrusor muscle
The muscle of the bladder wall contains fibres that run in all directions and allow the bladder to contract and expel urine

Trigone
A smooth, triangular area that lies between the two ureteric openings and the opening of the urethra

Internal urethral sphincter
Controls exit of urine from the bladder

Rugae
These ridges in the lining enable the bladder to expand as it fills with urine

Openings of ureters

Bladder neck

Urethra

Urethral orifice
In females, situated just in front of the entrance to the vagina

Differences in male and female anatomy

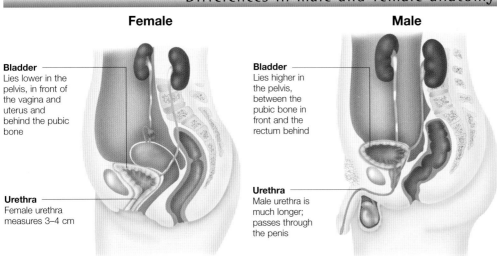

Female

Bladder
Lies lower in the pelvis, in front of the vagina and uterus and behind the pubic bone

Urethra
Female urethra measures 3–4 cm

Male

Bladder
Lies higher in the pelvis, between the pubic bone in front and the rectum behind

Urethra
Male urethra is much longer; passes through the penis

Owing to the presence of the reproductive organs the position of the bladder, and size, shape and position of the urethra vary between males and females:

■ In men the urethra is about 20 cm long, passing through the prostate gland and then running along the penis before opening at the external urethral orifice

■ In women the urethra is 3–4 cm in length and opens at the urethral orifice, which lies just in front of the vaginal opening.

The major difference in male and female urinary tract anatomy is the length of the urethra. An adult male urethra is five times the length of a female one.

The ureters

The ureters are tubular and propel the urine towards the bladder. Each ureter squeezes and contracts its muscles to encourage the free flow of urine.

The ureters are narrow, thin-walled muscular tubes which carry urine from the kidneys to the urinary bladder.

Each of the two ureters is 25–30 cm in length and about 3 mm wide. They originate at the kidney and pass down the posterior abdominal wall to cross the bony brim of the pelvis and enter the bladder by piercing its posterior wall.

PARTS OF THE URETER
Each ureter consists of three anatomically distinct parts:

■ Renal pelvis
This is the first part of the ureter, which lies within the hilum of the kidney. It is funnel-shaped as it receives urine from the major calyces and then tapers to form the narrow ureteric tube. The junction of this part of the ureter with the next is one of the narrowest parts of the whole structure.

■ Abdominal ureter
The ureter passes downwards through the abdomen and then slightly towards the midline until it reaches the pelvic brim and enters the pelvis. During its course through the abdomen the ureter runs behind the peritoneum, the membranous lining of the abdominal cavity.

■ Pelvic ureter
The ureter enters the pelvis just in front of the division of the large common iliac artery. It runs down the back wall of the pelvis before turning to enter the posterior wall of the bladder.

View of the ureters and bladder from behind

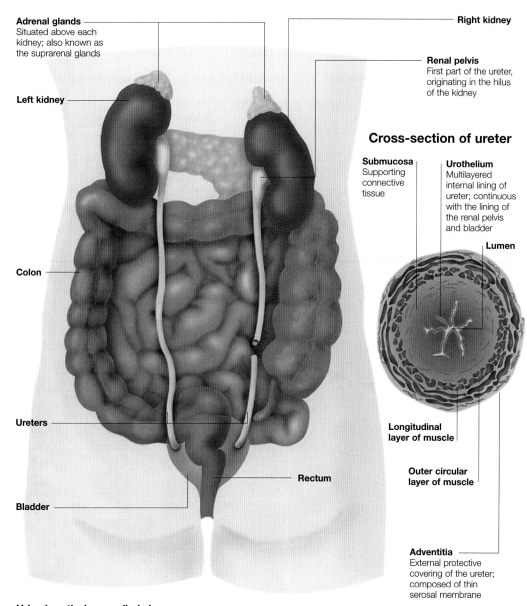

Adrenal glands
Situated above each kidney; also known as the suprarenal glands

Left kidney

Colon

Ureters

Bladder

Right kidney

Renal pelvis
First part of the ureter, originating in the hilus of the kidney

Rectum

Urine is actively propelled along the ureters to the bladder by contraction of the muscular walls. This is the action known as 'peristalsis'.

Cross-section of ureter

Submucosa
Supporting connective tissue

Urothelium
Multilayered internal lining of ureter; continuous with the lining of the renal pelvis and bladder

Lumen

Longitudinal layer of muscle

Outer circular layer of muscle

Adventitia
External protective covering of the ureter; composed of thin serosal membrane

Looking at the ureter on X-ray

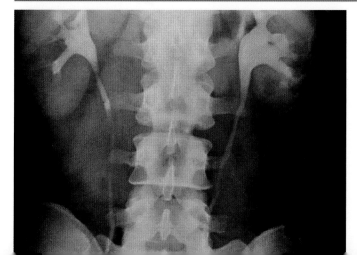

The ureter does not show up on a plain X-ray. However, calcium-rich renal stones may be seen on X-ray at one of the narrower points of the ureter.

UROGRAPHY
The kidneys, ureters and bladder may be outlined by performing an intravenous urogram.

This contrast X-ray of the urinary tract clearly shows two normal ureters. These muscular tubes pass down the entire length of the abdomen.

In this investigation a contrast dye, which shows up on X-ray, is injected intravenously and is then concentrated and excreted by the kidneys. Radiographs taken at intervals show the course of the ureters as they run from the kidneys down through the abdomen to the bladder.

The ureters appear to have constricted and dilated sections. This is due to the presence of waves of peristalsis, the muscular action by which the ureter propels urine towards the urinary bladder.

Chapter Seven

The Reproductive System

From the moment we are born, our bodies are equipped with the genital organs and tissues that make up the male and female reproductive systems. These remain inactive until we reach puberty, at which point hormone levels change and our bodies become sexually mature in preparation for procreation.

This chapter charts the anatomical makeup of our reproductive systems, and explores each stage of our sexual and reproductive journey – from the onset of puberty and our experience of sexual intercourse, to the development of the foetus, the process of childbirth and the effect of ageing on our reproductive organs.

LEFT: A child touches her pregnant mother's belly. It is important that older siblings form a bond with a baby as soon as possible.

Male reproductive system

The male reproductive system includes the penis, scrotum and the two testes (contained within the scrotum). The internal structures of the reproductive system are contained within the pelvis.

The structures constituting the male reproductive tract are responsible for the production of sperm and seminal fluid and their carriage out of the body. Unlike other organs it is not until puberty that they develop and become fully functional.

CONSTITUENT PARTS

The male reproductive system consists of a number of interrelated parts:

■ Testis – the paired testes lie suspended in the scrotum. Sperm are carried away from the testes through tubes or ducts, the first of which is the epididymis

■ Epididymis – on ejaculation sperm leave the epididymis and enter the vas deferens

■ Vas deferens – sperm are carried along this muscular tube en route to the prostate gland

■ Seminal vesicle – on leaving the vas deferens sperm mix with fluid from the seminal vesicle gland in a combined 'ejaculatory' duct

■ Prostate – the ejaculatory duct empties into the urethra within the prostate gland

■ Penis – on leaving the prostate gland, the urethra then becomes the central core of the penis.

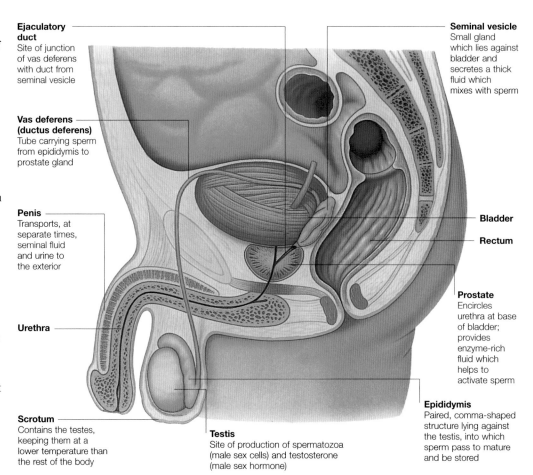

Ejaculatory duct
Site of junction of vas deferens with duct from seminal vesicle

Vas deferens (ductus deferens)
Tube carrying sperm from epididymis to prostate gland

Penis
Transports, at separate times, seminal fluid and urine to the exterior

Urethra

Scrotum
Contains the testes, keeping them at a lower temperature than the rest of the body

Testis
Site of production of spermatozoa (male sex cells) and testosterone (male sex hormone)

Seminal vesicle
Small gland which lies against bladder and secretes a thick fluid which mixes with sperm

Bladder

Rectum

Prostate
Encircles urethra at base of bladder; provides enzyme-rich fluid which helps to activate sperm

Epididymis
Paired, comma-shaped structure lying against the testis, into which sperm pass to mature and be stored

---

External genitalia

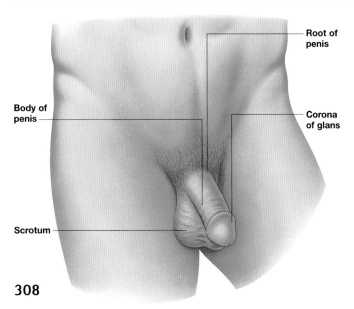

Root of penis

Body of penis

Scrotum

Corona of glans

The external genitalia are those parts of the reproductive tract which lie visible in the pubic region, while other parts remain hidden within the pelvic cavity.

Male external genitalia consists of:
■ The scrotum
■ The penis.
In adults, these are surrounded by coarse pubic hair.

SCROTUM

The scrotum is a loose bag of skin and connective tissue which

The external male genitalia consist of the scrotum and the penis, which are situated in the pubic area. In adults, pubic hair surrounds the root of the penis

holds the testes suspended within it. There is a midline septum, or partition, which separates each testis from its fellow.

Although it would seem unusual for the testes to be held in such a vulnerable position outside the protection of the body cavity, it is necessary for sperm production for them to be kept cool.

PENIS

Most of the penis consists of erectile tissue, which becomes engorged with blood during sexual arousal, causing the penis to become erect. The urethra, through which urine and semen pass, runs through the penis.

Prostate gland

The prostate gland forms a vital part of the male reproductive system, providing enzyme-rich fluid, and produces up to a third of the total volume of the seminal fluid.

About 3 cm in length, the prostate gland lies just under the bladder and encircles the first part of the urethra. Its base lies closely attached to the base of the bladder, its rounded anterior (front) surface lying just behind the pubic bone.

CAPSULE
The prostate is covered by a tough capsule made up of dense fibrous connective tissue. Outside this true capsule is a further layer of fibrous connective tissue, which is known as the prostatic sheath.

INTERNAL STRUCTURE
The urethra, the outflow tract from the bladder, runs vertically through the centre of the prostate gland, where it is known as the prostatic urethra. The ejaculatory ducts open into the prostatic urethra on a raised ridge, the seminal colliculus.

The prostate is said to be divided into lobes, although they are not as distinct as they may be in other organs:
■ Anterior lobe – this lies in front of the urethra and contains mainly fibromuscular tissue
■ Posterior lobe – this lies behind the urethra and beneath the ejaculatory ducts
■ Lateral lobes – these two lobes, lying on either side of the urethra, form the main part of the gland
■ Median lobe – this lies between the urethra and the ejaculatory ducts.

Location of the prostate gland

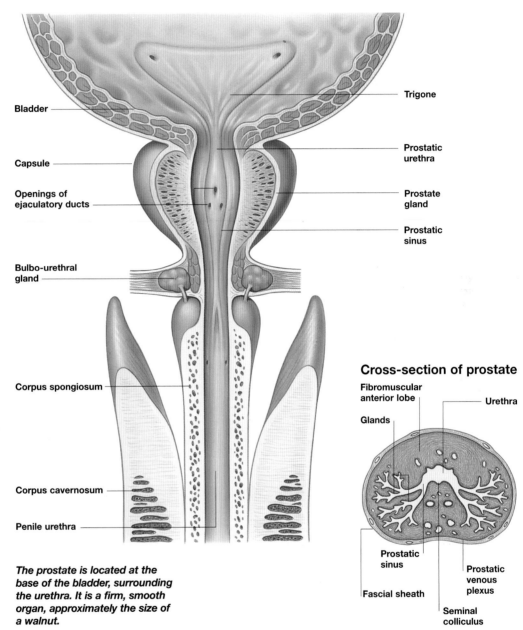

The prostate is located at the base of the bladder, surrounding the urethra. It is a firm, smooth organ, approximately the size of a walnut.

Labels (clockwise): Bladder, Capsule, Openings of ejaculatory ducts, Bulbo-urethral gland, Corpus spongiosum, Corpus cavernosum, Penile urethra, Trigone, Prostatic urethra, Prostate gland, Prostatic sinus

Cross-section of prostate
Labels: Fibromuscular anterior lobe, Glands, Urethra, Prostatic sinus, Fascial sheath, Prostatic venous plexus, Seminal colliculus

Seminal vesicles

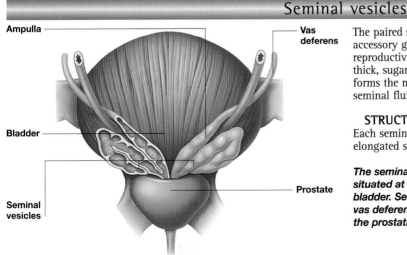

Labels: Ampulla, Vas deferens, Bladder, Seminal vesicles, Prostate

The paired seminal vesicles are accessory glands of the male reproductive tract and produce a thick, sugary, alkaline fluid that forms the main part of the seminal fluid.

STRUCTURE AND SHAPE
Each seminal vesicle is an elongated structure about the

The seminal vesicles are situated at the back of the bladder. Secretions pass into the vas deferentia, which empty into the prostatic urethra.

size and shape of a little finger and lies behind the bladder and in front of the rectum, the two forming a V-shape.

PROSTATE VOLUME
The prostate gland is sac-like, with a volume of approximately 10–15 millilitres. It consists internally of coiled secretory tubules with muscular walls.

The secretions leave the gland in the duct of the seminal vesicle, which joins with the vas deferens just inside the prostate to form the ejaculatory duct.

Testes, scrotum and epididymis

The testes, which lie suspended within the scrotum, are the sites of sperm production. The scrotum also contains the two epididymides – long, coiled tubes, which connect to the vas deferens.

The paired testes are firm, mobile, oval-shaped structures about 4 cm in length and 2.5 cm in width. The testes lie within the scrotum, a bag formed as an outpouching of the anterior abdominal wall, and are attached above to the spermatic cord, from which they hang.

TEMPERATURE CONTROL
Normal sperm can only be produced if the temperature of the testes is about three degrees lower than the internal body temperature. Muscle fibres within the spermatic cord and walls of the scrotum help to regulate the scrotal temperature by lifting the testes up towards the body when it is cold, and relaxing when the ambient temperature is higher.

EPIDIDYMIS
Each epididymis is a firm, comma-shaped structure which lies closely attached to the upper pole of the testis, running down its posterior surface. The epididymis receives the sperm made in the testis and is composed of a highly coiled tube which, if extended, would be six metres in length.

From the tail of the epididymis emerges the vas deferens. This tube will carry the sperm back up the spermatic cord and into the pelvic cavity on the next stage of the journey.

Sagittal section of the contents of the scrotum

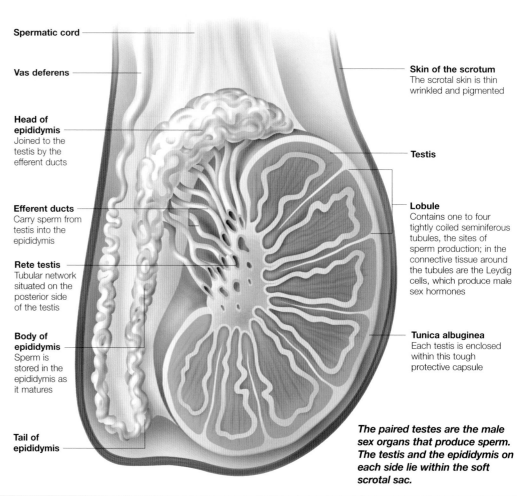

Spermatic cord

Vas deferens

Head of epididymis
Joined to the testis by the efferent ducts

Efferent ducts
Carry sperm from testis into the epididymis

Rete testis
Tubular network situated on the posterior side of the testis

Body of epididymis
Sperm is stored in the epididymis as it matures

Tail of epididymis

Skin of the scrotum
The scrotal skin is thin wrinkled and pigmented

Testis

Lobule
Contains one to four tightly coiled seminiferous tubules, the sites of sperm production; in the connective tissue around the tubules are the Leydig cells, which produce male sex hormones

Tunica albuginea
Each testis is enclosed within this tough protective capsule

The paired testes are the male sex organs that produce sperm. The testis and the epididymis on each side lie within the soft scrotal sac.

Walls of the scrotum

Cross-section of the scrotum

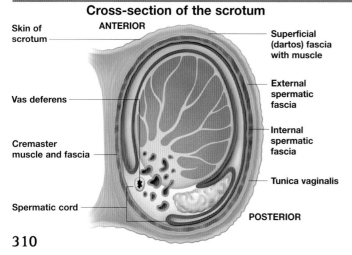

ANTERIOR

Skin of scrotum

Vas deferens

Cremaster muscle and fascia

Spermatic cord

Superficial (dartos) fascia with muscle

External spermatic fascia

Internal spermatic fascia

Tunica vaginalis

POSTERIOR

The walls of the scrotum have a number of layers, as would be expected from its origin as an outpouching of the multi-layered anterior abdominal wall.

LAYERS OF THE SCROTUM
The scrotum consists of:
■ Skin, which is thin, wrinkled and pigmented
■ Dartos fascia, a layer of

The scrotum contains the testes and hangs outside the body. It consists of an outer covering of skin, which surrounds several protective layers.

connective tissue with smooth muscle fibres
■ Three layers of fascia derived from the three muscular layers of the abdominal wall, with further cremasteric muscle fibres
■ Tunica vaginalis, a closed sac of thin, slippery, serous membrane, like the peritoneum in the abdomen, which contains a small amount of fluid to lubricate movement of the testes against surrounding structures.

Unlike the abdominal wall, there is no fat in the coverings around the testes, which is believed to help keep them cool.

Blood supply of the testes

The arterial blood supply of the testes arises from the abdominal aorta, and descends to the scrotum. Venous drainage follows the same route in reverse.

During embryonic life, the testes develop within the abdomen; it is only at birth that they descend into their final position within the scrotum. Because of this the blood supply of the testes arises from the abdominal aorta, and travels down with the descending testis to the scrotum.

TESTICULAR ARTERIES
The paired testicular arteries are long and narrow and arise from the abdominal aorta. They then pass down on the posterior abdominal wall, crossing the ureters as they go, until they reach the deep inguinal rings and enter the inguinal canal.

As part of the spermatic cord they leave the inguinal canal and enter the scrotum where they supply the testis, also forming interconnections with the artery to the vas deferens.

TESTICULAR VEINS
Testicular veins arise from the testis and epididymis on each side. Their course differs from that of the testicular arteries within the spermatic cord where, instead of a single vein, there is a network of veins, known as the pampiniform plexus.

Further up in the abdomen, the right testicular vein drains into the large inferior vena cava, while the left normally drains into the left renal vein.

The blood supply to the testes originates from high up in the abdominal blood vessels. These resulting long vessels allow for the testes' descent in early life.

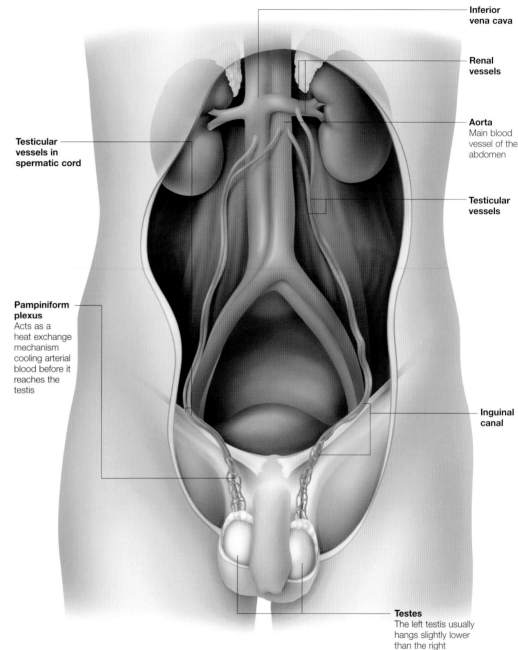

Inferior vena cava

Renal vessels

Aorta
Main blood vessel of the abdomen

Testicular vessels in spermatic cord

Testicular vessels

Pampiniform plexus
Acts as a heat exchange mechanism cooling arterial blood before it reaches the testis

Inguinal canal

Testes
The left testis usually hangs slightly lower than the right

Internal structure of the testis

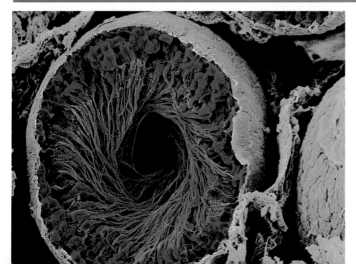

Each testis is enclosed within a tough, protective capsule, the tunica albuginea, from which numerous septa, or partitions, pass down to divide the testis into about 250 tiny lobules.

Each wedge-shaped lobule contains one to four tightly coiled seminiferous tubules, which are the actual sites of production of sperm.

This micrograph shows a sectioned seminiferous tubule. Developing sperm (red) are inside the tubule, which is surrounded by Leydig cells (green).

It has been estimated that there is a total of 350 metres of sperm-producing tubules in each testis.

TUBULES
Sperm are collected from the coiled seminiferous tubules into the straight tubules of the rete testis and from there into the epididymis.

Between the seminiferous tubules lie groups of specialized cells, the interstitial or Leydig cells, which are the site of production of hormones such as testosterone.

311

Penis

The penis is the male copulatory organ, which, when erect,
conveys sperm into the vagina during sexual intercourse.
To enable this, the penis is largely composed of erectile tissue.

The penis is mostly composed of three columns of sponge-like erectile tissue, the two corpora cavernosa and the corpus spongiosum. These are able to fill and become engorged with blood, causing an erection.

STRUCTURE OF THE PENIS

There is only a small amount of muscular tissue associated with the penis, and what there is lies in its root. The shaft and glans have no muscle fibres.

The main components of the penis are:

■ Root – this first part of the penis is fixed in position and is made up of the expanded bases of the three columns of erectile tissue covered by muscle fibres
■ Shaft – this hangs down in the flaccid condition and is made up of erectile tissue, connective tissue, and blood and lymphatic vessels
■ Glans – the tip of the penis, this is formed from the expanded end of the corpus spongiosum and carries the outlet of the urethra, the external urethral orifice
■ Skin – this is continuous with that of the scrotum and is thin, dark and hairless. It is attached only loosely to the underlying fascia and lies in wrinkles when the penis is flaccid.

At the tip of the penis the skin extends as a double layer which covers the glans; this is known as the prepuce, or foreskin.

The penis is anatomically divided into three parts: the root, the shaft and the glans, or head of the penis.

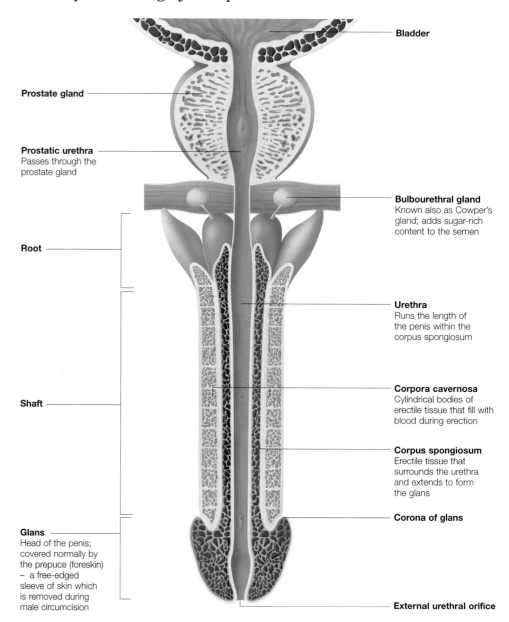

Bladder

Prostate gland

Prostatic urethra
Passes through the prostate gland

Bulbourethral gland
Known also as Cowper's gland; adds sugar-rich content to the semen

Root

Urethra
Runs the length of the penis within the corpus spongiosum

Corpora cavernosa
Cylindrical bodies of erectile tissue that fill with blood during erection

Shaft

Corpus spongiosum
Erectile tissue that surrounds the urethra and extends to form the glans

Corona of glans

Glans
Head of the penis; covered normally by the prepuce (foreskin) – a free-edged sleeve of skin which is removed during male circumcision

External urethral orifice

Cross-section through the penis

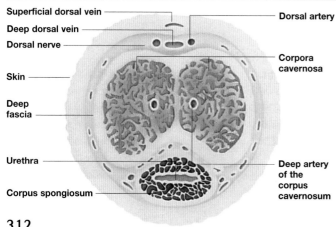

Superficial dorsal vein

Deep dorsal vein

Dorsal nerve

Skin

Deep fascia

Urethra

Corpus spongiosum

Dorsal artery

Corpora cavernosa

Deep artery of the corpus cavernosum

In a cross-section of the shaft of the penis, the relationship of erectile tissue, blood vessels and fascia can be seen more easily. The main bulk is made up of the three masses of erectile tissue, the smaller corpus spongiosum containing within its length the urethra. Each corpus cavernosum carries a central deep artery, which supplies the blood needed for erection.

The main body of the penis, the shaft, consists of three bodies of erectile tissue. These fill with blood during sexual stimulation, resulting in an erection.

CONNECTIVE TISSUE

A sleeve of connective tissue, the deep fascia, encloses the erectile tissue and the deep dorsal vein and dorsal arteries and nerves. Outside the deep fascia is a layer of loose connective tissue which contains the superficial veins. The skin which overlies this loose connective tissue layer is firmly attached to the underlying structures only at the glans.

Muscles associated with the penis

Several muscles are associated with the penis. Their fibres are confined to the root and structures around the penis, rather than to the shaft or glans.

These muscles are known collectively as the superficial perineal muscles, due to the fact that they lie in the perineum, the area around the anus and external genitalia.

There are three main muscles in this area:

■ Superficial transverse perineal muscle

This narrow, paired muscle lies just under the skin in front of the anus. It runs from the ischial tuberosity of the pelvic bone on each side right across to the midline of the body.

■ Bulbospongiosus

This muscle acts to compress the base of the corpus spongiosum, and thus the urethra, to help expel its contents. It originates in a central tendon or raphe, which unites the two sides and passes round to encircle the root of the penis.

■ Ischiocavernosus

This muscle originates from the ischial tuberosity of the pelvic bone to surround the crura or bases of the corpora cavernosa on each side. Contraction of this muscle helps to maintain erection of the penis.

The muscles near the penis are known as the superficial perineal muscles. They surround the base of the penis and help to maintain an erection.

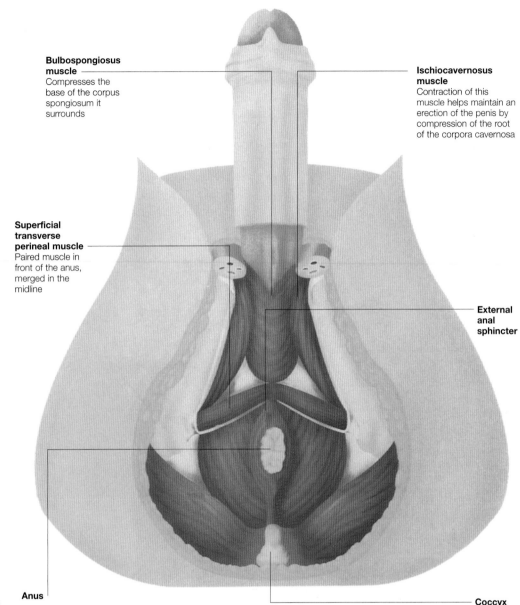

Bulbospongiosus muscle
Compresses the base of the corpus spongiosum it surrounds

Ischiocavernosus muscle
Contraction of this muscle helps maintain an erection of the penis by compression of the root of the corpora cavernosa

Superficial transverse perineal muscle
Paired muscle in front of the anus, merged in the midline

External anal sphincter

Anus

Coccyx

Blood supply of the penis

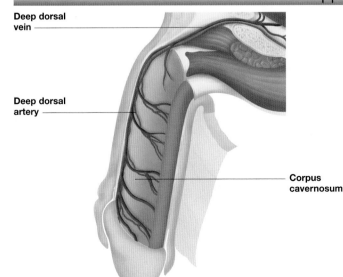

Deep dorsal vein

Deep dorsal artery

Corpus cavernosum

The arterial supply of the penis has two functions. As with any organ, it has to provide the necessary oxygenated blood for the tissues of the penis. It must also, however, provide an additional supply to allow engorgement of the spongy erectile tissues for erection.

ARTERIES

All the arteries supplying the penis originate from the internal pudendal arteries of the pelvis.

The blood supply of the penis originates from the internal pudendal arteries. The deep arteries supply the corpora cavernosa during an erection.

The dorsal arteries lie on each side of the midline deep dorsal vein, and supply connective tissue and skin.

The deep arteries run within the spongy tissue of the corpora cavernosa to supply tissue there and to allow flooding of that tissue during erection.

VENOUS DRAINAGE

The deep dorsal vein of the penis receives blood from the cavernous spaces while blood from the overlying connective tissue and skin is drained by the superficial dorsal vein.

Venous blood drains ultimately into the pudendal veins within the pelvis.

313

How sperm are produced

Sperm are the male sex cells, produced and stored in the testes. Owing to the process of meiosis, a specialized division of the cell nucleus, each cell contains a unique set of genes.

Sperm are mature male sex cells, vital to fertilization. They are produced in the testes, two walnut-sized organs located in the scrotum. The scrotum is the pouch that hangs below the penis, and is around two degrees cooler than the core temperature of the body, so providing the optimum temperature for the production of sperm.

In order to maintain this temperature, the scrotum can pull up closer to the body when the surrounding temperature is low, and can drop farther away as the temperature rises.

SEXUAL ORGANS
The testes are the primary producers of testosterone (the male sex hormone).

These specialized organs each contain around 1,000 seminiferous tubules, which are responsible for the manufacture and storage of sperm. The tubules are lined by small cells known as spermatogonia.

From puberty onwards spermatogonia cells start to divide to produce cells which

eventually develop into sperm.

Alternating with the spermatogonia are much larger cells, the Sertoli cells, which secrete nutrient fluid into the tubules.

SPERMATOGENESIS
Spermatogenesis (the formation of sperm) is a complex process, involving the constant proliferation of spermatogonia cells, to form primary spermatocytes. These cells have a full set of genes, identical to those in other body cells.

MEIOSIS
The primary spermatocytes then undergo a specialized division known as meiosis, in which they split twice to produce cells with a random half (haploid) set of genes. These cells, known as spermatids, develop and grow to produce mature, motile sperm.

The seminiferous tubules of the testes are lined with small cells, called spermatogonia. These cells divide to produce primary spermatocytes.

Formation of sperm

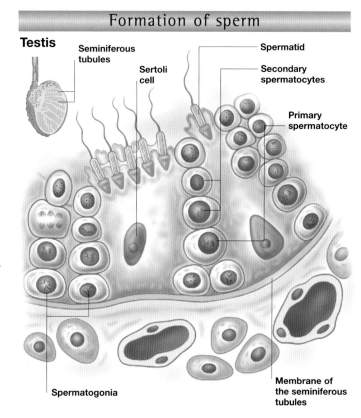

Testis
Seminiferous tubules
Sertoli cell
Spermatid
Secondary spermatocytes
Primary spermatocyte
Spermatogonia
Membrane of the seminiferous tubules

Division of genetic information

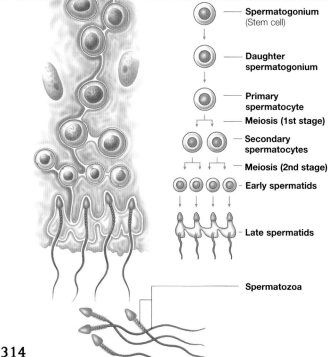

Spermatogonium (Stem cell)

Daughter spermatogonium

Primary spermatocyte

Meiosis (1st stage)

Secondary spermatocytes

Meiosis (2nd stage)

Early spermatids

Late spermatids

Spermatozoa

Each primary spermatocyte contains a set of chromosomes, arranged in 23 pairs (diploid set). It then undergoes the specialized process of meiosis by which it splits in two, to form a pair of spermatids, each containing only half a set of chromosomes (haploid set). This process occurs in two stages to produce four spermatids.

FIRST STAGE
During the first stage, the chromosomes within the spermatocyte's nucleus replicate (double up) and then pair off. The chromosomes exchange random blocks of genes within each pair.

As a result of meiosis, each spermatocyte divides into four spermatids. Each spermatid contains half the genetic material found in the spermatocyte.

This exchange is nature's way of 'shuffling' the gene pool and introducing variation within the offspring. The paired chromosomes separate as the cell divides, each cell receiving two copies of one member of each chromosome pair. The spermatocytes then divide again.

SECOND STAGE
During the second stage of meiosis, the 23 replicated chromosomes within each nucleus split up and the spermatocytes divide again.

The end result of meiosis is the production of spermatids that contain half the number of chromosomes of a spermatocyte. The genetic make-up of each resulting spermatid is unique, due to the mixing process and the chances of any two being identical are virtually zero.

Structure of sperm

Mature sperm cells are specially designed to facilitate the swimming movement used to propel them towards the female's egg.

The spermatids move towards the nearest Sertoli cell, where they receive nourishment in the form of glycogen, proteins, sugars and other nutrients. This provides them with energy and helps them to mature into spermatozoa.

Spermatozoa are among the most specialized cells in the body. Each sperm (spermatozoon) measures 0.05 mm in length and consists of a head, neck and tail.

SPERMATOZOA
The head of the sperm is shaped like a flattened teardrop, and contains a sac of enzymes known as the acrosome. These enzymes are vitally important to the sperm's ability to break down and penetrate the protective outer layer of the female's egg during fertilization.

Behind the acrosome is the cell nucleus, which contains a random half set of male genetic material (DNA) tightly coiled within 23 chromosomes. Thanks to the process of meiosis, each sperm possesses a unique set of genetic information.

The neck is a fibrous area where the middle part of the sperm joins the head. It is a flexible structure and allows the head to swing from side to side, facilitating the swimming movement.

TAIL STRUCTURE
The sperm tail consists of a pair of a long filaments surrounded by two rings each containing nine fibrils. At the front end of the tail are a further ring of outer dense fibres and also a protective tail sheath. The tail is divided into three sections:
■ The middle piece – the fattest part of the tail, due to an additional spiral layer full of energy-producing units known as mitochondria. These produce energy which fuels the sperm, allowing it to swim.
■ The principal piece – consists of the 20 filaments, along with the outer dense fibres and tail sheath.
■ The end piece – here the dense fibres and tail sheath thin out, with the result that this part of the tail is enclosed only by a thin cell membrane. This gradual tapering is what produces the sperm's characteristic whiplash-like swimming motion, driving the sperm towards the egg.

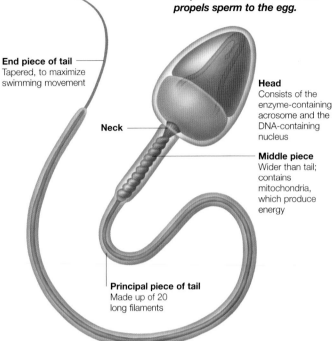

Each sperm cell consists of an enzyme-containing head, a middle piece and a tail. The whiplash movement of the tail propels sperm to the egg.

End piece of tail
Tapered, to maximize swimming movement

Neck

Head
Consists of the enzyme-containing acrosome and the DNA-containing nucleus

Middle piece
Wider than tail; contains mitochondria, which produce energy

Principal piece of tail
Made up of 20 long filaments

The tail consists of a central pair of filaments surrounded by an outer ring of nine paired filaments. At the front end of the tail is a further ring of outer dense fibres and a protective sheath.

Cross-section through tail

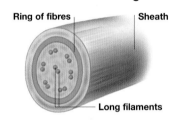

Ring of fibres

Sheath

Long filaments

Production of semen and ejaculation

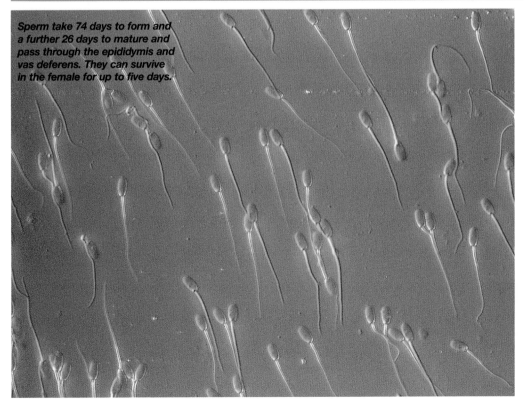

Sperm take 74 days to form and a further 26 days to mature and pass through the epididymis and vas deferens. They can survive in the female for up to five days.

Once their tails are fully developed the sperm are released by the Sertoli cell into the seminiferous tubule. As fluid is secreted into the tubule by the Sertoli cells, a current is produced which washes the sperm cells towards the epididymis. This is a long tube coiled against the testes in which the mature sperm are stored.

EJACULATION
The sperm are propelled from the epididymis during sexual stimulation and up the vas deferens via a wave of muscular contractions within the ducting system. They travel to the ejaculatory duct, through the prostate, and into the urethra. Here they are bathed in secretions from the prostate gland and seminal vesicles (small sacs that hold constituents of semen). The result is a thick, yellowish-white fluid, known as semen.

The average discharge of semen (ejaculate) contains approximately 300 million sperm.

Female reproductive system

The role of the female reproductive tract is twofold.
The ovaries produce eggs for fertilization, and the uterus nurtures
and protects any resulting fetus for its nine-month gestation.

The female reproductive tract is composed of the internal genitalia – the ovaries, uterine (Fallopian) tubes, uterus and vagina – and the external genitalia (the vulva).

INTERNAL GENITALIA

The almond-shaped ovaries lie on either side of the uterus, suspended by ligaments. Above the ovaries are the paired uterine tubes, each of which provides a site for fertilization of the oocyte (egg), which then travels down the tube to the uterus.

The uterus lies within the pelvic cavity and rises into the lower abdominal cavity as a pregnancy progresses. The vagina, which connects the cervix to the vulva, can be distended greatly, as occurs during childbirth when it forms much of the birth canal.

EXTERNAL GENITALIA

The female external genitalia, or vulva, is where the reproductive tract opens to the exterior. The vaginal opening lies behind the opening of the urethra in an area known as the vestibule. This is covered by two folds of skin on each side, the labia minora and labia majora, in front of which lies the raised clitoris.

The female reproductive system is composed of internal and external organs. The internal genitalia are T-shaped and lie within the pelvic cavity.

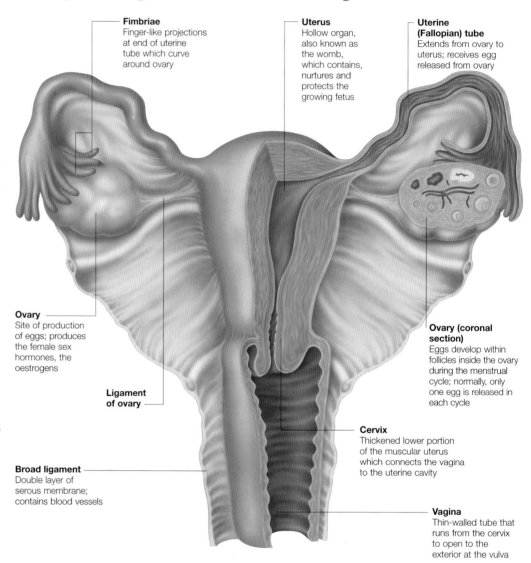

Fimbriae
Finger-like projections at end of uterine tube which curve around ovary

Uterus
Hollow organ, also known as the womb, which contains, nurtures and protects the growing fetus

Uterine (Fallopian) tube
Extends from ovary to uterus; receives egg released from ovary

Ovary
Site of production of eggs; produces the female sex hormones, the oestrogens

Ligament of ovary

Broad ligament
Double layer of serous membrane; contains blood vessels

Ovary (coronal section)
Eggs develop within follicles inside the ovary during the menstrual cycle; normally, only one egg is released in each cycle

Cervix
Thickened lower portion of the muscular uterus which connects the vagina to the uterine cavity

Vagina
Thin-walled tube that runs from the cervix to open to the exterior at the vulva

Position of the female reproductive tract

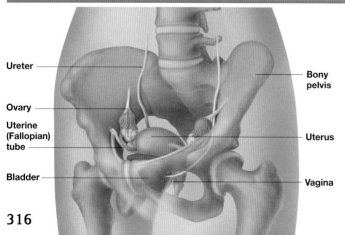

Ureter

Ovary

Uterine (Fallopian) tube

Bladder

Bony pelvis

Uterus

Vagina

In adult women the internal genitalia (which, apart from the ovaries, are basically tubular in structure) are located deep within the pelvic cavity. They are thus protected by the presence of the circle of bone which makes up the pelvis.

This is in contrast to the

The internal reproductive organs in adult women are positioned deep within the pelvic cavity. They are therefore protected by the bony pelvis.

pelvic cavity of young children, which is relatively shallow. A child's uterus, therefore, like the bladder behind which it sits, is located within the lower abdomen.

BROAD LIGAMENTS

The upper surface of the uterus and ovaries is draped in a 'tent' of peritoneum, the thin lining of the abdominal and pelvic cavities, forming the broad ligament which helps to keep the uterus in its position.

Blood supply of the internal genitalia

The female reproductive tract receives a rich blood supply via an interconnecting network of arteries. Venous blood is drained by a network of veins.

The four principal arteries of the female genitalia are:

■ **Ovarian artery** – this runs from the abdominal aorta to the ovary.
   Branches from the ovarian artery on each side pass through the mesovarium, the fold of peritoneum in which the ovary lies, to supply the ovary and uterine (Fallopian) tubes. The ovarian artery in the tissue of the mesovarium connects with the uterine artery

■ **Uterine artery** – this is a branch of the large internal iliac artery of the pelvis. The uterine artery approaches the uterus at the level of the cervix, which is anchored in place by cervical ligaments.
   The uterine artery connects with the ovarian artery above, while a branch connects with the arteries below to supply the cervix and vagina

■ **Vaginal artery** – this is also a branch of the internal iliac artery. Together with blood from the uterine artery, its branches supply blood to the vaginal walls

■ **Internal pudendal artery** – this contributes to the blood supply of the lower third of the vagina and anus.

VEINS
A plexus, or network, of small veins lies within the walls of the uterus and vagina. Blood received into these vessels drains into the internal iliac veins via the uterine vein.

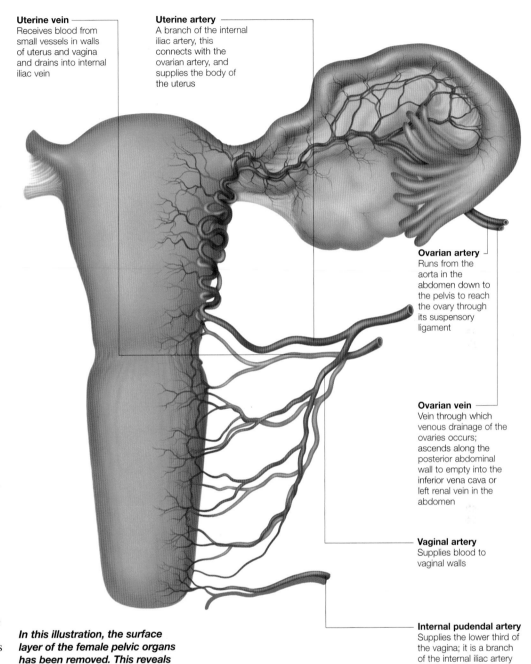

Uterine vein
Receives blood from small vessels in walls of uterus and vagina and drains into internal iliac vein

Uterine artery
A branch of the internal iliac artery, this connects with the ovarian artery, and supplies the body of the uterus

Ovarian artery
Runs from the aorta in the abdomen down to the pelvis to reach the ovary through its suspensory ligament

Ovarian vein
Vein through which venous drainage of the ovaries occurs; ascends along the posterior abdominal wall to empty into the inferior vena cava or left renal vein in the abdomen

Vaginal artery
Supplies blood to vaginal walls

Internal pudendal artery
Supplies the lower third of the vagina; it is a branch of the internal iliac artery

In this illustration, the surface layer of the female pelvic organs has been removed. This reveals the vasculature beneath.

Visualizing the female reproductive tract

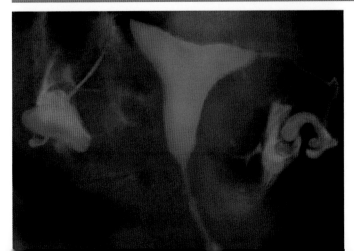

The tubal or hollow parts of the female reproductive tract can be outlined by performing a hysterosalpingogram.
   In this procedure a special radio-opaque dye is passed up into the uterus through the cervix, while X-ray pictures of the area are taken. The dye fills the uterine cavity, and enters

This hysterosalpingogram shows the uterine cavity (centre) filled with dye. Dye is also seen in the uterine tubes and emerging into the peritoneal cavity.

the uterine tubes. It then runs along their length until it flows into the peritoneal cavity at their far end.

ASSESSING TUBES
A hysterosalpingogram is sometimes carried out in the investigation of infertility to determine whether the uterine tubes are still patent (unobstructed). If the tubes have been blocked, as may happen after an infection, the dye will not be able to travel along their full length.

317

Uterus

The uterus, or womb, is the part of the female reproductive tract that nurtures and protects the fetus during pregnancy. It lies within the pelvic cavity and is a hollow, muscular organ.

During a woman's reproductive years, in the non-pregnant state, the uterus is about 7.5 cm long and 5 cm across at its widest point. However, it can expand enormously to accommodate the fetus during pregnancy.

STRUCTURE

The uterus is said to be made up of two parts:
■ The body, forming the upper part of the uterus – this is fairly mobile as it must expand during pregnancy. The central triangular space, or cavity, of the body receives the openings of the paired uterine (Fallopian) tubes
■ The cervix, the lower part of the uterus – this is a thick, muscular canal, which is anchored to the surrounding pelvic structures for stability.

UTERINE WALLS

The main part of the uterus, the body, has a thick wall which is composed of three layers:
■ Perimetrium – the thin outer coat which is continuous with the pelvic peritoneum
■ Myometrium – forming the great bulk of the uterine wall
■ Endometrium – the delicate lining, which is specialized to allow implantation of an embryo should fertilization occur.

The uterus resembles an inverted pear in shape. It is suspended in the pelvic cavity by peritoneal folds or ligaments.

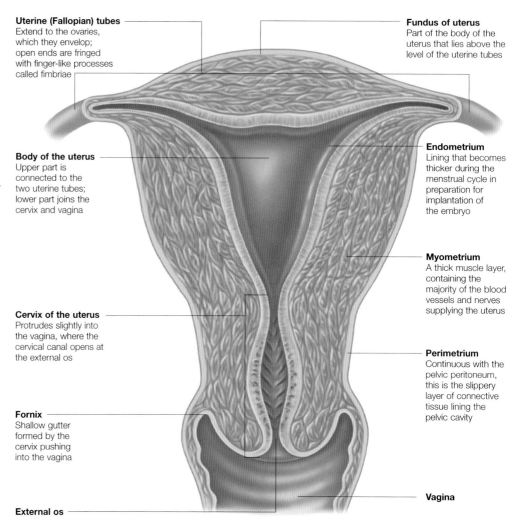

Uterine (Fallopian) tubes
Extend to the ovaries, which they envelop; open ends are fringed with finger-like processes called fimbriae

Body of the uterus
Upper part is connected to the two uterine tubes; lower part joins the cervix and vagina

Cervix of the uterus
Protrudes slightly into the vagina, where the cervical canal opens at the external os

Fornix
Shallow gutter formed by the cervix pushing into the vagina

External os

Fundus of uterus
Part of the body of the uterus that lies above the level of the uterine tubes

Endometrium
Lining that becomes thicker during the menstrual cycle in preparation for implantation of the embryo

Myometrium
A thick muscle layer, containing the majority of the blood vessels and nerves supplying the uterus

Perimetrium
Continuous with the pelvic peritoneum, this is the slippery layer of connective tissue lining the pelvic cavity

Vagina

Position of the uterus

Normal position of the uterus

Bladder

Vagina

Uterus in extreme retroverted position

Rectum

The uterus lies in the pelvis between the bladder and the rectum. However, its position changes with the stage of filling of these two structures and with different postures.

NORMAL POSITION

Normally the long axis of the uterus forms an angle of 90 degrees with the long axis of the vagina, with the uterus lying forward on top of the bladder. This usual position is known as anteversion.

In most women the uterus lies on the bladder, moving backwards as the bladder fills. However, it may lie in any position between the two extremes shown.

ANTEFLEXION

In some women, the uterus lies in the normal position, but may curve forwards slightly between the cervix and fundus, This is termed anteflexion.

RETROFLEXION

In some cases, however, the uterus bends not forwards but backwards, the fundus coming to lie next to the rectum. This is known as a retroverted uterus.

Regardless of the uterine position it will normally bend forwards as it expands in pregnancy. A pregnant retroverted uterus, however, may take longer to reach the pelvic brim, at which point it becomes palpable abdominally.

The uterus in pregnancy

In pregnancy the uterus must enlarge to hold the growing fetus. From being a small pelvic organ, it increases in size to take up much of the space of the abdominal cavity.

Pressure of the enlarged uterus on the abdominal organs pushes them up against the diaphragm, encroaching on the thoracic cavity and causing the ribs to flare out to compensate. Organs such as the stomach and bladder are compressed to such an extent in late pregnancy that their capacity is greatly diminished and they become full sooner.

After pregnancy, the uterus will rapidly decrease in size again although it will always remain slightly larger than one which has never been pregnant.

HEIGHT OF FUNDUS

During pregnancy the enlarging uterus can be accommodated within the pelvis for the first 12 weeks, at which time the uppermost part, the fundus, can just be palpated in the lower abdomen. By 20 weeks, the fundus will have reached the region of the umbilicus, and by late pregnancy it may have reached the xiphisternum, the lowest part of the breastbone.

WEIGHT OF UTERUS

In the final stages of pregnancy the uterus will have increased in weight from a pre-pregnant 45 g to around 900 g. The myometrium (muscle layer) grows as the individual fibres increase in size (hypertrophy). In addition, the fibres increase in number (hyperplasia).

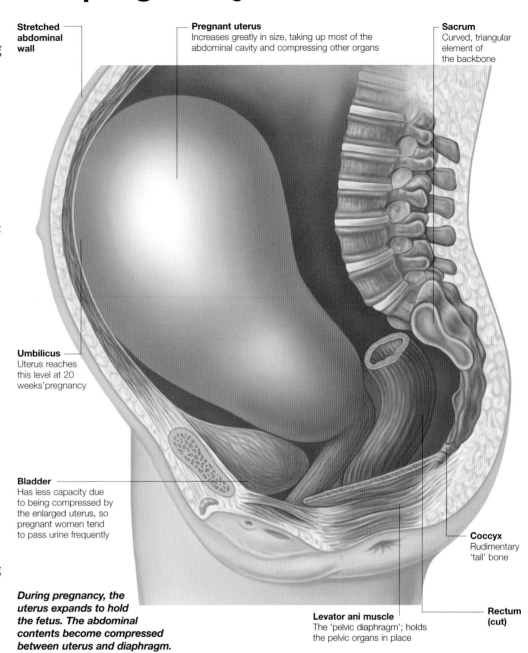

Stretched abdominal wall

Pregnant uterus
Increases greatly in size, taking up most of the abdominal cavity and compressing other organs

Sacrum
Curved, triangular element of the backbone

Umbilicus
Uterus reaches this level at 20 weeks' pregnancy

Bladder
Has less capacity due to being compressed by the enlarged uterus, so pregnant women tend to pass urine frequently

Coccyx
Rudimentary 'tail' bone

Rectum (cut)

Levator ani muscle
The 'pelvic diaphragm'; holds the pelvic organs in place

During pregnancy, the uterus expands to hold the fetus. The abdominal contents become compressed between uterus and diaphragm.

Lining of the uterus

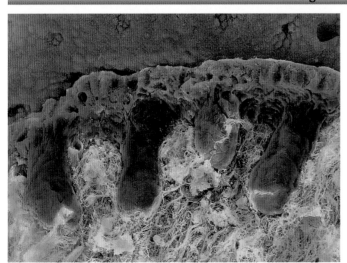

The endometrium is the name given to the lining of the uterus. It consists of a simple surface layer, or epithelium, overlying a thicker layer of highly cellular connective tissue, the lamina propria. Numerous tubular glands are also present within the endometrium.

MENSTRUAL CYCLE

Under the influence of sex hormones the endometrium undergoes changes during the

This enlarged section through the endometrium of the uterus shows the layer of epithelial cells (blue). Three tubular glands are also clearly visible.

monthly menstrual cycle which prepare it for the possible implantation of an embryo. It may vary in thickness from 1 mm to 5 mm before being shed at menstruation.

BLOOD SUPPLY

Arteries within the myometrium, the underlying muscle layer, send numerous small branches into the endometrium. There are two types: straight arteries, which supply the lower, permanent layer; and tortuous (twisted) spiral arteries, which supply the upper layer shed during menstruation. The tortuosity of the spiral arteries prevents excess bleeding during menstruation.

Anatomy of the placenta

The placenta is the organ which provides the developing fetus with all the nutrients it needs. It is a temporary structure formed in the uterus during pregnancy from fetal and maternal tissues.

The placenta takes on the role of the lungs and the intestine for the developing fetus. It achieves this by bringing the blood of the fetus close to the maternal blood within its internal structure, allowing the fetus to take up oxygen and nutrients, while waste products are carried away.

The placenta becomes detached at birth and is delivered after the baby in what is known as the third stage of labour. It is then examined to check that it is complete and shows no evidence of abnormality or disease which may have affected the fetus.

PLACENTAL APPEARANCE

At full term the placenta is a deep red, round or oval flattened organ. It normally weighs about 500 g, or one sixth of the weight of the fetus it nourishes.

There are two sides to the delivered placenta:
■ The maternal aspect (which is attached to the lining of the womb) – this shows subdivisions where the placental tissue is divided by fibrous bands (septa). It is deep red and feels spongy
■ The fetal aspect (from which the umbilical cord arises) – this is covered in fetal membranes. Its surface is shiny and smooth with large umbilical vessels.

Fetal aspect of placenta

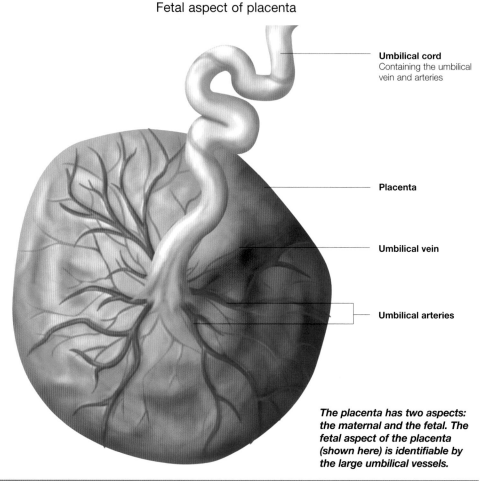

Umbilical cord
Containing the umbilical vein and arteries

Placenta

Umbilical vein

Umbilical arteries

The placenta has two aspects: the maternal and the fetal. The fetal aspect of the placenta (shown here) is identifiable by the large umbilical vessels.

Variations in the placenta

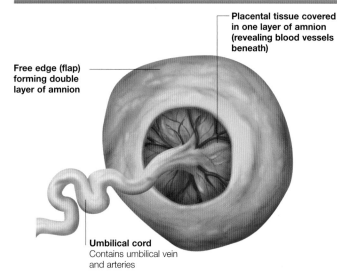

Placental tissue covered in one layer of amnion (revealing blood vessels beneath)

Free edge (flap) forming double layer of amnion

Umbilical cord
Contains umbilical vein and arteries

In a circumvallate placenta the amnion (the membranous sac containing the baby) folds in on itself, creating a double layer over most of the placenta.

There are a number of variations that can occur in the form or arrangement of the placenta. These are usually of little clinical significance, posing no threat to mother or developing fetus, although occasionally they may present problems.

PLACENTAL VARIATIONS

Possible variations include:
■ Succenturiate placenta – here there is an extra, or accessory, lobe of the placenta which lies within the fetal membranes a short distance away from the main placenta
■ Battledore placenta – this is the name given to a placenta in which the cord inserts at one edge rather than centrally, as is the normal situation
■ Velamentous insertion of the cord – this describes the unusual arrangement whereby the umbilical cord itself does not reach the placenta but inserts into the fetal membranes a little distance away. The umbilical vessels then divide on their way to the placenta
■ Circumvallate placenta – this occurs when there is extensive folding back of the membranes, which may be associated with bleeding during birth.

Inside the placenta

As the placenta develops, the fetal blood vessels form chorionic villi (finger-like projections) within it, to absorb nutrients and oxygen from the incoming maternal blood vessels. Waste is also passed back to the maternal blood.

The placenta provides the means by which the growing fetus can receive oxygen and nutrients from the maternal blood circulation and, at the same time, dispose of its waste products. To allow these transfers to take place, the placenta has a very rich blood supply from both the mother and the fetus.

A cross-section of the placenta reveals that this organ is made up partly from maternal tissue and partly from fetal tissue. The spiral arteries that arise from the maternal uterine arteries bring blood into the base of the placenta. This blood then leaves the arteries and fills wide 'pools' (intervillous spaces) in which the fetal villi are suspended. The maternal blood then returns to her circulation through numerous veins.

The fetal villi are finger-like projections that contain blood vessels connected to the fetus through the umbilical cord. They branch again and again to create the maximum amount of surface area for the transfer of oxygen, nutrients and waste substances to and from the maternal blood.

Although the two circulations come close to each other, maternal and fetal blood do not

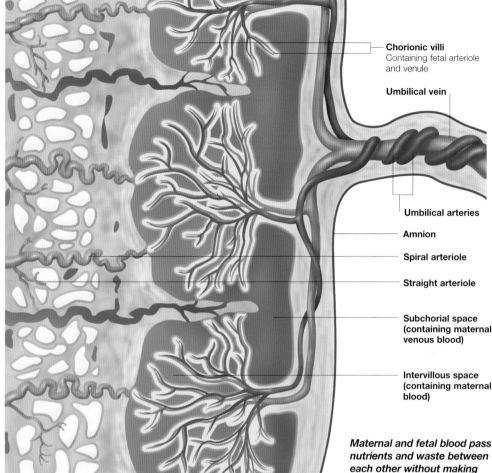

Chorionic villi
Containing fetal arteriole and venule

Umbilical vein

Umbilical arteries

Amnion

Spiral arteriole

Straight arteriole

Subchorial space
(containing maternal venous blood)

Intervillous space
(containing maternal blood)

Maternal and fetal blood pass nutrients and waste between each other without making direct contact, via the capillary network of the chorionic villi.

mix, being divided by the thin walls of the villi.

FUNCTIONS OF THE PLACENTA

The placenta has a number of functions, which are vital to fetal growth and development:
■ Respiration – fetal blood is supplied with oxygen from the maternal circulation via the placenta, which also carries away waste carbon dioxide
■ Nutrition – nutrients which circulate in the maternal bloodstream are passed to the fetus through the placenta
■ Excretion – waste products from the fetus are passed from the two umbilical arteries to the

villi, and ultimately the maternal circulation, for disposal
■ Hormone production – the placenta is an important source of hormones, especially oestrogen and progesterone. These hormones not only help to maintain the pregnancy, but also prepare the mother for the birth of her child.

Abnormal conditions of the placenta

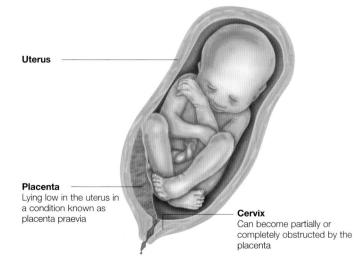

Uterus

Placenta
Lying low in the uterus in a condition known as placenta praevia

Cervix
Can become partially or completely obstructed by the placenta

There are several problems involving the placenta that may occur during pregnancy. One of the most well known is placenta praevia, a condition in which the placenta has implanted in an abnormally low position within the uterus.

Because of its low position, the placenta in these cases may actually come between the fetus and the cervix, making a vaginal delivery very difficult. Placenta praevia is often associated with

When the placenta implants in the lower segment of the uterus, it is known as placenta praevia. Its position between the fetus and the cervix is problematic.

bleeding in the later months of pregnancy.

PLACENTAL ABRUPTION

Placental abruption refers to the separation (partial or complete) of the placenta from the uterine wall. This is potentially a very serious condition in which bleeding occurs between the placenta and the uterine wall.

This bleeding may remain within the uterus, or may track down through the cervix to present as vaginal bleeding. Placental abruption can usually be managed to give a safe outcome, but the lives of both mother and baby are at risk from this condition.

Vagina and cervix

The vagina is the thin-walled muscular tube that extends from the cervix of the uterus to the external genitalia. The vagina is closed at rest but is designed to stretch during intercourse or childbirth.

The vagina is approximately 8 cm in length and lies between the bladder and the rectum. It forms the main part of the birth canal and receives the penis during sexual intercourse.

STRUCTURE OF THE VAGINA

The front and back walls of the vagina normally lie in contact with one another, closing the lumen (central space), although the vagina can expand greatly, as occurs in childbirth.

The cervix, the lower end of the uterus, projects down into the lumen of the vagina at its upper end. Where the vagina arches up to meet the cervix, it forms recesses known as the vaginal fornices. These are divided into anterior, posterior, right and left fornices, although they form a complete ring.

The thin wall of the vagina has three layers:
■ Adventitia – outer layer composed of fibroelastic connective tissue which allows distension when necessary
■ Muscularis – the central muscular layer of the vaginal wall
■ Mucosa – the inner layer of the vagina; this is thrown into many rugae (deep folds), and has a layered, stratified squamous (skin-like) epithelium (cell lining), which helps to resist abrasion during intercourse.

The vagina is a muscular, tubular organ designed to expand during sexual intercourse and childbirth. It is approximately 8 cm in length.

Coronal section through the vagina

Vaginal fornices

Adventitia

Muscularis
Muscular layer that stretches considerably in childbirth

Mucosa
Lined with layered epithelium, the vaginal mucosa has no glands; lubrication comes from secretions of the cervical glands above

Cervical os
The entrance to the uterine cavity

Vaginal artery
Supplies oxygenated blood

Vaginal lumen (central space)
Enclosed by epithelial lining of the vagina, which lies in a series of folds, called rugae

Hymenal caruncle
Remains of the hymen – a fold of mucosa covering the entrance to the vagina at birth; it divides the vagina from the vestibule

External genitalia

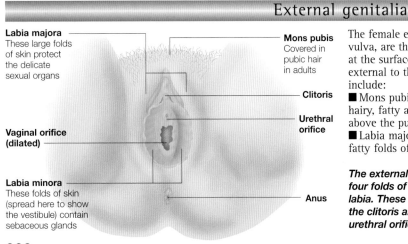

Labia majora
These large folds of skin protect the delicate sexual organs

Vaginal orifice (dilated)

Labia minora
These folds of skin (spread here to show the vestibule) contain sebaceous glands

Mons pubis
Covered in pubic hair in adults

Clitoris

Urethral orifice

Anus

The female external genitalia, or vulva, are those parts which lie at the surface of the body, external to the vagina. They include:
■ Mons pubis – the rounded, hairy, fatty area which lies above the pubic bone
■ Labia majora – the two outer fatty folds of skin, which lie

The external genitalia include four folds of skin, known as the labia. These cover and protect the clitoris and the vaginal and urethral orifices.

across the vulval opening
■ Labia minora – the two smaller folds of skin which lie inside the cleft of the vulva
■ Vestibule – area into which the urethra and vagina open
■ Clitoris – a structure composed of erectile tissue and containing a rich sensory nerve supply; it is analogous to the penis in males.

The vulval opening is partially closed off by a fold of mucosa, the hymen; this may rupture at first intercourse, with tampon use or during a pelvic examination.

The cervix

The cervix, or neck of the uterus, is the narrowed, lower part of the uterus which projects down into the upper vagina.

The cervix is fixed in position by the cervical ligaments, and so anchors the relatively mobile uterine body above.

CERVICAL STRUCTURE
The cervix has a narrow canal which is approximately 2.5 cm long in adult women. The walls of the cervix are tough, containing much fibrous tissue as well as muscle, unlike the body of the uterus, which is mainly muscular.

The central canal of the cervix is the downwards continuation of the uterine cavity which opens at its lower end, the external os, into the vagina. The canal is widest at its central point, constricting slightly at the internal os at the upper end and the external os below.

LINING OF THE CERVIX
The epithelium, or lining, of the cervix is of two types:
■ Endocervix – this is the lining of the cervical canal, inside the cervix. The epithelium is a simple, single layer of columnar cells which overlies a surface thrown into many folds containing glands.
■ Ectocervix – this covers the portion of the cervix which projects down into the vagina; it is composed of squamous epithelium and has many layers.

The cervix is located at the lower end of the uterus. It contains less muscle tissue than the uterus and is lined with two different types of epithelial cell.

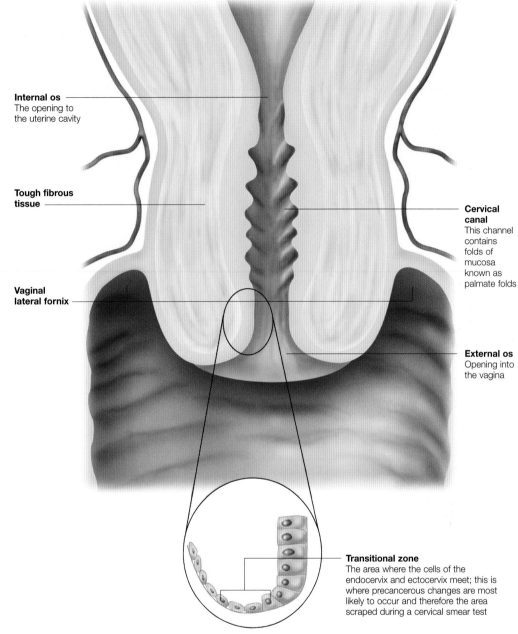

Internal os
The opening to the uterine cavity

Tough fibrous tissue

Vaginal lateral fornix

Cervical canal
This channel contains folds of mucosa known as palmate folds

External os
Opening into the vagina

Transitional zone
The area where the cells of the endocervix and ectocervix meet; this is where precancerous changes are most likely to occur and therefore the area scraped during a cervical smear test

Cervical os

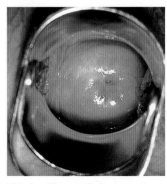

This healthy cervix is viewed through a metal speculum. The deeper pink lining of the inside of the cervix can be seen at the external cervical os.

The opening of the cervical canal into the upper vagina is known as the cervical os.

It may be necessary to look more closely at this area if, for example, some abnormal cells have been seen under the microscope during a routine cervical smear test. In this case a colposcope, a type of low-powered microscope, is used.

COLPOSCOPY
During colposcopy, the cervix is coated with a staining fluid that shows up any abnormal cells. A biopsy may be taken of any suspicious areas; further treatment may then be needed.

Nulliparous cervix

Parous cervix

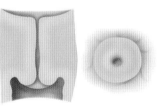

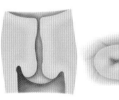

In a woman who has never given birth (nulliparous), the cervical os appears round in shape. The canal is also more tightly closed before childbirth.

After childbirth, the os becomes slit-like in appearance. The cervical canal is slightly looser, following the passage of the fetus.

Ovaries and uterine tubes

The ovaries are the site of production of oocytes, or eggs, which are fertilized by sperm to produce embryos. The uterine (or Fallopian) tubes conduct the oocytes from the ovaries to the uterus.

The paired ovaries are situated in the lower abdomen and lie on either side of the uterus. Their position may be variable, especially after childbirth, when the supporting ligaments have been stretched.

Each ovary consists of:
- Tunica albuginea – a protective layer of fibrous tissue
- Medulla – a central region with blood vessels and nerves
- Cortex – within which the oocytes develop
- Surface layer – smooth before puberty but becoming more pitted in the reproductive years.

BLOOD SUPPLY

The arterial supply to the ovaries comes via the ovarian arteries, which arise from the abdominal aorta. After supplying the uterine tubes also, the ovarian arteries overlap with the uterine arteries.

Blood from the ovaries enters a network of tiny veins, the pampiniform plexus, within the broad ligament, from which it enters the right and left ovarian veins. These ascend into the abdomen to drain ultimately into the large inferior vena cava and the renal vein respectively.

This cross-section shows the follicles situated in the cortex of the ovary. Each follicle contains an oocyte at a different stage of development.

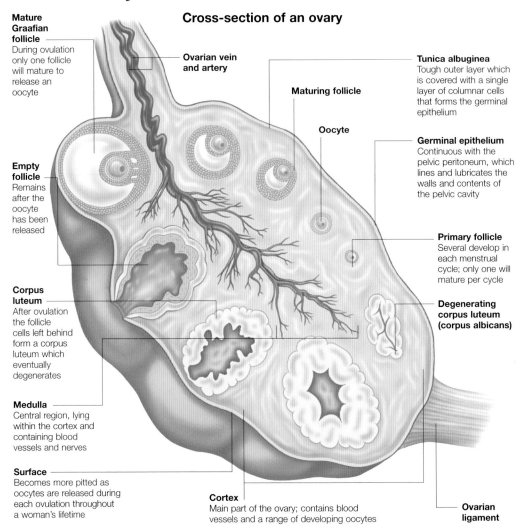

Cross-section of an ovary

Mature Graafian follicle
During ovulation only one follicle will mature to release an oocyte

Empty follicle
Remains after the oocyte has been released

Corpus luteum
After ovulation the follicle cells left behind form a corpus luteum which eventually degenerates

Medulla
Central region, lying within the cortex and containing blood vessels and nerves

Surface
Becomes more pitted as oocytes are released during each ovulation throughout a woman's lifetime

Ovarian vein and artery

Maturing follicle

Oocyte

Cortex
Main part of the ovary; contains blood vessels and a range of developing oocytes

Tunica albuginea
Tough outer layer which is covered with a single layer of columnar cells that forms the germinal epithelium

Germinal epithelium
Continuous with the pelvic peritoneum, which lines and lubricates the walls and contents of the pelvic cavity

Primary follicle
Several develop in each menstrual cycle; only one will mature per cycle

Degenerating corpus luteum (corpus albicans)

Ovarian ligament

Supporting ligaments

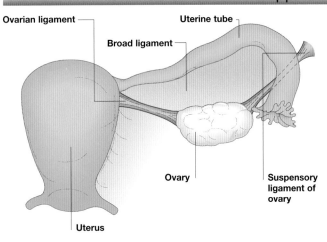

Ovarian ligament

Broad ligament

Uterine tube

Ovary

Suspensory ligament of ovary

Uterus

Each ovary is held in its position relative to the uterus and uterine tubes by several ligaments.

MAIN LIGAMENTS

These ligaments include the following:
- Broad ligament – the tent-like fold of pelvic peritoneum which hangs down on either side of the uterus, enclosing the uterine tubes and ovaries

Each ovary is suspended by several ligaments to hold it in position. However, the position varies, especially if the ligaments have stretched.

- Suspensory ligament of the ovary – that part of the broad ligament which anchors the ovary to the side wall of the pelvis and carries the ovarian vessels and lymphatics
- Mesovarium – the fold of the broad ligament within which the ovary lies.
- Ovarian ligament – attaches the ovary to the uterus and runs within the broad ligament.

These ligaments may become stretched in women following childbirth, which in many cases means that the position of the ovary may be more variable than before pregnancy.

The uterine tubes

The uterine, or Fallopian, tubes collect the oocytes released from the ovaries and transport them to the uterus. They also provide a site for fertilization of the oocyte by a sperm to take place.

Each uterine tube is about 10 cm long and extends outwards from the upper part of the body of the uterus towards the lateral wall of the pelvic cavity.

The tubes run within the upper edge of the broad ligament and open into the peritoneal cavity in the region of the ovary.

STRUCTURE
The tubes are divided anatomically into four parts which, from outer to inner are:

■ Infundibulum – the funnel-shaped outer end of the uterine tubes which opens into the peritoneal cavity
■ Ampulla – the longest and widest part and the most usual site for fertilization of the oocyte
■ Isthmus – a constricted region with thick walls
■ Uterine part – this is the shortest part of the tube.

BLOOD SUPPLY
The uterine tubes have a very rich blood supply which comes from both the ovarian and the uterine arteries; these overlap to form an arterial arcade.

Venous blood drains from the tubes in a pattern which mirrors the arterial supply.

Major parts of a uterine tube

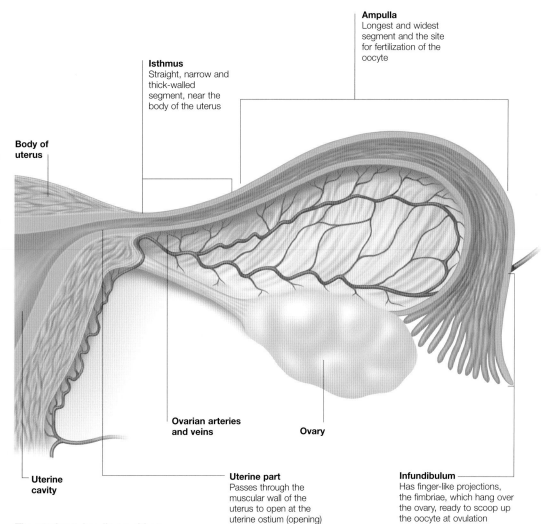

Ampulla
Longest and widest segment and the site for fertilization of the oocyte

Isthmus
Straight, narrow and thick-walled segment, near the body of the uterus

Body of uterus

Ovarian arteries and veins

Ovary

Uterine cavity

Uterine part
Passes through the muscular wall of the uterus to open at the uterine ostium (opening)

Infundibulum
Has finger-like projections, the fimbriae, which hang over the ovary, ready to scoop up the oocyte at ovulation

The uterine tubes lie on either side of the body. The outer part of each tube lies near the ovary, its end opening there into the abdominal cavity.

Wall of a uterine tube

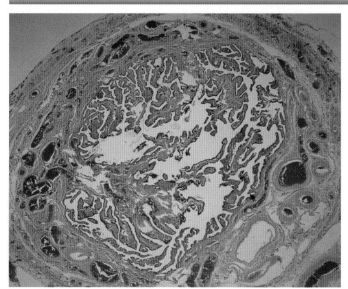

The uterine tube wall is lined with two types of cell: mucus-secreting and ciliated. These act to nourish and propel the oocyte along the length of the tube.

The structure of the wall of a uterine tube shows features which have developed to assist in the task of maintaining the oocyte and carrying it safely to the uterus for implantation:

■ A layer of smooth muscle fibres within the walls allows the uterine tubes to contract rhythmically, the waves of contraction passing towards the uterus.

■ The walls are lined with cells which bear cilia, tiny brush-like projections which beat to 'sweep' the oocyte inwards towards the uterus.

■ Non-ciliated cells in deep crypts in the lining of the uterine tubes produce secretions which keep the oocyte, and any sperm which may be present, nourished during their journey along the tube.

OVARIAN HORMONES
The lining of the uterine tubes is influenced by ovarian hormones, and so may vary in its activity according to the phase of the menstrual cycle. The hormone progesterone, for instance, increases the amount of mucous secretions that are produced.

Menstrual cycle

The menstrual cycle is the regular process by which an egg is released from an ovary in preparation for pregnancy. This occurs approximately every four weeks from the time of a woman's first period right up to the menopause.

The menstrual cycle is characterized by the periodic maturation of oocytes (cells that develop into eggs) in the ovaries and associated physical changes in the uterus. Reproductive maturity occurs after a sudden increase in the secretion of hormones during puberty, usually between the ages of 11 and 15.

CYCLE ONSET

The time of the first period, which occurs at about the age of 12, is called the menarche. After this, a reproductive cycle begins, averaging 28 days. This length of time may be longer, shorter or variable, depending on the individual. The cycle is continuous, apart from during pregnancy. However, women suffering from anorexia nervosa or athletes who train intensively may cease to menstruate.

MENSTRUATION

Each month, if conception does not occur, oestrogen and progesterone levels fall and the blood-rich lining of the uterus is shed at menstruation (menses). This takes place every 28 days or so, but the time can range from 19 to 36 days.

Menstruation lasts for about five days. Around 50 ml (about an eggcup) of blood, uterine tissues and fluid is lost during this time, but again this volume varies from woman to woman. Some women lose only 10 ml of blood, while others lose 110 ml.

Excessive menstrual bleeding is known as menorrhagia; temporary cessation of menstruation – such as during pregnancy – is called amenorrhoea. The menopause is the complete cessation of the menstrual cycle, and usually occurs between 45 and 55.

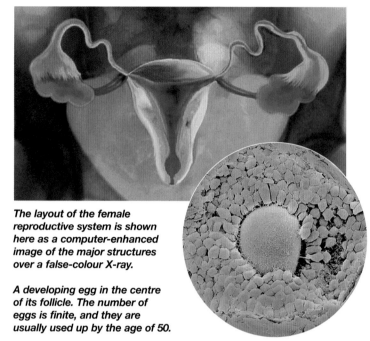

The layout of the female reproductive system is shown here as a computer-enhanced image of the major structures over a false-colour X-ray.

A developing egg in the centre of its follicle. The number of eggs is finite, and they are usually used up by the age of 50.

Monthly physiological changes

This diagram illustrates the ongoing changes during the cycle. Between days one and five, the lining is discharged, while another follicle is developing. The uterine lining thickens, and around day 14, the egg is released, at the point called ovulation.

Gonadotrophic hormones
Released by the pituitary gland to promote the production of the egg and of sex hormones in the gonads (ovaries)

Ovarian activity
Each month, one follicle develops to maturity, then releases an egg at ovulation; the surviving tissue in the ovary forms the corpus luteum, a temporary hormone-producing gland

Ovarian hormones
Secreted by the ovary to encourage the lining to grow; extra progesterone is produced by the corpus luteum after ovulation to prepare the uterus for pregnancy

Lining of uterus
Progressively thickens to receive the fertilized egg; if the egg does not implant, the lining is shed (menses) during the first five days of the cycle

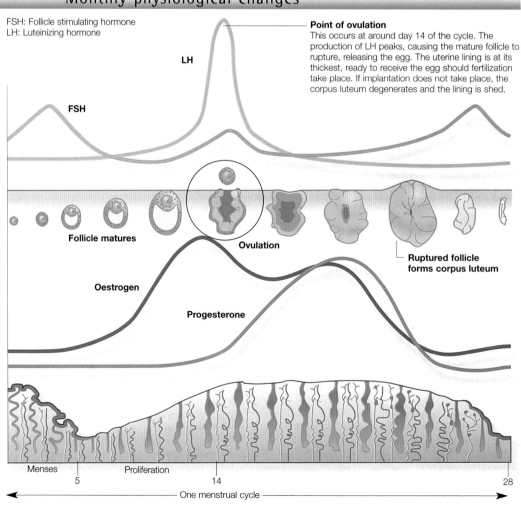

FSH: Follicle stimulating hormone
LH: Luteinizing hormone

Point of ovulation
This occurs at around day 14 of the cycle. The production of LH peaks, causing the mature follicle to rupture, releasing the egg. The uterine lining is at its thickest, ready to receive the egg should fertilization take place. If implantation does not take place, the corpus luteum degenerates and the lining is shed.

LH

FSH

Follicle matures

Ovulation

Ruptured follicle forms corpus luteum

Oestrogen

Progesterone

Menses   Proliferation

Days   5   14   28

One menstrual cycle

Egg development

The process of developing a healthy egg for release at ovulation takes around six months. It occurs throughout life until the stock of oocytes is exhausted.

Two million eggs (oogonia) are present at birth, distributed between the two ovaries, and 400,000 are left by the time of the first period. During each menstrual cycle, only one egg – from a pool of around 20 potential eggs – develops and is released. By the time menopause is reached, the process of atresia (cell degeneration) in the ovaries is complete and no eggs remain.

Eggs develop within cavity-forming secretory structures called follicles. The first stage of follicle development occurs when an oogonium becomes surrounded by a single layer of granulosa cells and is called a primordial (primary) follicle. The genetic material within the egg at this stage remains undisturbed – but susceptible to alteration –

until ovulation of that egg occurs, up to 45 years after it first developed. This helps to explain the increase in abnormal chromosomes in eggs and offspring of women who conceive later in life.

Primordial follicles develop into secondary follicles by meiotic (reductive) division and then into tertiary (or antral, meaning 'with a cavity') follicles. As many as 20 primary follicles will begin to mature, although 19 will eventually regress. If more than one follicle develops to maturity, twins or triplets may be conceived.

The follicles are located in the cortex of the ovary. This micrograph shows the follicle separated by connective tissue.

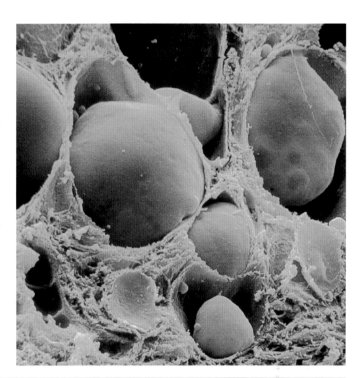

Ovulation

The final 14 day period of follicular development takes place during the first half of the menstrual cycle and depends on the precise hormonal interplay between the ovary, pituitary gland and the hypothalamus.

The trigger for selecting a

healthy egg for development at the start of each cycle is a rise in the secretion of follicle stimulating hormone (FSH) by the pituitary gland. This occurs in response to a fall in the hormones oestrogen and progesterone during the luteal

phase (second 14 days) of the previous cycle if conception has not occurred.

EGG SELECTION
At the time of the FSH signal, there are about 20 secondary follicles, 2–5 mm (0.1–0.2 in) in diameter, distributed between the two ovaries. A single follicle is selected from this pool, while the others undergo atresia. Once a follicle is selected, the development of further follicles is prevented. A typical 5 mm (0.2 in) secondary follicle will then require 10–12 days of sustained stimulation by FSH to grow to a diameter of 20 mm (0.8 in) before rupturing, releasing the

Under a light microscope, a secondary oocyte (mature egg) can be seen surrounded by the cells of the corona radiata that support it during development.

egg into the uterine (Fallopian) tube. As the follicle enlarges, there is a steady rise in oestrogen production, triggering a mid-cycle rise in luteinizing hormone (LH) by the pituitary, which in causes release and maturation of the egg. The interval between the LH peak and ovulation is relatively constant (about 36 hours). The ruptured follicle (corpus luteum) that remains after ovulation becomes a very important endocrine gland, secreting oestrogen and progesterone.

HORMONE REGULATION
Progesterone levels rise to a peak about seven days after ovulation. If fertilization takes place, the corpus luteum maintains the pregnancy until the placenta takes over at about three months' gestation. If no conception takes place, the gland has a lifespan of 14 days, and oestrogen and progesterone levels decline in anticipation of the next cycle.

In the first half of the cycle, oestrogen secreted by the developing follicle (stage before corpus luteum) enables the lining of the uterus (endometrium) to proliferate and increase in thickness ready to nourish the egg should it become fertilized. Once the corpus luteum is formed, progesterone converts the endometrium to a more compact layer in anticipation of an embryo implanting.

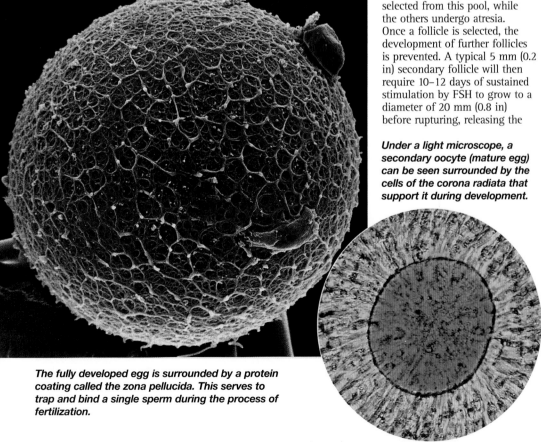

The fully developed egg is surrounded by a protein coating called the zona pellucida. This serves to trap and bind a single sperm during the process of fertilization.

How ovulation occurs

The total supply of eggs for a woman's reproductive years is determined before she is born. The immature eggs are stored in the ovary until puberty, after which one is released every month.

An ovum (egg) is the female gamete, or sex cell, which unites with a sperm to form a new individual. Eggs are produced and stored in the ovaries, two walnut-sized organs connected to the uterus via the uterine (Fallopian) tubes.

THE OVARY

Each ovary is covered by a protective layer of peritoneum (abdominal lining). Immediately below this layer is a dense fibrous capsule, the tunica albuginea. The ovary itself consists of a dense outer region, called the cortex, and a less dense inner region, the medulla.

GAMETE PRODUCTION

In females, the total supply of eggs is determined at birth. Egg-forming cells degenerate from birth to puberty and the timespan during which a woman can release mature eggs is limited from puberty until menopause.

The process by which ova are produced is known as oogenesis, which literally means 'the beginning of an egg'. Germ cells in the fetus produce many oogonia cells. These divide to form primary oocytes which are enclosed in groups of follicle cells (support cells).

GENETIC DIVISION

The primary oocytes begin to divide by meiosis (a specialized nuclear division) but this process is interrupted in its first phase and is not completed until after puberty. At birth, a lifetime's supply of primary oocytes, numbering between 700,000 and two million, will have been formed. These specialized cells will lie dormant in the cortical region of the immature ovary and slowly degenerate, so that by puberty only 40,000 remain.

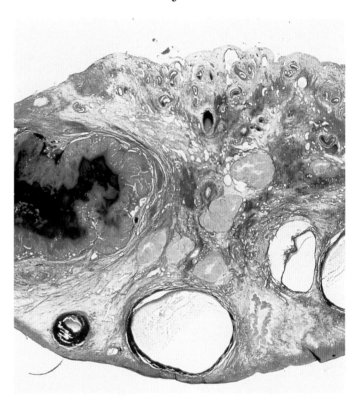

This micrograph shows an ovary with several large follicles (white). During ovulation, up to 20 follicles begin to develop, but only one matures to release an egg.

Egg development

How an egg develops

BEFORE BIRTH — Primordial follicle

CHILDHOOD

Zona pellucida — Follicular development arrested

AT PUBERTY — Primary follicle

Granulosa cells — Developing secondary follicle

Cumulus mass — Graafian follicle
Although several primary follicles develop with each menstrual cycle, only one Graafian follicle is formed; the other follicles regress

— Ruptured follicle

— Released egg

Follicular development begins in the fetus, stops during childhood and is stimulated to continue each month by the onset of the ovarian cycle at puberty.

Before puberty the primary oocyte is surrounded by a layer of cells (the granulosa cells), forming a primary follicle.

PUBERTY

With the onset of puberty, some of the primary follicles are stimulated each month by hormones to continue development and become secondary follicles:
■ A layer of clear viscous fluid, the zona pellucida, is deposited on the surface of the oocyte.
■ The granulosa cells multiply and form an increasing number of layers around the oocyte.
■ The centre of the follicle becomes a chamber (the antrum) that fills with fluid secreted by the granulosa cells.
■ The oocyte is pushed to one side of the follicle, and lies in a mass of follicular cells called the cumulus mass.

A mature secondary follicle is called a Graafian follicle.

Meiosis

The first meiotic division produces two cells of unequal size – the secondary oocyte and the first polar body. The secondary oocyte contains nearly all the cytoplasm of the primary oocyte. Both cells begin a second division; however, this process is halted, and is not completed until the oocyte is fertilized by a sperm.

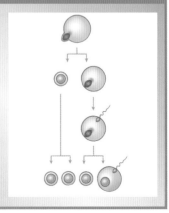

Meiosis, a specialized nuclear division, occurs in the ovaries, giving rise to a female sex cell and three polar bodies.

Egg release

Ovulation occurs when a follicle ruptures, releasing a mature oocyte into the uterine tube. It is at this stage in the menstrual cycle that fertilization may occur.

As the Graafian follicle continues to swell, it can be seen on the surface of the ovary as a blister-like structure.

HORMONAL CHANGES

In response to hormonal changes, the follicular cells surrounding the oocyte begin to secrete a thinner fluid at an increased rate, so that the follicle rapidly swells. As a result, the follicular wall becomes very thin over the area exposed to the ovarian surface, and the follicle eventually ruptures.

OVULATION

A small amount of blood and follicular fluid is forced out of the vesicle, and the secondary oocyte, surrounded by the cumulus mass and zona pellucida, is expelled from the follicle into the peritoneal cavity – the process of ovulation.

Women are generally unaware of this phenomenon, although some experience a twinge of pain in the lower abdomen. This is caused by the intense stretching of the ovarian wall.

FERTILE PERIOD

Ovulation occurs around the 14th day of a woman's menstrual cycle, and it is at this time that a woman is at her most fertile. As sperm can survive in the uterus for up to five days, there is a period of about a week when fertilization can occur.

In the event that the secondary oocyte is penetrated by a sperm cell and pregnancy ensues, the final stages of meiotic division will be triggered. If, however, the egg is not fertilized, the second stage of meiosis will not be completed and the secondary oocyte will simply degenerate.

The ruptured follicle forms a gland called the corpus luteum that secretes progesterone. This hormone prepares the uterine lining to receive an embryo.

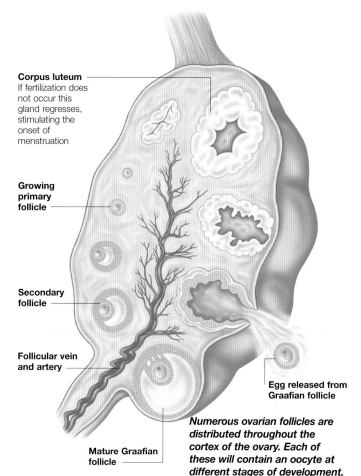

Corpus luteum
If fertilization does not occur this gland regresses, stimulating the onset of menstruation

Growing primary follicle

Secondary follicle

Follicular vein and artery

Mature Graafian follicle

Egg released from Graafian follicle

Numerous ovarian follicles are distributed throughout the cortex of the ovary. Each of these will contain an oocyte at different stages of development.

The menstrual cycle

This graph shows the fluctuation of anterior pituitary and ovarian hormones during the menstrual cycle, together with structural changes within the ovary and uterus.

Luteinizing hormone

Oestrogen

Follicle-stimulating hormone

Progesterone

Follicle matures

Ovulation

Corpus luteum forms

Menstrual flow
Uterine lining is shed at the beginning of the cycle

Uterine lining thickens
In readiness for fertilized egg

Menstruation
Begins again at day 28

Day 5

Day 14

Day 28

◄——— One menstrual cycle ———►

The oestrus, or menstrual cycle, refers to the cyclical changes which take place in the female reproductive system during the production of eggs.

These changes are controlled by hormones released by the pituitary gland and ovaries: oestrogen, progesterone, luteinizing hormone and follicle-stimulating hormone.

UTERINE CHANGES

Following menstruation the endometrium thickens and becomes more vascular under the influence of oestrogen and follicle-stimulating hormone.

During the first 14 days of the menstrual cycle a Graafian follicle matures. Ovulation occurs around day 14 when the secondary oocyte is expelled and swept into the uterine tube.

The ruptured follicle becomes a hormone-secreting body called the corpus luteum. This secretes progesterone, stimulating further thickening of the uterine lining (endometrium) in which the fertilized ovum will implant.

If fertilization does not occur, the levels of progesterone and oestrogen decrease. This causes the endometrium to break down and be excreted into the menstrual flow.

How orgasm occurs

Men and women undergo many physiological changes during orgasm – the climax of sexual intercourse. Male orgasm involves ejaculation, and female orgasm increases the likelihood of successful fertilization.

Sexual intercourse is the means by which the male sex cells (sperm) are transferred to the female reproductive tract.

During sexual intercourse, the man inserts his erect penis into the woman's vagina. Sexual stimulation causes semen to be pumped from the testicles out through the penis, causing ejaculation to occur.

STAGES OF AROUSAL

Sexual arousal occurs in a series of definite stages. During each of these phases, the body experiences different physical changes as it reaches different levels of arousal. After an initial period of desire, both men and women go through four phases:

■ Excitement
■ Plateau
■ Orgasm
■ Resolution.

Men and women exhibit different sexual responses, and these differ significantly from person to person. For both sexes, however, orgasm is the climax of sexual intercourse.

PHYSIOLOGICAL REASONS

The ejaculation of semen that accompanies male orgasm is a prerequisite for fertilization occurring, and it is believed that the female orgasm increases the chance of an egg being fertilized.

Orgasm also creates the urge for sexual intercourse in the first place; for many, it is the pursuit of this pleasurable sensation that is the driving force to copulate.

For orgasm to occur, men and women must be aroused both physically and mentally. The exact extent of the orgasm will differ from person to person.

Male sexual response

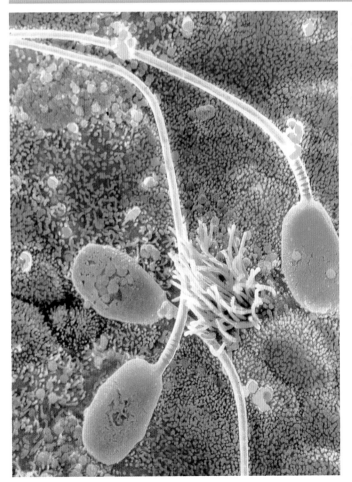

EXCITEMENT

When a man becomes aroused, there is a sudden increase in blood flow to his genitals that causes the penis to become erect. In addition, his heart rate, blood pressure and breathing rate will rise.

PLATEAU

As the penis continues to stiffen, it deepens in colour, and the tip may become lubricated by secretions from the bulbourethral glands (situated at the base of the penis). The testicles swell and contract towards the man's body.

Sperm is moved by a series of muscular contractions from the epididymides to the end of the vas deferens. Here, the sperm is mixed with the prostate and fluids from the seminal vesicles to produce semen. It is at this stage that a man experiences a sensation known as 'ejaculatory inevitability' so that even if stimulation to the penis is ceased, ejaculation will still occur.

The contractions during male orgasm are strong enough to propel semen into the woman's reproductive tract. A man usually has three to five main contractions during orgasm.

ORGASM

An orgasm is the climax of sexual excitement. The intense release of sexual tension built up during sexual stimulation and arousal is generally focused on the genitals, but it may also affect the rest of the body.

Orgasm in men is generally accompanied by simultaneous ejaculation. This occurs when intense contractions of the muscles in the urethra and around the base of the penis force semen out of the body. There are usually three to five main contractions at intervals of 0.8 seconds. The sensation of orgasm can be overwhelming, and many men may involuntarily thrust their pelvis forward, forcing their penis deeper into the woman's vagina.

Male orgasms tend to be shorter than the majority of female orgasms, generally lasting around seven or eight seconds. During orgasm, breathing, heart rate and blood pressure all reach a peak.

RESOLUTION

After orgasm, the penis and testicles return to their normal size. The man's breathing and heart rate slow down, and his blood pressure drops.

Female sexual response

The female orgasm is believed to aid the passage of sperm into the uterus during sexual intercourse, thereby maximizing the chance of fertilization. However, some women never experience an orgasm during intercourse, and are still able to conceive.

EXCITEMENT

During the female excitement phase the clitoris and vagina swell as a result of increased blood supply. The labia majora darken in colour, and the labia minora flatten and part.

One of the first signs of sexual arousal in women is wetness around the opening to the vagina. This is caused by stimulation of the secretory cells lining the vagina. This fluid lubricates the vagina, preparing it for penetration, which may or may not occur at a later stage.

The breasts become slightly enlarged, and the nipples become erect. The areolae (around the nipple) swell and darken. Blood pressure, heart rate, breathing rate and muscle tension all increase.

This stage of excitement can last a variable length of time. It may lead to the plateau phase or gently subside.

PLATEAU

If sexual excitement and stimulation continue, a woman will enter the plateau phase. This is characterized by increased blood flow to the entire genital area. The lower part of the vagina narrows to help grip the penis during intercourse. The upper vagina becomes enlarged and the uterus rises from the pelvic cavity, causing an expansion of the vaginal cavity and creating an area where semen can pool.

During this phase the labia minora deepen in colour, and the clitoris shortens and withdraws under the labial hood. A few drops of fluid may be secreted by

Physical arousal in women

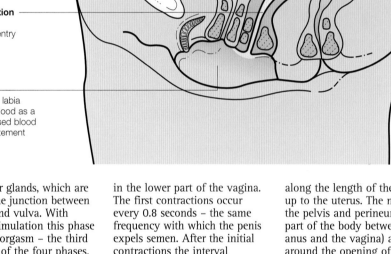

The contractions during orgasm may help to move sperm into the uterus and uterine (Fallopian) tubes.

Uterus
Rises from the pelvic cavity during the plateau phase; begins to contract rhythmically during orgasm

Upper vagina
Enlarges in the plateau phase, creating a space for semen to pool, maximizing chances of fertilization

Vaginal secretion
Lubricates the vagina, aiding entry of the penis

Genital area
The clitoris and labia engorge with blood as a result of increased blood flow in the excitement phase

the vestibular glands, which are situated at the junction between the vagina and vulva. With continued stimulation this phase may lead to orgasm – the third and shortest of the four phases.

ORGASM

Female orgasms can be intense, but rarely last for longer than 15 seconds. An orgasm begins with a wave of rhythmic contractions in the lower part of the vagina. The first contractions occur every 0.8 seconds – the same frequency with which the penis expels semen. After the initial contractions the interval becomes progressively longer. It is possible that a woman's contractions help to move sperm into the uterus and uterine (Fallopian) tubes.

Orgasmic contractions spread along the length of the vagina up to the uterus. The muscles of the pelvis and perineum (the part of the body between the anus and the vagina) and around the opening of the bladder and rectum also contract. Women usually experience 5 to 15 orgasmic contractions, depending on the intensity of the orgasm.

Muscles in the back and feet may also undergo involuntary spasms during orgasm, causing the back to arch and the toes to curl. The heart rate can rise to as much as 180 beats per minute and the breathing rate to as much as 40 breaths per minute. Blood pressure rises and the pupils and nostrils dilate. A woman may breathe rapidly or hold her breath for the duration of the orgasm.

RESOLUTION

Once the orgasm phase is complete, the resolution phase begins. During this time the woman's breasts return to their normal size, the body muscles relax, and her normal heart and breathing rates are restored.

Refractory period

After ejaculation, men experience a refractory period, during which time they are unable to achieve another orgasm. This latent period can last from about two minutes to several hours.

Women do not experience a refractory period, and some may experience multiple orgasms.

Men and women react differently after an orgasm. However, it is common for both men and women to feel relaxed and sleepy.

How conception occurs

Millions of sperm cells travel up the female reproductive tract in search of the oocyte (egg). It takes hundreds of sperm to break down the outer coating of the oocyte, but only one will fertilize it.

Fertilization occurs when a single male gamete (sperm cell) and a female gamete (egg or oocyte) are united following sexual intercourse. Fusion of the two cells occurs and a new life is conceived.

SPERM

Following sexual intercourse, the sperm contained in the man's semen travel up through the uterus. Along the way they are nourished by the alkaline mucus of the cervical canal. From the uterus the sperm continue their journey into the uterine (Fallopian) tube.

Although the distance involved is only around 20 cm, the journey can take up to two hours, since in relation to the size of the sperm the distance is considerable.

SURVIVAL

Although an average ejaculation contains around 300 million sperm cells, only a fraction of these (around 10,000) will manage to reach the uterine tube where the oocyte is located. Even fewer will actually reach the oocyte. This is because many sperm will be destroyed by the hostile vaginal environment, or become lost in other areas of the reproductive tract.

Sperm do not become capable of fertilizing an oocyte until they have spent some time in the woman's body. Fluids in the reproductive tract activate the sperm, so that the whiplash motion of their tails becomes more powerful.

The sperm are also helped on their way by contractions of the uterus, which force them upwards into the body. The contractions are stimulated by prostaglandins contained in the semen, and which are also produced during female orgasm.

THE OOCYTE

Once it has been ejected from the follicle (during ovulation) the oocyte is pushed towards the uterus by the wave-like motion of the cells lining the uterine tube. The oocyte is usually united with the sperm about two hours after sexual intercourse in the outer part of the uterine tube.

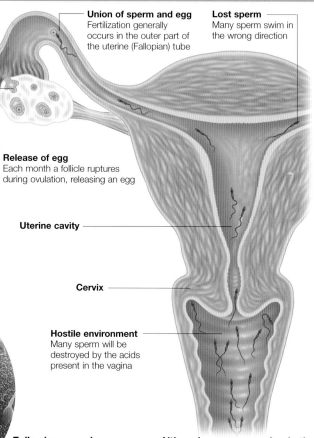

The path to fertilization

Union of sperm and egg
Fertilization generally occurs in the outer part of the uterine (Fallopian) tube

Lost sperm
Many sperm swim in the wrong direction

Release of egg
Each month a follicle ruptures during ovulation, releasing an egg

Uterine cavity

Cervix

Hostile environment
Many sperm will be destroyed by the acids present in the vagina

Following sexual intercourse, millions of sperm cells make their way up the reproductive tract in search of the oocyte.

Although many sperm begin the journey towards the oocyte, only a fraction reach the uterine tube. The majority are destroyed or become lost on the way.

Reaching the oocyte

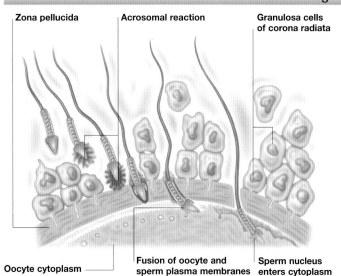

Zona pellucida

Acrosomal reaction

Granulosa cells of corona radiata

Oocyte cytoplasm

Fusion of oocyte and sperm plasma membranes

Sperm nucleus enters cytoplasm

On the journey towards the oocyte, secretions present in the female reproductive tract deplete the sperm cells' cholesterol, thus weakening their acrosomal membranes. This process is known as capacitation, and without it fertilization could not occur.

Once in the vicinity of the oocyte, the sperm are chemically attracted to it. When the sperm cells finally come in to contact with the oocyte, their acrosomal membranes are completely

When sperm cells reach the oocyte they release enzymes. These enzymes break down the protective outer layers of the ovum, allowing a sperm to enter.

stripped away, so that the contents of each acrosome (the enzyme-containing compartment of the sperm) are released.

PENETRATION

The enzymes released by the sperm cells cause the break-down of the cumulus mass cells and the zona pellucida, the protective outer layers of the oocyte. It takes at least 100 acrosomes to rupture in order for a path to be digested through these layers for a single sperm to enter.

In this way the sperm cells that reach the oocyte first sacrifice themselves, to allow penetration of the cytoplasm of the oocyte by another sperm.

Fertilization

When a single sperm has entered the oocyte, the genetic material from each cell fuses. A zygote is formed, which divides to form an embryo.

Once a sperm has penetrated the oocyte, a chemical reaction takes place within the oocyte, making it impossible for another sperm to enter.

MEIOSIS II
Entry of the sperm nucleus into the oocyte triggers the completion of nuclear division (meiosis II) begun during ovulation. A haploid oocyte and the second polar body (which degenerates) are formed.

Almost immediately, the nuclei of the sperm and oocyte fuse to produce a diploid zygote, containing genetic material from both the mother and father.

DETERMINATION OF SEX
It is at the point of fertilization that sex is determined. It is the sperm, and therefore the father, that dictates what sex the offspring will be.

Sex is determined by a combination of the two sex chromosomes, the X and the Y. The female will contribute an X chromosome, while a male may contribute either an X or a Y. Fertilization of the oocyte (X), will either be by a sperm containing an X or a Y to give a female (XX) or a male (XY).

CELL DIVISION
Several hours after fertilization the zygote undergoes a series of mitotic divisions to produce a cluster of cells known as a morula. The morula cells divide every 12 to 15 hours, producing a blastocyst comprised of around 100 cells.

The blastocyst secretes the hormone human chorionic gonadotrophin. This prevents the corpus luteum from being broken down, thus maintaining progesterone secretion.

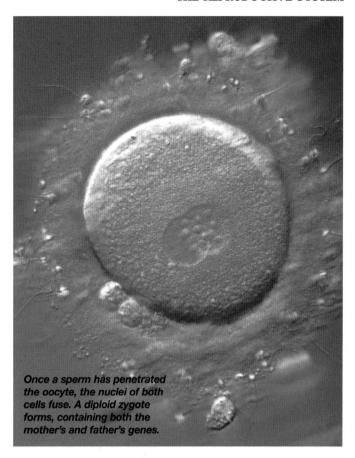

Once a sperm has penetrated the oocyte, the nuclei of both cells fuse. A diploid zygote forms, containing both the mother's and father's genes.

Implantation and development

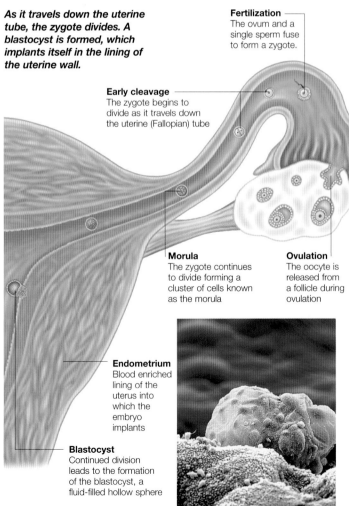

As it travels down the uterine tube, the zygote divides. A blastocyst is formed, which implants itself in the lining of the uterine wall.

Fertilization
The ovum and a single sperm fuse to form a zygote.

Early cleavage
The zygote begins to divide as it travels down the uterine (Fallopian) tube

Morula
The zygote continues to divide forming a cluster of cells known as the morula

Ovulation
The oocyte is released from a follicle during ovulation

Endometrium
Blood enriched lining of the uterus into which the embryo implants

Blastocyst
Continued division leads to the formation of the blastocyst, a fluid-filled hollow sphere

Around three days after fertilization, the blastocyst will begin its journey from the uterine (Fallopian) tube to the uterus.

Normally the blastocyst would be unable to pass through the sphincter muscle in the uterine tube. However, the increasing levels of progesterone triggered by fertilization cause the muscle to relax, allowing the blastocyst to continue its journey to the uterus.

A damaged or blocked uterine tube preventing the blastocyst from passing at this stage would result in an ectopic pregnancy in which the embryo starts to develop in the uterine tube.

MULTIPLE BIRTHS
In most cases a woman will release one oocyte every month from alternate ovaries.

Occasionally however, a woman may produce an oocyte from each ovary both of which are fertilized by separate sperm, resulting in the development of non-identical twins. In this case, each fetus will be nourished by its own placenta.

When the zygote reaches the uterus it will adhere to the endometrium. Nourished by the rich blood supply, it begins to develop.

Very occasionally a fertilized oocyte may split spontaneously in two to produce two embryos. This will result in identical twins that share exactly the same genes, and even the same placenta.

Siamese twins occur when there is an incomplete split of the oocyte several hours after fertilization.

IMPLANTATION
Once it has reached the uterus, the blastocyst will implant itself in the thickened lining of the uterine wall.

Hormones released from the blastocyst mean that it is not identified as a foreign body and expelled. Once the blastocyst is safely implanted, gestation will begin.

IMPERFECTIONS
About one third of fertilized oocytes fail to implant in the uterus and are lost.

Of those that do implant, many embryos contain imperfections in their genetic material, such as an extra chromosome.

Many of these imperfections will cause the embryo to be lost soon after implantation. This can occur even before the first missed period, so that a woman will not even have known that she was pregnant.

How childbirth occurs

Towards the end of pregnancy physiological changes occur in both mother and fetus. Hormonal triggers cause the muscles in the uterine wall to contract, expelling the baby and placenta.

Parturition – meaning 'bringing forth the young' – is the final stage of pregnancy. It usually occurs 280 days (40 weeks) from the last menstrual period.

The series of physiological events that lead to the baby being delivered from the mother's body are referred to collectively as labour.

INITIATION OF LABOUR

The precise signal that triggers labour is not known, but many factors which play a role in its initiation have been identified.

Before parturition, levels of progesterone secreted by the placenta into the mother's circulation reach a peak. Progesterone is the hormone responsible for maintaining the uterine lining during pregnancy and has an inhibitory effect on the smooth muscle of the uterus.

HORMONAL TRIGGERS

Towards the end of the pregnancy, there is increasingly limited space in the uterus and the fetus' limited oxygen supply becomes increasingly restricted (resulting from a more rapid increase in the size of the fetus than in the size of the placenta). This causes an increased level of adrenocorticotropic hormone (ACTH) to be secreted from the anterior lobe of the fetus' pituitary.

Consequently, the fetus' adrenal cortex is triggered to produce chemical messengers (glucocorticoids) which inhibit

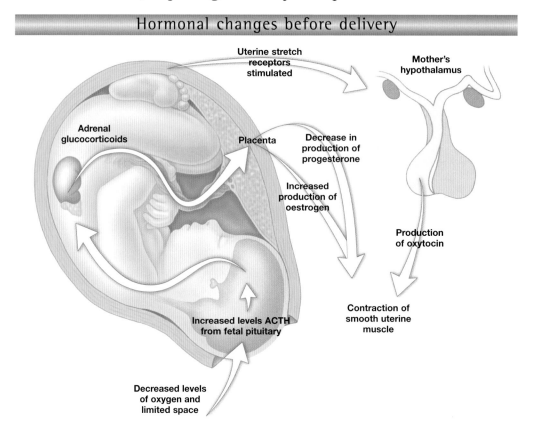

Hormonal changes before delivery

Uterine stretch receptors stimulated

Mother's hypothalamus

Adrenal glucocorticoids

Placenta

Decrease in production of progesterone

Increased production of oestrogen

Production of oxytocin

Increased levels ACTH from fetal pituitary

Contraction of smooth uterine muscle

Decreased levels of oxygen and limited space

progesterone secretion from the placenta.

Meanwhile the levels of the hormone oestrogen released by the placenta into the mother's circulation reach a peak. This causes the myometrial cells of the uterus to form an increased number of oxytocin receptors (making the uterus more sensitive to oxytocin).

CONTRACTIONS

Eventually the inhibitory influence of progesterone on the smooth muscle cells of the uterus is overcome by the stimulatory effect of oestrogen.

The inner lining of the uterus (myometrium) weakens, and the uterus begins to contract irregularly. These contractions, known as Braxton Hicks

As the pregnancy reaches full term a number of hormonal changes occur. These cause the lining of the uterus to weaken and contractions to commence.

contractions, help to soften the cervix in preparation for the birth and are often mistaken by pregnant mothers for the onset of labour.

Onset of labour

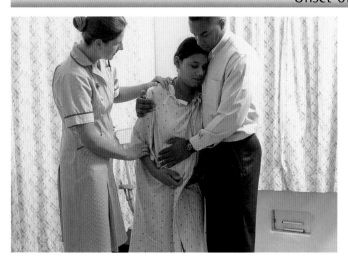

As the pregnancy reaches full term, stretch receptors in the uterine cervix activate the mother's hypothalamus (a region of the brain) to stimulate her posterior pituitary gland in order to release the hormone oxytocin. Certain cells of the fetus also begin to release this hormone.

Elevated levels of oxytocin trigger the placenta to release prostaglandins and together they stimulate the uterus to contract.

Oxytocin triggers uterine contractions that push the fetus against the cervix. Further stretch of the cervix stimulates more oxytocin to be released.

INTENSIFICATION OF CONTRACTIONS

As the uterus is weakened due to suppressed levels of progesterone and is more sensitive to oxytocin, the contractions become stronger and more frequent, and the rhythmic contractions of labour begin.

A 'positive feedback' mechanism is activated whereby the greater the intensity of the contractions the more oxytocin is released, which in turn causes the contractions to become more intense. The chain is broken when the cervix is no longer stretched after delivery and oxytocin levels drop.

Stages of labour

The birth can be divided in to three distinct stages: dilatation of the cervix, expulsion of the fetus and delivery of the placenta.

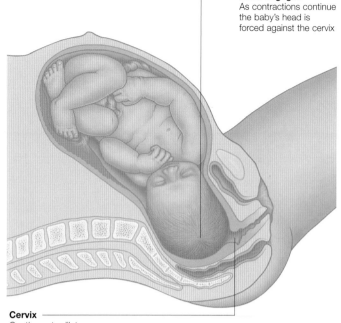

Head engaged
As contractions continue the baby's head is forced against the cervix

Cervix
Continues to dilate as contractions progress

DILATATION

In order for the baby's head to pass through the birth canal, the cervix and vagina must dilate to around 10 cm in diameter. As labour commences, weak but regular contractions begin in the upper part of the uterus.

These initial contractions are 15–30 minutes apart and last around 10–30 seconds. As the labour progresses, the contractions become faster and more intense, and the lower part of the uterus begins to contract as well.

The baby's head is forced against the cervix with each contraction, causing the cervix to soften, and gradually dilate.

Eventually the amniotic sac, which has protected the baby for the duration of the pregnancy, ruptures, and the amniotic fluid is released.

ENGAGEMENT

The dilatation stage is the longest part of labour and can last from 8 to 24 hours.

During this phase the baby begins to descend through the birth canal, rotating as it does so, until the head engages, entering the pelvis.

Dilatation is the longest stage of labour. It can take up to 24 hours for the cervix to dilate sufficiently to allow delivery.

Expulsion

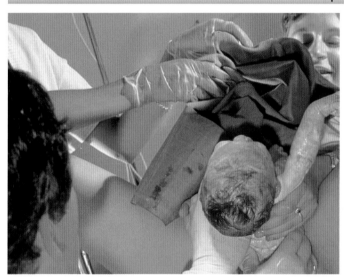

The second stage of labour, expulsion, lasts from full dilatation to the actual delivery of the child.

Usually by the time the cervix is fully dilated, strong contractions occur every 2–3 minutes and each lasts around a minute.

URGE TO PUSH

At this point the mother will have an overwhelming urge to push or bear down with the abdominal muscles.

Once the cervix is fully dilated the baby is ready to be delivered. The mother will feel a strong urge to push, expelling the baby through the cervix.

This phase can take as long as two hours, but is generally much quicker in subsequent births.

DELIVERY

Crowning takes place when the largest part of the baby's head reaches the vagina. In many cases the vagina will distend to such an extent that it tears.

Once the baby's head has exited, the rest of the body is delivered much more easily.

When the baby emerges head first, the skull (at its widest diameter) acts as a wedge to dilate the cervix. This head-first presentation allows the baby to breathe even before it is completely delivered from the mother.

Delivery of the placenta

The final stage of labour, when the placenta is delivered, can take place up to 30 minutes after the birth.

After the baby has been delivered the rhythmical uterine contractions continue. These act to compress the uterine blood vessels thus limiting bleeding. The contractions also cause the placenta to break away from the wall of the uterus.

AFTERBIRTH

The placenta and attached fetal membranes (the afterbirth) are then easily removed by pulling gently on the umbilical cord. All placental fragments must be removed to prevent continued uterine bleeding and infection after birth.

The number of vessels in the severed umbilical cord will be counted, as the absence of an umbilical artery is often associated with cardiovascular disorders in the baby.

HORMONE LEVELS

Blood levels of oestrogen and progesterone fall dramatically once their source, the placenta, has been delivered. During the four or five weeks after parturition the uterus becomes much smaller but remains larger than it was before pregnancy.

Contractions continue after the birth. This causes the placenta to detach from the uterine wall, and it can be removed with a gentle tug of the umbilical cord.

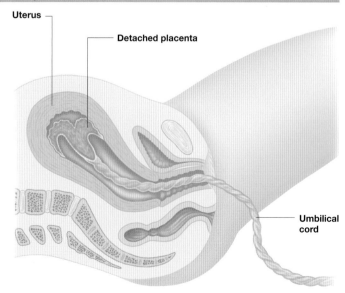

Uterus

Detached placenta

Umbilical cord

Chapter Eight

The Pelvis & Lower Limbs

The pelvis and lower limbs bear the entire weight of the upper body and therefore play an essential role in terms of balance, posture and stability. The ability to stand on two legs depends on a network of powerful muscles that connect the bones and joints of the lower body, enabling them to fulfil their supportive role and provide a secure base for the body to rest upon.

This chapter details the interconnecting bones, muscles and blood vessels of the lower body and explains the importance of the basin-like pelvis, both as a protective framework for the reproductive organs and bladder and as a secure link between the upper and lower body.

LEFT: Cross-country runners train off-road. The legs are one of the hardiest parts of human anatomy, capable of undergoing great stress and strain.

Bones of the pelvis

The basin-like pelvis is formed by the hip bones, sacrum and coccyx. The pelvic bones provide sites of attachment for many important muscles, and also help to protect the vital pelvic organs.

The bones of the pelvis form a ring which connects the spine to the lower limbs and protects the pelvic contents, including the reproductive organs and bladder.

The pelvic bones, to which many powerful muscles are attached, allow the weight of the body to be transferred to the legs with great stability.

STRUCTURE OF THE PELVIS
The basin-like pelvis consists of the innominate (hip) bones, the sacrum and the coccyx. The innominate bones meet at the pubic symphysis anteriorly. Posteriorly, these two bones are joined to the sacrum. Extending down from the sacrum at the back of the pelvis is the coccyx.

FALSE AND TRUE PELVIS
The pelvis can be said to be divided into two parts by an imaginary plane passing through the sacral promontory and the pubic symphysis:
■ Above the sacral promontory, the false pelvis flares out and supports the lower abdominal contents
■ Below this plane lies the true pelvis lies; in females, it forms the constricted birth canal through which the baby passes.

The bony structure of the pelvis is formed by the hip bones, sacrum and coccyx. The adult female pelvis, shown here, is adapted for childbirth.

Adult female pelvis from the front

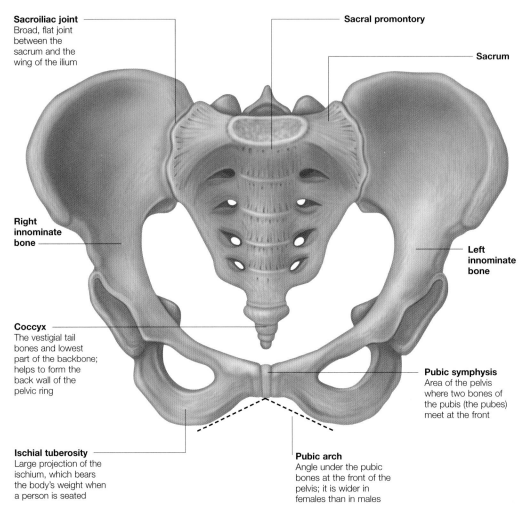

Sacroiliac joint
Broad, flat joint between the sacrum and the wing of the ilium

Sacral promontory

Sacrum

Right innominate bone

Left innominate bone

Coccyx
The vestigial tail bones and lowest part of the backbone; helps to form the back wall of the pelvic ring

Pubic symphysis
Area of the pelvis where two bones of the pubis (the pubes) meet at the front

Ischial tuberosity
Large projection of the ischium, which bears the body's weight when a person is seated

Pubic arch
Angle under the pubic bones at the front of the pelvis; it is wider in females than in males

Differences between male and female pelvis

Adult male pelvis from the front

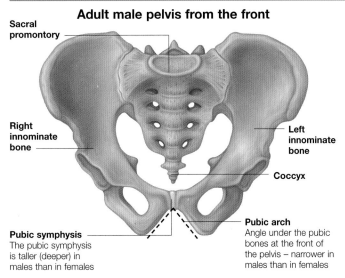

Sacral promontory

Right innominate bone

Left innominate bone

Coccyx

Pubic symphysis
The pubic symphysis is taller (deeper) in males than in females

Pubic arch
Angle under the pubic bones at the front of the pelvis – narrower in males than in females

The skeletons of men and women differ in a number of places, but nowhere is this more marked than in the pelvis.

PHYSICAL VARIATIONS
The differences between the male and female pelvis can be attributed to two factors: the requirements of childbirth and the fact that, in general, men are heavier and more muscular than women. Some of the more obvious differences are:
■ General structure – the male

The male pelvis differs from the female pelvis in being heavier, with thicker bones. The pubic arch is narrower and the pubic symphysis deeper in males.

pelvis is heavier, with thicker bones
■ Pelvic inlet – the 'way into' the true pelvis is a wide oval in females but narrower and heart-shaped in males
■ Pelvic canal – the 'way through' the true pelvis is roughly cylindrical in females, whereas in males it tapers
■ Pubic arch – the angle under the pubic bones at the front of the pelvis is wider in females (100 degrees or more) than in males (90 degrees or less).

These differences, together with other, more subtle, measurements, may be used by forensic pathologists and anthropologists to determine the sex of a skeleton.

The hip bone

The two hip bones are fused together at the front and join with the sacrum at the back. They each consist of three bones – the ilium, ischium and pubis.

The two innominate (hip) bones constitute the greater part of the pelvis, joining with each other at the front and with the sacrum at the back.

STRUCTURE

The hip bone is large and strong, due to its function of transmitting the forces between the legs and the spine. As with most bones, it has areas which are raised or roughened by the attachments of muscle or ligaments.

The hip bone is formed by the fusion of three separate bones: the ilium, the ischium and the pubis. In children, these three bones are joined only by cartilage. At puberty, they fuse to form the single innominate, or hip, bone on each side.

FEATURES

The upper margin of the hip bone is formed by the widened iliac crest. Further down the hip bone is the ischial tuberosity, a projection of the ischium.

The obturator foramen lies below and slightly in front of the acetabulum, the latter receiving the head of the femur (thigh bone).

This lateral view of the hip bone clearly shows its constituent parts of ilium, ischium and pubis. These three bones fuse together at puberty.

Right hip bone, lateral view

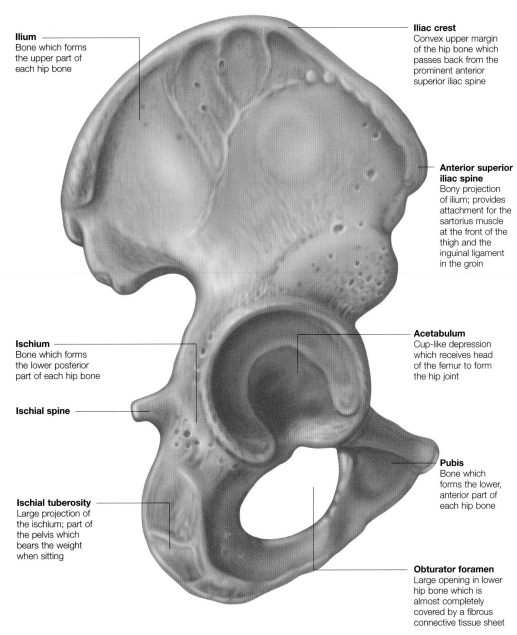

Ilium
Bone which forms the upper part of each hip bone

Iliac crest
Convex upper margin of the hip bone which passes back from the prominent anterior superior iliac spine

Anterior superior iliac spine
Bony projection of ilium; provides attachment for the sartorius muscle at the front of the thigh and the inguinal ligament in the groin

Ischium
Bone which forms the lower posterior part of each hip bone

Ischial spine

Acetabulum
Cup-like depression which receives head of the femur to form the hip joint

Pubis
Bone which forms the lower, anterior part of each hip bone

Ischial tuberosity
Large projection of the ischium; part of the pelvis which bears the weight when sitting

Obturator foramen
Large opening in lower hip bone which is almost completely covered by a fibrous connective tissue sheet

The female pelvic canal

Lateral view of right pelvis

Sacral promontory

Plane of pelvic inlet

Pubic symphysis
Front area of pelvis

Plane of pelvic outlet

In childbirth, the baby passes down into the pelvic canal, through the pelvic inlet and out through the pelvic outlet. The dimensions of the pelvic canal in women are therefore vital.

TRIANGULAR SHAPE

The pelvic canal is almost triangular in section, the short front wall being formed by the pubic symphysis. The much longer back wall is formed by the sacrum and coccyx.

The pelvic canal is defined by the pubic symphysis at the front, and the sacrum and coccyx at the back. The coccyx moves back out of the way in childbirth.

From front to back, the pelvic inlet usually has a diameter of about 11 cm, known as the obstetric conjugate. The inlet is slightly wider from side to side owing to its oval shape.

CHANGES IN CHILDBIRTH

The pelvic outlet is normally slightly larger than the inlet, especially at the end of pregnancy when the ligaments holding the pelvic bones together can stretch under the influence of hormones.

The joint between the coccyx and the sacrum also becomes looser, allowing the coccyx to move back out of the way during childbirth.

339

Ligaments and joints of the pelvis

The pelvic bones are connected by joints, which are bound together by the ligaments to form a solid structure. The ligaments of the pelvis are some of the strongest in the body.

The pelvis needs to be structurally strong in order to perform its functions of transferring weight to the legs, and supporting the abdominal contents.

The pelvic bones are themselves thick and strong but their overall stability is assured by the presence of a series of tough pelvic ligaments which bind them together.

STRUCTURE OF PELVIS
The pelvis is formed from the paired innominate (hip) bones together with the sacrum and coccyx. These bones have joints between them and the pelvic ligaments are arranged so that they hold these joints together while resisting forces that would otherwise pull them apart.

FRONT VIEW
The major ligaments of the pelvis are generally named after the two areas of bone that they link.

Those most visible on an anterior (frontal) view of the pelvis are:
■ Iliolumbar ligament
■ Anterior sacroiliac ligament
■ Sacrospinous ligament
■ Anterior longitudinal ligament.

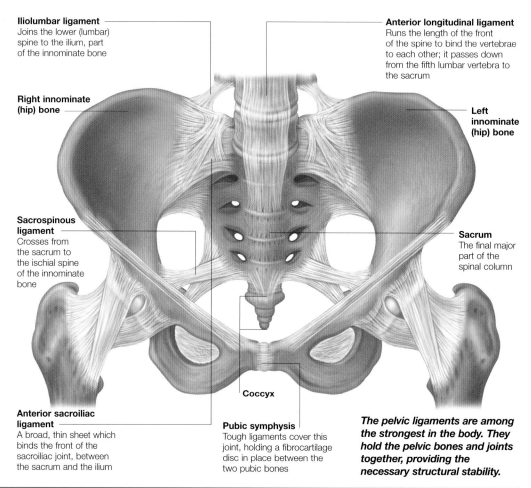

Iliolumbar ligament
Joins the lower (lumbar) spine to the ilium, part of the innominate bone

Right innominate (hip) bone

Sacrospinous ligament
Crosses from the sacrum to the ischial spine of the innominate bone

Anterior sacroiliac ligament
A broad, thin sheet which binds the front of the sacroiliac joint, between the sacrum and the ilium

Coccyx

Pubic symphysis
Tough ligaments cover this joint, holding a fibrocartilage disc in place between the two pubic bones

Anterior longitudinal ligament
Runs the length of the front of the spine to bind the vertebrae to each other; it passes down from the fifth lumbar vertebra to the sacrum

Left innominate (hip) bone

Sacrum
The final major part of the spinal column

The pelvic ligaments are among the strongest in the body. They hold the pelvic bones and joints together, providing the necessary structural stability.

Posterior pelvic ligaments

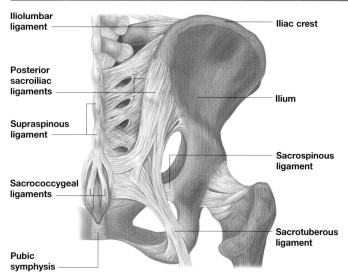

Iliolumbar ligament

Posterior sacroiliac ligaments

Supraspinous ligament

Sacrococcygeal ligaments

Pubic symphysis

Iliac crest

Ilium

Sacrospinous ligament

Sacrotuberous ligament

Viewing the pelvis from the rear shows the arrangement of ligaments which bind the bones at the back of the pelvis.

FUNCTION OF LIGAMENTS
The posterior sacroiliac ligament crosses the sacroiliac joint as it passes downwards and inwards from the ilium to the sacrum.

This ligament is stronger than the relatively thin anterior sacroiliac ligament, taking much of the strain of keeping the ilium

At the rear of the pelvis there are a number of ligaments. Each ligament strengthens the link between the different pelvic bones at the site of the joints.

connected to the sacrum on each side.

The large and powerful sacrotuberous ligament can be clearly seen as it passes from the sacrum down to the roughened ischial tuberosity. Together with the sacrospinous ligament, which lies just in front of it, the sacrotuberous ligament acts to resist the rotational force on the sacrum generated by the weight of the body.

Just as the anterior longitudinal ligament joins the front of the spinal column, the tough supraspinous ligament stabilizes the vertebrae from the rear, joining their spinous processes. It terminates by forming the sacrococcygeal ligaments.

Joints of the pelvis

The pelvis is a ring of bone on which the weight of the body is carried. Where bone meets bone, the pelvic joints are formed, bound together by the pelvic ligaments.

The pelvis is formed by the paired innominate bones and the sacrum with the coccyx attached. The pelvic joints which connect these bones, unlike the elbow or knee joints, are not designed to allow movement.

The joints of the pelvis are bound strongly together by the pelvic ligaments to form a single, solid structure.

THE SACROILIAC JOINT

The sacroiliac joint is the largest and most important joint of the pelvis, between the sacrum (part of the lower spine) and the ilium (part of the large innominate bone). This joint needs to be strong as it bears the weight of the body. The shape of the bones at the joint surface contributes to the stability of this joint, having irregular indentations which can partially interlock.

However, the main stabilizing factor is the presence of the very strong posterior sacroiliac ligaments and the interosseous ligaments. These ligaments 'suspend' the sacrum between the two iliac bones, bearing the weight of the upper body.

JOINT MOVEMENT

The sacroiliac joint belongs to the group of synovial, or fluid-filled, joints such as those of the elbow, shoulder and knee. However, unlike these joints there is very little movement allowed in the sacroiliac pelvic joint, a state of affairs which is necessary for its stability.

Section through the sacroiliac joints and ligaments

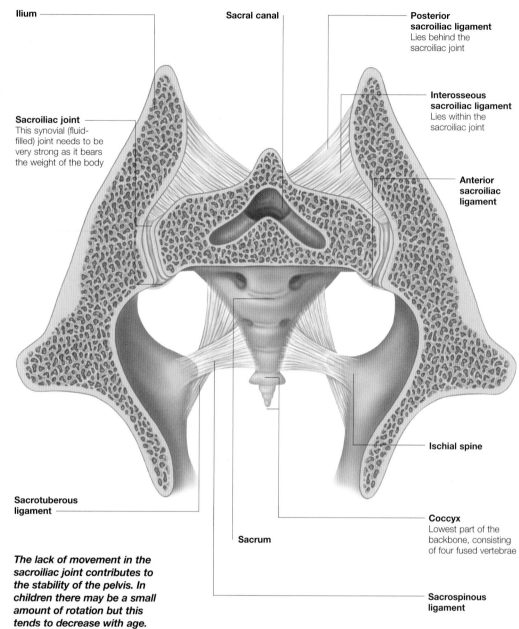

Ilium

Sacral canal

Posterior sacroiliac ligament
Lies behind the sacroiliac joint

Interosseous sacroiliac ligament
Lies within the sacroiliac joint

Sacroiliac joint
This synovial (fluid-filled) joint needs to be very strong as it bears the weight of the body

Anterior sacroiliac ligament

Ischial spine

Sacrotuberous ligament

Coccyx
Lowest part of the backbone, consisting of four fused vertebrae

Sacrum

Sacrospinous ligament

The lack of movement in the sacroiliac joint contributes to the stability of the pelvis. In children there may be a small amount of rotation but this tends to decrease with age.

Pubic symphysis

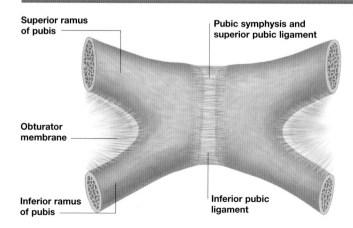

Superior ramus of pubis

Pubic symphysis and superior pubic ligament

Obturator membrane

Inferior ramus of pubis

Inferior pubic ligament

The pubic symphysis is the joint between the two pubic bones at the front of the pelvis. This is a very strong and stable joint, allowing almost no movement between the two pubic bones.

CARTILAGE

The pubic symphysis is a cartilaginous joint, the two bone surfaces being covered with a

The pubic symphysis connects the two pubic bones. The bones are held in place by ligaments, and the bone surface is lined with cartilaginous tissue.

layer of hyaline cartilage and connected by fibrous ligaments.

DISC

Between the two bones within the joint there is fibrocartilage, which has a small cavity in the midline. This tissue tends to be more extensive in women than in men.

The joint, and especially the springy fibrocartilage, serves to act as a 'shock absorber', helping to reduce the chance of fracture of the bones when the pelvis receives sudden forces either directly or from the legs.

Pelvic floor muscles

The muscles of the pelvic floor play a vital role in supporting the abdominal and pelvic organs. They also help to regulate the processes of defecation and urination.

The pelvic floor muscles play an important role in supporting the abdominal and pelvic organs. In pregnancy, these muscles help to carry the growing weight of the uterus, and in childbirth they support the baby's head as the cervix dilates.

MUSCLES

The muscles of the pelvic floor are attached to the inside of the ring of bone that makes up the pelvic skeleton, and slope downwards to form a rough funnel shape.

The levator ani is the largest muscle of the pelvic floor. It is a wide, thin sheet made up of three parts:
- Pubococcygeus – the main part of the levator ani muscle
- Puborectalis – joins with its counterpart on the other side to form a U-shaped sling around the rectum
- Iliococcygeus – the posterior fibres of the levator ani.

A second muscle, the coccygeus (or ischiococcygeus), lies behind the levator ani.

PELVIC WALLS

The pelvic cavity is described as having an anterior, a posterior and two lateral walls.

The anterior wall is formed by the pubic bones and their connection, the pubic symphysis. The posterior wall is formed by the sacrum and coccyx and the neighbouring parts of the iliac bones. The two lateral walls are formed by the obturator internus muscles overlying the hip bones.

Female pelvic diaphragm from above

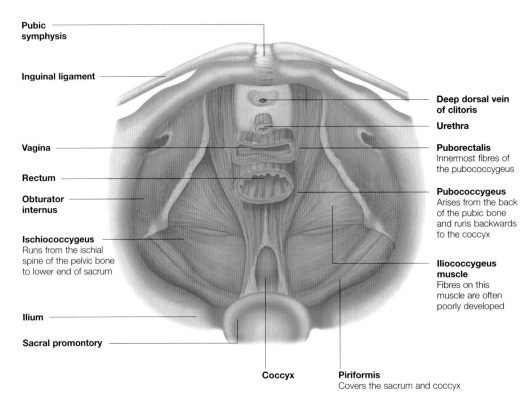

Pubic symphysis

Inguinal ligament

Vagina

Rectum

Obturator internus

Ischiococcygeus
Runs from the ischial spine of the pelvic bone to lower end of sacrum

Ilium

Sacral promontory

Deep dorsal vein of clitoris

Urethra

Puborectalis
Innermost fibres of the pubococcygeus

Pubococcygeus
Arises from the back of the pubic bone and runs backwards to the coccyx

Iliococcygeus muscle
Fibres on this muscle are often poorly developed

Coccyx

Piriformis
Covers the sacrum and coccyx

The pelvic floor muscles are known as the pelvic diaphragm. The levator ani is the most important muscle and is named for its action in lifting the anus.

Perineal body

Female pelvis

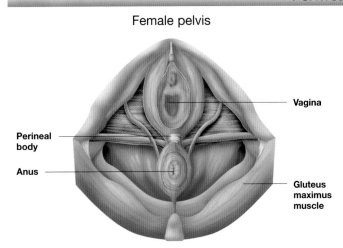

Perineal body

Anus

Vagina

Gluteus maximus muscle

The perineal body is a small mass of fibrous tissue that lies within the pelvic floor, just in front of the anal canal. This structure provides a site for the attachment of many of the pelvic floor and perineal muscles, so allowing paired muscles to pull against each other, normally one of the functions of bone. It also provides support for the internal organs of the pelvis.

Although the perineal body is small and tucked away, it is a very important structure. It supports the organs of the pelvis which lie above it.

EPISIOTOMY

The perineal body may become damaged during childbirth, either by stretching or tearing as the baby's head passes through the pelvic floor. Loss of the perineal body's support of the posterior vaginal wall may eventually lead to vaginal prolapse.

To prevent damage to the perineal body during childbirth, an obstetrician may perform an episiotomy. This deliberate incision into the muscle behind the vaginal opening enlarges this opening and avoids damage to the perineal body.

Openings of the pelvic floor

The pelvic floor resembles the diaphragm in the chest in that it forms a nearly continuous sheet, but does have openings to allow important structures to pass through it. There are two important openings situated in the pelvic floor region.

From below, the pelvic floor can be seen to assume a funnel shape. The muscles of the pelvic floor are so arranged that there are two main openings:

■ Anorectal hiatus – this opening, or hiatus, allows the rectum and anal canal to pass through the sheet of pelvic floor muscles to reach the anus beneath. The U-shaped fibres of the puborectalis muscle form the posterior edge of this hiatus

■ Urogenital hiatus – lying in front of the anorectal hiatus there is an opening in the pelvic floor for the urethra, which carries urine from the bladder out of the body. In females, the vagina also passes through the pelvic diaphragm within this opening, just behind the urethra.

FUNCTIONS OF THE PELVIC FLOOR MUSCLES
The functions of the pelvic floor include:
■ Supporting the internal organs of the abdomen and pelvis
■ Helping to resist rises in pressure within the abdomen, such as during coughing and sneezing, which would otherwise cause the bladder/bowel to empty
■ Assisting in the control of defecation and urination
■ Helping to fix and brace the trunk during forceful movements of the upper limbs, such as weight-lifting.

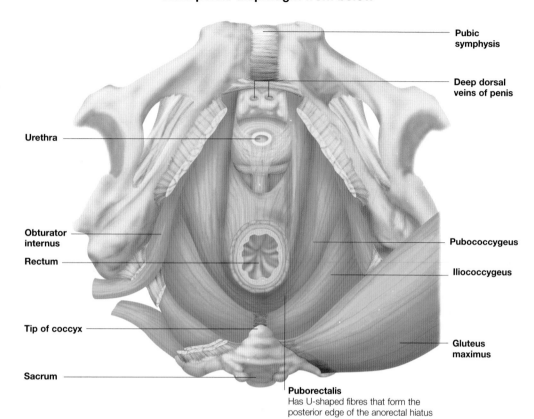

Male pelvic diaphragm from below

- Pubic symphysis
- Deep dorsal veins of penis
- Urethra
- Obturator internus
- Rectum
- Tip of coccyx
- Sacrum
- Pubococcygeus
- Iliococcygeus
- Gluteus maximus

Puborectalis
Has U-shaped fibres that form the posterior edge of the anorectal hiatus

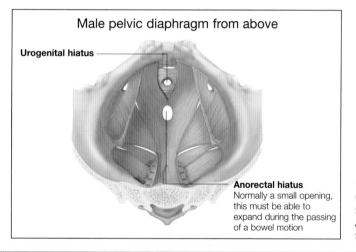

Male pelvic diaphragm from above

- Urogenital hiatus
- Anorectal hiatus
Normally a small opening, this must be able to expand during the passing of a bowel motion

The pelvic floor muscles play a vital supporting role. Without them, the internal organs of the abdomen and pelvis would sink through the bony pelvic ring.

Ischioanal fossae

Coronal section through pelvis

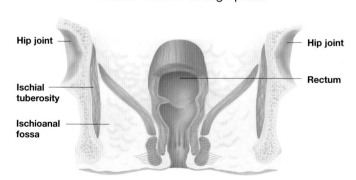

- Hip joint
- Hip joint
- Ischial tuberosity
- Rectum
- Ischioanal fossa

The ischioanal, or ischiorectal, fossae are spaces formed between the outside of the pelvic diaphragm and the skin around the anus.

The ischioanal fossae are filled with fat. This fat is divided into sections and supported by bands of connective tissue. The fat in the ischioanal fossae acts as a

The ischioanal fossae are wedge-shaped, being narrowest at the top and widest at the bottom. The fossae are filled with sections of fat.

soft packing material which accommodates changes in the size and position of the anus during a bowel movement.

INFECTION
Ischioanal fossae can become infected (ischioanal/ischiorectal abscess). Any area of the body with a poor blood supply is susceptible to infection and this is certainly the case with the fat within the ischioanal fossae. Infection may spread to the other side and infected areas may need to be surgically drained.

Muscles of the gluteal region

The gluteus maximus is the largest and heaviest of all the gluteal muscles and is situated in the buttock region. This strong, thick muscle plays an important part in enabling humans to stand.

The gluteal, or buttock, region lies behind the pelvis. The shape is formed by a number of large muscles which help to stabilize and move the hip joint. A layer of fat covers these muscles.

GLUTEUS MAXIMUS
This is one of the largest muscles in the body. It covers the other gluteal muscles with the exception of about one-third of the smaller gluteus medius. The gluteus maximus arises from the ilium (a part of the bony pelvis), the back of the sacrum and the coccyx. Its fibres run down and outwards at a 45 degree angle towards the femur. Most of the fibres then insert into a band (the iliotibial tract).

ACTIONS
The main function of the gluteus maximus is to extend (straighten) the leg as in standing from a sitting position. When the leg is extended, as in standing, the gluteus maximus covers the bony ischial tuberosity. This bears the weight of the body when sitting. However, we never sit on the gluteus maximus muscle itself as it moves up and away from the ischial tuberosity when the leg is flexed (bent forward).

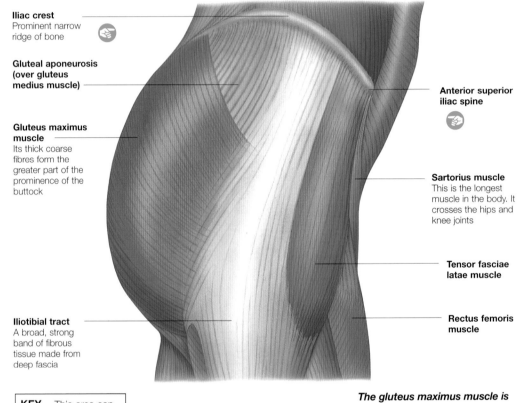

Iliac crest
Prominent narrow ridge of bone

Gluteal aponeurosis (over gluteus medius muscle)

Gluteus maximus muscle
Its thick coarse fibres form the greater part of the prominence of the buttock

Iliotibial tract
A broad, strong band of fibrous tissue made from deep fascia

Anterior superior iliac spine

Sartorius muscle
This is the longest muscle in the body. It crosses the hips and knee joints

Tensor fasciae latae muscle

Rectus femoris muscle

KEY This area can easily be felt under the skin

The gluteus maximus muscle is not very active during normal walking but comes into play during forceful actions such as running or walking upstairs.

Surface anatomy of the gluteal region

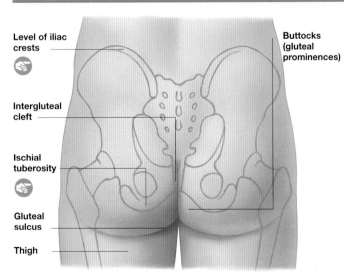

Level of iliac crests

Intergluteal cleft

Ischial tuberosity

Gluteal sulcus

Thigh

Buttocks (gluteal prominences)

The gluteal region overlies the back of the bony pelvis between the level of the iliac crests and the lower edge of the gluteus maximus muscle. Its shape is formed mainly by the mass of that muscle and by some fat.

FEATURES
There are a number of obvious features of the gluteal region:
■ The intergluteal or natal cleft which separates the buttocks
■ The gluteal fold is formed by the lower edge of the gluteus

Muscle and fat form the familiar shape of the gluteal region. The gluteal sulcus is the dividing line between the lower end of the buttock and the top of the thigh.

maximus muscle, which is usually covered by a layer of fat
■ The gluteal sulcus is the crease which lies beneath the gluteal fold. It is the line that separates the buttock and thigh.

BONY LANDMARKS
Except in very overweight people, some of the underlying bony protuberances of the gluteal region can be felt through the skin:
■ The iliac crests are usually palpable along their lengths
■ The ischial tuberosity can be felt in the lower part of the buttock, covered by the gluteus maximus when standing
■ The tip of the coccyx can be felt in the upper natal cleft.

Deeper muscles of the gluteal region

The muscles that lie deep to the gluteus maximus region play an important part in walking. They keep the pelvis level as each foot is lifted off the ground.

Beneath the gluteus maximus lie a number of other muscles which act to stabilise the hip joint and move the lower limb.

GLUTEUS MEDIUS AND MINIMUS

The gluteus medius and gluteus minimus muscles lie deep to the gluteus maximus. They are both fan-shaped muscles with fibres that run in the same direction.

Gluteus medius lies directly beneath gluteus maximus, with only about one-third of it not covered by this larger muscle. Its fibres originate from the external surface of the ilium (part of the pelvis) and insert into the greater trochanter, a protuberance of the femur.

Gluteus minimus lies directly beneath gluteus medius and is of a similar fan-like shape. Its fibres also originate from the ilium and insert into the greater trochanter.

ESSENTIAL ROLE

The gluteus medius and gluteus minimus together have an essential role in the action of walking. These muscles act to hold the pelvis level when one foot is lifted from the ground, rather than letting it sag to that side. This allows the non-weight-bearing foot to clear the ground before being swung further forward.

Quadratus femoris
This rectangular-shaped muscle extends laterally from the pelvis

Ischial tuberosity
The ischial tuberosities are the strongest parts of the hip bones

Gluteus maximus

Greater trochanter

Gluteus medius muscle
This is a thick muscle largely covered by the gluteus maximus

Gluteus minimus muscle
The smallest and deepest of the gluteal muscles

Piriformis muscle

Superior gemellus muscle

Obturator internus

Inferior gemellus muscle

KEY This area can easily be felt under the skin

A number of other muscles lie within this region, acting mainly to help certain movements of the lower limb at the hip. These include:
- Piriformis – this muscle, which is named for its pear shape, lies below gluteus minimus. It acts to rotate the thigh laterally, a movement which results in the foot turning outwards
- Obturator internus, superior and inferior gemelli – these three muscles together form a composite three-headed muscle which lies below the piriformis muscle. These muscles rotate the thigh laterally and stabilize the hip joint

The deep muscles of this group rotate the thigh laterally and stabilize the hip joint. The main muscles are the gluteus medius and gluteus minimus.

- Quadratus femoris – this short, thick muscle rotates the thigh laterally and helps to stabilize the hip joint.

Bursae of the gluteal region

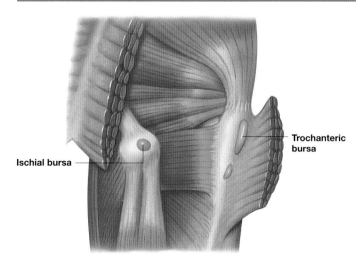

Ischial bursa

Trochanteric bursa

A bursa is a small fluid-filled sac rather like an underfilled water bottle. In many places in the body a bursa will be found where two structures, usually bone and tendon, regularly move against each other.

PROTECTION

The bursa lies between these structures, protecting them from wear and tear.

There are three main groups of bursae in the gluteal region:

The gluteal region contains three main groups of bursae. The bursae help to ease the movement of the bones and tendons upon each other.

- The trochanteric bursae
These large bursae lie between the thick, upper fibres of the gluteus maximus muscle and the greater trochanter of the upper femur (thigh bone)

- The ischial bursa
This bursa, if present, lies between the lower fibres of the gluteus maximus muscle and the ischial tuberosity, the part of the pelvis which bears our weight during sitting

- The gluteofemoral bursa
This bursa lies on the outer side of the leg, between the gluteus maximus and vastus lateralis muscles.

345

Hip joint

The hip joint is the strong ball-and-socket joint that connects the lower limb to the pelvis. Of all the body's joints, the hip is second only to the shoulder in the variety of movements it allows.

In the hip joint, the head of the femur (thigh bone) is the 'ball' that fits tightly into the 'socket' formed by the cup-like acetabulum of the hip bone of the pelvis.

The articular surfaces – the parts of the bone which come into contact with each other – are covered by a protective layer of hyaline cartilage, which is very smooth and slippery. The hip joint is a synovial joint, which means that movement is further lubricated by a thin layer of synovial fluid, which lies between these articular surfaces within the synovial cavity. The fluid is secreted by the synovial membrane.

ACETABULAR LABRUM

The depth of the socket formed by the acetabulum is increased by the presence of the acetabular labrum. This structure brings greater stability to the joint, allowing the almost spherical femoral head to rest deep within the joint.

The cartilage-covered articular surface of the acetabulum is not a continuous cup, or even a ring, but is horseshoe-shaped. There is a gap, the acetabular notch, at the lowest point, which is bridged by the complete ring of the acetabular labrum. The open centre of the 'horseshoe' is filled with a cushioning pad of fat.

The hip joint is the ball-and-socket joint between the head of the femur and the hip bone. The joint is capable of a wide range of movement.

Cross-section of the right hip joint

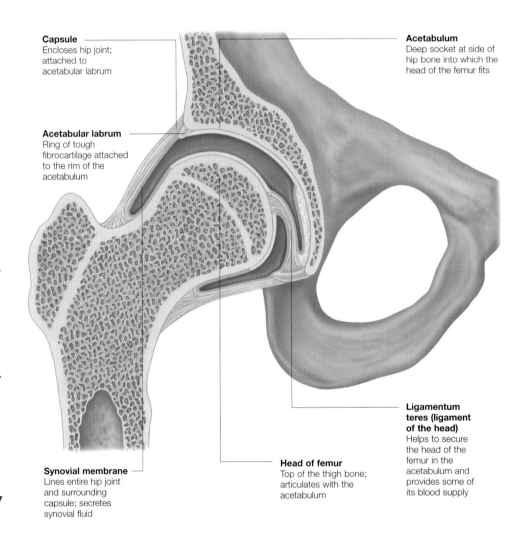

Capsule
Encloses hip joint; attached to acetabular labrum

Acetabular labrum
Ring of tough fibrocartilage attached to the rim of the acetabulum

Acetabulum
Deep socket at side of hip bone into which the head of the femur fits

Ligamentum teres (ligament of the head)
Helps to secure the head of the femur in the acetabulum and provides some of its blood supply

Synovial membrane
Lines entire hip joint and surrounding capsule; secretes synovial fluid

Head of femur
Top of the thigh bone; articulates with the acetabulum

Blood supply of the hip joint

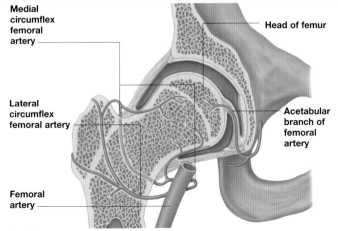

Medial circumflex femoral artery

Head of femur

Lateral circumflex femoral artery

Acetabular branch of femoral artery

Femoral artery

The hip joint receives its blood supply from two main sources:
■ The medial and lateral circumflex femoral arteries, – approach the femoral head from the femoral neck
■ The artery to the head of the femur – approaches the femoral head from the acetabulum, passing through the ligamentum teres (ligament of the head).

Blood is supplied to the hip joint from two main sources. These are the medial/lateral circumflex femoral arteries and the artery to the femoral head.

AVASCULAR NECROSIS
The origin of the blood supply of the hip is important clinically. In children especially a significant amount of the blood supplying the femoral head comes via the artery passing through the ligamentum teres.

If this artery is damaged then the femoral head may undergo avascular necrosis, (tissue death due to lack of an adequate blood supply). This damage may occur particularly in children aged three to nine years and may result in pain in the hip and knee.

Ligaments of the hip joint

The hip joint is enclosed and protected by a thick, fibrous capsule. The capsule is flexible enough to allow the joint a wide range of movements but is strengthened by a number of tough ligaments.

The ligaments of the hip joint are thickened parts of the joint capsule, which extends from the rim of the acetabulum down to the neck of the femur. These ligaments, which generally follow a spiral path from the hip bone to the femur, are named according to the parts of the bone to which they attach:
■ Iliofemoral ligament
■ Pubofemoral ligament
■ Ischiofemoral ligament.

MOVEMENT AND STABILITY
The ball-and-socket nature of the hip joint allows it great mobility, second only to the shoulder joint in its range of movement. Unlike the shoulder, however, it needs to be very stable as it is a major weight-bearing joint. It is capable of the following movements:

■ Flexion (bending forward, the knee coming up)
■ Extension (bending the leg back behind the body)
■ Abduction (moving the leg out to the side)
■ Adduction (bringing the leg back to the midline)
■ Rotation, which is greatest when the leg is flexed.

A fibrous capsule encloses the hip joint. This capsule is reinforced by a number of ligaments which spiral down from the hip bone to the femur.

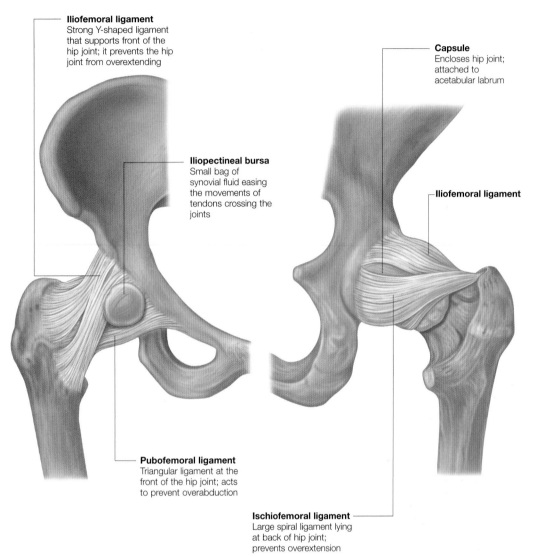

Anterior view of right hip

Iliofemoral ligament
Strong Y-shaped ligament that supports front of the hip joint; it prevents the hip joint from overextending

Iliopectineal bursa
Small bag of synovial fluid easing the movements of tendons crossing the joints

Pubofemoral ligament
Triangular ligament at the front of the hip joint; acts to prevent overabduction

Posterior view of right hip

Capsule
Encloses hip joint; attached to acetabular labrum

Iliofemoral ligament

Ischiofemoral ligament
Large spiral ligament lying at back of hip joint; prevents overextension

---

Artificial hip joints

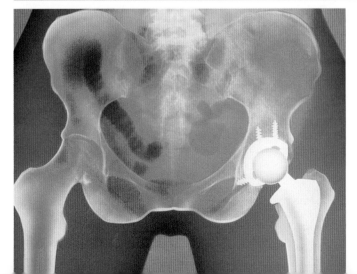

The hip was the first joint for which a successful prosthesis (artificial replacement) was developed, the first total hip replacement being carried out by Sir John Charnley in 1963.

Although the hip joint is generally very stable, it is susceptible to damage due to trauma, or to arthritis, which can be very disabling. Great improvements in mobility and a

This composite X-ray shows a left hip prosthesis. It consists of a plastic socket inserted into the pelvis, and a metal prosthesis cemented into the femur.

decrease in pain can be achieved by hip replacement surgery.

During surgery, the damaged femoral head and neck are replaced by a metal prosthesis, which is anchored into the femur with special bone cement. The acetabulum of the hip bone is replaced by a plastic socket which is cemented to the pelvis.

Currently, artificial hips have an expected lifespan of about 10 years and so are not really suitable for young, active people. It is hoped that, with research, the techniques and materials used will improve and more people will benefit from them.

Femur

The femur, or thigh bone, is the longest and heaviest bone in the body. Measuring approximately 45 cm in length in adult males, the femur makes up about one quarter of a person's total height.

The femur has a long, thick shaft with two expanded ends. The upper end articulates with the pelvis to form the hip joint while the lower end articulates with the tibia and patella to form the knee joint.

UPPER END

The femur upper end includes:
■ Head – this is the near-spherical projection which forms the 'ball' of the ball and socket hip joint
■ Neck – this is the narrowed area which connects the head to the body of the femur
■ Greater and lesser trochanters – projections of bone allowing the attachment of muscles.

SHAFT

The long central shaft of the femur is slightly bowed, being concave on its posterior surface. For much of its length the femur appears cylindrical, with a circular cross-section.

LOWER END

The lower end of the femur is made up of two enlarged bony processes, the medial and lateral femoral condyles. These carry the smooth, curved surfaces which articulate with the tibia and patella to form the knee joint. The shape of the femoral condyles is outlined when the leg is viewed with the knee bent.

The femur is the thigh bone, running from the hip joint to the knee. It is the longest bone in the body and is very strong.

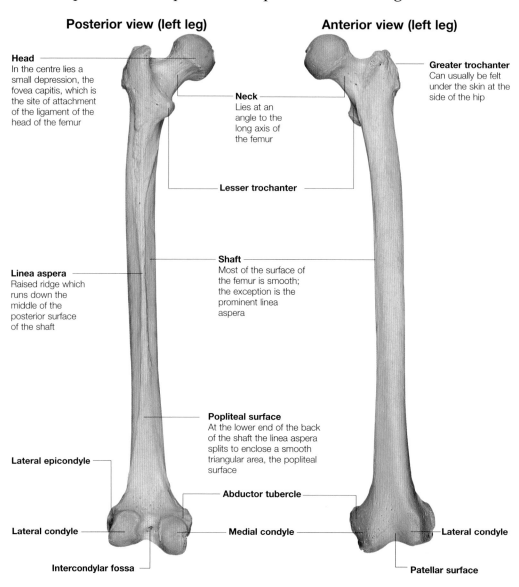

Posterior view (left leg)

Head
In the centre lies a small depression, the fovea capitis, which is the site of attachment of the ligament of the head of the femur

Neck
Lies at an angle to the long axis of the femur

Lesser trochanter

Linea aspera
Raised ridge which runs down the middle of the posterior surface of the shaft

Shaft
Most of the surface of the femur is smooth; the exception is the prominent linea aspera

Popliteal surface
At the lower end of the back of the shaft the linea aspera splits to enclose a smooth triangular area, the popliteal surface

Lateral epicondyle

Abductor tubercle

Lateral condyle

Medial condyle

Intercondylar fossa

Anterior view (left leg)

Greater trochanter
Can usually be felt under the skin at the side of the hip

Lateral condyle

Patellar surface

Internal structure of the femur

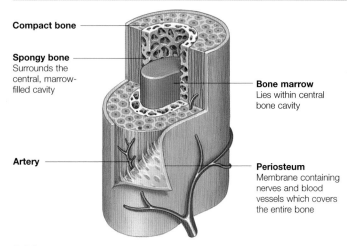

Compact bone

Spongy bone
Surrounds the central, marrow-filled cavity

Artery

Bone marrow
Lies within central bone cavity

Periosteum
Membrane containing nerves and blood vessels which covers the entire bone

The femur is one of the bones of the body classed as a long bone. Bones of this type have a relatively long shaft, or diaphysis, and two expanded ends, or epiphyses.

PERIOSTEUM

All the bone is covered with a protective membrane, the periosteum, nourished by tiny nutrient arteries in the bone.

The shaft of long bones is composed of layers of spongy and compact bone around a central cavity. The periosteum surrounds the outer layer.

DIAPHYSIS

The diaphysis of the femur is a tube composed of compact bone, which is strong and dense. This layer of compact bone encloses a core of yellow bone marrow which, in adults, is made up of fat cells.

EPIPHYSES

The expanded ends of the femur are made up of a surface layer of compact bone which surrounds a central area of spongy bone. This central area is much looser in structure and there is no marrow in the epiphyses.

Muscle attachments of the femur

The femur is a very strong bone which provides sites of attachment for many of the muscles of locomotion in the hip joint and legs.

MUSCLE ORIGINS
Some muscles, such as the powerful gluteus muscles, have their origins on the pelvic bones and so cross the hip joint to insert into the femur. When these muscles contract they cause the hip joint to move, allowing the leg to bend, straighten or move sideways.

Other muscles originate on the femur itself and pass down across the knee joint to insert on the tibia or the fibula, the two bones of the lower leg. These muscles allow the knee to bend or straighten.

Together these muscles bring about movements of the legs such as in climbing, or rising from a sitting position.

BONY PROCESSES
Where muscle is attached to bone it causes a projection, or bony process, to arise. If the muscle is powerful, or if a number of muscles attach at the same site, the bony process can be pronounced. This is the case in the femur. The surface of the bone at the site of muscle attachment can also become quite roughened, unlike the smooth bone surface in between.

The surface of the femur is roughened by projections to which muscles are attached. These muscles bring about movements in the legs and hips.

Anterior view (right leg)

Obturator internus and superior and inferior gemellus muscles

Piriformis muscle

Gluteus minimus muscle

Vastus lateralis muscle

Vastus medialis muscle

Iliopsoas muscle

Vastus intermedius muscle

Articularis genus muscle

Adductor magnus muscle

Posterior view (right leg)

Obturator externus muscle

Gluteus medius muscle

Quadratus femoris muscle

Iliopsoas muscle

Gluteus maximus muscle

Vastus lateralis muscle

Adductor magnus muscle

Adductor brevis muscle

Vastus intermedius muscle

Biceps femoris muscle

Adductor magnus muscle

Adductor longus muscle

Vastus lateralis muscle

Plantaris muscle

Gastrocnemius muscle (lateral head)

Gastrocnemius muscle (medial head)

Popliteus muscle

Pectineus muscle

Vastus medialis muscle

Adductor magnus muscle

Fractures of the femur

Fracture of femoral neck

Intertrochanteric fracture

Shaft fracture

Fracture of the neck of the femur, the narrowed segment which attaches the head to the body, is a fairly common event.

'BROKEN HIP'
Often referred to as a 'broken hip', a femoral neck fracture usually occurs in older people, especially over the age of 60 years and often as the result of a minor fall. It is more common in women due to the effects of osteoporosis, or thinning of the

Intertrochanteric and femoral neck fractures are common in elderly people who fall. However, a shaft fracture is usually caused by a violent direct injury.

bones, which can occur after the menopause.

INTERTROCHANTERIC FRACTURES
Intertrochanteric fractures, where the fracture line runs between the greater and lesser trochanters, are also fairly common in elderly women.

SHAFT FRACTURES
Fracture of the shaft of the femur is a much less common injury due to the great strength of the bone. Fractures here are usually the result of major trauma such as a road traffic accident, and may take many months to heal.

Tibia and fibula

The tibia and fibula together form the skeleton of the lower leg. The tibia is much larger and stronger than the fibula as it must bear the weight of the body.

Second only to the femur (thigh bone) in size, the tibia (shin bone) has the shape of a typical long bone, with an elongated shaft and two expanded ends. The tibia lies alongside the fibula, on the medial (inner) side, and articulates with the fibula at its upper and lower ends.

TIBIAL CONDYLES

The upper end of the tibia is expanded to form the medial and lateral tibial condyles, which articulate with the femoral condyles at the knee joint. The lower end of the tibia is less pronounced than the upper end. It articulates with both the talus (ankle bone) and the lower end of the fibula.

FIBULA

The fibula is a long, narrow bone which has none of the strength of the tibia. It lies next to the tibia, on its lateral (outer) side, and articulates with that bone. The fibula plays no part in the knee joint, but is an important support for the ankle.

The shaft of the fibula is narrow and bears the grooves and ridges associated with its major role as a site of attachment for leg muscles.

The tibia (shin bone) articulates with the thigh bone above, the ankle below and the fibula to the side. The thinner fibula helps to form the ankle joint.

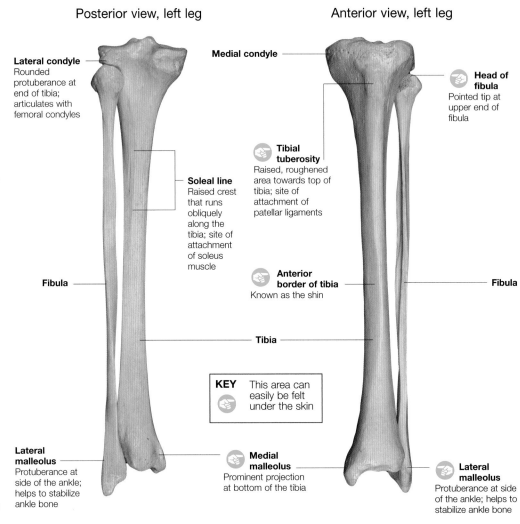

Posterior view, left leg

Lateral condyle
Rounded protuberance at end of tibia; articulates with femoral condyles

Soleal line
Raised crest that runs obliquely along the tibia; site of attachment of soleus muscle

Fibula

Lateral malleolus
Protuberance at side of the ankle; helps to stabilize ankle bone

Anterior view, left leg

Medial condyle

Head of fibula
Pointed tip at upper end of fibula

Tibial tuberosity
Raised, roughened area towards top of tibia; site of attachment of patellar ligaments

Anterior border of tibia
Known as the shin

Fibula

Tibia

Medial malleolus
Prominent projection at bottom of the tibia

Lateral malleolus
Protuberance at side of the ankle; helps to stabilize ankle bone

KEY This area can easily be felt under the skin

Cross-section of tibia and fibula

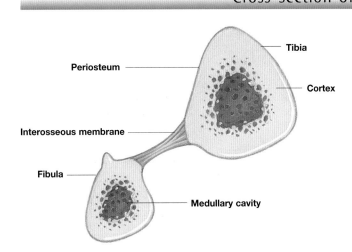

Periosteum

Interosseous membrane

Fibula

Tibia

Cortex

Medullary cavity

The shafts of the tibia and fibula are roughly triangular in cross section. The tibial shaft is much greater in diameter than that of the fibula as it is the main weight-bearing element of the lower leg. The fibula acts as a strut, increasing the stability of the lower leg under load.

LONG BONES

The tibia and fibula have a typical long bone structure, with a thick, tubular outer cortex surrounding a spongy medullary

In cross-section, the tibial and fibular shafts are triangular in shape. The two bones are anchored together by the interosseous membrane.

cavity. Their hollow structure provides maximal mechanical strength with minimal support material, namely dense, cortical bone.

The shape of the bones is genetically determined, but is modified by the pull of developing muscles during childhood and into adulthood. In this way the bony ridges, such as the soleal line and tuberosities, form.

The tibia and fibula are enveloped in the periosteum (a tough connective tissue layer). The periosteum from the lateral border of the tibia and the medial border of the fibula blends into the interosseous membrane.

Ligaments of the tibia and fibula

The ligaments which surround the tibia and fibula bind the two bones to each other and to the other leg bones with which they articulate.

Ligaments are the strong fibrous bands which bind bones together. There are a number of ligaments which surround the tibia and the fibula; they bind the two bones to each other and to other bones of the leg.

PROXIMAL (UPPER) END

Just under the knee is the upper joint between the head of the fibula and the underside of the lateral tibial condyle. The joint is surrounded and protected by a fibrous joint capsule, which is strengthened by the anterior and posterior tibiofibular ligaments.

The anterior ligament of the head of the fibula runs from the front of the fibular head across to the front of the lateral tibial condyle. The posterior ligament of the head of the fibula runs in a similar fashion behind the fibular head.

Other ligaments bind the bones of the lower leg to the femur. The strongest of these are the medial and lateral collateral ligaments of the knee joint, which run vertically down from the femur to the corresponding bone (tibia or fibula) beneath.

DISTAL (LOWER) END

The joint between the lower ends of the tibia and fibula allows no movement of one bone upon the other. Rather, the fibula is bound tightly to the tibia by fibrous ligaments in order to maintain the stability of

the ankle joint. The main ligaments concerned are the anterior and posterior inferior (lower) tibiofibular ligaments. Other ligaments around the ankle bind the tibia and fibula to the bones of the foot.

Anterior view of left leg with ligament attachments

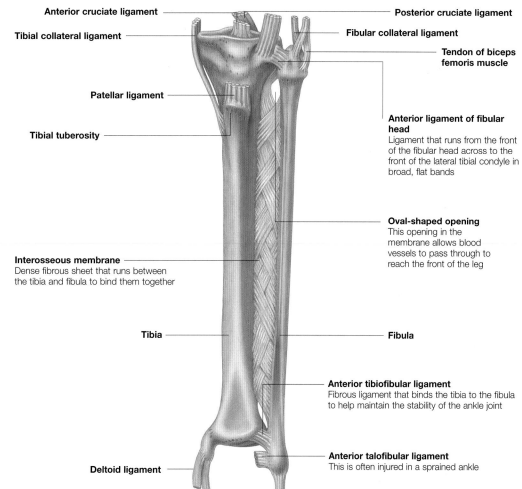

- Anterior cruciate ligament
- Tibial collateral ligament
- Patellar ligament
- Tibial tuberosity
- Posterior cruciate ligament
- Fibular collateral ligament
- Tendon of biceps femoris muscle
- **Anterior ligament of fibular head**
  Ligament that runs from the front of the fibular head across to the front of the lateral tibial condyle in broad, flat bands
- **Oval-shaped opening**
  This opening in the membrane allows blood vessels to pass through to reach the front of the leg
- **Interosseous membrane**
  Dense fibrous sheet that runs between the tibia and fibula to bind them together
- Tibia
- Fibula
- **Anterior tibiofibular ligament**
  Fibrous ligament that binds the tibia to the fibula to help maintain the stability of the ankle joint
- **Anterior talofibular ligament**
  This is often injured in a sprained ankle
- Deltoid ligament

INTEROSSEOUS MEMBRANE

The fibres of the dense interosseous membrane run obliquely from the sharp interosseous border of the tibia across to the front of the fibula, binding the two bones together.

The tibia and fibula of the lower leg are surrounded by a number of ligaments. Ligaments are tough fibrous bands of connective tissue that bind bones together where they articulate at a joint.

Fractures of the tibia and fibula

The tibia lies close to the skin surface throughout its length and is therefore the bone that is most susceptible to compound fractures. These are fractures in which the skin is torn and blood vessels are damaged as well as bone being broken.

SPORTS INJURIES

The weakest part of the tibia is at the junction of its middle and inferior thirds. Fractures may occur in this area during contact

The tibia and fibula can be fractured in a number of different places. Fractures usually occur as a result of sports or road traffic injuries.

sports or skiing, or as a result of road accidents.

The cortex (outer surface) of the tibia may also be the site of stress (or march) fractures which can occur after long walks, especially if a person is unfit.

FIBULAR FRACTURES

Fractures of the fibula may be associated with tibial fractures if the trauma is severe.

The most usual site for a fracture of the fibula is about 2 to 6 cm above the lateral malleolus at the lower end. Fractures of the fibula are frequently associated with fracture-dislocations of the ankle joint.

Knee joint and patella

The knee is the joint between the end of the thigh bone and the top of the tibia. In front of the knee is the patella (kneecap), the convex surface of which can readily be felt under the skin.

The knee is the joint between the lower end of the femur (thigh bone) and the upper end of the tibia (the largest bone of the lower leg). The fibula (the smaller of the two lower leg bones) plays no part in the joint.

STRUCTURE

The knee is a synovial joint – one in which movement is lubricated by synovial fluid which is secreted by a membrane lining the joint cavity.

Although the knee tends to be thought of as a single joint it is, in fact, the most complex joint in the body, being made up of three joints which share a common joint cavity. These three joints are:

■ The joint between the patella (kneecap) and the lower end of the femur. Classified as a plane joint, this allows one bone to slide upon the other

■ A joint on either side between the femoral condyles (the large bulbous ends of the femur) and the corresponding part of the upper tibia. These are said to be hinge joints as the movement they allow is akin to the movement of a door on its hinges.

STABILITY OF THE KNEE

Considering that there is not a very good 'fit' between the femoral condyles and the upper end of the tibia, the knee is actually a reasonably stable joint. It relies heavily on the surrounding muscles and ligaments for its stability.

Sagittal section of knee

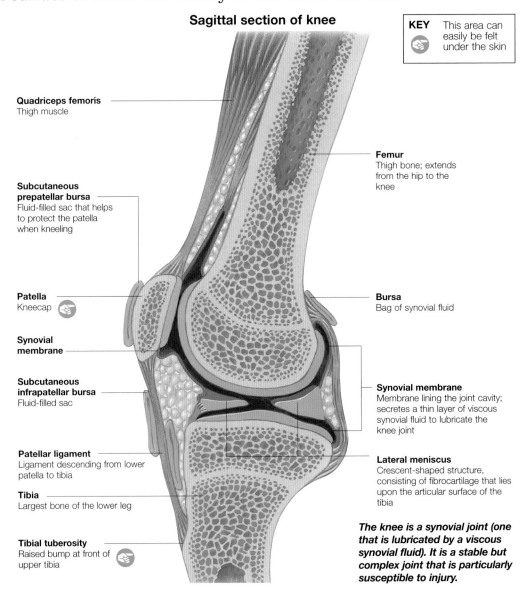

KEY This area can easily be felt under the skin

Quadriceps femoris
Thigh muscle

Subcutaneous prepatellar bursa
Fluid-filled sac that helps to protect the patella when kneeling

Patella
Kneecap

Synovial membrane

Subcutaneous infrapatellar bursa
Fluid-filled sac

Patellar ligament
Ligament descending from lower patella to tibia

Tibia
Largest bone of the lower leg

Tibial tuberosity
Raised bump at front of upper tibia

Femur
Thigh bone; extends from the hip to the knee

Bursa
Bag of synovial fluid

Synovial membrane
Membrane lining the joint cavity; secretes a thin layer of viscous synovial fluid to lubricate the knee joint

Lateral meniscus
Crescent-shaped structure, consisting of fibrocartilage that lies upon the articular surface of the tibia

The knee is a synovial joint (one that is lubricated by a viscous synovial fluid). It is a stable but complex joint that is particularly susceptible to injury.

Surface anatomy of the knee

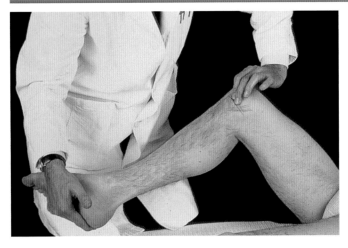

Many of the structures of the knee can be discerned by gently feeling the overlying skin, especially when the knee is flexed (bent).

PATELLAR LIGAMENT

The outline of the patella can be seen and its surface readily felt. Underneath the patella is the patellar ligament, the strong fibrous band which descends

A physical examination of the knee can reveal many of its components. The bony areas are particularly easy to feel as they lie just beneath the skin.

from the patella to the front of the tibia.

On either side of the patella, and just behind it, are the medial and lateral femoral condyles (the rounded ends of the femur). Below the patella the tibial tuberosity (a raised area at the front of the upper tibia) can easily be felt.

ARTERY

At the back of the knee lies a depression known as the popliteal fossa. Gentle pressure over this area, when the knee is flexed, reveals the pulsations of the large popliteal artery.

Inside the knee – the menisci

The menisci are crescent-shaped plates of tough fibrocartilage lying on the articular surface of the tibia. They act as 'shock absorbers' within the knee and prevent sideways movement of the femur.

Looking down on the upper surface of the tibia within the opened knee, the two c-shaped menisci can clearly be seen.

Named after the Greek word for 'crescent', the menisci are plates of tough fibrocartilage which lie upon the articular surface of the tibia, deepening the depression into which the femoral condyles fit.

SHOCK ABSORBERS

The menisci also have the function of acting as 'shock absorbers' within the knee and help to prevent the side-to-side rocking of the joint.

STRUCTURE OF THE MENISCI

The two menisci are wedge-shaped in cross-section, their external margins being widest. Centrally, they taper to a thin, unattached edge. Anteriorly, the two menisci are attached to each other by the transverse ligament of the knee, while the outer edges of the menisci are firmly attached to the joint capsule.

ATTACHMENT

Attachment of the medial meniscus to the tibial collateral ligament is of great clinical significance as the meniscus can be damaged when this ligament is injured during contact sports.

Superior view of knee (tibial plateau)

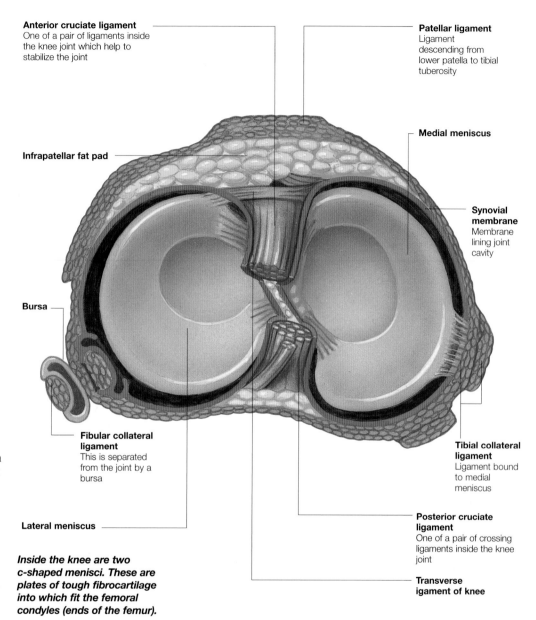

Anterior cruciate ligament
One of a pair of ligaments inside the knee joint which help to stabilize the joint

Infrapatellar fat pad

Bursa

Fibular collateral ligament
This is separated from the joint by a bursa

Lateral meniscus

Patellar ligament
Ligament descending from lower patella to tibial tuberosity

Medial meniscus

Synovial membrane
Membrane lining joint cavity

Tibial collateral ligament
Ligament bound to medial meniscus

Posterior cruciate ligament
One of a pair of crossing ligaments inside the knee joint

Transverse igament of knee

Inside the knee are two c-shaped menisci. These are plates of tough fibrocartilage into which fit the femoral condyles (ends of the femur).

The patella

Anterior view

Posterior view

Lying within the tendon of the powerful quadriceps femoris thigh muscle, the patella is the largest sesamoid bone in the body. A sesamoid bone is one which develops within the tendon of a muscle to protect that tendon from wear and tear where it passes over the end of a long bone.

STRUCTURE

The patella is flattened with a convex surface at the front

The patella (kneecap) is situated over the front of the knee. It has a flattened shape, and its convex outer surface can easily be felt beneath the skin.

which can readily be felt under the skin. Between the patella and the skin lies a bursa (fluid-filled sac), which helps to reduce friction and protect the bone when kneeling.

CARTILAGE

The posterior, or back, surface of the patella is covered with smooth cartilage and articulates with the lower end of the femur in a synovial joint.

Fibres from the strong quadriceps thigh muscle insert into the upper border of the patella, while the patellar ligament passes from the lower border of the patella down to the tibial tuberosity.

Ligaments and bursae of the knee

The knee joint is only partially enclosed in a capsule and relies on ligaments for its stability. Bursae are situated around the knee and allow smooth movement to take place.

Unlike the bones of the hip joint, the bones of the knee do not fit together in a particularly stable fashion. For this reason the stability of the knee joint depends to a great extent upon the ligaments and muscles that surround it.

The joint cavity of the knee is enclosed within a fibrous capsule. The ligaments that support the knee can be divided into two groups, depending on their relationship to this capsule.

EXTRACAPSULAR LIGAMENTS

The extracapsular ligaments lie outside the capsule and act to prevent the lower leg bending forward at the knee, or hyperextending. They include:

■ Quadriceps tendon – extends from tendon of the quadriceps femoris muscle. This supports the front of the knee (not shown)
■ Fibular (or lateral) collateral ligament – a strong cord which binds the lower end of the outer femur to the head of the fibula
■ Tibial (or medial) collateral ligament – a strong flat band, which runs from the lower end of the inner femur down to the tibia. It is weaker than the fibular collateral ligament and is more easily damaged
■ Oblique popliteal ligament – this strengthens the capsule at the back (not shown)
■ Arcuate popliteal ligament – also adds strength to the back of the knee (not shown).

Anterior view of flexed left knee

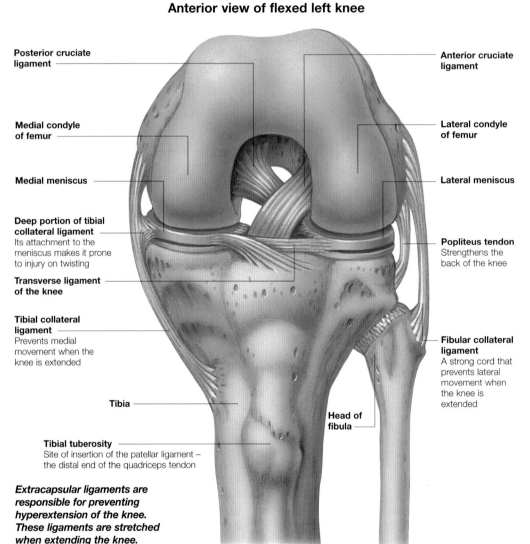

Posterior cruciate ligament

Medial condyle of femur

Medial meniscus

Deep portion of tibial collateral ligament
Its attachment to the meniscus makes it prone to injury on twisting

Transverse ligament of the knee

Tibial collateral ligament
Prevents medial movement when the knee is extended

Tibia

Tibial tuberosity
Site of insertion of the patellar ligament – the distal end of the quadriceps tendon

Anterior cruciate ligament

Lateral condyle of femur

Lateral meniscus

Popliteus tendon
Strengthens the back of the knee

Fibular collateral ligament
A strong cord that prevents lateral movement when the knee is extended

Head of fibula

Extracapsular ligaments are responsible for preventing hyperextension of the knee. These ligaments are stretched when extending the knee.

Intracapsular ligaments

Lateral view of intracapsular ligaments

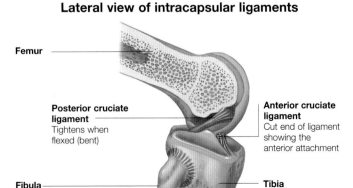

Femur

Posterior cruciate ligament
Tightens when flexed (bent)

Anterior cruciate ligament
Cut end of ligament showing the anterior attachment

Fibula

Tibia

Intracapsular ligaments connect the tibia to the femur within the centre of the knee joint and prevent forward and backward displacement of the knee.

CRUCIATES

The two main intracapsular ligaments are known as the

The intracapsular, or cruciate, ligaments form a cross. They act to prevent anterior–posterior displacement and to stabilize articulating bones.

cruciate ligaments, as they form the shape of a cross.

■ The anterior cruciate ligament – this is the weaker of the two cruciates, and is slack when the knee is flexed, taut when the knee is extended (straightened)
■ The posterior cruciate ligament – this ligament tightens during flexion (bending) and is very important for the stability of the knee when bearing weight in a flexed position (for example, when walking downhill).

Bursae of the knee

The bursae of the knee are small sacs filled with synovial fluid. They act to protect the structures inside the knee, reducing friction, as they slide over each other when the joint is moving.

Bursae are small fluid-filled sacs found between two structures, usually bone and tendon, that regularly move against each other. The bursae protect the structures from wear and tear.

There are a number of bursae around the knee that protect the tendons during movement or allow easy movement of the skin across the patella.

SUPRAPATELLAR BURSA
Some of the bursae around the knee joint are continuous with the joint cavity, the fluid-filled space between the articular surfaces. The suprapatellar bursa lies above the joint cavity between the lower end of the femur and the powerful quadriceps femoris muscle.

PREPATELLAR AND INFRAPATELLAR BURSAE
These bursae surround the patella and the patellar ligament. The prepatellar bursa allows the skin to move freely over the patella during movement. The superficial and deep infrapatellar bursae lie around the lower end of the patellar ligament where it attaches to the tibial tuberosity.

There are about a dozen bursae located around the knee joint. They allow the structures of the knee to move freely over one another, reducing friction.

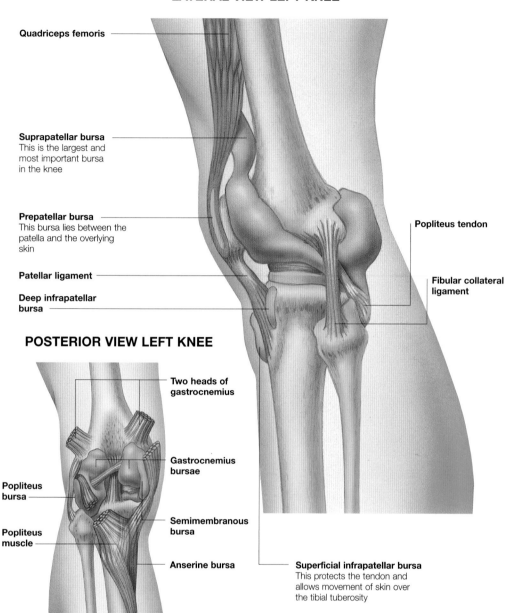

LATERAL VIEW LEFT KNEE

Quadriceps femoris

Suprapatellar bursa
This is the largest and most important bursa in the knee

Prepatellar bursa
This bursa lies between the patella and the overlying skin

Patellar ligament

Deep infrapatellar bursa

Popliteus tendon

Fibular collateral ligament

POSTERIOR VIEW LEFT KNEE

Two heads of gastrocnemius

Gastrocnemius bursae

Popliteus bursa

Semimembranous bursa

Popliteus muscle

Anserine bursa

Superficial infrapatellar bursa
This protects the tendon and allows movement of skin over the tibial tuberosity

Investigating the knee joint

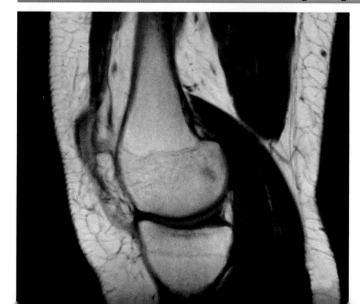

The knee joint is very susceptible to damage from trauma or osteoarthritis. To determine the extent of such damage involves clinical examination and often some further investigation.

X-RAYS
X-rays of the knee in different positions, perhaps with the injection of a dye into the joint space (arthrography), may be useful to show up abnormalities of the bones and menisci.

MRI is an effective technique for visualizing the complex anatomy of joints. In this scan, the bones of the knee joint and the tissues around them are clearly visible.

MRI INVESTIGATIONS
Magnetic resonance imaging (MRI) has greatly helped in the investigation of knee disorders. This type of non-invasive scanning can show up problems within the soft tissues around the knee as well as within the bone and so has largely taken over from arthrography.

ARTHROSCOPY
Another way of looking inside the knee is to use a tiny camera in an endoscope (arthroscopy). During this procedure, which is carried out under a general anaesthetic, the surgeon can often remove damaged tissue, thus avoiding further surgery.

Muscles of the thigh

The thigh is composed mainly of groups of large muscles which act to move the hip and the knee joint. Muscles that effect the movements of the thigh are among the strongest in the body.

The muscles of the thigh are divided into three basic groups; the anterior muscles lie in front of the femur, the posterior muscles lie behind, and the medial muscles (adductors) run between the inner femur (thigh) and the pelvis.

ANTERIOR MUSCLES

The muscles of the anterior compartment of the thigh flex or bend the hip, and extend or straighten the knee. These are the actions associated with lifting the leg up and bringing it forward during walking.

The muscles of this group include:

■ **Iliopsoas.** This large muscle arises partly from the inside of the pelvis and partly from the lumbar (lower) vertebrae. Its fibres insert into the projection of the upper femur known as the lesser trochanter. The iliopsoas is the most powerful of the muscles that flex the thigh, bringing the knee up and forwards

■ **Tensor fasciae latae.** This muscle inserts into the strong band of connective tissue that runs down the outside of the leg to the tibia below the knee

■ **Sartorius.** The longest muscle in the body, the sartorius runs as a flat strap across the thigh from the anterior superior iliac spine of the pelvis. It crosses both the hip and the knee joint before it inserts into the inner side of the top of the tibia.

■ **Quadriceps femoris.** A large four-headed muscle.

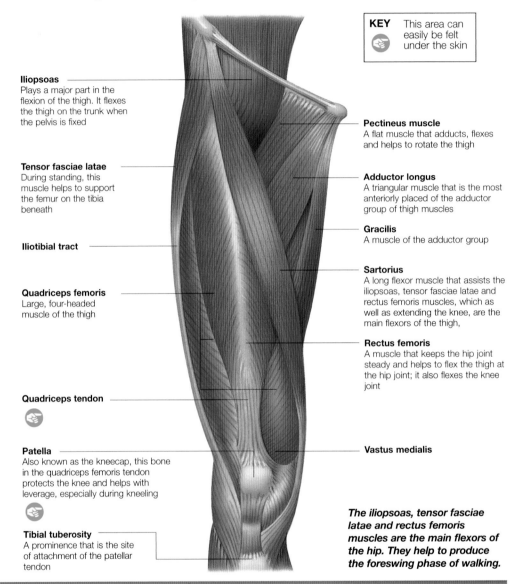

KEY This area can easily be felt under the skin

Iliopsoas
Plays a major part in the flexion of the thigh. It flexes the thigh on the trunk when the pelvis is fixed

Tensor fasciae latae
During standing, this muscle helps to support the femur on the tibia beneath

Iliotibial tract

Quadriceps femoris
Large, four-headed muscle of the thigh

Quadriceps tendon

Patella
Also known as the kneecap, this bone in the quadriceps femoris tendon protects the knee and helps with leverage, especially during kneeling

Tibial tuberosity
A prominence that is the site of attachment of the patellar tendon

Pectineus muscle
A flat muscle that adducts, flexes and helps to rotate the thigh

Adductor longus
A triangular muscle that is the most anteriorly placed of the adductor group of thigh muscles

Gracilis
A muscle of the adductor group

Sartorius
A long flexor muscle that assists the iliopsoas, tensor fasciae latae and rectus femoris muscles, which as well as extending the knee, are the main flexors of the thigh,

Rectus femoris
A muscle that keeps the hip joint steady and helps to flex the thigh at the hip joint; it also flexes the knee joint

Vastus medialis

The iliopsoas, tensor fasciae latae and rectus femoris muscles are the main flexors of the hip. They help to produce the foreswing phase of walking.

Quadriceps femoris

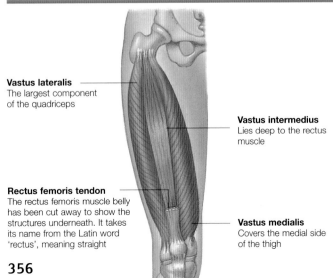

Vastus lateralis
The largest component of the quadriceps

Vastus intermedius
Lies deep to the rectus muscle

Rectus femoris tendon
The rectus femoris muscle belly has been cut away to show the structures underneath. It takes its name from the Latin word 'rectus', meaning straight

Vastus medialis
Covers the medial side of the thigh

This large, four-headed muscle makes up the bulk of the thigh and is one of the most powerful muscles in the body. It consists of four major parts whose tendons combine to form the strong quadriceps tendon. This inserts into the top of the patella and then continues down, as the patella tendon, to the front of the top of the tibia. The quadriceps femoris acts to straighten the knee.

The quadriceps femoris is an extensor that is used in running, jumping and climbing. It helps to straighten the knee when standing from a sitting position.

The four parts of the quadriceps femoris are:
■ **Rectus femoris** – a straight muscle, overlying the other parts. It helps to flex the hip joint and straighten the knee
■ **Vastus lateralis** – the largest part of the quadriceps muscle
■ **Vastus medialis** – lies on the inner side of the thigh
■ **Vastus intermedius** – lies centrally, underneath the rectus femoris muscle.

A few small slips of the vastus intermedius pass down to the joint capsule of the knee. They ensure that folds of the capsule do not become trapped when the knee is straightened.

Posterior thigh muscles

The three large muscles of the posterior thigh are commonly known as the hamstrings. These three muscles are the biceps femoris, semitendinosus and semimembranosus.

The hamstring muscles can both extend the hip and flex the knee. However, they cannot do both fully at the same time.

BICEPS FEMORIS

The biceps femoris has two heads. The long head arises from the ischial tuberosity of the pelvis and the short head arises from the back of the femur. The rounded tendon of the biceps femoris can easily be felt and seen behind the outer side of the knee, especially if the knee is flexed against resistance.

SEMITENDINOSUS

Like the biceps femoris muscle, the semitendinosus arises from the ischial tuberosity of the pelvis. It is named for its long tendon, which begins about two-thirds of the way down its course. This tendon attaches to the inner side of the upper tibia.

SEMIMEMBRANOSUS

This muscle arises from a flattened, membranous attachment to the ischial tuberosity of the pelvis. The muscle runs down the back of the thigh, deep to the semitendinosus. It inserts into the inner side of the upper tibia.

KEY This area can easily be felt under the skin

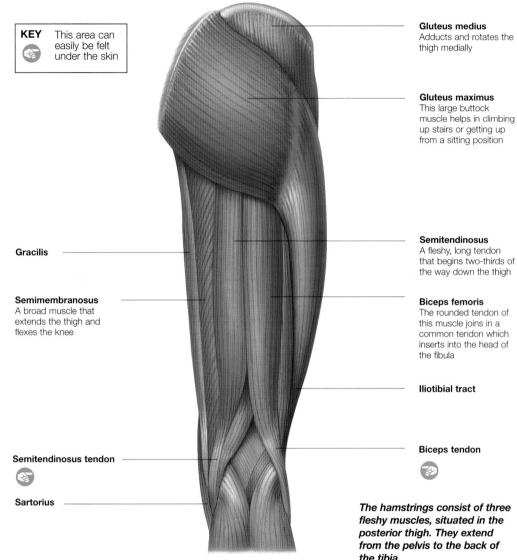

Gluteus medius
Adducts and rotates the thigh medially

Gluteus maximus
This large buttock muscle helps in climbing up stairs or getting up from a sitting position

Semitendinosus
A fleshy, long tendon that begins two-thirds of the way down the thigh

Biceps femoris
The rounded tendon of this muscle joins in a common tendon which inserts into the head of the fibula

Iliotibial tract

Biceps tendon

Gracilis

Semimembranosus
A broad muscle that extends the thigh and flexes the knee

Semitendinosus tendon

Sartorius

The hamstrings consist of three fleshy muscles, situated in the posterior thigh. They extend from the pelvis to the back of the tibia.

The adductors

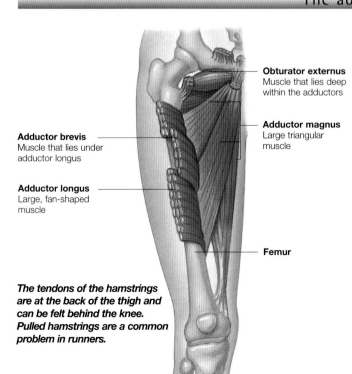

Adductor brevis
Muscle that lies under adductor longus

Adductor longus
Large, fan-shaped muscle

Obturator externus
Muscle that lies deep within the adductors

Adductor magnus
Large triangular muscle

Femur

The tendons of the hamstrings are at the back of the thigh and can be felt behind the knee. Pulled hamstrings are a common problem in runners.

The muscles of the inner thigh are known as adductors because they allow adduction of the thigh, which means moving the lower limb in towards the mid-line as when gripping the sides of a horse when riding. These muscles arise from the lower part of the pelvis and insert into the femur at various levels. The muscles of this group include:
■ **Adductor longus** – a large, fan-shaped muscle which lies in front of the other adductors and has a palpable tendon in the groin
■ **Adductor brevis** – a shorter muscle which lies under the adductor longus
■ **Adductor magnus** – a large triangular muscle which fulfils the function of both an adductor and a hamstring muscle
■ **Gracilis** – a strap-like muscle that runs vertically down the inner thigh (not shown)

■ **Obturator externus** – a small muscle which lies deeply within this group of adductors.

It is this adductor group of muscles which are active when gripping the horse during horse-riding and may become strained during sporting activities, leading to a groin injury.

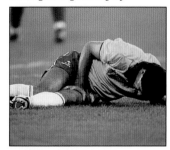

Footballers are at risk of groin injuries (pulled adductor muscles). This is due to the kicking action that moves the leg across the midline.

Muscles of the lower leg

There are three groups of muscles in the lower leg.
Depending where they lie, they support and flex the ankle and foot,
extend the toes and assist in lifting the body weight at the heel.

The muscles of the lower leg can be divided into three groups: the anterior group which lie in front of the tibia, the lateral group which lie on the outer side of the lower leg and the posterior group.

ANTERIOR MUSCLES
The anterior muscles of the lower leg include:
■ **Tibialis anterior** – this muscle can be felt under the skin alongside the edge of the tibia
■ **Extensor digitorum longus** – this muscle lies under the tibialis anterior and attaches to the outer four toes
■ **Peroneus (fibularis) tertius** – this muscle is not always present but, when it is, it may join the extensor digitorum longus muscle. It inserts into the fifth metatarsal bone near the little toe
■ **Extensor hallucis longus** – this thin muscle runs down to insert into the end of the hallux (big toe).

ACTION OF THE ANTERIOR MUSCLES
These muscles all have a similar action in that they are dorsiflexors of the foot. This means that when they contract they bend the ankle, bringing the toes up and the heel down.

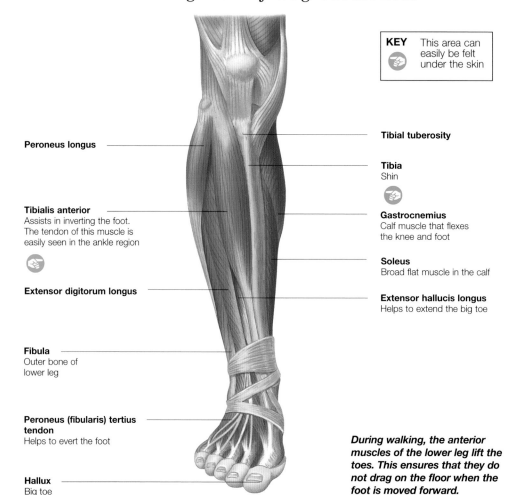

KEY This area can easily be felt under the skin

Peroneus longus

Tibialis anterior
Assists in inverting the foot. The tendon of this muscle is easily seen in the ankle region

Extensor digitorum longus

Fibula
Outer bone of lower leg

Peroneus (fibularis) tertius tendon
Helps to evert the foot

Hallux
Big toe

Tibial tuberosity

Tibia
Shin

Gastrocnemius
Calf muscle that flexes the knee and foot

Soleus
Broad flat muscle in the calf

Extensor hallucis longus
Helps to extend the big toe

During walking, the anterior muscles of the lower leg lift the toes. This ensures that they do not drag on the floor when the foot is moved forward.

Lateral muscles of the lower leg

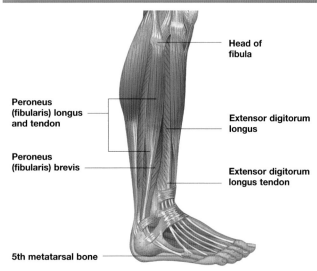

Head of fibula

Peroneus (fibularis) longus and tendon

Peroneus (fibularis) brevis

Extensor digitorum longus

Extensor digitorum longus tendon

5th metatarsal bone

The muscles of the lateral (outer) compartment lie alongside the smaller of the two lower leg bones, the fibula. There are two muscles in this group:
■ **Peroneus (fibularis) longus**
This muscle is the longer of the two and lies more superficially. It arises from the head and the upper portion of the narrow fibula and runs down to the sole of the foot
■ **Peroneus (fibularis) brevis**
As its name suggests, this is a

The lateral muscles protect the ankle by resisting inversion of the foot. When the foot is inverted, the ankle is in a very weak position.

short muscle, which lies underneath the peroneus longus muscle. It arises from the lower portion of the fibula and has a broad tendon that runs down to insert into the base of the fifth metatarsal bone of the foot.

ACTION OF THE LATERAL MUSCLES
These two muscles together cause the foot to plantar flex, when the toes point down, and to evert, which means bending so that the sole faces outwards. In practice, these muscles help to support the ankle by resisting the movement of inversion (the sole facing inwards), which is when the joint is weakest.

Posterior muscles of the lower leg

The posterior group of muscles of the lower leg form the mound of the calf. Together, these muscles are strong and heavy, enabling them to work together to flex the foot and to support the weight of the body.

The muscles that lie within the posterior compartment form the largest group of the lower leg. Also known as the 'calf muscles', this group can be further divided into superficial and deep layers.

SUPERFICIAL CALF MUSCLES

The superficial group of posterior muscles forms the bulk of the rounded calf, a feature peculiar to humans and which is due to our upright posture. These muscles include:

■ **Gastrocnemius** – this large fleshy muscle is the most superficial. It has a distinctive shape with two heads which arise from the medial and lateral condyles of the femur. Its fibres run mainly vertically, which allows for the rapid and strong contractions needed in running and jumping

■ **Soleus** – this is a large and powerful muscle which lies under the gastrocnemius. It takes its name from its shape, being flat like a sole (a type of flatfish). Contraction of the soleus muscle is important for maintaining balance when standing

■ **Plantaris** – this muscle is sometimes absent, and when it is present it is rather small and thin. Due to its relative unimportance in the lower leg, it

Plantaris muscle
This muscle is not always present

Gastrocnemius
A two-headed muscle, used in running and jumping

Soleus

Flexor hallucis longus

Flexor retinaculum

Soleus
Can be felt deep to the gastrocnemius when a person is standing on tiptoe

Calcaneal (Achilles) tendon
The largest tendon in the body; it is located at the back of the ankle and attached to the heel bone (calcaneus)

Calcaneal tuberosity

is sometimes used by surgeons to replace damaged tendons in the hand.

ACTIONS OF THE SUPERFICIAL MUSCLES

These muscles have the job of plantar flexing the foot, which means lifting the heel and

pointing the toes downwards. Strong muscles are needed for this job because, during walking, running and jumping, the heel needs to be lifted against the whole body weight.

Gastrocnemius and soleus have a single, common tendon that is known as the large and

Together, the gastrocnemius, soleus and plantaris muscles help to flex the foot at the ankle joint. The tiny plantaris is the weakest of the three muscles.

powerful Achilles tendon, which runs down from the lower edge of the calf to the heel.

Deep calf muscles

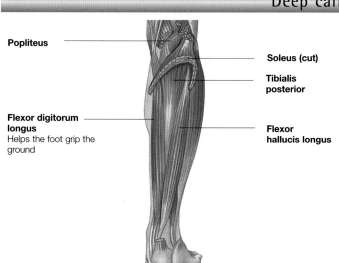

Popliteus

Flexor digitorum longus
Helps the foot grip the ground

Soleus (cut)

Tibialis posterior

Flexor hallucis longus

There are four muscles which together make up the deep group of calf muscles:

■ **Popliteus** – this is a thin, triangular muscle that lies at the back of the knee in the popliteal fossa. Popliteus has the particular role of 'unlocking' the knee joint by rotating it slightly to allow the straightened leg to be bent

■ **Flexor digitorum longus** – this muscle has long tendons

The actions of the deep muscles vary. The popliteus helps to unlock the knee joint, whereas the other muscles act on the ankle and foot joints.

which pass down to the outer four toes to allow them to curl under, or to flex

■ **Flexor hallucis longus** – although this muscle only runs to one toe, the big toe or hallux, it is a very powerful muscle. Its long tendon runs between the sesamoid bones located at the base of the big toe and it acts to give a 'push off' or 'spring' to the step during walking and running

■ **Tibialis posterior** – this is the deepest muscle in this group. It is the main provider of the action of inversion, in which the foot moves so that the sole faces inwards.

Deep fascia of the leg

Lying just below the subcutaneous tissue, the deep fascia of the leg forms a strong, circular sheath around the muscles, bone and blood vessels. The fascia partitions the leg into three compartments.

Fascia is the name given to sheets of connective tissue within the body that enclose and bind structures such as muscles.

Lying under the subcutaneous tissue, but above the muscles, the deep fascia of the leg is a membranous sheath that envelops the limb. It also forms partitions, which run down the leg until they meet bone, dividing the thigh and lower leg into a series of compartments.

ILIOTIBIAL TRACT

On the outer side of the thigh, the deep fascia is thickened and strengthened to form a tough vertical band known as the iliotibial tract. The tensor fasciae latae muscle inserts into this band, as does the greater part of the large and powerful gluteus maximus muscle.

SAPHENOUS OPENING

The saphenous opening is a gap in the deep fascia of the thigh. Usually about 3.75 cm in length and 2.5 cm wide, it allows the great saphenous vein to pass through in order to empty into the femoral vein.

VENOUS RETURN

One of the functions of the deep fascia of the leg is to aid the return of venous blood by acting as the 'muscle pump'. Being tough and relatively inelastic, the fascia resists the bulging of muscles as they contract. This causes them to put pressure upon the valved deep veins within the leg, so forcing venous blood back up to the body.

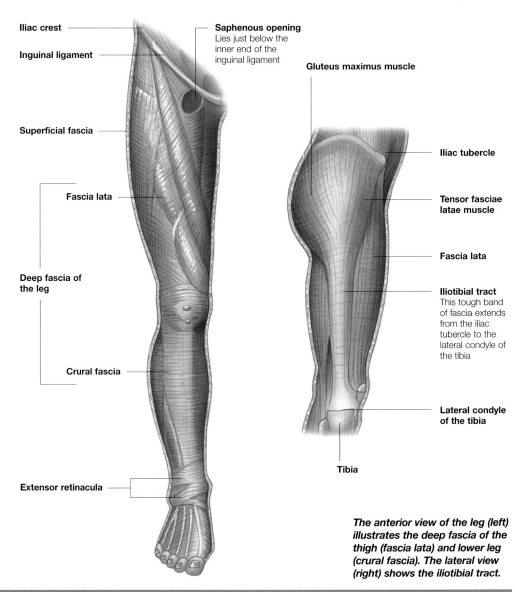

Iliac crest

Inguinal ligament

Saphenous opening
Lies just below the inner end of the inguinal ligament

Superficial fascia

Fascia lata

Deep fascia of the leg

Crural fascia

Extensor retinacula

Gluteus maximus muscle

Iliac tubercle

Tensor fasciae latae muscle

Fascia lata

Iliotibial tract
This tough band of fascia extends from the iliac tubercle to the lateral condyle of the tibia

Lateral condyle of the tibia

Tibia

The anterior view of the leg (left) illustrates the deep fascia of the thigh (fascia lata) and lower leg (crural fascia). The lateral view (right) shows the iliotibial tract.

Compartments of the thigh

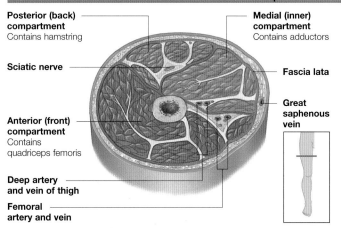

Posterior (back) compartment
Contains hamstring

Sciatic nerve

Anterior (front) compartment
Contains quadriceps femoris

Deep artery and vein of thigh

Femoral artery and vein

Medial (inner) compartment
Contains adductors

Fascia lata

Great saphenous vein

Partitions, or septa, arise from the fascia lata and run down to the femur. These divide the thigh into three compartments, each of which contain a group of muscles with similar actions and a similar nerve and blood supply:
■ The anterior compartment – this contains muscles that flex the hip and extend the knee, supplied mainly by the femoral

The fascia lata envelops the limb, encircling the muscles, bone and blood vessels. The limb also has sheets of connective tissue running down to the femur.

nerve and artery
■ The medial compartment – this is also known as the adductor compartment as it contains muscles that adduct the thigh (bring it in to the midline). These muscles are supplied by the obturator nerve and the profunda femoris and obturator arteries
■ The posterior compartment – this contains the powerful hamstring muscles that extend the hip and flex the knee. These muscles are supplied by the sciatic nerve and the profunda femoris artery.

Compartments of the lower leg

The deep fascia of the lower leg is also known as the crural fascia. It is a thick, membranous sheath, capable of binding muscles, arteries and veins within separate compartments: the anterior, lateral and posterior compartments.

The deep (crural) fascia of the lower leg attaches to the front and inner borders of the tibia below the knee, where it is continuous with its periosteum (the tough membrane that encloses bone). In the upper part of the lower leg, the crural fascia is quite thick and provides a site of attachment for muscle. Lower down it becomes thinner, except where it forms the horizontal retinacula at the ankle.

Partitions arise from the deep surface of the crural fascia and attach to the fibula beneath to divide the lower leg into three compartments: anterior, lateral and posterior.

ANTERIOR COMPARTMENT
The anterior compartment contains muscles that dorsiflex (point upwards) the foot and extend, or straighten, the toes.

LATERAL COMPARTMENT
The lateral (outer) compartment lies alongside the fibula bone. It contains two muscles which plantarflex (point down) the foot and evert the foot (the sole moves to face outwards).

POSTERIOR COMPARTMENT
The posterior compartment of the lower leg is subdivided by another partition, the transverse

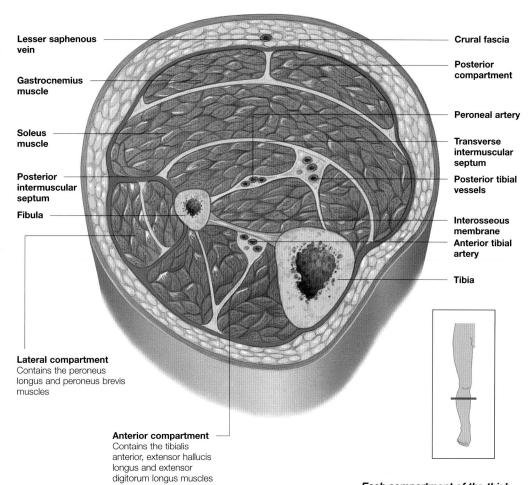

Lesser saphenous vein

Gastrocnemius muscle

Soleus muscle

Posterior intermuscular septum

Fibula

Crural fascia

Posterior compartment

Peroneal artery

Transverse intermuscular septum

Posterior tibial vessels

Interosseous membrane

Anterior tibial artery

Tibia

Lateral compartment
Contains the peroneus longus and peroneus brevis muscles

Anterior compartment
Contains the tibialis anterior, extensor hallucis longus and extensor digitorum longus muscles

Each compartment of the thigh houses particular muscles. The posterior compartment is further subdivided by the transverse intermuscular septum.

intermuscular septum, into deep and superficial layers. The powerful muscles here act to flex the foot and provide the main forward propulsive force in walking. These muscles are

supplied by the tibial nerve and the posterior tibial artery.

RETINACULA
Around the ankle the deep fascia forms thickened horizontal

bands known as the extensor retinacula. These are tough restraining bands which keep the underlying tendons in position against the ankle as the foot changes position.

Compartment syndrome

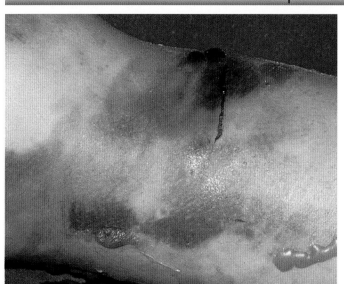

The deep fascia and its partitions are very strong and supportive. They are also relatively inelastic, which under normal conditions is a positive feature. However, when there is damage or trauma to the lower leg, such as a bone fracture, this inflexibility of the deep fascia becomes a hazard.

TRAUMA
Trauma causes soft tissues to swell and bleed inside the

Compartment syndrome develops as a result of tissue swelling in the limb following trauma. In severe cases, surgery is undertaken to relieve pressure.

compartments, which are unable to expand due to the tough fascial septa. As a result the pressure inside the compartment (normally the anterior) increases.

This increased pressure causes compression, making muscle contraction painful. As the pressure continues to rise, venous blood is unable to leave the area.

In the most severe cases, even pressurized arterial blood cannot pass and so there are no pulses in the foot. If necessary, a surgical procedure called a fasciotomy can be performed, in which the deep fascia in the affected area is cut open to release the pressure.

Arteries of the leg

The lower limb is supplied by a series of arteries which arise from the external iliac artery of the pelvis. These arteries pass down the leg, branching to reach muscles, bones, joints and skin.

A network of arteries supply the tissues of the lower limb with nutrients. The main arteries give off important and smaller branches to provide nourishment to various joints and muscles.

THE ARTERIES

■ **The femoral artery** – the main artery of the leg. Its main branch is the profunda femoris (deep femoral) artery. Small branches supply nearby muscles before the artery enters a gap in the adductor magnus muscle, the 'adductor hiatus', to enter the popliteal fossa (behind the knee)

■ **Profunda femoris (deep femoral artery)** – the main artery of the thigh. It gives off several branches including the medial and lateral circumflex femoral arteries and the four perforating arteries

■ **Popliteal artery** – a continuation of the femoral artery. It runs down the back of the knee, giving off small branches to nourish that joint, before dividing into the anterior and posterior tibial arteries

■ **Anterior tibial artery** – this supplies the structures within the anterior (front) compartment of the lower leg. It runs downwards to the foot and becomes the dorsalis pedis artery

■ **Posterior tibial artery** – this artery remains at the back of the lower leg and, together with the peroneal (fibular) artery, supplies the structures of the back and outer compartments.

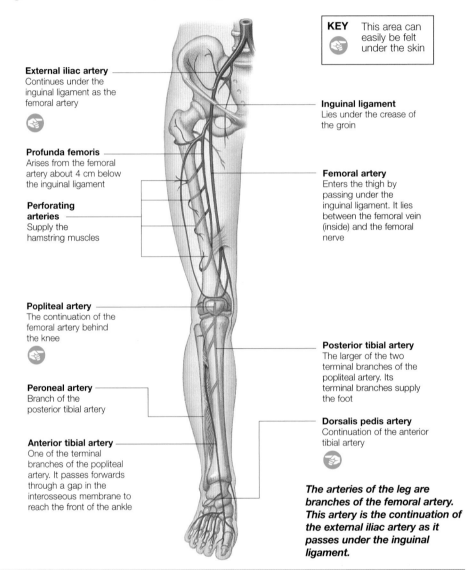

KEY This area can easily be felt under the skin

External iliac artery
Continues under the inguinal ligament as the femoral artery

Profunda femoris
Arises from the femoral artery about 4 cm below the inguinal ligament

Perforating arteries
Supply the hamstring muscles

Popliteal artery
The continuation of the femoral artery behind the knee

Peroneal artery
Branch of the posterior tibial artery

Anterior tibial artery
One of the terminal branches of the popliteal artery. It passes forwards through a gap in the interosseous membrane to reach the front of the ankle

Inguinal ligament
Lies under the crease of the groin

Femoral artery
Enters the thigh by passing under the inguinal ligament. It lies between the femoral vein (inside) and the femoral nerve

Posterior tibial artery
The larger of the two terminal branches of the popliteal artery. Its terminal branches supply the foot

Dorsalis pedis artery
Continuation of the anterior tibial artery

The arteries of the leg are branches of the femoral artery. This artery is the continuation of the external iliac artery as it passes under the inguinal ligament.

Arteries around the knee

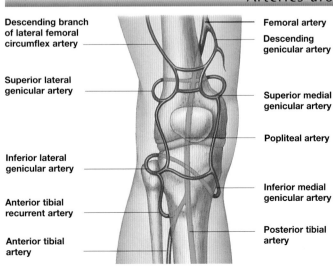

Descending branch of lateral femoral circumflex artery

Superior lateral genicular artery

Inferior lateral genicular artery

Anterior tibial recurrent artery

Anterior tibial artery

Femoral artery

Descending genicular artery

Superior medial genicular artery

Popliteal artery

Inferior medial genicular artery

Posterior tibial artery

Behind the knee, the popliteal artery gives off a number of small branches which surround the knee joint and form connections, or anastomoses, with other small branches of the femoral and the anterior and posterior tibial arteries. Together they form an arterial network through which blood can bypass the normal route via the main popliteal artery. This may be of importance when the knee is held bent for a considerable time

The arterial supply around the knee forms a network of vessels that connect the femoral artery to the terminal branches of the popliteal artery.

or if the main artery is narrowed or blocked.

POPLITEAL PULSE

Just like the pulsations of the femoral artery, which can be felt in the groin, those of the popliteal artery can be felt behind the knee. However, since the popliteal artery lies so deeply within the tissues behind the knee, it may be difficult to feel. It is often necessary to examine the leg when flexed (bent) to loosen the fascia and muscle in the popliteal fossa so making the pulse easier to feel. If the pulse is weak or absent this means that the femoral artery is narrowed or blocked.

Arteries of the foot

In a pattern similar to that in the hand, the small arteries of the foot form arches which interconnect, giving off branches to each side of the toes. Branches of the arteries give the sole of the foot a particularly rich blood supply.

The arterial supply of the foot is provided by the terminal branches of the anterior and posterior tibial arteries.

TOP OF THE FOOT
As the anterior tibial artery passes down in front of the ankle it becomes the 'dorsalis pedis' artery. This then runs down across the top of the foot towards the space between the first and second toes, where it gives off a deep branch that joins the arteries on the sole of the foot. Branches of the dorsalis pedis on the top of the foot join to form an arch which gives off branches to the toes.

The dorsalis pedis pulse can be felt by an examining doctor on the top of the foot next to the tendon of tibialis anterior. As the artery lies just under the skin the pulse should be fairly easy to feel if the blood vessels are healthy.

SOLE OF THE FOOT
The sole of the foot has a rich blood supply which is provided by branches of the posterior tibial artery. As the artery enters the sole, it divides into two parts to form the medial and lateral plantar arteries.

Plantar aspect of foot (sole)

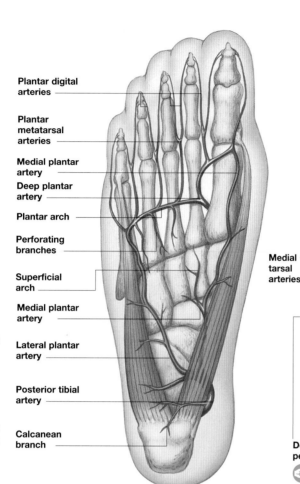

- Plantar digital arteries
- Plantar metatarsal arteries
- Medial plantar artery
- Deep plantar artery
- Plantar arch
- Perforating branches
- Superficial arch
- Medial plantar artery
- Lateral plantar artery
- Posterior tibial artery
- Calcanean branch

Dorsum of foot (top)

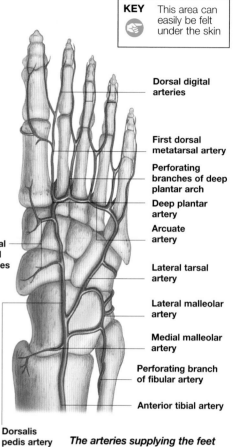

KEY This area can easily be felt under the skin

- Dorsal digital arteries
- First dorsal metatarsal artery
- Perforating branches of deep plantar arch
- Deep plantar artery
- Arcuate artery
- Lateral tarsal artery
- Lateral malleolar artery
- Medial malleolar artery
- Perforating branch of fibular artery
- Anterior tibial artery

- Medial tarsal arteries
- Dorsalis pedis artery

The arteries supplying the feet branch in a similar fashion to those of the hands. The sole of the foot has a particularly rich blood supply.

■ **The medial plantar artery –** this is the smaller of the two branches of the posterior tibial artery. It provides blood for the muscles of the big toe and sends tiny branches to the other toes.

■ **The lateral plantar artery –** this artery is much larger than the medial plantar artery and curves around under the metatarsal bones to form the deep plantar arch.

The deep branch of the dorsalis pedis artery joins the inner end of this arch so making a connection between the arterial supply of the top of the foot and the sole.

Arteriograms

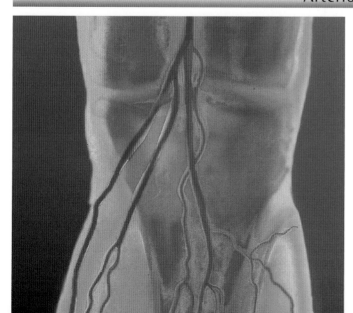

It is possible to examine an artery by using arteriography. This involves the injection of a dye into an artery followed by a series of X-ray pictures which show how the dye travels onward through the arterial system. If dye is injected into the femoral artery high in the thigh a study can be made of the arterial system of the leg. This can show up areas of blockage or narrowing, or the bulging outline of a popliteal aneurysm.

Arteriography may be used if

Arteriography is used to identify the exact location of a blockage in the artery. Radio-opaque dye is injected into the bloodstream, making vessels visible on X-ray.

surgery is being contemplated but there are less invasive ways to investigate blood flow in the leg. These include Doppler ultrasound scanning and MRI, which provide information with less discomfort and risk.

ATHEROSCLEROSIS
One of the more common conditions that affects the blood flow in the leg is atherosclerosis, or hardening of the arteries, which is often due to smoking. This may cause cramping pain in the calf muscles during exercise, which settles down after a few minutes rest. Such pain is due to decreased blood supply to the muscles caused by narrowed arteries.

Veins of the leg

The lower limb is drained by a series of veins which can be divided into two groups, superficial and deep. The perforating veins connect the two groups of veins.

Lying within the subcutaneous (beneath the skin) tissue, there are two main superficial veins of the leg, the great and small saphenous veins.

GREAT SAPHENOUS VEIN

The great saphenous vein is the longest vein in the body and is sometimes used during surgical procedures to replace damaged or diseased arteries in areas such as the heart. It arises from the medial (inner) end of the dorsal venous arch of the foot and runs up the leg towards the groin.

On its journey, the great saphenous vein passes in front of the medial malleolus (inner ankle bone), tucks behind the medial condyle of the femur at the knee and passes through the saphenous opening in the groin to drain into the large femoral vein.

SMALL SAPHENOUS VEIN

This smaller superficial vein arises from the lateral (outer) end of the dorsal venous arch and passes behind the lateral malleolus (outer ankle bone) and up the centre of the back of the calf. As it approaches the knee, the small saphenous vein empties into the deep popliteal vein.

TRIBUTARIES

The great and small saphenous veins receive blood along the way from many smaller veins and also intercommunicate freely, or 'anastomose', with each other.

Superficial veins of leg, anterior view

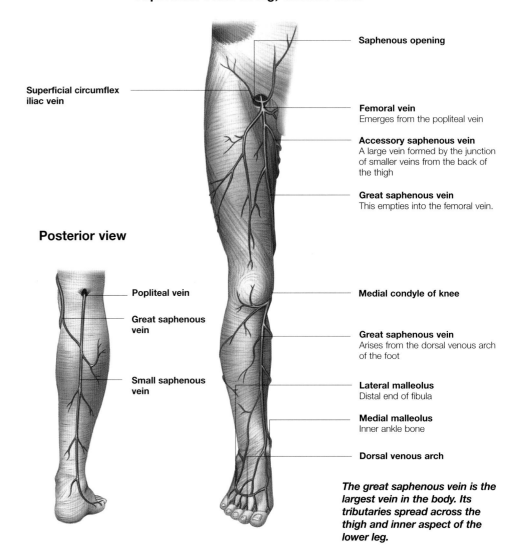

Saphenous opening

Superficial circumflex iliac vein

Posterior view

Popliteal vein

Great saphenous vein

Small saphenous vein

Femoral vein
Emerges from the popliteal vein

Accessory saphenous vein
A large vein formed by the junction of smaller veins from the back of the thigh

Great saphenous vein
This empties into the femoral vein.

Medial condyle of knee

Great saphenous vein
Arises from the dorsal venous arch of the foot

Lateral malleolus
Distal end of fibula

Medial malleolus
Inner ankle bone

Dorsal venous arch

The great saphenous vein is the largest vein in the body. Its tributaries spread across the thigh and inner aspect of the lower leg.

Valves and venous pump

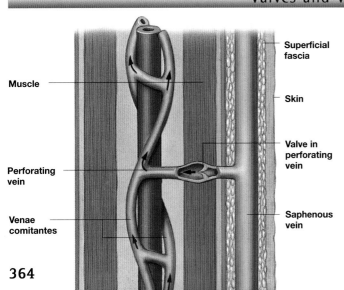

Muscle

Perforating vein

Venae comitantes

Superficial fascia

Skin

Valve in perforating vein

Saphenous vein

The arrangement of blood vessels in the leg means that blood flows from the superficial veins through the perforating veins to the deep veins. Venous blood is then pumped back up to the body mainly by the calf muscles which surround these deep veins (the venous pump).

Unlike arteries, veins contain tiny valves which prevent the backflow of blood within them. These valves are of great

The valved perforating veins play a key role in helping the venous pump to function. The valves enable the blood to make its way towards the heart.

importance in the veins of the leg as they ensure that, when the calf muscles contract, the blood is pushed up the vein towards the heart rather than back out into the superficial veins.

VARICOSE VEINS

If the valves in the perforating veins become damaged, then there can be backflow into the relatively low pressure superficial veins, which then become distended and tortuous. Causes of these varicose veins include hereditary factors, pregnancy, obesity and thrombosis (abnormal clotting) of the deep leg veins.

Deep veins of the leg

The deep veins of the leg follow the pattern of the arteries, which they accompany along their length. As well as draining venous blood from the tissues of the leg, the deep veins receive blood from the superficial veins via the perforating veins.

Although the deep leg veins are referred to and illustrated as single veins they are usually, in fact, paired veins which lie either side of the artery. These veins are known as venae comitantes and they are common throughout the body.

DEEP VEINS
The main deep veins comprise:
■ **The posterior tibial vein** – this is formed by the joining together of the small medial and lateral plantar veins of the sole of the foot. As it approaches the knee it is joined from its lateral side by the fibular (peroneal) vein before joining with the anterior tibial vein to form the large popliteal vein
■ **The anterior tibial vein** – this is the continuation of the dorsalis pedis vein on the top of the foot. It passes up the front of the lower leg
■ **The popliteal vein** – this lies behind the knee and receives blood from the small veins which surround the knee joint
■ **The femoral vein** – this is the continuation of the popliteal vein as it passes up the thigh. The large femoral vein receives blood from the superficial veins and continues up into the groin to become the external iliac vein of the pelvis.

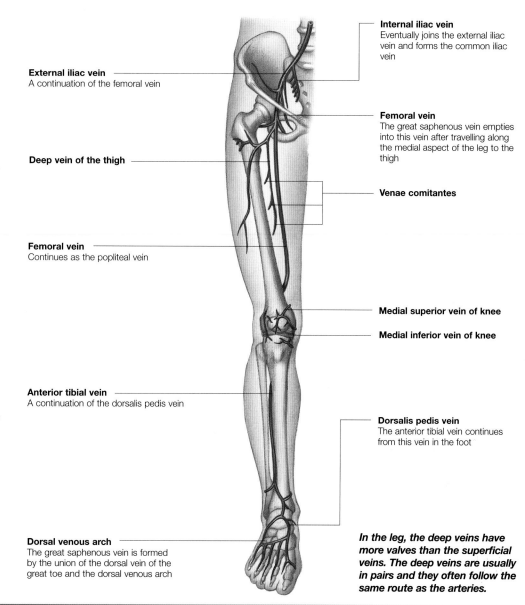

Deep veins of leg, anterior view

External iliac vein
A continuation of the femoral vein

Deep vein of the thigh

Femoral vein
Continues as the popliteal vein

Anterior tibial vein
A continuation of the dorsalis pedis vein

Dorsal venous arch
The great saphenous vein is formed by the union of the dorsal vein of the great toe and the dorsal venous arch

Internal iliac vein
Eventually joins the external iliac vein and forms the common iliac vein

Femoral vein
The great saphenous vein empties into this vein after travelling along the medial aspect of the leg to the thigh

Venae comitantes

Medial superior vein of knee

Medial inferior vein of knee

Dorsalis pedis vein
The anterior tibial vein continues from this vein in the foot

In the leg, the deep veins have more valves than the superficial veins. The deep veins are usually in pairs and they often follow the same route as the arteries.

Deep vein thrombosis

Thrombosis of the blood within the deep veins of the leg is a relatively common disorder. It is usually associated with sluggish blood flow in these vessels, which may have a number of causes, including:
■ Prolonged bed rest – this increases the risk of developing a deep vein thrombosis (DVT). It is for this reason that post-operative patients and women who have just given birth are encouraged to be up and about

Sitting in cramped conditions on a long-haul flight is believed to contribute to DVT formation. Passengers are encouraged to carry out in-flight leg exercises.

as soon as possible
■ Inactivity for long periods (for example, on long-haul air flights) – this is known to predispose to DVT formation
■ Fractures of the leg bones – this increases the likelihood of developing a DVT
■ Pregnancy, or the presence of abnormal abdominal masses – this can impede the return of blood from the leg and cause sluggish flow in the deep veins.

The main clinical importance of DVTs is that they may lead to a pulmonary embolus, in which a piece of the blood clot breaks off and is carried to the lungs. In some cases, a pulmonary embolism is fatal.

Nerves of the leg (1)

The main nerve of the leg – the sciatic nerve – is the largest
nerve in the body. Its branches supply the muscles of the hip, many
of the thigh and all of the muscles of the lower leg and foot.

The sciatic nerve is made up of
two nerves, the tibial nerve and
the common peroneal (or fibular)
nerve. These are bound together
by connective tissue to form a
wide band that runs the full
length of the back of the thigh.

ORIGIN AND COURSE
The sciatic nerve arises from a
network of nerves at the base of
the spine, called the sacral
plexus. From here, it passes out
through the greater sciatic
foramen and then curves
downwards through the gluteal
region under the gluteus
maximus muscle (midway
between the bony landmarks of
the greater trochanter of the
femur and the ischial tuberosity
of the pelvis).

The sciatic nerve leaves the
gluteal region by passing under
the long head of the biceps
femoris muscle to enter the thigh
and runs down the centre of the
back of the thigh, branching off
into the hamstring muscles (a
collective name for the biceps
femoris, semitendinosus and
semimembranosus muscles). It
then divides to form two
branches, the tibial nerve and
the common peroneal nerve just
above the knee.

HIGHER DIVISION
In a few cases the sciatic nerve
divides into two at a much
higher level. In this situation the
common peroneal nerve may
pass above or even through the
piriformis muscle.

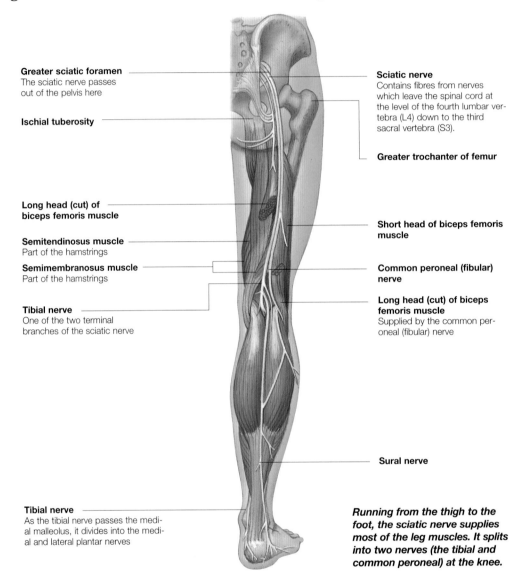

Greater sciatic foramen
The sciatic nerve passes
out of the pelvis here

Ischial tuberosity

**Long head (cut) of
biceps femoris muscle**

Semitendinosus muscle
Part of the hamstrings

Semimembranosus muscle
Part of the hamstrings

Tibial nerve
One of the two terminal
branches of the sciatic nerve

Tibial nerve
As the tibial nerve passes the medial malleolus, it divides into the medial and lateral plantar nerves

Sciatic nerve
Contains fibres from nerves
which leave the spinal cord at
the level of the fourth lumbar vertebra (L4) down to the third
sacral vertebra (S3).

Greater trochanter of femur

**Short head of biceps femoris
muscle**

**Common peroneal (fibular)
nerve**

**Long head (cut) of biceps
femoris muscle**
Supplied by the common peroneal (fibular) nerve

Sural nerve

*Running from the thigh to the
foot, the sciatic nerve supplies
most of the leg muscles. It splits
into two nerves (the tibial and
common peroneal) at the knee.*

The sciatic nerve and intramuscular injections

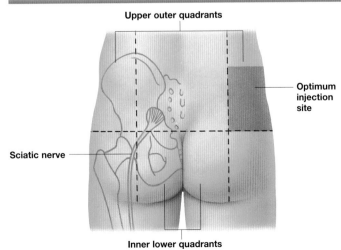

Upper outer quadrants

**Optimum
injection
site**

Sciatic nerve

Inner lower quadrants

The buttock is often chosen as a
site for intramuscular injections
as it has a large muscle mass.

It is vitally important to have
an accurate knowledge of the
position and course of the sciatic
nerve when administering
intramuscular injections in this
area. If the injection site
impinges on the sciatic nerve
there is a high risk of damaging
it, with severe consequences for
the future function of the leg.

*Intramuscular injections are
often given in the gluteal area.
The sciatic nerve can be safely
avoided if injections are given in
the upper outer quadrant.*

If each gluteal (buttock) region
is divided into four quadrants it
can be seen that the sciatic
nerve lies within the lower
quadrants. The only safe place
for an intramuscular injection in
the buttock, therefore, is in the
upper outer quadrant.

ALTERNATIVE SITE
Anyone responsible for giving
injections will have been taught
the importance of injecting only
in the upper outer quadrant. In
many cases people choose to
inject the outer side of the thigh
in preference to the buttock as
there is much less risk of
damage to important structures.

Terminal branches of the sciatic nerve

The sciatic nerve divides into two terminal branches: the common peroneal (fibular) nerve and the tibial nerve. The common peroneal nerve supplies the front of the leg, while the tibial nerve supplies the muscles and skin at the back.

The common peroneal nerve leaves the sciatic nerve in the lower third of the thigh and runs down around the outer side of the lower leg before dividing into two just below the knee.

NERVE BRANCHES
The two branches of the peroneal nerve comprise:
■ The superficial branch of the peroneal nerve – supplies the lateral (outer) compartment of the lower leg in which it lies. It sub-divides into smaller branches to supply the muscles around it
■ The deep peroneal nerve – runs in front of the interosseous membrane between the tibia and the fibula, and then passes over the ankle into the foot.
   These two branches also supply the knee joint and the skin over the outer side of the calf and the top of the foot.

DAMAGE
As the common peroneal nerve passes around the outer side of the lower leg, it lies just under the skin and very close to the head of the fibula. It is very vulnerable to damage, especially if the fibula suffers a fracture, and is the most commonly damaged nerve in the leg.

Common peroneal (fibular) nerve
Branches off the sciatic nerve and runs around the outer side of the lower leg before dividing into deep and superficial branches

Common peroneal nerve
Point of maximum vulnerability

Peroneus longus muscle (cut)
Supplied by the superficial branch of the peroneal nerve

Superficial peroneal nerve
Supplies the fibularis longus and fibularis brevis muscles

Peroneus longus muscle
Supplied by the superficial branch of the peroneal nerve

Peroneus brevis muscle
Supplied by the superficial branch of the peroneal nerve

Extensor digitorum brevis muscle
Supplied by the lateral branch of the deep peroneal nerve

Medial branch of deep peroneal (fibular) nerve
Supplying the skin between the first and second toes

Articular branch of common peroneal nerve

Head of the fibula
Lies underneath the peroneal muscles

Tibia
Larger inner bone of the lower leg

Deep peroneal nerve
Supplies the tibialis anterior and extensor digitorum longus muscles

Tibialis anterior muscle
Supplied by the deep peroneal nerve

Extensor digitorum longus muscle
Supplied by the deep peroneal nerve

Extensor hallucis longus muscle
Supplied by the deep peroneal nerve

Extensor hallucis brevis muscle
Supplied by the lateral branch of the deep peroneal nerve

The common peroneal nerve splits into two branches to supply the inner and outer lower leg. Close to the skin at points, it is vulnerable to damage.

The tibial nerve

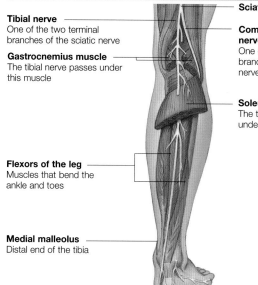

Tibial nerve
One of the two terminal branches of the sciatic nerve

Gastrocnemius muscle
The tibial nerve passes under this muscle

Flexors of the leg
Muscles that bend the ankle and toes

Medial malleolus
Distal end of the tibia

Sciatic nerve

Common peroneal nerve
One of the two terminal branches of the sciatic nerve

Soleus muscle
The tibial nerve passes under this muscle

The tibial nerve is the larger of the two terminal branches of the sciatic nerve. It supplies the flexors of the leg: those muscles which bend, rather than straighten, the joints.

PATH DOWN THE LEG
The tibial nerve arises in the lower third of the thigh, where it supplies the hamstring muscles. It then separates from the common peroneal nerve before following a course down the back of the leg:

The tibial nerve splits off from the sciatic nerve to course down the back of the lower leg. Branches supply the muscles and skin with sensation.

■ It passes through the popliteal fossa (a space behind the knee) alongside the popliteal artery
■ It then descends under the large gastrocnemius and soleus muscles
■ It reaches the posterior compartment of the lower leg where it gives off branches to the flexor muscles found there
■ At the ankle it passes behind the medial malleolus, before dividing into the medial and lateral plantar nerves of the foot.

BRANCHES
The tibial nerve has two cutaneous branches which supply areas of skin: the sural nerve (in the calf) and the medial calcaneal nerve (heel).

367

Nerves of the leg (2)

Cutaneous nerves provide the skin with sensation. There are a number of these within the leg, many of which branch off from the main nerves of the leg: the sciatic, femoral and tibial nerves.

Cutaneous nerves supply the skin. Those of the leg lie within the subcutaneous tissue and often branch off larger nerves serving muscles and joints.

NERVES OF THE THIGH
Cutaneous nerves include:
■ Ilioinguinal nerve – supplies an area of the front/inner thigh
■ Genitofemoral nerve – supplies a small area of skin under the centre of the inguinal ligament
■ Lateral cutaneous nerve – supplies the outer thigh
■ Obturator nerve – branches of this nerve supply the inner thigh
■ Medial/intermediate cutaneous nerves – supply front of the thigh not supplied by the ilioinguinal
■ Posterior cutaneous nerve – supplies the back of the thigh and the popliteal fossa.

LOWER LEG AND FOOT
The following nerves branch off the sciatic and femoral nerves:
■ Saphenous nerve – supplies the front and inner lower leg
■ Lateral cutaneous nerve – supplies the upper front and outer side of the lower leg
■ Superficial peroneal (fibular) nerve – supplies the lower outer calf and the top of the foot
■ Sural nerve – supplies the lower outer part of the back of the lower leg and the lateral border of the foot and little toe
■ Medial and lateral plantar nerves – supply the sole
■ Tibial nerve – supplies the heel.

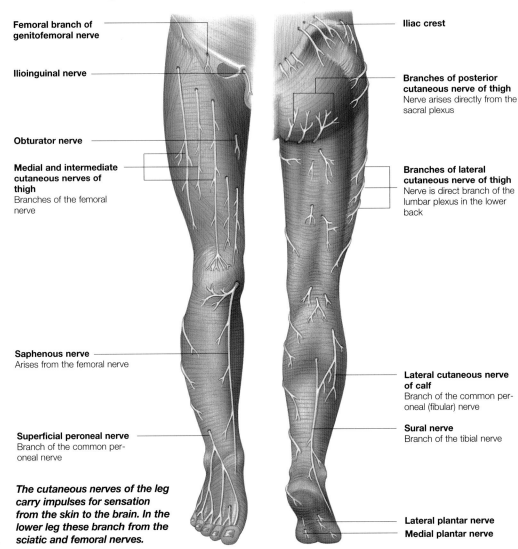

Femoral branch of genitofemoral nerve

Ilioinguinal nerve

Obturator nerve

Medial and intermediate cutaneous nerves of thigh
Branches of the femoral nerve

Saphenous nerve
Arises from the femoral nerve

Superficial peroneal nerve
Branch of the common peroneal nerve

Iliac crest

Branches of posterior cutaneous nerve of thigh
Nerve arises directly from the sacral plexus

Branches of lateral cutaneous nerve of thigh
Nerve is direct branch of the lumbar plexus in the lower back

Lateral cutaneous nerve of calf
Branch of the common peroneal (fibular) nerve

Sural nerve
Branch of the tibial nerve

Lateral plantar nerve
Medial plantar nerve

The cutaneous nerves of the leg carry impulses for sensation from the skin to the brain. In the lower leg these branch from the sciatic and femoral nerves.

Dermatomes

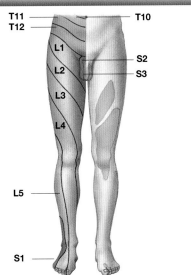

T11
T12
T10
L1
L2
L3
L4
L5
S1
S2
S3

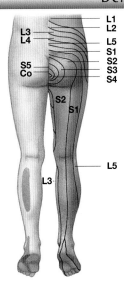

L1
L2
L3
L4
L5
S1
S2
S3
S4
S5
Co
S2
S1
L5
L3

A dermatome is an area of skin that receives its sensory nerve supply from a single spinal nerve (and therefore a single segment of the spinal cord). The nerve supply may actually be taken to the skin in two or more cutaneous branches.

PATTERN
The dermatome pattern in the limbs can be more easily understood if the human body is imagined to be assuming an ''animal-like' all-fours position.

The skin of the leg can be mapped into well-defined areas of sensitivity, where each area relates exactly to a single segment of the spinal cord.

The spinal nerves from the lumbar region (L1–L5) mostly serve the front of the leg.

Those from the sacral region (S1–S5), which is further down, supply the back of the leg and the buttocks.

The lowermost spinal nerves, emerging from the region of the coccyx (Co), supply an area that corresponds to where the tail would be.

TESTING
Doctors can test the sensation in these areas to see if the spinal nerves at a particular level are intact. There is usually a degree of overlap between dermatomes, however, as well as a variation between individuals.

The femoral nerve

The femoral nerve branches off from lumbar spinal nerves L2, L3 and L4. It runs through the pelvis and down the front of the thigh, supplying the powerful quadriceps muscles and the skin of the front and inner aspects of the leg.

The femoral nerve is a large nerve arising from the lumbar plexus, a network formed by the spinal nerves which emerge from the lumbar vertebrae.

The nerve enters the front of the thigh by passing under the inguinal ligament lateral to the femoral artery and vein.

However, unlike those vessels, it does not lie within the protective femoral sheath.

BRANCHES

The femoral nerve divides into its terminal branches around 3–4 cm below the inguinal ligament. These smaller nerves provide a supply to a number of structures within the lower limb:

■ The anterior thigh muscles – the femoral nerve has a very important role in the provision of nerve stimuli to the powerful muscles of the front of the thigh. The four muscles that make up the large quadriceps femoris group are all supplied by the femoral nerve, as are the pectineus and sartorius muscles

■ The hip and knee joints – the femoral nerve sends articular branches to these two large joints, between which it lies

■ The skin over the front of the thigh – the cutaneous branches of the femoral nerve provide this area with sensation

■ The skin below the knee – this

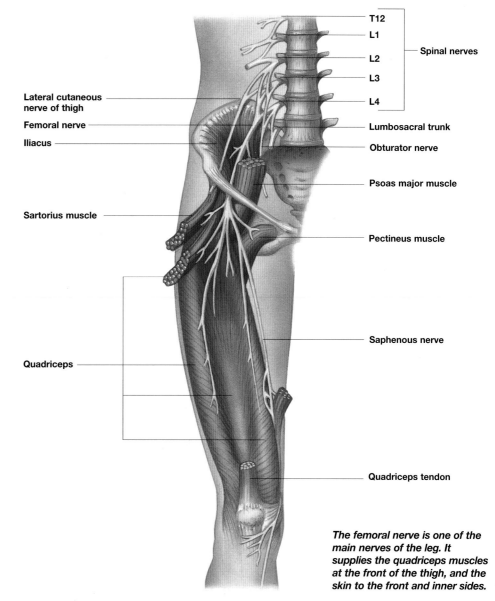

The femoral nerve is one of the main nerves of the leg. It supplies the quadriceps muscles at the front of the thigh, and the skin to the front and inner sides.

is supplied by another large cutaneous branch, the saphenous nerve. This passes down the leg with the femoral artery from the knee to the toes.

When the spinal roots of the lumbar region are compressed, such as in the case of a herniated (or slipped) disc, the structures supplied by the femoral nerve may be affected.

As the nerve supplies muscles that move both the hip and the knee, this can have a serious effect upon walking. Numbness (paraesthesia) of the skin over the front of the thigh may also develop.

The obturator nerve

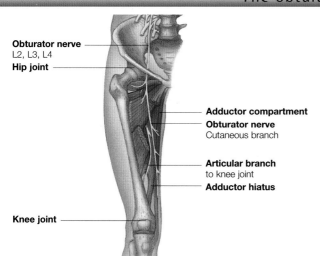

Like the femoral nerve, the obturator nerve arises from the lumbar plexus.

It is formed within the psoas major muscle and passes down through the obturator foramen of the pelvis alongside the obturator artery and vein.

From here, the obturator nerve enters the inner adductor compartment of the thigh. This contains the muscles which

The obturator nerve supplies the hip joint, knee joint and the thigh's adductor muscles. These are the muscles used by horse riders for gripping the saddle.

adduct the legs (pull them in towards the centre of the body). It lies within the inner aspect of the thigh.

From within this compartment, the obturator nerve supplies:

■ All the adductor muscles except the lower part of the adductor magnus

■ The skin on the inner aspect of the lower thigh, via a cutaneous branch

■ The hip and the knee joints. The small branch that supplies the knee joint descends through the adductor hiatus, a gap in the adductor magnus muscle, to reach the joint.

369

Ankle

The ankle is the joint between the lower ends of
the tibia and fibula, and the upper surface of the large foot
bone, the talus. It is an example of a hinge joint.

At the ankle, a deep socket is
formed by the lower ends of the
tibia and fibula, the bones of the
lower leg. Into this socket fits
the pulley-shaped upper surface
of the talus. The shape of the
bones and the presence of strong
supporting ligaments mean that
the ankle is very stable. This is
an important feature for such a
major weight-bearing joint.

THE JOINT
The articular surfaces of the
ankle joint – those parts of the
bone which move against each
other – are covered with a layer
of smooth hyaline cartilage. This
cartilage is surrounded by a thin
synovial membrane that secretes
a viscous fluid and helps to
lubricate the joint.

The articular surfaces of the
ankle joint consist of the:
■ Inside of the lateral malleolus,
the expanded lower end of the
fibula. This carries a facet
(depression) that articulates with
the outer side of the upper
surface of the talus
■ Undersurface of the lower end
of the tibia. This forms the roof
of the socket, which articulates
with the talus
■ Inside of the medial malleolus,
the projection at the lower end
of the tibia. This moves against
the inner side of the upper
surface of the talus
■ Trochlea of the talus. Named
for its pulley shape, this upper
part of the talus fits into the
ankle joint, and articulates with
the lower ends of the tibia and
fibula.

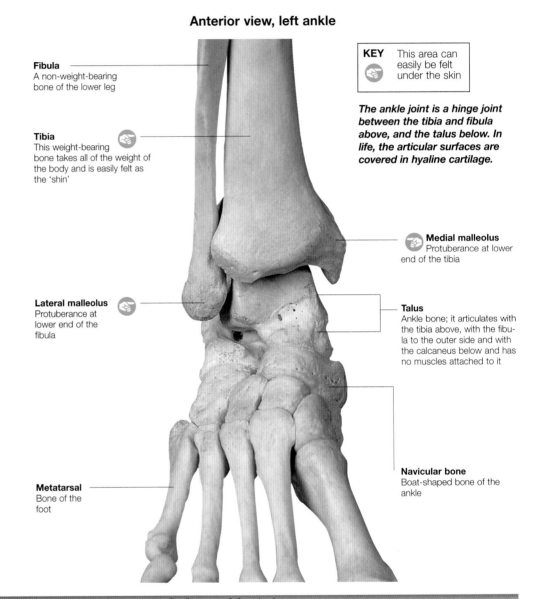

Anterior view, left ankle

Fibula
A non-weight-bearing
bone of the lower leg

Tibia
This weight-bearing
bone takes all of the weight of
the body and is easily felt as
the 'shin'

Lateral malleolus
Protuberance at
lower end of the
fibula

Metatarsal
Bone of the
foot

KEY This area can
easily be felt
under the skin

*The ankle joint is a hinge joint
between the tibia and fibula
above, and the talus below. In
life, the articular surfaces are
covered in hyaline cartilage.*

Medial malleolus
Protuberance at lower
end of the tibia

Talus
Ankle bone; it articulates with
the tibia above, with the fibula to the outer side and with
the calcaneus below and has
no muscles attached to it

Navicular bone
Boat-shaped bone of the
ankle

Movements of the ankle joint

Plantarflexion

Dorsiflexion

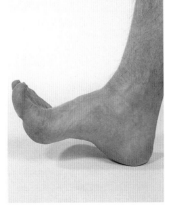

Although the foot is capable of a
variety of movements, much of
this flexibility is due to the
presence of other joints within
the foot and below the ankle. The
ankle joint itself acts only as a
hinge joint, allowing the talus to
rotate in one plane only. In this
respect, it is rather like the elbow
joint.

Movement of the foot at the
ankle is thus limited to:

*As a hinge joint, the ankle allows
movement in one plane only. In
dorsiflexion, the toes are pulled
upwards and in plantarflexion,
they are pushed downwards.*

■ Dorsiflexion. This is the term
that describes the movement of
the foot upwards, the heel
pointing down and the toes up.
The action of dorsiflexion is
partly limited by the pull of the
calcaneal (or Achilles) tendon at
the back of the ankle
■ Plantarflexion. This is the
opposite movement to
dorsiflexion; the toes point down.
This action is limited by the pull
of muscles and ligaments at the
front of the ankle.

Due to the shape of the bones
and ligaments of the ankle, the
joint is much more stable in
dorsiflexion than plantarflexion.

Ligaments of the ankle

The ankle is supported by strong ligaments which help to stabilize this important weight-bearing joint.

The ankle joint needs to be stable as it bears the weight of the body. The presence of a variety of strong ligaments around the ankle helps to maintain this stability, while still allowing the necessary freedom of movement.

Like most joints, the ankle is enclosed within a tough fibrous capsule. Although the capsule is quite thin in front and behind, it is reinforced on each side by the strong medial (inner) and lateral (outer) ankle ligaments.

MEDIAL LIGAMENT

Also known as the deltoid ligament, the medial ligament is a very strong structure which fans out from the tip of the medial malleolus of the tibia. It is usually described in three parts, each named for the bones that they connect:

■ Anterior and posterior tibiotalar ligaments. Lying close against the bones, these parts of the medial ligament connect the tibia to the medial sides of the talus beneath

■ Tibionavicular ligament. More superficially, this part of the ligament runs between the tibia and the navicular, one of the bones of the foot

■ Tibiocalcaneal ligament. This strong ligament runs just under the skin from the tibia to the sustentaculum tali, a projection of the calcaneus (large heel bone).

Together, these parts of the medial ligament support the

Lateral view (right foot)

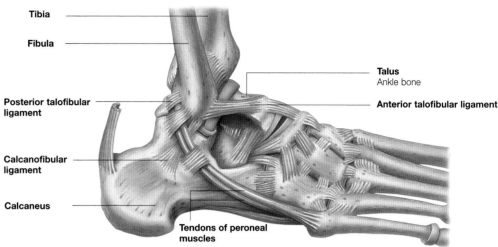

Tibia

Fibula

Posterior talofibular ligament

Calcanofibular ligament

Calcaneus

Talus
Ankle bone

Anterior talofibular ligament

Tendons of peroneal muscles

Medial view (right foot)

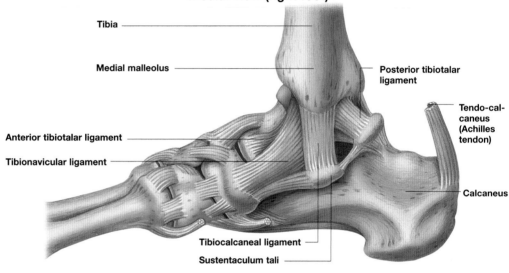

Tibia

Medial malleolus

Anterior tibiotalar ligament

Tibionavicular ligament

Posterior tibiotalar ligament

Tendo-cal-caneus (Achilles tendon)

Calcaneus

Tibiocalcaneal ligament

Sustentaculum tali

ankle joint during the movement of eversion (where the foot is turned out to the side).

LATERAL LIGAMENT

The lateral ligament is weaker than the medial ligament, and is

made up of three distinct bands:
■ Anterior talofibular ligament. This runs forward from the lateral malleolus of the fibula to the talus
■ Calcanofibular ligament. This passes down from the tip of the

lateral malleolus to the side of the talus
■ Posterior talofibular ligament. This is a thick, stronger band which passes back from the lateral malleolus to the talus behind.

Injuries to the ankle

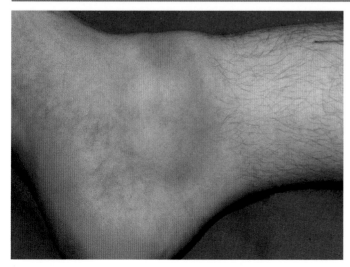

Injuries to the ankle are not uncommon; of all the major joints, it is the most likely to be damaged.

Most injuries to the ankle joint are sprains, in which one or more of the ligaments are stretched to such a degree that some of their fibres tear. A sprained ankle is a common occurrence in the sporting world. It is usually the result of sudden and unexpected twisting of the weight-bearing foot, in which

If the foot is twisted outwards, a Pott's fracture can occur. This causes the talus to be twisted, the fibula to fracture and the medial ligament to tear.

the foot becomes inverted (foot facing inwards). Ankle sprains most commonly affect the lateral ligament on the outer side of the joint, as it is by far the weakest ligament.

POTT'S FRACTURE

A Pott's fracture may occur when the foot is suddenly and forcibly twisted outwards. In this injury, the twisting of the talus causes the fibula to fracture, and the strong medial ligament is stretched until it tears away from the medial malleolus of the tibia. In extreme cases, the damage may be so great that the end of the tibia itself may be sheared off.

Bones of the foot

The human foot has 26 bones in total: seven larger,
irregular tarsal bones; five metatarsals running the length of the
foot; and 14 phalanges forming the skeleton of the toes.

The tarsal bones in the foot are equivalent to the carpal bones in the wrist, but there are seven tarsals as opposed to eight wrist bones. The tarsal bones differ somewhat from the wrist in terms of their arrangement, reflecting the different functions of the hand and the foot.

TARSAL BONES

The tarsal bones consist of:
■ The talus – articulates with the tibia and fibula at the ankle joint. It bears the full weight of the body, transferred down from the tibia. Its shape is such that it can then spread this weight by passing this force backwards and downwards, and forwards to the front of the foot
■ The calcaneus – the large heel bone
■ The navicular – a relatively small bone, named for its boat-like appearance. It has a projection, the navicular tuberosity which, if too large, may cause foot pain as it rubs against the shoe
■ The cuboid – a bone roughly the shape of a cube. It lies on the outer side of the foot, and has a groove on its under surface to allow passage of a muscle tendon
■ The three cuneiforms – bones named according to their positions: medial, intermediate and lateral. The medial cuneiform is the largest of these three wedge-shaped bones.

Tarsal bones

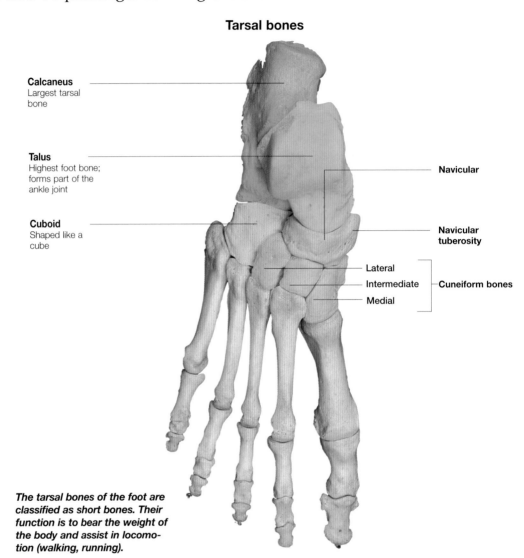

Calcaneus
Largest tarsal bone

Talus
Highest foot bone; forms part of the ankle joint

Cuboid
Shaped like a cube

Navicular

Navicular tuberosity

Lateral
Intermediate — **Cuneiform bones**
Medial

The tarsal bones of the foot are classified as short bones. Their function is to bear the weight of the body and assist in locomotion (walking, running).

Calcaneus (heel bone)

View from above the calcaneus bone

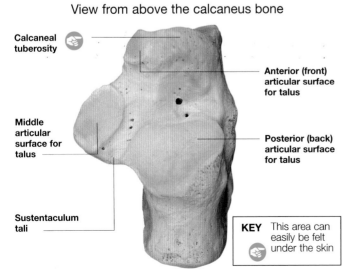

Calcaneal tuberosity

Middle articular surface for talus

Sustentaculum tali

Anterior (front) articular surface for talus

Posterior (back) articular surface for talus

KEY This area can easily be felt under the skin

The calcaneus is the largest bone in the foot and it can easily be felt under the skin as the prominence of the heel. It needs to be a large bone with great strength, as it has the important role of transmitting the weight of the body from the talus to the ground.

ARTICULAR SURFACES

This large, irregular bone has several articular surfaces where it forms joints with the talus above, and the cuboid in front.

The calcaneus bone has several articulating surfaces. These surfaces are where the calcaneus moves against the talus and cuboid bones.

The inner surface of the calcaneus bears a projection, the sustentaculum tali, which supports the head of the talus. On the underside of this projection is a groove for the passage of a long muscle tendon.

POSTERIOR SURFACE

The back of the calcaneus has a roughened prominence, the calcaneal tuberosity, the medial process of which comes into contact with the ground when standing.

Halfway up the posterior surface of the calcaneus is a ridge which indicates the site of attachment of the powerful Achilles tendon.

Metatarsals and phalanges

The metatarsals and phalanges in the foot are miniature long bones, consisting of a base, shaft and head.

Like the metacarpals in the hand, there are five metatarsals in the foot. While the individual bones tend to resemble the metacarpals in structure, their arrangement is slightly different. This is mainly due to the fact that the big toe lies in the same plane as the other toes and is not opposable like the thumb.

METATARSALS
Each metatarsal has a long shaft with two expanded ends, the base and the head. The bases of the metatarsals articulate with the tarsal bones in the middle of the foot. The heads articulate with the phalanges of the corresponding toes.

The metatarsals are numbered from 1 to 5 starting with the most medial, which lies behind the big toe. The first metatarsal is shorter and more sturdy than the rest. It articulates with the first phalanx of the big toe.

PHALANGES
The phalanges of the toe resemble the phalanges of the fingers. There are 14 phalanges in the foot, the big toe (hallux) having just two while the other four toes have three each.

The base of the first phalanx of each toe articulates with the head of the corresponding metatarsal. The phalanges of the big toe are thicker than those of the other toes.

Lateral view of the foot

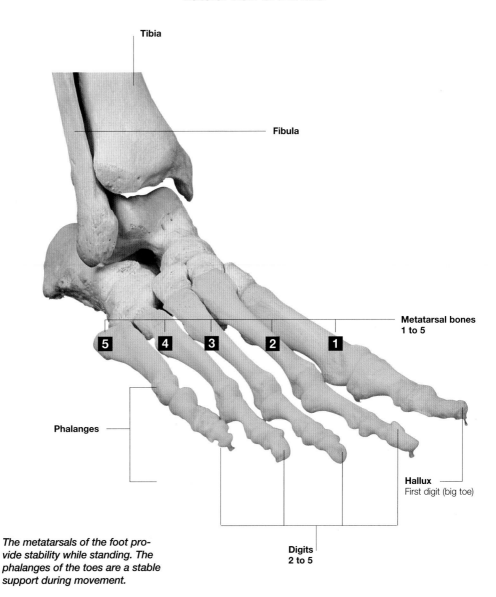

Tibia

Fibula

Metatarsal bones 1 to 5

5   4   3   2   1

Phalanges

Hallux
First digit (big toe)

Digits 2 to 5

The metatarsals of the foot provide stability while standing. The phalanges of the toes are a stable support during movement.

Sesamoid bones of the foot

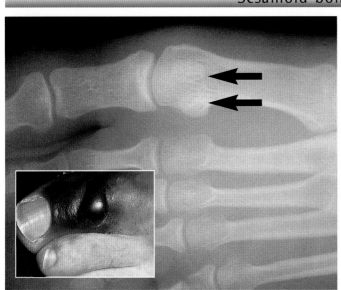

The foot is one of the sites in the body that has sesamoid bones.

PROTECTIVE ROLE
A sesamoid bone is one which develops within the tendon of a muscle to protect that tendon from wear and tear where it passes over the end of a long bone.

POSITION OF BONES
The two sesamoid bones in the foot lie under the head of the first metatarsal within the two heads of the flexor hallucis

The sesamoid bones of the foot can be seen in this X-ray (arrows). These tiny bones are often injured when objects are dropped on to the big toe (inset).

brevis muscle and bear the weight of the body, especially as the toe pushes off in walking.

Additional sesamoid bones may also be found elsewhere in the foot within other flexor tendons of the digits.

BONE DEVELOPMENT
The sesamoid bones develop before birth and gradually begin to ossify (become bony) during late childhood. Once ossified these bones can be seen clearly on an X-ray of the foot – the sesamoid bones are seen to overlap with the head of the first metatarsal bone.

These small bones may be damaged by a crushing injury to the foot, such as when a heavy weight falls on the big toe.

Ligaments and arches of the foot

The bones of the foot are arranged in such a way that they form bridge-like arches. These bones are supported by the presence of a number of strong ligaments.

The main supportive ligaments of the foot lie on the plantar (under) surface of the bones. The three most prominent ligaments are:

■ The plantar calcaneonavicular, or spring ligament – stretches forward from the sustentaculum tali, a projection of the calcaneus (heel bone), to the back of the navicular (boat-shaped) bone. This ligament is important in helping to maintain the longitudinal arch of the foot

■ The long plantar ligament – runs forward from the underside of the calcaneus to the cuboid (outer) bone and to the bases of the metatarsals (foot bones). It helps to maintain the arches of the foot

■ The plantar calcaneocuboid, or short plantar, ligament – lies under the long plantar ligament and runs from the front of the undersurface of the calcaneus forward to the cuboid.

OTHER LIGAMENTS

Many other ligaments support and bind together the long metatarsals and the phalanges (toe bones). The metatarsals are bound to the tarsals and to each other by ligaments running across the foot on both their dorsal and plantar surfaces.

Ligaments of foot (plantar view)

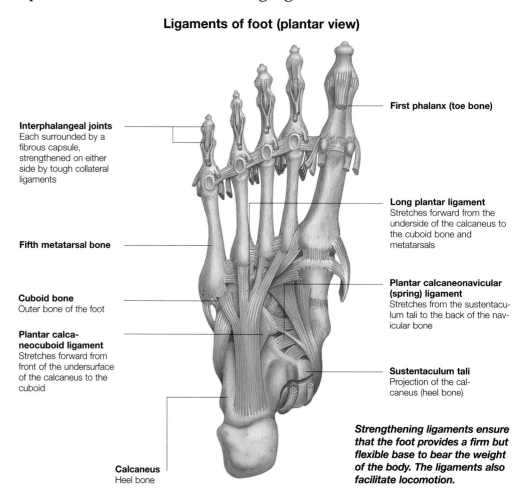

Interphalangeal joints
Each surrounded by a fibrous capsule, strengthened on either side by tough collateral ligaments

Fifth metatarsal bone

Cuboid bone
Outer bone of the foot

Plantar calca-neocuboid ligament
Stretches forward from front of the undersurface of the calcaneus to the cuboid

Calcaneus
Heel bone

First phalanx (toe bone)

Long plantar ligament
Stretches forward from the underside of the calcaneus to the cuboid bone and metatarsals

Plantar calcaneonavicular (spring) ligament
Stretches from the sustentaculum tali to the back of the navicular bone

Sustentaculum tali
Projection of the calcaneus (heel bone)

Strengthening ligaments ensure that the foot provides a firm but flexible base to bear the weight of the body. The ligaments also facilitate locomotion.

Joints of the foot

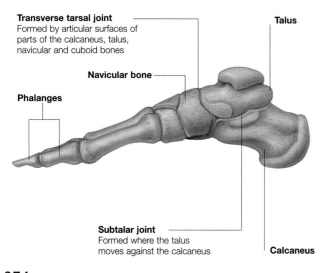

Transverse tarsal joint
Formed by articular surfaces of parts of the calcaneus, talus, navicular and cuboid bones

Navicular bone

Phalanges

Talus

Subtalar joint
Formed where the talus moves against the calcaneus

Calcaneus

The ankle joint allows the foot to move up and down only. Other movements of the foot, such as eversion, where it faces outwards, or inversion, where it faces inwards, take place further down the foot at two joints: the transverse tarsal and the subtalar joint:

■ Transverse tarsal joint. This complicated joint is formed by the adjoining articular surfaces of parts of the calcaneus, talus (ankle bone), navicular and cuboid. It is across this joint that

Joints between the bones of the foot allow movement between the hind- and forefoot. Such movements are necessary when walking on rough ground.

amputation of the foot is performed, when necessary

■ Subtalar joint. This is formed where the talus moves against the calcaneus.

OTHER JOINTS OF THE MID-FOOT

There are many other small synovial (fluid-filled) joints located within the foot where bone meets bone. However, these joints are generally held tightly together by tough ligaments and so little movement is possible.

The joints between the phalanges allow movement of the toes, although the range of movement is less than that for the fingers.

Arches of the foot

A distinctive feature of the human foot is that the bones within it are arranged in bridge-like arches. This allows the foot to be flexible enough to cope with uneven ground, while still being able to bear the weight of the body.

The arched shape of the foot can be illustrated by looking at a footprint. Only the heel, the outer edge of the foot, the pads under the metatarsal heads and the tips of the toes leave an impression. The rest of the foot is lifted away from the ground.

THREE ARCHES

The foot has two longitudinal arches (medial and lateral) running along its length, and a transverse arch lying across it:

■ Medial longitudinal arch. This is the higher and more important of the two longitudinal arches. The bones involved are the calcaneus, the talus, the navicular bone, the three cuneiform bones and the first three metatarsals. The head of the talus supports this arch

■ Lateral longitudinal arch. This arch is much lower and flatter, the bones resting on the ground when standing. The lateral arch is formed by the calcaneus, the cuboid and the fourth and fifth metatarsal bones

■ Transverse arch. This arch runs across the foot, supported on either side by the longitudinal arches, and is made up of the bases of the metatarsal bones, the cuboid and the three cuneiform bones.

Bones forming medial longitudinal arch of foot

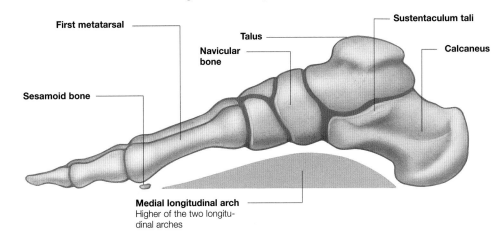

First metatarsal

Sesamoid bone

Navicular bone

Talus

Sustentaculum tali

Calcaneus

Medial longitudinal arch
Higher of the two longitudinal arches

Bones forming lateral longitudinal arch of foot

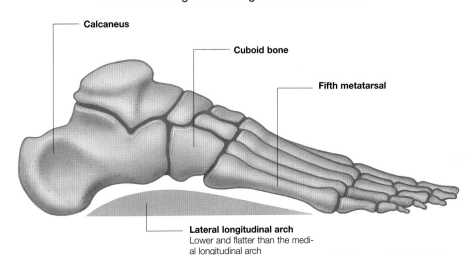

Calcaneus

Cuboid bone

Fifth metatarsal

Lateral longitudinal arch
Lower and flatter than the medial longitudinal arch

The bones of the foot form bridge-like arches. These are maintained by the shape of the bones and the strength of the ligaments and muscle tendons.

Weight bearing in the foot

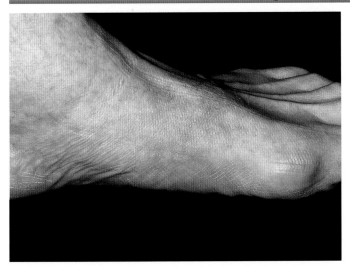

The weight of the body is transmitted down to the talus from the tibia (shin bone). The force is then passed down and backwards to the calcaneus and forwards to the heads of the second to fifth metatarsals and the tiny sesamoid bones underlying the first metatarsal. Between these points the weight is absorbed by the stretching of the 'elastic' longitudinal and transverse arches of the foot, which act as shock absorbers.

A person with flat feet has collapsed medial longitudinal arches, so that the sole lies flat upon the ground. Treatment is only required if pain is caused.

METATARSAL HEADS

It used to be thought that the body's weight was supported on a 'tripod' formed by the heel and the heads of the first and fifth metatarsals. It is now known that all the metatarsal heads are involved in weight bearing and, indeed, long marches may cause a 'stress' fracture of the head of the second metatarsal.

In the condition known as pes planus (flat feet), the medial longitudinal arches collapse until the head of the talus comes down between the navicular and calcaneus. The footprint of an affected person shows the whole foot to be in contact with the ground.

Muscles of the upper foot

Many of the muscles which move the foot lie in the lower leg, rather than in the foot itself. This allows them to be more powerful than if they were contained within the small space of the foot.

To have an effect upon the bones and joints of the foot, the leg muscles have long tendons. To reach the bones of the foot these tendons must first cross the ankle joint, where they are held in place by a series of retaining bands, or retinacula. If these bands were not present, the tendons would run straight to their attachments like a bowstring rather than following the contours of the ankle joint.

RETINACULA OF THE FOOT

There are four main retinacula in this area:

■ **Superior extensor retinaculum.** Lies just above the ankle joint and retains the long tendons of the extensor muscles

■ **Inferior extensor retinaculum.** Lies beneath the ankle joint. It also retains extensor muscles

■ **Peroneal retinaculum.** Lies on the outer side of the ankle. It is in two parts, upper and lower, and retains the long peroneal muscle tendons

■ **Flexor retinaculum.** Lies on the inner side of the ankle and retains the long flexor tendons as they pass under the medial malleolus to reach the sole of the foot.

The leg muscles have long tendons which connect to the foot bones like puppet strings. Retinacula are fibrous bands that hold tendons in place.

Lateral view

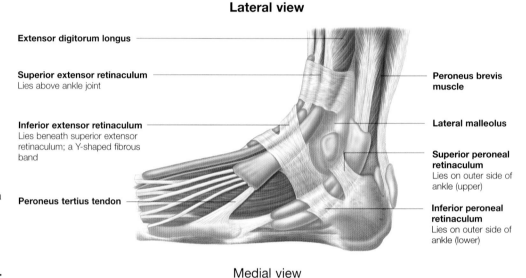

Extensor digitorum longus

Superior extensor retinaculum
Lies above ankle joint

Inferior extensor retinaculum
Lies beneath superior extensor retinaculum; a Y-shaped fibrous band

Peroneus tertius tendon

Peroneus brevis muscle

Lateral malleolus

Superior peroneal retinaculum
Lies on outer side of ankle (upper)

Inferior peroneal retinaculum
Lies on outer side of ankle (lower)

Medial view

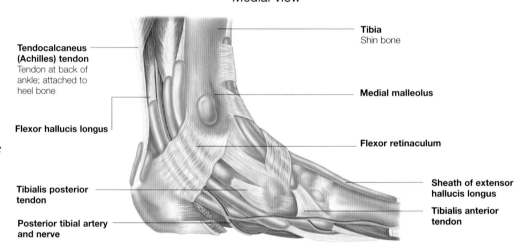

Tendocalcaneus (Achilles) tendon
Tendon at back of ankle; attached to heel bone

Flexor hallucis longus

Tibialis posterior tendon

Posterior tibial artery and nerve

Tibia
Shin bone

Medial malleolus

Flexor retinaculum

Sheath of extensor hallucis longus

Tibialis anterior tendon

Long tendons around the ankle joint

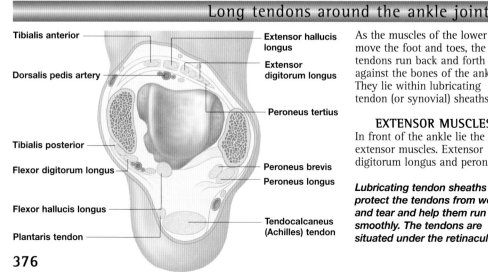

Tibialis anterior

Dorsalis pedis artery

Tibialis posterior

Flexor digitorum longus

Flexor hallucis longus

Plantaris tendon

Extensor hallucis longus

Extensor digitorum longus

Peroneus tertius

Peroneus brevis

Peroneus longus

Tendocalcaneus (Achilles) tendon

As the muscles of the lower leg move the foot and toes, the long tendons run back and forth against the bones of the ankle. They lie within lubricating tendon (or synovial) sheaths.

EXTENSOR MUSCLES

In front of the ankle lie the long extensor muscles. Extensor digitorum longus and peroneus

Lubricating tendon sheaths protect the tendons from wear and tear and help them run smoothly. The tendons are situated under the retinacula.

tertius share a common synovial sheath. The long flexor tendons, which bend the foot down or the toes under, lie behind the bony medial malleolus. Behind the lateral malleolus lie the long tendons of the peroneal muscles, while the tendocalcaneus tendon inserts into the heel bone.

Blood vessels and nerves must also cross the ankle joint. It is important for a doctor to know where these structures lie in relation to the ankle because it is a common site for fractures, sprains and dislocations, which may cause surrounding damage.

Muscles of the top of the foot

Although they are not particularly powerful, the muscles that lie over the top of the foot play an important part in helping to extend the toes. The extensor digitorum brevis muscle tends to be used when the foot is already pointing upwards.

Most of the muscles which lie within the foot, the intrinsic muscles, are in the sole. The top, or dorsal surface, of the foot has just two muscles: the extensor digitorum brevis and the extensor hallucis brevis.

MUSCLES OF THE DORSAL SURFACE

■ **Extensor digitorum brevis.** As its name suggests, this is a short muscle which extends (straightens or pulls upwards) the toes. It arises from the upper surface of the calcaneus, or heel bone, and the inferior extensor retinaculum. This muscle divides into three parts, each with a tendon that joins the corresponding long extensor tendon to insert into the second, third and fourth toes

■ **Extensor hallucis brevis.** This short muscle is really part of the extensor digitorum brevis. It runs down to insert into the big toe, or 'hallux', from which its name derives

ACTION OF THE MUSCLES

Together these two muscles assist the long extensor tendons in

extending the first four toes. Although they do not have a particularly powerful action, they are useful in extending the toes when the foot itself is already pointing up, or dorsiflexed, as in this position the long extensors are unable to act further.

Extensor digitorum longus muscle — Superior extensor retinaculum

Inferior extensor retinaculum

Extensor hallucis brevis

Peroneus tertius tendon — This lies over the extensor digitorum brevis

Extensor digitorum brevis — Extensor hallucis longus tendon

CLINICAL RELEVANCE

The top of the foot is one of the sites in the body where excess tissue fluid (oedema) may accumulate and be visible to an examining doctor. The position of the muscle bellies of these two short extensor muscles must

The muscles that lie at the top of the foot help to extend the toes. They assist the long extensors when the foot is dorsiflexed.

be known to prevent them being mistaken for such oedema.

Surface anatomy of the foot

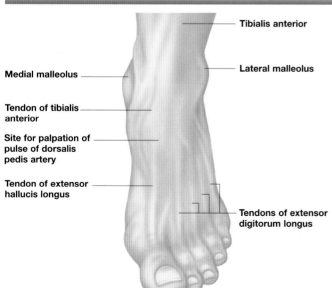

Tibialis anterior

Lateral malleolus

Medial malleolus

Tendon of tibialis anterior

Site for palpation of pulse of dorsalis pedis artery

Tendon of extensor hallucis longus

Tendons of extensor digitorum longus

Surface, or living, anatomy is the study of the live, intact body at rest and during movement. The foot is a good example as the relative lack of subcutaneous fat, together with the presence of numerous bony landmarks and prominent tendons, means there are many points of interest.

BONY LANDMARKS

The most obvious bony landmarks in the area are the medial (inner) and lateral (outer) malleoli, projections on either side of the ankle joint. In the foot itself, the most prominent

The skin is thinner on the top of the foot than on the sole of the foot. This means that the bony landmarks are easy to locate and to study.

landmark is the tuberosity of the navicular bone which can be felt on the inner aspect of the foot.

TENDONS

Many of the long tendons can be seen and felt as they pass across the ankle joint and along the foot. The most obvious are the extensor tendons on the top of the foot, which are prominent when the foot is dorsiflexed.

PULSES

An important landmark on top of the foot is the site where the pulse of the dorsalis pedis artery can be felt. This is usually midway between the two malleoli at the front of the ankle. A doctor might feel for a pulse here to check the circulation to the foot.

Muscles of the sole of the foot

Many of the movements of the bones and joints of the feet are brought about by muscles in the lower leg. However, there are also many small 'intrinsic' muscles which lie entirely within the foot.

The sole of the foot has four layers of intrinsic muscles, which work with the extrinsic muscles to meet the varying demands placed upon the foot during standing, walking, running and jumping. They also help to support the bony arches of the foot and to allow us to stand on sloping or uneven ground.

FIRST MUSCLE LAYER
The first layer of sole muscles is the most superficial, lying just under the thick plantar aponeurosis. The muscles of this layer include:

■ **Abductor hallucis** – this muscle lies along the medial (inner) border of the sole. It acts to abduct the big toe, or 'hallux', which means moving it away from the mid-line. It also flexes, or bends down, the big toe

■ **Flexor digitorum brevis** – this fleshy muscle lies down the centre of the sole and inserts into each of the lateral four toes. Contraction of this muscle causes those toes to flex

■ **Abductor digiti minimi** – lying along the lateral (outer) border of the sole within this first layer, this muscle acts to abduct and flex the little toe.

These muscles are similar to the corresponding muscles in the hand, but their individual function is less important because the toes do not have such a wide range of movement as the fingers.

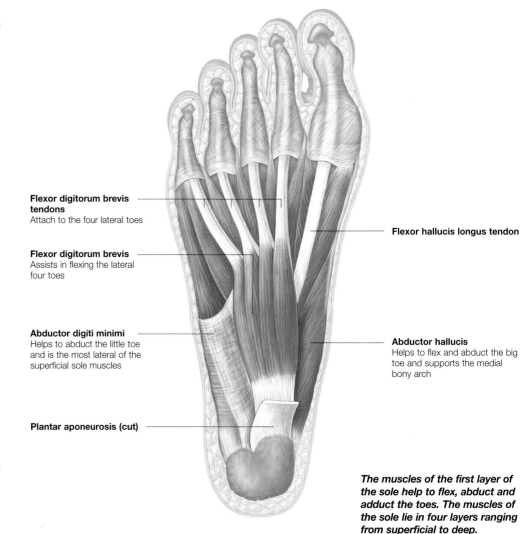

Flexor digitorum brevis tendons
Attach to the four lateral toes

Flexor digitorum brevis
Assists in flexing the lateral four toes

Abductor digiti minimi
Helps to abduct the little toe and is the most lateral of the superficial sole muscles

Plantar aponeurosis (cut)

Flexor hallucis longus tendon

Abductor hallucis
Helps to flex and abduct the big toe and supports the medial bony arch

The muscles of the first layer of the sole help to flex, abduct and adduct the toes. The muscles of the sole lie in four layers ranging from superficial to deep.

Plantar aponeurosis

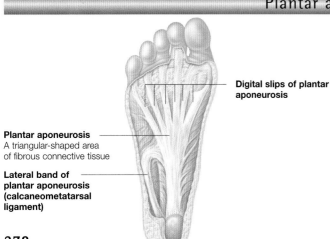

Digital slips of plantar aponeurosis

Plantar aponeurosis
A triangular-shaped area of fibrous connective tissue

Lateral band of plantar aponeurosis (calcaneometatarsal ligament)

The skin of the sole is thick and overlies a layer of shock-absorbing fat pads. Under this layer lies a sheet of tough connective tissue called the plantar aponeurosis.

The plantar aponeurosis is the thickened central portion of the plantar fascia, the connective tissue which surrounds and

The plantar aponeurosis is a strong sheet of connective tissue. 'Plantar' refers to the sole of the foot just as 'palmar' refers to the palm of the hand.

encloses the muscles of the sole. The plantar aponeurosis consists of bands of strong fibrous tissue that run the length of the sole and insert into each of the toes. It also attaches to the skin above it and to the deeper tissues that lie below.

ACTION
The plantar aponeurosis acts to hold together the parts of the foot and helps to protect the sole of the foot from injury. It also helps to support the bony arches of the foot.

Deeper muscle layers of the sole

The muscles of the sole of the foot are made up of four different layers; three of these layers lie under the top layer of the sole of the foot. All of these muscles act together, to help to keep the bony arches of the feet stable.

Beneath the superficial layer of intrinsic muscles of the sole lie three further layers. These all have a contribution to make to the stability and flexibility of the foot, both at rest and in motion.

Although the deep muscles each have individual actions, their main role is to act together to maintain the stability of the bony arches of the feet.

SECOND MUSCLE LAYER OF THE SOLE

The second muscle layer of the sole of the foot includes some tendons from the extrinsic muscles, as well as some smaller intrinsic muscles.

Muscles and tendons that are included within this second layer of the sole are:

■ **Quadratus plantae (or flexor accessorius) muscle –** this wide, rectangular muscle arises from two heads on either side of the heel.

It inserts into the edge of the tendon of flexor digitorum longus where it acts by pulling backwards on this tendon and so stabilizing it while it flexes the toes

■ **Tendons of flexor hallucis**

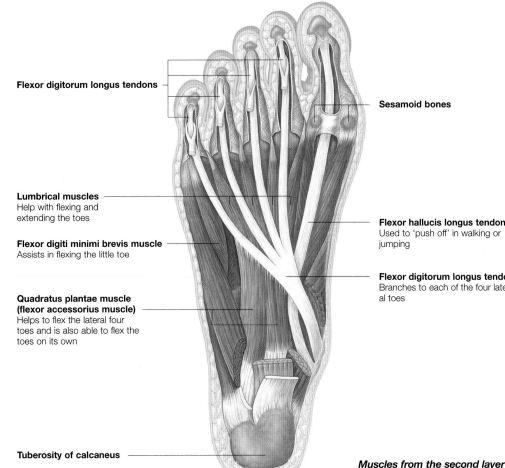

Flexor digitorum longus tendons

Sesamoid bones

Lumbrical muscles
Help with flexing and extending the toes

Flexor hallucis longus tendon
Used to 'push off' in walking or jumping

Flexor digiti minimi brevis muscle
Assists in flexing the little toe

Flexor digitorum longus tendon
Branches to each of the four lateral toes

Quadratus plantae muscle (flexor accessorius muscle)
Helps to flex the lateral four toes and is also able to flex the toes on its own

Tuberosity of calcaneus

Muscles from the second layer of the sole of the foot help to extend and flex the toes. They also help to stabilize the tendons during flexion of the toes.

longus and flexor digitorum longus – these tendons enter the second muscular layer of the sole after winding around the medial malleolus (inner 'ankle bone')

■ **The four lumbrical muscles –** named for their worm-like appearance, these four muscles arise from the tendons of flexor digitorum longus. They are similar to the lumbrical muscles

in the hand. These muscles act to extend (straighten) the toes while the long tendons are flexing them, which helps to prevent the toes 'buckling under' when walking or running.

Third and fourth muscle layers

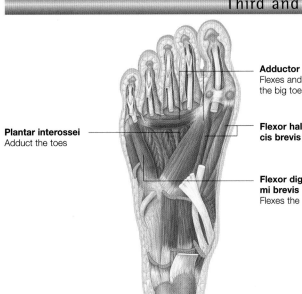

Plantar interossei
Adduct the toes

Adductor hallucis
Flexes and adducts the big toe

Flexor hallucis brevis

Flexor digiti minimi brevis
Flexes the little toe

Lying deep to the long flexor tendons, the third muscle layer of the sole is made up of three small muscles:

■ **Flexor hallucis brevis.** This is a short muscle, which flexes the big toe. It arises from the cuboid and lateral cuneiform bones and then splits into two parts. Each of these two parts has a tendon that inserts into the base of the big toe. The two sesamoid bones of the foot lie within these tendons

■ **Adductor hallucis.** This

The three small muscles in the deep muscle layer of the sole help to flex the toes. Even deeper, between the bones, lie the seven interossei muscles.

muscle arises from two heads; an oblique head and a transverse head. They join to insert into the base of the big toe

■ **Flexor digiti minimi brevis.** This small muscle runs along the outer border of the foot to the little toe, which it helps to flex.

FOURTH MUSCLE LAYER OF THE SOLE

The muscles of the fourth, and deepest layer of the sole are called the 'interossei' muscles which literally means 'between the bones'. Unlike the hand, which has eight, there are only seven interossei in the foot. The four dorsal interossei muscles (not shown) abduct the toes, whereas the three plantar interossei adduct them.

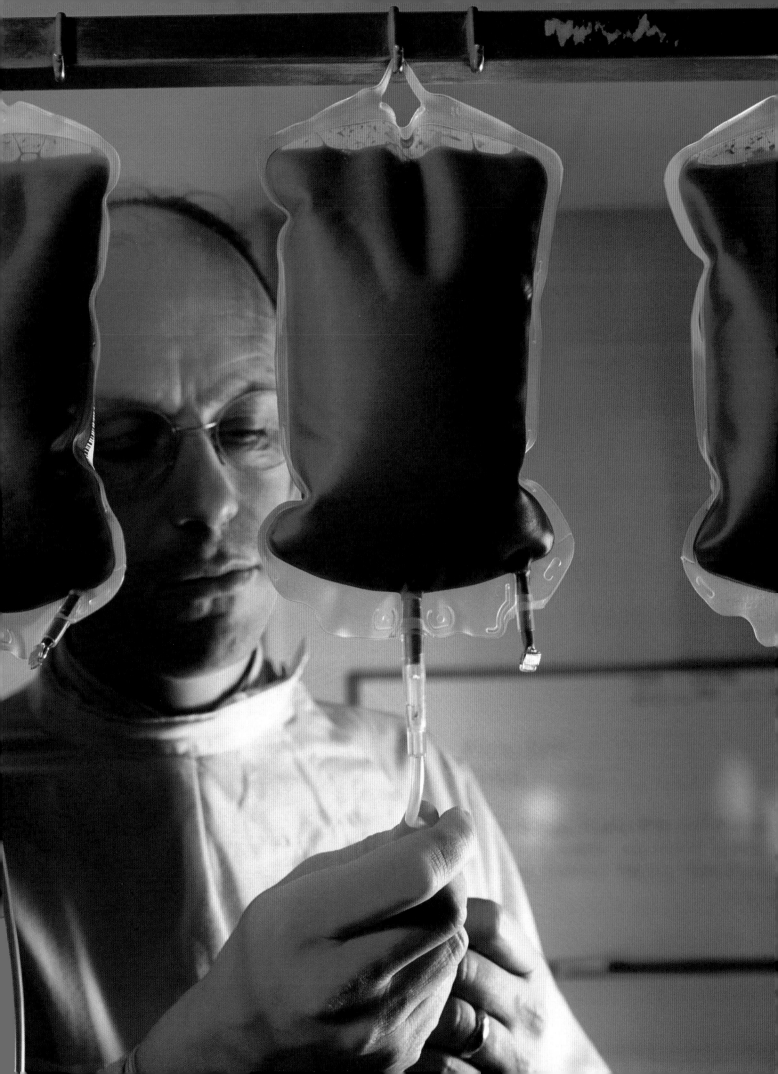

Chapter Nine

The Blood Stream

As blood circulates around the body, it provides a remarkably efficient transport system. All the elements essential for our survival – namely oxygen, vital nutrients and hormones – are carried via the blood to each organ and tissue, while harmful waste products are picked up and eliminated from the body.

Here you can explore the body's vast network of blood vessels and discover the important role of the component's of blood in protecting us from disease. This chapter also explores the body's built-in mechanisms for controlling blood pressure and enabling blood to clot, both of which ensure our survival.

LEFT: A laboratory technician checks blood bags stored in a blood bank.

Overview of blood circulation

There are two blood vessel networks in the body. The pulmonary circulation transports blood between the heart and lungs; the systemic circulation supplies blood to all parts except the lungs.

The blood circulatory system can be divided into two parts:
■ Systemic circulation – those vessels that carry blood to and from all the tissues of the body
■ Pulmonary circulation – the vessels that carry blood through the lungs to take up oxygen and release carbon dioxide.

SYSTEMIC ARTERIAL SYSTEM

The systemic arterial system carries blood away from the heart to nourish the tissues. Oxygenated blood from the lungs is first pumped into the aorta via the heart. Branches from the aorta pass to the upper limbs, head, trunk and the lower limbs in turn. These large branches give off smaller branches, which then divide again and again. The tiniest arteries (arterioles) feed blood into capillaries.

PULMONARY CIRCULATION

With each beat of the heart, blood is pumped from the right ventricle into the lungs through the pulmonary artery (this carries deoxygenated blood). After many arterial divisions, the blood flows through the capillaries of the alveoli (air sacs) of the lung to be reoxygenated. The blood eventually enters one of the four pulmonary veins. These pass to the left atrium, from where the blood is pumped through the heart to the systemic circulation.

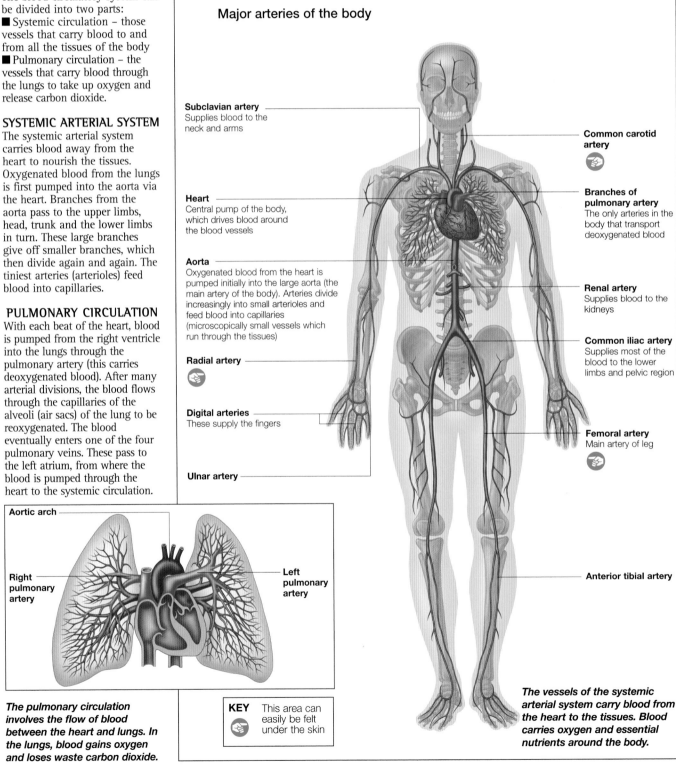

Major arteries of the body

Subclavian artery
Supplies blood to the neck and arms

Heart
Central pump of the body, which drives blood around the blood vessels

Aorta
Oxygenated blood from the heart is pumped initially into the large aorta (the main artery of the body). Arteries divide increasingly into small arterioles and feed blood into capillaries (microscopically small vessels which run through the tissues)

Radial artery

Digital arteries
These supply the fingers

Ulnar artery

Common carotid artery

Branches of pulmonary artery
The only arteries in the body that transport deoxygenated blood

Renal artery
Supplies blood to the kidneys

Common iliac artery
Supplies most of the blood to the lower limbs and pelvic region

Femoral artery
Main artery of leg

Anterior tibial artery

The vessels of the systemic arterial system carry blood from the heart to the tissues. Blood carries oxygen and essential nutrients around the body.

Aortic arch

Right pulmonary artery

Left pulmonary artery

The pulmonary circulation involves the flow of blood between the heart and lungs. In the lungs, blood gains oxygen and loses waste carbon dioxide.

KEY This area can easily be felt under the skin

The venous system

The systemic venous system carries blood back to the heart from the tissues. This blood is then pumped through the pulmonary circulation to be reoxygenated before entering the systemic circulation again.

Veins originate in tiny venules that receive blood from the capillaries. The veins converge upon one another, forming increasingly large vessels until the two main collecting veins of the body, the superior and inferior vena cavae, are formed. These then drain into the heart. At any one time, about 65 per cent of the total blood volume is contained in the venous system.

DIFFERENCES

The systemic venous system is similar in many ways to the arterial system. However, there are some important differences:

■ Vessel walls – arteries tend to have thicker walls than veins to cope with the greater pressure exerted by arterial blood.

■ Depth – most arteries lie deep within the body to protect them from injury, but many veins lie superficially, just under the skin.

■ Portal venous system – the blood that leaves the gut in the veins of the stomach and intestine does not pass directly back to the heart. It first passes into the hepatic portal venous system, which carries the blood through the liver tissues before it can return to the systemic circulation.

■ Variations – while the pattern of systemic arteries tends to be the same from person to person, there is far greater variability in the layout of the systemic veins.

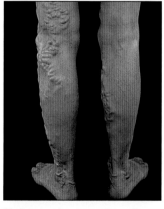

Varicose veins are enlarged or twisted superficial veins, those of the leg being most commonly affected. They are caused by defective valves in the veins.

Major veins of the body

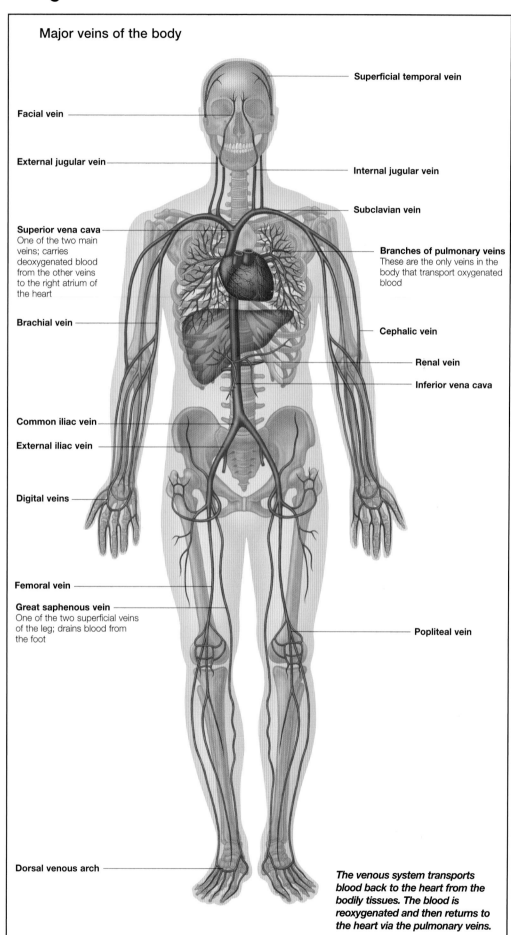

Superficial temporal vein

Facial vein

External jugular vein

Internal jugular vein

Subclavian vein

Superior vena cava
One of the two main veins; carries deoxygenated blood from the other veins to the right atrium of the heart

Branches of pulmonary veins
These are the only veins in the body that transport oxygenated blood

Brachial vein

Cephalic vein

Renal vein

Inferior vena cava

Common iliac vein

External iliac vein

Digital veins

Femoral vein

Great saphenous vein
One of the two superficial veins of the leg; drains blood from the foot

Popliteal vein

Dorsal venous arch

The venous system transports blood back to the heart from the bodily tissues. The blood is reoxygenated and then returns to the heart via the pulmonary veins.

The function of blood

Blood transports life-giving oxygen and all the vital nutrients which the cells of our bodies need in order to function. It also carries away the waste products which are produced by our tissues.

Blood makes up about eight per cent of the weight of the human body. The average adult man has around 5 litres (8.8 pints) of it, although the volume varies a great deal, depending mainly on the size of the person.

The blood volume of the average adult woman is about 4 litres (7 pints); a six-year-old child will have around 1.6 litres and a newborn baby will have only about 0.35 litres.

BLOOD CIRCULATION
Blood circulates inside a closed system of blood vessels, made up of arteries, capillaries and veins. This complex network transports blood to and from all tissues and organs of the body.

At any one time in the average man, the amount of blood in the various parts of the circulation is approximately as follows:
- Arteries   1,200 ml
- Capillaries   350 ml
- Veins   3,400 ml

Therefore, most of the blood in circulation is actually in our veins, and very little is in the capillaries.

The blood in the veins (venous blood, returning to the heart) is much darker in colour than arterial blood because it contains relatively little oxygen. Oxygenated blood from the heart, which is found in the arteries, is strikingly scarlet. Capillary blood – which we see when we cut ourselves – has a slightly less bright red colour than arterial blood.

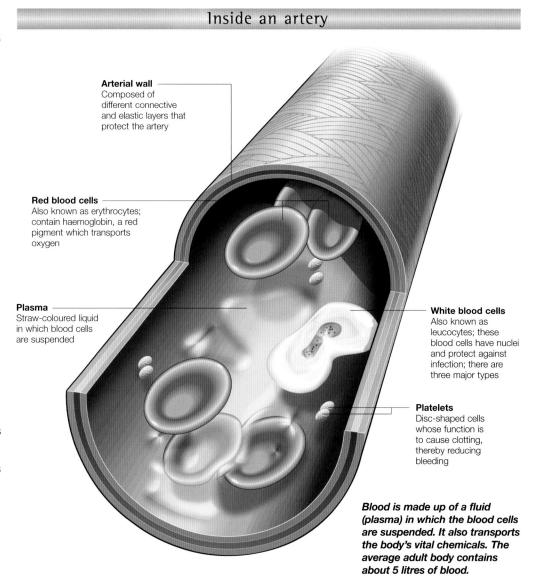

Inside an artery

Arterial wall
Composed of different connective and elastic layers that protect the artery

Red blood cells
Also known as erythrocytes; contain haemoglobin, a red pigment which transports oxygen

Plasma
Straw-coloured liquid in which blood cells are suspended

White blood cells
Also known as leucocytes; these blood cells have nuclei and protect against infection; there are three major types

Platelets
Disc-shaped cells whose function is to cause clotting, thereby reducing bleeding

Blood is made up of a fluid (plasma) in which the blood cells are suspended. It also transports the body's vital chemicals. The average adult body contains about 5 litres of blood.

How blood is made

Blood cells are mostly manufactured in the bone marrow – the soft tissue found in the centre of bones – and this process is called erythropoiesis. Some blood cells are also made in the spleen, a large organ located in the top left-hand corner of the abdomen.

In children, blood cells are mainly manufactured in the marrow of the long bones – the bones of the arms and legs. In adults they are mostly produced in the flatter bones of the body, such as those of the pelvis.

Blood production goes on at an astonishing rate. Literally billions of new red cells are turned out by bone marrow every 24 hours. The reason for this massive rate of manufacture is simply the fact that the cells of the blood are very rapidly worn out; the average red blood cell only lasts between 80 and 120 days, and approximately two million die every second.

This false-coloured micrograph shows immature red and white blood cells in bone marrow. All the cells derive from a single ancestral cell type by a process called haemopoiesis.

Components of blood

The blood that circulates around our bodies is not a single substance, but consists of several important ingredients. Suspended in plasma are red and white blood cells and platelets; each type of cell has a specific purpose.

Blood is made up of various cells suspended in a pale yellow liquid called plasma. Plasma is a sticky fluid containing various chemicals which are in transit from one part of the body to another. Its constituents include:

■ Proteins   7 per cent
■ Salt   0.9 per cent
■ Glucose   0.1 per cent

The main proteins in blood plasma are called albumin,

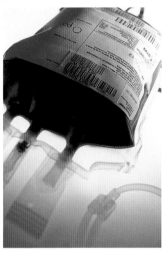

Donated blood can be used whole for transfusions during surgery or after trauma. Sometimes the red cells can be separated out and concentrated.

globulin and fibrinogen. They help to supply nutrition to tissues of the body, and are also important in protecting against infectious diseases. Fibrinogen plays a vital role in blood clotting – it turns into fibrin, a mesh-like material, which helps to stop bleeding after an injury.

Glucose – a form of sugar – is the body's principal fuel, and salt is the body's most important mineral. Its presence is the reason why blood tastes salty.

RED BLOOD CELLS

There are three types of cells in blood: red cells, white cells and platelets. Red cells (also known as red corpuscles or erythrocytes) are by far the commonest cells in the blood. Red cells contain the pigment haemoglobin. This is the iron-containing chemical which takes up oxygen in the lungs.

Major blood elements

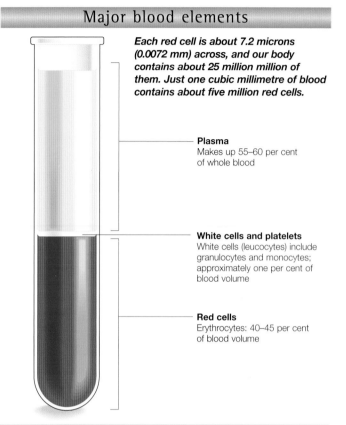

Each red cell is about 7.2 microns (0.0072 mm) across, and our body contains about 25 million million of them. Just one cubic millimetre of blood contains about five million red cells.

Plasma
Makes up 55–60 per cent of whole blood

White cells and platelets
White cells (leucocytes) include granulocytes and monocytes; approximately one per cent of blood volume

Red cells
Erythrocytes: 40–45 per cent of blood volume

White cells and platelets

White cells (also known as white corpuscles or leucocytes) are far fewer in number than red cells. Children have about 10,000 of them in a cubic millimetre of blood, but adults have much less than this.

White cells are vital in protecting against disease. They are divided into various types:
■ Neutrophils: combat bacterial and fungal infection
■ Eosinophils: help defend the body against parasites, and also

in allergic reactions
■ Lymphocytes: involved in creating immunity to infection
■ Monocytes: capable of engulfing invading particles in the bloodstream
■ Basophils: can also engulf invaders, but little is known about them.

Platelets (also known as thrombocytes) are very small cells involved in the process of blood clotting. In a cubic millimetre of blood, there are

about a quarter of a million of them. When a blood vessel is cut or damaged, platelets – which are very sticky – immediately adhere to the injured spot, and to each other, and so (along with fibrin) help to plug the gap and stop the bleeding.

This white blood cell is a T-lymphocyte, or T-cell, covered by characteristic microvilli (hair-like structures). These cells are important for immunity.

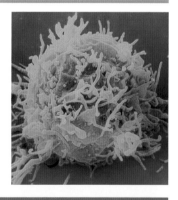

What happens when we bleed

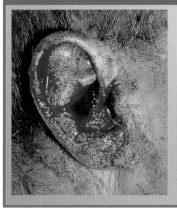

If our skin is cut, we immediately start bleeding. Most cuts are small, and only involve slight loss from the capillaries. The blood flow soon stops, particularly if firm pressure is applied to the wound.

The main reason why the bleeding ceases is the blood's

Bleeding from the ear may be serious, as it can indicate brain trauma; alternatively, it may be a superficial cut to the ear tissue. A doctor would have to assess the wound and treat accordingly.

natural ability to clot. Threads of a substance called fibrin form a mesh-like plug at the bleeding point, and this helps stop the blood loss.

If a wound involves a vein or an artery, however, it is rather more serious. Veins are quite large tubes – as can be seen just under the skin – and if they are sliced through, they tend to ooze quite large quantities of blood over a long period. Pressure on the spot may stop this, but surgical stitching may be needed.

Even more serious is a cut through an artery, because very large amounts of blood can pump forcibly out of it in a very short time. If firm pressure is not applied promptly, the person can bleed to death in a few minutes.

The reason why heavy blood loss can rapidly cause death is that the body – and particularly the brain – needs a constant supply of blood to function. If there is insufficient blood, then there is insufficient oxygen, and as a result our cells soon die.

How the blood circulates

The circulation transports blood to and from every tissue in the body, maintaining an optimal environment for cell survival and function. It also allows the transport of hormones around the body.

The function of the circulation is to supply blood to every bodily tissue, carrying fuel, nutrition and oxygen to the cells. It also carries waste products away from the tissues, transporting them to the kidneys or lungs for excretion.

Circulation is achieved by the heart pumping blood forcefully, in a series of 'jets', through the arterial system. The arteries divide into increasingly smaller branches, and the smallest arteries (arterioles) deliver blood into microscopic capillaries. The capillaries pass through the tissues and anastomose (join) with the smallest veins (venules).

The venules join up to form veins, which take blood back to the heart again. The blood on returning to the heart is then pumped to the lungs to be re-oxygenated.

The arteries and veins are linked by a meshwork of capillaries. Over 150,000 km long, this network allows the exchange of oxygen and nutrients between the arterial and venous systems.

The circulatory system is the branching network of blood vessels. The arteries carry oxygenated blood (red) to the tissues, and the veins return the de-oxygenated blood (blue) to

Circulatory system

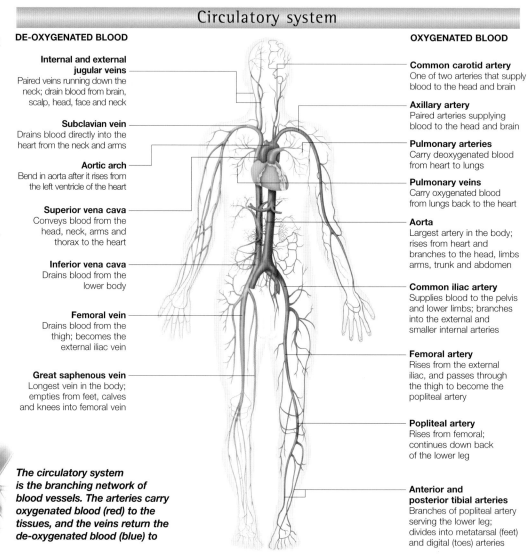

DE-OXYGENATED BLOOD

Internal and external jugular veins
Paired veins running down the neck; drain blood from brain, scalp, head, face and neck

Subclavian vein
Drains blood directly into the heart from the neck and arms

Aortic arch
Bend in aorta after it rises from the left ventricle of the heart

Superior vena cava
Conveys blood from the head, neck, arms and thorax to the heart

Inferior vena cava
Drains blood from the lower body

Femoral vein
Drains blood from the thigh; becomes the external iliac vein

Great saphenous vein
Longest vein in the body; empties from feet, calves and knees into femoral vein

OXYGENATED BLOOD

Common carotid artery
One of two arteries that supply blood to the head and brain

Axillary artery
Paired arteries supplying blood to the head and brain

Pulmonary arteries
Carry deoxygenated blood from heart to lungs

Pulmonary veins
Carry oxygenated blood from lungs back to the heart

Aorta
Largest artery in the body; rises from heart and branches to the head, limbs arms, trunk and abdomen

Common iliac artery
Supplies blood to the pelvis and lower limbs; branches into the external and smaller internal arteries

Femoral artery
Rises from the external iliac, and passes through the thigh to become the popliteal artery

Popliteal artery
Rises from femoral; continues down back of the lower leg

Anterior and posterior tibial arteries
Branches of popliteal artery serving the lower leg; divides into metatarsal (feet) and digital (toes) arteries

Blood pressure

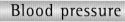

Blood pressure is force per unit area exerted by the blood in the arterial system. It is measured in millimetres of mercury (mm Hg – UK and USA), or kilopascals (other European countries).

Blood pressure is expressed as two figures, for instance 150/110. The first, or upper figure represents the pressure in the arteries when the heart is contracting (systole) – the systolic pressure. The second, or lower figure represents the

pressure in the arteries while the heart is relaxing (diastole) – the diastolic pressure.

The diastolic pressure is often considered to be more clinically important, especially when assessing high blood pressure – because the systolic pressure is so readily affected by factors such as anxiety. Blood pressure is generally measured by placing an inflatable cuff, connected to a measuring device, around the upper part of the arm.

Hypertension (high blood pressure) affects millions of people; in most cases, its cause is unknown. But it is important to detect and to treat, because its presence increases the risk of heart attack or stroke.

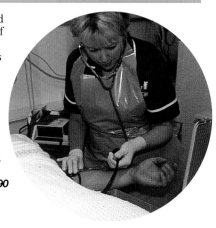

A doctor takes a blood pressure reading in his patient's upper arm. Ideally, pressure in this part of the body should be below 140/90 mm Hg.

Blood flow through the body

Blood flow is the volume of blood flowing through the circulation, an organ of the body or an individual blood vessel in a given period of time.

The flow of blood through a blood vessel is determined by a combination of the pressure difference between the two ends of the vessel and the resistance to blood flow through the vessel.

Blood pressure is greatest in the vessels nearest the pump, that is, in the aorta and pulmonary artery. As the blood flows away from the heart, the pressure falls. However, of the two parameters – pressure and resistance – it is resistance that has the greater influence on

blood flow. The total blood flow in the the circulation of an adult at rest is about five litres per minute; this is referred to as the cardiac output.

The blood flow to individual tissues is almost precisely controlled in relation to the tissue's needs. When tissues are active, they may require up to 20 or 30 times more blood flow than when they are at rest. However, cardiac output cannot increase more than about four to seven times.

Since the body cannot simply increase total blood flow, local blood flow to specific tissues is control by internal monitoring mechanisms. Blood is distributed according to the specific tissues' needs, and redirected away from tissues that do not require nutrients or oxygen at that time.

VENOUS BLOOD FLOW
The 'drive' produced by the heartbeat is not carried on through the tiny capillaries. Therefore, there is no pulse in the veins. However, blood flows back through veins towards the heart by a combination of mechanisms: the contraction of the leg and arm muscles; the presence of efficient valves in the veins; and the simple process of breathing, which helps to 'suck' the blood through the veins, towards the chest.

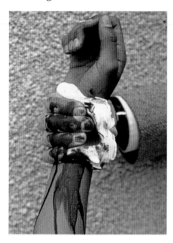

When an artery is cut, blood spurts out of the wound because aterial blood is pumped under pressure. Venous blood, however, is not pressurized and so flows out more slowly.

Distribution of blood

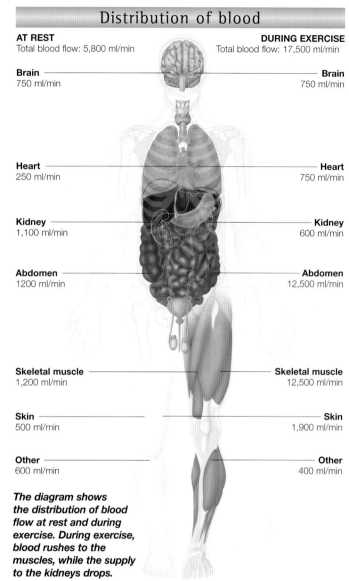

AT REST
Total blood flow: 5,800 ml/min

DURING EXERCISE
Total blood flow: 17,500 ml/min

	AT REST	DURING EXERCISE
Brain	750 ml/min	750 ml/min
Heart	250 ml/min	750 ml/min
Kidney	1,100 ml/min	600 ml/min
Abdomen	1200 ml/min	12,500 ml/min
Skeletal muscle	1,200 ml/min	12,500 ml/min
Skin	500 ml/min	1,900 ml/min
Other	600 ml/min	400 ml/min

The diagram shows the distribution of blood flow at rest and during exercise. During exercise, blood rushes to the muscles, while the supply to the kidneys drops.

Distribution of blood volume

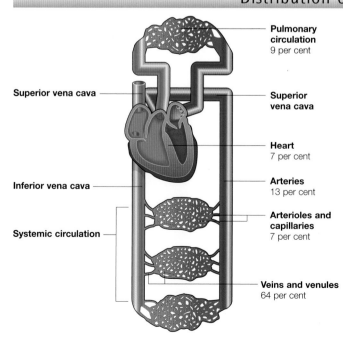

- **Pulmonary circulation** 9 per cent
- **Superior vena cava**
- **Heart** 7 per cent
- **Arteries** 13 per cent
- **Arterioles and capillaries** 7 per cent
- **Veins and venules** 64 per cent

Superior vena cava

Inferior vena cava

Systemic circulation

The circulatory system can be divided into two main portions: the pulmonary (lungs) and systemic (entire body). The diagram shows how blood is distributed around these areas.

The circulation moves blood around the body in two networks, starting and finishing with the heart.

SYSTEMIC CIRCULATION
The systemic circulation contains the greatest proportion of the circulating blood volume – about 84 per cent. However, only 7 per cent of the blood volume is in the capillary beds, where the essential exchange of cell nutrients and waste products take place. In these tiny blood vessels, the blood is in intimate contact with the tissues for the first time. The capillaries have

permeable walls, allowing chemical molecules to pass out of the blood into the tissues. Similarly, chemicals that have been formed in the tissues can diffuse through the capillary walls into the blood so that they can be carried away.

PULMONARY CIRCULATION
The pulmonary circulation allows the discharge of waste products from the blood into the lungs, and the uptake of oxygen from the air. Blood returning from the major veins of the body to the right side of the heart is pumped out again via the pulmonary artery to the lungs. Here, the artery divides into tiny arterioles and then capillaries, which traverse the tissues of the lung. The pulmonary veins then take the oxygen-rich blood back to the heart.

How blood is transported

Blood vessels are the tubes that carry blood around the body.
Arteries carry blood from the heart to the body's tissues. From there
veins carry the deoxygenated blood back to the heart.

TYPES OF BLOOD VESSEL

Blood vessels vary in size according to the amount of blood they carry; thus the largest vessels are found nearest the heart. Blood destined for body tissues leaves the heart via the aorta, which arches over and behind the heart and carries blood down the trunk. From the aorta, smaller arteries lead to the main organs of the body, where they branch into smaller vessels.

The smallest arteries, or arterioles, deliver blood to the capillaries, from which oxygen and nutrients are absorbed into the tissues, and into which carbon dioxide and waste materials are taken up. Blood leaving the tissues collects into veins, which feed blood into larger and larger vessels, the largest of which, the two vena cavae, deliver blood back to the heart. From the heart, the blood is pumped to the lungs where it is reoxygenated for circulation.

Structure of a typical artery

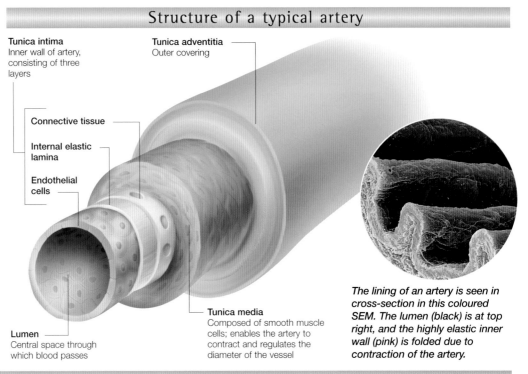

Tunica intima
Inner wall of artery, consisting of three layers

Tunica adventitia
Outer covering

Connective tissue

Internal elastic lamina

Endothelial cells

Tunica media
Composed of smooth muscle cells; enables the artery to contract and regulates the diameter of the vessel

Lumen
Central space through which blood passes

The lining of an artery is seen in cross-section in this coloured SEM. The lumen (black) is at top right, and the highly elastic inner wall (pink) is folded due to contraction of the artery.

Arteries and arterioles

Blood leaves the heart under pressure, so arteries have thick, muscular walls made up of several layers (tunicae). Surrounding the central canal (lumen) is the tunica intima which consists of a lining of endothelial cells, a layer of connective tissue and a layer of tissue called the internal elastic lamina. The middle layer (tunica media) is made up of smooth muscle cells and sheets of elastic tissue known as elastin. The outer layer (tunica adventitia) is a tough outer coat of fibrous connective tissue.

The largest arteries lead directly from the heart. They are known as elastic, or conducting, arteries because they contain a relatively high proportion of elastic tissue. This allows them to expand as they fill with blood and then contract again, forcing the blood onwards towards the smaller arteries.

ARTERIOLES

Arteries with a diameter of between 0.3 mm and 0.01 mm are called arterioles. The largest of these possess all three tunicae, but the tunica media contains only scattered elastic fibres. The smallest have no outer coat and consist only of an endothelial lining surrounded by a single layer of spiralling muscle cells. The flow of blood from arterioles into capillaries is controlled by sympathetic nerves, which cause the muscle cells to contract, thus constricting or dilating the lumen of the arterioles.

Pulse

When the heart beats, the impact of the blood being forced into the aorta from the left ventricle causes a pressure wave to travel down all the arteries of the body. Where an artery lies close to the skin, this pressure wave can be felt as a pulse. The easiest points at which a pulse can be felt are at the radial artery in the wrist and the common carotid artery in the neck.

Doctors usually feel for a pulse at the patient's wrist. The pulse corresponds to the heart rate, and the average for a healthy adult at rest is 60–80 beats per minute.

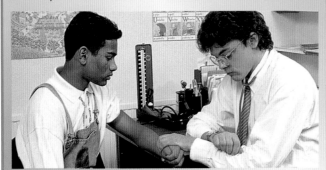

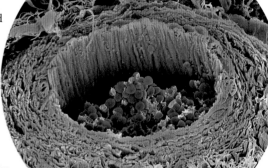

Red blood cells are visible travelling through the lumen (centre) of this arteriole. The vessel is surrounded by connective tissue (yellow).

Veins and capillaries

Veins are the vessels that move deoxygenated blood from around the body to the heart. The capillaries make up the network between veins and arteries in all the tissues.

Veins

The structure of veins is very similar to that of arteries, but veins are generally larger and have thinner walls, and contain less muscle and less elastic and collagenous tissue, so they can be compressed or distended. Venules – the smallest veins – collect blood from the capillaries, and it is then collected into increasingly larger veins. Blood from the lower body arrives at the inferior vena cava and drains into the right atrium of the heart. Blood from the upper body is collected by the superior vena cava and also drains into the right atrium.

Most veins have a system of one-way valves that allow blood to flow in one direction only. The valves are semi-lunar, formed of two half-circles of tissue, and are prevalent in veins of the lower limbs.

The blood pressure in veins is low. Movement of the blood is helped by the skeletal muscle pump, in which the contraction of surrounding skeletal muscles squeezes the vein and forces the blood along. In veins of less than one mm diameter and in regions where muscular activity is more or less continuous, such as the chest and abdominal cavities, there are no valves, as the blood flow is maintained by muscle pressure alone.

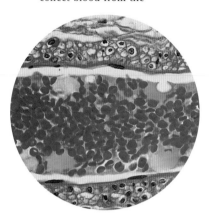

Erythrocytes (red blood cells) can be seen in the lumen of this vein. They contain haemoglobin, which transports oxygen.

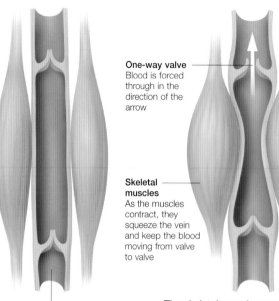

Muscles relaxed

Muscles contracted

One-way valve
Blood is forced through in the direction of the arrow

Skeletal muscles
As the muscles contract, they squeeze the vein and keep the blood moving from valve to valve

Vein
Each vein is divided into segments by non-return valves to prevent the blood flowing backwards

The skeletal muscle pump moves blood through the veins back to the heart. Muscles contract against the flexible vein, forcing the valves to open.

Types of capillary

There are at least three different kinds of capillary:
■ Continuous capillaries are made up of a single long endothelial cell curved round to form a tube
■ Fenestrated capillaries are made up of two or more endothelial cells that have a number of pores (fenestrations), especially near the junctions of the cells
■ Discontinuous capillaries, also called sinusoids or vascular sinuses, are made up of a number of cells with large fenestrations.

Continuous capillaries are the least permeable, and liquids are transferred to and from the surrounding tissues by exocytosis and endocytosis, processes by which vesicles containing the liquids are moved across the endothelial cells.

In fenestrated capillaries and sinusoids, chemicals pass more easily through the thin membranes that cover the pores. Fenestrated capillaries are common in the endocrine glands and kidneys; sinusoids are found in the liver and spleen.

Structure of a fenestrated capillary

Endothelial cell
Inner wall of capillary is just one cell thick

Lumen
Wide enough to allow red blood cells to pass singly

Nucleus of endothelial cell

Intercellular cleft
Allows passage of fluids

Fenestrations
Pores in the cells which allow rapid transfer of materials into the tissues

Basement membrane
Surrounds endothelial layer

Fainting

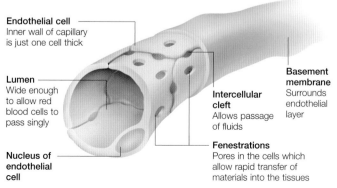

Fainting (syncope) is a temporary loss of consciousness due to a decrease in the supply of blood to the brain. This can be caused by a stuffy atmosphere, standing up suddenly, standing for a long time in one position, an obstruction to the neck arteries on moving the head suddenly or by an emotional reaction to shock. It can also be caused by poor output from the heart, due to a heart attack, arrhythmia (disturbance of the heartbeat) or disease of the heart valves.

Fainting can affect anyone at any age, irrespective of health or fitness, but it is more common in the elderly. Before losing consciousness, the victim may complain of light-headedness and nausea, and their skin may be pale and clammy to the touch.

Faintness from standing for long periods is due to blood collecting in the legs. Blood flow can be restored by flexing the leg muscles.

389

How blood clots

Blood makes a complete circuit of the body every minute, and injury to the vascular bed must therefore be plugged quickly in order to prevent excessive blood loss. This process is called haemostasis.

Blood flows freely in intact blood vessels due partly to an excess of naturally occurring anticoagulants. However, if the blood vessel wall breaks, a series of chemical reactions are initiated to stop the bleeding (haemostasis). Without these haemostatic processes, even the smallest cut could cause a person to bleed to death.

Haemostasis involves many blood coagulation factors, which are present in the plasma, as well as chemicals released from platelets and injured cells.

THE STAGES OF HAEMOSTASIS

Haemostasis can be broken down into three main stages, which occur in rapid succession after an injury:
■ Vasoconstriction – the first stage involves the constriction of the damaged blood vessel; this can significantly reduce blood loss in the short term
■ Platelet plug formation – damage to the blood vessel causes platelets, which are present within the plasma, to become sticky and adhere to each other and to the damaged vessel wall
■ Coagulation (blood clotting) – next, the platelet plug is reinforced with a meshwork of fibrin fibres. This fibrin net traps red and white blood cells to form a secondary haemostatic plug, or blood clot.

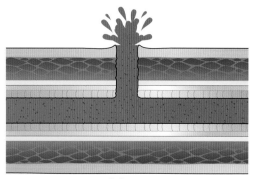
Injury
When a blood vessel is damaged, blood escapes the circulation, reducing blood volume. Excessive blood loss is prevented by haemostasis.

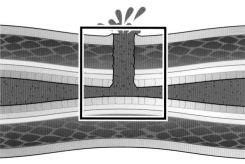
Stage 1
The first stage of haemostasis is vasoconstriction; the damaged blood vessel constricts to reduce the amount of blood flowing through it.

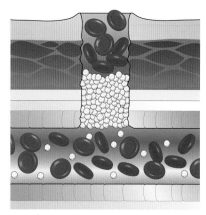
Stage 2
The second stage is the formation of a platelet plug. Platelets (white) stick to one another to temporarily seal the hole in the vessel wall.

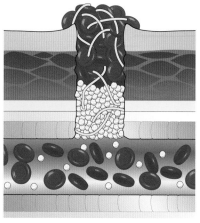
Stage 3
Finally, a blood clot is formed; blood cells are trapped in a fibrin mesh (yellow strands) which seals the hole until it can be permanently repaired.

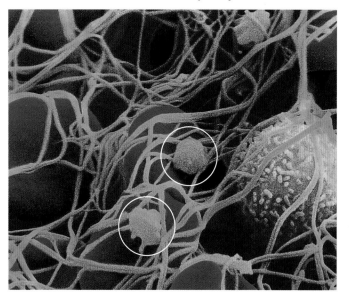

A web of fibrin strands can be seen trapping red blood cells during clot formation. This micrograph also reveals a white blood cell (yellow) and platelets (circled) within the clot.

How blood clots form

The formation of a blood clot is a very complicated process involving over 30 different chemicals. Some of these chemicals, called coagulation factors, enhance clot formation, whereas others, called anticoagulants, inhibit clotting.

Clotting is initiated by a complex cascade of biochemical reactions involving 13 coagulation factors. The end result is the formation of a complex chemical called prothrombin activator. This compound catalyses the conversion of a plasma protein called prothrombin into a smaller protein called thrombin. Thrombin, in turn, catalyses the joining together of fibrinogen molecules present in the plasma to produce a fibrin mesh. It is this mesh that traps blood cells in the hole in the blood vessel wall.

The large number of chemical steps involved in the clotting process means that coagulation must be tightly controlled. This is important because unnecessary clotting can be very dangerous, especially if it blocks a blood vessel supplying a major organ.

Clot contraction and repair

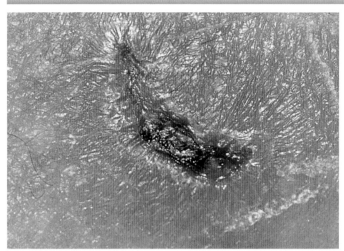

About 30–60 minutes after a blood clot has formed, the platelets within the clot contract – like muscle, platelets contain two contractile proteins called actin and myosin. This contraction pulls on the fibrin strands, bringing the edges of the injured tissue closer together and helping the wound to close.

The blood clot is temporary; at the same time as the clot is contracting, surrounding tissues divide to repair the vessel wall.

A healing dog bite on the shin can be seen here. Scar tissue is forming at either end of the laceration.

FIBRINOLYSIS
Once the tissue has healed (after about two days), the fibrin mesh which holds the clot together is dissolved. This process, called fibrinolysis, is catalysed by the enzyme plasmin, which is produced from the plasma protein plasminogen.

Plasminogen molecules are incorporated into the blood clot during its formation, where they lie dormant until activated by the healing process. As a result, most of the plasmin is restricted to the clot.

Normally, a balance between coagulation and fibrinolysis is maintained in the body.

Platelets

Platelets are cytoplasmic fragments that are able to survive in the circulation for up to 10 days. They are formed in bone marrow by extremely large cells called megakaryocytes. Strictly speaking, they are not cells as they do not have a nucleus and so cannot divide.

Electron microscopy reveals three platelet zones:
1 The outer membrane consists of a glycoprotein surface coat which causes it to adhere only to injured tissues. The membrane also contains large numbers of phospholipids which play a number of roles in the blood clotting process.
2 The cytosol (solution inside the cellular membrane) contains contractile proteins (including actin and myosin), microfilaments and microtubules. These are important for clot contraction.

3 Platelet granules contain a variety of haemostatically active compounds which are released when platelets are activated. These compounds are potent aggregating agents that attract more platelets to the wound site. Thus, the formation of the platelet plug is a self-perpetuating process.

This electron micrograph shows activated blood platelets grouping on the surface of a damaged blood vessel wall.

A single activated blood platelet is seen in this micrograph. In this activated state, platelets develop extensions (pseudopodia) from the cell wall, which are also visible in this image.

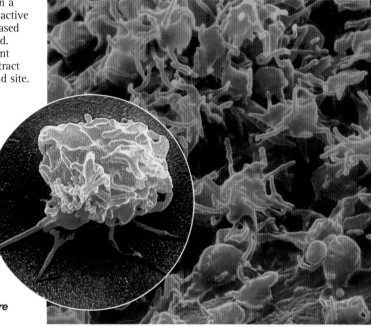

Anticoagulant drugs

The main clinical use of anti-coagulants is to prevent the formation of a blood clot (thrombus) in an undamaged blood vessel. A large clot could potentially block a blood vessel, leading to the death of the tissues that it supplies.

CLINICAL USES
Anticoagulants, such as heparin, are given by intravenous injection (parenterally), while others drugs, such as warfarin, are administered orally. The two types of drugs have different modes of action; whereas warfarin takes 48 to 72 hours to take effect, the effect of heparin is immediate.

Heparin is the anticoagulant most often used clinically, particularly for cardiac operations and for patients receiving blood transfusions. Warfarin is predominantly used in patients at risk of suffering arrhythmia (irregular heart rate).

Aspirin blocks platelet aggregation and platelet plug formation. A dose of 75-150 mg per day is used in the secondary prevention of thrombotic cerebrovascular (stroke) or cardiovascular disease.

Warfarin has been widely used as a rat poison. Rats who eat food laced with warfarin die from blood loss because their blood is unable to coagulate.

Haemophilia

Haemophilia is a group of inherited bleeding disorders caused by a lack of one of the clotting factors. The most common (85 per cent of cases) is haemophilia A, which is caused by a deficiency of clotting factor VIII. The disease is characterized by painful spontaneous bleeding into joints and muscles. The most famous case of haemophilia A is Queen Victoria's family, where many of the males fell victim to the disease.

The condition is treated by replacing the missing factor obtained from human plasma. Genetically engineered versions are also available for patients who cannot produce factors VIII or IX (which results in Christmas disease).

How blood protects us from disease

As well as carrying nutrients to and waste products from all tissues of the body, blood contains components that are a vital part of the human immune response to infection.

The blood is the great defensive fluid of our bodies. It is constantly present in the circulatory (cardiovascular) system, ready to respond to any microbial threat which may present.

BONE MARROW

All blood cells begin life in the bone marrow – the jelly-like substance contained within the cavities of bones. All types of blood cells are derived from a single type of cell called a stem cell, which may go on to form red blood cells, platelets or the white blood cells of the immune system.

Cells may migrate to other regions, such as the spleen or thymus (in the neck) where they mature in to other cell types.

THE LYMPHATIC SYSTEM

The functioning of the immune system is facilitated by the lymphatic system. The lymphatic system circulates a liquid mixture called lymph around the whole of the body, but is different to the circulatory system, which carries blood. Importantly, the lymphatic system carries white blood cells around the body.

In capillaries – the smallest of the blood vessels – pressure causes fluid and small molecules to be forced into the spaces between cells. This is called interstitial fluid, which bathes and feeds surrounding tissues. This is subsequently drained into the lymphatic system, where it circulates and eventually drains back into the bloodstream. It is not actively pumped, but relies on the vessels being squeezed by surrounding muscles.

When the body is infected by bacteria, chemical signals are released. These cause white blood cells, called leucocytes, to leave the capillaries and attack the invading bacteria.

Defending against infection

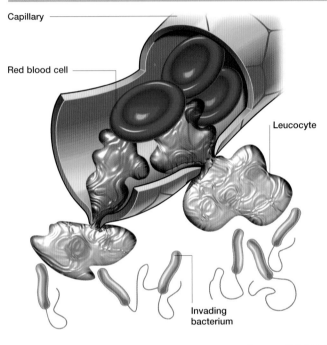

Capillary

Red blood cell

Leucocyte

Invading bacterium

Viruses

Because viruses are so small (only 0.00001 mm in diameter), they are very efficient at entering the respiratory and gastro-intestinal tracts. Blood is able to fight viruses by delivering antibodies to the affected area.

The rhinovirus, shown here, is one of the causes of the common cold. Blood defends against such viruses by carrying antibodies.

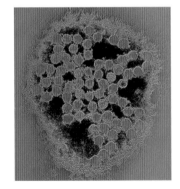

Single-cell invaders

Bacteria and protozoans are sought out, ingested (engulfed by phagocytosis) and killed by white blood-cell phagocytes.

Invading microbes cause the production of factors that attract phagocytes to the infected area; they are then coated with antibodies and ingested.

E. coli bacteria are associated with food poisoning. Phagocytes in the blood are capable of ingesting such microbes.

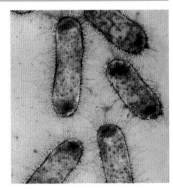

Multi-cellular invaders

Helminths are parasitic worms, commoner in warmer countries. The blood attacks them with specialized white cells called eosinophils – so named because they stain red when exposed to eosin laboratory dye.

Parasites, such as this hookworm, are often found in the intestines. Eosinophils in the bloodstream are capable of attacking some of these invaders.

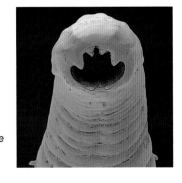

Fungi

Fungal organisms are very effective at invading moist, warm areas of the human body, such as between the toes. The body tries to fight back against these invasions by bringing antibodies to the site, via the blood, as part of the immune response.

The body responds to many fungal infections by producing antibodies. These are carried in the blood to the relevant area.

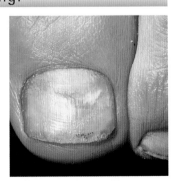

Defensive components of blood

Although some infections can overcome our defences, the various components of the blood successfully fight back against most invaders.

The components of the blood which combat infections are:

■ Phagocytes. If a microbe enters the body it will almost certainly encounter specialized white blood cells: neutrophil polymorphs and monocytes. Their function is to engulf (phagocytose) invading particles and break them up through a process of intracellular digestion.

Phagocytes do not live exclusively in the blood. Instead, they spread out from the blood vessels and into the tissues where they are best placed to attack invading microbes.

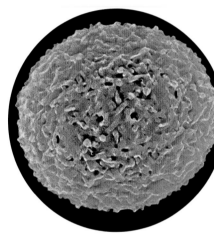

Neutrophil polymorphs – the commonest type of white blood cell – attack invading organisms by phagocytosis.

Of the two types of phagocyte, polymorphs are relatively short-lived, while monocytes are longer-lasting and turn into another group of cells, termed macrophages. Macrophages create a zone of inflammation around microbes, helping to limit their spread. Where possible, they engulf them.

■ Lymphoid cells. These white cells come in three forms:

T-lymphocytes. These are very effective at attacking viruses. Virologists classify them into various groups (helper T cells; suppressor T-cells; cytotoxic T-cells; and hypersensitivity-mediating T-cells), which all combine to attempt to destroy viruses.

B-lymphocytes. These are involved in the production of antibodies against microbes.

Killer cells and natural killer (NK) cells. These are often able to recognize human cells that have been taken over by viruses as intracellular 'factories' and destroy them.

■ **Interferons.** These are chemical agents which are produced by cells which have been infected by viruses and T-lymphocytes. Interferons flow through the bloodstream, activating NK cells and providing defence against viruses.

■ Complement. This blood

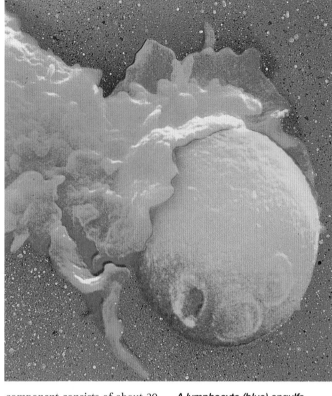

component consists of about 20 proteins. When infection occurs, they work together to attack bacteria and organize inflammation around the infected area.

■ Acute phase proteins. These are blood proteins with the ability to attach to certain bacteria and disable them in the early stages of an infection.

■ Eosinophils. These are specialized white blood cells

A lymphocyte (blue) engulfs a yeast spore (yellow) by phagocytosis. Lymphocytes normally attack invaders with enzymes, rather than by phagocytosis.

which play a role in fighting off infection by helminths. They are capable of inactivating some of these parasites by binding to them and releasing a toxic protein.

Blood antibodies

Antibodies are vital components of the blood. They are complex molecules called immunoglobulins, which are formed in response to infection. There are various types of immunoglobulin:

■ IgG makes up about three-quarters of the immunoglobulin in normal blood.

It is very effective in neutralizing the toxins (poisons) produced by certain microbes.

■ IgM makes up about one-fourteenth of the serum immunoglobulins. It activates complement so that it can attack foreign cells.

■ IgA makes up about a fifth of the blood's immunoglobulin load,

and it is mainly delivered to areas such as the mouth, air passages and intestine, where germs are likely to attack. It acts as an antiseptic secretion, helping to keep microbes from penetrating the mucous surfaces of the body.

■ IgE is thought to play a part in defending the body against helminths, by creating defensive inflammations. Unfortunately, it is often produced in vastly excessive amounts in people who have allergies. In these individuals, it causes inappropriate inflammation, and this is associated with the symptoms of asthma, hay fever and allergic skin reactions.

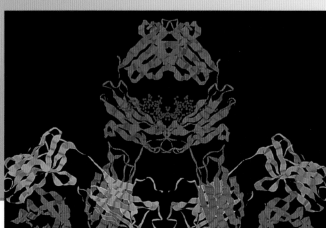

The structure of an antibody is shown on this computer-generated image. Antibodies are able to bind to foreign cells or toxins and neutralize them.

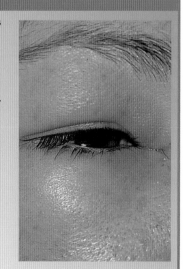

In some cases, excessive amounts of an antibody called IgE are produced. This can result in the symptoms of an allergic reaction.

What is blood pressure?

The heart must pump out blood at sufficient pressure to supply the body's tissues with both oxygen and nutrients. Blood pressure is closely monitored by the body and maintained at its optimal level.

Blood leaves the heart in a pulsatile fashion: each time the heart contracts, about 70 millilitres of blood are ejected from it. However, despite this discontinuous and choppy flow of blood through the root of the aorta, the blood flow through the capillaries is smooth and continuous.

ELASTIC ARTERIES

Continuous flow happens because arteries are not rigid cylinders. Rather, they have elastic walls that can expand or recoil like an elastic band. Thus during systole (when the heart contracts), blood enters the arteries quicker than it leaves the capillary beds; the increased volume of blood present in the arteries forces the arterial walls to expand.

In contrast, during diastole (when the heart is relaxed and

Heart contracting

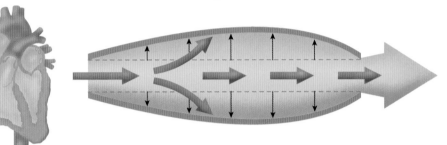

Heart relaxing

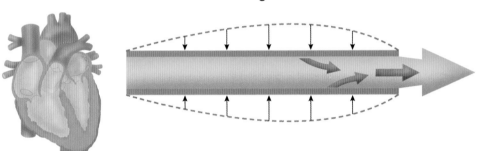

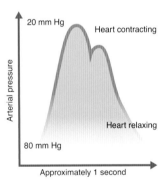

Arterial pressure changes during each heart beat from about 80 (diastolic) to 120 (systolic) millimetres of mercury (mm Hg). The difference (40 mm Hg) is called the pulse pressure.

no blood is ejected from it) the blood stored in the arteries is propelled towards the capillaries by the recoil of the expanded arterial walls.

SMOOTH BLOOD FLOW

This elasticity causes the blood to flow more smoothly as it travels down the vascular tree; while the arterial pressure fluctuates with each heartbeat, it would be much more pulsatile if the arteries were rigid, inflexible tubes (an analogy to this would be the flow of water from a garden hose if the tap were to be intermittently turned on and

During systole (top) blood is forced into elastic arteries, which expand. During diastole (bottom), they recoil, propelling the blood smoothly onwards.

off). The smooth flow of blood through the capillaries is advantageous because large changes in pressure would damage the capillaries, whose walls are only one cell thick.

How blood pressure is measured

Doctors measure blood pressure using a sphyngmomanometer:
1. A cuff is wrapped around the patient's upper arm and inflated to block the flow of blood through the brachial artery.
2. Air is gradually let out of the cuff while the doctor listens with a stethoscope placed 'downstream' from the cuff. The pressure at which the blood can be heard running through the artery is noted (the systolic pressure).
3. As more air is let out of the cuff, blood flows smoothly through the artery again and the sound of blood rushing through the artery disappears. This pressure is called the diastolic pressure.

The brachial artery is blocked using an inflatable cuff. When the cuff is deflated, the sound of blood flowing back through the vessel can be heard.

1 No sound heard through stethoscope

2 Pulse heard as 'banging' sound

3 Sound of pulse disappears

What determines blood pressure?

At its simplest level, blood pressure is the product of two factors: cardiac output and total peripheral resistance.

■ Cardiac output is the amount of blood that the heart pumps around the body each minute. For example, in a healthy adult man, the heart beats around 70 times per minute, with each ventricular contraction pumping out around 70 millilitres (ml) of blood (called the stroke volume). Thus the cardiac output would be 4,900 millilitres (70 ml multiplied by 70 ml equals 4,900 ml) per minute.

■ Total peripheral resistance (TPR) is the resistance that the blood encounters as it flows around the body. Resistance is very sensitive to the diameter of the vessel that the blood is flowing through; halving the diameter of a vessel increases its resistance by 16 times.

IMPORTANCE OF BLOOD VOLUME

However, the circulation is a closed system – blood is returned to the heart by the veins and does not drain out of the body after each contraction. Thus the volume of circulating blood also determines blood pressure. This can be important in severe haemorrhage (blood loss), for example.

Blood pressure **=** **Cardiac output** **X** **Total peripheral resistance**

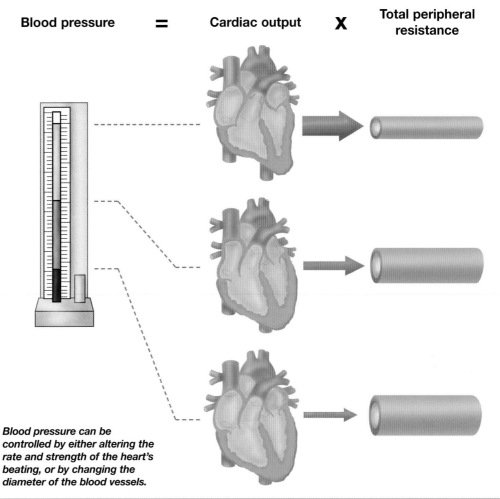

Blood pressure can be controlled by either altering the rate and strength of the heart's beating, or by changing the diameter of the blood vessels.

How is blood pressure controlled?

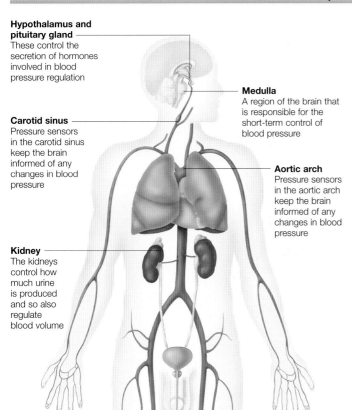

Hypothalamus and pituitary gland
These control the secretion of hormones involved in blood pressure regulation

Carotid sinus
Pressure sensors in the carotid sinus keep the brain informed of any changes in blood pressure

Kidney
The kidneys control how much urine is produced and so also regulate blood volume

Medulla
A region of the brain that is responsible for the short-term control of blood pressure

Aortic arch
Pressure sensors in the aortic arch keep the brain informed of any changes in blood pressure

The body has three ways of regulating blood pressure: it can alter the cardiac output by changing the force or rate of the heart's contraction; it can alter the diameter and elasticity of the blood vessels to regulate TPR; or it can change the circulating volume of blood.

SHORT-TERM CONTROL

There are two mechanisms that control blood pressure in the short term:

■ **Nervous control**
Blood pressure sensors in the arteries send information, via nerves, to a region of the brain called the medulla, which calculates whether blood pressure needs to be corrected. If so, the medulla in turn sends nerve signals to the heart, to modify its rate and strength of beating, and to the blood vessels to modify their diameter.

■ **Chemical control**

A large number of structures are involved in the regulation of blood pressure, both in the short and the long term.

A large number of blood-borne chemicals can either constrict or dilate blood vessels.

LONG-TERM CONTROL

In the long term, blood volume is controlled by chemicals that act on the kidneys. If blood pressure falls, the kidneys conserve water by producing a more concentrated amount of urine, thus increasing blood volume.

Excessive stress may contribute to the development of high blood pressure by causing a region of the brain called the medulla to malfunction.

How the brain controls blood pressure

The medulla, a region of the brain situated just above the spinal cord, constantly monitors arterial pressure. It corrects changes in pressure by sending nervous signals to the heart and blood vessels.

The pressure of blood in the arteries is constantly measured by specialized pressure sensors called baroreceptors (baro- is a prefix for pressure). Baroreceptors are nerve endings contained within the walls of an artery which are able to detect even the smallest distension of the arterial wall. These pressure sensors are mainly found in the aortic arch and the carotid sinuses.

BARORECEPTOR NERVES

The baroreceptor endings are part of nerve fibres which travel up to a region of the brain called the medulla.

The afferent (from the Latin 'afferere' – to carry towards) fibres of the aortic baroreceptors form the aortic nerve, which joins the vagus (10th cranial) nerve before entering the medulla, where they terminate in a region called the nucleus tractus solitarii (NTS).

The afferent fibres of the carotid baroreceptors form the carotid sinus nerve, which joins the glossopharyngeal nerve (ninth cranial nerve) before also terminating in the NTS.

Anatomy of the baroreceptor reflex

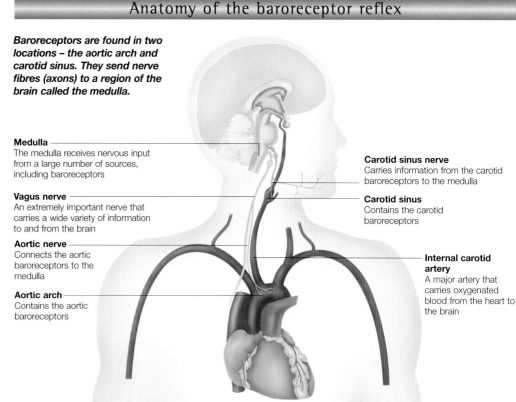

Baroreceptors are found in two locations – the aortic arch and carotid sinus. They send nerve fibres (axons) to a region of the brain called the medulla.

Medulla
The medulla receives nervous input from a large number of sources, including baroreceptors

Vagus nerve
An extremely important nerve that carries a wide variety of information to and from the brain

Aortic nerve
Connects the aortic baroreceptors to the medulla

Aortic arch
Contains the aortic baroreceptors

Carotid sinus nerve
Carries information from the carotid baroreceptors to the medulla

Carotid sinus
Contains the carotid baroreceptors

Internal carotid artery
A major artery that carries oxygenated blood from the heart to the brain

Response of baroreceptors

The response of baroreceptor nerves to increasing pressures

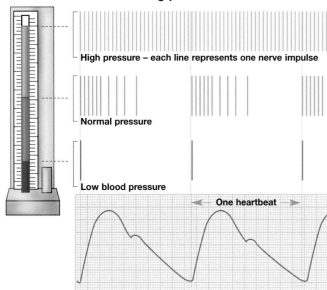

High pressure – each line represents one nerve impulse

Normal pressure

Low blood pressure

One heartbeat

Since blood moves through the arteries via pulsatile, rather than constant flow, the baroreceptor nerves do not 'fire' at a uniform rate.

This is because during systole (when the heart contracts and the pressure is highest) the arterial walls are distended causing the baroreceptor nerves to fire a volley of nervous impulses which travel up to the medulla. However, during diastole (when the heart is relaxed and the pressure is lowest) the arterial walls are not stretched, and this causes the baroreceptors to fall silent.

Arterial wall stretching is transformed into electrical activity in the baroreceptor nerve fibres. When pressure rises, nerve activity increases.

Importantly, many baroreceptors will be active at normal pressures; this allows them to inform the medulla when the pressure falls (by slowing the rate of nervous impulses), which would be impossible if the nerves were silent at rest.

BARORECEPTOR PROPERTIES

Not all baroreceptors have the same properties:
■ Some are responsive at low pressures, whereas others fire only when the arterial pressure has reached very high levels
■ The range of pressure over which they are sensitive also varies considerably
■ Baroreceptors vary in their sensitivity to the rate of change of arterial pressure – this parameter is thought to be very important, as it allows the brain to pre-empt changes in pressure.

Role of medulla

The baroreceptor nerves project to, and terminate in, a region of the medulla called the nucleus tractus solitarii (NTS). The NTS plays an important role in the control of autonomic (unconscious) functions, including, but not restricted to, the control of blood pressure. If it is damaged, for example following a stroke, the consequences can be fatal.

ROLE OF THE NTS
The NTS receives information not just from baroreceptors, but also from a large number of other sources including receptors found in the heart, gastro-intestinal tract, lungs, oesophagus and tongue. The NTS neurones do not act as a simple relay station for this diverse afferent input. Rather, they calculate what the correct blood pressure should be after taking into account information obtained from all the other sources.

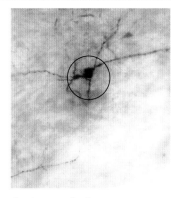

A micrograph of a neurone located in the NTS, which receives input from baroreceptors. The cell body, which contains the nucleus, is the dark oval (circled).

The top trace shows the electrical activity of a neurone located in the NTS. The neurone's rate of firing increases when arterial pressure is raised (bottom).

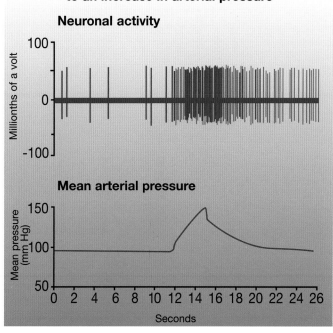

The response of an NTS neurone to an increase in arterial pressure

Neuronal activity

Mean arterial pressure

The baroreceptor reflex pathway

The baroreceptor reflex pathway

Medulla

NTS

Baroreceptors
Increased blood pressure

Spinal cord

Arteries dilate

Heart
Decreased heart rate and decreased force of contraction

The baroreceptor reflex corrects a rise in arterial pressure by reducing the rate and strength of the heartbeat, as well as causing the arteries to relax, so lowering blood pressure.

During periods of intense stress, the baroreceptor reflex is strongly suppressed by nerves originating from the hypothalamus. This may be one of the causes of hypertension.

If arterial pressure rises, baroreceptors respond to the distension of the arterial wall by sending a volley of nervous impulses to the NTS.

Under normal conditions, the NTS will try to correct this increase in pressure by sending nervous impulses to the heart, telling it to reduce its rate and strength of contraction; and to the arteries, telling them to become more elastic. This will have the effect of reducing both the cardiac output (the amount of blood that the heart pumps out each minute) and the resistance to blood flow in the arteries. These combined effects will act to lower blood pressure.

RESETTING THE BARORECEPTOR REFLEX
The baroreceptor reflex acts to maintain blood pressure at what physiologists call the 'set-point'.

An analogy to the set-point is the temperature setting of a central heating thermostat; the set-point of the baroreceptor reflex can be altered in the same way as a thermostat. The body does this either by affecting the threshold pressure at which the baroreceptors fire (peripheral resetting), or by altering the sensitivity of the neurones within the medulla (central resetting).

PERIPHERAL RESETTING
If pressure is maintained at a raised level for many minutes, the baroceptors become accustomed to the new pressure and 'think' that it is the correct level. Thus baroreceptors cannot accurately inform the brain about blood pressure levels over the long-term.

CENTRAL RESETTING
When we are exposed to a stressful situation, the neurones within the NTS which mediate the baroreceptor reflex are strongly suppressed, allowing blood pressure to rise. This was advantageous to our ancestors because it prepared them either to fight or to run away from their aggressors. However, this neural mechanism could be responsible for the high incidence of hypertension seen in modern Western society. The stress that we experience in our day-to-day lives, could, in some people at least, raise the set-point and so cause hypertension.

Chapter Ten

Body Systems

The human body is comparable to a hugely complex machine, run with amazing efficiency by a range of systems. Each system has its own unique role and function, but all are interconnected, working together to ensure a smoothly running whole. When all of our systems are working successfully, we are provided with a stable internal environment that enables us to carry out everyday tasks and activities and encourages growth and development.

This chapter explains how each individual system operates while at the same time illustrating the interdependent nature of all parts of the human body.

LEFT: This microscopic view of a section of muscle fibre magnified 27000 times shows the fine longitudinal myofibrils.

The skeleton

The skeleton is made up of bone and cartilage, and it accounts for one-fifth of the body's weight. Over 200 bones form a living structure, superbly designed to support and protect the body.

The human skeleton provides a stable yet flexible framework for the other tissues of the body. Cartilage is more flexible than bone and is found in the places where movement occurs.

FUNCTIONS OF BONE
The bones of the skeleton have a number of vital functions:
■ Support – bones support the body when standing, and hold soft internal organs in place
■ Protection – the brain and spinal cord are protected by the skull and vertebral column, while the rib cage protects the heart and lungs
■ Movement – throughout the body, muscles attach to bones to give them the leverage to bring about movement
■ Storage of minerals – calcium and phosphate ions are stored in bone to be drawn upon when necessary
■ Blood cell formation – the marrow cavity of some bones, such as the sternum, is a site of production of red blood cells.

Formation

The bony skeleton is formed in fetal life, but grows throughout childhood. A fetus of six weeks has a skeleton made of fibrous membranes and hyaline cartilage, which converts into bone during pregnancy. After birth, and until the end of adolescence, the skeleton grows in weight and length as well as being remodelled.

This image of a fetus shows early bone development. The dark ends of the bone are primary ossification centres, which produce new bone cells.

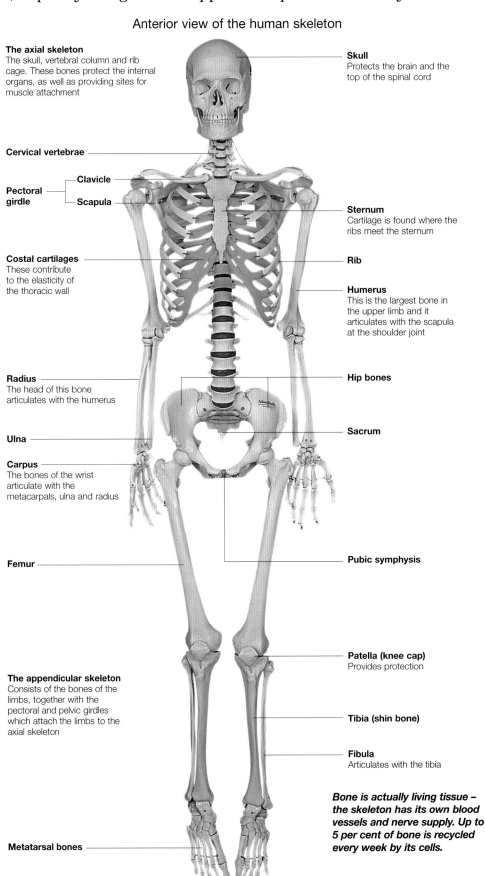

Anterior view of the human skeleton

The axial skeleton
The skull, vertebral column and rib cage. These bones protect the internal organs, as well as providing sites for muscle attachment

Cervical vertebrae

Pectoral girdle — Clavicle — Scapula

Costal cartilages
These contribute to the elasticity of the thoracic wall

Radius
The head of this bone articulates with the humerus

Ulna

Carpus
The bones of the wrist articulate with the metacarpals, ulna and radius

Femur

The appendicular skeleton
Consists of the bones of the limbs, together with the pectoral and pelvic girdles which attach the limbs to the axial skeleton

Metatarsal bones

Skull
Protects the brain and the top of the spinal cord

Sternum
Cartilage is found where the ribs meet the sternum

Rib

Humerus
This is the largest bone in the upper limb and it articulates with the scapula at the shoulder joint

Hip bones

Sacrum

Pubic symphysis

Patella (knee cap)
Provides protection

Tibia (shin bone)

Fibula
Articulates with the tibia

Bone is actually living tissue – the skeleton has its own blood vessels and nerve supply. Up to 5 per cent of bone is recycled every week by its cells.

Bone markings and features

Each bone of the skeleton is shaped to fulfil its own functions. Bones bear marks, ridges and notches which relate to other structures with which they come into contact.

Over the years, anatomists have given names to the various types of feature which can be found on bones. Using these names, a bone can be described quite clearly and accurately, something which can be of importance clinically.

PROJECTIONS

Projections on the surface of a bone often occur where muscles, tendons or ligaments are attached or where a joint is formed. Examples include:
■ Condyle – rounded projection at a joint (such as the femoral condyle at the knee)
■ Epicondyle – the raised area above a condyle (such as on the lower humerus, at the elbow)
■ Crest – prominent ridge of bone (such as the iliac crest of the pelvic bone)
■ Tubercle – small raised area (such as the greater tubercle at the top of the humerus)
■ Line – long, narrow raised ridge (such as the soleal line at the back of the tibia).

Depressions and grooves

Depressions, holes and grooves are usually found where blood vessels and nerves must pass through or around bones.

FEATURES

Examples include:
■ Fossa – shallow, bowl-like depression (such as the infraspinous fossa of the shoulder blade or the iliac fossa, which is a depression found on the ilium)
■ Foramen – a hole in a bone to allow the passage of a particular vessel or nerve, (such as the jugular foramen in the skull which allows the internal jugular vein to leave)
■ Notch – indentation that is found at the edge of a bone (such as the greater sciatic notch, which is partly formed by the ilium)
■ Groove – a furrow or elongated depression that marks the route of a vessel or nerve along a bone (such as the oblique radial groove at the back of the humerus).

Posterior view of the human skeleton

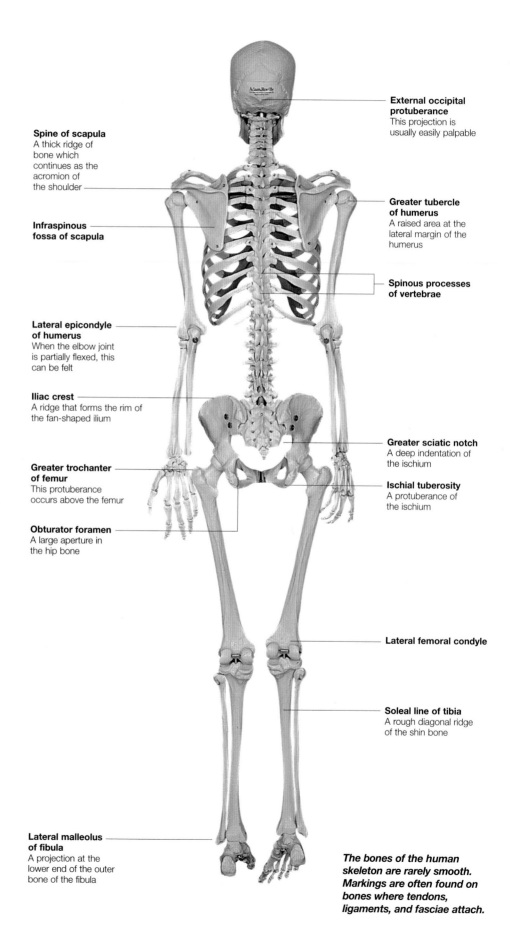

Spine of scapula
A thick ridge of bone which continues as the acromion of the shoulder

Infraspinous fossa of scapula

Lateral epicondyle of humerus
When the elbow joint is partially flexed, this can be felt

Iliac crest
A ridge that forms the rim of the fan-shaped ilium

Greater trochanter of femur
This protuberance occurs above the femur

Obturator foramen
A large aperture in the hip bone

Lateral malleolus of fibula
A projection at the lower end of the outer bone of the fibula

External occipital protuberance
This projection is usually easily palpable

Greater tubercle of humerus
A raised area at the lateral margin of the humerus

Spinous processes of vertebrae

Greater sciatic notch
A deep indentation of the ischium

Ischial tuberosity
A protuberance of the ischium

Lateral femoral condyle

Soleal line of tibia
A rough diagonal ridge of the shin bone

The bones of the human skeleton are rarely smooth. Markings are often found on bones where tendons, ligaments, and fasciae attach.

How bones are formed

Bones are living tissue, and are in a constant state of renewal. They form the basis of the skeleton, and are responsible for locomotion as well as containing bone marrow and vital minerals.

Bones are the rigid body tissues that are the basis of the human skeleton. They are a living tissue, constantly being renewed and shaped by the process of growth and reabsorption.

BONE MATRIX

Bone is composed of a calcified matrix in which bone cells are embedded. The matrix is made up of flexible collagen fibres in which crystals of hydroxyapatite (a calcium salt) are deposited. Three principal bone cell types are found within this matrix:

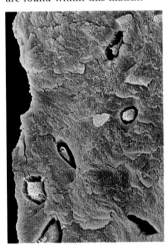

■ Osteoblasts – cells responsible for forming bone
■ Osteoclasts – bone-eating cells
■ Osteocytes – bone cells that have fully matured.

Bone-forming and bone-eating cells permit the constant turnover of bone matrix that occurs throughout life.

SKELETAL SUPPORT

Linked together at joints by ligaments and moved by attached muscles, bones form levers vital to locomotion.

The intricate arrangement of the bones making up the skeleton provides cages which protect the soft, delicate parts of the body, while still allowing for great flexibility and movement.

In addition, bones contain bone marrow, the soft fatty substance that produces most of the body's blood cells.

Bones also act as a reservoir for the minerals calcium and phosphorus, vital to many body processes.

Osteoblasts are bone-forming cells. This micrograph shows osteoblast cells (irregular ovals) surrounded by the bone matrix they have created.

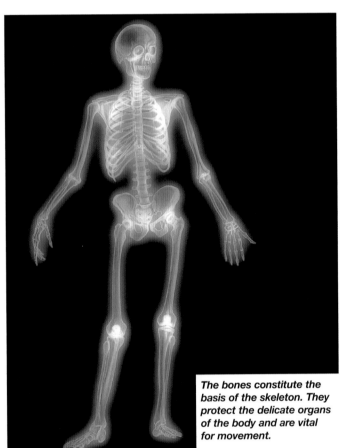

The bones constitute the basis of the skeleton. They protect the delicate organs of the body and are vital for movement.

Structure of bone tissue

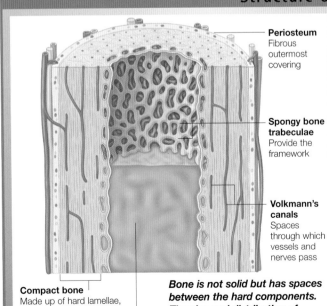

Periosteum
Fibrous outermost covering

Spongy bone trabeculae
Provide the framework

Volkmann's canals
Spaces through which vessels and nerves pass

Compact bone
Made up of hard lamellae, interspersed with lacunae

Medullary cavity
Filled with bone marrow

Bone is not solid but has spaces between the hard components. The size and distribution of these spaces dictates whether bone is compact or spongy.

Bone tissue exists in two forms: compact (or cortical) bone and spongy (or cancellous) bone.

Compact bone

Compact bone makes up the outer covering of all bones and is thickest in the places that receive the greatest stress. It is made up of a series of canals and passageways; these provide a route for the nerves, blood vessels and lymphatic vessels that extend through each bone.

The structural units of compact bones (osteons) are elongated cylinders which lie parallel to the long axis of the bone. Osteons are composed of a group of lamellae (hollow tubes) of bone matrix arranged concentrically.

The lamellae are organized in such a way that the collagen fibres in adjacent lamellae run in opposite directions; this is intended to reinforce the bone against twisting forces. Each

osteon is nourished by blood vessels and served by nerve fibres which run throughout its centre, known as the Haversian canal.

The Volkmann's canals connect the blood vessels and nerve supplies of the periosteum (membrane around the bone) to those in the central canals and medullary cavity (which contains bone marrow).

Mature bone cells (osteocytes) are located in the small cavities (lacunae) between each lamella.

Spongy bone

Spongy bone makes up the inner part of most bones and is much lighter and less dense than compact bone. This is due to the fact that it contains a number of cavities which are filled with marrow. Spongy bone is strengthened by a criss-cross network of boney supports, known as trabeculae.

Formation of bone

Bone formation begins in the embryo and continues throughout the first 20 years of life. Development takes place from a number of ossification centres and, once these are fully calcified, no further elongation can occur.

The skeleton is made up of a variety of different bones, ranging from the flat bones found in the skull to the long bones of the limbs. Each bone is designed for a different function.

LONG BONES
The longest bones within the body are those of the upper and lower limbs. Each long bone consists of three main components:
- Diaphysis – a hollow shaft, composed of compact bone
- Epiphysis – at each end of the bone; site of articulation between bones
- Epiphyseal (growth) plate – composed of spongy bone and the site of bone elongation.

PROTECTIVE MEMBRANE
The entire bone is covered by the two-layered periosteum.

The outer layer of this membrane consists of fibrous connective tissue. The inner layer of the periosteum contains osteoblasts and osteoclasts, the cells that are responsible for the constant replenishment of the bone.

The humerus, a typical 'long bone', is found in the upper arm. The bone is divided into a diaphysis (shaft), with epiphyses (heads) at either end.

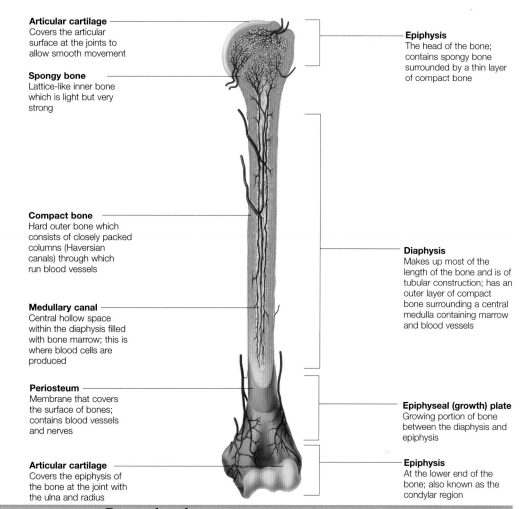

Articular cartilage
Covers the articular surface at the joints to allow smooth movement

Spongy bone
Lattice-like inner bone which is light but very strong

Compact bone
Hard outer bone which consists of closely packed columns (Haversian canals) through which run blood vessels

Medullary canal
Central hollow space within the diaphysis filled with bone marrow; this is where blood cells are produced

Periosteum
Membrane that covers the surface of bones; contains blood vessels and nerves

Articular cartilage
Covers the epiphysis of the bone at the joint with the ulna and radius

Epiphysis
The head of the bone; contains spongy bone surrounded by a thin layer of compact bone

Diaphysis
Makes up most of the length of the bone and is of tubular construction; has an outer layer of compact bone surrounding a central medulla containing marrow and blood vessels

Epiphyseal (growth) plate
Growing portion of bone between the diaphysis and epiphysis

Epiphysis
At the lower end of the bone; also known as the condylar region

Bone development

Long bone of a newborn

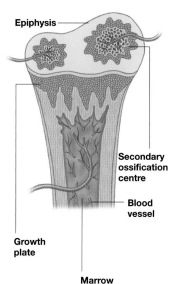

Epiphysis (bone end)

Growth plate

Blood vessel

Marrow cavity

Diaphysis (shaft)

In a newborn baby, the shaft is mostly bone, while the bone ends consist of cartilage. In a child, new bone forms from secondary ossification centres in the bone ends.

Long bone of a child

Epiphysis

Secondary ossification centre

Blood vessel

Growth plate

Marrow cavity

Skeletal development begins in the embryo and continues for around two decades. It is a complex process under genetic control, and is modulated by endocrine, physical and biological processes.

A template of the skeleton forms in the embryo from the primitive embryonic tissue. As the embryo develops, this tissue becomes recognisable as cartilage (soft, elastic connective tissue) and individual 'bones' begin to be seen.

OSSIFICATION
Normal bone then forms within these templates by a process known as ossification. This takes place either directly around the early bone-forming cells of the fetus (intramembranous ossification) or by replacing a cartilage model with bone (endochondral ossification).

The formation of compact bone commences at sites in the bone shafts known as primary ossification centres. Osteoblasts within the cartilage secrete a gelatinous substance called osteoid, which is hardened by mineral salts to form bone. The cartilage cells die and are replaced by further osteoblasts.

Ossification of long bones continues until only a thin strip of cartilage remains at either end. This cartilage (the epiphyseal plate) is the site of secondary bone growth up to late adolescence.

The sequence of formation of ossification centres follows a prescribed pattern, allowing experts to age skeletons by the extent of ossification.

MATURE BONE
Once the bone has reached full length, the shaft, growth plate and epiphyses are all ossified and fuse to form continuous bone. No further elongation can take place after this time.

How bone repairs itself

Although bones cease to grow after late adolescence, bone
is a very dynamic tissue. Bone is continually being reabsorbed
and regenerated as its structure is constantly changing.

One of the most amazing features of bone is its ability to reshape itself. This process, known as remodelling, occurs during growth and continues throughout life.

BONE REMODELLING

During bone formation, bone is deposited in a random pattern by a process known as ossification. Remodelling continually occurs, organizing the bone into orderly units that enable the bone mass to best withstand mechanical forces. Old bone is removed by osteoclasts (bone-eating cells), and osteoblasts (bone-forming cells) deposit new bone.

BONE REABSORPTION

Osteoclasts secrete enzymes that break down the bone matrix, as well as acids which convert the resulting calcium salts into a soluble form (which can enter the bloodstream).

Osteoclast activity takes place behind the epiphyseal growth zone to reduce expanded ends to the width of the lengthening shaft. Osteoclasts also act within the bone in order to clear the long tubular spaces that will accommodate bone marrow.

HORMONAL REGULATION

While the osteoclasts reabsorb bone, osteoblasts make new bone to maintain the skeletal structure. This process is regulated by hormones, growth factors and vitamin D.

During childhood, bone formation outweighs bone destruction, resulting in gradual growth. After skeletal maturity has been reached, however, the two processes occur in equilibrium so that growth proceeds more gradually.

LONG BONES

The process of remodelling is especially important for the long bones which support the limbs. These are wider at each end than in the middle, providing extra strength at the joint.

As osteoclasts destroy the old epiphyseal swellings of the bone, osteoblasts within the growth zone create a new epiphysis.

Within each of the tubular spaces cleared by osteoclasts inside the bone, the osteoblasts follow along, laying down a layer of new bone.

RATES OF REMODELLING

Bone remodelling is not a uniform process; it takes place at different rates throughout the skeleton. Bone formation tends to take place in areas where the bone undergoes the greatest stress. This means that bones which receive the most stress are subject to much remodelling. The femur for example (one of the load-bearing bones of the leg), is effectively replaced every five to six months.

A bone that is under-used, such as a leg that is immobilized after injury, will be prone to reabsorption however, as bone destruction outweighs formation.

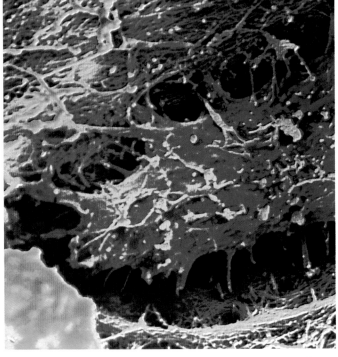

Osteoblasts (the orange cell pictured) secrete a substance called osteoid which hardens to become bone. This bone may be reabsorbed by osteoclasts as remodelling occurs.

Bone that is subject to increased stress is constantly remodelled. The femur, for example, is effectively replaced every six months.

Remodelling gives rise to the distinctive shape of the long bones. These are wider at each end than in the middle.

Calcium regulation

Bone remodelling not only alters the structure of the bone, but also helps to regulate the levels of calcium ions in the blood. Calcium is necessary for healthy nerve transmission, the formation of cellular membranes and the ability of the blood to clot.

Bone contains about 99 per cent of the body's calcium. When body fluid calcium levels fall too low, parathyroid hormone stimulates osteoclast activity and calcium is released into the bloodstream. When body fluid calcium levels become too high, calcitonin hormone inhibits reabsorption, restricting the release of calcium from the bones.

Bone repair

If bone is subjected to a force beyond its strength it will fracture. New bone must be formed and remodelled for the fracture to heal.

One of the processes which is dependent on the remodelling of bone is the repair mechanism that takes place after a fracture.

BONE FRACTURES

Fractures occur when a bone experiences a force greater than its resistance or strength.

These can occur as the result of a spontaneous force, or after years of continued stress upon a bone. Bones are particularly susceptible to fractures later in life when they are less elastic and bone mineral density declines. Bone repair takes place in four main stages.

A plaster cast aids the healing of fractured bone by immobilizing the limb. This is important to ensure that the ends of the broken bone realign correctly.

Blood clot formation

1 A fracture of the bone causes the blood vessels in the area (mainly those of the periosteum, the protective covering of the bone) to rupture.

As these vessels bleed, a clot is formed at the site of the fracture giving rise to the characteristic swelling that often accompanies a broken bone. Very soon, bone cells deprived of nutrition begin to die and the site becomes extremely painful.

Blood vessels at the site of the fracture rupture, causing a blood clot to form. The nerves lining the periosteum are also severed, causing much pain.

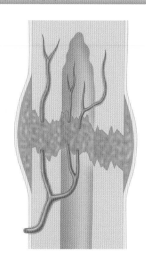

Fibrocartilage callus formation

2 Several days after the injury, blood vessels and undifferentiated cells from surrounding tissues invade the area. Some of these cells develop into fibroblasts, which produce a network of collagen fibres between the bone fragments. Other cells form chondroblasts, which secrete cartilage matrix.

This zone of tissue repair between the two ends is known as a fibrocartilage callus.

Blood vessels and cells invade the site of the fracture. The cells produce a matrix of collagen fibres and cartilage, forming a fibrocartilage callus.

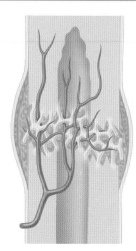

Bony callus formation

3 Osteoblasts and osteoclasts migrate towards the affected area multiplying rapidly within the fibrocartilage callus.

Osteoblasts within the callus secrete osteoid, converting it into a bony callus.

This bony callus is composed of two portions: an external callus located around the outside of the fracture and an internal callus located between the broken bone fragments.

Osteoblasts and osteoclasts multiply within the fibrous callus. Osteoblasts secrete a substance known as osteoid, which hardens, forming a bony callus.

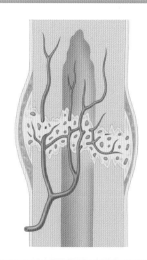

Bone remodelling

4 Bone formation is usually complete within four to six weeks of injury.

Once the new bone has been formed it will slowly be remodelled to form compact and spongy bone.

Total healing may require up to several months depending on the nature of the fracture and the specific function of the limb – weight-bearing limbs take longer to repair.

As the new bone is formed it is remodelled by osteoclasts. In this way the bony callus is smoothed out, and the bone regains its original structure.

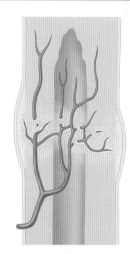

Bone injury

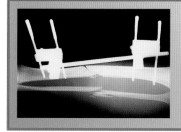

In some cases the extent of a fracture can be so severe that the normal process by which the

In cases where damage to the bone is very severe it may fail to heal. The use of orthopaedic pins to hold the bone in place may therefore be necessary.

bone repairs itself cannot occur.

Examples include shattered bones, or fractures in which fragments of bone are lost, so that the gap between severed ends is too great to heal.

The bones may need to be fixed in place with the use of orthopaedic screws, pins, plates or wires, in order to encourage the bone's repair mechanisms to to take place.

Bone chips can be transplanted from other parts of the patient's skeleton in order to aid bone formation. In cases where there is massive injury, amputation may be necessary.

Types of joints (1)

A joint is formed where two or more bones meet. Some allow movement and so give mobility to the body while others protect and support the body by holding the bones rigid against each another.

The joints of the body can be divided into three main structural groups, according to the tissues that lie between the bones. These groups are fibrous, cartilaginous and synovial.

FIBROUS JOINTS

Where two bones are connected by a fibrous joint, they are held together with collagen (a protein). Collagen fibres allow little, if any, movement. Fibrous joints are located in the body where the movement of one bone upon the other should be prevented, such as in the skull.

CARTILAGINOUS JOINTS

The ends of the bones in a cartilaginous joint are covered with a thin layer of hyaline (glass-like) cartilage, with the bones being connected by tough fibrocartilage. The whole joint is covered by a fibrous capsule.

Cartilaginous joints do not allow much movement but they can 'relax' under pressure, so giving flexibility to structures such as the spinal column.

SYNOVIAL JOINTS

Most joints of the body are synovial, and allow easy movement between the bones. In a synovial joint, the bones are covered by hyaline cartilage and separated by fluid. The joint cavity is lined by a synovial membrane and the whole joint is enclosed by a fibrous capsule.

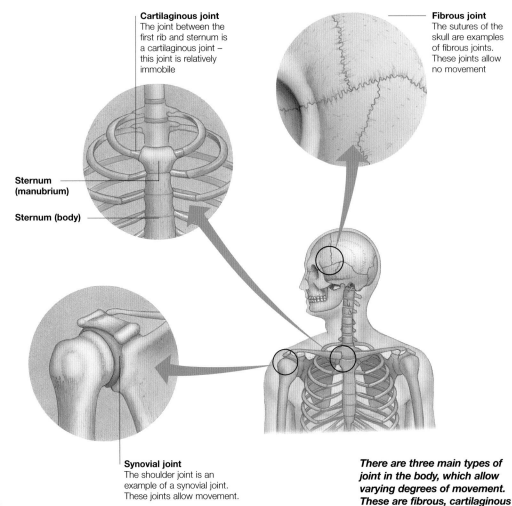

Cartilaginous joint
The joint between the first rib and sternum is a cartilaginous joint – this joint is relatively immobile

Sternum (manubrium)

Sternum (body)

Fibrous joint
The sutures of the skull are examples of fibrous joints. These joints allow no movement

Synovial joint
The shoulder joint is an example of a synovial joint. These joints allow movement.

There are three main types of joint in the body, which allow varying degrees of movement. These are fibrous, cartilaginous and synovial joints.

Functional groups of joints

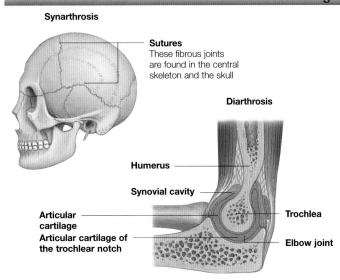

Synarthrosis

Sutures
These fibrous joints are found in the central skeleton and the skull

Diarthrosis

Humerus

Synovial cavity

Articular cartilage

Articular cartilage of the trochlear notch

Trochlea

Elbow joint

The elbow is a diarthrotic joint, which allows flexibility. The articular capsule allows plenty of freedom for extending the elbow joint.

The classification of joints shown above is based on the structure of the tissues which make up the joint.

Joints can also be grouped according to their function. Perhaps the most important function of a joint is to allow or prevent movement. On this basis, there are three groups:
■ Synarthroses – joints which allow no movement. These joints lie predominantly within the axial skeleton (the central skeleton, excluding the limbs), where bones are more likely to fulfil the functions of support and protection than mobility. An example is the fibrous joints (sutures) of the skull
■ Amphiarthroses – joints that allow slight movement. These are found in areas where some flexibility is needed but greater degrees of movement would be unsuitable. Examples include the vertebral joints or the fibrous interosseous membrane in the forearm
■ Diarthroses – joints which allow free movement. These predominate in the limbs, where mobility and movement are the prime functions. Some examples are the hip, shoulder and elbow joints.

Fibrous and cartilaginous joints

Fibrous and cartilaginous joints have an important role to play in the human skeleton. Unlike the more widespread synovial joints, which are designed to allow mobility, fibrous and cartilaginous joints help to maintain stability of the body's frame.

The bones of a fibrous joint are connected solely by long collagen fibres; there is no cartilage, and no fluid-filled joint cavity. Because of its structure, a fibrous joint does not allow much real movement of the bones against each other. What little movement there is, is determined by the length of the collagen fibres.

GROUPS OF FIBROUS JOINTS

Fibrous joints can be further subdivided into three groups:
■ Sutures – literally meaning 'seams', sutures are the tough fibrous joints between the interlocking bones of the skull. Short collagen fibres allow no side-to-side movement of these bones upon each other although there may be some slight 'springing' of the bones if pressure is applied. The presence of fixed fibrous joints in the skull gives great protection to the vulnerable brain tissue that lies beneath
■ Syndesmoses – here, the bones are connected by a sheet of fibrous connective tissue, and the length of the fibres varies from joint to joint. These may also be known as interosseous membranes and are a feature of the forearm and the lower leg,

Cranial bone

Suture
Skull sutures consist of dense, fibrous connective tissue. The fixed nature of these joints gives greater protection

When a person becomes an adult, the fibrous tissue hardens and the skull bones become a single unit. The sutures are then known as synostoses.

where two bones lie side by side, acting as a unit. Syndesmoses tend to have longer fibres than sutures and so allow a little more movement
■ Gomphoses – this is a very

specialized type of fibrous joint with only one example in the human body, the tooth socket. In a gomphosis, a peg-like process sits in a depression, or socket, and is held in place by fibrous

tissue, in this case the periodontal ligament. Movement is generally abnormal but micromovement is essential to eating to allow adjustment of the pressure of the bite.

Cartilaginous joints

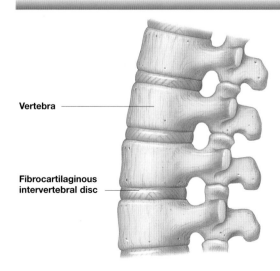

Vertebra

Fibrocartilaginous intervertebral disc

The fibrocartilaginous discs are in the joints in the vertebrae and act as shock absorbers. These tough joints allow a small amount of movement.

In a cartilaginous joint, the bone ends are covered by hyaline cartilage. In some cases, there is a plate of tough fibrocartilage between the bones. The joint is usually enclosed within a fibrous capsule.

There are said to be two types of cartilaginous joints:
■ Primary cartilaginous joints – those where two ends of bone are connected by a plate of hyaline cartilage. They are found

in the growing long bones of children and, in the adult, between the first rib and the top of the sternum (breastbone)
■ Secondary cartilaginous joints or 'symphyses' – joints where a plate of tough fibrocartilage lies between the bones. These are strong, slightly movable joints that often perform the function of shock absorbers. An example of this type of cartilaginous joint is found in the vertebral column, in which the individual vertebrae are covered with hyaline cartilage and are connected to each other by resilient fibrocartilaginous intervertebral discs.

Types of joints (2)

Synovial joints allow greater movement than fibrous and cartilaginous joints. They are classified into six groups according to their structure and movement.

Synovial joints, the most common type of joint in the body, have a fluid-filled joint cavity between the articulating bones. This arrangement allows a great deal of free movement, unlike the restriction of fibrous and cartilaginous joints, and is the reason why all of the many joints of the limbs are synovial.

FEATURES

There are a number of features common to all synovial joints:

■ Joint cavity – the distinctive feature of a synovial joint is the presence of a joint cavity between the articulating bones. This is filled with synovial fluid

■ Synovial membrane – this is a layer of connective tissue that is rich in blood vessels. It lines the joint cavity except where cartilage covers the bone ends

■ Synovial fluid – this fills the joint cavity and is produced by the synovial membrane. In a healthy joint there is only a fine layer of this thick and viscous fluid, which is ideal for lubricating the joint

■ Articular cartilage – this is hyaline (transparent), flexible and slightly spongy so that it can act as a shock absorber for the bones, helping to reduce wear and tear and friction

■ Articular capsule – this tough, fibrous capsule surrounds and protects the joint. It is continuous with the periosteum (protective covering) of the bones above and below the joint.

Schematic representation of a synovial joint

Bone

Synovial membrane
This layer of connective tissue lines the joint cavity and secretes synovial fluid

Joint cavity
Filled with synovial fluid

Bone

Articular capsule
A tough, fibrous capsule surrounding and protecting the joint, which is rich in blood vessels and nerves

Articular cartilage
This lines the articular surfaces (the surfaces of the bone that move against each other)

Synovial joints are the most common type of joint in the body. Their design allows great flexibility, but degeneration occurs over time.

Stability of synovial joints

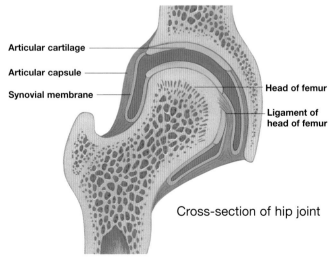

Articular cartilage

Articular capsule

Synovial membrane

Head of femur

Ligament of head of femur

Cross-section of hip joint

To prevent instability and possible dislocation, a synovial joint depends upon three factors – the shape of the bone surfaces, the presence of ligaments and the tone of the surrounding muscle:

■ Articular surfaces – in some cases the shape of the articular surfaces is important in the stability of a joint. The ball and deep socket arrangement of the hip, for example, greatly adds to this joint's stability

The free movement of synovial joints means they rely for stability on the surrounding ligaments and muscles. The joint shape also plays a part.

■ Ligaments – these support and strengthen the articular capsule of synovial joints and help to stabilize them by preventing excessive movement. However, the ligaments can be damaged by over-stretching if the joint is not also stabilized by other means

■ Muscle tone – the tone in the joint's surrounding muscles is the most important factor in its stabilization. The gentle background contraction of muscle fibres, even when the muscle is relaxed, acts to provide stability. The muscle both holds the joint ends together and brings them back into alignment after movement.

Types of synovial joints

Although all synovial joints have many structural features in common, they can be divided into six distinct groups. The distinctions rest on the shape of their articular surfaces and the type of movement the joints allow.

PLANE JOINTS
In plane joints the articular surfaces are flat and usually allow movement in one plane only. Examples include the acromioclavicular joint between the shoulder blade and the collarbone, and the joints between the articular processes of the vertebrae in the spine.

HINGE JOINTS
Hinge joints act like the hinge on a door: the articular surfaces can move in only one plane around one axis. The best example of a hinge joint is the elbow where only flexion (bending) and extension (straightening) are allowed.

PIVOT JOINTS
In a pivot joint a rounded or conical process of one bone inserts into a sleeve or ring of another. Rotation is the only permitted movement, such as is illustrated by the atlas and axis vertebrae, which allow the head to be turned from side to side.

BALL AND SOCKET JOINTS
In a ball and socket joint one articular surface is rounded and sits in a cup-shaped socket. This is the most mobile of joint types, allowing movement in all directions. Examples include the hip and the shoulder joints.

Plane joint

Plane joints allow movement in one plane only

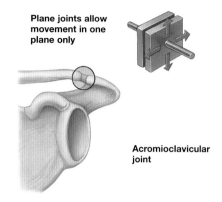

Acromioclavicular joint

Hinge joint

Hinge joints move in only one plane around one axis

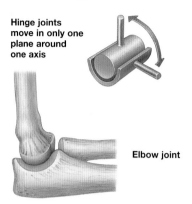

Elbow joint

Pivot joint

Pivot joints only allow a rotating movement

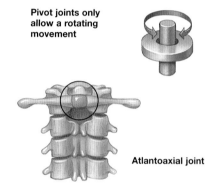

Atlantoaxial joint

Ball and socket joint

Ball and socket joints allow movement in all directions

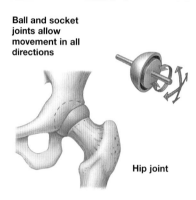

Hip joint

Each type of synovial joint allows a particular kind of movement. This ranges from movement in one plane to that over several.

SADDLE JOINTS
The saddle-shaped articular surfaces of this joint allow movement in two different planes, such as occurs at the base of the thumb where the first metacarpal articulates with the trapezium of the wrist.

CONDYLOID JOINTS
Condyloid joints have oval articular surfaces which allow a range of movements including flexion, extension, side-to-side movement and the circular movement of circumduction. One example is the 'knuckles', or metacarpophalangeal joints.

Degeneration of synovial joints

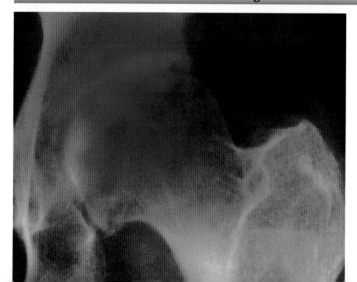

The structure of synovial joints makes them capable of withstanding considerable physical stress over the course of a lifetime. However, in many cases they eventually suffer damage and degenerate over time. This degenerative change in synovial joints is known as osteoarthritis and is one of the most common disabling conditions of modern life.

The degeneration that occurs in osteoarthritis is clearly shown in this X-ray of a hip. The head of the femur is almost touching the pelvic socket (highlighted pink).

OSTEOARTHRITIS
In osteoarthritis the protective cartilage covering the joint's articular surfaces gradually roughens and wears away. As the cartilage thins and erodes it becomes less effective as a shock absorber and lubrication is lost. This leads to stiffness and pain on movement. Repeated stress on a joint and previous disease or injury increase the risk of acquiring the disease.

Although ageing is not the only precipitating factor, the condition is more common in the elderly and occurs most often in weight-bearing joints.

Types of muscle

There are three main types of muscle in the body – skeletal muscle is used for voluntary movement, smooth muscle controls internal organs and cardiac muscle keeps the heart beating.

The most familiar muscles in the body are the skeletal muscles, (also known as striated, or voluntary, muscles), many of which are visible under the skin. Voluntary muscles can be under conscious control, and can also contract in a reflex action such as when the knee straightens when the patellar tendon is tapped (the knee jerk reflex).

STRUCTURE
The muscle fibres of each skeletal muscle are bound together by connective tissue (epimysium), and divided into groups or fascicles by a sheath (perimysium). Within these fascicles, each muscle fibre is surrounded by an endomysium. The whole muscle is attached to bone by a tough fibrous band, the muscle tendon.

FUNCTION
Skeletal muscle can be very adaptable. These muscles can contract powerfully, exerting a great deal of force, such as in lifting a heavy object. Alternatively, they can exert a small force to perform a delicate action such as picking up a feather. Another feature of skeletal muscle, which becomes obvious after performing exercise, is that it tires easily. Whereas the heart can beat all day, every day, without ceasing, skeletal muscle needs a period of rest after a contraction.

Connective tissue sheaths of skeletal muscle

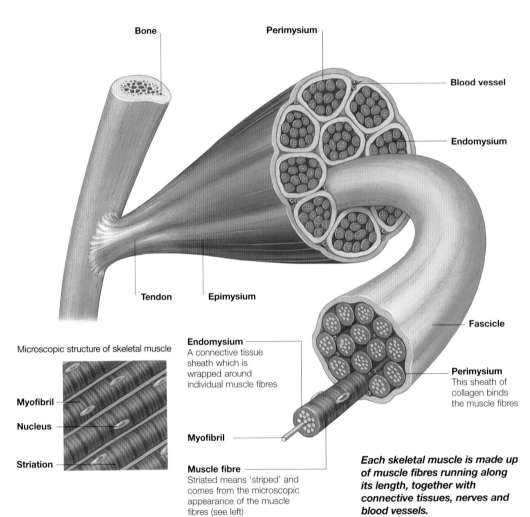

Bone

Perimysium

Blood vessel

Endomysium

Tendon

Epimysium

Fascicle

Perimysium
This sheath of collagen binds the muscle fibres

Microscopic structure of skeletal muscle

Myofibril

Nucleus

Striation

Endomysium
A connective tissue sheath which is wrapped around individual muscle fibres

Myofibril

Muscle fibre
Striated means 'striped' and comes from the microscopic appearance of the muscle fibres (see left)

Each skeletal muscle is made up of muscle fibres running along its length, together with connective tissues, nerves and blood vessels.

Smooth (involuntary) muscle

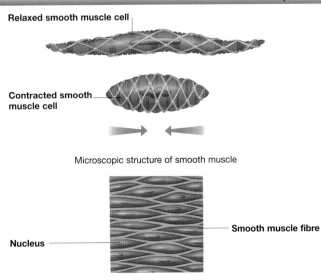

Relaxed smooth muscle cell

Contracted smooth muscle cell

Microscopic structure of smooth muscle

Nucleus

Smooth muscle fibre

Smooth muscle cells contract in a gradual, synchronized manner. These contractions are much slower than those of the skeletal muscles.

Smooth muscle is named for the lack of striations, or stripes, when viewed under the microscope. It is also known as involuntary muscle as its actions do not come under a person's conscious control.

LOCATION OF SMOOTH MUSCLE
Smooth muscle is found in the walls of hollow structures within the body, such as the gut, blood vessels and the bladder. Here, it acts to regulate the size of the lumen (central space) as well as causing wave-like peristalsis in some organs (such as the gut and ureters). Smooth muscle is also found in the skin, where it acts upon hairs, and in the eyeball, where it determines the thickness of the lens and size of the pupil.

NERVOUS SYSTEM
Smooth muscle is controlled by the autonomic nervous system, the part of the nervous system that is concerned with regulation of the internal environment of the body as well as the response to stress. Unlike skeletal muscle, smooth muscle can keep up a steady contraction for a long period of time.

Shapes of skeletal muscle

Although all skeletal muscles are made up of fascicles, or groups of muscle fibres, the arrangement of these fascicles may vary. This variation leads to a number of different muscle shapes throughout the body.

There are several ways of describing the various shapes of muscle, including:

■ Flat – muscles, such as the external oblique in the abdominal wall, may be flat, yet fairly broad. They may cover a wide area and sometimes insert into an aponeurosis (a broad sheet of connective tissue)

■ Fusiform – many muscles are of this 'spindle-shaped' form, where the rounded belly tapers at each end. Examples include the biceps and triceps muscles of the upper arm which have more than one head

■ Pennate – these muscles are named for their similarity to a feather (the word 'penna' means feather). They may be described as being unipennate (for example, extensor digitorum longus), bipennate (such as rectus femoris) or multipennate (for example, the deltoid). Multipennate muscles resemble a number of feathers placed next to one another

■ Circular – these muscles, also known as sphincteral muscles, surround body openings. Contraction of these muscles, where the fibres are arranged in concentric rings, closes the opening. Circular muscles within

Fascicle arrangement in relation to muscle structure

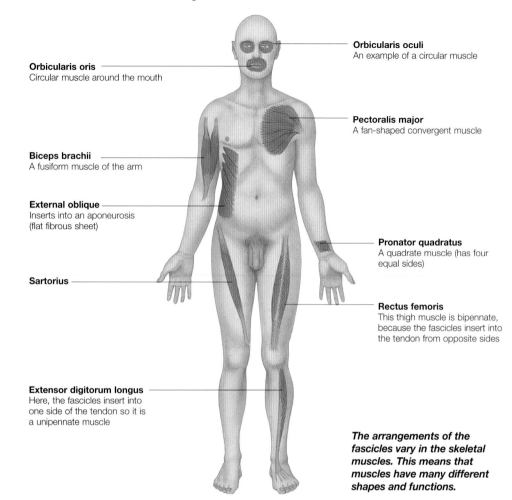

Orbicularis oris
Circular muscle around the mouth

Biceps brachii
A fusiform muscle of the arm

External oblique
Inserts into an aponeurosis (flat fibrous sheet)

Sartorius

Extensor digitorum longus
Here, the fascicles insert into one side of the tendon so it is a unipennate muscle

Orbicularis oculi
An example of a circular muscle

Pectoralis major
A fan-shaped convergent muscle

Pronator quadratus
A quadrate muscle (has four equal sides)

Rectus femoris
This thigh muscle is bipennate, because the fascicles insert into the tendon from opposite sides

The arrangements of the fascicles vary in the skeletal muscles. This means that muscles have many different shapes and functions.

the face include the orbicularis oculi, which closes the eye

■ Convergent – these muscles are fan-shaped and the muscle fibres arise from a wide origin and converge on a narrow tendon. In some cases, these muscles take on a triangular shape. Examples include the large pectoral muscles.

FUNCTION
The arrangement of the fascicles within a muscle influences that muscle's action and power. When muscle fibres contract, they shorten to about 70 per cent of their relaxed length. If the muscle is long with parallel fibres, such as the sartorius muscle in the leg, it can shorten a great deal but has

little strength.

If the degree of shortening is not as important as the power it can produce, the muscle may have numerous fibres packed tightly together and converging on a single point. This is the arrangement in multipennate muscles such as the deltoid in the shoulder.

Cardiac muscle

Microscopic structure of cardiac muscle

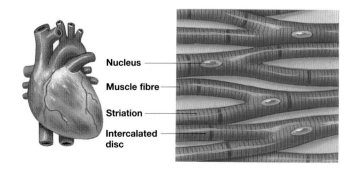

Nucleus

Muscle fibre

Striation

Intercalated disc

Cardiac muscle is a specialized form of striated muscle which is found only in the heart and walls of the great vessels adjoining it, such as the aorta and superior vena cava.

This type of muscle makes up almost all the mass of the thick heart walls, the myocardium. Here, the fibres are arranged in a distinctive spiral pattern which causes the blood to be squeezed through the heart as a wave of contraction spreads.

The function of the cardiac muscle is to pump blood from the heart. This involuntary muscle contracts rhythmically and spontaneously.

Although cardiac muscle is striated, it is not under conscious control like skeletal muscle, but is controlled by the autonomic nervous system. The fibres in cardiac muscle are unusual in that they branch and have specialized junctions called intercalated discs.

RATE OF CONTRACTION
Cardiac muscle has the ability to contract spontaneously, without an external signal from a nerve although, in a healthy heart, the rate of contraction is controlled by the heart's nerve supply. Even when removed from the body, the heart will continue to contract for a short time.

How muscles contract

Muscle tissue accounts for about half of the body's total mass,
and is constantly at work, whether articulating the skeleton, enabling the
heart to beat or passing food through the gut.

Muscle is tissue that is capable of contracting. The two main types of muscle are voluntary and involuntary muscle. The contraction of voluntary – or skeletal – muscle can be consciously controlled, and this type of muscle is linked to parts of the skeleton to produce physical movement.

INVOLUNTARY MUSCLE

Involuntary muscle is not under the brain's conscious control. It is controlled automatically by a special part of the nervous system and is found in non-skeletal parts of the body. The heart, for example, is made up of involuntary muscle, beating without conscious effort.

VOLUNTARY MUSCLE

Muscle that moves bones is known as striated muscle due to its striped appearance under the microscope. It consists of bundles of fibres bound tightly together, with each fibre made up of a single long, multi-nucleated cell that stretches from one end of the muscle to the other. Each fibre consists of many long thin strands, known as myofibrils. These are made up of two kinds of tiny, overlapping protein filaments made of actin and myosin, giving the myofibril a banded appearance. The bands of neighbouring myofibrils line up so that the whole fibre appears to be striped.

Muscle structure

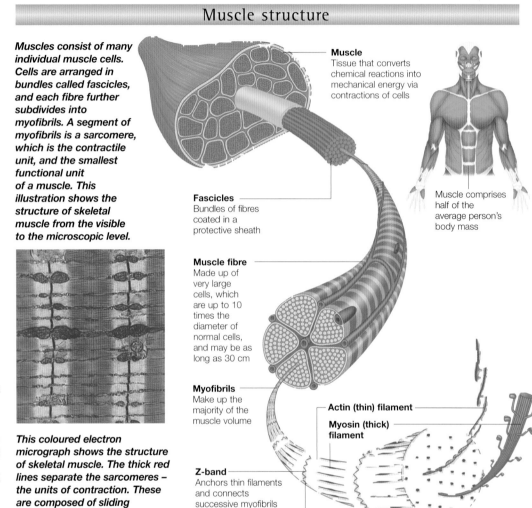

Muscles consist of many individual muscle cells. Cells are arranged in bundles called fascicles, and each fibre further subdivides into myofibrils. A segment of myofibrils is a sarcomere, which is the contractile unit, and the smallest functional unit of a muscle. This illustration shows the structure of skeletal muscle from the visible to the microscopic level.

This coloured electron micrograph shows the structure of skeletal muscle. The thick red lines separate the sarcomeres – the units of contraction. These are composed of sliding filaments: myosin (pink) and actin (yellow).

Muscle
Tissue that converts chemical reactions into mechanical energy via contractions of cells

Fascicles
Bundles of fibres coated in a protective sheath

Muscle comprises half of the average person's body mass

Muscle fibre
Made up of very large cells, which are up to 10 times the diameter of normal cells, and may be as long as 30 cm

Myofibrils
Make up the majority of the muscle volume

Actin (thin) filament

Myosin (thick) filament

Z-band
Anchors thin filaments and connects successive myofibrils

Sarcomere

Muscle contraction

A muscle contracts when it is stimulated by a nerve impulse, which causes complex chemical changes to take place in the muscle fibres. Each group of filaments lies in a small chamber (sarcomere) in which the thin actin filaments are attached to each end. The thick myosin filaments lie between the actin filaments in the middle of the sarcomere.

When provided with energy – usually obtained from glycogen ('animal starch') stored in the muscle – they form chemical bonds with the actin filaments, and these bonds are repeatedly broken and remade further along. In this way, the myosin filaments work their way along the actin

filaments like ratchets, with the result that the whole sarcomere becomes shorter and fatter.

When the muscle is no longer being stimulated, the chemical action ceases. The bonds between the filaments are no longer formed and the muscle relaxes.

Contraction of an opposing muscle stretches the filaments apart, and this is triggered by a chemical called acetylcholine, which is released by the nerve endings and alights on special receptive areas in the muscle. As long as acetylcholine is present in these areas, the muscle remains contracted.

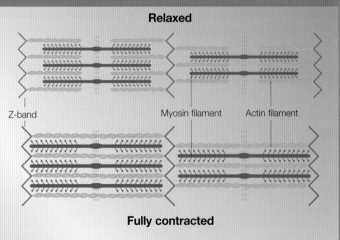

Relaxed

Z-band

Myosin filament   Actin filament

Fully contracted

Contraction is achieved by myosin fibres rapidly breaking and reforming bonds with the actin fibres, in a ratchet-like manner.

How involuntary muscles move

The body contains two types of involuntary muscle (muscle not under the conscious control of the brain). Smooth muscle can focus the eye and pass food along the digestive tract; cardiac muscle causes the heart to beat.

Smooth muscle

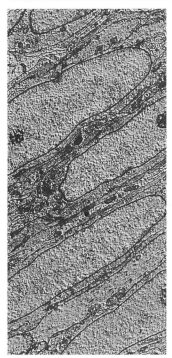

The smooth muscle cells enclosing the inner wall of the uterus are shown in this false coloured electron micrograph. They are responsible for muscular contractions during labour and childbirth.

Smooth muscle and cardiac muscle are both able to contract involuntarily, without conscious control. They are controlled by nerve impulses from the autonomic (unconscious) nervous system.

Smooth muscle is found in many parts of the body, notably the gut, but also in such places as the lungs, bladder and sex organs. It consists of spindle-shaped cells, whose average length is only a fraction of a millimetre.

CELL ARRANGEMENT

The cells are tapered at both ends, have single nuclei and are arranged in bundles held together by a substance that acts as a cement. These bundles are grouped into larger bundles or flattened bands, held together by connective tissue. The arrangement of the cells is much looser than the regular pattern found in striped muscle, but the contraction of smooth muscle still results from the movement of filaments, which are found in the walls of the cells.

Contraction of smooth muscle is generally slower than that of striated muscle, and contraction does not necessarily take place throughout the whole muscle.

An action typical of smooth muscle is found in the intestines, where a band of muscle usually contracts over a certain part of its length, then relaxes while another part contracts, thus producing waves of contraction down the muscle, called peristalsis. This enables food to be passed down the digestive tract, into the stomach and through the intestines.

Smooth muscle surrounds hollow body structures, such as the oesophagus, bladder, uterus and blood vessels. The rate of contraction of its cells is relatively slow, but they are more energy-efficient and able to maintain contraction for a longer time.

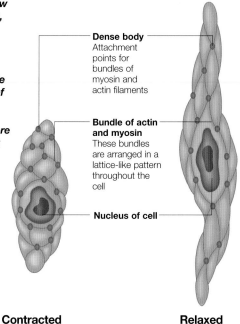

Dense body
Attachment points for bundles of myosin and actin filaments

Bundle of actin and myosin
These bundles are arranged in a lattice-like pattern throughout the cell

Nucleus of cell

Contracted   **Relaxed**

Cardiac muscle

Cardiac muscle is only found in the heart, and its structure is somewhere between that of striated muscle and smooth muscle. It has a striped appearance when viewed through a microscope, but the cells of which it is made are shorter and more box-shaped than striated muscle fibres. Most of the cells are divided at the ends, and the subdivisions form connections with cells that lie alongside. In this way, a resilient network of fibres is formed with the ability to act in unison, and it is this structure that gives heart muscle its toughness.

Cardiac muscle has to be immensely strong for the demanding job it has to do. In an average lifetime, the heart beats over two billion times and pumps some 550,000 tonnes of blood. In order to keep the heart contracting steadily and regularly throughout this time, the heartbeat is controlled by electrical impulses.

Cardiac muscle cells are less elongated than skeletal muscle. Adjacent cells are closely attached to each other by proteins called intercalated discs. Structures called desmosomes form junctions, allowing electrical signals to be transmitted between cells.

Sliding filaments
Thick and thin filaments of actin and myosin

Cardiac cell

Intercalated disc
Connects cardiac cells together, both physically and electrically

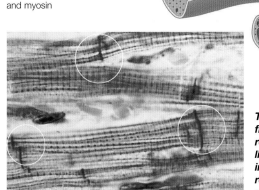

This light micrograph shows the individual fibres which make up cardiac muscle. The round bodies are cell nuclei and the dark lines (circled) at right angles to the fibres are intercalated discs. These have a low electrical resistance which allows contractions to spread rapidly throughout the muscle.

Skin and nails

The skin, together with the hair and nails, makes up
the integumentary system. Functions of the skin include heat
regulation and defence against microbial attack.

The skin covers the entire
human body and has a surface
area of about 1.5–2 m². It
accounts for about 7 per cent of
the weight of the body and
weighs around 4 kg.

TWO LAYERS
Skin is composed of two layers –
the epidermis and dermis.
■ Epidermis – this is the thinner
of the two layers of skin and
serves as a tough protective
covering for the underlying
dermis. It is made up of
numerous layers of cells, the
innermost of which consist of
living cube-shaped cells that
divide rapidly, providing cells for
the outer layers.

By the time these cells reach
the outer layers, they have died
and become flattened, before
being 'sloughed off' by abrasion.
The epidermis has no blood
supply of its own, and depends
upon diffusion of nutrients from
the plentiful supply of blood to
the dermis below
■ Dermis – this is the thicker
layer of skin, which lies
protected under the epidermis. It
is composed of connective tissue
which has elastic fibres to keep
it stretchy, and collagen fibres
for strength. The dermis contains
a rich supply of blood vessels as
well as numerous sensory nerve
endings. Lying within this layer
are the other important
structures of the integumentary
system, including hair follicles
and oil (sebaceous) and sweat
glands.

Cross section of skin

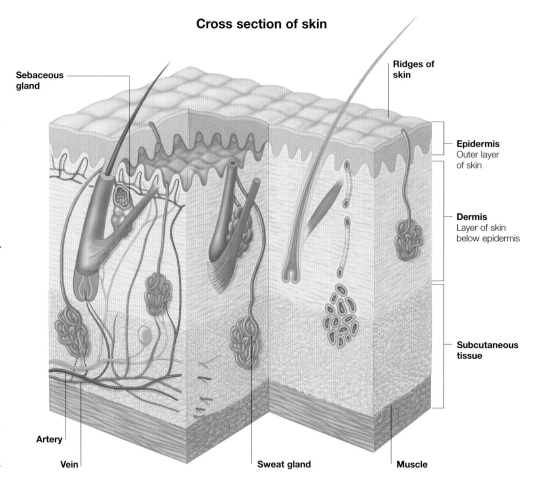

Sebaceous gland

Ridges of skin

Epidermis
Outer layer
of skin

Dermis
Layer of skin
below epidermis

Subcutaneous tissue

Artery

Vein

Sweat gland

Muscle

*The skin has been described as
the largest organ in the body. It
helps to regulate temperature
through narrowing and widening
of blood vessels in the dermis.*

Skin colour

The colour of skin varies greatly,
not just between different races,
but also between individuals of
the same race.

THREE PIGMENTS
Skin colour is determined by
three pigments: melanin,
carotene and haemoglobin.
Melanin, which ranges in colour
from red to brown to black, is
made in specialized cells called
melanocytes that lie within the

*Skin colour varies greatly
between individuals, especially
between those of different racial
groups. This is due to varying
levels of the three skin pigments.*

lower layers of the epidermis. All
humans of all racial groups have
the same number of melanocytes
even though skin colour varies
so widely. The melanocytes of
dark-skinned people produce
more and darker melanin than
those of light-skinned people.

Carotene is an orange
pigment, absorbed from
vegetables such as carrots. It
accumulates in the outermost
layer of the epidermis and is
most noticeable on the palms
and soles. Haemoglobin within
the blood vessels of the dermis
gives the skin a pinkish hue,
especially if there is little
melanin present in the skin.

Nails

Human nails are the equivalent of the hooves or claws of other animals. They form a hard protective covering for the vulnerable fingers and toes, and they provide a useful tool for scratching or scraping when this is required.

Nails lie on the dorsal (back) surfaces of the ends of the fingers and toes, overlying the terminal phalanx, or final bone, of each.

CONSTITUENT PARTS
The parts of the nail include:
■ Nail plate – each nail is composed of a plate of hard keratin (the same substance as is found in hair) which is continuously produced at its root
■ Nail folds – except for the free edge of the nail, at its furthermost end, the nail is surrounded and overlapped by folds of skin (nail folds)
■ Free edge – the nail separates from the underlying surface at its furthermost point to form a free edge. The extent of this nail at the free edge depends upon personal preference and wear and tear
■ Root, or matrix – this lies at the base of the nail beneath the nail itself and the nailfold. This part of the nail is closest to the skin, and it is here that the hard keratin of the nail is produced by cell division. If the root of the nail is destroyed the nail cannot grow back
■ Lunula – the paler, crescent-shaped area located at the base

Cross section of nail

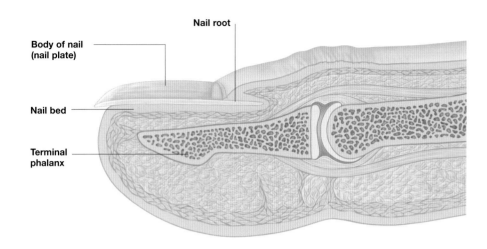

Body of nail (nail plate)

Nail root

Nail bed

Terminal phalanx

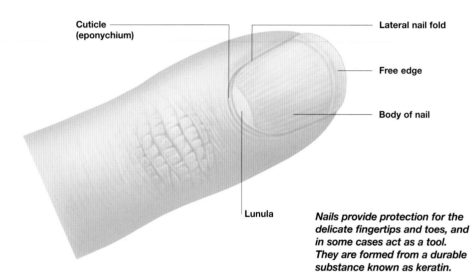

Cuticle (eponychium)

Lateral nail fold

Free edge

Body of nail

Lunula

Nails provide protection for the delicate fingertips and toes, and in some cases act as a tool. They are formed from a durable substance known as keratin.

of the nail where the matrix is visible through the nail
■ Cuticle (eponychium). This covers the proximal (near) end of the nail and extends over the nail plate to help protect the matrix from infection by invading micro-organisms.

GROWTH
Fingernails grow much more quickly than toenails. A mark made over the lunula of a fingernail will take three months to reach the free edge, whereas the corresponding time for a toenail may be up to two years. For a normal rate of growth, and

to produce normal, pink, healthy nails, there needs to be a good blood supply to the root of the nail; nails look pink because of the large number of blood vessels in the dermis. Nails grow at a rate of about 0.1mm a day, but when there is injury to a nail, the growth speeds up.

Psoriasis

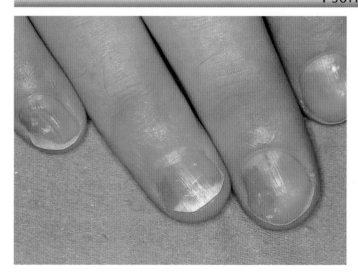

Psoriasis is a troublesome skin condition that occurs in about 2 per cent of the population. Its cause is unknown, although there seems to be an inherited factor in some cases. The onset tends to occur in adolescence and further attacks may be triggered by stress or infection.

CELL BUILD-UP
The main feature of psoriasis is a very rapid proliferation of the cells at the base of the

Psoriasis causes certain characteristic changes to the nails. The nails may be pitted, thick and ridged, and may be loosened from the nail bed.

epidermis, the outermost layer of the skin. This causes a build up of cells in the epidermis which then form red, scaly plaques.

For many people, psoriasis is no more than a nuisance which recurs from time to time. For some, however, it is a severe debilitating disease that can affect other parts of the body as well, such as the joints.

ABNORMAL NAILS
The nails are often affected in psoriasis. Separation of the nail plate from the nail bed at its distal end (onycholysis) can occur as well as general thickening and ridging of the nails (dystrophy).

How skin protects the body

The skin is a remarkable organ, covering the entire surface area of the body. Skin plays a number of important roles in protecting the body and also helps to control body temperature.

The skin is the largest organ of the body. It can weigh from around 2.5 to 4.5 kg and covers an area of about two square metres.

ANATOMY OF THE SKIN
The skin is composed of two distinct layers: the epidermis and the dermis.

The epidermis, or cuticle, is the outer protective layer of the skin. The outermost layer of the epidermis (the stratum corneum, or horny layer) accounts for up to three quarters of the epidermal thickness.

KERATIN
Cells of the epidermis produce keratin (a fibrous protein also found in the hair and nails) and are progressively pushed outward by dividing cells beneath them.

As the cells move outwards they become enriched with keratin, flatten out and die. These dead cells are constantly

shed, the epidermis thus being effectively replaced every few weeks. In fact, the average person sheds around 18 kg of skin (in the form of dandruff or dry skin flakes) in a lifetime.

SKIN THICKNESS
The epidermis is thickest on the parts of the body that receive the greatest wear, for example the soles of the feet and the palms of the hands.

DERMIS
The innermost layer of the skin is the dermis. This fibrous layer comprises a network of collagen and elastic fibres.

The dermis also contains blood vessels, nerves, fat lobules, hair roots, sebaceous glands and sweat glands.

The skin is composed of two main layers: the epidermis and the dermis. The epidermis is nourished indirectly by the blood vessels in the dermis.

Anatomy of the skin

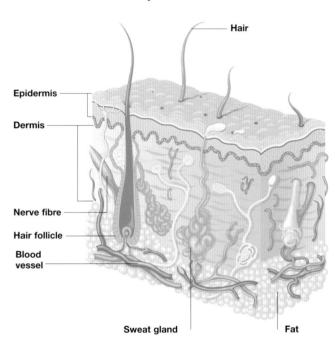

- Hair
- Epidermis
- Dermis
- Nerve fibre
- Hair follicle
- Blood vessel
- Sweat gland
- Fat

Role of skin

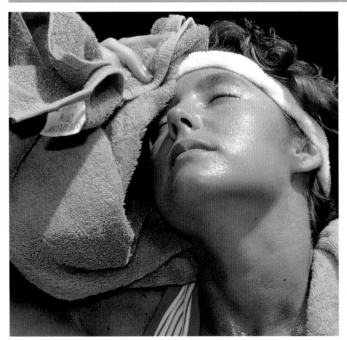

The skin plays an important role in the regulation of temperature. Sweat glands produce a salty solution which cools the body as it evaporates.

The skin plays a number of important roles. These include:

■ Protection – the collagen fibres of the dermis give the skin strength and resistance, preventing any object from penetrating the body.

■ Regulation of temperature – through vasoconstriction (narrowing) and vasodilation (widening) of blood vessels in the dermis. The production of sweat also helps to cool the body.

■ Barrier against bacterial infection – large numbers of micro-organisms are naturally present on the surface of the skin. These compete with harmful bacteria, preventing them from invading the body.

■ Sensitivity to touch and pain – the dermis contains a dense network of nerve endings sensitive to pain and pressure.

These nerves provide the brain with vital information about the body in relation to its environment, allowing it to act accordingly, for example retracting the hand when something hot is touched.

■ Prevention of unregulated water loss – the sebaceous glands of the dermis secrete an oily substance known as sebum. This coats the skin, making it effectively waterproof. Collagen fibres within the dermis also hold water.

■ Protection against ultraviolet (UV) radiation – the pigment melanin (produced by melanocytes in the epidermis) acts as a filter to the harmful ultraviolet radiation produced by the sun.

■ Manufacture of vitamin D – this is produced in response to sunlight and helps to regulate the metabolism of calcium.

Skin colour

Skin colour largely depends on the presence of melanin. Production of this pigment protects the skin from harmful radiation produced by the sun.

The colour of the skin depends on a combination of factors, such as skin thickness, blood flow, and pigment concentration.

PIGMENTS

In areas where the skin is very thin and blood flow is good, it will appear much darker (such as over the lips) due to the red colour of the pigment

Skin colour is mainly dependent upon the number of melanin-producing cells. People with albinism have no such cells, making them very pale skinned.

haemoglobin in the blood.

In general, the production of melanin will determine how dark the skin is. This pigment is produced by melanocyte cells present in the epidermal layer.

Dark-skinned people have a high proportion of melanocytes, and hence greater concentrations of melanin in their skin.

SUN EXPOSURE

Skin responds to ultraviolet rays in sunlight by producing greater amounts of melanin.

As levels of melanin increase, the skin darkens forming a filter against the harmful radiation produced by the sun.

Freckles are another example of the skin's reaction to the sun,

representing concentrated areas of melanin-producing cells.

SUN BURN

If sun exposure does not take place gradually however, the skin is unable to produce melanin fast enough to filter out the suns harmful rays.

As a result the skin burns, becoming inflamed and very tender. Prolonged exposure to UV radiation can permanently

In response to sunlight, melanin-producing cells become more active. As levels of melanin increase the skin darkens, filtering out harmful radiation.

damage the skin cells leading to premature ageing of the skin and, sometimes, skin cancer.

Skin cancer tends to be less common in dark-skinned people, which is testament to the protective role of melanin.

Skin repair

When skin is cut, for example by surgical incision, the sides of the wound will automatically grow back together if they are held in place with stitches.

Where there is tissue loss however, a remarkable process

occurs whereby new skin is regenerated.

Skin cells adjacent to the wound break away from the cells below, migrate to the wounded area and enlarge.

Other cells surrounding the

wound multiply rapidly to replace the cells lost.

Eventually from all sides of the wound the migrating cells meet. Once the wound is entirely covered, cell migration stops.

The wound will continue to

heal as epithelial cells multiply, and normal thickness is restored.

When skin is wounded, the surrounding cells move to the site of the wound, and multiply until the area is covered.

Skin grafts

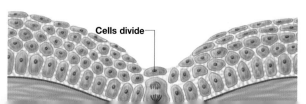

In cases where the skin is severely damaged, such as with third degree burns, medical intervention may be required. This is because the damaged area may be too large for the skin to regenerate itself before infection sets in.

Transplantation of skin
A process known as skin grafting involves the removal of a fine sheet of skin from elsewhere on the body, usually a fleshy area such as the thigh or buttock.

Sometimes damage to the skin is so bad that a skin graft is required. This procedure involves transplanting skin from elsewhere on the body.

This skin is then transplanted on to the wound. With time these new skin cells proliferate, join together and heal the area.

New techniques are being developed, which involve the culturing of skin cells in a laboratory. Using this technique, skin can be grown specifically for transplantation.

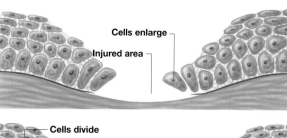

Cells enlarge
Injured area

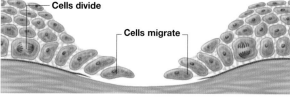

Cells divide
Cells migrate

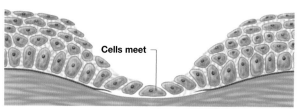

Cells meet

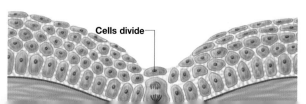

Cells divide

How nails grow

The nails are extensions of the outer skin layer, and grow continuously throughout life. Apart from their protective function, they can also give a good indication of a person's state of health.

Like the hair, the nails are a derivative of skin, and form part of the outer covering of the body. Each nail is a scale-like extension of the epidermis (outer layer of skin) that covers the end of the finger and toe.

ANATOMY OF THE NAIL

The nails are flattened, elastic structures that begin to grow on the upper surface of the tips of the fingers and toes in the third month of fetal development.

Each nail consists of the following parts:
■ Body – otherwise known as the nail plate. This is the main, exposed part of the nail
■ Free edge – this is the part of the nail that tends to grow beyond the fingertip
■ Lateral nail fold – this is the bulge of skin that grows on either side of the nail. Folds arise at the boundary of the epidermis and nail because the epidermal cells divide more quickly than those of the nail and cause the skin to bulge over the nail
■ Eponychium (cuticle) – this is

a fold of cornified (dead) skin that partially covers the nail and also protects the growing area of the nail
■ Lunula – this is the slightly opaque area of the nail, which is crescent shaped (lunula means 'little moon' in Latin). This area of the nail may be partially obscured by the cuticle
■ Hyponchium – this is the area of skin attached just below the free border of the nail. The hyponchium has a very rich nerve supply, which is why it can be very painful if foreign bodies such as a splinter of wood penetrate it
■ Root – otherwise known as the matrix, this is the proximal part of the nail (closest to the skin) and is implanted in a groove beneath the cuticle
■ Nail bed – this is the area underlying the entire nail.

The nails consist of curved plates of hard keratin. Beneath the lunula area lies the nail matrix – this is responsible for the growth of the nail.

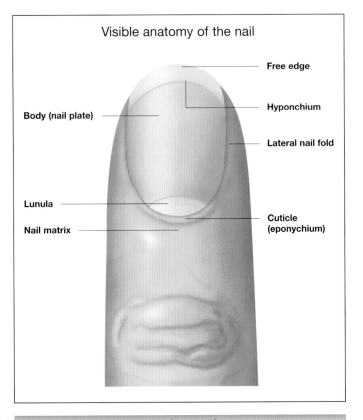

Visible anatomy of the nail

- Body (nail plate)
- Lunula
- Nail matrix
- Free edge
- Hyponchium
- Lateral nail fold
- Cuticle (eponychium)

The role of nails

Despite the fact that the nails are not as strong as those of our ancestors, they still serve a number of important roles.

PROTECTIVE ROLE

Like skin and hair, the nails are composed of keratin, a tough protein. This acts as a shock absorber, protecting the tips of the fingers and toes.

In addition, the fingernails are useful tools for tasks such as undoing a shoelace, picking up

The nails are well designed for enhancing the movements of the fingers, such as scratching. They also protect the sensitive tips of the fingers and toes.

small objects or scratching an itch.

Despite the fact that the nails lack nerves, they also serve as excellent 'antennae', since they are embedded in sensitive tissue that detects any impact when the nail touches an object.

Brittle nails

The nails are very porous and they can hold up to 100 times as much water as the equivalent weight of skin. In this way, the nails limit the amount of water entering the tissues of the fingertips.

Water taken up by the nails is eventually lost through evaporation as they dry out and

resume their normal size.

Frequent immersion in water and then drying can cause the structure of the nail to become weakened, resulting in nails becoming brittle and splitting.

In addition, the use of nail varnish, and its removal with solvents, can cause the nails to become brittle.

After death

It is a common misconception that the fingernails continue to grow for a short time after death.

It is understandable how this myth came to be believed, however, since after death the skin surrounding the nails dries up and shrivels away from the nail plates. This phenomenon creates the impression that the nails have actually grown in length, in the same way that pushing the cuticles back makes the nails seem longer.

In reality, however, every cell in the body ceases to grow after death, including those of the nails.

It is often suggested that nails and hair continue to grow after death. Once death has occurred, however, every cell in the body ceases to function.

Growth rate of nails

It can take up to six months for a nail to grow from the root to the tip of the finger. At certain times, nail growth is accelerated, for instance in warm weather.

There are two areas of the nail in which growth occurs:
■ The germinal matrix – this is the area beneath the root of the nail. Here, epidermal cells divide, and become enriched with keratin, which thickens to become nail
■ The nail bed – this is the area underneath the nail plate; it provides a surface over which the growing nail divides.

RATE OF GROWTH
On average, it takes around three to six months for a nail to grow from its base to the tip of the finger. The average fingernail grows at a rate of about 0.5 mm a week, with faster growth taking place in the summer. It is thought that blood circulates faster in the summer so that cell division is more rapid. Fingernails grow around four times faster than toenails; the reason for this is unknown.

Interestingly, if a person is right handed, the nail of the right thumb grows faster than that of the left thumb.

▶ *In most people, finger and toe nails are kept short due to abrasion or cutting. Without this, the nails are capable of growing to a great length.*

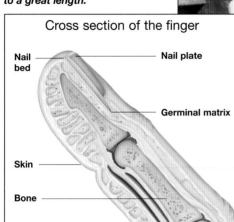

Cross section of the finger

Nail bed — Nail plate
— Germinal matrix
Skin —
Bone —

▲ *Injury to a nail tends to accelerate its growth until it has recovered. However, if the root of the nail is destroyed, the nail will cease to grow.*

Nail disorders and damage

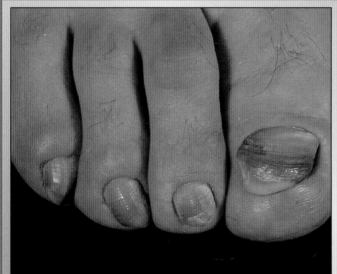

The nails can reveal much about the health of a person.

Blood supply
Normally appearing pink in colour due to the rich blood supply in the skin beneath it, the nail acts as an indicator of oxygen supply to anaesthetists during surgery. This is the reason why women are always asked to remove any nail varnish before undergoing an

In yellow nail syndrome, the nails become thickened and yellow. It is associated with swelling of the feet or may be due to thyroid disease.

operation. If the nail were to become pale in colour or even to turn blue, this would alert the anaesthetist to the fact that the patient was not receiving enough oxygen.

Nail disorders
The state of the nails can be useful in helping to diagnose a number of disorders.

Grooves running across all the nails may reveal that a person has suffered a serious illness several months before. This is because illness slows down the rate of nail growth, causing ridges to develop in the nail root. These ridges are then pushed outwards as the nail grows.

Similarly, misshapen nails which are bent backwards may indicate anaemia (iron deficiency).

The colour of the fingernails is also very revealing. For example, white opaque nails may indicate cirrhosis of the liver, while white bands on the nails may be a clue that mild arsenic poisoning has taken place.

Nail damage
More severe changes in the nail such as the nail turning blue or falling off altogether are commonly due to the nail bed becoming damaged through injury. As long as the nail root is not destroyed, the nail will eventually replace itself and continue to grow.

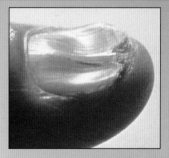

Brittle, spoon-shaped nails indicate that a person has koilonychia. This is a sign of anaemia and occurs because of a lack of iron in the cells.

Ingrowing toenails are caused by cutting nails too close at the edges. This causes the nail to grow into the flesh, resulting in inflammation and infection.

Hair

There are two main types of human hair: vellus and terminal. Only terminal hair, which is found mainly on men, has a central core or medulla and responds to the male sex hormone, testosterone.

The surface of the human body is covered with millions of hairs. They are most noticeable on the head, around the external genitalia and under the arm. The only regions of the body without hairs are the lips, nipples, parts of the external genitalia, the palms of the hands and the soles of the feet.

Although hair does not really serve to keep us warm, as it does in other mammals, it has a number of other functions:
■ Sensing small objects or insects that approach the skin
■ Protecting/insulating the head
■ Shielding the eyes
■ Sexual signalling.

STRUCTURE OF A HAIR

Hair is composed of flexible strands of the hard protein, keratin. It is produced by hair follicles within the dermis (the inner layer of the skin) but arises from an 'inpouching' of the epidermis (the outer layer).

Each hair follicle has an expanded end – the hair bulb – which receives a knot of capillaries to nourish the root of the growing hair shaft. The shape of the hair shaft determines whether the hair is straight or curly: the rounder the shaft in cross section, the straighter the hair.

Each hair is made up of three concentric layers:
■ The medulla
■ The cortex
■ The cuticle.

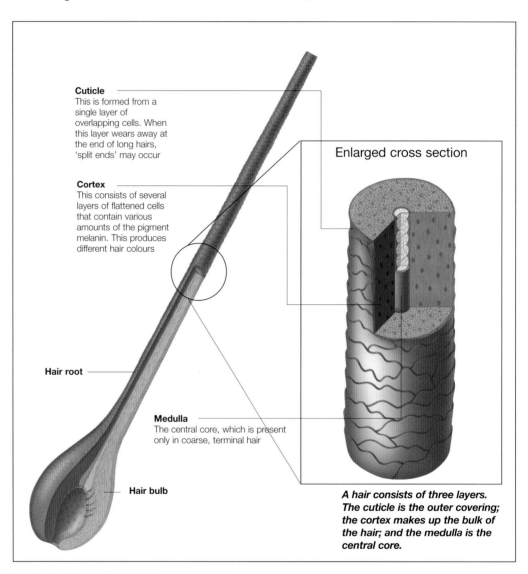

Cuticle
This is formed from a single layer of overlapping cells. When this layer wears away at the end of long hairs, 'split ends' may occur

Cortex
This consists of several layers of flattened cells that contain various amounts of the pigment melanin. This produces different hair colours

Enlarged cross section

Hair root

Medulla
The central core, which is present only in coarse, terminal hair

Hair bulb

A hair consists of three layers. The cuticle is the outer covering; the cortex makes up the bulk of the hair; and the medulla is the central core.

Types of hair and their distribution

Although it seems as though there are many different types of human hair, it can be divided into just two main groups:
■ Vellus hair
■ Terminal hair.

VELLUS HAIR

Vellus hair is the name given to the soft hair that covers most of the body in women and children. It is short, fine and usually light in colour, making it much less noticeable than

Eyelashes are one of the few examples of terminal hair to be found on men, women and children. They prevent foreign bodies from entering the eye.

terminal hair. Vellus hair shafts do not have a central medulla.

TERMINAL HAIR

Terminal hair is much coarser than vellus hair. It occurs on top of the head, as eyelashes and eyebrows, as pubic and axillary (armpit) hair, and it makes up most of the body hair of adult men. Terminal hair does have a central medulla within its shaft.

Terminal hairs develop and grow in response to the presence of male sex hormones, such as testosterone. In medical conditions where women have too much of these hormones, unwanted male pattern hair growth (hirsutism) occurs.

The hair follicle

Hairs are produced within hair follicles, which are present on most of the skin surface. A number of other structures are associated with these follicles, including sebaceous glands, nerve endings and tiny muscles that pull the hair erect.

The root of each hair sits in a follicle and is buried about 4–5 mm in the skin. Hair is kept lubricated with oil produced in the sebaceous glands.

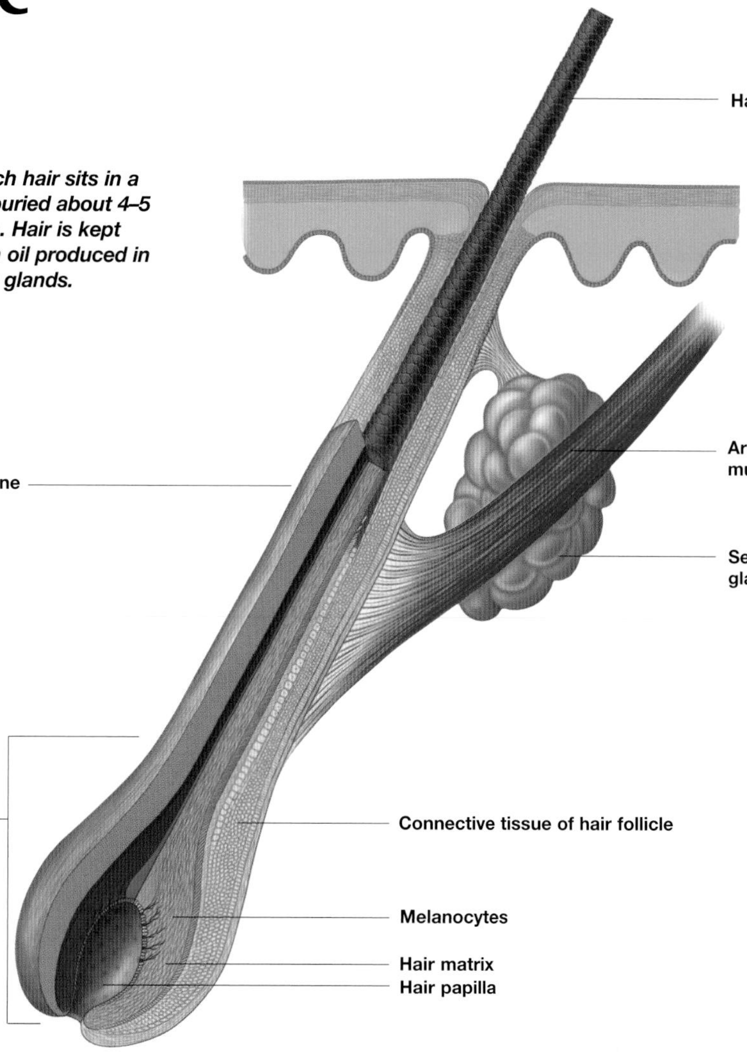

Sebaceous, or oil, glands lie alongside hair follicles wherever they are on the surface of the body. They produce an oily substance, known as sebum, which drains out of the gland through a sebaceous duct into the hair follicle. The sebum then passes out around the emerging hair shaft to reach the surface of the body.

The amount of sebum produced depends upon the size of the sebaceous gland, which in turn depends upon the levels of circulating hormones, especially androgens (male sex hormones). The largest sebaceous glands are found on the head, neck, and back and front of the chest.

The function of sebum is to soften and lubricate the skin and hair, and to prevent the skin from drying out. It also contains substances that kill bacteria, which might otherwise cause infection of the skin and hair follicle.

NERVE ENDINGS
A network of tiny nerve endings lie around the bulb of the hair follicle. These nerves are stimulated by any movement of the base of the hair. If the hair is bent by pressure somewhere along its shaft, these nerve

Labels on diagram:
Hair shaft
Arrector pili muscle
Sebaceous gland
Connective tissue of hair follicle
Melanocytes
Hair matrix
Hair papilla
Hyaline membrane
Hair bulb

endings will fire, sending signals to the brain. This is what happens, for instance, when an insect alights on the skin; the slight bending of hairs it causes sets off a chain of events, resulting in a reflex action to remove it before it stings. In this way hair contributes to our sense of touch.

ARRECTOR PILI MUSCLE
Each hair follicle is attached to a tiny muscle called an arrector pili, which literally means 'raiser of hair'. When this muscle contracts, it causes the hair to move from its normal, angled position to a vertically erect one.

When this occurs within many hair follicles, we see (and feel)

the condition known as goosepimples, which is commonly stimulated by either cold or fear.

The action of these muscles is more important in furry mammals, as it allows them to trap a large amount of air within their fur for insulation from the cold.

Hair thinning and baldness

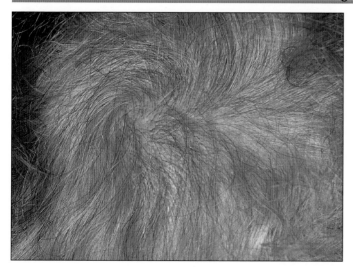

Hair growth is fastest between childhood and early adulthood. After about the age of 40 this high rate of growth starts to fall as the hair follicles begin to age.

Hairs are not replaced as rapidly when they fall out, leading to general thinning, and some degree of baldness in both men and women. Thinning of the hair is also caused by the replacing of coarse terminal hairs with less noticeable, softer vellus hair.

After the age of around 40, hair follicles start to age and hair is not replaced as quickly as it falls out. Thicker terminal hair is also replaced by thinner vellus hair.

ONSET OF BALDNESS
True baldness, which is usually known as male pattern baldness, is a different condition, linked to a number of factors. These include:
■ Family history
■ Levels of androgens (male sex hormones)
■ Increasing age.

It is believed to be due to a gene that only 'switches on' in adult life and somehow alters the response of the hair follicle to circulating hormones.

Abnormal hair thinning or loss may also be linked to a wide variety of medical conditions and treatments of which doctors should be aware.

How hair grows

Hair is a derivative of the skin, and is composed of keratin – a strong structural protein. Hair plays an important role in protecting the body, particularly the scalp, where it is most dense.

Hair is a distinguishing characteristic of mammals and in humans plays a role in the protection of the body from trauma, heat loss and sunlight.

HAIR STRUCTURE

Hair is a complex structure comprised of keratin fibres – keratin is a strong structural protein also found in the nails and outer layer of the skin.

Each hair is made up of three concentric (circular) layers of dead keratinized (keratin-containing) cells: the medulla, cortex and cuticle.

The medulla (the central core) consists of large cells containing soft keratin, partially separated by air spaces. The cortex, the bulky layer surrounding the medulla, consists of several layers of flattened, hard keratin-containing cells.

PROTECTIVE LAYER

The cuticle is the outermost layer, and is composed of a single layer of hard keratin cells that overlap one another like roof tiles.

This outer layer of the hair contains the most keratin, and strengthens and protects the hair, helping to keep the inner layers compacted. The cuticle tends to wear away as the hair becomes older or is damaged, allowing the keratin fibrils, or small fibres, in the cortex and medulla to escape, giving rise to the common phenomenon of 'split ends'.

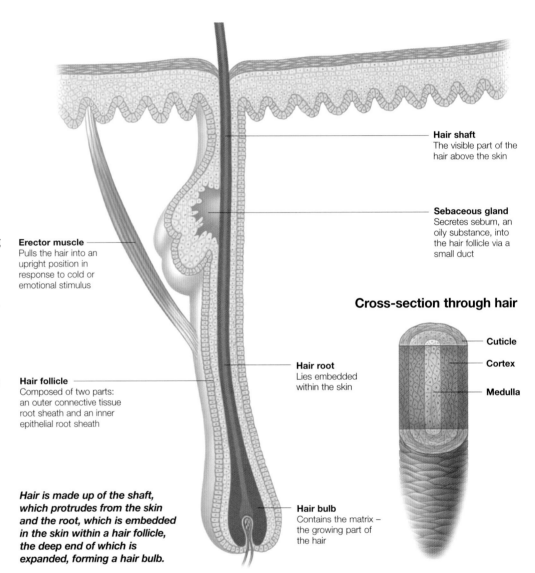

Hair shaft
The visible part of the hair above the skin

Sebaceous gland
Secretes sebum, an oily substance, into the hair follicle via a small duct

Erector muscle
Pulls the hair into an upright position in response to cold or emotional stimulus

Hair follicle
Composed of two parts: an outer connective tissue root sheath and an inner epithelial root sheath

Hair root
Lies embedded within the skin

Hair bulb
Contains the matrix – the growing part of the hair

Cross-section through hair

Cuticle

Cortex

Medulla

Hair is made up of the shaft, which protrudes from the skin and the root, which is embedded in the skin within a hair follicle, the deep end of which is expanded, forming a hair bulb.

What causes hair to grow?

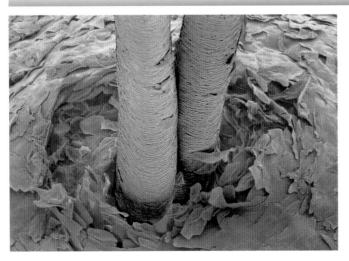

This electron micrograph shows hairs on the scalp. There are two shafts of hair emerging from follicles located in the epidermis of in the skin.

Each hair is divided into the shaft (the visible part) and the root. The root of each hair is enclosed within a hair follicle, below the surface of the skin. At its base the hair follicle is expanded to form the hair bulb.

PRODUCTION OF HAIR

The hair bulb encloses a mass of undifferentiated epithelial cells (the hair matrix), which divide to produce hair. The hair bulb is nourished by a dense network of capillaries which are supplied by the dermal papilla (a projection of the dermis).

STIMULATION OF GROWTH

Chemical signals from the papilla stimulate the adjacent matrix cells to divide and produce hair. As new hair cells are produced by the matrix, the older cells are pushed upwards and fuse together. They become increasingly keratinized and die. Thus the hair that extends from the scalp is no longer living, but due to the active cell division at its root, grows at a rate of around 0.3 mm every day.

Stages of growth

Hair is produced in different stages. Any factors, such as stress or certain drugs, that upset this balance can lead to hair thinning and baldness.

Hair is produced in cycles that involve a growth phase, and a resting phase. During the growth phase the hair is formed and extends as cells are added at the base of the root. This phase can last from around two to six years. As hair grows approximately 10 cm a year, any individual hair is unlikely to grow more than one metre long.

RESTING PHASE

Eventually, cell division pauses (the resting phase) and growth of the hair stops. The hair follicle shrinks to one sixth of its normal length, and the dermal papilla, responsible for the nourishment of new hair cells, breaks away from the root bulb. During this phase the dead hair is held in place. It is these hairs which seem to come out in handfuls when hair is washed or brushed. Eventually a new cycle begins, and the hair is shed from the hair follicle as the production of a new hair begins.

DIFFERENT HAIR TYPES

The length of each phase depends on the type of body hair: scalp hairs tend to grow for a period of three years and rest for one or two years, while eyelash hair, which is much shorter, will grow for around 30 days, and rest for 105 days before being shed. At any one time around 90 per cent of scalp hairs will be in the growing stage, and there is a normal loss of around 100 scalp hairs per day.

Hair does not grow at a constant rate; individual hairs pass through a growth phase and a resting phases, before falling out and being replaced.

Hair loss

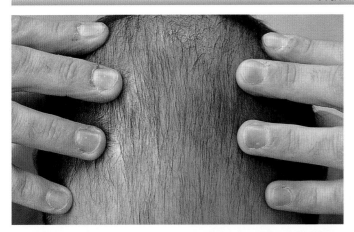

As we age the rate at which our hair grows declines. This can mean that hairs are not replaced as quickly as they are shed and there is an overall thinning, with balding in places (alopecia) often occurring, especially in men.

PREMATURE HAIR LOSS

The physiological changes which bring about male pattern

Male pattern baldness is a common hereditary condition. The growth stage of each hair is so short that they are shed before emerging from the scalp.

baldness are different to those occurring with alopecia. Male pattern baldness is a genetically determined condition and is thought to be caused by changes in the response of the hair follicles to testosterone. The growth cycles of each hair follicle become so short that many hairs never emerge from their follicles before they are shed, and those that do are only very fine.

Hair thinning and loss may also result from factors such as stress that upset the normal hair loss and replacement cycle.

Hair colour and texture

The colour of hair depends upon the presence of the pigment melanin, produced by melanocytes in the bulb of the hair follicle, and then transferred to the cortex.

Dark hair contains true melanin like that found in the skin, while blond and red hair results from types of melanin that contain sulphur and iron. Grey or white hair results from decreased melanin production (genetically triggered) and from the replacement of melanin by air bubbles in the shaft.

Adults have around 120,000 hairs on their head. Redheads tend to have fewer hairs, while blonds have more.

The exact composition of the keratin produced by the body is determined by our genes and differs among individuals. Since keratin is responsible for the texture of the hair shaft, this can vary greatly.

A smooth, cylindrical hair shaft will produce straight hair, while an oval hair shaft will produce wavy hair. A hair shaft that has a kidney-shaped appearance will produce curly hair.

Hair colour and texture are genetically determined and can vary greatly. Colour is determined by melanin content while texture depends upon the exact composition of keratin.

Peripheral nervous system

The peripheral nervous system includes all the body's nerve tissue that is not in the brain and spinal cord. Its principal anatomical components are the cranial and spinal nerves.

The nervous system of the human body is divided into two parts: the central nervous system (CNS) and the peripheral nervous system (PNS).

The major components of the PNS are:

■ Sensory receptors – specialized nerve endings which receive information about temperature, touch, pain, muscle stretching, and taste

■ Peripheral nerves – bundles of nerve fibres which carry information to and from the CNS

■ Motor nerve endings – specialized nerve endings which cause the muscle on which they lie to contract in response to a signal from the CNS.

ARRANGEMENT

Peripheral nerves are of two types:

■ **Cranial nerves**

These emerge from the brain and are concerned with receiving information from, and allowing control of, the head and neck. There are 12 pairs of cranial nerves

■ **Spinal nerves**

These arise from the spinal cord, each containing thousands of nerve fibres, to supply the rest of the body. Many of the 31 pairs of spinal nerves enter one of the complex networks, such as the brachial plexus which serves the upper limb, before becoming part of a large peripheral nerve.

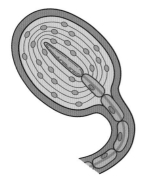

Sensory nerve endings are either free endings or encapsulated. This 'Pacinian corpuscle' is an example of an encapsulated nerve ending.

Major nerves of the peripheral nervous system

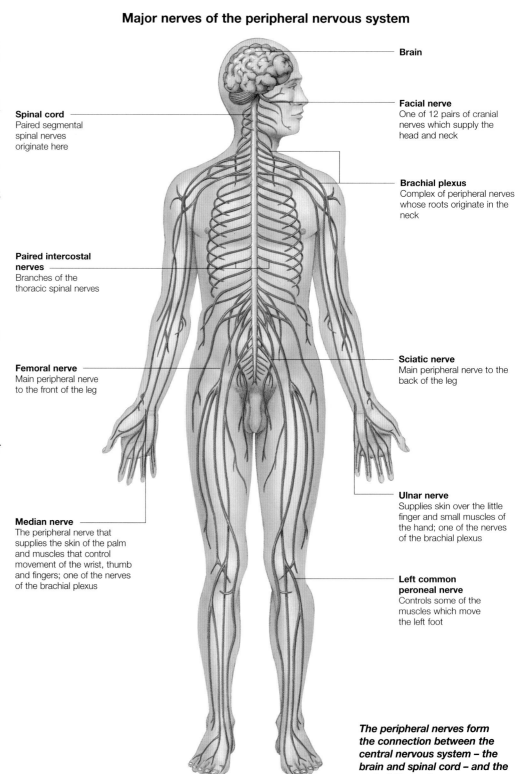

Brain

Facial nerve
One of 12 pairs of cranial nerves which supply the head and neck

Brachial plexus
Complex of peripheral nerves whose roots originate in the neck

Sciatic nerve
Main peripheral nerve to the back of the leg

Ulnar nerve
Supplies skin over the little finger and small muscles of the hand; one of the nerves of the brachial plexus

Left common peroneal nerve
Controls some of the muscles which move the left foot

Spinal cord
Paired segmental spinal nerves originate here

Paired intercostal nerves
Branches of the thoracic spinal nerves

Femoral nerve
Main peripheral nerve to the front of the leg

Median nerve
The peripheral nerve that supplies the skin of the palm and muscles that control movement of the wrist, thumb and fingers; one of the nerves of the brachial plexus

The peripheral nerves form the connection between the central nervous system – the brain and spinal cord – and the rest of the body.

Structure of a peripheral nerve

Each peripheral nerve consists of separate nerve fibres, some with an insulating layer of myelin, enclosed within connective tissue.

The greater part of the final bulk of a peripheral nerve is made up of three protective connective tissue coverings, without which the fragile nerve fibres would be vulnerable to injury.

■ Endoneurium
The endoneurium is a layer of delicate connective tissue that surrounds the smallest unit of the peripheral nerve, the axon. This layer may also enclose an axon's myelin sheath.

■ Perineurium
The perineurium is a layer of connective tissue that encloses a group of protected nerve fibres, called fascicles, that are tied together in bundles.

■ Epineurium
Nerve fascicles are bound together by a tough connective tissue coat, the epineurium, into a peripheral nerve. The epineurium also encloses blood vessels which help to nourish the nerve fibres and their connective tissue coverings.

NERVE FUNCTION
Most peripheral nerves carry information to and from the central nervous system (sensory and motor functions respectively), and thus are known as 'mixed' nerves.

Nerves that are either purely sensory or purely motor are very rare within the body.

Peripheral nerve fibres are grouped together in bundles, called fascicles. These carry both sensory (afferent) and motor (efferent) fibres.

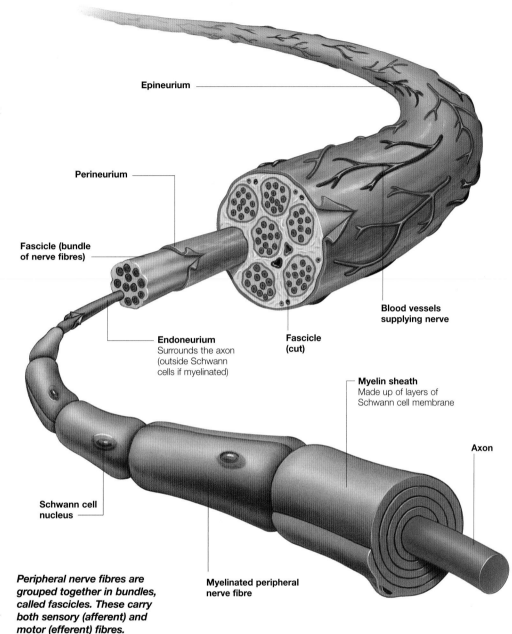

Epineurium

Perineurium

Fascicle (bundle of nerve fibres)

Endoneurium Surrounds the axon (outside Schwann cells if myelinated)

Fascicle (cut)

Blood vessels supplying nerve

Myelin sheath Made up of layers of Schwann cell membrane

Axon

Schwann cell nucleus

Myelinated peripheral nerve fibre

Motor nerve endings

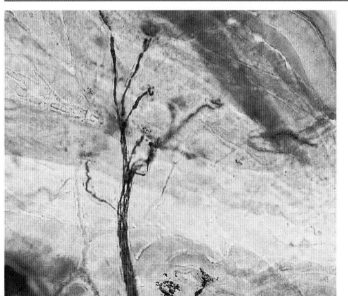

A neuromuscular junction is shown on this micrograph. The connections between the nerve fibre and voluntary muscle can be seen at the top of the image.

Motor nerve endings are specialized nerve fibres that lie on muscle fibres and secretory cells. They receive signals from the central nervous system via peripheral nerves and pass them on to cause muscles to contract or cells to secrete their products. In this way the CNS is able to control each part of the body.

JUNCTION
The neuromuscular junction is where a motor nerve ending of a peripheral nerve fibre connects with the voluntary muscle (also known as striated or skeletal muscle) which it supplies.

At the junction, the axon of a motor nerve fibre divides and branches several times, like a tree, to produce many tiny endings, which lie against a small muscle fibre.

TRANSMITTING A SIGNAL
When an electrical signal is sent down the nerve fibre to the neuromuscular junction it is transmitted to the muscle fibre by chemicals (neurotransmitters) released by the motor nerve endings. The muscle then contracts in response.

Autonomic nervous system

The autonomic nervous system provides the nerve supply to those parts of the body which are not consciously directed. It can be subdivided into the sympathetic and parasympathetic systems.

The autonomic nervous system is divided into two parts: the sympathetic system and the parasympathetic system. Both systems generally supply the same organs, but with opposing effects. In each system two neurones (nerve cells) make up the pathway from the central nervous system (CNS) to the organ which is being supplied.

SYMPATHETIC NERVOUS SYSTEM

The effects upon the body of stimulation by the sympathetic nervous system are often referred to as the 'fight or flight' response. In exciting or dangerous situations, the sympathetic nervous system becomes more active, causing the heart rate to increase and the skin to become pale and sweaty as blood is diverted to muscle.

STRUCTURE

The cell bodies of the neurones of the sympathetic nervous system lie within a section of the spinal cord. Fibres from these cell bodies exit the spinal column at the ventral root and pass through the white rami communicantes to reach the paravertebral sympathetic chain.

Some of the fibres which enter the sympathetic chain connect there with the second cell of their pathway. Fibres then exit through the grey rami communicantes to join the ventral spinal nerve.

Anatomy of a sympathetic trunk

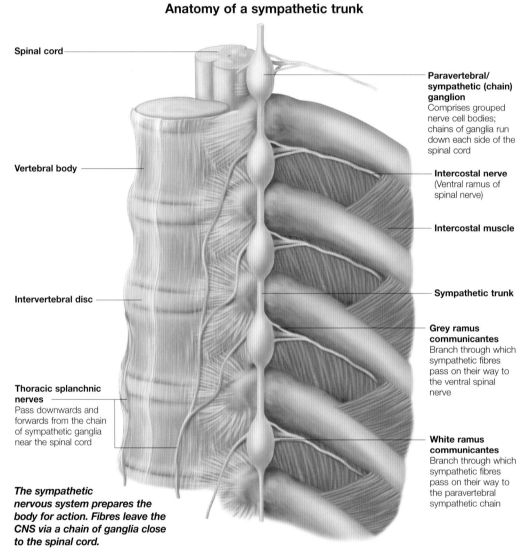

Spinal cord

Vertebral body

Intervertebral disc

Thoracic splanchnic nerves
Pass downwards and forwards from the chain of sympathetic ganglia near the spinal cord

Paravertebral/ sympathetic (chain) ganglion
Comprises grouped nerve cell bodies; chains of ganglia run down each side of the spinal cord

Intercostal nerve
(Ventral ramus of spinal nerve)

Intercostal muscle

Sympathetic trunk

Grey ramus communicantes
Branch through which sympathetic fibres pass on their way to the ventral spinal nerve

White ramus communicantes
Branch through which sympathetic fibres pass on their way to the paravertebral sympathetic chain

The sympathetic nervous system prepares the body for action. Fibres leave the CNS via a chain of ganglia close to the spinal cord.

Adrenal medulla

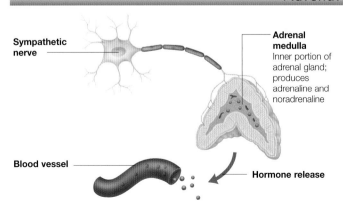

Sympathetic nerve

Blood vessel

Adrenal medulla
Inner portion of adrenal gland; produces adrenaline and noradrenaline

Hormone release

In its role as the mediator of the 'fight or flight' response, the sympathetic nervous system also stimulates the adrenal medulla, the inner portion of the adrenal gland.

The adrenal medulla, in turn, releases the hormones adrenaline and noradrenaline into the blood

As a reaction to stress, the adrenal medulla is stimulated to release hormones into the bloodstream. These hormones prepare the body for action.

stream. These hormones act upon many parts of the body to amplify the effects of the sympathetic nervous system.

The innervation of the adrenal medulla by the sympathetic nervous system is unique in the body in that there is only one neurone in the pathway from the CNS to the gland, rather than two. The adrenal medulla itself seems to act as a sympathetic ganglion, and indeed is derived embryologically from the same tissue.

Parasympathetic nervous system

The parasympathetic system is the part of the autonomic nervous system which is most active during periods of rest.

The structure of the parasympathetic nervous system is simpler than that of the sympathetic nervous system.

LOCATION OF CELL BODIES
The cell bodies of the first of the two neurones in the pathway are located in only two places:

■ The brainstem – fibres from the parasympathetic cell bodies in the grey matter of the brainstem leave the skull as part of a number of cranial nerves. Together, these fibres make up what is known as the cranial parasympathetic outflow

■ The sacral region of the spinal cord – the sacral outflow arises from parasympathetic cell bodies which lie within part of the spinal cord. Fibres leave through the ventral root.

Because of the locations of the origins of parasympathetic fibres, the parasympathetic system is sometimes known as the craniosacral division of the autonomic nervous system; the sympathetic system is known as the thoracolumbar division.

DISTRIBUTION
The cranial outflow provides parasympathetic innervation for the head, and the sacral outflow supplies the pelvis. The area between (the majority of the abdominal and thoracic internal organs) is supplied by part of the cranial outflow which is carried within the vagus (tenth cranial nerve).

Organs controlled by the parasympathetic nervous system

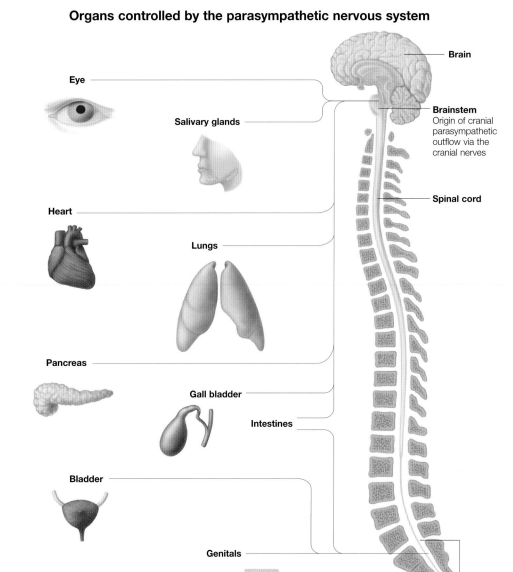

Eye

Salivary glands

Heart

Lungs

Pancreas

Gall bladder

Intestines

Bladder

Genitals

Brain

Brainstem
Origin of cranial parasympathetic outflow via the cranial nerves

Spinal cord

Origin of sacral parasympathetic outflow

The parasympathetic nervous system is most active when the body is at rest. Fibres leave the CNS from the brain and sacral region of the spinal cord.

Opposing effects

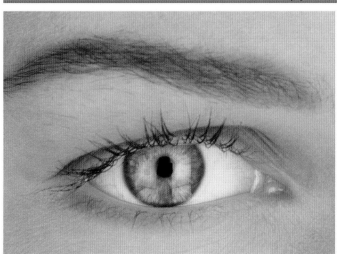

The sympathetic nervous system prepares the body in times of stress or danger, while the parasympathetic system helps the body to rest, digest food and conserve energy. As these tasks are in many ways mutually exclusive, the two systems often have opposite effects upon the body, some of which are:

■ Heart – the sympathetic system increases the rate and the strength of the heartbeat; the parasympathetic system

The sympathetic and parasympathetic nervous systems have opposing effects on the eye. The former dilates the pupil; the latter constricts it.

decreases them

■ Digestive tract – the sympathetic system inhibits digestion and reduces the blood supply; the parasympathetic system stimulates them

■ Liver – the sympathetic system encourages the breakdown of glycogen (a carbohydrate) in the liver to provide energy; the parasympathetic system encourages its formation

■ Salivary glands – the sympathetic system reduces the production of saliva, which also becomes thicker; the parasympathetic system promotes a free flow of watery saliva.

How reflexes work

Bodily actions that can occur independently of conscious control are called reflexes. They are especially important when a rapid involuntary response is required.

The central nervous system is able to perform highly complex tasks, and not all of these require conscious thought. Those actions that are involuntary in nature are called reflexes, pre-programmed and predictable responses to a specific sensory stimulus.

SOMATIC REFLEXES
Somatic reflexes result in the movement of a muscle, or the secretion of a chemical from a gland.

For example, if you were to touch a hot oven, pain receptors in your hand would send nerve impulses to neurones in the spinal cord. These in turn would communicate with the appropriate muscles in your arm telling them to withdraw the hand instantly. Only after the hand was withdrawn, however, would your brain become aware of what had happened.

AUTONOMIC REFLEXES
We are not conscious of the outcome of all the reflexes that occur in our bodies. For example, the baroreceptor reflex corrects a rise in arterial blood pressure without us being aware that it is doing so.

A simple reflex arc

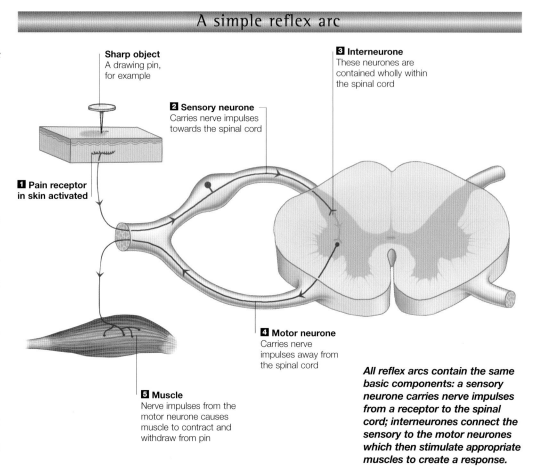

Sharp object
A drawing pin, for example

2 Sensory neurone
Carries nerve impulses towards the spinal cord

3 Interneurone
These neurones are contained wholly within the spinal cord

1 Pain receptor in skin activated

4 Motor neurone
Carries nerve impulses away from the spinal cord

5 Muscle
Nerve impulses from the motor neurone causes muscle to contract and withdraw from pin

All reflex arcs contain the same basic components: a sensory neurone carries nerve impulses from a receptor to the spinal cord; interneurones connect the sensory to the motor neurones which then stimulate appropriate muscles to create a response.

The patellar reflex

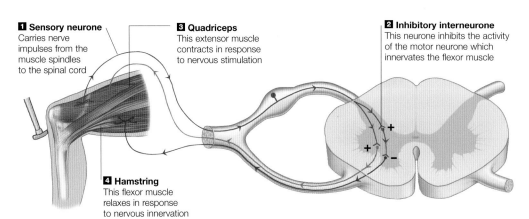

1 Sensory neurone
Carries nerve impulses from the muscle spindles to the spinal cord

3 Quadriceps
This extensor muscle contracts in response to nervous stimulation

2 Inhibitory interneurone
This neurone inhibits the activity of the motor neurone which innervates the flexor muscle

4 Hamstring
This flexor muscle relaxes in response to nervous innervation

The patellar reflex is tested by doctors after a patient has suffered a traumatic injury to determine whether their lower spine has been damaged.

Babies up to about one year old exhibit the Babinski reflex when the sole of the foot is rubbed. This disappears as their nervous system develops.

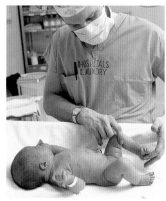

Clinicians often use the 'patellar reflex' to test the integrity of a patient's lower spine. The patient sits high up so that the legs hang freely. The doctor then lightly taps the patellar tendon (situated just below the knee-cap) and looks for a response.

MUSCLE SPINDLES
In a healthy person, the knock to the tendon stretches the quadriceps muscle. This stretch is detected by structures in the muscle called muscle spindles. These send nervous signals to neurones in the spinal cord,

which in turn send impulses to the quadriceps muscle telling it to contract (to counteract the initial stretch). This causes the foot to spring forward. At the same time the antagonistic muscle, the hamstrings, are inhibited.

Complex reflexes

Although some spinal reflexes, such as the patellar reflex, are relatively simple and involve only a few nerve cells, the spinal cord is capable of carrying out more complicated functions without needing to involve the brain.

If you were to step on a sharp object, such as a drawing pin, with your right foot, a complex reflex (the crossed extensor reflex) would be initiated in order to withdraw the foot and to shift the body's weight onto the left leg.

Initially, the drawing pin stimulates pain receptors in the skin of the right foot, causing them to send nerve impulses, via afferent nerve fibres (from the Latin 'afferre' – to carry towards), to the right hand side of the spinal cord. Neurones in this half of the spinal cord send nerve signals away from the cord via efferent nerve fibres to tell the extensor muscles to relax and the flexors to contract.

TRANSFER OF WEIGHT

These events result in the injured leg being moved away from the drawing pin. However, unless the body's weight is transferred to the other leg, you will fall over.

Thus neurones from the right-hand side of the spinal cord cross over to the left-hand side and synapse with motor neurones which innervate muscles in the left leg. These motor neurones inform the extensor muscles in the left leg to contract and the flexors to relax, causing the leg to be extended so that it can carry the body's weight.

When a bare foot treads on a sharp object, it is rapidly withdrawn and the body's weight is transferred to the other leg.

The crossed extensor reflex

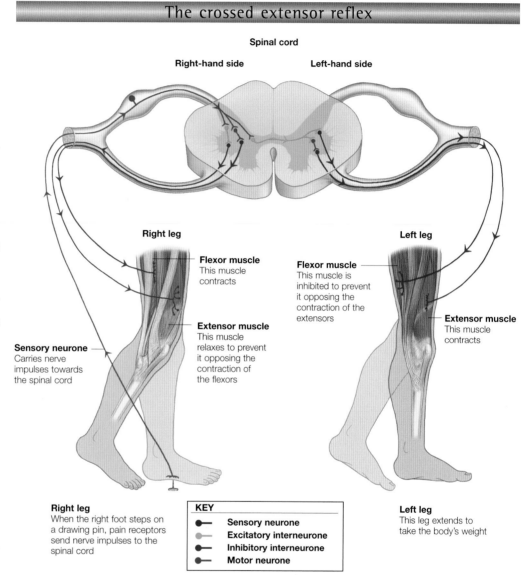

Spinal cord

Right-hand side   Left-hand side

Right leg

Flexor muscle
This muscle contracts

Flexor muscle
This muscle is inhibited to prevent it opposing the contraction of the extensors

Left leg

Extensor muscle
This muscle relaxes to prevent it opposing the contraction of the flexors

Extensor muscle
This muscle contracts

Sensory neurone
Carries nerve impulses towards the spinal cord

Right leg
When the right foot steps on a drawing pin, pain receptors send nerve impulses to the spinal cord

Left leg
This leg extends to take the body's weight

KEY	
●	**Sensory neurone**
●	**Excitatory interneurone**
●	**Inhibitory interneurone**
●	**Motor neurone**

Learned reflexes

The reflexes that have been discussed so far are 'hard-wired' into the nervous system.

However, while babies are born with the innate ability to learn how to walk, we have to make a conscious effort to learn to drive a car, ride a bicycle or play the piano.

With time, new movements can become as automatic as walking. For example, although learning to drive is a relatively difficult experience for most people, after a while the

Pianists are able to read music and then play the appropriate note without thinking about what they are doing. This is an example of a learned reflex.

movements become automatic and the driver no longer has to consciously think what he or she is doing.

Similarly, touch-typists do not need to think where their fingers are on a keyboard, and as a result many can type at up to 80 words per minute. Assuming that the average word contains around six letters, a fast typist can make up to eight keystrokes every second!

It is thought that during the learning process the neurones involved in controlling the movement change the way that they are connected to each other. Important connections between cells are reinforced and unnecessary synapses are lost.

How the body feels pain

Pain is not just a signal that certain tissues in the body have been damaged – it also alerts the sufferer to danger. Painkillers can bring relief, but the body also has its own built in pain inhibition system.

Any event that causes a degree of damage to the tissues of the body – be it mechanical (from pressure or a wound), chemical (exposure to acid, for example) or thermal (extreme heat or cold) – brings about the release of large amounts of chemicals, such as serotonin and histamine.

As well as producing reactions within the tissues, such as swelling and redness, these chemicals are detected by special sensory cells, called free nerve endings, which are found in the superficial layers of skin as well as in some of the internal organs. They are also known as nociceptors, because they react to noxious substances.

PAIN IMPULSES

In response to the chemical changes within the tissues, the sensory cells send nerve impulses to relay stations in the spinal cord. From here, they are passed through further relays in the lower part of the brain in the brain stem and the thalamus, and so on to the higher levels of the brain. There, the information is analysed and perceived as pain. In most circumstances, a person will withdraw from the source of the pain.

Receptors in the skin

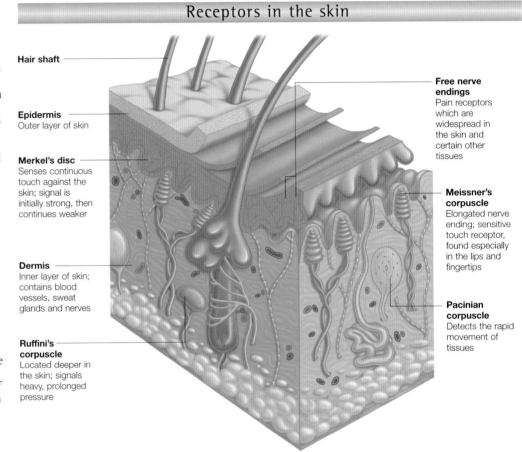

Hair shaft

Epidermis
Outer layer of skin

Merkel's disc
Senses continuous touch against the skin; signal is initially strong, then continues weaker

Dermis
Inner layer of skin; contains blood vessels, sweat glands and nerves

Ruffini's corpuscle
Located deeper in the skin; signals heavy, prolonged pressure

Free nerve endings
Pain receptors which are widespread in the skin and certain other tissues

Meissner's corpuscle
Elongated nerve ending; sensitive touch receptor, found especially in the lips and fingertips

Pacinian corpuscle
Detects the rapid movement of tissues

Classifying pain

There are two types of pain, distinguished according to the speed with which the sensations are felt. The first, which is felt as soon as tissue damage is sensed, is sharp and stabbing and is known as acute pain. Its impulses travel extremely quickly to the brain along special nerve fibres, called A-fibres, that have myelin sheaths to speed the impulses along.

The purpose of acute pain is to bring about an immediate, subconscious, reaction, to remove the body from the danger; A-fibre impulses cause a hand to be moved out of a flame, for example.

After some time, acute pain dies down and is replaced by the second type: the dull, throbbing, aching, persistent feeling that characterizes chronic pain. The impulses of chronic pain come from sensory receptors deeper in the tissues, and they travel 10 times more slowly that those of acute pain along unmyelinated nerve fibres called C-fibres.

People suffering from serious chronic pain, such as this cancer patient, may need intravenous painkilling drugs. These work by suppressing the C-fibre impulses.

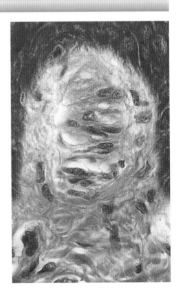

Meissner's corpuscles, one of which is shown here, transmit signals along myelinated nerve fibres. Acute pain signals also travel along myelinated fibres.

Pain inhibition

The body has three pain relief systems: each depends on
preventing nerve impulses from reaching the higher levels of the brain
by blocking them at the spinal relays or lower brain levels.

The first, and most simple, pain relief system is best summed up as 'rubbing it better'. However, this phrase disguises a complex sequence of events.

Two nerves join at the relay station in the spine, the junction of the two being called a synapse. One nerve carries signals from the sensory nerve endings, and the other carries them up the spine to the brain. Neurologists think of the synapse as a gate: normally it is shut, but strong impulses, as in acute pain can force it open.

However, the synapse is only open to one type of pathway at a time. This is why A-fibre impulses, which travel faster, reach the synapse before C-fibre

impulses and block them out until they have themselves died down. But if a painful area is rubbed vigorously, A-fibre impulses are generated, and again they reach the synapse first, blocking out the slower C-fibre impulses. As a result, the aching, chronic pain is relieved.

CHEMICAL BLOCK

The second system depends on blocking the passage of nerve impulses by chemical means. In response to pain signals, the brain produces chemicals called endorphins. These are the body's own painkillers, and they block receptors in the brain stem and thalamus, and block the gates in the spinal relays. Heroin and morphine are painkillers because they block the same receptors.

SUPPRESSION

Finally, the brain can send impulses down the spinal cord to suppress pain signals at the spinal relay. This is most apparent when pain is extreme, when, for example, a soldier is fighting for his life or an athlete is pushed to the limits.

PAIN TOLERANCE

How much pain is felt is determined by the quantity of endorphins (pain-relieving chemicals in the brain). Exercise increases endorphin levels, as does relaxation, a positive mental outlook and

sleep. In contrast, fear, depression, anxiety, lack of exercise and concentrating on pain all reduce endorphin levels. The fewer endorphins there are, the more pain is felt.

This second-degree burn was caused by boiling fat. Pain from such injuries is acute at first, becoming chronic for several days afterwards.

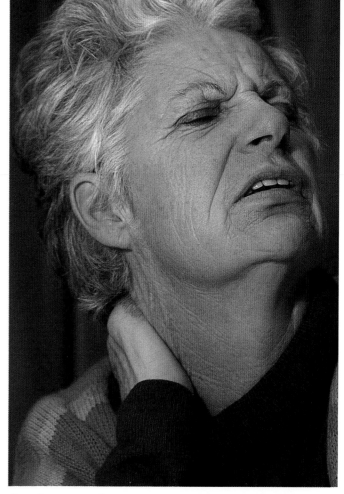

A natural and subconscious response is to rub a painful area, particularly when muscles are affected. Physiologically, the action of rubbing works effectively to ease discomfort.

Referred pain

Sometimes pain is felt in an area that is not in fact the source of the pain, and in such cases the sensation is called referred pain. Examples of this include pain from the area of the diaphragm, which can be felt at the tip the shoulder, and pain from the heart – as in angina – which is felt across the chest, in the neck and along the inner side of the arm.

There are two explanations for this phenomenon. First, tissues which originate

Referred pain that affects the ear is very common. The cause is often found to be tooth-related, such as abscesses or impaction, or associated with the larynx or pharynx (tonsillitis, for example).

from the same embryological building block – that is, they come from the same area of basic tissue in a fetus as it develops inside the uterus – often share the same spinal relays, so activity in one part of the relay triggers activity in another part of the same relay. Second, there can be so many nerve impulses from an internal organ that they flood the pathways normally reserved for other areas of the body.

Doctors often check for referred pain as part of the diagnosis of a disorder that affects the internal organs. This is often somewhat of a surprise to the patient, who perhaps cannot understand why the main source of their discomfort (that is, the source of the pain) is being ignored during an investigation.

Lymphatic system

The lymphatic system consists of a network of lymph vessels and organs and specialized cells throughout the body. It is an essential part of the body's defence against invading micro-organisms.

The lymphatic system is the lesser known part of the circulatory system, working together with the cardiovascular system to transport a fluid called lymph around the body. The lymphatic system plays a vital role in the defence of the body against disease.

LYMPH FLUID

Lymph is a clear, watery fluid containing electrolytes and proteins which is derived from blood and bathes the body's tissues. Lymphocytes – specialized white blood cells involved in the body's immune system – are found in lymph. They attack and destroy foreign micro-organisms, thereby maintaining the body's health. This is known as an immune response.

Although the vessels of the lymphatic system carry lymph, the fluid is not pumped around the body as blood is; instead, contractions of muscles surrounding the lymph vessels move the fluid along.

CONSTITUENT PARTS OF THE LYMPHATIC SYSTEM

The lymphatic system is made up of a number of interrelated parts:
■ Lymph nodes – lie along the routes of the lymphatic vessels and filter lymph
■ Lymphatic vessels – small capillaries leading to larger vessels that eventually drain lymph into the veins
■ Lymphoid cells (lymphocytes) – cells through which the body's immune response is mounted
■ Lymphoid tissues and organs – scattered throughout the body, these act as reservoirs for lymphoid cells and play an important role in immunity.

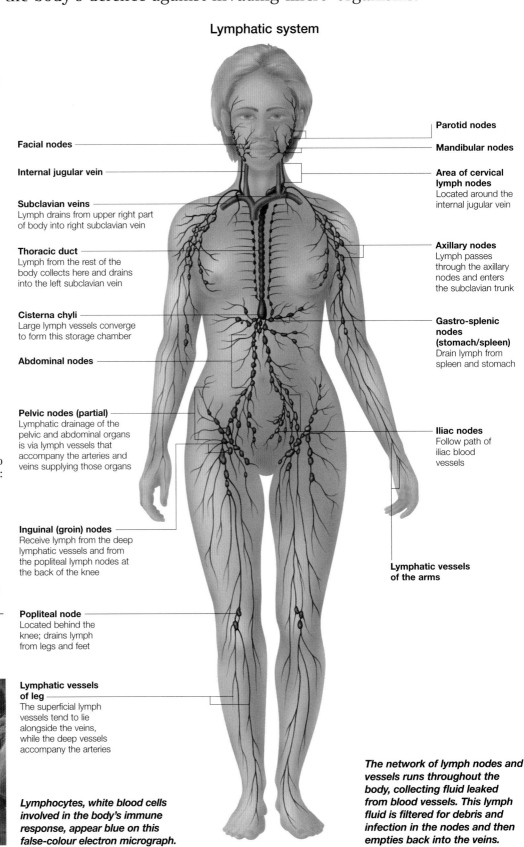

Lymphatic system

Facial nodes

Internal jugular vein

Subclavian veins
Lymph drains from upper right part of body into right subclavian vein

Thoracic duct
Lymph from the rest of the body collects here and drains into the left subclavian vein

Cisterna chyli
Large lymph vessels converge to form this storage chamber

Abdominal nodes

Pelvic nodes (partial)
Lymphatic drainage of the pelvic and abdominal organs is via lymph vessels that accompany the arteries and veins supplying those organs

Inguinal (groin) nodes
Receive lymph from the deep lymphatic vessels and from the popliteal lymph nodes at the back of the knee

Popliteal node
Located behind the knee; drains lymph from legs and feet

Lymphatic vessels of leg
The superficial lymph vessels tend to lie alongside the veins, while the deep vessels accompany the arteries

Parotid nodes

Mandibular nodes

Area of cervical lymph nodes
Located around the internal jugular vein

Axillary nodes
Lymph passes through the axillary nodes and enters the subclavian trunk

Gastro-splenic nodes (stomach/spleen)
Drain lymph from spleen and stomach

Iliac nodes
Follow path of iliac blood vessels

Lymphatic vessels of the arms

The network of lymph nodes and vessels runs throughout the body, collecting fluid leaked from blood vessels. This lymph fluid is filtered for debris and infection in the nodes and then empties back into the veins.

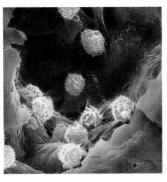

Lymphocytes, white blood cells involved in the body's immune response, appear blue on this false-colour electron micrograph.

Lymph nodes

Lymph nodes lie along the route of the lymphatic vessels. They filter the lymph for invading micro-organisms, infected cells and other foreign particles.

Lymph nodes are small, rounded organs that lie along the course of the lymphatic vessels and act as filters of the lymph. Lymph nodes vary in size, but they are mostly bean-shaped, 1–25 mm in length, surrounded by a fibrous capsule and usually embedded in connective tissue.

LYMPH NODE FUNCTION

As well as fluid, the tiny lymphatic vessels in the tissues may pick up other items, such as parts of broken cells, bacteria and viruses. Within the lymph node, fluid slows and comes into contact with lymphoid cells which ingest any solid particles and recognize foreign micro-organisms. To prevent these particles from entering the bloodstream – and to allow the body to mount a defence against invading organisms – lymph is filtered through a number of lymph nodes before draining into the veins.

Some lymph nodes are grouped together in regions and given names according to their position, the region in which they are found (for example, the axillary nodes in the axilla, or armpit), the blood vessels they surround (such as the aortic nodes around the large central artery of the body, the aorta), or the organ they receive lymph from (pulmonary nodes in the lungs).

Structure of a lymph node

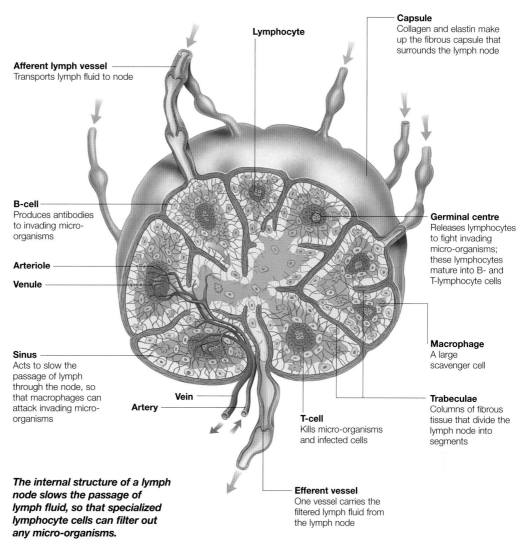

Lymphocyte

Capsule
Collagen and elastin make up the fibrous capsule that surrounds the lymph node

Afferent lymph vessel
Transports lymph fluid to node

B-cell
Produces antibodies to invading micro-organisms

Arteriole

Venule

Germinal centre
Releases lymphocytes to fight invading micro-organisms; these lymphocytes mature into B- and T-lymphocyte cells

Macrophage
A large scavenger cell

Sinus
Acts to slow the passage of lymph through the node, so that macrophages can attack invading micro-organisms

Vein

Artery

Trabeculae
Columns of fibrous tissue that divide the lymph node into segments

T-cell
Kills micro-organisms and infected cells

Efferent vessel
One vessel carries the filtered lymph fluid from the lymph node

The internal structure of a lymph node slows the passage of lymph fluid, so that specialized lymphocyte cells can filter out any micro-organisms.

Lymph vessels

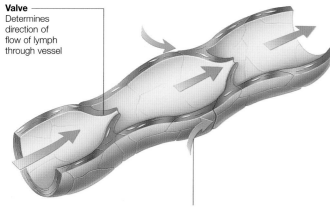

Valve
Determines direction of flow of lymph through vessel

The fluid circulating around the cells in tissues drains into lymph capillaries. From here, it flows through valves in these vessels to the lymph nodes.

Entry point for interstitial fluid
Lymph fluid is called interstitial fluid before it has drained into the lymph capillaries

Arteries supply blood to the body's tissues under pressure. This has the effect of causing fluid and proteins to leak out of the tiny capillaries and into the spaces around the cells of those tissues.

Much of this leaked fluid will pass back into the capillaries, which gradually converge to form veins that carry blood back to the heart for further circulation. However, some of the fluid – and the proteins – remain behind and would accumulate in the tissues were it not for the network of tiny lymphatic vessels in the tissue spaces.

The lymph fluid travels up the converging lymphatic vessels, which eventually join to form the main lymphatic trunks. These unite to form the two large lymphatic ducts – the thoracic duct and the right lymphatic duct. These drain into the large veins above the heart, returning the retrieved fluid and proteins to the bloodstream.

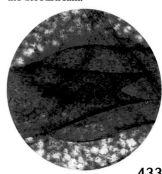

A valve within a lymph vessel is seen in this light micrograph. The valve allows lymph fluid to pass in one direction only.

Lymphoid cells and lymph drainage vessels

Lymphoid cells are divided into B-lymphocytes, which produce antibodies, and T-lymphocytes, which kill infected cells. The whole lymph network eventually drains into the venous system.

Scattered throughout the body are discrete groups of lymphoid tissue, which have an important role in the immune system:

■ The spleen – provides a site for the cells of the immune system to proliferate and monitor the blood for foreign or damaged cells

■ The thymus – a small gland which lies in the chest just behind the upper part of the sternum (breastbone). It receives newly formed lymphocytes from the bone marrow, which mature into T-lymphocytes, an important group of lymphoid cells

■ Lymphoid tissue of the gastro-intestinal tract – lymphoid tissue lying beneath the lining of the gut generally, the ring of lymphoid tissue at the back of the mouth and some discrete clumps of lymphoid nodules known as 'Peyer's patches', found in the walls of the last part of the small intestine. These are thought to be the site of maturation of B-lymphocytes, another important set of lymphocytes.

The large amount of lymphoid tissue in the gut wall helps to protect against infection by organisms entering through the mouth.

Lymphoid tissues and organs

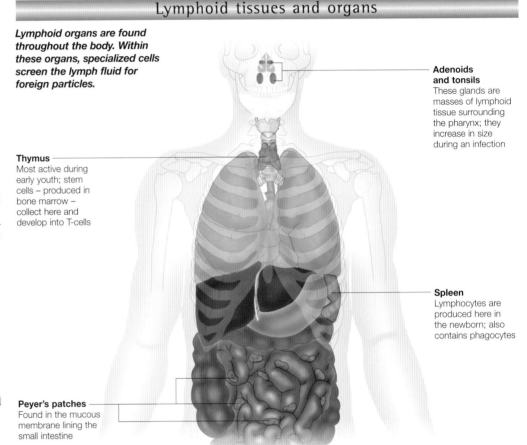

Lymphoid organs are found throughout the body. Within these organs, specialized cells screen the lymph fluid for foreign particles.

Adenoids and tonsils
These glands are masses of lymphoid tissue surrounding the pharynx; they increase in size during an infection

Thymus
Most active during early youth; stem cells – produced in bone marrow – collect here and develop into T-cells

Spleen
Lymphocytes are produced here in the newborn; also contains phagocytes

Peyer's patches
Found in the mucous membrane lining the small intestine

The role of lymphocytes

'Natural killer cells' are a type of lymphocyte. They are able to destroy cancer cells and cells infected with viruses.

The cells of the immune system, lymphocytes, can recognize foreign proteins, such as those found on the surface of invading micro-organisms or on the cells of transplanted organs.

In response, the lymphocyte cells multiply and mount an immune response, some (T-cells) by directly attacking the foreign cells and some (B-cells) by manufacturing antibodies which attach to the foreign proteins, allowing them to be found and destroyed.

Lymphocytes are made in the bone marrow and circulate freely in the bloodstream. As they circulate, they can quickly mount a response to infections.

Lymph drainage vessels

The lymphatic vessels form a network that runs through the tissues. These vessels converge and empty into the veins.

DRAINAGE OF THE CHEST

Of the lymph nodes that lie in the chest, the most important clinically are the internal mammary nodes on either side of the sternum. They receive 25 per cent of the lymph from the breast and may be a site for spread of breast cancer. Within the chest, the largest group of lymph nodes lie around the base of the trachea (windpipe) and the bronchi. Other lymph node groups within the chest lie alongside the major blood vessels.

UPPER AND LOWER LIMBS

In the limbs, there are superficial and deep lymph vessels; the superficial vessels tend to lie alongside the veins while the deep vessels accompany the arteries. The axillary (armpit) group of nodes receives lymph from the whole of the upper limb, the trunk above the umbilicus and the breast. The inguinal (groin) lymph nodes receive lymph from the superficial vessels and the deep lymph vessels that run alongside the arteries. Lymph travels up from the inguinal nodes to the nodes alongside the aorta and eventually join the lumbar lymph trunks.

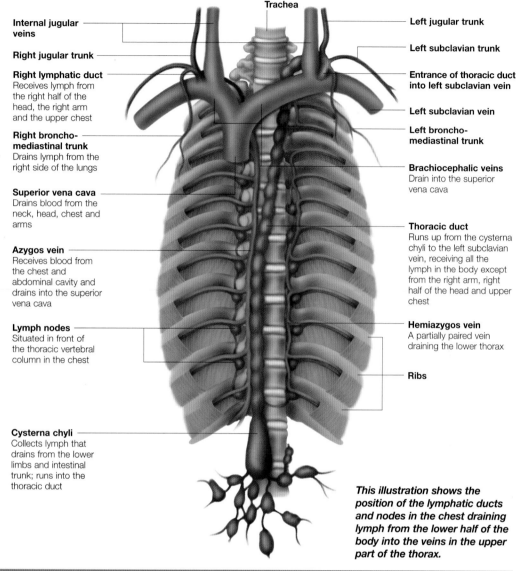

Trachea

Internal jugular veins

Right jugular trunk

Right lymphatic duct
Receives lymph from the right half of the head, the right arm and the upper chest

Right broncho-mediastinal trunk
Drains lymph from the right side of the lungs

Superior vena cava
Drains blood from the neck, head, chest and arms

Azygos vein
Receives blood from the chest and abdominal cavity and drains into the superior vena cava

Lymph nodes
Situated in front of the thoracic vertebral column in the chest

Cysterna chyli
Collects lymph that drains from the lower limbs and intestinal trunk; runs into the thoracic duct

Left jugular trunk

Left subclavian trunk

Entrance of thoracic duct into left subclavian vein

Left subclavian vein

Left broncho-mediastinal trunk

Brachiocephalic veins
Drain into the superior vena cava

Thoracic duct
Runs up from the cysterna chyli to the left subclavian vein, receiving all the lymph in the body except from the right arm, right half of the head and upper chest

Hemiazygos vein
A partially paired vein draining the lower thorax

Ribs

This illustration shows the position of the lymphatic ducts and nodes in the chest draining lymph from the lower half of the body into the veins in the upper part of the thorax.

Disorders of the lymphatic system

As lymph is carried from the tissues back to the bloodstream in the lymphatic vessels, it passes through a series of lymph nodes. These act as filters, removing cells and micro-organisms. Lymph from each area of the body drains through a particular set of lymph nodes and this pattern of drainage is of great clinical importance in the diagnosis and treatment of cancer and infection.

In cancer, the lymph nodes draining the affected area may enlarge and become firmer or even hard and may be felt by a doctor. Finding such enlarged lymph nodes may allow the doctor to suspect a secondary tumour and will give an indication of the site of the primary tumour. Knowledge of lymphatic drainage also allows a surgeon to remove the associated lymph nodes when he is removing a tumour to check for, or help prevent, secondary spread.

Bacterial infection of the skin can lead to a condition known as lymphangitis where the lymphatic vessels themselves become infected and inflamed. Where the pathway of these affected lymphatic vessels lies just under the skin, it can be seen as a series of red lines which are painful and tender to the touch. Lymphangitis, along with painful enlargement of the associated lymph nodes, is a feature of infection with *Streptococcus* bacteria.

The red line along the inside of this man's arm is caused by lymphangitis – an infection of the lymphatic vessels.

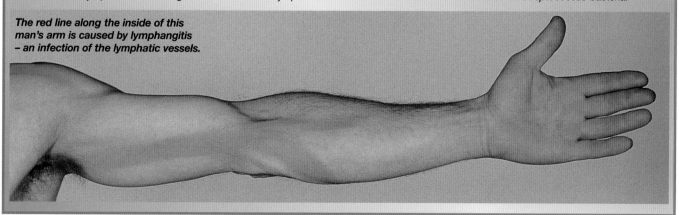

Regional lymphatic drainage

Lymph from every part of the body returns to the bloodstream via a series of lymph nodes. An understanding of the lymph drainage pattern is vital in monitoring the spread of cancers or infection.

Lymph is the fluid present within the vessels of the lymphatic system. The main function of the lymphatic vessels is to collect excess tissue fluid and return it to the blood circulation.

Lymph from each part of the body follows a specific path on its way back to rejoin the blood circulation, passing through lymph node groups – which have a filtering role – on the way.

HEAD AND NECK NODES

The lymph node groups of the structures of the head and neck are named according to their positions. The important lymph node groups include the:
- Occipital
- Mastoid, or retroauricular (behind the ear)
- Parotid
- Buccal
- Submandibular (under the jaw)
- Submental (under the chin)
- Anterior cervical
- Superficial cervical
- Deep within the neck lie other groups of nodes which surround and drain the pharynx, larynx and trachea.

DEEP CERVICAL NODES

These lymph nodes all drain ultimately into the deep cervical group of nodes which lie in a chain alongside the major blood vessels of the neck.

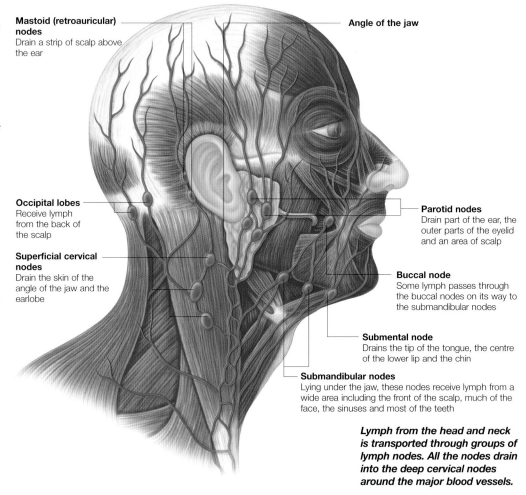

Mastoid (retroauricular) nodes
Drain a strip of scalp above the ear

Angle of the jaw

Occipital lobes
Receive lymph from the back of the scalp

Superficial cervical nodes
Drain the skin of the angle of the jaw and the earlobe

Parotid nodes
Drain part of the ear, the outer parts of the eyelid and an area of scalp

Buccal node
Some lymph passes through the buccal nodes on its way to the submandibular nodes

Submental node
Drains the tip of the tongue, the centre of the lower lip and the chin

Submandibular nodes
Lying under the jaw, these nodes receive lymph from a wide area including the front of the scalp, much of the face, the sinuses and most of the teeth

Lymph from the head and neck is transported through groups of lymph nodes. All the nodes drain into the deep cervical nodes around the major blood vessels.

Lymphatic drainage of the tongue

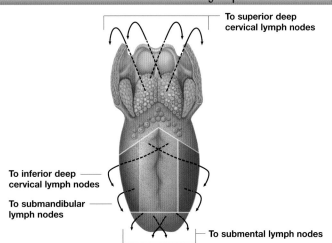

To superior deep cervical lymph nodes

To inferior deep cervical lymph nodes

To submandibular lymph nodes

To submental lymph nodes

Surgeons often face the problem of managing malignant ulcers of the tongue. Understanding the drainage pattern of the lymph vessels of the tongue is very helpful in gaining information about the spread of disease.

DRAINAGE PATTERN

Lymph drains from the following areas:
- The tip of the tongue – lymph

The lymph vessels of the tongue have their own drainage pattern. Studying this system helps in the treatment of malignant disease, often caused by smoking.

from both sides of this area of the tongue drains into the submental group of lymph nodes, under the chin
- The sides of the tongue – lymph drains from each side to the submandibular group of nodes
- The central part of the tongue – this area drains to the inferior (lower) deep cervical nodes which lie alongside the internal jugular vein, deep within the neck
- The back of the tongue – lymph from both sides of this area drains into the superior (upper) deep cervical lymph nodes.

Lymph drainage of the intestines

The lymph vessels and nodes that make up the lymphatic drainage of the gastrointestinal system follow the general pattern of the arteries which supply the gut with blood. Lymph from the small intestine transports fats absorbed from food into the bloodstream.

Much of the gut is enclosed and suspended within a fold of connective tissue, known as a mesentery. The blood vessels that supply the gut lie within this mesentery, forming arcades that connect with each other to reach all parts of this lengthy structure.

SITE OF NODES

The lymph nodes which initially receive lymph from the intestine are found within the mesentery in a number of places:
■ By the wall of the intestine
■ Among the arterial arcades
■ Alongside the large superior and inferior mesenteric arteries.

These mesenteric groups of nodes are, in some cases, named according to their positions in relation to the intestine or to the artery they accompany. From the intestinal wall, lymph drains through these nodes in turn to eventually reach the pre-aortic nodes, which lie next to the large central artery, the aorta.

ABSORPTION OF FAT

In addition to its normal function, the lymph which leaves the small intestine has a further role – that of transporting the fats absorbed from food.

The lining of the small intestine bears numerous microvilli. These tiny projections of the mucous membrane greatly increase the surface area of the intestine to help absorption.

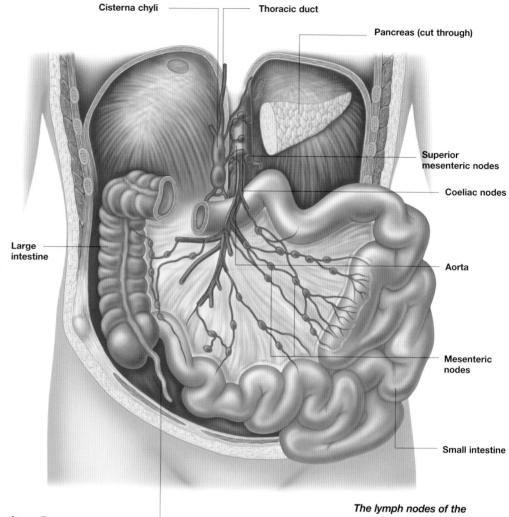

Cisterna chyli

Thoracic duct

Pancreas (cut through)

Superior mesenteric nodes

Coeliac nodes

Large intestine

Aorta

Mesenteric nodes

Small intestine

Appendix
This has its own lymph node

The lymph nodes of the gastrointestinal system are located in the mesentery. This is a fold of membrane that encloses a large part of the gut.

CENTRAL VESSELS

Within each microvillus lies a central lymph vessel, called a lacteal. The function of the lacteals is to carry away fat particles absorbed from food which are too big to enter the blood capillaries.

These fats travel through the lymphatic system to be delivered into the bloodstream with the rest of the lymph.

Lymphatic drainage of the stomach

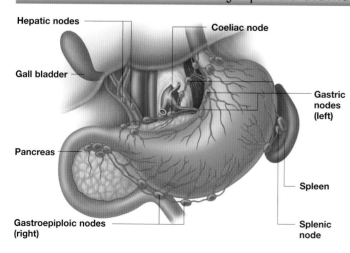

Hepatic nodes

Coeliac node

Gall bladder

Gastric nodes (left)

Pancreas

Spleen

Gastroepiploic nodes (right)

Splenic node

Like the intestine, the lymphatic drainage of the stomach tends to follow the pattern of the arterial blood supply.

FOUR GROUPS

The nodes which receive lymph from the stomach comprise four main groups:
■ The left and right gastric nodes receive lymph from the area supplied by the left and

There are four main groups of lymph nodes within the stomach. These comprise the gastric, splenic, gastroepiploic and coeliac nodes.

right gastric arteries, respectively. They lie along the lesser curve of the stomach
■ The splenic nodes lie at the hilum (hollow) of the spleen on the left side of the stomach. These nodes receive lymph from the area of the stomach supplied by the short gastric arteries
■ The left and right gastroepiploic nodes lie along the greater curve of the stomach and receive lymph from areas supplied by the corresponding gastroepiploic arteries.

All the lymph received from the stomach by these groups travels on to drain into the coeliac nodes.

How the body produces sweat

Sweat is secreted from the sweat glands during physical exercise, stress and in excessive heat. It is produced in two different types of glands, both of which are located in the dermis of the skin.

The body constantly produces sweat. This process is the body's main way of ridding itself of excess heat.

The amount of sweat the body produces depends upon the state of emotion and physical activity. Sweat can be produced in response to stress, high air temperature and exercise.

SWEAT GLANDS

Sweat is manufactured in the sweat glands. These are located in the dermis of the skin, along with nerve endings and hair follicles. On average, each person has around 2.6 million sweat glands, which are distributed over the entire body, with the exception of the lips, nipples and genitals.

Sweat glands consist of long, coiled, hollow tubes of cells. The coiled portion in the dermis is where sweat is produced. The long portion is a duct that connects the gland to tiny openings (pores) located on the outer surface of the skin. Nerve cells from the sympathetic nervous system (a division of the autonomic nervous system) connect to the sweat glands.

TYPES OF SWEAT GLAND

There are two types of gland:
- Eccrine – these are the most numerous type of sweat gland, found all over the body, particularly on the palms of the hands, soles of the feet and forehead. Eccrine glands are active from birth
- Apocrine – these sweat glands are mostly confined to the armpits and around the genital area. Typically, they end in hair follicles rather than pores. These are larger than eccrine glands, and only become active once puberty has begun.

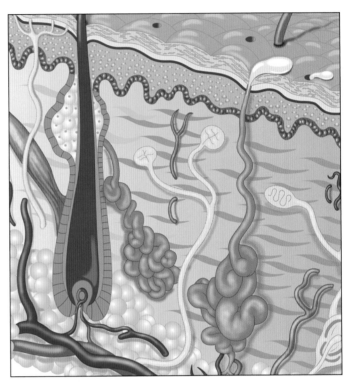

Sweat is produced in sweat glands, located in the dermis. These glands comprise long, coiled tubes of cells that connect to pores on the skin surface.

Sweat production

Stimulation of an eccrine gland causes the cells lining the gland to secrete a fluid that is similar to plasma, but without the fatty acids and proteins. This is mostly water with high concentrations of sodium and chloride (salts) and a low concentration of potassium.

This fluid originates in the spaces between cells (interstitial spaces), which are provided with fluid by the blood vessels (capillaries) in the dermis.

The fluid passes from the coiled portion and up through the straight duct. What happens to this fluid when it reaches the straight portion of the sweat duct depends upon the rate of sweat production.

- Low sweat flow – at rest and in a cool environment, the sweat glands are not stimulated to produce much sweat. The cells of the straight duct have time to reabsorb most of the water and salts, so not much fluid actually reaches the surface of the skin as sweat.

The composition of this sweat is different from that of its primary source: it contains less sodium and chloride, and more potassium.
- High sweat flow – this occurs in higher temperatures or during exercise. Cells in the straight portion of the sweat duct do not have time to reabsorb all the water, sodium and chloride from the primary secretion. As a result, alot of sweat reaches the surface of the skin, and its composition is similar to that of the primary secretion.

APOCRINE SWEAT

Sweat is produced in the apocrine glands in a similar way, but apocrine differs from eccrine sweat in that it contains fatty acids and proteins. For this reason, apocrine sweat is thicker and milky-yellow in colour.

ODOUR

Sweat itself has no odour, but when bacteria present on the hair and skin metabolize the proteins and fatty acids present in apocrine sweat, an unpleasant odour is produced. Deodorants are designed to eliminate this distinctive body odour.

High Flow | Low Flow

Na^+ Na^+
Cl^- Cl^-
H_2O H_2O
H_2O
H_2O Na^+
Cl^-
Na^+ Cl^-
K^+ K^+

H_2O Water
K^+ Potassium
Na^+ Sodium
Cl^- Chloride

The constituents of sweat vary according to temperature and activity. If sweat production is minimal, then the sweat contains less salts.

The role of sweat

When sweat evaporates, it takes excess body heat with it. In a very hot climate, the sweat glands can produce up to three litres of sweat an hour.

The role of sweat is to cool the body. Sweat on the surface of the skin evaporates into the atmosphere, taking with it excess body heat.

VAPORIZATION HEAT

Heat loss from sweating is governed by a basic rule of physics. Heat is required to convert water from a liquid to a vapour (gas); when sweat evaporates this heat is taken from the body.

However, not all of the sweat evaporates and much runs off the skin and is absorbed by items of clothing. Not all heat energy produced by the body is lost through sweat; some is directly radiated from the skin to the air, and some is lost through breathing.

EVAPORATION RATE

Humidity affects the rate at which sweat evaporates. If the air is humid, for example, then it already has water vapour in it and might not be able to take more (near-saturation). If this is the case, then sweat does not evaporate and cool the body as it does when the air is dry.

When the water in sweat evaporates, it leaves the salts (sodium, chloride and potassium) behind on the skin, which is why the skin can taste salty.

DEHYDRATION

A body that is not acclimatized to very hot temperatures can easily produce one litre of sweat per hour. In fact, the maximum amount that the body can produce appears to be around two to three litres per hour.

The loss of excessive water and salts from the body can lead to dehydration, causing circulatory problems, kidney failure and heat stroke. It is important therefore to drink plenty of fluids when exercising or in high temperatures.

Specialized drinks are also available for people taking part in sports – these contain vital salts to replace those lost through sweating.

In areas of high humidity, such as tropical rain forests, the air is already saturated with water. Thus, reduced evaporation of sweat prevents body cooling.

Other causes of sweating

Sweating can also occur as a result of nervous activity, or as the sign of a disorder.

Nervous sweating
Sweating responds to the emotional state. If a person is nervous, afraid or anxious, there is an increase in sympathetic nerve activity, and an increase in adrenaline secretion from the adrenal gland.

Adrenaline acts on the sweat glands, particularly those on the palms of the hands and armpits, causing them to produce sweat. This phenomenon is often referred to as a 'cold sweat' and is a factor exploited in the use of lie detector tests. This is because

People in stressful situations can sweat in the absence of a high temperature. This is due to an adrenaline surge that stimulates the sweat glands.

the increased sympathetic nerve activity in the skin changes its electrical resistance.

Excessive sweating
Diaphoresis or hyperhidrosis is a condition in which excessive sweating occurs. The exact cause of this embarrassing condition is not known, although it may be due to the following:
■ Overactive thyroid gland – the thyroid hormone increases body metabolism and heat production
■ Certain foods and medications
■ Overactivity of the sympathetic nervous system
■ Hormonal imbalances – for example the menopause.
If the problem of sweating becomes severe, surgery to remove the sympathetic nerve trunk may be performed – this procedure is know as a sympathectomy.

How body temperature is controlled

Body temperature is regulated by a part of the brain called the hypothalamus. If the external temperature rises or falls, the body uses various mechanisms to ensure it maintains a comfortable equilibrium.

Endothermic (warm-blooded) animals, such as birds or mammals, maintain their bodies at a more or less constant temperature using internal control mechanisms. In comparison, ectothermic (cold-blooded) animals, such as fish and reptiles, have no such internal mechanisms and are dependent to a great extent on the surrounding temperature.

CONTROLLING BODY HEAT

Humans, like all warm-blooded animals, produce heat as a result of metabolism. All of the body's tissues produce heat, but the most heat is produced by the tissues that are most active, such as the liver, heart, brain and endocrine glands.

Muscles also produce heat – about 25 per cent of body heat is produced by inactive muscles. Active muscles may produce up to 40 times more heat than the rest of the body, which is why the body warms during exercise.

HOMEOSTASIS

Humans have a fairly constant body temperature that is, under normal conditions, maintained independently of their external surroundings. This maintenance of a constant internal environment, in spite of variations in the outside environment, is known as homeostasis.

One of the advantages of maintaining a constant body temperature is that the danger of overheating is greatly reduced. Extreme cases of overheating can result in convulsions and death, as nerve pathways are suppressed and the activities of vital proteins are affected.

A thermogram (heat image) shows the distribution of heat around the body after exercise. The hottest parts are white, followed by yellow and purple; the coldest parts are shown as red, blue and black.

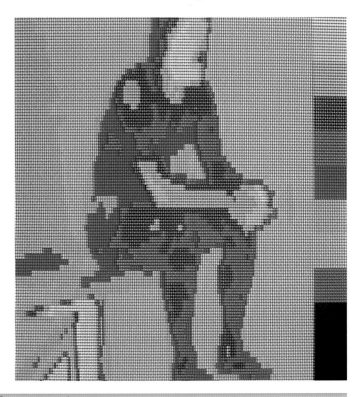

Mechanisms for warming up

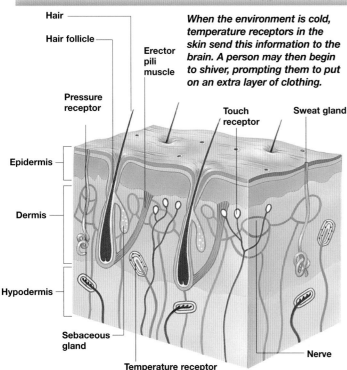

Hair
Hair follicle
Erector pili muscle
Pressure receptor
Touch receptor
Sweat gland
Epidermis
Dermis
Hypodermis
Sebaceous gland
Temperature receptor
Nerve

When the environment is cold, temperature receptors in the skin send this information to the brain. A person may then begin to shiver, prompting them to put on an extra layer of clothing.

The normal temperature of the human body varies between 35.6 °C and 37.8 °C. In order to maintain this degree of constancy, the temperature is monitored by a part of the brain called the hypothalamus. This operates using a feedback mechanism, similar to that used by the thermostat of a domestic central heating system.

When the external environment starts to cool the body down, temperature sensors in the skin send this information to the hypothalamus and the person starts to feel cold. This information is then passed to other parts of the brain, which initiate physiological responses designed to increase body heat and reduce heat loss.

Some reactions to feeling cold are conscious, such as jumping up and down, putting on extra clothing or moving to a warmer place. Other reactions occur spontaneously. Shivering occurs when body muscles contract and

relax very rapidly, giving out four or five times as much heat as they do in their resting state. At the same time, adrenaline production is increased, which increases the body's metabolic rate – the rate at which energy, stored in the form of glucose, is used. As a result more heat is generated inside the whole body.

REGULATING HEAT LOSS

To reduce heat loss from the body's surface, the capillaries near the surface of the skin become constricted, resulting in a reduced blood flow to the skin and a paler complexion. At the same time, the tiny muscles attached to the hair follicles contract, resulting in the hairs on the skin becoming erect. In most mammals, this has the effect of trapping a layer of warmer air near the skin, but because skin hair is sparse in humans, this pilo-erection has very little effect on heat loss, other than causing 'goose-pimples'.

Temperature control mechanisms

Our skin is equipped with thousands of receptors that monitor the overall
temperature of the body. These sensors detect changes in the external environment and
alert the brain, which in turn stimulates shivering or sweating to maintain homeostasis.

VASODILATION

Vasodilation is a key mechanism
for conserving and losing heat.
At high temperatures, the blood
vessels dilate (widen), allowing
heat to be lost and giving a
flushed appearance. The degree
of dilation of the blood vessels is
controlled by nerves called
vasomotor fibres, which are in
turn controlled by the brain.

VASOCONSTRICTION

At low temperatures, arterioles
(branches of arteries) leading to
capillaries in upper skin layers
may constrict (vasoconstriction).
This reduces blood flow to the
skin, and reduces heat loss.

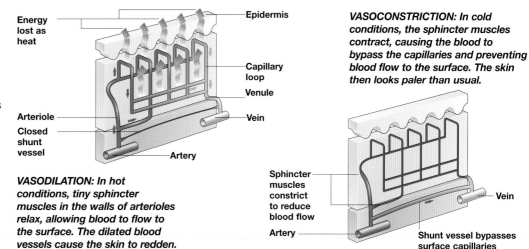

*VASOCONSTRICTION: In cold
conditions, the sphincter muscles
contract, causing the blood to
bypass the capillaries and preventing
blood flow to the surface. The skin
then looks paler than usual.*

*VASODILATION: In hot
conditions, tiny sphincter
muscles in the walls of arterioles
relax, allowing blood to flow to
the surface. The dilated blood
vessels cause the skin to redden.*

Mechanisms for cooling down

The body's temperature is
normally higher than that of the
surrounding air. Therefore, heat
is lost to the surrounding
environment by radiation and
convection, as currents of
moving air pass over the surface
of the skin.

If, however, the body starts to
become too warm, due to either
a high external temperature or
an internal fever, heat sensors
send nerve impulses to the
hypothalamus, and the brain
initiates cooling measures.

The blood capillaries near the
surface of the skin become
dilated so that blood flow
increases and more heat is lost

through the skin to the outside.
Sweating also increases heat
loss: as liquid produced by the
sweat glands evaporates, it has
a cooling effect on the skin.

In dry air, sweating works
very effectively: a person can
tolerate temperatures of up to
65 °C for several hours in dry
conditions. However, if the air
is moist, sweat cannot evaporate
easily and the body becomes
overheated more rapidly.

*A coloured electron micrograph
shows droplets of sweat (blue)
on human skin. Sweat, mostly in
the form of dissolved salts, cools
the body down.*

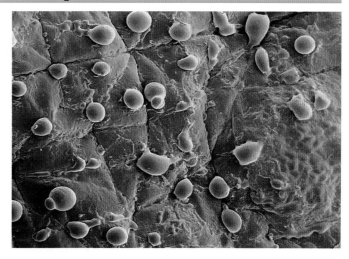

Fever and hypothermia

A fever is a raised body
temperature that may occur
as a result of infection. Chemical
substances called cytokines are
released by white blood cells and
damaged tissue cells. These

*Symptoms of hypothermia
include lethargy, muscle
stiffness and a confused mental
state. If untreated, it results in
unconsciousness, brain damage
and ultimately death.*

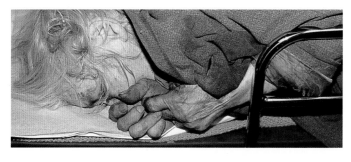

chemicals cause the
hypothalamus to produce
prostaglandins (hormones that
dilate blood vessels), which in
turn 'reset' the thermostatic
control mechanism of the
hypothalamus to a higher
temperature. The result is that
heat-producing mechanisms are
triggered; even though the body
temperature may rise to 40 °C,
the patient still feels a chill.

Body temperature remains

high until the infection is
cleared. At this point, the normal
setting of the hypothalamus is
restored and cooling
mechanisms are initiated. The
patient sweats and becomes
flushed as the blood vessels in
the skin dilate. Research has
shown that fever both boosts
the body's immune system and
inhibits the growth of micro-
organisms.

Hypothermia occurs when
the core body temperature falls
below 35 °C. It results from the
body being exposed to cold
conditions, rendering it unable
to maintain normal body
temperature. Newborn babies,
the elderly and those suffering
from illness are most susceptible.
Hypothermia is usually the result
of a combination of inadequate
food, clothing and heating in
cold conditions.

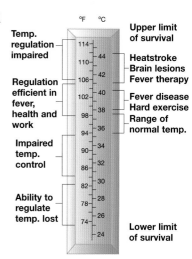

	°F	°C	
Temp. regulation impaired	114	44	Upper limit of survival
	110	42	Heatstroke / Brain lesions / Fever therapy
Regulation efficient in fever, health and work	106		
	102	40	Fever disease
	98	38	Hard exercise / Range of normal temp.
	94	36	
Impaired temp. control	90	34	
	86	32	
	82	30	
Ability to regulate temp. lost	78	28	
	74	26	Lower limit of survival
		24	

*Extremes of body temperature –
whether too high or too low –
have a devastating effect
on mental and physical health.*

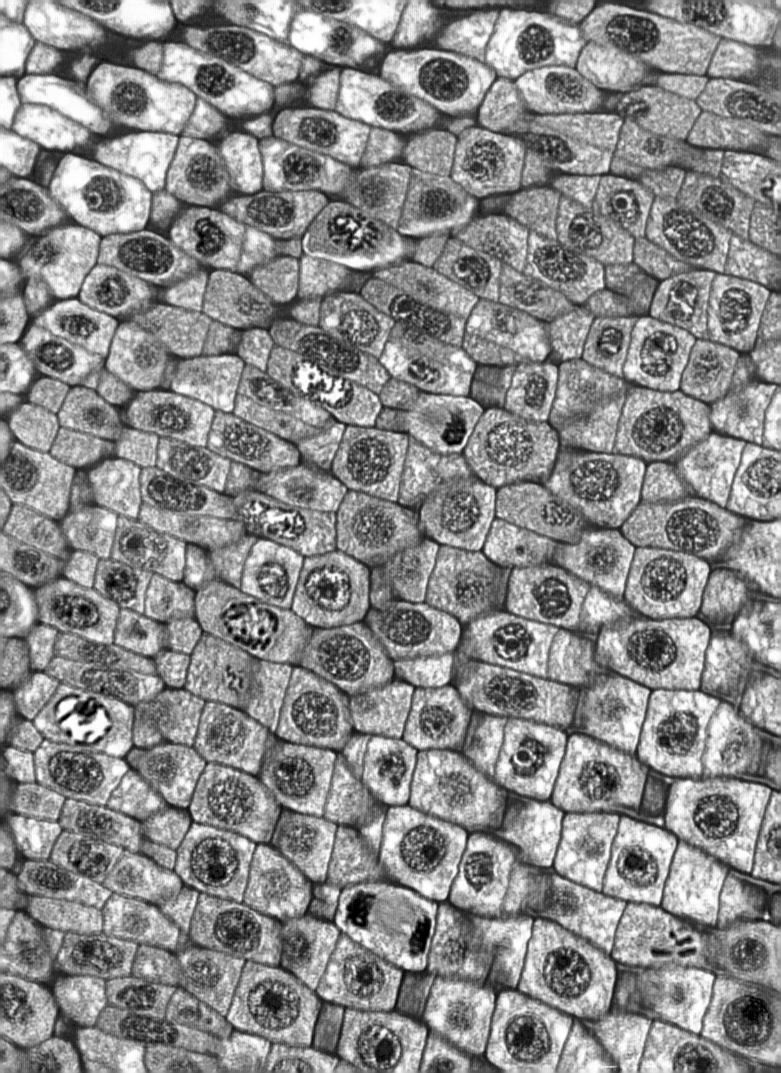

Chapter Eleven

Cellular &
Chemical Structure

Our bodies are made up of billions of cells, each with its own specialized function. Cells are the smallest units within the body and the fundamental building blocks for the muscles, bones, tissues, blood, nerves and skin that make up our body systems.

At the centre of each cell is the nucleus. This contains the DNA that holds the cell's genetic code and which is the blueprint for making proteins, essential for the development and growth of all body structures.

This chapter details the structure and function of human cells, explaining how cells communicate and work together and how the genes carried within cells dictate the unique characteristics of each individual.

LEFT: This microscopic view of human skin cells has been magnified by 400 times. The purple blobs are sweat pores.

Neurone

A neurone is a specialized cell of the nervous system.
The main function of neurones is to carry information in the form
of electrical impulses from one part of the body to the other.

The tissues of the nervous system are made up of two types of cells: neurones, or nerve cells, which transmit information in the form of electrical signals; and the smaller supporting cells (glial cells) which surround them.

COMMON FEATURES

Neurones are the large, highly specialized cells of the nervous system, whose function is to receive information and transmit it throughout the body. Although variable in structure, neurones have some features in common:
■ Cell body – the neurone possesses a single cell body from which a variable number of branching processes emerge
■ Dendrites – these are thin, branching processes of the neurone, which are in fact extensions of the cell body
■ Axon – each neurone has an axon carrying electrical impulses away from the cell body.

CHARACTERISTICS

Neurones have several other special characteristics:
■ Neurones cannot divide and so cannot replace themselves if damaged or lost
■ Neurones live for a very long time; as they cannot replace themselves they need to last for a lifetime
■ Neurones have very high energy requirements and so cannot survive for more than a few minutes without oxygen or glucose from the blood.

Structure of a motor neurone

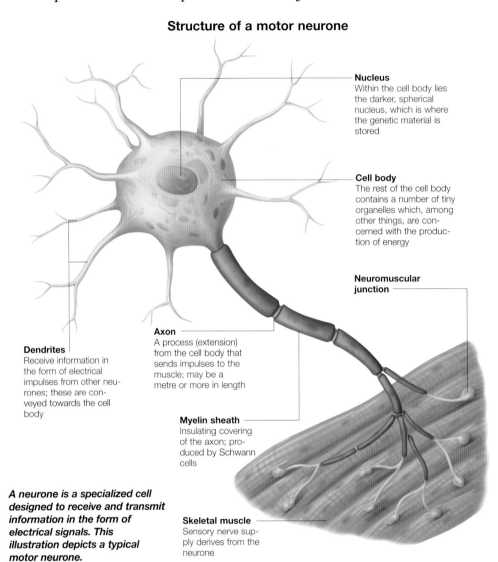

Nucleus
Within the cell body lies the darker, spherical nucleus, which is where the genetic material is stored

Cell body
The rest of the cell body contains a number of tiny organelles which, among other things, are concerned with the production of energy

Neuromuscular junction

Dendrites
Receive information in the form of electrical impulses from other neurones; these are conveyed towards the cell body

Axon
A process (extension) from the cell body that sends impulses to the muscle; may be a metre or more in length

Myelin sheath
Insulating covering of the axon; produced by Schwann cells

Skeletal muscle
Sensory nerve supply derives from the neurone

A neurone is a specialized cell designed to receive and transmit information in the form of electrical signals. This illustration depicts a typical motor neurone.

Structural types of neurone

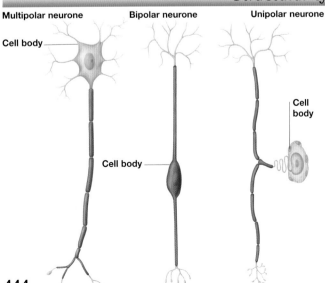

Multipolar neurone

Cell body

Bipolar neurone

Cell body

Unipolar neurone

Cell body

There are three major groups of neurones, based on the number of processes extending from their cell bodies:
■ Multipolar neurones – have many processes extending from the cell body, all except one of which (the axon) are dendrites. This is the most common form of neurone, especially within the central nervous system (CNS). Sometimes the axon is absent
■ Bipolar neurones – have only two processes: a single dendrite

The arrangement of cell processes from the cell body of a neurone fall into three categories. The structural type of the cell is related to function.

and an axon. This type of neurone is unusual within the body, and they are found in special sense organs, such as the retina of the eye
■ Unipolar neurones – have a single process, which is divided into a peripheral process that receives information, often from a sense receptor, and a central process which enters the CNS.

NEURONE FUNCTION

Neurones may also be classified according to their functions into sensory (or afferent) neurones and motor (or efferent) neurones. Most sensory neurones are unipolar while motor neurones are multipolar.

The myelin sheath

The speed of an electrical signal along a neurone's axon is increased by the presence of a myelin sheath – a layer of fatty insulation.

The myelin sheath is formed differently according to where it is located:
■ In the peripheral nervous system (those nerves lying outside the brain and spinal cord), the myelin sheath is produced by specialized Schwann cells. These wrap themselves around the axon of a nerve cell to form a sheath of concentric circles of their cell membranes
■ In the central nervous system, neurones are given their myelin sheath by cells known as oligodendrocytes, which can myelinate more than one nerve axon at a time.

APPEARANCE
Nerve fibres with myelin sheaths tend to look whiter than unmyelinated ones, which have a grey tinge. The 'white matter' of the brain is composed of dense collections of myelinated nerve fibres, whereas the 'grey matter' is made up of nerve cell bodies and unmyelinated fibres.

FUNCTION
Each Schwann cell lies adjacent to, but not touching, the next. The gap between the cells, where there is no myelin, is known as the node of Ranvier. As an electrical signal passes down the nerve it must 'hop' from one node to the other, which makes it travel faster overall than if no myelin sheath were present.

Insulation of a peripheral nerve

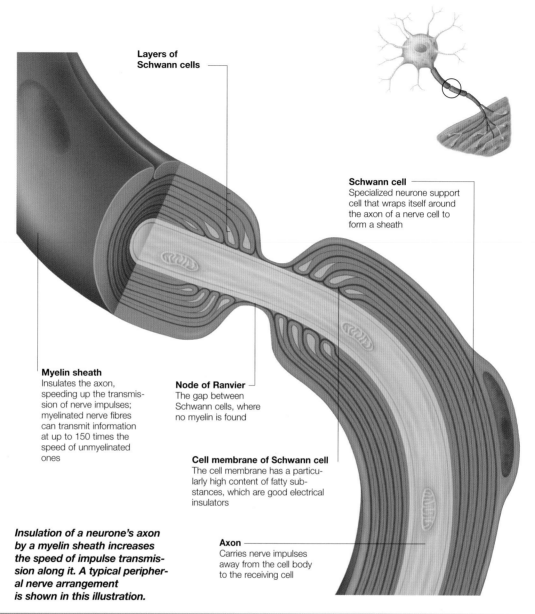

Layers of Schwann cells

Schwann cell
Specialized neurone support cell that wraps itself around the axon of a nerve cell to form a sheath

Myelin sheath
Insulates the axon, speeding up the transmission of nerve impulses; myelinated nerve fibres can transmit information at up to 150 times the speed of unmyelinated ones

Node of Ranvier
The gap between Schwann cells, where no myelin is found

Cell membrane of Schwann cell
The cell membrane has a particularly high content of fatty substances, which are good electrical insulators

Axon
Carries nerve impulses away from the cell body to the receiving cell

Insulation of a neurone's axon by a myelin sheath increases the speed of impulse transmission along it. A typical peripheral nerve arrangement is shown in this illustration.

Supporting cells of the central nervous system

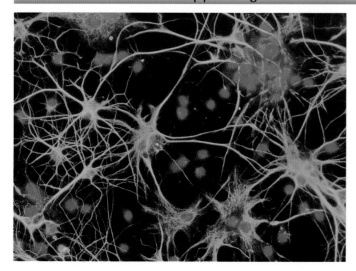

Astrocytes are star-shaped cells in the central nervous system. Their numerous branches of connective tissue provide support and nutrition for neurones.

Neurones are surrounded by neuroglia, a collective name given to the group of small support cells which make up about half the bulk of the central nervous system.

Neuroglial cells outnumber neurones by about 10 to 1 and have a variety of functions:
■ Astrocytes – the most abundant neuroglial cells; they are star-shaped. They anchor the neurones to their blood supply and determine what substances can pass between the blood and the brain (the so-called blood–brain barrier)
■ Microglia – like similar cells in other parts of the body, these small oval cells are specialized to ingest, or phagocytose, any invading micro-organisms or dead tissue
■ Oligodendrocytes – these cells provide the myelin sheath for neurones of the CNS
■ Ependymal cells – lining the fluid-filled ventricles of the CNS, these cells may be of a variety of shapes, from flat to columnar. They have tiny brush-like cilia on their surfaces which beat to maintain circulation of the cerebrospinal fluid.

How nerve cells work

Nerve cells generate nerve impulses, electrical messages which travel from one end of a nerve cell to the other. This ability is essential for us to interact successfully with the world around us.

The human central nervous system contains at least two hundred billion neurones (nerve cells); on average, each neurone communicates with thousands of other nerve cells. This complexity allows the brain to interpret the rich sensory input that it receives from the five senses and to react accordingly.

NEUROANATOMY

Although nerve cells from different regions of the nervous system can look very dissimilar, they all contain the same three basic elements: dendrites, a cell body and an axon:

■ Dendrites (from the Greek 'dendros', meaning tree) are branch-like protrusions of the cell membrane which provide a large surface area to receive neurotransmitters released from other neurones. The dendrites transduce (convert) this chemical information into small electrical impulses, which are then conveyed to the cell body.

■ The greater part of a neurone is made up of the cell body which, like the majority of the body's cells, contains a nucleus. A region of the cell body called the axon hillock collates all the small nervous impulses generated by the many dendrites and initiates action potentials (nerve impulses) accordingly.

■ The axon of a neurone carries nerve impulses from the cell body to the synaptic terminals – specialized endings which release neurotransmitters to communicate with other neurones.

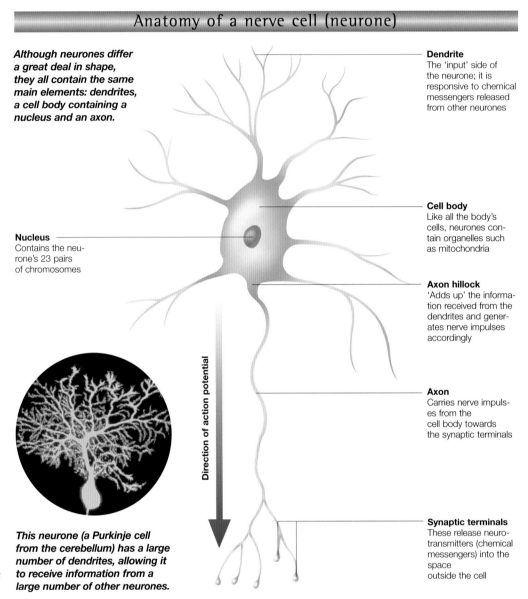

Anatomy of a nerve cell (neurone)

Although neurones differ a great deal in shape, they all contain the same main elements: dendrites, a cell body containing a nucleus and an axon.

Nucleus
Contains the neurone's 23 pairs of chromosomes

Direction of action potential

Dendrite
The 'input' side of the neurone; it is responsive to chemical messengers released from other neurones

Cell body
Like all the body's cells, neurones contain organelles such as mitochondria

Axon hillock
'Adds up' the information received from the dendrites and generates nerve impulses accordingly

Axon
Carries nerve impulses from the cell body towards the synaptic terminals

Synaptic terminals
These release neurotransmitters (chemical messengers) into the space outside the cell

This neurone (a Purkinje cell from the cerebellum) has a large number of dendrites, allowing it to receive information from a large number of other neurones.

What makes neurones different from other cells in the body?

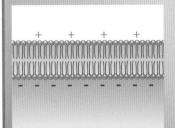

Unlike neurones, the majority of the body's cells do not have protein pores which can open and close in response to a predetermined signal.

The chemical composition of the fluid inside a cell (called the cytosol) is different from the composition of the fluid outside the cell (extracellular fluid).

Compared to the extracellular fluid, the cytosol has fewer positive charges and more negative charges; this means that the inside of the cell is slightly negative compared to the outside of the cell. This electrical charge across the membrane is called the membrane potential, and in most cells is about minus 70 millivolts (thousandths of a volt).

What makes neurones so special is that they can alter the electrical charge across their membrane to generate nerve impulses. They can do this because their cell membrane contains gated protein pores which allow electrically charged ions (sodium, potassium, calcium or chloride ions) to cross it transiently, so altering the neurone's membrane potential. Other cells do not have these protein pores and so their membrane potential stays relatively constant.

Neurones are able to generate nerve impulses because their membrane contains gated channels that respond either to chemical messengers (left) or to a change in voltage (right).

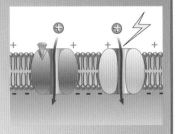

How nerve impulses are generated

Neurones generate nerve impulses by altering the charge across their membrane. If this is reduced (e.g. by cooling) production of impulses is decreased.

A neurone can only generate an action potential when it has been adequately stimulated. Neurotransmitters (chemical messengers), released from nearby neurones, cause receptor proteins in the dendritic membrane to open. This allows positive sodium ions to flow into the dendrite, causing the membrane potential to become slightly less negative.

THE 'UPSHOOT' OF AN ACTION POTENTIAL
If sufficient sodium ions enter the neurone to raise the membrane potential to the 'threshold potential', other voltage-dependent protein pores open, which allows even more positive sodium ions to enter the neurone.

An electroencephalogram (EEG) records the electric fields generated by the billions of action potentials that the brain generates every second.

Action potentials are not graded in amplitude (that is, they do not vary in 'strength'). Rather, when the threshold potential is reached the membrane potential suddenly rockets to its maximum level.

By analogy, in order to flush a toilet, sufficient pressure has to be applied to the handle to open the valve that connects the cistern with the toilet bowl. However, once the water starts flowing, it is not possible to stop the cistern emptying.

ACTION POTENTIAL RECOVERY
When the membrane potential reaches its maximum level, the sodium channels close and other channels, which are permeable to positive potassium ions, open in response to the high membrane potential (the potassium channels only open in response to a high voltage). The positive potassium ions flow out of the cell, bringing the membrane potential back towards its resting value.

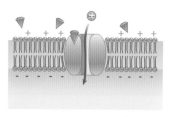

1 *Some neurotransmitters open sodium channels in the dendritic membrane, thereby allowing positive sodium ions to flow into the cell.*

2 *When the voltage reaches a threshold value, voltage-activated sodium channels open, allowing even more positive ions to enter the cell.*

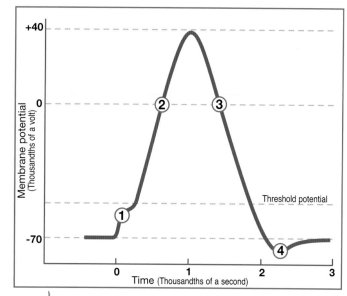

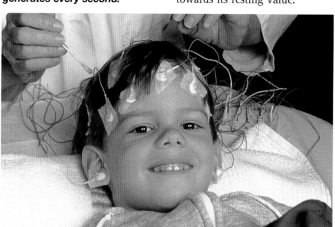

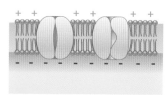

3 *Sodium channels close and potassium channels open, allowing positive potassium ions to leave the cell; both these events act to lower the voltage.*

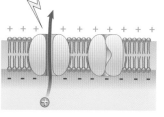

4 *Eventually, both the sodium and potassium channels inactivate – at this point the neurone is at rest and no further action potentials can occur.*

Speed of nerve impulses

Each axon transmits nerve impulses at a constant speed. However, there is a wide degree of variability in the speed at which different axons conduct action potentials.

NERVE DIAMETER
For example, conduction speeds can vary between about 0.5 and 120 metres per second. The speed depends on the diameter of the nerve (nerves with large diameters conduct quicker than nerves with small diameters) as well as the degree to which the nerve is insulated; conduction speed is increased in those

nerve fibres that are wrapped in a fatty insulating substance called myelin.

EFFECT OF TEMPERATURE
Furthermore, the speed of nerve impulse transmission varies according to temperature. For example, cooling a twisted ankle with an ice pack dulls the pain because it reduces the number of action potentials along the nerve.

Placing an ice pack on a swollen ankle eases the pain by slowing down the transmission of the nerve impulses.

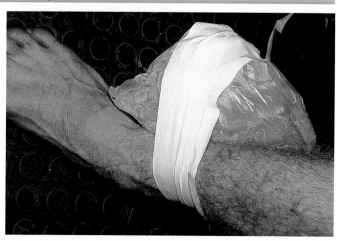

How nerve cells communicate

Nerve cells communicate with each other by releasing chemical messengers called neurotransmitters. Both therapeutic and illicit drugs act by altering the effectiveness of these transmitter molecules.

Nerve cells do not make direct contact with each other. Rather, there is a very small gap, called the synaptic gap, which separates the nerve cell sending the information (the pre-synaptic neurone) from the neurone receiving the information (the post-synaptic neurone).

This gap means that an electrical nerve impulse cannot flow directly from one neurone to the next. Instead, when a

This transmission electron micrograph shows a presynaptic neurone (left) containing vesicles (blue) in synaptic contact with a post-synaptic neurone (right).

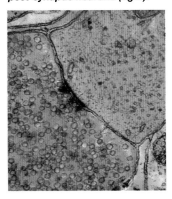

nerve impulse reaches the synaptic terminals, the sudden change in voltage causes calcium ions to flow into the pre-synaptic cell.

RELEASE OF NEUROTRANSMITTERS

Calcium ions cause vesicles (small membrane-bound sacs containing chemical messengers called neurotransmitters) to move towards, and dock with, the pre-synaptic cell membrane, releasing their contents into the synaptic gap.

The neurotransmitter molecules diffuse across to the post-synaptic cell and activate receptor proteins located within its membrane. This can have the effect of either exciting or inhibiting the post-synaptic cell (depending on the neurotransmitter and its associated receptor), increasing or decreasing the likelihood of an action potential being generated respectively.

Neurotransmitter molecules diffuse across the synaptic gap and bind to receptors in the post-synaptic membrane.

1 Voltage change across membrane causes influx of calcium ions

Vesicles containing neurotransmitter molecules

Synaptic gap

Pre-synaptic neurone

2 Calcium ions cause vesicles to release neurotransmitter

3 Neurotransmitter opens pores to either excite or inhibit the post-synaptic neurone

Post-synaptic neurone

Nervous control of muscle

Some nerves project from the spinal cord to serve muscle. When a nerve impulse arrives at a 'neuromuscular junction' it causes a neurotransmitter called acetylcholine to be released from the nerve endings.

A micrograph showing a neuromuscular junction. The nerve (black) can be seen innervating the pink muscle.

Acetylcholine diffuses across the synaptic gap and binds with receptors in the muscle tissue. This initiates a sequence of events that results in the contraction of the muscle fibres.

In this way, the central nervous system controls which muscles contract at a given moment. This is essential for complicated movements such as walking.

Neurotransmitter effects

After a neurotransmitter has bound with, and activated, its receptor on the post-synaptic membrane, it rapidly disengages and is either broken down by enzymes floating in the synaptic gap, or is taken up into the pre-synaptic terminal, where it is repackaged into another vesicle. This ensures that the effect of the neurotransmitter on a receptor molecule is short-lived.

Some illegal drugs, such as cocaine, as well as some prescription drugs, work by preventing the neurotransmitter

(dopamine in the case of cocaine) from being reabsorbed; this prolongs the time that the neurotransmitter can activate receptors in the post-synaptic membrane leading to a much greater stimulatory effect.

Cocaine works by inhibiting the uptake of the neurotransmitter dopamine. This allows the dopamine molecules to activate their receptors for longer.

Neural processing

The brain is an incredibly complex structure; each of its neurones is connected to thousands of others located throughout the nervous system.

Since nerve impulses do not vary in strength, information is encoded in the frequency of nerve impulses (that is, the number of action potentials that a neurone generates per second), in a similar way to Morse code.

One of the big problems that neuroscientists face today is to try to understand how this relatively simple encoding system produces; for example, the emotional responses that we feel when a friend or relative dies, or the ability to throw a ball with such accuracy that it hits a target 20 metres away.

In this regard, it is clear that information is not transmitted from one neurone to another in a linear fashion. Rather, a single neurone is likely to receive synaptic inputs from many other neurones (called convergence) and be able to influence a large number (up to 100,000) of other neurones (called divergence).

Indeed, it has been calculated that the number of possible routes for nerve impulses to take through this vast neural network is greater than the number of sub-atomic particles contained in the entire universe!

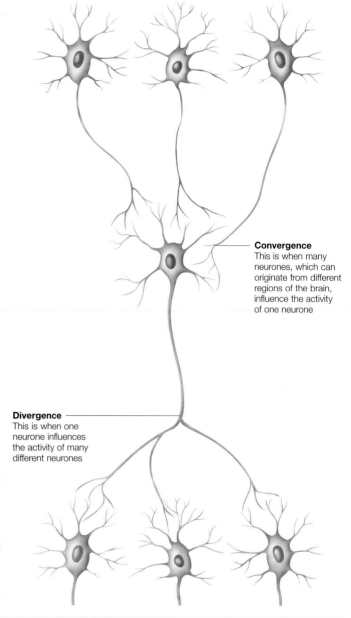

Convergence
This is when many neurones, which can originate from different regions of the brain, influence the activity of one neurone

Divergence
This is when one neurone influences the activity of many different neurones

This scanning electron micrograph shows many pre-synaptic neurones (blue) synapsing with a post-synaptic neurone (orange).

Information transfer does not occur in a linear fashion. A single neurone can thus influence and be influenced by thousands of other such cells.

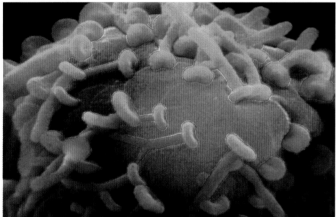

Types of synapse

Axo-somatic synapse
Inhibitory synapses are often of this type – they reduce the likelihood of a nerve impulse being generated

Axo-dendritic synapse
The vast majority of synapses are of this type

Axon of post-synaptic neurone
This carries nerve impulses away from the cell body towards the synaptic terminals

Synapses are named based on their constituent parts. For example, in an axo-dendritic synapse an axon makes synaptic contact with a dendrite.

Axo-axonic synapse
This type of synapse is relatively rare

There are two main types of synapse: those that cause the post-synaptic neurone to become excited, and those that cause it to become inhibited (this depends to a large degree on the type of neurotransmitter that is released). A neurone will only fire a nerve impulse when the excitatory inputs outweigh the inhibitory ones.

STRENGTH OF SYNAPSES
Each neurone receives a large number of both excitatory and inhibitory inputs. Each of the synapses present will have a greater or lesser effect in determining whether an action potential is initiated.

For example, synapses that have the most powerful effect are generally those close to the nerve impulse-initiating zone in the cell body (soma).

How cells work

All the living tissue in the body is made up of cells – microscopic membrane-bounded compartments filled with a concentrated solution of chemicals. Cells are the smallest living unit in the body.

Every tissue in the body is made up of groups of cells performing specialized functions, linked by intricate systems of communication. There are over 200 different types of cells in the body. Although enormously complex, the final structure of the human body is generated by a limited repertoire of cell activities. Most cells grow, divide and die while performing functions particular to their tissue type, such as the contraction of muscle cells.

Typically, cells contain structural elements called organelles, which are involved in the cell's metabolism and life cycle. This includes the uptake of nutrients, cell division and synthesis of proteins – the molecules responsible for most of the cell's enzymatic, metabolic and structural functions.

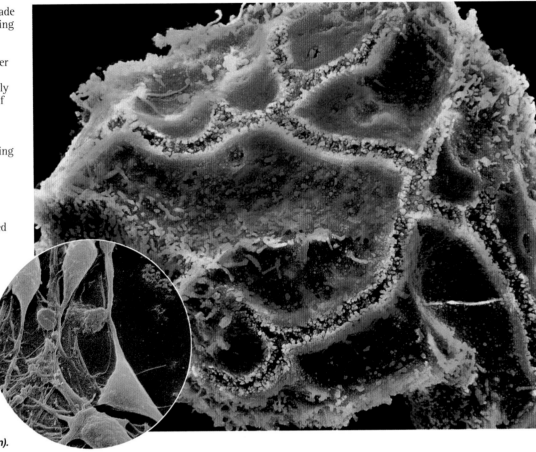

A hepatocyte is seen on a micrograph (far right). This is a specialized liver cell that performs several functions. Neurones – nerve cells – of the cerebral cortex are shown in the inset (green).

Immortal cells

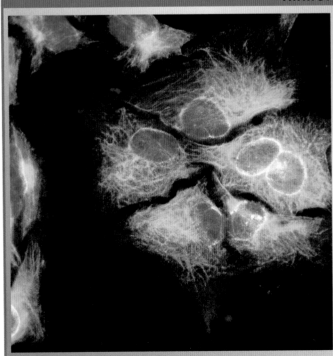

Most cells, when grown in a laboratory, can divide only about 50 times before they die. Immortal cells are cells that can be grown in Petri dishes indefinitely, and such cells are extremely useful in research.

In 1951, Henrietta Lacks, a 31-year-old American woman, was found to have a small lesion on her cervix, and a biopsy was taken to determine if the cells were malignant (cancerous). The sample of cells sent to the laboratory were indeed malignant, and despite treatment, she died eight months later from cervical cancer.

The sample of cells ended up in the laboratory of George Gey, a pioneer of tissue culture, and after working with the cells for several

weeks, he concluded that they divided faster than any cells he had ever seen before.

The cells, now called HeLa cells, proved to be robust and immortal, and because they grow so rapidly and reliably, they were eventually made available to other researchers and have been used extensively in biological research ever since. The polio vaccine was developed in under a year thanks to their use.

Unfortunately, HeLa cells have the ability to contaminate and subvert other cells growing in the same laboratory, and there were instances of experiments performed on one particular type of cell, unknowingly being performed on HeLa cells instead.

HeLa cells are still maintained in laboratory cultures. Such colonies have been maintained for the 40 years since the tumour they were cultured from was removed from Henrietta Lacks' cervix.

HeLa cells, unlike normal cells, continue to divide indefinitely. They have been used in research worldwide because they are so easily cultured.

Structure of a cell

The cell structure can be divided into the outer membrane, the DNA-containing nucleus and the structures called organelles within the cell. Each component of a cell has a specific function, such as energy production, storage or the synthesis of proteins.

THE PLASMA MEMBRANE

The plasma membrane surrounds each cell and separates it from its external environment, which includes other cells. Contained inside the membrane is a solution of proteins, electrolytes and carbohydrates called the cytosol, as well as membrane-bound subcellular structures called organelles. Spanning across the membrane are proteins responsible for communication with the external environment and for transport of nutrients and waste.

THE NUCLEUS

The nucleus is in the centre of the cell, and contains the cell's DNA arranged into chromosomes, as well as structural proteins for coiling and protecting the DNA. The nucleus is surrounded by a membrane with large pores in it, allowing for movement of molecules between the nucleus and the cytosol, while retaining the chromosomes inside the nucleus.

The shape of each type of cell varies according to function. Many organelles found in most cells are seen in this cut-away.

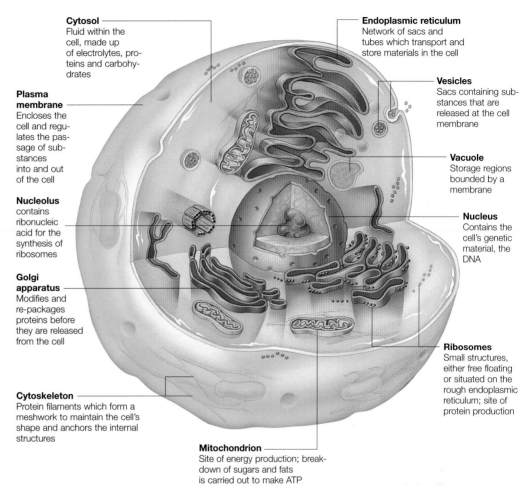

Cytosol
Fluid within the cell, made up of electrolytes, proteins and carbohydrates

Plasma membrane
Encloses the cell and regulates the passage of substances into and out of the cell

Nucleolus contains ribonucleic acid for the synthesis of ribosomes

Golgi apparatus
Modifies and re-packages proteins before they are released from the cell

Cytoskeleton
Protein filaments which form a meshwork to maintain the cell's shape and anchors the internal structures

Endoplasmic reticulum
Network of sacs and tubes which transport and store materials in the cell

Vesicles
Sacs containing substances that are released at the cell membrane

Vacuole
Storage regions bounded by a membrane

Nucleus
Contains the cell's genetic material, the DNA

Ribosomes
Small structures, either free floating or situated on the rough endoplasmic reticulum; site of protein production

Mitochondrion
Site of energy production; breakdown of sugars and fats is carried out to make ATP

Inside the cell – the cytoplasm

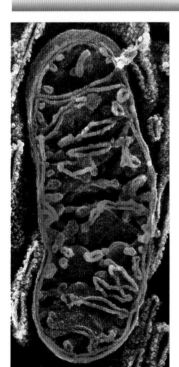

A single mitochondrion is seen coloured pink in this high-powered micrograph. These are the 'powerhouses' of the cell, where respiration occurs.

The cytoplasm is the inner contents of the cell, not including the nucleus, which is made up of fluid (the cytosol) and large numbers of organelles. The organelles include:

■ **Mitochondria**
Responsible for energy production. Nutrients in the form of sugars and fats are broken down in the presence of oxygen to make ATP (adenosine triphosphate), a source of energy used by a cell.

■ **Ribosomes**
Ribosomes carry out the production of proteins, using the blueprint recorded in the genetic material of the cell.

■ **Endoplasmic reticulum**
This is a vast network of tubes, sacs and sheets of membrane

that runs throughout the cell. It allows for the transport and storage of molecules.

■ **Golgi apparatus**
The Golgi apparatus is a stack of flattened sacs, critical in the modification, packaging and sorting of large molecules in the cell.

■ **Vesicles and vacuoles**
Vesicles are membrane-bounded areas within a cell for specialized processes or storage. Vacuoles appear as 'holes' under the microscope, and are typically

regions of storage or digestion surrounded by a membrane.

■ **Cytoskeleton**
The cytoskeleton is the fine meshwork of protein filaments used to maintain the cell's shape, to anchor components in place and to provide a basis for the cell's movements.

This electron micrograph shows a section through the rough endoplasmic reticulum of an animal cell (red lines). Attached to the surface are ribosomes.

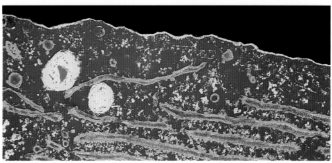

How cells divide

The vast majority of cells that make up the human body divide on a regular basis. This occurs not only during periods of growth, but also when worn out cells need to be replaced.

All tissues are made up of cells, microscopic membrane-bounded compartments. New cells are made by cell division, during which a cell replicates its genetic material and then separates its contents into two daughter cells. The process of cell division occurs continuously throughout the body, both during the development of the fetus and throughout adulthood.

WHY DO CELLS DIVIDE?

Cells divide when body tissue is growing, or when the cells in that tissue wear out and need to be replaced. Division is carefully regulated and must occur in accordance with the needs of the surrounding tissue, as well as in time with the internal cell growth cycle.

Cells that divide out of control can become cancers. Most chemotherapy is based around regimens that kill dividing cells, but which have less of an effect on non-dividing cells.

EMBRYONIC CELLS

The most prolific cell division occurs during early embryonic development; in nine months, a fertilized egg (one cell) develops into an embryo and, subsequently, a fetus of over 10 thousand million cells.

As development proceeds, many cells switch from dividing to performing a specialized function (such as becoming pacemaker cells in the heart), a process called differentiation.

In almost all tissues there are stem cells, which are cells that are not fully differentiated, but which can divide and differentiate in response to stimuli or wounding.

Once an egg cell is fertilized, it divides progressively; this human embryo is at the four cell stage.

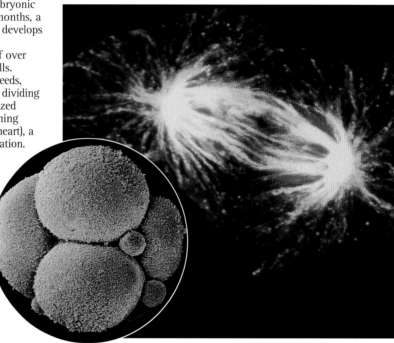

During the process of cell division, the chromosomes, which contain genetic material, separate into each of the two new (daughter) cells.

The life cycle of a cell

The cell division cycle is the process by which one cell doubles its genetic material and then divides into two identical daughter cells. The cycle is divided into two main stages: interphase, when the cell's components are replicated, and mitosis, when the cell divides into two.

Interphase is divided into two gap phases (G_1 and G_2) and a synthesis (S) phase.

During the first gap phase (G_1), the cell produces carbohydrates, lipids and proteins. Slow-growing cells, such as liver cells, may remain in this phase for years, whereas fast-growing cells, such as those in bone marrow, spend only 16–24 hours in the G_1 phase.

If a cell is not actively dividing, it exits the cell cycle during G_1 and enters a state called G_0. For example, in adults many highly specialized cells, such as neurones (nerve cells) and heart muscle cells, do not divide and remain in phase G_0. This makes healing and regeneration in these tissues slow and sometimes impossible.

REPLICATING CHROMOSOMES

The next period of interphase – which is known as the S phase – sees the replication of the chromosomes so that the cell temporarily has 92 chromosomes instead of the normal 46. Proteins are also synthesized during the S phase, including those that form the spindle structures that pull the chromosomes apart. In most human cells, the S phase lasts between 8 and 10 hours.

Additional proteins are synthesized in the second gap phase, G_2.

The cell divides into two daughter cells during a process called mitosis. Mitosis is subdivided into four phases: prophase, metaphase, anaphase and telophase.

The duration of a complete cell cycle varies from a day to a year, depending on the type of cell concerned. Examples of the rates of replacement for different cell types are:

- Liver cells: 12 months
- Red blood cells: 80–120 days
- Skin cells: 14–28 days
- Intestinal mucosa: 3–5 days.

Cell division is divided into two stages: interphase (purple), when the cell contents are replicated, and mitosis (orange), when the cell divides. Interphase is subdivided into G_1, S and G_2 phases. Mitosis is subdivided into prophase, metaphase, anaphase and telophase.

INTERPHASE

S Phase

G_1 Phase

G_2 Phase

G_0 Phase

Telophase

Anaphase Metaphase

Prophase

MITOSIS

Cell Division

The four stages of mitosis

1 Prophase
During prophase, the DNA condenses into recognizable chromosomes, the nucleus disbands and the nuclear contents enter the cytoplasm.

This scanning electron micrograph (SEM) shows the condensed chromosomes (red), nuclear membrane (orange) and cytoplasm (green).

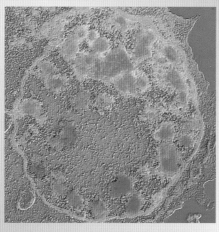

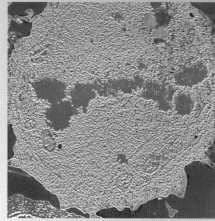

2 Metaphase
During metaphase the chromosomes attach to the mitotic apparatus, a series of specially synthesized protein filaments anchored to opposite sides of the cell.

Here, the cell is in late metaphase: the nuclear membrane has disappeared and the chromosomes (red) are aligned along the centre of the cell.

3 Anaphase
During anaphase, the chromosomes are pulled away from each other by the mitotic apparatus in the cell. Half of the chromosomes go to each side of the cell.

This SEM shows the first stages of anaphase, when the chromosomes separate and the cell membrane becomes indented.

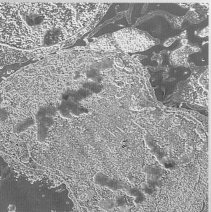

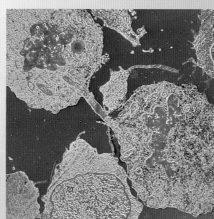

4 Telophase
During telophase, the nuclear membranes re-form, the cell's contents are redistributed and the membrane 'pinches off' to form two cells.

The cell is in late telophase in this SEM: the two newly-formed cells are still joined by a narrow bridge containing elements of the mitotic apparatus.

Cell death and cell suicide

There are two ways that cells can die: they can be killed by injurious agents, a process called necrosis; or they can be induced to 'commit suicide', a mechanism that scientists call apoptosis.

NECROSIS

When the body is exposed to mechanical or chemical damage, cells may die simply because they are no longer able to function properly. This process, called necrosis, happens when the cell's integrity is violated, or when molecules or structures essential to the cell's survival are no longer available or functional. For example, after a person dies, nutrients and oxygen are unavailable to all of the cells of the body, which then undergo necrosis and die. Gangrene is another example of necrosis – the dead tissue turns black due to the action of certain bacteria on haemoglobin, which is broken down to produce dark iron sulphide deposits.

APOPTOSIS

The majority of cells have an in-built programme which makes them commit suicide. Scientists believe that this programme is as intrinsic to the cell as mitosis.

There are two main reasons why a cell would commit suicide. First, programmed cell death is often needed for the proper development of the human body. For example, the formation of the fingers and toes of the fetus requires the removal of the tissue between them by apoptosis.

Second, cell suicide may be needed to destroy cells that represent a threat to the organism. For example, defensive T-lymphocyte cells kill virus-infected cells by inducing apoptosis in them.

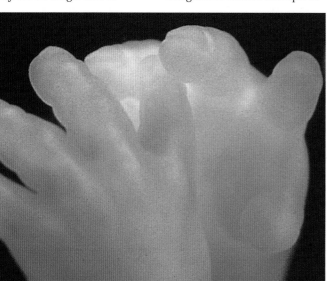

In the early stages of fetal development, fingers are connected, giving them a webbed appearance. This webbing disappears as the fetus develops, due to apoptosis.

Meiosis

Meiosis is a special form of cell division that only occurs during the formation of sperm and eggs. During meiosis, there are two cycles of division, but only one duplication of chromosomes, so that sperm and eggs end up with only 23 chromosomes. Meiosis is unique, since 'crossing-over' occurs between paired chromosomes. As a result, the chromosomes that are in the sperm and eggs are not identical to the chromosomes from the parent.

Meiosis is the process that creates both sperm and eggs. Unlike all the other cells in the body, these gametes (sex cells) contain only 23 – rather than 46 – chromosomes

453

How cells communicate

For the body to act in a co-ordinated manner, it is essential that cells communicate with each other. They do this either by releasing chemical messengers or by electrically exciting neighbouring cells.

The human body contains a total of around 10,000,000,000,000 (10 trillion) cells, made from just over 200 different cell types. However, the benefits of having specialized cells can only be realized if this multicellular organization acts in a co-ordinated manner.

■ Internal stimuli
The body must be able to respond to changes in its internal environment. For example, cells in the pancreas detect the rise in blood glucose concentration after a meal; they release a hormone – insulin – which makes the cells of other tissues absorb glucose from the blood to provide energy.

■ External stimuli
Similarly, the body must also be able to detect and respond to external stimuli. For example, it would be no good having eyes with which to see a predator if this visual information could not be relayed to the rest of the body to prepare it to fight with, or run away from, an aggressor.

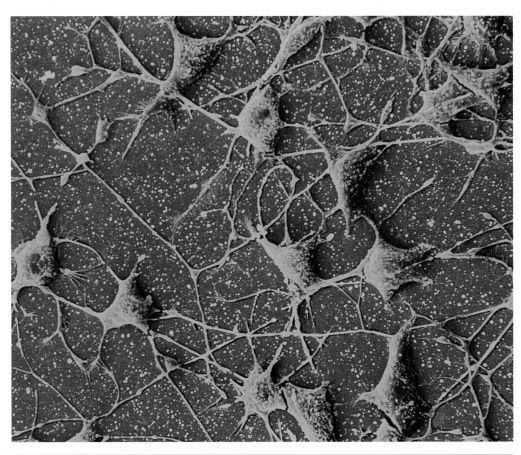

Nerve cells communicate by releasing chemical messengers which affect the electrical excitability of neighbouring cells.

Electrical and chemical communication between cells

Heart cells communicate electrically

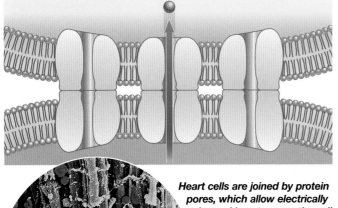

Heart cells are joined by protein pores, which allow electrically charged ions to cross the cell membrane. These allow a wave of electrical excitation to travel through the heart.

Heart cells (green) communicate with each other electrically. Chemicals (for example, adrenaline) released by distant tissues can affect their behaviour, however.

Both internal and external stimuli are detected by specialized chemicals (normally proteins) called 'receptors', which transduce (convert) information into a form that can be relayed to other cells within the body. Broadly speaking, communication between the body's cells is accomplished using either chemical messengers or electrical currents.

ELECTRICAL COMMUNICATION
Most electrical messages are carried by nerve cells (though heart cells also communicate electrically) which are specially adapted for carrying nerve impulses from one region of the body to another. For example, some nerve fibres can be up to a metre in length.

The main advantage of electrical communication is the speed at which information can be transmitted; some nerves are able to propagate nerve impulses at rates of 120 metres per second. Furthermore, because the 'wiring' of neurones is very precise, information can be delivered to very specific locations.

CHEMICAL COMMUNICATION
In contrast, by virtue of the fact that many chemical messengers, such as hormones, are released into the bloodstream, these molecules can affect a wide number of cells, but can do so only relatively slowly. For example, when a person is exposed to a stressful situation, the adrenaline rush does not 'kick in' for 15–30 seconds. This is because the adrenaline molecules have to diffuse from the adrenal gland (located just above the kidneys) into the bloodstream, which then carries them around the body to the target organs (such as the heart, increasing both its rate and strength of beating).

Types of chemical communication

Chemical messengers can be classified into three groups based on the type of cell that releases the chemical and how the chemical messenger reaches its site of action. These groups include hormones, paracrine and autocrine factors and neurohormones.

Hormones

Hormones are chemicals that are released by a gland into the bloodstream, which then carries them to distant sites throughout the body. They may have a specific site of action or may affect a wide variety of different cells, simultaneously regulating a large number of different bodily processes.

For example, adrenaline is released into the blood from the adrenal medulla, the central region of each of the adrenal glands, which lie above the kidneys. Adrenaline has a wide number of actions, which include constriction of the blood vessels, increased cardiac activity, dilatation of the pupils in the eye and inhibition of the gastro-intestinal tract.

HORMONE SPECIFICITY

Since all the body's cells are in close proximity to passing blood vessels, one might expect a hormone to be able to affect every cell in the body. However, this is not the case. For a hormone to affect the internal biochemistry of a cell (a cell's 'behaviour'), the cell must have an appropriate protein receptor embedded within its cell membrane; by way of an analogy, a front door must contain a letterbox to allow the postman to deliver a letter.

Hormones are chemical messengers that are released by a gland into the bloodstream which then carries the hormone to distant tissues.

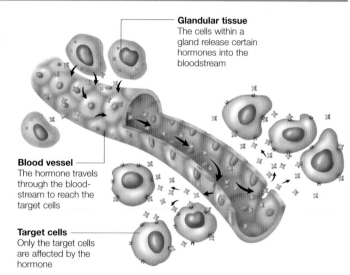

Glandular tissue
The cells within a gland release certain hormones into the bloodstream

Blood vessel
The hormone travels through the bloodstream to reach the target cells

Target cells
Only the target cells are affected by the hormone

Paracrine and autocrine factors

The second group of chemical messengers differ from hormones in that they are not transported by the bloodstream to their target cells.

Rather, these chemicals are released into the watery space that lies between the cells to affect either the same type of cell that released them (autocrine factors – 'auto' meaning 'self') or different, though nearby, cells (paracrine factors). It should be noted, however, that a chemical can be both an autocrine and paracrine factor.

PARACRINE FACTORS

One of the most common paracrine factors is the chemical histamine. Histamine is released

from specialized cells called mast cells that are present in most tissues. It is involved in allergic reactions and in some of the inflammatory chemical pathways that are initiated when a tissue is damaged. Anti-histamines work by preventing mast cells from releasing this paracrine factor.

AUTOCRINE FACTORS

Autocrine factors affect the same type of tissue that released them. For example, most cells release autocrine factors which inhibit their own cell division and that of similar nearby cells. Cancerous cells are thought to either not release, or not respond to, these inhibitors resulting in cell division proceeding unabated.

Autocrine factors are chemical messengers which only affect the same type of cell that originally released them.

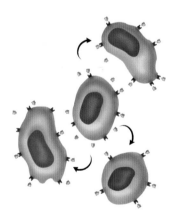

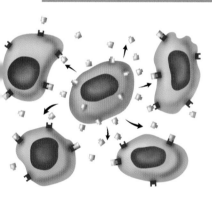

Paracrine factors are released into the water-filled space between the cells. They affect cells of a different type to the one that released them.

Neurohormones

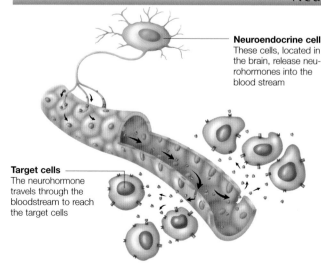

Neuroendocrine cell
These cells, located in the brain, release neurohormones into the blood stream

Target cells
The neurohormone travels through the bloodstream to reach the target cells

Most neurones communicate with each other by releasing a chemical messenger which diffuses between the gap (called a synapse) that separates them.

However, some neurones do not synapse with another nerve cell. Rather, their synaptic terminals are located near blood vessels; when these neurones are stimulated they release a neurohormone into the

Neurohormones are released by specialized nerve cells called neuroendocrine cells. These chemicals are carried in the bloodstream to the target cells.

bloodstream, which is then carried to distant target organs in much the same way that a hormone is released from a gland.

OXYTOCIN

Oxytocin is a neurohormone which is released into the bloodstream by neuroendocrine cells located in the hypothalamus. This occurs in response to the stimulation of sensory nerves in the mother's nipple by a suckling infant. The blood carries the neurohormone to the mammary gland where it causes milk to be ejected from the nipple.

Structure of the cell membrane

The cell membrane separates the inside from the outside of the cell. Since it is only permeable to certain molecules, the internal environment of the cell can be tightly controlled.

Every cell is covered in a membrane, made mainly of phospholipids (phosphate-containing fat molecules) and proteins, which acts as a barrier between the inside and outside of the cell. Some molecules can pass freely through it, while others have either restricted or no access.

A membrane surrounds the cell as a whole (also called the plasma membrane), as well as the organelles (the sub-cellular components) which are contained within the membrane.

The cell membrane is much more than a simple protective covering; by determining which chemicals are allowed to pass into and out of the cell, the cell is able to tightly control its internal environment, as well as communicate with other cells.

CHEMICAL CONSTRUCTION
The cell membrane is composed of four groups of chemicals: phospholipids (25 per cent), proteins (55 per cent), cholesterol (15 per cent) and carbohydrates and other lipids (5 per cent).
■ Phospholipid molecules are arranged in two layers (known as a 'bilayer'). They act as a very thin yet impenetrable barrier to water, and to water-soluble molecules such as glucose.

How the cell membrane is made up

The cell membrane is a complex structure that separates the inside from the outside of the cell.

Carbohydrate
Carbohydrates are only found on the outside surface of the cell membrane

Cholesterol
Cholesterol molecules are embedded within the cell membrane

The cell

Phospholipid bilayer

Protein
Some membrane proteins connect the inside to the outside of the cell

Cytoskeleton
The internal 'scaffolding' of the cell is called the cytoskeleton

However, fat-soluble molecules, such as oxygen, carbon dioxide and steroids, can pass through it freely.

The phospholipid portion of the membrane is extremely thin – if a human were shrunk to the height of the membrane, a passing red blood cell would appear to be about a mile wide

■ Proteins provide a means by which water-soluble molecules can enter and leave a cell. They also allow cells to communicate with, recognize and adhere to each other
■ Cholesterol molecules are, in a sense, 'dissolved' within the phospholipid bilayer. Cholesterol reduces the fluidity of the

membrane by interfering with the lateral movement of the phospholipid tails
■ Carbohydrates are attached to proteins (glycoproteins) and to lipids (glycolipids). They invariably protrude on the outside surface of the membrane and are important for cell adhesion and communication.

Specializations of the membrane

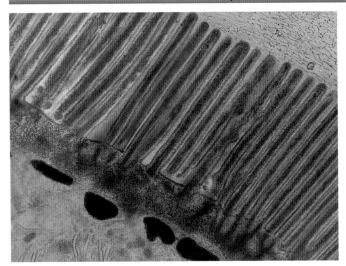

Microvilli (purple) increase the surface area available for absorption of nutrients from the gut lumen (yellow).

Cell membranes are not the same in all cells. This is because of the huge variety of functions that cells perform in different parts of the body.

MICROVILLI
Microvilli are specialized infoldings of the cell membrane which greatly increase its total surface area. This is especially useful in cells whose main role is to absorb chemicals from the outside to the inside of the cell. For example, around 1,000

microvilli are found in each intestinal epithelial cell, which act to absorb nutrients from the gastro-intestinal tract. Each of these microvilli is about one-thousandth of a millimetre in length, and so increases the surface area available for uptake of nutrients by up to 20 times.

ADHESION BETWEEN CELLS
While some cells, such as blood cells and sperm, are independent entities with some degree of movement, the majority of the body's cells are knitted together to form tissues; the body's cells are joined together by specialized membrane junctions.

The role of membrane proteins

Proteins embedded in the cell membrane play an important role in many
cellular functions. Some span the cell membrane, connecting the inside to the outside
of the cell, and this allows cells to communicate with each other chemically.

Membrane proteins are responsible for most of the specialized functions of the cell membrane. They can be broken down into two main groups:
■ Integral proteins – while some integral proteins protrude through the cell membrane on one side only, the vast majority cross the membrane and so are exposed to both the inside and outside of the cell.

These 'trans-membrane' proteins often allow substances to be exchanged between the internal and external environments, either by providing a pore through the membrane, or by physically ferrying the molecules across.

Integral proteins also provide binding sites for chemicals released by other cells; this allows cells, including neurones (nerve cells), to communicate with each other.
■ Peripheral proteins – these are not imbedded in the phospholipid bilayer. Rather, they are usually attached to the internal side of integral proteins. They may act as enzymes, which speed up chemical reactions inside the cell, or may be involved in changing the cell's shape; for example, during cell division.

Functions of membrane proteins

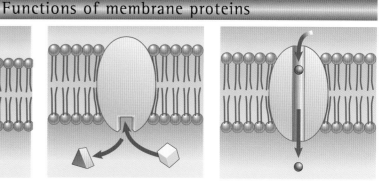

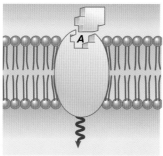

The external surface of some proteins provides a 'binding site' (A) for chemical messengers released from other cells.

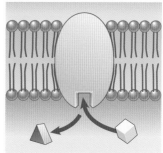

The internal surface of some proteins acts as an enzyme, speeding up chemical reactions that occur inside the cell.

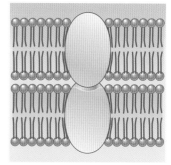

Transport proteins span the membrane, providing a pore for chemicals to travel either into, or out of, the cell.

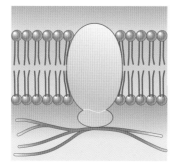

The internal scaffolding of the cell (cytoskeleton – red strands) attaches to the internal surface of membrane proteins.

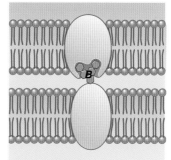

Some glycoproteins (molecules made of proteins joined to carbohydrates) act as 'identification tags' (B).

Membrane proteins of adjacent cells may join together, providing various kinds of junctions between the two cells.

Why are phospholipids so important?

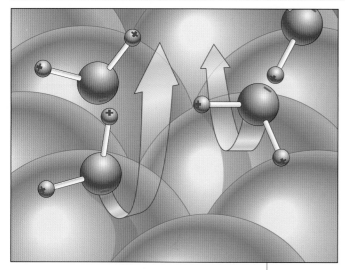

Water molecules are made of a slightly negative oxygen atom (red), attached to two slightly positive hydrogen atoms (blue). Since the phospholipid 'tails' are non-polar, the cell membrane is impermeable to water.

Cell membrane

WATER MOLECULES

Water molecules are composed of two hydrogen atoms attached to one oxygen atom (hence the chemical formula H_2O). Although they have no net electrical charge (a water molecule as a whole is electrically neutral), the oxygen at one end of the molecule tends to be slightly negative and the two hydrogens at the other tend to be slightly positive.

As a result of this chemical property, water is said to be a 'polar' molecule since, like a magnet, it has two electrical 'poles'.

This results in water molecules interacting with each other electrically: the negative oxygen in one water molecule is attracted to the positive hydrogens of neighbouring water molecules. The degree of this attraction, which is dependent on the surrounding temperature, determines whether the molecules form ice, water or steam.

PHOSPHOLIPIDS AND WATER

Water also interacts with other polar molecules, such as glucose (hence glucose is 'water soluble'). However, non-polar molecules, which include fats, are insoluble in water.

Phospholipids are ideal building blocks for the cell membrane, which is designed to separate the internal contents of the cell from the outside environment, by virtue of its special chemical structure: a phospholipid molecule is composed of a phosphorus-containing, 'water-loving' head attached to a lipid-containing, 'water-hating' tail. This means that when phospholipids are mixed with water, the 'water-loving' heads mix with, while the 'water-hating' tails avoid, the surrounding water molecules. Thus water molecules can only pass through the cell membrane by travelling through pores in proteins which are embedded within it.

How chemicals cross the cell membrane

Cells in the body need to control their internal environment carefully in order to function properly. The cell membrane provides a barrier that regulates which chemicals can pass into and out of the cell.

Each cell in the body is surrounded by a membrane. This is an important barrier which separates the cells' internal contents from their external environment. This is vital because the contents within cells need to be tightly regulated in order for cells to function properly.

SEMI-PERMEABLE MEMBRANES

The cell membrane is not an impenetrable barrier. Rather, it allows some substances to pass freely across it, while restricting, or totally preventing, the passage of other chemicals; hence the term 'semi-permeable membrane'.

For example, glucose, an essential molecule which provides the body with energy, is able to move easily through the cell membrane. However, in order to prevent unused glucose leaking back out of the cell, glucose is converted into a

chemical called glucose-6-phosphate, which is unable to travel back across the cell membrane.

Other molecules that can pass through the cell membrane easily include oxygen, which is used in the metabolism of glucose; and carbon dioxide, a waste-product which diffuses out of the cell.

PROTEIN PORES

Other particles, such as sodium ions or amino acids, can only pass through the membrane via 'pores' provided by specific membrane proteins; many of these act like gates – opening or closing only in response to a predetermined chemical signal.

For example, some protein pores only open when a chemical released from another cell (for example a hormone) binds to its external surface. Other protein pores open in response to a change in electrical voltage.

Passive transport

Simple diffusion

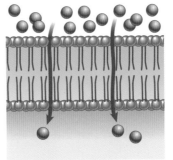

Some molecules, such as steroids, can travel freely across the cell membrane (green) into the cell.

Protein pores

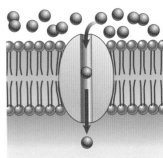

Some proteins provide 'pores' which allow small chemicals, such as sodium atoms, to cross the cell membrane.

Facilitated transport

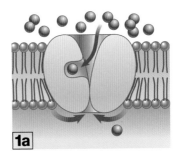

1a

Larger molecules, such as glucose, are 'ferried' across the membrane. These differ from pores as they are not permanently open.

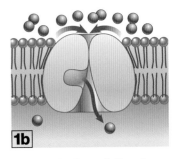

1b

The protein channel alters its shape slightly after the chemical has 'docked' with it, allowing the chemical to be released on the inside of the cell membrane.

Receptor-mediated opening of protein pore

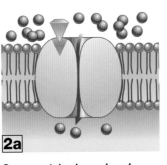

2a

Some protein channels only open when a messenger molecule (often released by another cell) binds to the outer surface of the protein, like a key.

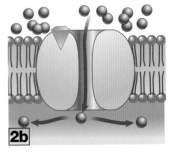

2b

These 'receptor proteins' are especially important in neurones (nerve cells), as they allow one neurone to influence the internal environment of another.

Osmosis

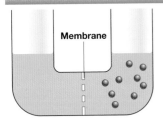

A U-shaped tube is separated into two compartments, one containing pure water, the other a concentrated sugar solution.

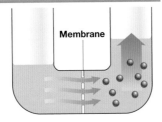

The membrane is permeable to water, but not sugar molecules; therefore water travels across the membrane to the sugar solution.

Osmosis is the movement of water molecules across a membrane from where they are at high concentration to where they are at low concentration.

This process can be illustrated by filling one half of a two compartment vessel with pure water and the other half with a concentrated sugar solution. Separating the compartments is a semi-permeable membrane whose small pores allow the passage of water but not sugar molecules.

Water molecules will travel across the membrane from the side containing the pure water into the sugar solution.

IMPORTANCE OF OSMOSIS

Osmosis plays a very important role in the human body. For example, blood volume can be controlled by altering the concentration of sodium in the urine; water flows into the urinary tract as a result of osmosis and is then excreted, thus lowering the blood volume.

Active transport

Some molecules are unable to cross the cell membrane unaided. This may be because they are insoluble in the cell membrane or too large to pass through the protein pores.

Like the osmotic action of water, other molecules have a natural tendency to move from a high to a low concentration (like a ball rolling downhill) until they are evenly dispersed in the space available. This is termed 'diffusing down the concentration gradient'.

If a molecule has to travel against its concentration gradient ('uphill') to pass into or out of a cell, it is carried by a process called active transport. However, active transport requires the cell to use energy.

There are two main types of active transport: membrane pumps and vesicular transport. Membrane pumps are proteins that run through the membrane which are able to propel a small number of molecules across the membrane. Vesicular transport, in contrast, involves the transport of many molecules.

Membrane pumps

Simple active transport

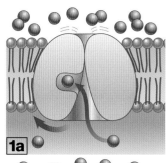
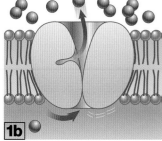

One way of carrying a chemical against its concentration gradient (in this case from inside to outside a cell) is for a protein to ferry it; this process requires energy.

Co-transport

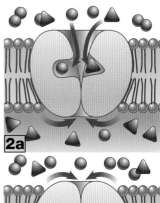
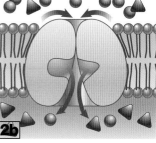

If one molecule is diffusing down its concentration gradient (red spheres), another molecule (purple triangles) can 'hitch a ride' and so travel against its concentration gradient.

Counter-transport

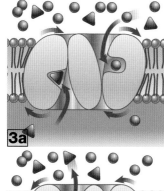
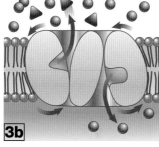

Counter-transport is very similar to co-transport in that the 'downhill' movement of one chemical provides the energy for another chemical to be transported 'uphill'.

Vesicular transport – exocytosis

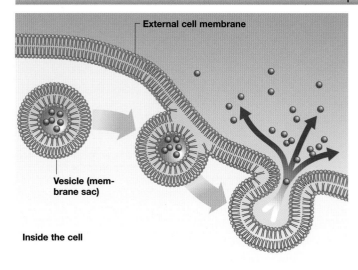

External cell membrane

Vesicle (membrane sac)

Inside the cell

Exocytosis is a process by which large quantities of a substance are transported from the inside to the outside of the cell.

Like membrane pumps, exocytosis requires energy. This mode of transport is responsible for the secretion of hormones from endocrine glands and for the release of neurotransmitters from nerve cells. As a result, exocytosis plays a major role in allowing communication between cells.

The chemical to be released from the cell is packaged in a membrane sac (vesicle). This sac fuses with the cell membrane, releasing its contents.

THE PROCESS OF EXOCYTOSIS

The substance to be released from the cell is first packaged within a sac, called a 'vesicle', composed of phospholipids and proteins, just like those found in the plasma membrane. The vesicle then moves to the cell membrane. Proteins on the vesicle recognize, and bind with, proteins on the membrane causing the two membranes to fuse and finally rupture, spilling the contents of the vesicle outside of the cell. The cell membrane does not increase in size as more of these vesicles dock with it, instead these vesicles are constantly recycled.

Vesicular transport – endocytosis

Endocytosis is in many ways the complete opposite of exocytosis; it is a process which allows substances to be taken into the cell from the external environment. There are three main types of endocytosis:
■ Phagocytosis (literally 'cell eating'): large, solid material (for example a bacterium) is engulfed by the cell's membrane and taken into the cell where it is digested

■ Pinocytosis (literally 'cell drinking'): droplets of fluid from outside the cell, containing dissolved molecules, are engulfed
■ Receptor-mediated endocytosis: when only very specific molecules are engulfed.

A cell (brown) is seen eating a Clostridium bacterium (blue). The cell's membrane engulfs the bacterium by a process called phagocytosis.

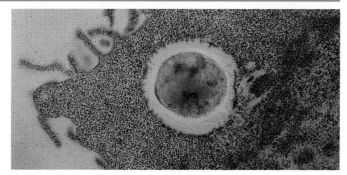

How DNA works

DNA is the genetic material of all organisms, located in the nucleus of every cell. The discovery of its chemical structure revolutionized the biological sciences and our understanding of human genetics.

The chemical properties of DNA (deoxyribonucleic acid) allow it to carry out two very important functions:
■ It provides the body's cells with the 'recipes' needed to build proteins from the 20 essential amino acids found in proteins
■ It is able to make copies of itself, and so provides the means by which these protein 'recipes' can be transmitted from one generation to the next; this means that characteristics such as eye colour or facial features can be passed from parent to child.

In humans, DNA is packaged within 23 pairs of chromosomes. These 'X' shaped structures replicate during cell division.

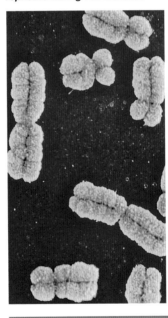

CHROMOSOMES
The vast majority of human DNA is packaged into 23 pairs of chromosomes, which are stored within the cell's nucleus; one set of 23 chromosomes is inherited from the father and one set is inherited from the mother.

The exceptions to this rule are sperm cells and egg cells, which contain only one set of 23 chromosomes; and red blood cells, which contain no chromosomes at all.

GENES
Useful DNA (as opposed to so-called 'junk' DNA – see below) is packaged within the chromosomes into what are known as genes, of which there are thought to be around 100,000 in the human body. Each of these genes provides the 'recipe' which tells the cell how to make a specific protein.

However, while each of the body's cells contains a copy of every protein recipe, not all of these recipes are 'switched on'. This is what differentiates a heart cell from, for example, a liver cell – each produces its own set of proteins.

'JUNK' DNA
Most DNA is so-called 'junk' DNA, which serves no known purpose in the human body. Much of this DNA is inherited from our distant ancestors and their parasites, dating back to when life on earth began around four billion years ago.

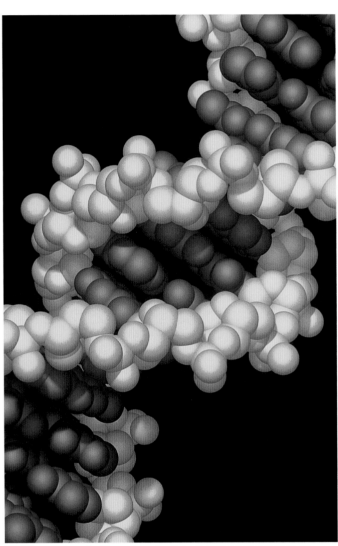

DNA is composed of two strands of nucleotides (shown here as yellow and blue) which curl around each other in a helical manner (called a double helix).

Mutations in DNA

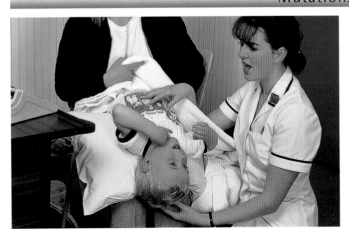

Cystic fibrosis sufferers have a mutated gene that produces a faulty protein. The result is that lungs become clogged with mucus.

If the set of instructions (that is, the DNA sequence) detailing how to make a specific protein becomes altered in even the smallest way, a 'mutation' is said to have occurred; this may result in a defective protein, or no protein at all, being made.

The consequences of this can be quite severe; for example cystic fibrosis is the result of a mutation at a single point in one of the DNA molecules in an affected individual.

Mutations are not necessarily harmful, though, and occur spontaneously throughout our lifetimes. However, certain chemicals increase the rate of mutation, one example being agent orange, a defoliant used in the Vietnam war. Nuclear radiation can also have a mutating effect. If a large number of mutations occurs, the chances of one being harmful are statistically higher, so these mutagens are harmful to humans.

Structure of DNA

The ability of DNA to make copies of itself is the direct result of its chemical structure. A molecule of DNA consists of two interlinked strands of nucleotides, which are exact

James Watson (left) and Francis Crick (right) were awarded a Nobel Prize for their contribution to the discovery of DNA structure.

mirror images of each other, and which run in opposite directions.

These two strands are each made up of a sugar-phosphate 'backbone', to which are attached specialized molecules, called bases. These bases are: adenine, guanine, cytosine and thymine (abbreviated to the letters A, G, C and T). What gives DNA its special properties is that the four bases will only pair off in the following combinations: A with T and G with C. Thus a strand of DNA with the sequence of 'TGATCG' will only bind with a complementary strand with the sequence 'ACTAGC'.

DNA is made of two strands, which run in opposite directions. The two strands are joined by special molecules called bases.

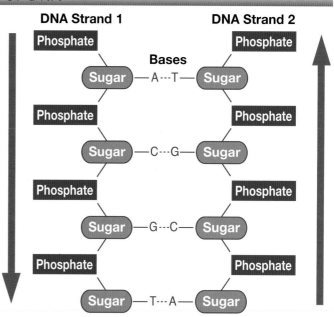

How DNA makes copies of itself

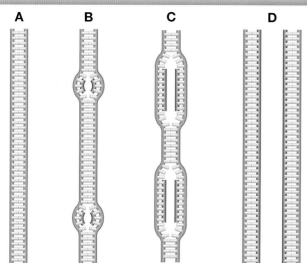

A **B** **C** **D**

DNA makes a copy of itself during a process called replication. First, the original double helix 'parent' DNA is unwound, so separating and exposing the base pairs. Since each base will only bind with

DNA (A) is simultaneously unwound at a number of points (B). A new strand (red) is built on each parent strand (B and C) forming two DNA molecules (D).

The two strands of a DNA molecule are separated from each other and a new strand is built onto each one. Thus, two identical strands can be made from the original.

one of the three others (such as A with T, but not with G or C), a complementary strand can then be built up on each of the parent strands. Thus one DNA double helix becomes two identical double helices.

DNA molecule

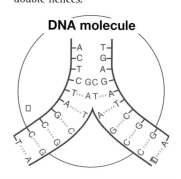

DNA as a recipe book for proteins

In many ways DNA is similar to a language; however, unlike the English language there are only 64 three-letter 'words' that can be formed using the four letters A, G, C and T. Geneticists call these 'words' codons as each codes for a specific amino acid, the building blocks of proteins.

Proteins are made on structures called ribosomes in the cytoplasm. However, since DNA is unable to leave the nucleus, first of all one strand of DNA is 'transcribed' onto a single-stranded messenger molecule, with a very similar structure to a DNA strand, called messenger RNA (mRNA), which can cross the nuclear membrane. The mRNA is then 'translated' on the ribosomes in the cytoplasm so that the correct amino acids are joined together in the correct order.

Proteins are built from amino acids according to a template copied from DNA. This process occurs in subcellular structures called ribosomes, which are found inside the cell.

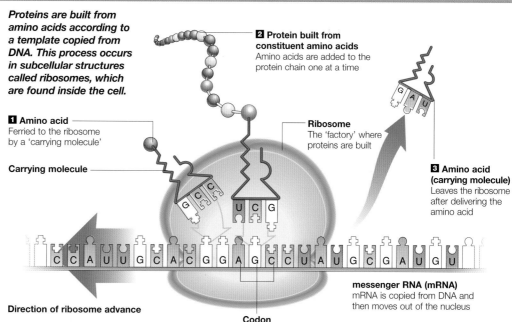

1 Amino acid
Ferried to the ribosome by a 'carrying molecule'

Carrying molecule

2 Protein built from constituent amino acids
Amino acids are added to the protein chain one at a time

Ribosome
The 'factory' where proteins are built

3 Amino acid (carrying molecule)
Leaves the ribosome after delivering the amino acid

messenger RNA (mRNA)
mRNA is copied from DNA and then moves out of the nucleus

Direction of ribosome advance

Codon

461

How our genes can affect us

Faulty genes do not always lead to disease. It is possible for people who are normal to carry abnormal genes. Such people often only learn of the problem when they have an affected child by another, equally normal carrier.

An observable characteristic in a person is called a phenotype. This could be a disease, a blood group, eye colour, nose shape or any other such attribute. The genetic information that gives rise to a phenotype is known as the genotype.

A gene locus is the site on a chromosome at which a gene for a particular trait lies. The different forms of a gene that may be present at a gene locus are called alleles. If there are two alleles – 'A' and 'a' – for a given gene locus, then three genotypes may be formed. These are: 'AA', 'Aa' or 'aa'. 'Aa' is known as a heterozygote and 'aa' and 'AA' are homozygotes.

If the 'A' allele is dominant, it will mask the effects of the

In these family trees, the parents have a dominant allele 'A' and recessive allele 'a'. If 'A' represents the phenotype of brown eyes, and 'a' represents blue eyes, only those with the 'aa' genotype will have blue eyes. Otherwise, the dominant 'A' allele dictates the phenotype.

recessive 'a' allele, and produce a recognizable phenotype in a heterozygous (Aa) individual. Recessive alleles only result in a recognizable phenotype in the homozygous state (aa).

If both alleles are recognized in the heterozygous state (Aa), they are designated co-dominant. The expression of the ABO blood groups is a example of the effects of co-dominance.

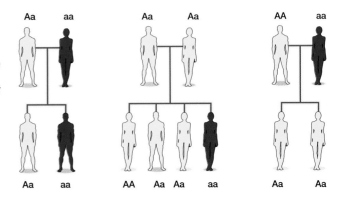

Autosomal dominant conditions

Affected individuals (male or female) who carry a dominant abnormal gene have a 50 per cent chance of producing affected offspring with a normal partner. Only people who inherit a copy of the gene will be affected. Achondro-plasia (dwarfism) is an autosomal dominant condition.

KEY

Normal male and female

Affected male and female

An individual is either normal or affected by an autosomal dominant condition. When their partner is unaffected there is a 50 per cent chance that a child will be affected.

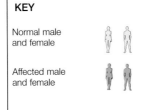

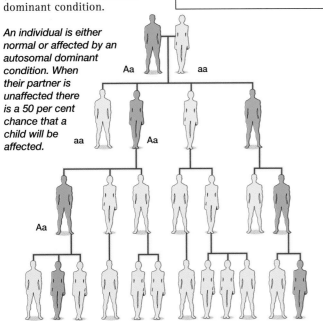

Conditions such as achondroplasia (pictured here) result from the inheritance of an affected gene from either the mother or father.

Autosomal recessive conditions

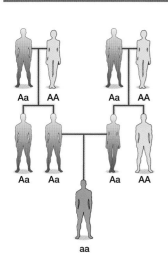

Unaffected individuals of either sex may be carriers. When two carriers (Aa; heterozygotes) have a child, there is a one in four (25 per cent) risk that the child will be affected. An example of an autosomal recessive condition is sickle-cell disease, a disease of the blood that mainly affects people of African ancestry.

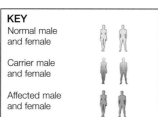

KEY
Normal male and female

Carrier male and female

Affected male and female

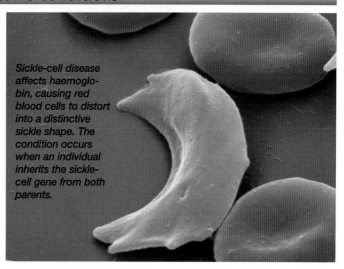

Sickle-cell disease affects haemoglobin, causing red blood cells to distort into a distinctive sickle shape. The condition occurs when an individual inherits the sickle-cell gene from both parents.

Sex-linked conditions

In these conditions, the abnormal trait is carried on the sex chromosomes (X and Y). Males only have one X chromosome, therefore all daughters inherit their father's X chromosome. They will also inherit one of the two X chromosomes from their mother. A son will inherit his father's Y chromosome and one of his mother's two X chromosomes.

If one of the two X chromosomes from the mother contains a gene that can give rise to a disorder, she is referred to as a 'carrier'. Half the sons of a carrier female are likely to be affected. Half the daughters will be carriers for the gene. Males are clinically affected because they carry only one X chromosome, while females are unaffected because they have two X chromosomes. Affected males can only inherit

the gene through a female line.

A well known example of an X-linked disorder is the haemophilia that affected Queen Victoria's family line. Baldness can also be X-linked.

Y-linked traits include genes for sex determination and male development. Father-to-son transmission is only possible in Y-liked traits as the Y chromosome is only inherited by sons.

This family tree shows the sex-linked recessive inheritance of haemophilia in the royal families of Europe. All affected individuals can trace their inheritance back to Queen Victoria in the 19th century, who was a carrier of the disease. The current British royal family is unaffected as they are descended from an unaffected individual (Victoria's son, King Edward VII).

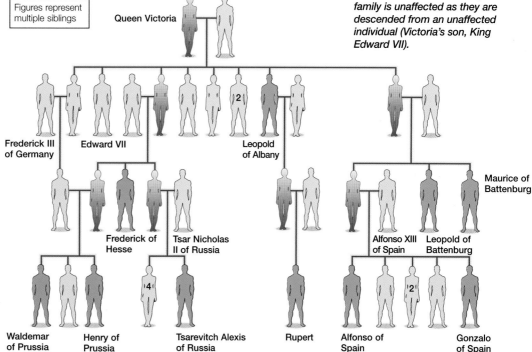

Figures represent multiple siblings

Queen Victoria

Frederick III of Germany | Edward VII | Leopold of Albany | Maurice of Battenburg

Frederick of Hesse | Tsar Nicholas II of Russia | Alfonso XIII of Spain | Leopold of Battenburg

Waldemar of Prussia | Henry of Prussia | Tsarevitch Alexis of Russia | Rupert | Alfonso of Spain | Gonzalo of Spain

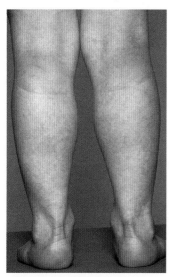

Sex-linked (Duchenne) muscular dystrophy is an example of a severe X-linked recessive condition. It produces progressive muscle weakness, beginning in early childhood.

Sporadic mutations

Mistakes can be made during the normal process of DNA duplication. These are known as new mutations. Only a few of the mutations that occur are in areas of the DNA that lead to an alteration of phenotype. Most occur in regions that are not concerned with gene function.

Achondroplasia (dwarfism) has been shown to arise through new mutations in some families. The mutation then continues to be inherited in subsequent generations in an autosomal dominant fashion.

Achondroplasia occurring as a new mutation is subsequently inherited by offspring in an autosomal dominant manner. That is, a child has a 50 per cent chance of being affected if one parent has the condition.

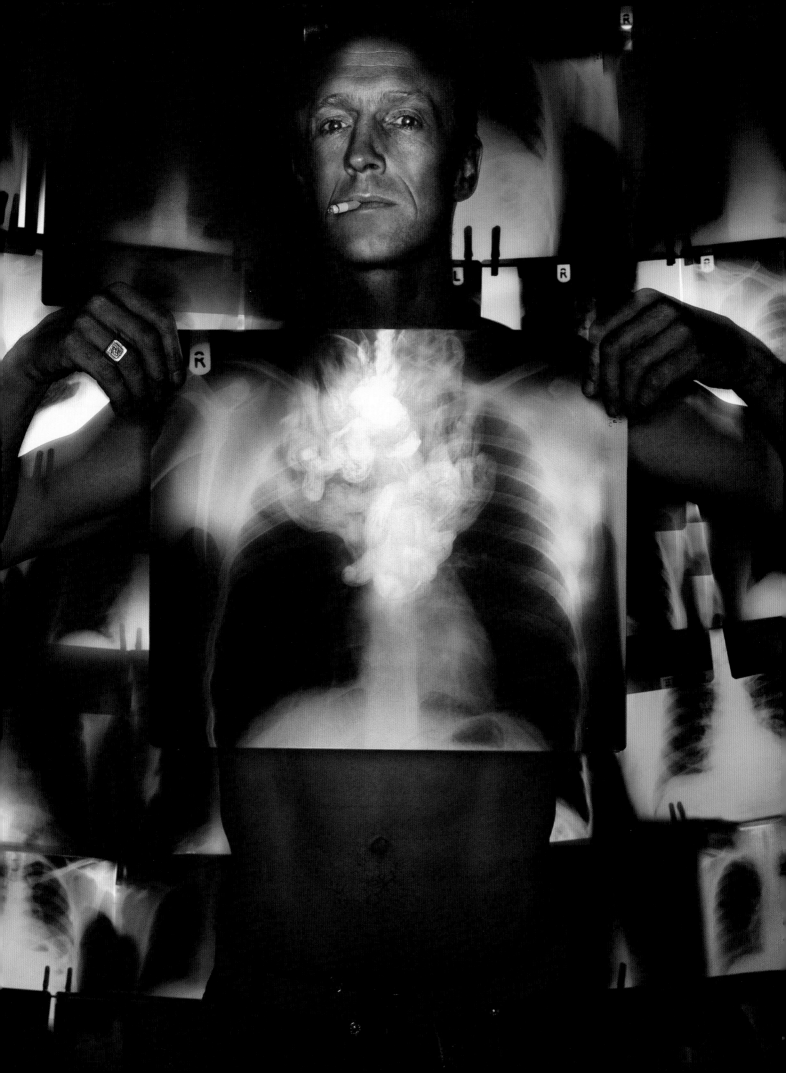

Chapter Twelve

Development & Body Cycles

Throughout life, our bodies are exposed to numerous influences that can alter our sense of equilibrium and have a detrimental effect on health, from substances that cause an allergic reaction, to alcohol, cigarettes, stress and medicine.

This chapter charts the effects of these influences on our bodies, and uncovers the built-in response mechanisms that enable us to cope with certain situations and restore balance. In addition, you can discover the fascinating role of biorhythms, or natural biological cycles, in providing the body with its own internal clock to regulate certain physiological functions.

LEFT: A doctor demonstrates the potential damage caused by cigarette smoke to the human lungs in graphic fashion.

How biorhythms occur

Many of the of the body's important physiological processes take place in cycles, known as biorhythms. These cycles occur at specific intervals, and are controlled by an internal biological clock.

Many of the physiological processes that take place within the body are timed to occur at specific intervals. The onset of puberty, for example, is a physiological event triggered by a form of timing mechanism to occur in early adolescence.

Many of the body's physiological processes are controlled by hormones that fluctuate in cycles known as biorhythms.

MONTHLY CYCLE

One example of a biorhythm is the female menstrual cycle. The uterine lining develops, degenerates and is shed in a cycle that begins approximately every 28 days.

This cycle suggests that the hormones responsible for menstruation are controlled by a form of internal clock.

Different types of biorhythms occur over varying intervals of time. They include:
■ Pulses – hormones may be secreted in spurts every few minutes, for example insulin, or over every hour
■ Circadian rhythms – regulated over a 24-hour period, for example the hormones that control the sleep-wake cycle
■ Monthly cycles – for example, the fluctuations of the hormones that control the menstrual cycle
■ Seasonal – levels of thyroid hormones decrease in winter while melatonin levels increase.

Many physiological processes occur in cycles. These biorhythms appear to be synchronized by external factors such as light and dark.

Circadian rhythms

Many of the biorhythms exhibited by humans appear to be linked to environmental rhythms.

Biorhythms that occur in 24-hour cycles (roughly corresponding with the solar or light-dark cycle) are referred to as circadian rhythms (literally meaning 'about a day').

SLEEP-WAKE CYCLE
An obvious example of a circadian biorhythm is the sleep-wake cycle. In general, adults tend to wake around 7 am and become sleepy around 10 pm.

Likewise, body temperature fluctuates over a 24-hour period; it is at its lowest in the middle of the night, and tends to reach a peak during mid-afternoon.

The levels of many hormones are seen to correspond with this circadian pattern.

HORMONES
The production of the hormone cortisol by the adrenal glands is one such hormone. If levels of cortisol are monitored over a 24-hour period, a distinct pattern can be seen. Production of cortisol rises when we wake, peaking at around 9 am. Levels of this hormone reach their lowest around midnight.

THYROID-STIMULATING HORMONE
The production of thyroid-stimulating hormone from the pituitary also follows a circadian rhythm. Thyroid hormones act directly on almost all the body cells, controlling their rate of metabolism. Levels of this hormone reach a peak around 11 pm, and are at their lowest around 11 am.

The production of other hormones, such as endorphins and sex hormones, also vary in a circadian manner.

The sleep-wake cycle is one example of a circadian rhythm. Our levels of alertness appear to be synchronized with the 24-hour light-dark cycle.

Light-dark cycle

Studies reveal that when subjects are placed in isolation chambers (with no indication of time) they continue to follow a regular sleep-wake cycle, although it is closer to 25 hours. With time, therefore, subjects lose synchronization with day and night.

Synchronization
If subjects are once again exposed to light-dark cycles the body soon reverts to the circadian cycle.

This demonstrates that the natural period of the body clock is not caused by the day-night cycle, but is simply synchronized by it.

In the absence of external cues, humans continue to follow a regular sleep-wake cycle. Our natural cycle is slightly longer than 24 hours however.

The biological clock

Biorhythms appear to be regulated by a self-sustaining timing mechanism or biological clock. Research indicates that this clock is synchronized with the light-dark cycle through interaction between the hypothalamus and pineal gland.

Light entering the eye reaches the retina (the densely innervated area at the back of the eye) stimulating the visual cortex of the brain.

Some of the retinal nerve fibres however, are connected to the suprachiasmatic nuclei, two tiny structures located within the hypothalamus of the brain.

PINEAL GLAND

Stimulation of the suprachiasmatic nuclei causes the hypothalamus to send signals to the pineal gland (a small oval gland located immediately above the brainstem). This gland is often referred to as the 'third eye' since it is triggered by levels of light and dark.

MELATONIN

The pineal gland secretes the hormone melatonin in response to the signals it receives from the retina (mediated by the hypothalamus).

In darkness, the pineal gland secretes melatonin, while the presence of light suppresses it. Research shows that melatonin influences the activity of a number of endocrine glands.

Melatonin plays an important role in the regulation of the sleep cycle, as increased levels lead to sleepiness and fatigue.

Research indicates that there is a relationship between the activity of the pineal gland and seasonal affective disorder (SAD).

The presence of melatonin also reduces the activity of the suprachiasmatic nuclei.

LOST RHYTHMS

It has long been known that damage to the suprachiasmatic nuclei, for example following surgery to remove a brain tumour, causes circadian rhythm to be lost. Likewise, diseases of the hypothalamus cause disorders of normal patterns such as sleep.

RESEARCH

Although research continues, it certainly seems that synchronization of the body with the day-night cycle involves the stimulation of the suprachiasmatic nuclei to activate the clock, and the action of the pineal gland to turn it off.

Together these specialized areas of the brain regulate the timing of events such as sleeping, waking, times of eating, and body temperature.

The pineal gland is located immediately above the brainstem. This gland secretes melatonin in response to signals from the hypothalamus.

The pineal gland

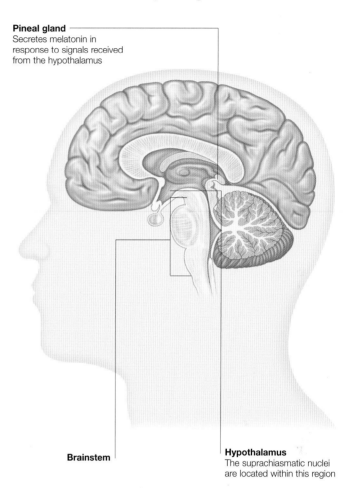

Pineal gland
Secretes melatonin in response to signals received from the hypothalamus

Brainstem

Hypothalamus
The suprachiasmatic nuclei are located within this region

The effects of jet lag

Rapid travel across time zones gives rise to a phenomenon commonly referred to as jet lag.

Disrupted biorhythms
As the body is thrown into a different time zone, its biological clock does not corresponds with that of the time zone it is actually in, with the result that circadian rhythms no longer synchronize with the light-dark cycle.

Consequently, the normal circadian rhythms of sleeping, waking, eating and drinking become disrupted. This can lead to insomnia, fatigue during the day, light-headedness, malaise and reduced mental and physical performance.

Travel across time zones, can interfere with the body's natural cycles. Circadian rhythms, such as eating patterns, can become very disrupted.

These effects do not only occur as a result of air travel, but are also experienced by people translocated to extreme environments, as in space travel, or expeditions to the arctic or antarctic, where light-dark cycles are markedly different.

Travelling east
The effects of jet lag tend to be worse when travelling east due to the hours lost. As the body's natural cycle is around 25 hours long, it is easier for the body to adjust to a lengthened day when flying west.

Interestingly, if a person were to travel the entire circumference of the globe in a day and return to their original time zone, they would not suffer from jet lag.

Research into the effects of melatonin in resynchronizing the body with the 24-hour cycle after long distance travel is underway.

How puberty occurs

During puberty, both boys and girls undergo enormous physical and emotional change. This is due to the production of sex hormones, triggering development essential to fertility.

Puberty is the period of physical change that occurs during adolescence and results in sexual maturity. In girls it tends to occur between the ages of 10 and 14, while in boys it is likely to start between 10 and 14 and continue until about 17.

SECONDARY SEXUAL CHARACTERISTICS

The physical changes that take place during puberty are manifested in the appearance of secondary sexual characteristics, such as a deepening of the voice in boys and the growth of the breasts in girls.

ACCELERATED GROWTH

During puberty a striking growth spurt occurs, at around the age of 10 in girls and 12 in boys. A rate of growth of around 8–10 cm per year is attained. Since boys do not reach full maturity until later than girls, their growth period is extended, with the result that they tend to be significantly taller.

This accelerated growth spurt affects different parts of the body at any one time, so the body may appear to be out of proportion during this period.

Growth acceleration tends to affect the feet first, followed by the legs and the torso. Finally, the face, particularly the lower jaw, undergoes development.

Body weight may almost double during this time. In girls this is largely due to increased fat deposition in response to changing hormone levels, while in boys it is due to an increase in muscle bulk.

TRENDS

Studies have shown that the menarche (the onset of menstruation) appears to be occurring at an increasingly early age in girls, at a rate of four to six months earlier every decade. This is thought to be due to improved nutrition. It is likely that boys are also maturing at an earlier age.

Girls tend to start puberty at different ages. However, most girls will have reached the same level of sexual maturity by the time they are 16.

Hormonal triggers

In addition to undergoing major physical changes, teenagers also suffer the emotional consequences of hormonal fluctuations.

Puberty is triggered by the production of a gonadotrophin-releasing hormone from a region of the brain known as the hypothalamus.

It is not clear what triggers the release of this hormone. There is speculation that it may be controlled by the interaction between the pineal gland and the hypothalamus, acting as a biological clock.

SEX GLAND STIMULATION

Gonadotrophin-releasing hormone stimulates a small gland in the brain known as the pituitary. This triggers the release of a group of hormones known as gonadotrophins (sex gland stimulators), at around the age of 10–14 years.

Gonadotrophins stimulate the ovaries to secrete oestrogens, and the testes to produce testosterone. It is these hormones that are responsible for the development of the secondary sexual characteristics during puberty.

EMOTIONAL CHANGES

The many physical changes that take place during puberty are accompanied by a number of emotional changes.

The main reasons for this are as follows:
■ The individual may have difficulties in coming to terms with the many physical changes taking place in the body. The onset of menstrual periods in girls and the deepening of the voice in boys, for example, can be extremely distressing and cause great self-consciousness
■ The fluctuating levels of hormones during puberty can seriously affect mood, with the result that pubescent individuals are prone to mood swings, aggression, tearfulness and loss of confidence.

Physical changes during puberty

Testosterone is a key hormone during the period of puberty, causing both complex and profound change in both boys and girls.

Boys start to go through puberty between the ages of 10 and 14. The physical changes that occur during this time are brought about by the male sex hormone, testosterone. This is a growth-promoting hormone that is produced by cells within the testes.

SPERM PRODUCTION

Before puberty, the testes contain numerous solid cords of cells. With the onset of puberty, the cells at the centre of the cords die, so that the cords become the hollow tubes called seminiferous tubules, in which sperm cells develop.

The production of testosterone within the testes in turn triggers:
■ The onset of sperm production. Large numbers of sperm cells are produced – around 300-600 per gram of testicle every second
■ Growth of the testes, scrotum and penis
■ Spontaneous erections; present since birth, these can now be psychologically induced
■ Maturation of the sperm-carrying ducts and enlargement of the seminal vesicles (sperm-storing sacs)
■ Enlargement of the prostate gland, which starts to secrete fluid that makes up part of the seminal fluid
■ Ejaculation – first occurs around a year after the penis undergoes accelerated growth.

OTHER CHANGES IN BOYS

Changes continue until the age of around 17. The voice box enlarges, the vocal cords lengthen, and the voice deepens and becomes more resonant. Body hair begins to grow in the pubic region, in the armpits and on the face, chest and abdomen.

Testosterone also accelerates muscular development.

The release of testosterone triggers puberty in boys. It leads to growth of sexual organs and body hair, and an increase in muscle mass.

Male physical changes

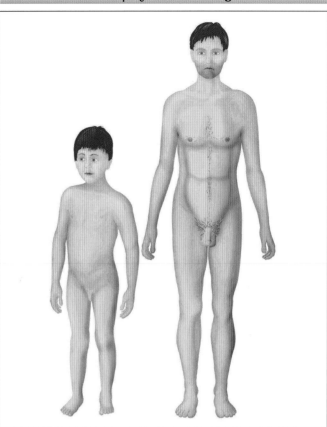

Puberty in girls

Puberty in girls tends to start between the ages of 10 and 14, but varies from person to person, and so some girls reach sexual maturity before others.

By the age of 16, however, most girls will have reached the same level of sexual maturity. This period of puberty is characterized by considerable body growth, alterations in body proportions and major changes in the the sexual and reproductive organs.

BREAST BUDDING

The first sign of puberty in girls is usually breast budding. Hormones trigger the nipples to enlarge and the breast tissue to grow, as milk glands and ducts develop. After this time, breast growth is very rapid.

ADRENAL GLANDS

During puberty the adrenal glands start producing male sex hormones, such as testosterone. These key hormones:
■ Cause a sudden surge of physical growth
■ Alter the development of hair, causing pubic and underarm hair to develop for the first time.

Menstruation usually begins around a year after these hormones are released.

HIP DEVELOPMENT

Changes take place in the bones of the pelvis, making it wider in relation to the rest of the skeleton. These changes occur in conjunction with increased deposits of fat around the breasts, hips and buttocks, and create a more curvaceous and womanly appearance.

Puberty is deemed complete when menstruation assumes a regular pattern. This means that ovulation is taking place on a monthly basis and that conception is possible.

Female physical changes

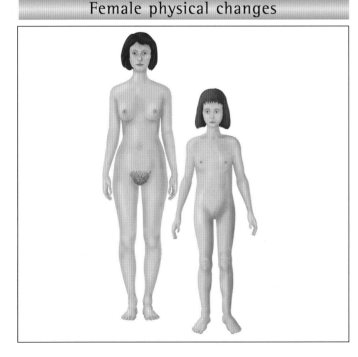

Girls undergo huge physical change in puberty. This includes the menarche, breast budding, pelvic bone widening, selective fat deposits and hair growth.

Abnormal puberty

Abnormal changes in the hypothalamus or adrenal glands, such as a tumour, can cause the process of puberty to occur at a much earlier age. This rare phenomenon is known as precocious puberty, and can result in full sexual development in young children.

Puberty in both sexes can be delayed by malnutrition or constant physical exertion. Many athletes and gymnasts do not develop sexual characteristics until they have more relaxed training regimes.

A number of genetic disorders (such as cystic fibrosis) can also affect puberty.

How the body ages

Ageing is the gradual degeneration of the body with time.
Biological processes, such as cardiovascular system functions,
become less efficient until they can no longer fulfil their role.

Ageing is the term used to describe the physiological changes that take place in the body as it slowly degenerates with time. This process occurs gradually over a number of years, beginning in the third decade of life (age 20–30).

LIFE EXPECTANCY
The longest lifespan recorded by the *Guinness Book of Records* is currently 122 years. With advances in lifestyle, medicine and sanitation, however, this record is likely to be broken. On average in the UK, life expectancy is 74 years for men and 79 years for women.

'STOPPING THE CLOCK'
Extensive research has been carried out into the biological mechanisms behind ageing, in an attempt to delay its effects, and even reverse the process.

Though much progress has been made in our understanding of the ageing process, what holds true is that ageing is an inevitable biological state and is as much a part of life as infancy, childhood and adolescence.

Ageing occurs gradually, over a period of many years. Medical advances have helped to ensure that people are living longer today than ever before.

Cell ageing

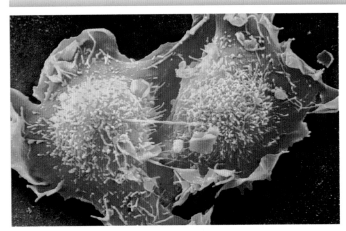

In order to understand the ageing process, it is necessary to investigate the biological mechanisms that occur at the cellular level. Cells are the individual building blocks, functioning together, that form the tissues which make up the body. These cells are replenished through the process of replication (cell division).

Cells divide a finite number of times before dying. It is likely therefore that cellular genes are programmed to stop functioning at a predetermined time.

CELL DEATH
Research has shown that cells divide a finite number of times before undergoing apoptosis (programmed cell death). In addition, the remaining cells may not function as efficiently as those in the young.

Certain cellular enzymes may be less active, so more time may be required for chemical reactions to occur that are essential to the basic functioning of the cell. As the cells fail to reproduce, the organ becomes less efficient, until it can no longer fulfil its biological role.

External changes

Ageing is most commonly characterized by the external changes that take place in the body.

HAIR CHANGES
Perhaps the most obvious change that takes place during the ageing process is the alteration in hair colour. Around the age of 30, grey or white hairs often begin to appear as the hair follicles lose their source of pigmentation. This greying becomes increasingly obvious as pigmented hairs are shed and replaced by grey hairs.

In both sexes, the hair thins considerably, and many men may experience balding.

SKIN CHANGES
The skin loses its elasticity with age, and wrinkles develop. This is due to changes in collagen (a structural protein) and elastin (the protein that gives the skin its elastic quality).

CHANGES IN STATURE
Middle age is often associated with an increase in weight as the metabolism slows, followed by a significant decrease in weight as old age progresses.

Muscle tissue may be replaced by fat, particularly around the trunk, while the arms and legs generally become thinner.

Older people tend to shrink in height, owing to compression of the spinal vertebrae.

As the body ages, the hair loses its colour and turns grey. The skin becomes less elastic and wrinkles develop due to changes in collagen and elastin.

Internal changes

Physiological studies reveal that the performance of many of the body's vital organs – such as the heart, kidneys and lungs – declines with age.

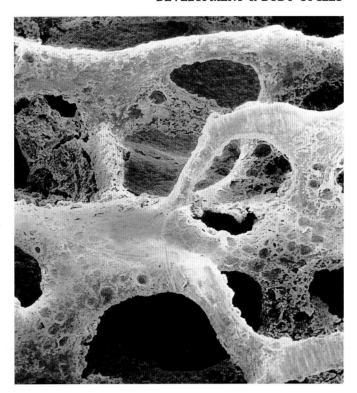

Changes associated with ageing also take place internally. Many internal organs, such as the liver, kidneys, spleen, pancreas, lungs and liver, shrink in size and function less efficiently, as the cells which comprise them gradually degenerate.

The circulation of blood by the heart is also affected by ageing. The heart's pumping action is greatly reduced and the body's response to exercise or stress by increasing the heart rate is much more extreme. The blood vessels (veins, arteries and capillaries) throughout the body lose some elasticity and tend to become convoluted.

The bones become more brittle, as the calcium content decreases, making older people more liable to fractures, even after relatively minor falls.

There is a general decline in the body's regulating mechanisms, resulting in the body being less adaptable to external changes. Older people are more sensitive to extremes of temperature, and may take longer to recover from illness.

The gradual decline of the immune system also means that older people are more vulnerable to infection and disease.

Calcium and protein are progressively lost from the bones with age. This can lead to the condition osteoporosis, in which the bones become brittle.

Changes in the nervous system

The ageing brain undergoes a gradual loss of neurones (brain cells), which are not replaced.

MENTAL ABILITY
However, although the number of brain cells decreases throughout life, this represents

only a small percentage of the total number of cells in the brain. There is no conclusive evidence that intelligence deteriorates with age, rather that it is closely associated with education and lifestyle.

AGE-RELATED DISEASE
Brain cells are extremely sensitive to oxygen deficiency. It is likely that, when deterioration of the brain does occur, it is caused not by ageing itself, but by age-related diseases, such as arteriosclerosis.

Such diseases affect the cardiovascular system and

Mental stimulation plays a part in countering the effects of ageing on the brain. Activities such as doing crosswords help to keep the brain active.

reduce oxygen supply to the brain. The efficiency of the brain is therefore reduced, and there may be a decline in intellectual performance. Logic, mental agility and the ability to grasp new ideas can all be affected.

BRAIN FUNCTION
Functions related to the brain also lose their efficiency. Reflexes and physical movements become slower and the memory may deteriorate, especially for recent events.

SENILE DEMENTIA
In severe cases, this can lead to senile dementia. This condition is characterized by a loss of memory, childlike behaviour, incoherent speech and a lack of awareness.

Senses

There is a gradual deterioration of the senses with age:
■ Vision – beyond the age of around 20, there is a decline in visual acuity, which deteriorates at an even greater rate after age 50. The size of the pupil also reduces with age, with the result that night vision is affected. The eyes are also increasingly susceptible to disease
■ Hearing – there is a gradual reduction in the ability to hear tones at higher frequencies. This can interfere with the identification of individuals by their voices, and with following group conversations
■ Taste – the number of taste buds is gradually reduced, and the sense of taste is dulled
■ Smell – this may deteriorate with age, also affecting the sense of taste.

Genetics of ageing

Medical advances have served to increase the average life expectancy, although they have not yet extended the maximum lifespan.

Laboratory research has shown that cells replicate themselves a certain number of times before dying, and that the quality of each cell gradually deteriorates. This suggests that human beings may

The rate at which a person ages is determined by both genetic and environmental factors. Taking regular exercise can help to delay the effects of time.

be programmed to age and die at some pre-determined point and that genes may carry instructions to cease functioning at a certain time.

Environmental factors
In reality, how a person ages is determined not solely by genes, but by environmental factors such as lifestyle and diet.

A person who smokes, has a generally poor diet and does not take any exercise is more likely to age more quickly, become ill and die before their genetically determined time.

How the body responds to stress

When we perceive a threat our sympathetic nervous system triggers a widespread response known as 'fight or flight'. The role of this reflex is to enable the body to react effectively to danger.

The autonomic nervous system regulates the body's basic processes (such as heart rate and breathing) in order to maintain homeostasis (normal functioning of internal bodily processes).

Humans have no voluntary control over this aspect of the nervous system, although certain events, such as emotional stress or fear, can bring about a change in the level of autonomic activity.

OPPOSING EFFECTS
The autonomic nervous system is divided into two parts: the sympathetic and parasympathetic nervous systems. Both generally serve the same organs, but cause opposite effects. In this way, the two divisions counterbalance each other's activities to keep the body's systems operating smoothly.

Under normal circumstances, the parasympathetic nervous system stimulates activities such as digestion, defecation and urination, as well as slowing heart rate and respiration.

The sympathetic nervous system, on the other hand, functions to produce localized adjustments (such as sweating) and reflex adjustments of the cardiovascular system (such as an increase in heart rate).

'FIGHT OR FLIGHT'
Under conditions of stress – such as those caused by fear or rage – however, the entire sympathetic nervous system is activated. This produces an immediate, widespread ('fight or flight') response. The overall effect is to prepare the body to react effectively to danger, whether to defend itself or to flee from a dangerous situation.

The sympathetic nervous system exerts control over a number of organs. In stressful conditions, all of these organs are stimulated simultaneously.

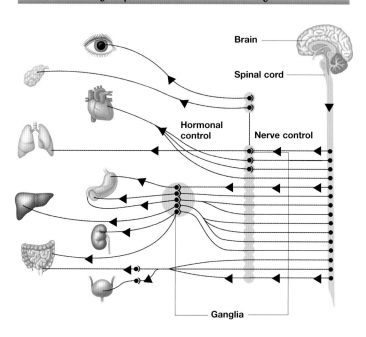

The sympathetic nervous system

Brain

Spinal cord

Hormonal control

Nerve control

Ganglia

Role of chemical messengers

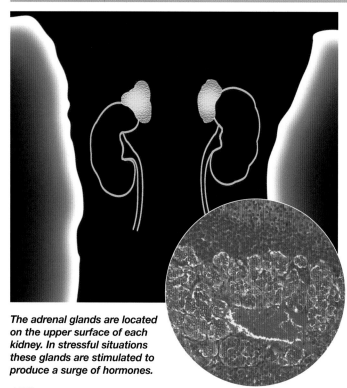

The adrenal glands are located on the upper surface of each kidney. In stressful situations these glands are stimulated to produce a surge of hormones.

The medulla of the adrenal gland, seen on this micrograph, secretes adrenaline and noradrenaline. These hormones are vital for 'fight or flight'.

The sympathetic nervous system exercises control over the organs via a series of nerves which extend to ganglia (collections of nerve cells) on either side of the spinal cord.

Nerve cells from the ganglia project to target tissues such as glands, smooth muscles or cardiac muscle.

NORMAL RESPONSE
Under normal circumstances, nerve impulses from the brain stimulate the ends of the sympathetic nerve fibres to secrete the chemical messengers adrenaline and noradrenaline.

These hormones stimulate the target organs, and in this way act as chemical mediators for conveying the nerve impulses to the target organs.

STRESSFUL STIMULI
In stressful situations, the whole of the sympathetic nervous system is activated at once. Adrenaline and noradrenaline are immediately secreted from the adrenal medulla (inner portion of the adrenal gland). These hormones are carried in the bloodstream, reinforcing the effects of the sympathetic nervous system.

Meanwhile, the hypothalamus (part of the forebrain) stimulates the pituitary gland to secrete adrenocorticotrophic hormone (ACTH). This triggers the adrenal cortex (outer portion of the adrenal gland) to release the hormone cortisol into the bloodstream.

Cortisol prepares the body for danger, by stabilizing membranes and increasing blood sugar. Stored amino acids are rapidly transported to the liver and converted into glucose, the fuel necessary for the production of energy.

Response to fear

The sympathetic nervous system triggers the characteristic symptoms of fear. These enable the body to achieve a heightened performance under stress.

The surge in levels of adrenaline and noradrenaline from both the nerve endings and the adrenal medulla causes an immediate reaction throughout the body, giving rise to a number of responses which are characteristic of fear.

The aim of these responses is to enable the body to respond effectively to danger, whether it be to run, see better, think more clearly or to stay and fight.

BODY RESPONSES

Fear responses include:
■ Rapid, deep breathing – the airways enlarge and breathing becomes more efficient to allow an increased intake of oxygen into the body

Under stressful situations, such as during an examination, there is an increase in blood flow to the brain. This enables us to think more clearly.

■ Pounding heart – the heart beats harder and faster and blood pressure rises considerably.
   Vasodilatation (increase in diameter of blood vessels) occurs within the vessels of those organs essential for emergency reaction, such as the brain, heart and limbs. This allows more blood to reach the organs, providing more oxygen and essential nutrients necessary for heightened performance
■ Pale skin – the effects of the sympathetic nervous system cause vasoconstriction (contraction in the walls of the blood vessels supplying the skin). As a result, blood flow is greatly reduced. This means that blood loss from superficial wounds is decreased should the body be required to fight. It also explains why people can literally go white with fear
■ A surge in energy – the body's metabolism is increased by as much as 100 per cent in order to maintain heightened responses. To compensate, the liver produces more glucose, which is rapidly respired to produce extra energy. This explains why a cup of sweet tea

Stressful stimuli, such as a confrontation with danger, activates the entire sympathetic nervous system. This triggers a number of fear responses.

may be helpful after a stressful event
■ Increased physical strength – as a result of increased blood flow and energy levels, the strength of muscular contraction increases. This is the reason why people can perform great feats of strength when they are in danger, for example lifting a very heavy weight, such as a human body
■ Resistance to pain – the secretion of endorphins (natural painkillers) from the brain increases the body's resistance to pain, enabling an individual to remain active despite injury
■ Hair shafts – hairs stand on end as part of a primitive reflex, similar to hair ruffling in cats and dogs
■ Pupils dilate – this sharpens the vision
■ Sweating of the skin – perspiration increases to keep the body cool
■ 'Butterflies' in the stomach – this is caused by decreased blood flow to the stomach (in favour of the vital organs). Urinary tract mobility is also suspended as blood is diverted away from the kidneys.

Effects of long-term stress

The fear responses are designed to help the body in threatening situations, such as those of immediate and physical danger.

Relaxation
As soon as a threat has passed, the body gradually reverts back to normal as the parasympathetic nervous system is activated.
   The muscles begin to relax, heart rate and blood pressure decrease, breathing becomes more regular and deeper, and the stomach relaxes as blood flow returns. The emotional state changes from one of anger and

Prolonged stress can be damaging to health. The effects can render individuals more susceptible to infection and stress-related illness.

fear to a more calm and peaceful condition.

Prolonged stress
However, in stressful conditions that are socially generated, such as those caused by a heavy workload or financial worries, the fear response can exist on a long-term basis – in other words, there is no relaxation in the body's response to stress.
   If there is no outlet for this tension, the effects of stress can have a detrimental effect on the body. An individual may suffer symptoms such as headaches, abdominal pain, tissue wasting (due to the constantly raised metabolic rate), fatigue and high blood pressure, which may lead to damage to the heart, blood vessels and kidneys).

How the body responds to exercise

During exercise, the body's physiological needs change in certain characteristic ways. Exercising muscle requires an increase in the supply of oxygen and energy, which must be met by the body.

The body needs energy for everyday activities. This energy is produced as the body burns food. However, when exercising, the muscles of the body require more energy than at rest.

To exercise for a brief period, a sprint to the bus stop for example, the body is capable of increasing the supply of energy to the muscles quickly. The body can do this because it has a small amount of stored oxygen and is able to respire anaerobically (producing energy without using oxygen).

The need for energy increases when exercising for a longer period. Muscles must be supplied with more oxygen to allow aerobic respiration (producing energy by using oxygen).

CARDIAC ACTIVITY

Our heart beats about 70 to 80 times a minute at rest; this can rise to 160 beats per minute after exercise, the heart also beating with a greater force. Thus, a normal person can increase their cardiac output a little over fourfold, while a trained athlete can increase output about sixfold.

VASCULAR ACTIVITY

At rest, blood flows through the heart at a rate of about five litres a minute; during exercise, blood circulates at a rate of 25 or even 30 litres a minute.

This bloodflow is directed towards active muscle, which needs it most. This is achieved by reducing the supply of blood to areas of the body with less need, and by widening blood vessels to allow greater bloodflow to active muscles.

RESPIRATORY ACTIVITY

Circulating blood must be fully oxygenated (saturated with oxygen), so the rate of breathing must also increase. The lungs fill with more oxygen, which can be passed to the blood.

During exercise, the rate at which air enters the lungs increases to as high as 100 litres a minute. This is vastly greater than the six litres per minute which we breathe at rest.

A marathon runner achieves 40 per cent greater cardiac output than an untrained person. Through training, heart mass and chamber size increase.

Changes in the heart's activity

Effects of exercise on the heart

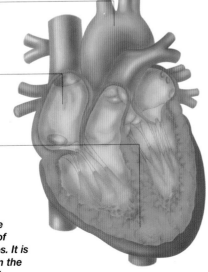

Aorta
Supplies blood to all muscles; supply to heart muscle must also increase

Right atrium
Volume and pressure of blood returning to the heart from the veins increases

Ventricular muscle
Stimulated via nerves to the heart's pacemaker to pump more rapidly

Strenuous exercise causes a number of circulatory changes. It is very demanding on the heart muscle itself.

During exercise, the heart rate (beats per minute) and cardiac output (volume pumped per minute) increase. This is due to increased activity of the nerves that supply the heart causing the heart to beat more quickly.

INCREASED VENOUS RETURN

The amount of blood returning to the heart increases due to:
■ Reduced resistance of vessels in the muscle bed due to dilation
■ Muscle action (contraction and relaxation), which pumps more blood back to the heart
■ Chest movements of rapid

Much research has been done into circulatory changes during exercise. This research shows that the more we exercise, the greater these changes will be.

breathing, which also have a pumping effect
■ Constriction of the veins, forcing blood back to the heart.

As the ventricles in the heart become increasingly full, the muscular walls of the heart are stretched and work with greater power. Thus, a higher volume of blood is expelled from the heart.

Circulatory changes

When we exercise, our body experiences an increase in the bloodflow to the muscles. This ensures a ready supply of oxygen and other essential nutrients.

Even before muscles begin to contract in exercise, the bloodflow to them can be increased through brain signals.

DILATION OF VESSELS

Nerve signals, carried by the sympathetic nervous system, cause blood vessels in the muscle bed to dilate (widen), allowing more blood to flow to muscle cells. However, to keep the vessels dilated, after this initial change, local changes occur. These include decreased levels of oxygen and a rise in carbon dioxide and other waste products of respiration in muscle tissue.

Increased temperature, caused by excessive heat produced by active muscles, also leads to vessel dilation.

CONSTRICTION OF VESSELS

In addition to these changes in the muscle bed, blood is diverted from other tissues and organs of the body which have less need for blood during exercise activity.

Nervous impulses cause vasoconstriction (narrowing of the blood vessels) in these areas, particularly the gut. This causes blood to be redirected to those areas where it is most needed, making it available to be fed into muscles in the next cycle of circulation.

The increased bloodflow to muscles during exercise is particularly dramatic in fit young adults. The increase can be more than 20-fold.

Respiratory changes

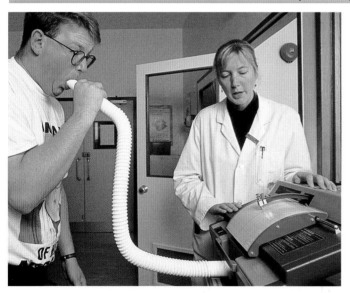

During exercise the body uses far more oxygen than it normally would, and the respiratory system must respond to the need by increasing the rate of ventilation. Although our breathing rate increases rapidly with the onset of exercise, the precise mechanism is uncertain.

As the body uses up more oxygen and produces more carbon dioxide, receptors in our bodies, which can detect changes in blood gas levels, may stimulate breathing. However, our reaction occurs

In order to meet the demands of increased muscular activity, the body will require more oxygen. For this reason, exercise leads to an increased breathing rate.

far earlier than any chemical change that can be detected. This indicates that it is a learned response which causes us to send a signal to our lungs to increase the rate of breathing, whenever we begin exercise.

RECEPTORS

Some experts suggest that the small increase in temperature which occurs almost as soon as our muscles start working, is responsible for triggering more rapid and deep breathing. However, fine control of breathing, which allows us to match our breathing with the amount of oxygen needed by our muscles, is controlled by chemical receptors in the brain and main arteries.

Body heat during exercise

In order to dissipate the heat produced during exercise, the body uses mechanisms similar to those it would employ on a hot day in order to cool down.

These mechanisms include:
■ Vasodilation at the skin – to allow loss of heat from the blood to the environment
■ Increased sweating – sweat evaporates on the skin using heat energy to do so
■ Increase in ventilation – this acts to dissipate heat via exhaled warm air from the lungs.

Oxygen consumption by the body can increase as much as 20-fold in a well-trained athlete, and the amount of heat liberated

in the body is almost exactly proportional to the oxygen consumption.

If the sweating mechanism cannot eliminate heat on a hot and humid day, a dangerous and sometimes even lethal condition called heatstroke can develop easily in an athlete. Under these conditions, the main aim should be to reduce the body temperature as rapidly as possible by artificial means.

The body employs several mechanisms to cool itself during exercise. An increase in sweating and ventilation helps to rid the body of excess heat.

How alcohol affects the body

Alcohol is appreciated in modern society for the pleasurable effects it can have on the body. In excess however, alcohol is an intoxicating substance and may be detrimental to health.

Alcohol (otherwise known as ethyl alcohol, or ethanol) has long been exploited for its pleasurable effects on the body. Historical archives record the use of alcohol in ancient civilizations, in a number of religious and social rituals.

FERMENTATION
Alcohol is an organic substance, produced by a natural process known as fermentation.

Sugars present in fruits or grains undergo a reaction with enzymes to form alcohol – a process exploited artificially by breweries and distilleries worldwide.

ALCOHOL CONCENTRATION
The concentration of alcohol varies with different drinks, from around four per cent in most beers and 12 per cent in wine, to 40 per cent in spirits such as vodka or whisky.

Today the consumption of alcohol plays a major role in society, and still features in many religious practices.

DETRIMENTAL EFFECTS
Ever since ancient times however, the perils of this intoxicating substance have been preached about, and strict laws exist in order to regulate alcohol consumption.

Although in moderation the effects of alcohol on the body are negligible, it is an addictive substance and excessive intake, particularly for prolonged periods, can have a serious impact upon health.

Alcohol plays an important role in social gatherings. People drink together in pubs and bars, enjoying the relaxing effects that alcohol can have.

Path of alcohol through the body

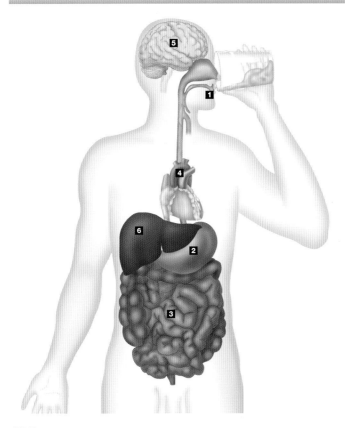

The route taken by alcohol during its passage through the body includes the alimentary canal and several organs, in the following order:

1 Mouth – alcohol may be diluted by saliva before it is swallowed

2 Stomach – alcohol passes via the oesophagus into the stomach where it is further diluted by gastric juices. Some alcohol is absorbed into the bloodstream here, but most passes into the small intestine. The rate of absorption will depend upon the strength of the alcohol and the presence of food in the stomach

3 Small intestine – this is supplied by a dense network of small blood vessels, and is the site of most of the absorption of alcohol into the bloodstream

4 Bloodstream – once in the bloodstream, alcohol is circulated around the body and

Alcohol is absorbed into the bloodstream as it passes down the alimentary canal. Once it reaches the liver, alcohol is metabolized to release energy.

taken up by the cells of various tissues

5 Brain – once alcohol reaches the brain it has an immediate intoxicating effect. Alcohol acts on many sites of the central nervous system including the reticular formation (responsible for consciousness), the spinal cord, cerebellum and cerebral cortex

6 Liver – absorbed alcohol quickly passes to the liver, where it is metabolized to water, carbon dioxide and energy at a rate of around 16 grams of alcohol (two units; for example two small glasses of wine) per hour. This rate varies however, depending upon the build of the individual.

OTHER SITES OF EXCRETION
A small proportion of alcohol goes to the lungs, and is excreted in exhaled air (allowing levels of intoxication to be calculated by the use of a breathalyzer). Some alcohol is disposed of in the urine, and a tinier amount still is excreted in the sweat.

Effects of alcohol

Once absorbed into the bloodstream, alcohol has an immediate effect on the central nervous system. This results in symptoms characteristic of drunkenness.

In general, alcohol reaches the bloodstream within five minutes of ingestion.

IMPAIRED JUDGEMENT
The most immediate effect of alcohol is that drinkers become relaxed and more sociable.

After a single unit of alcohol, the activity of the brain is slowed down, with the result that judgement may be impaired and reaction times slower.

LOSS OF CO-ORDINATION
Muscle co-ordination is increasingly reduced as the relevant control centres of the brain become intoxicated. This can result in clumsiness, staggering and slurred speech.

As the levels of alcohol in the blood rise, the pain centre of the brain is numbed, and the body becomes desensitized.

If the individual continues to drink, their vision may become blurred as the visual cortex is affected.

DRUNKEN BEHAVIOUR
A person is said to be 'drunk' when they no longer have control over their actions.

If sufficient alcohol is consumed the individual may fall into a deep sleep, or even lose consciousness. Extreme quantities of alcohol effectively anaesthetize certain centres of the brain, causing breathing or heart beat to cease, resulting in death.

MEMORY LOSS
Excessive measures of alcohol can affect the short-term memory, and thus actions carried out when drunk may not be recalled the following day.

As blood alcohol levels rise, the brain becomes increasingly intoxicated. The drinker may lose consciousness as certain brain centres are affected.

Long-term effects

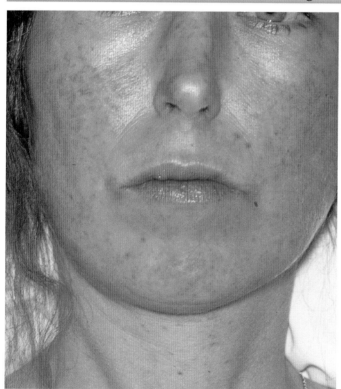

An excessive intake of alcohol can lead to skin changes. This woman is jaundiced (yellowed skin) and has tiny broken blood vessels on her face and neck.

If the body is subjected to excessive alcohol intake for prolonged periods, the effects can be extremely serious. These include:
■ Tissue damage – as an irritant, alcohol, especially in purer forms, damages the tissues of the mouth, throat, gullet and stomach, causing increased susceptibility to cancer
■ Loss of appetite – large quantities of alcohol affect the stomach and appetite; thus heavy drinkers tend to neglect their diet. Alcohol is calorific but it does not contain any useful nutrients or vitamins
■ Liver damage – excessive quantities of alcohol damage the liver, causing it to shrink and become defective (cirrhotic). Eventually the organ will be unable to carry out its detoxifying role
■ Brain damage – as alcohol destroys brain cells, prolonged use permanently reduces mental ability, leading to dementia. At low concentrations alcohol has a stimulatory effect on the brain, but as concentrations increase it has a more depressant effect
■ Weight gain – alcohol is rich in calories, causing heavy drinkers to become bloated and overweight, thus putting a strain on the heart
■ Skin damage – alcohol causes the small blood vessels in the skin to dilate, resulting in increased blood flow to the skin surface. This will give the individual a flushed appearance, and a false feeling of being hot. The capillaries in the skin eventually rupture, giving the skin a permanently ruddy and unsightly appearance
■ Accidental injury – fatal injury is more likely in heavy drinkers. Alcoholics are seven times as liable to be victims of serious accidents as non-alcoholics.

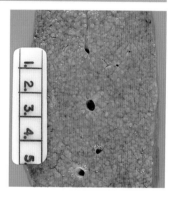

Alcohol is an addictive substance. Long-term abuse can lead to serious health problems such as cirrhosis of the liver (shown here).

Withdrawal
Heavy drinking may be followed by a headache, nausea and fatigue, otherwise known as a hangover. This is due to the dehydrating effect of alcohol, which effectively starves the body's cells of water.

Prolonged alcohol abuse can lead to dependence, and so withdrawal may result in DTs (or delirium tremens), causing shaking, loss of appetite, inability to digest food, sweating, insomnia and seizures. In severe cases, people may hallucinate.

How smoking affects the body

Tobacco contains a number of harmful compounds which are drawn in to the lungs during smoking. Tens of thousands of people die every year in the UK from this addictive and deadly habit.

The practice of inhaling smoke produced by burning tobacco leaves was introduced to the Western world by European explorers in the early 17th century. They had observed this custom in native Indians, who used tobacco in a number of rituals and believed it possessed medicinal properties.

ADVERSE EFFECTS

Before long, smoking became a fashionable pastime. Lung cancer, once comparatively rare, began to increase dramatically in the 20th century and research into the effects of smoking on the body began.

Today, despite the fact that a clear correlation has been made between smoking and a range of diseases, the number of people who smoke continues to increase. In developed countries, smoking leads to around three million deaths a year, and is the main cause of death in people under 65.

Smoking is favoured by many for its stress-relieving properties. In reality nicotine is a stimulant and has harmful effects.

GAS COMPOSITION

When a cigarette is lit, the burning tobacco gives off a pungent smoke, which is drawn into the lungs through inhalation. Cigarette smoke consists of both a gas and a particle phase. The particle phase (the smoke we see) consists of around 4000–5000 different particles of unburnt tobacco. Among these are chemicals that can cause cancer, poison cells, alter cell structure, suppress the immune system and alter neural activity in the brain.

The gas phase consists mainly of carbon dioxide, carbon monoxide and nicotine.

Carbon monoxide (the noxious gas in car exhaust fumes) combines with the blood pigment haemoglobin that is responsible for the transport of

oxygen to vital organs and tissues. This means that less oxygen can be carried in the blood, reducing the availability of oxygen to the tissues.

Nicotine affects the central

nervous system, constricting blood vessels and increasing heart rate and blood pressure. Many smokers are addicted to nicotine and have withdrawal symptoms when they stop.

Effect on the cardiovascular system

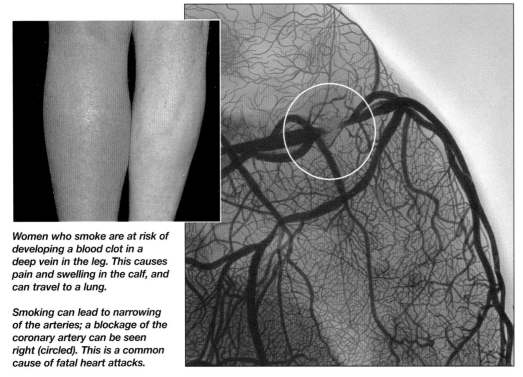

Women who smoke are at risk of developing a blood clot in a deep vein in the leg. This causes pain and swelling in the calf, and can travel to a lung.

Smoking can lead to narrowing of the arteries; a blockage of the coronary artery can be seen right (circled). This is a common cause of fatal heart attacks.

Smoking is a greater cause of death and disability than any single disease.

In particular, smoking has a grave impact upon the cardiovascular system and has been linked to around one in four deaths from cardiovascular disease.

NARROWING OF ARTERIES

Nicotine and carbon monoxide present in cigarette smoke encourage the narrowing of arteries, a disorder known as atherosclerosis. This increases the risk of stroke and other cardiovascular disorders.

Coronary artery disease is one example of a cardiovascular disorder, whereby the blood supply to the heart is restricted, increasing the risk of a fatal heart attack.

Women who smoke are also at a far greater risk of developing deep vein thrombosis and stroke, particularly if they take the oral contraceptive pill.

The effect of smoking on the lungs

With time, the effects of smoking reduce the capacity of the lungs and impair their defence mechanisms, exposing the body to attack by diseases.

As well as having a serious effect on the cardiovascular system, smoking is detrimental to the lungs.

THE LUNGS

The two lungs lie beneath the rib cage, and surround the heart. They function like bellows, drawing air into the airways, so that oxygen can pass from the lungs into the blood. Oxygen is then delivered throughout the body, and waste products such as carbon dioxide are returned to the lungs and exhaled.

In order to prevent foreign bodies, such as dust or pollen, from entering the lungs, the airways are are lined with specialized cells covered in cilia (hair-like projections). These cells maintain a constant wave-like motion, so that any potentially harmful particles are wafted up the airways and out of the lungs into the throat. The mechanism of coughing also serves to remove any foreign particles from the lungs.

IMPAIRED FUNCTION

Smoking inhibits the lungs' protective mechanisms. First, it reduces the body's response to smoke, so that people do not cough when smoking a cigarette as they normally would when inhaling pungent smoke.

Secondly, the ciliated cells beat much more slowly as they become paralysed by toxins in the tobacco. For this reason harmful substances contained in cigarettes are able to settle in the lungs, reducing the overall capacity of these vital organs, and compromising the entire body.

As harmful substances settle, the mucous membrane of the lungs produces more and more mucus (otherwise known as phlegm). Tar, ash and phlegm accumulate in the tiny air sacs of the lungs, reducing their capacity, and causing severe shortness of breath.

IMMUNE RESPONSE REDUCED

Smoking also damages the white blood cells that would normally scavenge and remove dirt and bacteria from the lungs. This means that the lungs are more prone to infection.

In this way, smoking causes the body to be exposed to a greater number of harmful foreign bodies, while the normal defence mechanisms for combating disease are severely impaired.

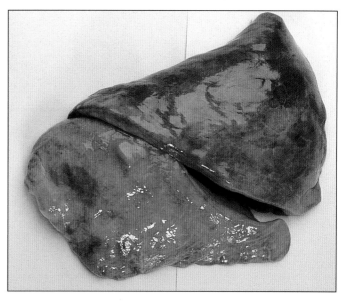
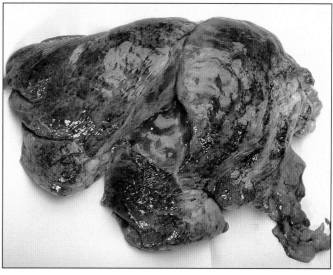

Shown here is a healthy lung (top) and the diseased lung of a smoker (below). Tar contained in tobacco smoke has severely discoloured the smoker's lung.

Nicotine addiction

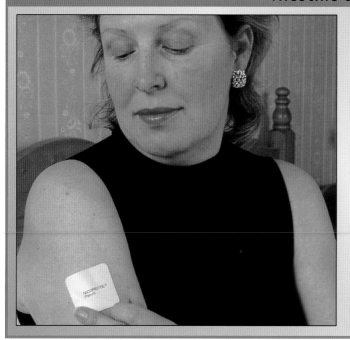

Everybody is aware of the hazards of smoking, but most smokers fail to give up. This is partly due to habituation to the stimulatory effects of nicotine, but also due to routine and social convention.

Stimulatory effect

Nicotine stimulates neurons in the brain, increasing attentiveness, decreasing appetite and irritability, and relaxing muscles. Indeed, many smokers find that smoking regulates their mood and they associate cigarettes with a pleasant sensation. In reality the body has no physiological need for nicotine.

Nicotine patches allow nicotine to be absorbed into the blood. This reduces the craving for cigarettes, avoiding the harmful effects of smoking.

Tobacco substitutes

There are a number of products available that act as a substitute for cigarettes, supplying nicotine without the effects of smoking.

These include nicotine patches, inhalers and gum. The idea is that if the body is receiving nicotine the smoker will no longer crave cigarettes. Eventually the dose of nicotine is reduced until it is no longer required by the body.

A drug known as Zyban has also been developed in recent years, which removes cravings for nicotine by interfering with the same chemical messengers in the brain that nicotine disrupts.

Acupuncture and hypnotherapy are also reputed to be helpful in smoking cessation. It is worth bearing in mind that the risks of developing smoking-related diseases decreases dramatically with time after giving up smoking.

How caffeine affects the body

Caffeine, found in a range of products, has a powerful stimulant action on the body. As a result, it can adversely affect sleep patterns and may become addictive in the long term.

Caffeine is the most widely used drug in the world – most people consume it in some form every day. Although caffeine is usually associated with coffee, it actually occurs naturally in many plants, including tea leaves and cocoa nuts, and is contained in many drinks. In fact, many people consume up to a gram of caffeine a day without even knowing it.

SOURCES

The most common sources of caffeine include:
■ Fresh coffee – a single cup can contain up to 200 mg
■ Tea – a cup of tea can contain as much as 70 mg of caffeine
■ Cola – a can may contain around 50 mg
■ Chocolate – milk chocolate can contain up to 6 mg per 28 g. Dark chocolate contains more cocoa and therefore more caffeine
■ Painkillers – certain headache tablets can contain as much as 200 mg per tablet.

STIMULANT

Recreationally, caffeine is enjoyed by many people who find that it gives them an energy boost and a feeling of heightened alertness. Many people drink coffee to help them wake up in the morning or to remain alert during the day.

Medically, caffeine (or trimethylxanthine) is used as a heart stimulant and a diuretic (it increases urine production).

Coffee beans are the seeds of coffee fruits. Coffee contains caffeine, which is a stimulant, boosting energy and increasing mental activity.

Short-term effects

Adenosine is a chemical secreted by the brain. Levels of this chemical build up throughout the day and bind to specialized adenosine receptors in the brain, causing nervous activity to slow down, blood vessels to dilate and drowsiness to occur.

In terms of its chemical make-up, caffeine looks very similar to adenosine. Caffeine is thus able to bind to the same receptors in place of adenosine. However, caffeine does not slow down the activity of a nerve cell as adenosine would, but has the opposite effect, causing nerve cell activity to speed up.

Moreover, because caffeine blocks the ability of adenosine to dilate blood vessels, the blood vessels of the brain constrict. This is the reason why some headache tablets contain caffeine (constriction of the brain's blood vessels can help to relieve certain types of headache).

ADRENALINE

When a person takes caffeine, the pituitary gland responds to the increased brain cell activity. It does this by acting as though the body were facing an emergency situation, and releases hormones that stimulate the adrenal glands to produce adrenaline.

The release of adrenaline (the 'fight or flight' hormone) has the following effects, explaining why, after a cup of coffee, a person might experience tense muscles, cold, clammy hands

A dose of caffeine has an instant effect on the nervous system. Brain cell activity is increased, resulting in release of the hormone adrenaline.

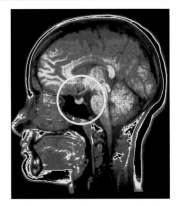

Caffeine stimulates the pituitary gland in the brain, shown on this scan circled. The gland triggers the releases of 'fight and flight' hormones from the adrenals.

and a feeling of excitement:
■ Pupils and airways dilate
■ Heart rate increases
■ Blood pressure rises, as blood vessels close to the surface of the skin constrict
■ Blood flow to the stomach is reduced
■ The liver releases sugar into the bloodstream to provide extra energy.

Addictive properties

Many people become addicted to caffeine. Not only is it a stimulant, but it raises the levels of dopamine in the brain, which increases feelings of pleasure.

Caffeine is an addictive drug. It belongs to a drug group known as stimulants, so called because of their excitatory effect on the brain. Other stimulants include amphetamines and cocaine.

BRAIN CHANNELS
Although the effects of caffeine are less powerful than those of other stimulants, the drug operates in a similar way and, because it manipulates the same channels in the brain, it is just as addictive.

In the short term, caffeine is a harmless substance, but the long-term consumption of caffeine can be a problem. Once the adrenaline released by ingesting caffeine wears off, a person may feel tired and mildly depressed and so may reach for

another cup of coffee.

It is in this way that many people become addicted to caffeine without even realizing it. It is not healthy for the body to be in a constant state of emergency and many people become jumpy and irritable.

PLEASURE
Like other stimulants, caffeine raises levels of dopamine, a neurotransmitter that activates the pleasure centre of the brain. It is suspected that this action is a contributory factor to the addictive nature of caffeine.

For years, scientists have studied the behaviour of neuro-transmitters – chemicals in the brain. Caffeine raises the level of the neurotransmitter dopamine.

Effect on sleep

Caffeine has a significant effect on sleep. It takes around 12 hours for caffeine to leave the body's system. This means that if a person has a cup of coffee containing 200 mg of caffeine at around 4 pm, then by 10 pm there will still be 100 mg of caffeine in their bloodstream.

LACK OF DEEP SLEEP
Although the person may be able to fall asleep, they will not be able to attain the deep sleep that the body requires. As a result, they will wake feeling tired, and may instinctively pour themselves a cup of coffee to help wake them up. And so the

cycle continues.

If a person tries to break this cycle, they may find that they feel very tired and mildly depressed. They may also experience headaches due to dilation of the blood vessels in the brain.

▶ *Caffeine may hinder deep sleep. A person is therefore likely to wake feeling tired and repeat the cycle by drinking coffee to help them wake up.*

◀ *It takes around 12 hours for caffeine to leave the body after consumption. If caffeine remains in the bloodstream, it can adversely affect sleep patterns.*

Decaffeinated drinks

With increasing awareness of the harmful effects that caffeine can have on the body, decaffeinated drinks are growing in popularity. These provide the taste of coffee, tea and cola, but without the detrimental effects.

Filtration
Decaffeination of coffee involves treating the coffee beans with a solvent that absorbs the caffeine. This is then filtered from the solution, leaving only the coffee oils (vital to flavour). This solution is then added back to the coffee beans, which are roasted and processed as normal.

Research has shown that

Decaffeinated coffee, which has no harmful effects, is very popular. Removing caffeine from coffee beans, however, is a complex process.

people suffering from hypertension benefit from cutting caffeine out of their diets.

How drugs work on the body

Drugs used to prevent or treat diseases work by causing biochemical or physiological changes in the body, or by alleviating symptoms. Some drugs affect specific cells while others act on the whole body.

TYPES OF DRUGS

Drugs exert their action in a variety of ways. A drug's effects may be described according to the changes that it causes, or with reference to the clinical symptoms it is intended to relieve or prevent. In general, drugs may be classified into those that:

- Artificially moderate or regulate the activity of specific body cells, tissues or organs
- Combat virulent organisms invading the body (bacteria that cause infections, for example)
- Act in place of substances occurring naturally within the body
- Have an effect on abnormal or malignant cells or tissues.

CELL-REGULATING DRUGS

Certain drugs affect the activity of the body's cells by influencing the maintenance of normal cell function. Cell-regulating drugs (artificial moderators) can act either on cells throughout the entire body (systemically) or only upon those located within certain tissues or organs.

Some of these artificial moderators enhance or inhibit substances in the cell necessary for energy production, synthetic reactions or other normal cell functions.

Such drugs often work on enzyme (biological catalyst) activity, either by inhibiting or enhancing it. One specific example is allopurinol, which is used to treat gout because it prevents the formation of uric acid. Gout occurs when joints become painfully swollen because uric acid salts collect around them.

Drugs acting at the cellular level may be specific to certain organs or they may have wide-ranging (systemic) effects. For example, the drug hydralazine is an antihypertensive, used to lower blood pressure. It causes the small arteries around the body to widen, increases the heart-rate and elevates cardiac output.

Therefore, hydralazine works at two levels: the effects on blood circulation are exerted at the cellular level, while those on the heart are exerted at the organ function level.

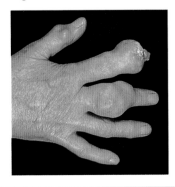

Gout is an inflammation of the joints caused by a failure in the meabolism of uric acid. Drugs prevent uric acid from forming and then collecting in joints.

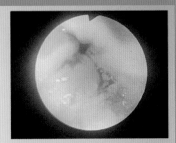

A Novopen is used to administer a metered doses of insulin to a patient with diabetes. The insulin is needed to regulate the patient's glucose metabolism.

Multi-drug management

The drug treatment for any particular disease is not necessarily restricted to one approach, but may utilize several drugs acting in different ways.

A good example is the management of peptic ulcers. The symptoms of this condition are exacerbated by the secretion of gastric acid. These can be alleviated, and ulcer healing promoted, by a range of drugs acting locally in the stomach (using antacids), or by reducing gastric acid secretion (perhaps with an H_2-receptor antagonist such as ranitidine), or by enhancing mucosal protection (for example, with carbenoxolone). In addition, since the bacterium *Helicobacter pylori* has been implicated in gastric ulceration, antibiotic treatment may also be initiated. Any combination of these approaches may be adopted.

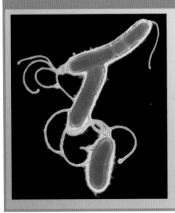

Helicobacter pylori bacteria are found in the stomach lining of patients with peptic ulcers. They respond well to antibiotics.

This endoscope image shows a peptic ulcer. Multi-drug therapy can reduce gastric acid and act against the bacteria involved.

Anti-infective drugs

Some drugs exert their effects on invading organisms (infections) or abnormal body cells (such as cancers). The specific activity of anti-infectives, including antibiotics, antifungals, and antimalarials is generally due to the intrinsic differences between the cells of the infecting agent (for example, the bacterium) and those of the host. An effective and safe antimicrobial preparation is one that is toxic to the infecting organism, but not to the host.

Some of these drugs merely arrest the growth of the susceptible organism, whereas others kill the organism, but these effects may depend on the dose taken.

The means by which anti-infective drugs exert their toxicity are diverse, and are frequently related to the drug's chemical structure. Some anti-infectives (such as gentamicin and erythromycin) interfere with the synthesis of bacterial protein, while others (penicillins, for example) interfere with bacterial cell wall synthesis, or cell function.

The fungus Candida albicans can cause oral candidiasis on the tongue. Antifungal preparations may treat the condition by killing the fungus or slowing its growth.

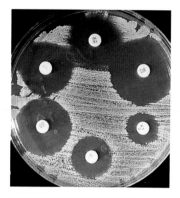

Antibiotic drugs are tested on a culture of a strain of bacteria to see which is the most effective. The drug must do as little harm as possible to human cells.

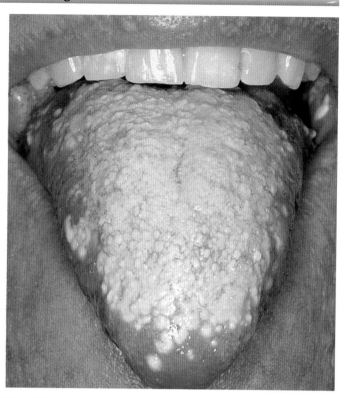

Drugs that replace natural substances

The treatment or prevention of some diseases involves the administration of a substance that, under normal circumstances, occurs naturally within the body.

An example is the use of insulin in the treatment of Type 1 (insulin-dependent) diabetes mellitus. In patients with Type 1 diabetes, the administered insulin compensates for the lack of insulin secreted by the pancreas cells and acts by enhancing the transport of glucose into cells, thus restoring normal function.

Similarly, oestrogenic and progestogenic hormones are used as hormone replacement therapy in post-menopausal women, to alleviate post-menopausal symptoms and to prevent osteoporosis. These hormones exert actions comparable to the actions of naturally occurring hormones prior to the menopause.

Oral contraceptive pills contain oestrogen and progestogen preparations. These suppress ovulation by mimicking the effects of the woman's own hormones on the pituitary and hypothalamus.

Other substances that act in place of naturally occurring substances are vitamin and mineral preparations used to treat or prevent deficiency states.

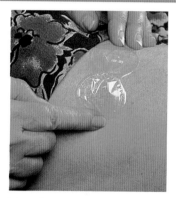

Transdermal patches, as used in hormone replacement therapy, deliver hormones through the skin. They release a set amount of the drug every hour.

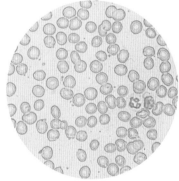

Anaemia (iron deficiency) is evident in this blood smear. Two white blood cells (purple) serve to fight infection. Anaemia may be treated by iron supplements.

Antimalignancy drugs

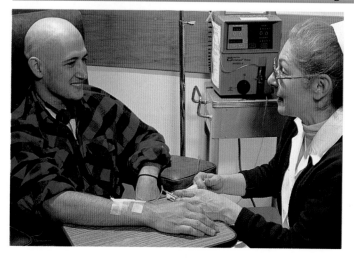

Cytotoxic (or antineoplastic) drugs are used in the treatment of malignant disease, either in instead of or as well as surgery or radiotherapy.

The actions of such drugs are often not specific to the cancerous cells, and they therefore affect healthy cells in the body. However, the ability of such drugs in targeting malignant cells is generally attributable to the different

Cytotoxic drugs used in cancer chemotherapy may be given intravenously. They often act to prevent tumour cells dividing, but may also affect normal cells.

characteristics shown by malignant and normal cells. For example, malignant cells undergo cell division at a more rapid rate than is usually seen in normal cells. Alkylating agents (such as cisplatin) exploit this feature, and arrest cell division in rapidly dividing cells.

Some normal body cells (including bone marrow cells) divide rapidly and are therefore prone to toxicity from these drugs. Some newer anticancer treatments aim for greater specificity for malignant cells by using antibodies that bind selectively to the malignant cells and not normal cells.

How anaesthetics work

Anaesthetic drugs work by stopping the conduction of pain sensations through the body's nervous system. There are different types of drug that can be given by different routes, but they all act to affect nerve conduction.

THE NERVE NETWORK

Nerve cells (or neurones) form a comprehensive network throughout the body relaying information from sensory receptors to the brain. The information is processed in the brain (itself a collection of many neurones) which then sends appropriate information, in the form of electrical impulses, via motor neurones to move muscles. Anaesthetics work by interfering with the transmission of these impulses between nerve cells.

NERVE CELL STRUCTURE

Unlike other cells, neurones can be very long – up to 100 cm – in order to conduct electrical impulses over long distances. The longest projection from a neurone is called the axon.

The nervous system forms an elaborate circuit throughout the body; however, nerve cells are not physically connected to each other. Instead, nerves connect to other nerves (or muscles) via gaps called synapses.

When an impulse has reached the end of a nerve cell (the synaptic bulb), chemicals called neurotransmitters are actively transported across the synapse. When they bind to receptors on the adjoining cell, they trigger an impulse which can then continue along the length of the adjoining cell.

Cell membranes are largely made up of adjoining layers of lipid (fat) molecules. Proteins embedded in the membrane act as channels, specifically controlling entry and exit of chemicals into and out of the cells.

The synapse (circled) is the point of contact between two nerve cells, where the synaptic bulb of one meets the axon, dendrite or cell body of another.

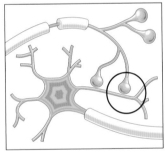

Normal synapse action

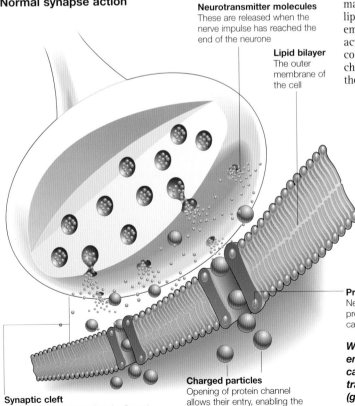

Neurotransmitter molecules
These are released when the nerve impulse has reached the end of the neurone

Lipid bilayer
The outer membrane of the cell

Synaptic cleft
Gap between nerve cell and adjacent cell (which may be a muscle, gland, or another nerve cell)

Charged particles
Opening of protein channel allows their entry, enabling the nerve impulse to continue

Protein channel
Neurotransmitter chemicals bind to protein channels on adjoining cells, causing them to open

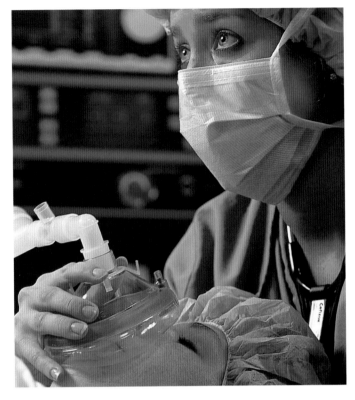

During major operations, the anaesthetist's role is vital. Gaseous anaesthetic mixed with oxygen may be administered via a mask (as here) or an endotracheal tube, which is inserted into the windpipe.

Anaesthetic effect at the synapse

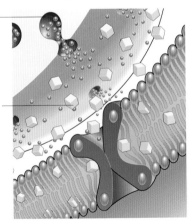

Synaptic vesicle
Sac which contains the neurotransmitter molecules

Anaesthetic agent

When an impulse reaches the end of a nerve cell, chemicals called neurotransmitters are transported across the synapse (gap) where they bind to the adjoining cell. This allows the entry of charged particles which continue the impulse.

It is believed that anaesthetics block the protein channel in the cell membrane or alter its ability to open as normal. Other researchers believe that anaesthetics can act on other sites, as the affected sites vary between different anaesthetics.

Where anaesthetics operate

Although the exact effect of anaesthetic agents is unclear, it is known that they act in the region of the synapse. This is the gap through which impulses are transferred between nerve cells, and between nerve cells and muscle fibres.

NERVE IMPULSES

The conduction of nerve impulses along the axon is achieved by the rapid entry and exit of ions through the protein channels, causing a small electrical current that spreads down the nerve. If these ions are prevented from passing through the protein channels, nerve conduction is impaired.

The exact mechanism by which drugs cause anaesthesia is still unknown, but because different types of molecules can all cause anaesthesia, it is thought that several molecular sites may be involved.

MOLECULAR SITE OF ACTION

Early studies suggested that the site of action of anaesthetic drugs is within the membrane of the cell, because the potency of inhaled anaesthetic drugs was proportional to their solubility in oil, a substance which closely resembles the membrane lipids. The assumption is that by inserting into the lipid bilayer, anaesthetic drugs may change the properties of the membrane – the membrane is a fluid structure and embedded structures are able to move freely within it. If the membrane was less fluid, this would affect the conduction of impulses.

Further studies suggested that anaesthetic drugs in the lipid bilayer cause the cell membrane to expand. Once a critical volume is reached, nerve conduction is impaired. A rise in pressure is thought to reduce the expansion of the cell membrane.

Types of anaesthetic

■ **Local anaesthetic**
Used for minor operations, such as stitching a wound, when a specific area (a local nerve) needs to be numbed. This may be done by injection, topical cream or eye-drops.

■ **Regional anaesthetic**
Numbs larger areas (often a limb), and operates in a similar way: a series of local anaesthetic injections is made around a nerve or a number of nerves to render them unresponsive to pain.

■ **General anaesthetic**
Renders the patient totally unconscious by the injection of drugs into the bloodstream, or by inhalation of gas; often both are combined. The drugs affect the brain to cause loss of consciousness and to prevent the sensation of pain.

Other drugs may be given during general anaesthetic to control post-operative pain, and in some cases to induce paralysis (neuro-muscular blockers) so that the muscles are relaxed during surgery.

As well as keeping the patient fully anaesthetized, the anaesthetist is responsible for monitoring their state throughout the operation. Equipment will measure blood pressure, heart rate and respiration.

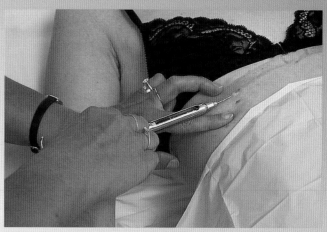

A local anaesthetic injection is used prior to the removal of a malignant melanoma (tumour of the skin's pigment cells). The patient is fully aware of what is going on but feels no pain.

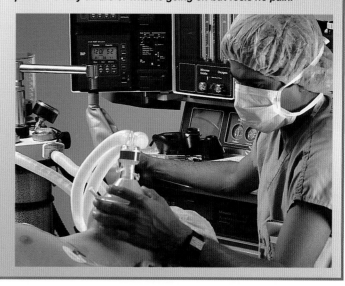

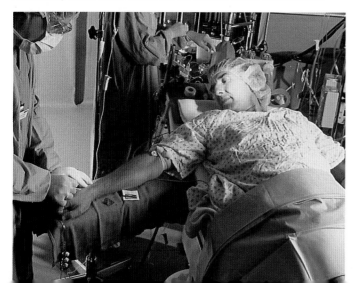

Premedication is commonly administered before major surgery to sedate the patient. Often, another drug is given to control lung secretions, which could otherwise be inhaled under anaesthesia.

Other sites of anaesthetic action

As well as the actions on the nerve cell membrane and within the lipid bilayer, anaesthetic drugs may well affect other sites involved with nerve impulse conduction.

SYNAPSES AND AXONS

When the nerve impulse arrives at the end of the nerve cell, this causes specific channels to open. These channels allow the passage of calcium ions into the nerve cell. This in turn causes neurotransmitter chemicals to be released – in sacs called vesicles – into the synapse.

Anaesthetic drugs may affect these calcium channels, preventing their normal opening and reducing the release of the neurotransmitter vesicles.

There is also evidence that some anaesthetic drugs bind to proteins on the surface of the adjoining nerve cell, impairing the binding of acetylcholine, an important neurotransmitter. In theory, this would reduce the nerve impulse that is triggered.

HIGHER NEURONAL CIRCUITS

The reticular activating system is a region of the brain involved in regulating consciousness.

General anaesthetic drugs may cause loss of consciousness by blocking the processing of sensory information as it passes through this area.

Common allergies

Allergies can be caused by anything from peanuts and bee stings, to penicillin and jewellery. Immunologists have divided these allergic, or hypersensitive, responses into four types.

Type I – Immediate, allergic responses

Allergy to peanuts is an increasingly recognised problem, that can lead to life-threatening anaphylactic shock.

Hay fever, an allergic reaction to pollen grains, is the most common example of a Type I, atopic allergy.

The faeces of the house-dust mite, which lives in bedding, carpets and furniture, are a common allergen.

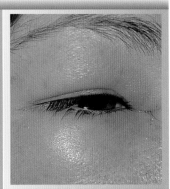

This young boy has had a severe anaphylactic reaction to a bee sting, causing oedema – fluid accumulation around the eye.

Type I hypersensitivity is an immediate response, beginning within seconds of exposure.

The commonest examples are hay fever, childhood eczema and extrinsic asthma. About 10 per cent of the population have a tendency to develop such a reaction, called atopy.

Upon encountering an allergen, instead of making a normal immune response, the body produces a class of antibody

molecule called IgE. These bind to mast cells, which are especially prevalent in the skin, respiratory passages and gastrointestinal tract, and cause the release of a number of inflammatory chemicals, including histamine.

Histamine causes blood vessels to dilate and become 'leaky', and is the main cause of typical allergic reactions - runny nose, watery eyes and itchy, red skin. Symptoms also depend upon

where the allergen enters the body. An inhaled allergen causes the airways to constrict, causing asthma symptoms; if ingested, symptoms include cramp, vomiting and diarrhoea.

A second, more dramatic, reaction can occur if an allergen enters the bloodstream. This is called anaphylactic shock. The airways constrict (and the tongue may swell) making breathing difficult, and the sudden dilation of

blood vessels and loss of fluid may cause circulatory collapse. This is typically triggered in susceptible individuals by bee stings and spider bites, injection of a foreign substance (for example penicillin, or other drugs) or certain foods, such as peanuts. Susceptible individuals may have to carry syringes of adrenaline (epinephrine) to administer in an emergency. Fortunately, reactions such as these are rare.

Type II – Reactions against 'foreign' cells

Type II hypersensitivity is caused by the binding of antibodies to 'self' molecules on the surface of cells. This does not generally cause damage, but may trigger a number of further responses.

One example of this may occur in mismatched blood transfusions. Another involves incompatibility between blood groups. All blood is either rhesus positive (Rh$^+$) or negative (Rh$^-$), depending on whether a certain protein is present on the surface of a person's blood cells. If a Rh$^-$ woman is pregnant with a Rh$^+$ fetus, it is possible for fetal blood to enter the mother's bloodstream during delivery, or following abortion.

Upon a subsequent pregnancy with a Rh$^+$ fetus, antibodies may cross the placenta, enter the fetal bloodstream and cause a number of detrimental effects. Injection of antibodies shortly after birth of an incompatible child will destroy fetal red cells in the mother's circulation.

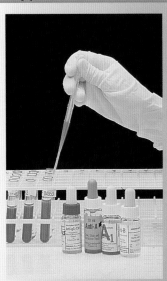

Accurate blood typing is crucial in preventing serious immune reactions in transplants. This occurs in both of the two blood matching systems.

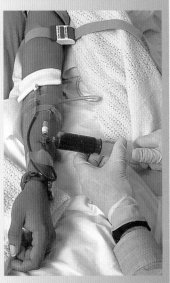

If a mismatch occurs in a blood transfusion (such as a Rh+ patient receiving Rh-), host defences cause destruction of the 'foreign' blood.

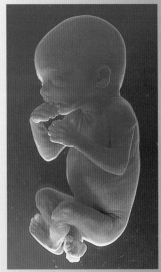

A mother may form antibodies to her own fetus's blood, if she comes into contact with it. This can lead to immune reactions in subsequent pregnancies.

Type III – Reactions against antibody-antigen complexes

Type III hypersensitivity results when allergens are distributed throughout the body. The body produces antibodies, which form insoluble antibody-antigen complexes. The body is unable to clear these, and a large inflammatory response develops.

Examples of such allergies include farmer's lung, which is caused by the inhalation of mould growing on hay, and mushroom grower's lung, caused by inhaling the spores produced by mushrooms.

A number of microorganisms can trigger immune complexes. Streptococcal throat infection may be exacerbated by the formation of these immune complexes, as can the organisms that cause malaria, syphilis and leprosy. Drugs can also have the same effect.

These responses are also involved in autoimmune disorders, when the body's defences attack host tissue. Examples are systemic lupus erythematosus (SLE) and rheumatoid arthritis.

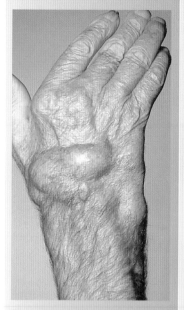

Rheumatoid arthritis is an autoimmune disorder, in which the body's defences attack host tissues. In this case, it is the lining of the joints that are affected, causing erosion and damage, leading to deformity.

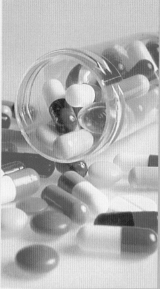

A number of drugs are known to cause allergic responses. For instance, penicillin in the body can bind to the protein albumin (the protein also present in egg white) and provoke a significant immune reaction.

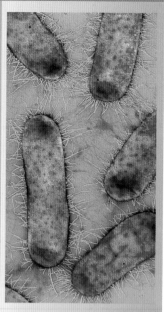

In microbial infections causing malaria, syphilis and leprosy, amongst others, the surface of the microorganism can trigger a Type III response. The complex of antibodies and bacteria can be harmful.

Type IV – Delayed reactions

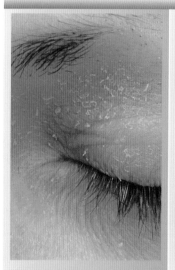

This type IV reaction is actually a reaction to nail varnish. These allergic reactions can occur some distance from the site of the original allergen – in this example, dermatitis has occurred on the eyelid.

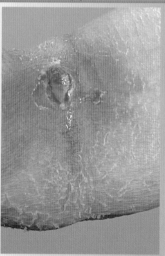

This sore has been caused by an allergic response to sticking plaster on the skin, used to cover a wound. Such reactions are caused by the release of chemicals called lymphokines from T-cell white blood cells.

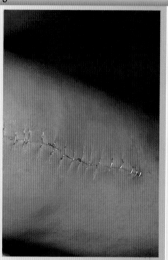

Here a patient has suffered a large wound, running across his knee. The red allergic patches which surround the trauma result from a hypersensitivity to the metal surgical sutures used to close up the wound.

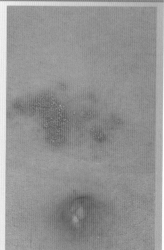

Contact dermatitis in an 18-year-old woman, caused by a reaction to nickel in jewellery. The nickel is absorbed into the skin, where it binds to body proteins and becomes 'foreign' to the immune system.

Type IV reactions are known as delayed hypersensitivities. They appear much more slowly and are caused by the actions of a number of white blood cells. The main effects are caused by a class of immune cell called T-cells. Inflammatory responses are caused by the release of chemicals from T-cells called lymphokines. Therefore antihistamines are not effective against these allergies.

A well-known manifestation of a Type IV reaction is allergic contact dermatitis. This results from skin contact with, for example, nettles, poison ivy, heavy metals (such as lead and mercury), cosmetics and deodorants. These substances are often too small to evoke an immune response, but upon absorption through the skin, they bind to body proteins and become recognized as 'foreign' (this is utilized in the Heaf test for tuberculosis, in which the bacterial proteins are 'punched' beneath the skin surface).

Nickel and copper in jewellery may cause contact dermatitis, and in these cases, the cause is obvious. A wide number of potential allergens exist, and careful questioning of the patient's circumstances and relevant patch tests can establish the cause. Rashes can be chronic (long-term), patchy and some distance from the allergen. For example, allergy to nail varnish may manifest itself as a rash on the face or neck.

How allergies occur

An allergy is an inappropriate response by the body's immune system to a normally harmless substance. Allergies vary from hay fever and asthma to life-threatening anaphylactic shock.

An allergy is a hypersensitivity of the body to a particular substance. If the body comes into contact with this substance, unpleasant and even life-threatening symptoms may occur.

IMMUNE REACTION

Allergies occur when the immune system – the body's defence against infection – misidentifies an innocuous substance as being harmful and overreacts to it. This can result in mildly inconvenient symptoms such as a rash or runny nose or, in some cases, life-threatening shock. Allergies can be caused by anything, but typical allergens are pollen, wasp sting venom, penicillin, latex, peanuts and shellfish.

The main components of the body's immune system are lymphocytes (white blood cells). B-cells are a form of lymphocyte that are able to identify foreign particles (antigens) and form appropriate antibodies

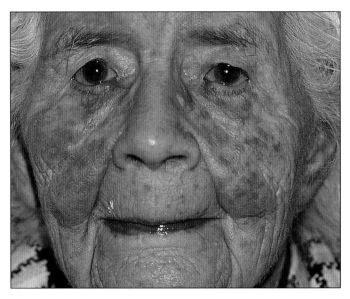

(immunoglobulins) specifically engineered to fight them. There are five basic types of antibodies: IgA, IgD, IgE, IgG and IgM. The immunoglobulin responsible for allergic reaction is IgE.

Allergies tend to be inherited, whereby the gene responsible for producing the protein that enables lymphocytes to distinguish between threatening and non-threatening proteins is faulty. This means that in a

Skin allergies are usually caused by direct contact with an allergen. Here, an elderly woman has developed a rash in response to a particular bubble bath.

person allergic to shellfish, for example, a B-cell is unable to recognize that a protein ingested as part of a meal containing shellfish is not invading the body. As a result, the B-cell produces large quantities of IgE antibodies.

SENSITIZATION

These antibodies subsequently attach themselves to basophils (a type of white blood cell) and mast cells (found in connective tissue) in the body, causing the body to become sensitized to the allergenic protein.

Basophils and mast cells both produce histamine, an important weapon in the body's defence against infection. When released in extreme quantities, histamine can have a devastating effect upon the body.

The allergic cascade

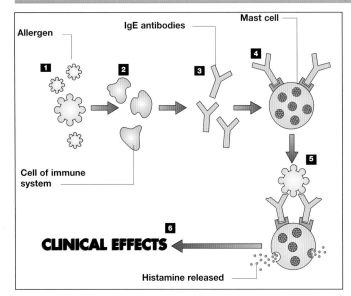

When an individual develops an allergic response to a substance, a domino reaction occurs. A chain of events is set in motion, known as the allergic cascade.

3) The IgE antibodies within the body are alerted
4) The IgE antibodies, bound to the surface of the mast cells and basophils, recognize the allergen by the specific protein markers on its surface
5) The IgE antibodies, still attached to the mast cells and basophils, attach to the surface proteins of the allergen. The healthy mast cells and basophils are destroyed (degranulation). Histamine is released, which causes the surface blood vessels to dilate, leading to a drop in blood pressure; the spaces between surrounding cells fill with fluid
6) Depending on the allergen, and where the reaction occurs, this may result in immediate symptoms. For example, if the reaction occurs in the mucous membrane of the nose, it may cause symptoms of hay fever, such as sneezing.

NON-ALLERGIC REACTION

In a normal person, the allergic cascade fails to progress because the allergen is destroyed. A group of around 20 proteins that are present in the blood bind, one by one, to the allergen/antibody site. When the string of proteins is complete, the allergen is destroyed.

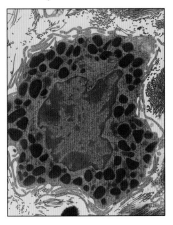

Mast cells are large cells found in connective tissue. Histamine (which helps the body to fight infection) is produced in the cells' granules (shown in black).

Over a period of around 10 days from initial exposure to the allergen, all the body's basophils and mast cells are primed with IgE antibodies and the body becomes sensitized to that allergen. If the body then comes into contact with the allergen for a second time, it will be prepared to attack immediately and a cascade

reaction occurs, in which a domino effect is triggered.

ALLERGIC CASCADE

The allergic cascade occurs as follows:
1) The body and the allergen come into contact
2) The cells of the immune system are stimulated

Anaphylaxis

Anaphylactic shock is an extreme allergic reaction that affects the whole body. Without treatment with adrenaline, the condition may be fatal.

In some cases, an allergic reaction can involve the entire body; this is known as a systemic reaction. During this reaction, histamine is released throughout the body, causing capillaries in many tissues to dilate. Anaphylaxis occurs when the reaction is so severe that the blood pressure becomes dangerously low. In extreme cases, the blood pressure drops so low that the body goes into shock. This is known as anaphylactic shock, and is often a fatal condition.

SEVERE REACTION

Anaphylaxis develops very suddenly and presents in a number of ways. A person may rapidly develop a rash and the throat may swell as cells release fluid into surrounding tissue, causing breathing difficulties. A dangerous and rapid drop in blood pressure accompanies this as the blood vessels throughout the body dilate. The brain and other vital organs become

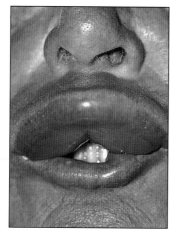

A severe allergic reaction can cause localized swelling, known as oedema, in the tissues. This man has been stung by a bee on his lip resulting in inflammation.

starved of oxygen and, within a matter of minutes, the person may die. Even if the victim survives this form of allergic reaction, the brain and kidneys may be permanently damaged.

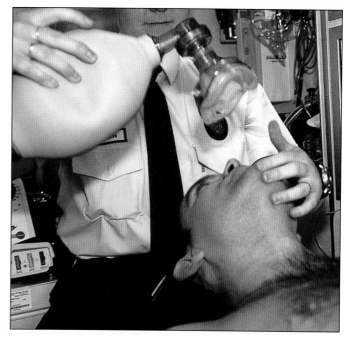

Anaphylactic shock can be a life-threatening event. In extreme cases, a person can suffer a respiratory and cardiac arrest and need resuscitation.

ADRENALINE

The only effective treatment for anaphylaxis is an intramuscular injection of adrenaline, a hormone naturally produced by the adrenal glands.

Adrenaline counteracts the symptoms caused by excess histamine by constricting the body's blood vessels and opening the airways. It is vital that the injection is administered correctly at the onset of symptoms for it to be effective.

People who are aware of a serious allergy usually carry an injection for self administration.

Treatment for allergies

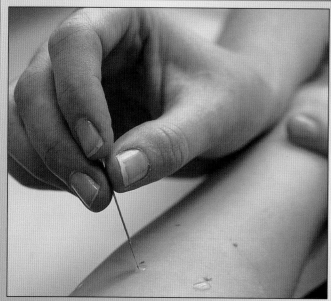

If a person suspects that they have an allergy, they can request tests to determine its exact nature. The scratch test is a common means of determining the cause of an allergy. This involves applying a diluted extract of a possible allergen to the skin (usually of the forearm) and then scratching the skin under the allergen with a needle. If swelling

or redness develops in the scratched area, it indicates that IgE antibodies to that allergen are present.

Blood tests may also be used to diagnose an allergy, especially in young children, since exposing a child to even minute amounts of allergen during a scratch test could trigger an anaphylactic reaction.

Scratch tests are often used to identify allergens responsible for allergic conditions. Allergens include pollen, fungal spores and dust.

Neither technique is one hundred per cent accurate, but a combination of both tests along with a patient's medical history can aid diagnosis and the formulation of a treatment plan.

Managing the allergy

Once identified, many allergens, such as dog hair or shellfish, can simply be avoided. However, some allergens, such as pollen, mould or dust, that are present in the environment cannot be avoided. The resulting allergies are kept in check using antihistamines, decongestants, corticosteroids and, in the case of anaphylaxis, adrenaline.

Immunotherapy

For people with severe allergies that cannot be avoided or managed with medication, immunotherapy may be their only hope of leading a normal life. This involves a number of injections of the specific allergen, starting with a very weak dilution and building

up to a higher dose that can be maintained over time.

These injections allow the immune system to adjust and desensitize to the allergen over time, so that it produces fewer IgE antibodies. Immunotherapy also stimulates the production of IgG antibodies, which block the effects of IgE. The treatment is expensive, time-consuming and entails risk (such as severe allergic reaction).

Some people may choose to consult a homeopath. Here, a vegetative reflex test for allergies is measuring substances in the body.

How infection occurs

Although the body is a natural host to a huge number of
bacteria, infection does not usually arise unless the body's defences
are damaged. Infections are generally acquired from other people.

The body is exposed to a myriad of micro-organisms every day. In fact, it plays host to millions of bacteria, all living in a state of co-existence. Most bacteria are harmless as long as they stay in protected places such as the surface of the skin, intestines, nose, mouth or vagina. However, if these surfaces become damaged through injury or disease, and micro-organisms are allowed to enter the normally sterile internal tissues of the body, infection can occur. The large intestine, for example, is home to numerous bacteria which do not usually cause any harm, but if they were to enter the abdominal cavity, serious infection would occur.

PROTECTIVE BARRIERS
The body, fortunately, has a number of protective barriers, which act as a first line of defence against infection, including:

■ The skin – this provides a physical barrier to pathogens (disease-causing organisms), helping to maintain a sterile internal environment
■ The nose – this contains sticky mucus and hairs to trap potentially harmful micro-organisms, while the sneezing mechanism expels any irritants
■ Saliva – this contains antibodies that combat pathogens
■ Tears – these contain antibodies to prevent infection of the eyes
■ Throat – this is protected by the reflex reaction of coughing
■ Stomach – this produces a strong acid that destroys any ingested pathogens.

Coughing is a reflex action by which micro-organisms in the airway are expelled. This is one of the body's defence mechanisms against infection.

Local infection

If pathogens manage to breach the body's first line of defence, they can multiply in the tissues, causing infection. The body's response to this is to produce inflammation, an important reaction that prevents the spread of infection.

REDNESS
If a sufficient number of pathogens invade the body, they will release harmful toxins or cause enough damage to cells for local blood vessels to dilate, resulting in an increased blood flow to the affected area. This gives rise to the redness and warmth typical of an inflamed area. In addition, a watery fluid leaks out of the blood vessels, causing the surrounding area to swell visibly.

The increase in blood flow enables cells of the immune system, including phagocytes (a

In some infections, a wall of fibrous tissue may form around the inflamed area. Pus can then build up within this wall to form an abscess.

form of white blood cell that engulfs and destroys pathogens) to reach the area and attack the organisms present. This is usually sufficient to prevent infection spread, and the swelling eventually subsides as the pathogens are destroyed.

If the infection is particularly severe, then the body will also form a wall of fibrous tissue around the infected area. This wall serves to keep the infection localized while it is brought under control by the immune system. Within the fibrous walls, a build-up of pus may occur; this contains dead white cells, body cells and bacteria, and cell debris.

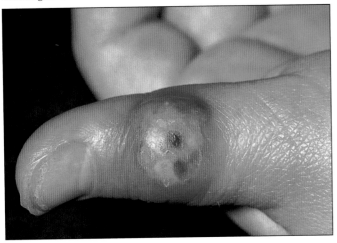

Incubation period

After a pathogen has invaded the body, there is an interval of time before there is any evidence of disease. This is because all pathogens undergo an incubation period during which they multiply. Once there are sufficient pathogens, they will cause noticeable effects or symptoms in the patient.

Variable length
Incubation periods vary greatly, from only a few hours to some years. Cholera, for example, can develop within a couple of hours of drinking contaminated water, but AIDS may not develop until many years after the HIV virus has been acquired.

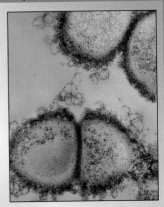

Pathogens go through an incubation period in which they multiply in the body. Here, Streptococcus bacteria are dividing (lower centre).

Systemic infection

Some micro-organisms enter the bloodstream and spread quickly to engulf the whole body. Common signs of systemic infection are a fever or rash.

In some cases, infecting organisms, or the toxins they produce, enter the bloodstream and rapidly spread throughout the entire body. This is known as systemic infection and can lead to some characteristic symptoms, such as fever or rash.

FEVER

Fever occurs when the immune system cells are damaged by invading pathogens, causing them to release substances called cytokines. These affect the body's 'thermostat' (controlled by the brain), effectively adjusting it to a higher setting. As a result, the normal body temperature is perceived by the brain to be too low, causing

shivering to occur, which automatically produces extra heat. This causes the body temperature to rise to a level that is fatal to most invading micro-organisms.

RASH

Skin rashes in systemic infection are caused by multiple areas of skin damage as a result of the micro-organisms or their toxins. Rashes indicate that similar damage may be occurring within the body.

Sometimes, infections affect the whole body. Such systemic infections are serious and cause characteristic symptoms, such as fever or a rash.

Spread of infection

The majority of infections are acquired, directly or indirectly, from other people, and may be spread in the following ways:
■ Skin-to-skin contact – if the dose of micro-organisms is large or virulent enough, skin infections may be spread by contact. Some organisms, such as staphylococci, penetrate the sweat glands and hair follicles, causing pustules and boils. For example, impetigo, a bacterial skin infection, can very easily be spread through contact with infected skin
■ Transfer to eye – pathogens may be spread from the fingers to the eye, causing infections such as conjunctivitis. This may be spread from one eye to the next, and may even be transmitted by infected towels or make-up products

If the water supply is contaminated, infection spreads easily. Washing cooking pots in polluted rivers can lead to diseases such as typhoid.

■ Transfer to nose – pathogens are often picked up by the fingers and spread to the nose through rubbing. In fact, the rhinoviruses that cause the common cold are more readily transmitted by hand-shaking than by sneezing
■ Inhalation – a number of infections are spread through the inhalation of airborne droplets released during coughing or sneezing. Some infective agents are inhaled in the form of dried spores contained in dust, for example Q fever (an influenza-type infection)
■ Ingestion – although stomach

acid destroys the majority of ingested pathogens, some manage to survive and pass through to the intestines. The consumption of contaminated food or water can spread infection in this way, such as in gastroenteritis. Food poisoning can also be caused by food contaminated with material from the infected hands of food handlers. This can contain a virulent toxin produced by staphylococci bacteria, which causes severe illness
■ Faecal contamination – this is a common cause of infection, since faeces can contain pathogens that are transmitted to food prepared by a person who has not washed their hands (such as in salmonella poisoning). Certain viruses (enteroviruses) can be spread by the ingestion of faecal traces, for example the viruses that cause polio and hepatitis A
■ Pregnancy – infection can be spread directly from mother to baby during pregnancy via the placenta, for example toxoplasmosis. During birth, babies may also contract infections such as herpes or syphilis through contact with an infected vagina
■ Blood – pathogens in blood can be transmitted by the use of an infected syringe, or through

The saliva of a particular female mosquito contains the parasite responsible for malaria. This disease is one of a number spread by insects or animals.

tattooing and ear-piercing with unsterilized needles. HIV can be spread in this way
■ Sexually-transmitted infection – some diseases (such as herpes) can be spread during sexual activity due to intimate contact and exchange of bodily fluids.

ANIMAL CONTACT

A few infections are contracted through contact with animals and insects. Some, such as rabies, may be acquired from infected animals; others, such as malaria, may be picked up from insects which act as vectors for a disease, but do not actually have it themselves.

Adapting to changes in atmospheric pressure

Changes in atmospheric pressure are experienced when we are above or below sea level. The body can adapt, within limits, to changes in the concentration of oxygen when pressure is raised or lowered.

The body depends upon oxygen, a major component of air, for survival. Oxygen (O_2) is transferred by red blood cells, from the lungs to body tissues, where it is exchanged for the waste product carbon dioxide (CO_2), which is exhaled. This process is essential to the production of energy necessary for the body to function.

ATMOSPHERIC PRESSURE
Oxygen accounts for 20.96 per cent of volume of the air. Atmospheric pressure determines how dense the air is and thus the amount of oxygen in the air we inhale.

Humans are designed to thrive at around sea level, at which pressure the air is dense enough to ensure that oxygen is present in adequate concentrations in every breath taken.

PHYSIOLOGICAL CHANGES
The further we move away from sea level, for example when climbing a mountain, or deep-sea diving, the atmospheric pressure changes. To survive, the body must adapt by undergoing physiological changes. This is known as acclimatization, or acclimation.

The amount of oxygen inhaled with each breath depends on ambient pressure. As pressure changes, the body must adapt itself or be artificially aided.

Surviving high pressure under water

Water represents a high-pressure medium to which humans are not adapted. The main obstacle to survival is, of course, an inability to extract oxygen from the water for long-term survival.

In addition, gaseous exchange within the lungs is compromised by the increased ambient pressure that occurs with increasing depth.

DIVING REFLEX
Although humans are poorly adapted to an aquatic environment, they do have some reflexes to prevent drowning and to conserve oxygen. These include inhibition of breathing, slowing down of the heart (bradycardia), constriction of the peripheral blood vessels, and reduced peripheral blood flow.

ADAPTATIONS
Experienced divers are able to exploit these reflexes, enabling them to remain underwater for longer periods of time.

Through practice, lung capacity is increased so that a greater amount of oxygen can be stored before surfacing. Also, the employment of techniques such as hyperventilation, whereby a greater concentration of oxygen is taken into the lungs, enables

Self-contained underwater breathing apparatus enables divers to breathe underwater. Increased pressure can have adverse effects on the body.

divers to remain below the surface for longer than would normally be possible.

DEEP-SEA DIVING
There are limitations to the body's ability to stay underwater, however, and for deeper diving an artificial supply of oxygen is required.

Modern equipment provides divers with a supply of oxygen, the pressure of which is constantly equalized to that of the lungs. This allows divers to remain underwater for greater lengths of time, reaching profound depths, although this in itself can be hazardous.

Decompression sickness

Although nitrogen (which forms 79 per cent of air) normally has little effect on the body, prolonged exposure to high pressure can cause it to concentrate in the body tissues, giving a narcotic effect. As a result, divers can become dizzy and appear drunk.

With gradual ascent, the dissolved nitrogen gas dissipates slowly. If a diver surfaces too quickly, however, the rapid decrease in pressure causes the dissolved nitrogen to form bubbles of gas in the blood which can develop emboli (clots) that may be fatal or cause paralysis (when bubbles migrate to the brain) and evoke acute musculoskeletal pain (commonly referred to as 'the bends').

Treatment should be carried out immediately, which involves recompression (hyperbaric therapy) of the body in a compression chamber, before gradual decompression.

Coping with low pressure

At high altitudes, atmospheric pressure becomes lower and oxygen more scarce.
The body is able to compensate for the lack of oxygen using certain mechanisms;
however, altitude sickness may occur if a climber ascends too high or too quickly.

With an increase in altitude there is a decrease in pressure and the air becomes less and less dense as the oxygen molecules spread out. As a result, each breath taken will contain less oxygen.

DECREASED OXYGEN

Under these challenging conditions there are noticeable effects on respiration. The body therefore adapts to the decrease in partial pressure of oxygen in the lungs by employing a number of compensatory mechanisms.

In the short term, the decrease in available oxygen will be compensated for by an increase in the rate and volume of air inspired. The respiratory centre of the brain causes deeper breaths to be taken in order to inhale greater volumes of air and therefore more oxygen.

The scarcity of oxygen at high altitudes also stimulates an increase in the production of haemoglobin and red blood cells, helping to increase the blood's oxygen-carrying capacity. In addition, heart rate and blood pressure are increased to maximize the amount of oxygen transported throughout the body.

For prolonged periods at high altitudes body tissues develop more blood vessels, increasing the efficiency of gaseous exchange. In addition the size of muscle fibres decreases, shortening the diffusion path of oxygen.

ACCLIMATIZATION

These acclimatizing physiological changes are effective, but not spontaneous, and acclimatization must occur progressively. Ascending to altitude too quickly or climbing too high will result in the body not being able to adapt quickly enough, if at all, and the body will be unable to cope with the oxygen depletion.

Most aircraft are pressurized to counteract low atmospheric pressure at high altitudes. Oxygen masks are available in case of emergency.

The Sherpas are people renowned for their ability to survive in the high Himalayan mountains. They are adapted to living at a very low atmospheric pressure.

Altitude sickness

Altitude sickness occurs when the altitude is simply too high for the body to cope with, or the decrease in pressure is too rapid.

Because there is so little oxygen, the body must work harder to pass more air through the lungs, while the increased breathing rate requires even more energy expenditure. Breathing becomes laboured and irregular, and as oxygen concentrations reaching the body tissues become inadequate, a state of hypoxia is reached. The climber experiences confusion, light-headedness, headache and nausea.

Treatment for this condition is gradual descent and, in some cases, drug therapy. Severe forms of altitude sickness are extremely dangerous and can give rise to brain haemorrhage and fluid accumulation in the lungs.

The body is unable to function unaided beyond 6,400 metres (21,000 feet). Mount Everest is 8,840 m (29,000 ft) high, meaning that climbers normally require oxygen to complete the ascent.

At extreme heights, or if ascent is too rapid, the body is unable to cope with the low pressure. Altitude sickness occurs due to an inadequate oxygen supply.

The effects of space travel on the body

The body adapts to life in space, but there are limitations.
Decreased bone and muscle mass, a weak heart and anaemia are just
a few of the problems that astronauts may experience.

Since Yuri Gagarin was launched into space in 1961, a great deal has been discovered about the effect of space travel on the body.

CHANGE OF ENVIRONMENT

All of the body's physiological processes are finely tuned to the Earth's environment. A drastic change of atmosphere and conditions, such as an individual encounters in space, has a number of dramatic effects on the body.

There are three key differences between space and Earth that affect the human body:
- Atmosphere
- Radiation
- Gravity.

Combined, these factors have an important effect on the body; incredibly, the body is able to detect these changes, automatically adjusting to its new environment by a series of integrated responses.

SURVIVAL IN SPACE

In the past, many people doubted whether human beings could even survive a journey into space. Over the years, however, we have learned that while survival is possible with the proper protection, there are limitations to the body's capabilities.

For humans to colonize space in the future, the effects that space travel has on the body need to be understood and counteracted.

Space is a very different environment to Earth. Within days, however, the human body begins to recognize and adjust to its new surroundings.

Short-term changes

Once in orbit, the body enters a state of weightlessness or microgravity. In fact, the body is in a state of free-fall, just like jumping out of a plane, except that the body is actually moving horizontally at high speed, orbiting the Earth.

MICROGRAVITY

When a person stands on bathroom scales, both the body and scales are forced down by gravity. However, because the scales are on the floor, they push back up (resist) with an equal force which is then recorded as weight. If a person were to stand on the scales in free-fall, without meeting any resistance, both the body and scales would be pulled down equally by gravity, and the weight would be recorded as zero.

ORIENTATION

On Earth the body learns to process combined signals from the eyes, ears and touch receptors in the skin, which inform the brain of the body's orientation in relation to its environment.

In space, the senses cannot determine any natural 'up' or 'down'. The lack of gravity, and therefore weight, means that astronauts are initially unable to grasp the orientation of their bodies at all; when the arms and legs are weightless, it is difficult to sense where they are.

MOVEMENT

As the body moves, the brain processes sensory information in order to determine the body's orientation at any point of movement.

The vestibular organ (within the inner ear) measures, and informs the brain, how fast the body is moving and in what direction in relation to gravity. In addition, the brain takes into account information from receptors in the muscles and joints.

However, in space this sensory information is conflicting. When the brain tries to process this information it becomes 'confused', and does not allow for the lack of gravity in space. Consequently, the body's responses are not appropriate for the new environment.

SIDE EFFECTS

As a result, astronauts develop a condition known as space motion sickness and suffer headaches, loss of appetite, abdominal pain and nausea. Within a few days, however, the brain adapts to the new environment and the symptoms subside. When the astronaut returns to Earth, the brain has to repeat the readjustment process.

Microgravity causes the body's sensory organs to become confused. As a result, the brain responds inappropriately and astronauts become disoriented.

Long-term changes

Over time, the body begins to make significant adjustments to cope with the requirements of space travel. These can have serious long-term implications.

While the body's balance and movement mechanisms soon adapt to being in space, other body functions do not adjust as quickly. The changes in some physiological processes are lasting and can cause serious long-term problems, especially on return to Earth.

CARDIOVASCULAR SYSTEM

In space, the body no longer experiences the pull of gravity that distributes blood and other fluids to the lower body.

Instead, body fluids make what is called a 'headward' shift, in which they are redistributed towards the upper part of the body and away from the lower extremities. This has some interesting effects.

In space, astronauts actually appear different, as the increased fluid in the upper body fills the facial cavities, making the face puffy. In addition, as body fluids move away from the lower body, the legs become smaller in circumference, giving them a 'bird-like', wasted appearance.

FLUID LOSS

When the body enters microgravity, the increase in arterial pressure is detected by baroreceptors (specialized sensory nerve endings that monitor fluid levels). These send information along nerve impulses to

Due to a lack of gravity, red blood cell production is decreased in space. Astronauts must monitor their haemoglobin levels closely to avoid anaemia.

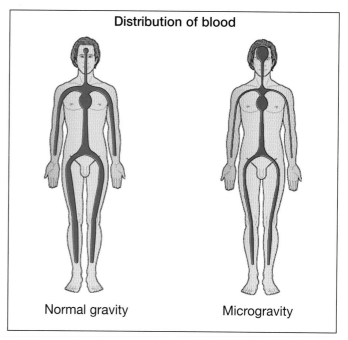

Distribution of blood

Normal gravity   Microgravity

Without the force of gravity, the distribution of blood and body fluids in the body changes. Blood is diverted upwards to the head, away from the legs.

the brain, which in turn stimulates the kidneys to rid the body of this excess fluid.

As a result, astronauts tend to urinate more frequently. At the same time, the pituitary gland within the brain decreases the secretion of anti-diuretic hormone (ADH), making the astronaut less thirsty.

These two factors combine to reduce fluid levels in the head and chest, achieving lower levels within a matter of days than is normal on Earth.

HEART DECONDITIONING

As there is less fluid to pump around the body, and less energy is required to combat the effects of gravity, the heart no longer has to work as hard. As a result, it shrinks.

SPACE ANAEMIA

When the astronaut returns to Earth, not only will the heart be weakened, but the blood will be severely deficient in red blood cells (anaemic).

This is because the kidneys reduce their secretion of erythropoietin, a hormone that stimulates red blood cell production. This decrease in red blood cells means that, in space, the percentage of blood volume occupied by red blood cells is normal, but on return to Earth, and a resultant increase in blood volume, astronauts are anaemic.

Musculoskeletal system

Another important effect of microgravity is that astronauts no longer require the full strength of the skeletal and muscular systems to sustain an upright posture.

ATROPHY

In space, the body assumes a fetal position: slightly crouched, with the arms and legs bent in front. In this position the muscles are fairly lax, particularly those that are normally used to maintain an upright position. If the astronaut stays in space for some time, muscle mass decreases, muscle fibres change and atrophy increases day by day.

BONE DETERIORATION

Bone mass also decreases in space. Bone mass and size is regulated by the rate at which bone is made by bone cells

(osteoblasts) and broken down by other cells (osteoclasts).

As the bones are used less in space, particularly the weight-bearing bones such as the hips, thighs and lower back, the rate at which osteoblasts lay down new bone is reduced, while the rate at which osteoclasts absorb bone stays the same. As a result, bone size and mass decreases at a rate of around one per cent a month. The bones become weak and brittle and are more likely to break upon return to Earth.

Research into these long-term implications is continuing, alongside work to counteract these physiological changes.

Bone mass decreases in space as the bones are not required to resist gravity. As a result, they weaken and are more likely to fracture on return to Earth.

Counteracting the effects of space travel

Space travel has some dramatic effects on the body. In order to counteract these effects, astronauts need to undergo specialized training, and rely on increasingly sophisticated equipment.

From the minute the space shuttle takes off, the astronaut's body is subject to dramatic environmental changes.

ADJUSTING TO SPACE

Immediately, the gravitational pull of the shuttle taking off puts the body under immense pressure. As the shuttle accelerates, the *g*-forces increase to three times normal gravity, causing chest compression, breathing difficulties and an extreme feeling of heaviness. Within minutes, the shuttle is in orbit and weightlessness occurs.

The body will then have to adjust all over again on return to Earth. In order to cope with all the extreme changes brought on by space travel, astronauts have to undergo rigorous training and must be in excellent health.

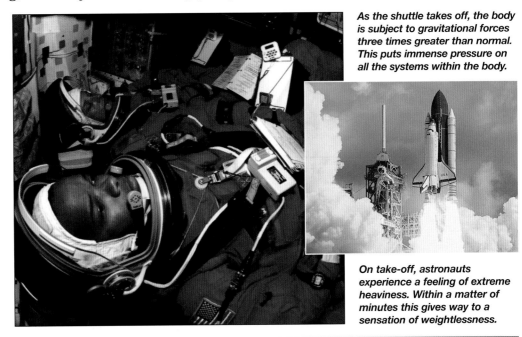

As the shuttle takes off, the body is subject to gravitational forces three times greater than normal. This puts immense pressure on all the systems within the body.

On take-off, astronauts experience a feeling of extreme heaviness. Within a matter of minutes this gives way to a sensation of weightlessness.

In-flight precautions

When in space, it is essential that astronauts follow a strict exercise programme (up to two hours a day) to counteract the wasting of muscle, bone and the heart.

Without this exercise, the body would be too weak to survive a return to the Earth's atmosphere.

The combination of confined space and weightlessness means that astronauts have to rely on exercise machines such as treadmills to prevent the muscles from wasting. They also use large rubber bands (bungee cords) and shoulder weights to hold themselves down, producing a sensation similar to weight.

Fluid loss

One way to deal with fluid loss is the use of a lower body negative pressure (LBNP) device. This involves the use of a vacuum-cleaner-like suction below the waist to keep fluids in the legs.

The LBNP device might be attached to an exercise machine, such as a treadmill. By spending around 30 minutes every day using this device, astronauts can keep their circulatory system in near-Earth condition.

Just before returning to Earth astronauts drink large volumes of water or electrolyte solutions to help replace the fluids lost. Without this measure the astronaut would be likely to faint when first standing up on return to Earth.

Monitoring

Constant monitoring of body changes on each mission is an

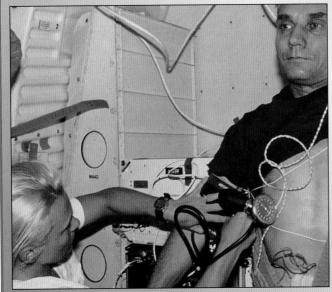

Astronauts constantly monitor body changes in space. This provides vital data for research into the effect of space travel on the body.

Astronauts exercise every day on machines such as treadmills. This counteracts the muscle wasting that would otherwise occur in space.

extremely important part of the astronaut's role. These measurements are essential since they allow any abnormal changes to be detected, as well as providing vital data for research into the effect of space travel on the body.

Creating a safe micro-environment

Space represents a hostile environment for the body. Beyond the spacecraft, astronauts would perish within seconds without the aid of a spacesuit.

Although astronauts have successfully landed and walked on the moon, this would be impossible without extremely specialized equipment.

HOSTILE ENVIRONMENT

If an astronaut were to leave the security of the spacecraft without a spacesuit, they would instantly perish for a number of reasons:
■ A lack of oxygen means they would lose consciousness within 15 seconds
■ There is little or no air pressure in space, which would cause the blood and other body fluids to boil instantly
■ Extreme temperatures ranging from 120°C in sunlight, to –100°C in the shade would be fatal
■ The body would be exposed to deadly levels of radiation from cosmic rays and charged particles emitted from the sun.

In addition, astronauts face the hazard of fast-moving rock particles and satellite debris bombarding the site, creating a dangerous environment.

For this reason, increasingly sophisticated equipment is required to create a safe micro-environment.

Spacesuits provide astronauts with a temperature- and pressure-controlled environment. They also protect them from radiation and space debris.

Spacesuits

Spacesuits allow astronauts to leave the safety of the spacecraft by providing the following:

■ A pressurized atmosphere – this is vital for keeping the body's fluids in a liquid state. Spacesuits operate below normal atmospheric pressure, while the space cabin operates at normal air pressure. For this reason, an airlock is operated between the cabin and the exterior so that pressure can be reduced before astronauts put on their suits, to prevent nitrogen building up in the blood (causing the 'bends').

■ Oxygen supply – spacesuits provide pure oxygen for breathing, supplied either from the spacecraft (via an 'umbilical cord') or from the astronaut's specialized backpack. As the shuttle has a normal air mixture (simulating Earth's atmosphere), astronauts must breathe pure oxygen for some time before putting on their spacesuits.

This eliminates nitrogen from the astronaut's blood and body tissues, minimizing the risk of it entering the blood and causing the 'bends'. The spacesuit is also designed to eliminate carbon

Astronauts enter an airlock compartment before putting on their spacesuits. This allows their body to adjust to the lower atmospheric pressure.

dioxide, which would otherwise build up and poison the body.

■ Insulation – spacesuits are designed to maintain the optimum temperature for the body, despite strenuous activity, and prevent exposure to extreme temperatures.

Spacesuits are heavily insulated with layers of sophisticated fabrics that allow the body to breathe, but maintain temperature. Heat produced by the body during strenuous activity is dissipated by a fan or water cooler, to prevent excessive sweating and subsequent dehydration. It has been known for astronauts to lose several pounds during a single space walk due to fluid loss.

■ Protection – spacesuits are made up of many layers of durable fabric which protect the body from flying debris and prevent the suit from tearing.

■ Defence against radiation – spacesuits offer only limited protection from radiation, so

Space is a hostile environment for human life. Spacesuits create optimum conditions for the body, enabling astronauts to explore new frontiers.

spacewalks are always planned during periods of low solar activity.

■ Easy mobility – joints in the fabric of the spacesuit enable astronauts to move easily.

■ Clear vision – visors are made of clear material, designed to reflect sunlight and reduce glare. Fitted lights enable the astronaut to see in the shadows

■ Communication – spacesuits are equipped with radio transmitters and receivers to allow communication.

Index

Picture Credits

Apart from those images listed below, all of the pictures contained in this book were originally sourced for the partwork *Inside the Human Body*, produced by Bright Star Publishing plc.

Corbis: 88 (Jens Nieth), 248 (Howard Sochurek), 336, 380 (Andrew Brookes),
  398 (Ron Boardman; Frank Lane Picture Agency), 442 (Jim Zuckerman)
Dreamstime: 8, 9, 11(t), 12, 14, 15
Getty Images: Front Cover (3D4Medical), 10 (Roger Ressmeyer), 16 (3D4Medical),
  130 (Patrik Giardino), 156 (Adam Smith), 210 (PM Images), 306 (Stephanie Rausser),
  464 (George Logan)
IStockphoto: 11(b), 13